The Kaleidoscope of Gender

To Kay's spouse, Paul J. Burgett, for sharing love, knowledge, and adventure

To Joan's children, their spouses, and grandchildren, for having the strength to seek out new ways of doing gender

The Kaleidoscope of Gender

Prisms, Patterns, and Possibilities
Second Edition

Joan Z. Spade
SUNY College at Brockport

Catherine G. Valentine
Nazareth College

SAGE Publications
Los Angeles • London • New Delhi • Singapore

For information:

Pine Forge Press
A Sage Publications Company
2455 Teller Road
Thousand Oaks, California 91320
E-mail: order@sagepub.com

Sage Publications Ltd.
1 Oliver's Yard
55 City Road
London EC1Y 1SP
United Kingdom

Sage Publications India Pvt. Ltd.
B 1/I 1 Mohan Cooperative Industrial Area
Mathura Road, New Delhi 110 044
India

Sage Publications Asia-Pacific Pte. Ltd.
33 Pekin Street #02-01
Far East Square
Singapore 048763

Printed in the United States of America.

Library of Congress Cataloging-in-Publication Data

The kaleidoscope of gender : prisms, patterns, and possibilities/[edited by] Joan Z. Spade, Catherine G. Valentine.—2nd ed.
p. cm.
Includes bibliographical references and index.
ISBN-13: 978-1-4129-5146-3 (pbk.)
1. Sex role. 2. Sex differences (Psychology) 3. Gender identity. 4. Man-woman relationships. 5. Interpersonal relations. I. Spade, Joan Z. II. Valentine, Catherine G.

HQ1075.K35 2008
305.3—dc22 2007031676

This book is printed on acid-free paper.

08 09 10 11 10 9 8 7 6 5 4 3 2

Acquisitions Editor:	Benjamin Penner
Editorial Assistant:	Nancy Scrofano
Production Editor:	Karen Wiley
Copy Editor:	Gretchen Treadwell
Typesetter:	C&M Digitals (P) Ltd.
Proofreader:	Dorothy Hoffman
Indexer:	Sheila Bodell
Cover Designer:	Edgar Abarca
Marketing Manager:	Jennifer Reed

CONTENTS

Preface

The second edition *of The Kaleidoscope of Gender: Prisms, Patterns, and Possibilities* provides an overview of the cutting edge literature and theoretical frameworks in the sociology of gender and related fields for understanding the social construction of gender. Although not ignoring classical contributions to gender research, this book focuses on where the field is moving and the changing paradigms and approaches to gender studies. *The Kaleidoscope of Gender* uses the metaphor of a kaleidoscope and three themes—prisms, patterns, and possibilities—to unify topic areas. It focuses on the prisms through which gender is shaped, the patterns which gender takes, and the possibilities for social change through a deeper understanding of ourselves and our relationships with others, both locally and globally.

The book begins, in the first part, by looking at gender and other social prisms that define gendered experiences across the spectrum of daily lives. We conceptualize prisms as social categories of difference and inequality that shape the way gender is defined and practiced, including culture, race/ethnicity, social class, sexuality, age, and ability/disability. Different as individuals' experiences might be, there are patterns to gendered experiences. The second part of the book follows this premise, and examines these patterns across a multitude of arenas of daily life. From here, the last part of the book takes a proactive stance, exploring possibilities for change. Basic to the view of gender as a social construction is the potential for social change. Students will learn that gender transformation has occurred and can occur, and, consequently, that it is possible to alter the genderscape. Because prisms, patterns, and possibilities themselves intersect, the framework for this book is fluid, interweaving topics and emphasizing the complexity and ever-changing nature of gender.

We had multiple goals in mind as we first developed this book, and the second edition reaffirms these goals:

1. Creating a book of readings that is accessible, timely, and stimulating in a text whose structure and content incorporate a fluid framework with gender presented as an emergent, evolving, complex pattern, not one fixed in traditional categories and topics;
2. Selecting articles that creatively and clearly explicate what gender is and is not, and what it means to say that gender is socially constructed by incorporating provocative illustrations and solid scientific evidence of the malleability of gender and the role of individuals, groups, and social institutions in the daily performance and transformation of gender practices and patterns;
3. Including readings that untangle and clarify the intricate ways in which gender is embedded in and defined by the prisms of culture, race/ethnicity, class, sexuality, age, ability/disability, and cultural patterns of identities, groups, and institutions;

4. Integrating articles with a crosscultural focus to illustrate that gender is a continuum of categories, patterns, and expressions whose relevance is contextual and continuously shifting, and that gender inequality is not a universal and natural social pattern, but rather one of many systems of oppression, none of which can be generalized;

5. Assembling articles that offer students useful cognitive and emotional tools for making sense of the shifting and contradictory genderscape they inhabit, its personal relevance, its implications for relationships both locally and globally, and possibilities for change.

These goals shaped the revisions in the second edition of *The Kaleidoscope of Gender.* The twenty-seven new selections in this edition provide an even stronger emphasis on intersectional analysis and a deeper understanding of masculinities. Across the chapters, readings examine the individual, situational, and organizational bases for gendered patterns in relationships, behaviors, and beliefs. Additionally, many readings skillfully illuminate how other prisms of difference and inequality, such as race and social class, create an array of patterns of gender—distinct, but sometimes similar to the idealized patterns in a culture. Additional adjustments that heighten this new addition include a sharper focus on work in Chapter 7; new readings on domestic violence, including an article on rape on campuses in Chapter 8; and an increased emphasis on readings about social movements for equality in Chapter 10. Throughout the book, every chapter introduction has been updated with the most recent statistics, new concepts, and references. We have also updated "Topics for Further Examination" in each chapter.

As in the first edition, reading selections include theoretical, classical, and review articles; however, the emphasis continues to be on contemporary contributions to the field. We incorporate classical and theoretical arguments throughout the book to provide a broader framework with which to understand gender. Chapter introductions and the introduction to the book contextualize the literature in gender, introduce the readings in the chapters and illustrate how they relate to analyses of gender, and develop the kaleidoscope metaphor as a tool for viewing gender. Introductions and questions for consideration precede each reading to help students to focus on and grasp the key points of the selections. Additionally, each section ends with questions for students to consider and topics for students to explore.

It is possible to use this book alone, as a supplement to a text, or in combination with other articles or monographs. It is designed for undergraduate audiences and the readings are appropriate for a variety of courses focusing on the study of gender such as sociology of gender, gender and social change, and women's studies. The book may be used in departments of sociology, anthropology, psychology, and women's studies.

We would like to thank those reviewers whose valuable suggestions and comments helped us to develop the second edition: Heather Laube, University of Michigan, Flint; Laura Kramer, Montclair State University; Linda Grant, University of Georgia; Patti Giuffre, Texas State University, San Marcos; Todd Migliaccio, California State University, Sacramento; Wendy Simonds, Georgia State University; Kristen Myers, Northern Illinois University; Debbie Storrs, University of Idaho; Minjeong Kim, University at Albany, SUNY; and Elroi Waszkiewicz, Georgia State University.

Finally, we would like to thank generations of students in our sociology of gender courses for challenging us to think about new ways to teach our courses and making us aware of arenas of gender that are not typically the focus of gender studies books. Our special thanks go to the students in Kay's fall 2001 and 2002 Gender and Society classes and Joan's spring 2002 and 2003 Gender and Social Change classes for their insightful feedback on the material for the first edition. Joan would like to thank Gloria Condoluci and Sue Smithson, Gloria's successor as sociology department secretary at SUNY Brockport, for their assistance with the project and her many colleagues who gave suggestions and comments along the way, including Ronnie Steinberg and Michael Ames.

INTRODUCTION

This book is an invitation to you, the reader, to enter the fascinating and challenging world of gender studies. Gender is briefly defined as the meanings, practices, and relations of femininity and masculinity that people create as they go about their daily lives in different social settings. Although we discuss gender throughout this book, it is a very complex term to understand. While a more detailed discussion of what gender is and how it is related to biological maleness and femaleness is provided in Chapter 1, in thinking about the complexity of the meaning of gender from a sociological viewpoint, we find the metaphor of a kaleidoscope useful.

THE KALEIDOSCOPE OF GENDER

A real kaleidoscope is a tube containing an arrangement of mirrors or prisms that produces different images and patterns. When you look through the eyepiece of a kaleidoscope, light is typically reflected from the mirrors or prisms through object cells containing glass pieces, seashells, and the like to create ever-changing patterns of design and color (Baker, 1999). In this book, we use the kaleidoscope metaphor to help us grasp the complex and dynamic meaning and practice of gender as it interacts with other social prisms—such as race, ethnicity, age, sexuality, and social class—to create complex patterns of identities and relationships. Three themes then emerge from the metaphor of the kaleidoscope: prisms, patterns, and possibilities.

Part I of the book focuses on prisms. A prism in a kaleidoscope is an arrangement of mirrors that refracts or disperses light into a spectrum of patterns (Baker, 1999). We use the term *social prism* to refer to socially constructed categories of difference and inequality through which our lives are reflected or shaped into patterns of daily experiences. In addition to gender, when we discuss social prisms, we consider other socially constructed categories such as race, ethnicity, age, physical ability, social class, and sexuality. Culture is also conceptualized as a prism in this book, as we examine how gender is shaped across groups and societies. The concept of social prisms helps us to understand that gender is not a universal or static entity, but rather it is continuously created within the social parameters of individual and group life. Looking at the interactions of the prism of gender with other social prisms helps us to see the bigger picture—gender practices and meanings are a montage of intertwined social divisions and connections that both pull us apart and bring us together.

Part II of the book examines the patterns of gendered expressions and experiences created by the interaction of multiple prisms of difference and inequality. Patterns are regularized, prepackaged ways of thinking, feeling, and acting in society, and gendered patterns are present in almost all aspects of daily life. In the United States, examples of gendered patterns include the association of the color pink with girls and blue with boys, and the disproportionate numbers of female nurses and male engineers (see Table 7.1 in this book). However, these patterns of gender are experienced and expressed in different ways depending on the other social prisms that shape our identities and life chances. For example, as you take a closer look at who is an engineer and who is a nurse (as discussed in Chapter 7), you will note that it is predominately White men who are engineers and White women who are nurses. Consequently, the patterns of gender are a result of the complex interaction of multiple prisms.

Part III of the book concerns possibilities for gender change. Just as the wonder of the kaleidoscope lies in the ever-evolving patterns it creates, gendered patterns are always in flux. Each life and the world we live in can be understood as a kaleidoscope of unfolding growth and continual change (Baker, 1999). This dynamic aspect of the kaleidoscope metaphor represents the opportunity we have, individually and collectively, to transform gendered patterns that can be harmful to women and men. Although the theme of gender change is prominent throughout this book, it is addressed specifically in Chapter 10 and the Epilogue.

One caveat must be presented before we take you through the kaleidoscope of gender. A metaphor is a figure of speech in which a word ordinarily used to refer to one thing is applied to better understand another thing. A metaphor should not be taken literally. It does not directly represent reality. We use the metaphor of the kaleidoscope as an analytical tool to aid us in grasping the complexity, ambiguity, and fluidity of gender. However, unlike the prisms in a real kaleidoscope, the meaning and experience of social prisms (e.g., gender, race, ethnicity, social class, sexuality, and culture) are socially constructed and change in response to the patterns in the larger society. Thus, although the prisms of a real kaleidoscope are static, the prisms of the gender kaleidoscope are shaped by the patterns of society.

As you step into the world of gender studies, you'll need to develop a capacity to see what is hidden by the cultural blinders that we all wear at least some of the time. This capacity to see into the complexities of human relationships and group life has been called sociological imagination or, to be hip, "sociological radar." It is a capacity that is finely honed by practice and training both in and out of the classroom. A sociological perspective enables us to see through the cultural smokescreens that conceal the patterns, meanings, and dynamics of our relationships.

Gender Stereotypes

What will the sociological perspective help you to understand about gender that you don't already know? It will, for example, help you to debunk *gender stereotypes*, which are rigid, oversimplified, exaggerated beliefs about femininity and masculinity that misrepresent most women and men (Walters, 1999). To illustrate, let's analyze one gender stereotype that many people in American society believe—women talk more than men (Anderson & Leaper, 1998; Wood, 1999).

Social scientific research is helpful in documenting whether women actually talk more than men, or whether this belief is just another gender stereotype. To arrive at a conclusion, social scientists study the interactions of men and women in an array of settings and count how often men speak compared to women. They almost always find that, on average, men talk more in mixed-gender groups (Wood, 1999). Researchers also find that men interrupt more and tend to ignore topics brought up by women (Anderson & Leaper, 1998; Wood, 1999). In and of itself, these are important findings—the stereotype turns reality on its head.

So, why does the stereotype continue to exist? First, we might ask how it is that people believe something to be real, such as the stereotype that women talk more than men, when, in general, it isn't true? Part of the answer lies in the fact that culture, defined as the way of life of a group of people, shapes what we experience as reality (see Chapter 3 for a more detailed discussion). As Allan Johnson (1997) aptly puts it, "Living in a culture is somewhat like participating in the magician's magic because all the while we think we're paying attention to what's 'really' happening, alternative realities unfold without even occurring to us" (p. 55).

In other words, we don't usually reflect on our own culture; we are mystified by it without much awareness of its bewildering effect on us. The power of beliefs, including gender beliefs, is quite awesome. By providing stereotypes along which perceptions proceed, beliefs shape reality.

A second question we need to ask about gender stereotypes is: What is their purpose? For example, do they set men against women and contribute to the persistence of a system of inequality that disadvantages women? Certainly, the stereotype that many Americans hold of women as nonstop talkers is not a positive one. The stereotype does not assume that women are assertive, articulate, or captivating speakers. Instead it tends to depict women's talk as trivial gossip or irritating nagging. In other words, the stereotype devalues women's talk while, at the same time, elevating men's talk as thoughtful and worthy of our attention. One of the consequences of this stereotype is that both men and women take men's talk more seriously (Wood, 1999). This pattern is reflected in the fact that the voice of authority in many areas of American culture, such as television and politics, is almost always a male voice. The message that is communicated is clear—women are less important than men. In other words, gender stereotypes help to legitimize role and power differences between men and women.

However, stereotypical images of men and women are not universal in their application because they are complicated by the kaleidoscopic nature of people's lives. Prisms, or social categories, such as race/ethnicity, social class, and age intersect with gender to produce stereotypes that differ in symbolic meaning and functioning. For example, the prisms of gender, race, and age interact for African American and Hispanic men, who are stereotyped as dangerously "macho," which is not reflected in reality, as you will read about in Coltrane, Parke, and Adams's article in Chapter 8. These variations in gender stereotypes act as controlling images that maintain complex systems of domination and subordination in which some individuals and groups are dehumanized and disadvantaged in relationship to others (see Collins and other readings in Chapter 2).

The Evolution of the Concept of Gender

Just a few decades ago, social scientists described gender as two "sex" roles—male/masculine and female/feminine—that they believed were tied to innate personality characteristics and to biological sex characteristics such as hormones and reproductive functions (Kimmel, 2004; Tavris, 1992). For example, women were thought to be naturally more nurturing because of their capacity to bear children, and men were seen as natural leaders because they were unencumbered by pregnancy and childcare. This definition of masculine and feminine was one-dimensional, relatively static, and ethnocentric, and it is not supported by biological, psychological, sociological, or anthropological research.

The definition of gender expanded as social scientists conducted research that questioned the simplicity and accuracy of the "sex roles" perspective. First, social scientists debunked the notion that biological sex characteristics cause differences in men's and women's behaviors (Tavris, 1992). For example, testosterone does not cause aggression in men (see Sapolsky in Chapter 1), and the menstrual cycle does not cause women to be more "emotional" than men (Tavris, 1992).

Second, social scientific research demonstrated that men and women are far more physically, cognitively, and emotionally alike than different. Then and now, what we perceive as natural differences are actually rooted in the asymmetrical and unequal life experiences, resources, and power of women compared to men (Tavris, 1992). For example, it was, and for many still is, a commonly held belief that biological sex is related to physical ability and that women are athletically inferior to men. That belief has been challenged by the outcomes of a recent series of legal interventions that opened the doors to the world of competitive sport to girls and women. Once legislation such as Title IX was implemented in 1972, the expectation that women could not be athletes began to change as girls and young women received the same training and support for athletic pursuits as men. Not surprisingly, the gap in physical strength and skills between women and men decreased dramatically.

Today, women athletes regularly break records and perform physical feats thought impossible for women just a few decades ago.

Third, social scientists documented patterns of gender inequality within the economy, family, religion, and other social institutions that benefit men as a group. To illustrate, in the 1970s when researchers began studying gender inequality, they found that women made between 60 and 70 cents for every dollar men made. Things are not much better today. In 2005, the median salary for women was 76.7 percent of that of men (U.S. Census, 2005).

Fourth, social scientists documented the fact that gender is "humanly-made" (Schwalbe, 2001). The theory that argues that gender is a human invention is called social constructionism. Understanding gender as a social construction means seeing the social processes by which gender is defined into existence and maintained or changed by human actions and interactions. Social constructionism will be discussed in more detail later in this introduction and throughout the book.

One of the most important arguments in support of the idea that gender is socially constructed is derived from crosscultural studies. The variations and fluidity in the definitions and expressions of gender across cultures illustrate that the American gender system is not universal. For example, some cultures have more than two genders (see Nanda in Chapter 1). Other cultures define men and women as similar, not different (see Helliwell in Chapter 3). Still others view gender as flowing and changing across the life span (Herdt, 1997).

Fifth, as social scientists examine gender variations through the prism of culture, their research has challenged the notion that masculinity and femininity are defined and experienced in the same way by all people, even within the subcultures of a particular society. For example, the meaning of femininity in orthodox, American religious subcultures is not the same as femininity outside those communities (Rose, 2001). The differences are expressed in a variety of ways, including the clothing of women. Typically, orthodox religious women adhere to modesty rules in dress, covering their heads, arms, and legs.

Elaborating on the idea of multiple masculinities and femininities, the Australian sociologist, R. W. Connell, coined the terms *hegemonic masculinity* and *emphasized femininity* to understand the relations between and among masculinities and femininities in patriarchal societies; that is, societies that are dominated by men. According to Connell (1987), hegemonic masculinity is the idealized pattern of masculinity in patriarchal societies, while emphasized femininity is the vision of femininity that is held up as the model of womanhood in those societies. In Connell's definition, hegemonic masculinity is "the pattern of practice (i.e., things done, not just a set of role expectations or an identity) that allow men's dominance over women to continue" (2005, p. 832). Key features of hegemonic masculinity include the subordination of women, the exclusion and debasement of gay men, and the celebration of toughness and competitiveness (Connell, 2000). However, hegemony does not mean violence per se. It refers to "ascendancy achieved through culture, institutions, and persuasion" (Connell, 2005, p. 832). Emphasized femininity, in contrast, is about women's subordination, with its key features being sociability, compliance with men's sexual and ego desires, and acceptance of marriage and childcare (Connell, 1987). Both hegemonic masculinity and emphasized femininity patterns are "embedded in specific social environments" and are, therefore, dynamic as opposed to fixed (p. 846).

Interestingly, according to Connell, hegemonic masculinity and emphasized femininity are not necessarily the most common gender patterns. They are, however, the versions of manhood and womanhood against which other patterns of masculinity and femininity are measured and found wanting (Connell, 2005; Kimmel, 2004). For example, hegemonic masculinity produces marginalized masculinities, which are, according to Connell (2000), characteristic of exploited groups such as racial and ethnic minorities. These marginalized forms of masculinity may share features with hegemonic masculinity, such as "toughness," but are socially debased (see Coltrane, Parke, & Adams in Chapter 8).

In patriarchal societies, the culturally idealized form of femininity, emphasized femininity, is produced in relation to male dominance. Emphasized femininity insists on compliance, nurturance, and empathy as ideals of womanhood to which all women should subscribe (Connell, 1987). Connell does not use the term *hegemonic* to refer to emphasized femininity because, he argues, emphasized femininity is always subordinated to masculinity. The reading by Pyke and Johnson in Chapter 2 describes the lives of young, second-generation Asian women and their attempts to balance two cultural patterns of gender, in which "White" femininity, they argue, is hegemonic, or the dominant form of femininity. Connell, however, argues emphasized femininity is constructed in the context of patriarchy and therefore cannot be hegemonic. As examples, Connell (1987) points to the resistant and marginalized patterns of femininity in the experiences of lesbians, prostitutes, so-called spinsters, and women who do highly masculinized work such as manual labor. Connell's work is important in helping us to understand that multiple masculinities and femininities are associated with complex systems of domination and subordination across and within gender categories.

Sixth, a major factor associated with multiple masculinities and femininities is the relationship of gender to other social categories of difference and inequality. Allan Johnson (2001) points out that:

> [C]ategories that define privilege exist all at once and in relation to one another. People never see me solely in terms of my race, for example, or my gender. Like everyone else's, my place in the social world is a package deal—white, male, heterosexual, middle-aged, married, . . .—and that's the way it is all the time. . . . It makes no sense to talk about the effect of being in one of these categories—say, white—without also looking at the others and how they're related to it. (p. 53)

Seeing gender through multiple social prisms is critical, but it is not a simple task, as you will discover in the readings in Chapters 2 and 3. We need be aware of how other social prisms alter life experiences and chances. For example, although an upper-class, African American woman is privileged by her social class category, she will face obstacles related to her race and gender. Or, consider the situation of a middle-class White man who is gay; he might lose some of the privilege attached to his class and race because of his sexual orientation.

Think about what happens when we look at the difference in men and women's incomes if we include the prisms of race and ethnicity. As compared to the 76.7 cent difference in men and women's incomes mentioned earlier, White non-Hispanic women who graduated from college and worked full-time, year around in 2005 made 73.0 cents for every dollar their male counterparts made (using the median yearly income for both; see U.S. Census, 2005). The situation for African American women is slightly better at 85.9 cents for every dollar African American men made; however, this is because African American, college-educated men who work full-time, year around earn considerably less than their White counterparts (median income $34,433 compared to $44,850 for White men, whereas African American, college-educated women earn $29,588 compared to their White counterparts who earn $33,237; see U.S. Census, 2005). The ratio of female-to-male earnings for Hispanics is also better than that for non-Hispanic Whites at 89.3, but Hispanics earn less than either Whites or African Americans ($27,380 for men and $24,451 for women; see U.S. Census, 2005). We will explore many other examples of patterns of gender inequities that are a consequence of the interaction of multiple social prisms in this book. We also include theoretical frameworks to help you understand why these inequalities occur and what we need to do to change them.

Theories of Gender

Numerous theories explain how and why gender differentiation and inequality exist in most societies. We cannot describe them all, but we will briefly mention some of the theoretical approaches you will encounter in this book.

Theories help us better understand gender; however, we want to emphasize that theories are not reality. Theories are explanations that exist in a conceptual world and allow us to make predictions and generalizations about human behavior. As such, we encourage you to apply more than one theory to explain the same gender phenomenon to achieve a more complete understanding. As with the definition of gender, theories explaining why gender differences and inequalities exist in many societies have evolved over time, and continue to evolve. Many of the readings in this book expand upon or criticize the theories that follow. A primary reason for criticism is that many of these theories ignore the interaction among gender, race, ethnicity, class, and other social categories, a point we will return to later.

Individual

At the individual level, social psychologists and sociologists study the social categories and stereotypes that individuals use to identify themselves and label others. Gendered expectations, including stereotypes, are incorporated into how we define ourselves, and shape how we act and react as well as how others perceive us. Considerable research has studied how boys and girls (later researchers also included adults) are socialized into gender identities and practices (see Chapter 4 and the selection on Halloween costumes by Nelson in Chapter 5). While, ultimately, gendered identities are the result of larger societal and cultural forces, they are deeply felt, and individuals act upon them. Gender ideals and images are translated into self-identities (Howard & Alamilla, 2001). However, as noted earlier, gender socialization does not produce static outcomes for individuals or societies.

Interactional

Socialization does not happen in a vacuum, but occurs as part of social interaction. We do not become gendered alone. Theories such as symbolic interactionism and social constructionism help us to understand how developing a self-identity is a social process that involves incorporating the views of others into our own perspectives. According to West and Zimmerman (1987), gender is "done" in interaction with others in specific situations. They argue that the very process of interaction involves the presentation of a gendered self, the responses of the persons we are interacting with, and our reactions to their anticipated or actual responses. As such, gender is an ongoing activity that is carried out in interaction with other people, and people vary their gender presentations as they move from situation to situation. For example, an aggressive macho attitude might work on the football field, but not generally in an interview for a job. Lots of makeup, a strapless dress, and stiletto heels are okay at a formal dance, but not on a physician in the operating room of a hospital. The point is that sociologically speaking, people make gender happen through what they do and don't do in relation to others, and what they think is appropriate for different situations (see Lucal in Chapter 1).

Institutional

Some theorists use social constructionism to explain gender at the institutional level. Acker (1992) argues that institutions are gendered because "gender is present in the processes, practices, images and ideologies, and distributions of power in the various sectors of social life" (p. 567). Institutions maintain gender inequality through processes that exclude women (see Acker in Chapter 7). There are many examples of these processes, including outright exclusion as in the Roman Catholic priesthood, subtle exclusion through linguistic conventions that privilege masculinity (e.g., chairman, mankind), and the interactional exclusion of women, such as men dominating decision-making processes (West & Zimmerman, 1987).

Like Acker, Judith Lorber (1994, 2001) also combines a social constructionist perspective with the concept of social institutions. She argues that gender be viewed "as a society-wide institution that is built into all major organizations of society:" family, schools, mass media, and so on (2001, p. 180). By viewing gender as an institution, Lorber sees it as a basis for inequality in society because it is through gender that resources, power, and privilege are distributed. Gender, according to Lorber, is "a process of creating distinguishable social statuses for the assignment of rights and responsibilities" (1994, p. 32). As such, the structures of American society are built upon and reinforce the idea that women and men are distinct and unequal human types (Lorber, 1994). Without the socially constructed idea of clear differences between men and women, the entire social system of inequality and power would, according to Lorber, be in jeopardy.

Major Schools of Thought

Historically, conflict and functionalist theories explained gender at a macro-level of analysis, with these theories having gone through many transformations since first proposed around the turn of the twentieth century. Scholars at that time were trying to sort out massive changes in society resulting from the industrial and democratic revolutions. However, currently, feminist, postmodernist, and queer theories provide more nuanced explanations of gender. These more recent theories frame their understanding of gender in the lived experiences of individuals, rather than focusing on a macrolevel analysis of society wherein gender does not vary in form or function across groups or contexts.

Functionalism

Functionalism attempts to understand how all parts of a society (e.g., institutions such as family, education, economy, and the polity or state) fit together to form a smoothly running social system. According to this theoretical paradigm, parts of society tend to complement each other to create social stability (Durkheim, 1933). Translated into gender relationships, Talcott Parsons and Robert Bales (1955), writing after World War II, saw distinct and separate gender roles in the heterosexual nuclear family as a functional adaptation to a modern, complex society. Women were thought to be more "functional" if they were socialized and aspired to raise children. And men were thought to be more "functional" if they were socialized and aspired to support their children and wives. However, as Michael Kimmel (2004) notes, this "sex-based division of labor is functionally anachronistic," and if there ever was any biological basis for specific tasks being assigned to men or to women, it has been eroded (p. 55).

Conflict Theories

Karl Marx and later conflict theorists, however, did not see social systems as functional or benign. Instead, Marx and his colleague Friedrich Engels described industrial societies as systems of oppression in which one group, the dominant social class, uses its control of economic resources to oppress the working class. The economic resources of those in control are obtained through profits gained from exploiting the labor of subordinate groups. Marx and Engels predicted that the tension between the "haves" and the "have nots" would result in an underlying conflict between these two groups. Most early Marxist theories focused on social class oppression; however, Engels (1942, 1970) wrote an important essay on the oppression of women as the earliest example of oppression of one group by another.

Feminist Theories

Feminist theorists expanded upon the ideas of Marx and Engels, turning attention to the causes of women's oppression. One group, socialist feminists, continued to emphasize the role of capitalism in interaction with a patriarchal family structure as the basis for the exploitation of women. These theorists argue that economic and power benefits accrue to men who dominate women in capitalist societies. Another group, radical feminists, argues that patriarchy—the domination of men over women—is the fundamental form of oppression of women. Both socialist and radical feminists call for far-reaching changes in all institutional arrangements and cultural forms, including the dismantling of systems of oppression such as sexism, racism, and classism; replacing capitalism with socialism; developing more egalitarian family systems; and making other structural changes (e.g., Bart & Moran, 1993; Daly, 1978; Dworkin, 1987; MacKinnon, 1989).

Not all feminist theorists call for deep, structural, and cultural changes. Liberal feminists are inclined to work toward a more equitable form of democratic capitalism. They argue that policies such as Title IX and affirmative action laws opened up opportunities for women in education and increased the number of women professionals, such as physicians. These feminists strive to achieve gender equality by removing barriers to women's freedom of choice and equal participation in all realms of life, eradicating sexist stereotypes, and guaranteeing equal access and treatment for women in both public and private arenas (e.g., Reskin & Roos, 1990; Schwartz, 1994; Steinberg, 1982; Vannoy, Hiller, & Philliber, 1989; Weitzman, 1985).

Postmodern and Queer Theories

Postmodernism focuses on the way knowledge about gender is constructed, not on explaining gender relationships themselves. To postmodernists, knowledge is never absolute—it is always situated in a social reality that is specific to a historical time period. Postmodernism is based on the idea that it is impossible for anyone to see the world without presuppositions. From a postmodernist perspective, then, gender is socially constructed through discourses, which are the "series of stories" that we use to explain our world (Andersen, 2004). Postmodernists attempt to "deconstruct" the discourses or stories used to support a group's beliefs about gender (Andersen, 2004; Lorber, 2001). For example, Jane Flax argues that in order to fully understand gender in Western cultures, we must deconstruct the meanings in Western religious, scientific, and other discourses relative to "biology/sex/gender/nature" (in Lorber, 2001, p. 199). As you will come to understand from the readings in Chapters 1 and 3 (e.g., Nanda and Helliwell), the association between sex and gender in Western scientific (e.g., theories and texts) and nonscientific (e.g., film, newspapers, media) discourses is not shared in other cultural contexts. Thus, for postmodernists, gender is a product of the discourses within particular social contexts that define and explain gender.

Queer theories borrow from the original meaning of the word queer to refer to that which is "outside ordinary and narrow interpretations" (Plante, 2006, p. 62). Queer theorists are most concerned with understanding sexualities in terms of the idea that (sexual) identities are flexible, fluid, and changing, rather than fixed. In addition, queer theorists argue, identity and behavior must be separated. Thus, we cannot assume that people are what they do. From the vantage point of this theory, gender categories, much like sexual categories, are simplistic and problematic. Real people cannot be lumped together and understood in relationship to big cultural categories such as men and women, heterosexual and homosexual (Plante, 2006).

Intersectional or Prismatic Theories

A major shortcoming with many of the theoretical perspectives just described is their failure to recognize how gender interacts with other social categories or prisms of difference and inequality within societies, including race/ethnicity, social class, sexuality, age, and ability/disability (see Chapter 2). A growing number of social scientists are responding to the problem of incorporating multiple social categories in their research by developing a new form of analysis, called multiracial theory or intersectional analysis, which is referred to as prismatic analysis in this book. Chapter 2 explores these theories of how gender interacts with other prisms of difference and inequality to create complex patterns. Without an appreciation of the interactions of socially constructed categories of difference and inequality, or what we call prisms, we end up with not only an incomplete, but also an inaccurate explanation of gender.

As you read through the articles in this book, consider the basis for the authors' arguments. What observations, data, or works of other social science researchers do these authors use to support their claims? Use a critical eye to examine the evidence as you reconsider the assumptions about gender that guide your life.

The Kaleidoscope of Gender: Prisms, Patterns, and Possibilities

Before beginning the readings that take us through the kaleidoscope of gender, let us briefly review the three themes that shape the book's structure: prisms, patterns, and possibilities.

Part I: Prisms

Understanding the prisms that shape our experiences provides an essential basis for the book. Chapter 1 explores the meanings of the pivotal prism, gender, and its relationship to biological sex. Chapter 2 presents an array of prisms or socially constructed categories that interact with gender in many human societies, including race/ethnicity, social class, sexuality, age, and ability/disability. Chapter 3 focuses on the prism of culture/nationality, which alters the meaning and practice of gender in surprising ways.

Part II: Patterns

The prisms of the kaleidoscope create an array of patterned expressions and experiences of femininity and masculinity. Part II of this book examines some of these patterns. We look at how people learn, internalize, and "do" gender (Chapter 4); how gender is exploited by capitalism (Chapter 5); how gender acts upon bodies, sexualities, and emotions (Chapter 6); how gendered patterns are reproduced and modified in work (Chapter 7); how gender is created and transformed in our intimate relationships (Chapter 8); and how conformity to patterns of gender is enforced and maintained (Chapter 9).

Part III: Possibilities

In much the same way as the colors and patterns of kaleidoscopic images flow, gendered patterns and meanings are inherently changeable. Chapter 10 examines the shifting sands of the genderscape and reminds us of the many possibilities for change. Finally, in the Epilogue, we examine changes we have seen and encourage you to envision future changes.

We use the metaphor of the gender kaleidoscope to discover what is going on under the surface of a society whose way of life we don't often penetrate in a nondefensive, disciplined, and deep fashion. In doing so, we will expose a reality that is astonishing in its complexity, ambiguity, and fluidity. With the kaleidoscope, you never know what's coming next. Come along with us as we begin the adventure of looking through the kaleidoscope of gender.

References

Acker, J. (1992). From sex roles to gendered institutions. *Contemporary Sociology, 21* (5), 565–570.

Andersen, M. L. (2004). *Thinking about women: Sociological perspectives on sex and gender* (6th ed.). Boston: Allyn & Bacon.

Anderson, K. J., & Leaper, C. (1998). Meta-analysis of gender effects on conversational interruption: Who, what, when, where, and how. *Sex Roles, 39* (3–4), 225–252.

Baker, C. (1999). *Kaleidoscopes: Wonders of wonder.* Lafayette, CA: C & T Publishing.

Bart, P. B., & Moran, E. G. (Eds.). (1993). *Violence against women: The bloody footprints.* Newbury Park, CA: Sage.

Connell, R. W. (1987). *Gender and power: Society, the person and sexual politics.* Stanford, CA: University of Stanford Press.

Connell, R. W. (2000). *The men and the boys.* Berkeley: University of California Press.

Connell, R. W. (2005). Hegemonic masculinity: Rethinking the concept. *Gender & Society, 19*(6), 829–859.

Daly, M. (1978). *Gyn/ecology, the metaethics of radical feminism.* Boston: Beacon Press.

Durkheim, E. (1933). *The division of labor in society.* Glencoe, IL: Free Press.

Dworkin, A. (1987). *Intercourse.* New York: Free Press.

Engels, F. (1942, 1970). *Origin of the family, private property, and the state.* New York: International Publishers Company, Inc.

Herdt, G. (1997). *Same sex, different cultures.* Boulder, CO: Westview Press.

Hill Collins, P. (1990). *Black feminist thought: Knowledge, consciousness, and the politics of empowerment.* New York: Routledge.

Howard, J. A., & Alamilla, R. M. (2001). Gender and identity. In D. Vannoy (Ed.), *Gender mosaics: Social perspectives* (pp. 54–64). Los Angeles: Roxbury Publishing Company.

Johnson, A. (1997). *The gender knot.* Philadelphia: Temple University Press.

Johnson, A. (2001). *Privilege, power, and difference.* Mountain View, CA: Mayfield Publishing.

Kimmel, M. S. (2004) *The gendered society* (2nd ed.). New York: Oxford University Press.

Lorber, J. (1994). *Paradoxes of gender.* New Haven, CT: Yale University Press.

Lorber, J. (2001). *Gender inequality: Feminist theories and politics.* Los Angeles, CA: Roxbury Publishing Company.

MacKinnon, C. A. (1989). *Toward a feminist theory of the state.* Cambridge, MA: Harvard University Press.

Parsons, T., & Bales, R. F. (1955). *Family, socialization, and interaction process.* Glencoe, IL: Free Press.

Plante, R. F. (2006). *Sexualities in context: A social perspective.* Cambridge, MA: Westview.

Reskin, B. F., & Roos, P. A. (1990). *Job queues, gender queues: Explaining women's inroads into male occupations.* Philadelphia: Temple University Press.

Rose, D. R. (2001). Gender and Judaism. In D. Vannoy (Ed.), *Gender mosaics: Social perspectives* (pp. 415–424). Los Angeles: Roxbury Publishing.

Schwalbe, M. (2001). *The sociologically examined life: Pieces of the conversation* (2nd ed.). Mountain View, CA: Mayfield Publishing.

Schwartz, P. (1994). *Love between equals: How peer marriage really works.* New York: Free Press.

Steinberg, R. J. (1982). *Wages and hours: Labor and reform in twentieth century America.* New Brunswick, NJ: Rutgers University Press.

Tavris, C. (1992). *The mismeasure of woman.* New York: Simon & Schuster.

U.S. Census Bureau. (2005). *Income, Earnings, and Poverty Data from the 2005 American Community Survey.* Retrieved March 22, 2007 from http://www.census.gov/prod/2006pubs/acs-02.pdf

Vannoy Hiller, D., & Philliber, W. W. (1989). *Equal partners: Successful women in marriage.* Newbury Park, CA: Sage.

Walters, S. D. (1999). Sex, text, and context: (In)between feminism and cultural studies. In M. M. Ferree, J. Lorber, & B. B. Hess (Eds.), *Revisioning gender* (pp. 193–257). Thousand Oaks, CA: Sage.

Weitzman, L. J. (1985). *The divorce revolution: The unexpected social and economic consequences for women and children in America.* New York: Free Press.

West, C., & Zimmerman, D. H. (1987). Doing gender. *Gender & Society, 1* (2), 125–151.

Wood, J. T. (1999). *Gendered lives: Communication, gender, and culture* (3rd ed.). Belmont, CA: Wadsworth.

PART I

PRISMS

1

The Prism of Gender

In the metaphorical kaleidoscope of this book, gender is the pivotal prism. It is central to the intricate patterning of social life, and encompasses power relations, the division of labor, symbolic forms, and emotional relations (Connell, 2000). The shape and texture of people's lives are affected in profound ways by the prism of gender as it operates in their social worlds. Indeed, our ways of thinking about and experiencing gender, and the related category of sex, originate in our society.

As we noted in the Introduction, gender is very complex. In part, the complexity of the prism of gender in North American culture derives from the fact that it is characterized by a marked contradiction between people's beliefs about gender and real behavior. Our real behavior is far more flexible, adaptable, and malleable than our beliefs would have it. To put it another way, contrary to the stereotypes of masculinity and femininity, there are no gender certainties or absolutes. Real people behave in feminine, masculine, and nongendered ways as they respond to situational demands and contingencies (Glick & Fiske, 1999; Tavris, 1992).

Two questions are addressed in this chapter to help us think more clearly about the complexity of gender: (1) how does Western culture condition us to think about gender, especially in relation to sex? and (2) how does social scientific research challenge Western beliefs about gender and sex?

Western Beliefs About Gender and Sex

Most people in Western cultures grow up learning that there are two and only two sexes, male and female, and two and only two genders, feminine and masculine (Bem, 1993; Wharton, 2005). We are taught that a real woman is feminine, a real man is masculine, and that any deviation or variation is strange or unnatural. Most people also learn that femininity and masculinity flow from biological sex characteristics (e.g., hormones, secondary sex characteristics, external and internal genitalia). We are taught that testosterone, a beard, big muscles, and a penis make a man, while estrogen, breasts, hairless legs, and a vagina make a woman. Many of us never question what we have learned about sex and gender, so we go through life assuming that gender is a relatively simple matter: a person who wears lipstick, high heel shoes, and a skirt is a feminine female, while a person who plays rugby, belches in public, and walks with a swagger is a masculine male (Lorber, 1994; Ridgeway & Correll, 2004).

The readings we have selected for this chapter reflect a growing body of social scientific research that challenges and alters the Western model of sex and gender. Overall, the readings are critical of the American tendency to explain virtually every human behavior in individual and biological terms. As Jodi O'Brien (1999) points out, Americans tend to assume that answers to the complex workings of social relationships can be found in the bodies and psyches of individuals, rather than in culture or in the interaction between bodies and the environment.

Americans habitually overemphasize biology and underestimate the power of social facts to explain sex and gender (O'Brien, 1999). For instance, Americans tend to equate aggression with biological maleness and vulnerability with femaleness; natural facility in physics with masculinity and natural facility in childcare with femininity; lace and ribbons with girlness and rough and tumble play with boyness (Glick & Fiske, 1999; Ridgeway & Correll, 2004). The notions of natural sex and gender difference, duality, and even opposition and inequality permeate our thinking, color our labeling of people and things in our environment, and affect our practical actions (Bem, 1993; Wharton 2005).

We refer to the American tendency to assume that biological sex differences cause gender differences as "the pink and blue syndrome." This syndrome is deeply lodged in our minds and feelings, and reinforced through everyday talk, performance, and experience. It's everywhere. Any place, object, discourse, or practice can be gendered. Children's birthday cards come in pink and blue. Authors of popular books assert that men and women are from different planets. People love PMS and alpha male jokes. In "The Pink Dragon Is Female" (see Chapter 5), Adie Nelson's research reveals that even children's fantasy costumes tend to be gendered as masculine and feminine. The "pink and blue syndrome" is so embedded within our culture, and consequently within individual patterns of thinking and feeling, that most of us cannot remember when we learned gender stereotypes and expectations or came to think about sex and gender as natural, immutable, and fixed. It all seems so simple and natural. Or is it?

What is gender? What is sex? How are gender and sex related? Why do most people in our society believe in the "pink and blue syndrome"? Why do so many of us attribute one set of talents, temperaments, skills, and behaviors to women and another set to men? These are the kinds of questions social scientists in sociology, anthropology, psychology, and other disciplines have been asking and researching for almost fifty years. Thanks to decades of good work by an array of scientists, we now understand that gender and sex are not so simple. Social scientists have discovered that the gender landscape is complicated, shifting, and contradictory. Among the beliefs that have been called into question by research are:

- the notion that there are two and only two sexes and, consequently, two and only two genders
- the assumption that men and women are the same everywhere and all the time
- the belief that biological factors cause the "pink and blue syndrome"

Using Our Sociological Radar

Before we look at how social scientists answer questions such as "What is gender," let's do a little research of our own. Try the following: relax, turn on your sociological radar, and examine yourself and the people you know carefully. Do all the men you know fit the ideal of masculinity all the time, in all relationships, and in all situations? Do all the women in your life consistently behave in stereotypical feminine fashion? Do you always fit into one as opposed to the other culturally approved gender box? Or are most of the people you know capable of "doing" both masculinity and femininity, depending on the interactional context? Our guess is that none of the people we know are aggressive all the time, nurturing all the time, sweet and submissive all the time, or strong and silent all the time. Thankfully, we are complex and creative. We stretch and grow and develop as we meet the challenges, constraints, and opportunities of different and new situations and life circumstances. Men can do mothering; women can "take care of business." Real people are not stereotypes.

Yet even in the face of real gender fluidity and complexity, the belief in gender dichotomy and opposition continues to dominate almost every aspect of the social worlds we inhabit. For example, recent research shows that even though men's and women's roles have changed and blended, the tendency of Americans to categorize and stereotype people based on the simple male/female dichotomy persists (Glick & Fiske, 1999). As Glick and Fiske (1999) put it, "we typically categorize people by sex effortlessly, even nonconsciously, with diverse and profound effects on social interactions" (p. 368). To reiterate, many Americans perceive humankind as divided into mutually exclusive, nonoverlapping groups: males/masculine/men and females/feminine/women (Bem, 1993; Wharton, 2005). The culturally created image of gender, then, is nonkaleidoscopic: no spontaneity, no ambiguity, no complexity, no diversity, no surprises, no elasticity, and no unfolding growth.

Social Scientific Understandings of Sex and Gender

Modern social science offers us a very different image of gender. It opens the door to the richness and diversity of human experience, and it resists the tendency to reduce human behavior to single factors. Research shows that the behavior of real women and men depends on time and place, and context and situation, not on fixed gender differences (Lorber, 1994; Tavris, 1992). For example, just a few decades ago in the United States, cheerleading was a men's sport because it was considered too rigorous for women (Dowling, 2000), women were thought to lack the cognitive and emotional "stuff " to pilot flights into space, and medicine and law were viewed as too intellectually demanding for women. As Carol Tavris (1992) says, research demonstrates that perceived gender differences turn out to be a matter of "now you see them, now you don't" (p. 288).

If we expand our sociological examination of gender to include cultures outside North America, the real-life fluidity of gender comes fully alive. (See Chapter 3 for detailed discussion.) In some cultures (for example, the Aka hunter-gatherers) fathers as well as mothers suckle infants (Hewlett, 2001). In other cultures, such as the Agta Negritos hunter-gatherers, women as well as men are the hunters (Estioko-Griffin & Griffin, 2001). As Serena Nanda notes in her reading in this chapter, extraordinary gender diversity was expressed in complex sex/gender systems in many precontact Native American societies.

Context, which includes everything in the environment of a person's life such as work, family, social class, race—and more—is the real source of gender definitions and practices. Gender is flexible, and "in its elasticity it stretches and unfolds in manifold ways" so that depending upon its contexts, including the life progress of individuals, we see it and experience it differently" (Sorenson, 2000, p. 203). Most of us "do" both masculinity and femininity, and what we do is situationally dependent and institutionally constrained.

Let's use sociological radar again and call upon the work of social scientists to help us think more precisely and "objectively" about what gender and sex are. It has become somewhat commonplace to distinguish between gender and sex by viewing sex as a biological fact, meaning that it is noncultural, static, scientifically measurable and unproblematic, while we see gender as a cultural attribute, a means by which people are taught who they are, how to behave, and what their roles will be (Sorenson, 2000). However, this mode of distinguishing between sex and gender has come under criticism, largely because new studies have begun to reveal the cultural dimensions of sex itself. That is, the physical characteristics of sex cannot be separated from the cultural milieu in which they are labeled and given meaning. For example, Robert Sapolsky's chapter reading debunks the widely held myth that testosterone causes males to be more aggressive and domineering than females. He ends his article by stating firmly that "our behavioral biology is usually meaningless outside the context of the social factors and environment in which it occurs." In other words, the relationship between biology and behavior is reciprocal, both inseparable and intertwined (Yoder, 2003).

Sex, as it turns out, is not a clear-cut matter of DNA, chromosomes, external genitalia, and the like, factors which produce two and only two sexes—females and males. First, there is considerable biological variation. Sex is not fixed in two categories. There is overlap. For example, all humans have estrogen, prolactin, and testosterone, but in varying and changing levels (Abrams, 2002). Think about this. In our society, people tend to associate breasts and related phenomena, such as breast cancer and lactation, with women. However, men have breasts. Indeed, some men have bigger breasts than some women, some men lactate, and some men get breast cancer. Also, in our society, people associate facial hair with men. What's the real story? All women have facial hair and some have more of it than some men. Indeed, recent hormonal and genetic studies (e.g., Abrams, 2002; Beale, 2001) are revealing that, biologically, women and men are far more similar than distinct.

In fact, variations in and complexities of sex development produce *intersexed* people whose bodies do not fit the two traditionally understood sex categories (Fausto-Sterling, 2000; Fujimora, 2006). Fujimora (2006) examined recent research on sex genes and concluded that "there is no single pathway through which sex is genetically determined" and that we might consider sex variations, such as intersex, as resulting from "multiple developmental pathways that involve genetic, protein, hormonal, environmental, and other agents, actions, and interactions" (p. 71). Lorber and Moore (2007) argue that intersexed people are akin to multiracial people. They point out that just as scientists have demonstrated through DNA testing that almost all of us are genetically interracial, similarly, "if many people were genetically sex-typed, we'd also find a variety of chromosomal, hormonal, and anatomical patterns unrecognized" in our rigid, two sex system (p. 138). Sharon E. Preves's reading in this chapter offers exciting insights into the meanings and consequences of contemporary and historical responses to individuals who are intersexed.

Biology is complicated business, and that should come as no surprise. The more we learn about biology, the more elusive and complex sex becomes. What seemed so obvious—two, opposite sexes—turns out to be an oversimplification. Humans are not unambiguous, clearly demarcated, biologically distinct, nonoverlapping, invariant groups of males and females.

So, again, what is gender? First, gender is not sex. Biological sex characteristics do not cause specific gender behaviors or activities. As discussed earlier, biological sex is virtually meaningless outside the social context in which it develops and is expressed (Yoder, 2003). Second, gender is not an essential identity. It "does not have a locus nor does it take a particular form" (Sorenson, 2000, p. 202). In other words, individuals do not possess a clearly defined gender that is the same everywhere and all the time. At this point, you may be thinking, what in the world are these authors saying? We are saying that gender is a human invention, a means by which people are sorted (in our society, into two genders), a basic aspect of how our society organizes itself and allocates resources (e.g., certain tasks assigned to people called women and other tasks to those termed men), and a fundamental ingredient in how individuals understand themselves and others ("I feel feminine." "He's manly." "You're androgynous.").

One of the fascinating aspects of gender is the extent to which it is negotiable and dynamic. In effect, masculinity and femininity exist because people believe that women and men are distinct groups and, most important, because people "do gender," day in and day out. The chapter reading by Betsy Lucal illustrates vividly how gender is a matter of attribution and enactment. Some social scientists call gender a performance, while others term it a masquerade. The terms "performance" and "masquerade" emphasize that it is through the ways in which we present ourselves in our daily encounters with others that gender is created and recreated.

We even do gender by ourselves and sometimes quite self-consciously. Have you ever tried to make yourself look and act more masculine or feminine? What is involved in "putting on" femininity or masculinity? Consider *transvestism* or cross-gender dressing. "Cross-dressers know that successfully being a man or a woman simply means convincing others that you are what you appear to be" (Kimmel, 2000, p. 104). Think about the emerging communities of *transgender* people who are "challenging, questioning, or changing gender from that assigned at birth to a chosen gender" (Lorber & Moore, 2007, p. 139). Although most people have deeply learned gender and view the gender box they inhabit as natural or normal, intersex and transgender activists attack the boundaries of "normal" by refusing to choose a traditional sex, gender, or sexual identity (Lorber & Moore, 2007). In so doing, cultural definitions of sex and gender are destabilized and expanded.

You may be wondering why we have not used the term *role*, as in *gender role*, to describe "doing gender." The problem with the concept of roles is that typical social roles, such as those of teacher, student, doctor, or nurse, involve situated positions and identities. However, gender, like race, is a status and identity that cuts across many situations and institutional arenas. In other words, gender does not "appear and disappear from one situation to another" (Van Ausdale & Feagin, 2001, p. 32). People are always doing gender. They rarely let their guard down. In part, this is a consequence of the pressures that other people exert on us to "do gender" no matter the situation in which we find ourselves. Even if an individual would like to "give up gender," others will define and interact with that individual in gendered terms. If you were a physician, you could "leave your professional role behind you" when you left the hospital or office and went shopping or vacationing. Gender is a different story. Could you leave gender at the office and go shopping or vacationing? What would that look like, and what would it take to make it happen?

So far, we have explored gender as a product of our interactions with others. It is something we do, not something we inherit. Gender is also built into the larger world we inhabit, including its institutions, images and symbols, organizations, and material objects. For example, jobs, wages, and hierarchies of dominance and subordination in workplaces are gendered. Even after decades of substantial increase in women's workforce participation, occupations continue to be allocated by gender (e.g., secretaries are overwhelmingly women; men dominate construction work) and a wage gap between men and women persists (Bose & Whaley, 2001; Steinberg, 2001; Introduction to this book; Introduction to Chapter 7). In addition, men are still more likely to be bosses and women to be bossed. The symbols and images with which we are surrounded and by which we communicate are another part of our society's gender story. Our language speaks of difference and opposition in phrases such as "the opposite sex" and the absence of any words, except awkward medical terms (e.g., transsexual) or epithets (e.g., pervert), to refer to sex and gender variants. In addition, the swirl of gendered images in the media is almost overwhelming. Blatant gender stereotypes still dominate TV, film, magazines, and billboards (Lont, 2001). Gender is also articulated, reinforced, and transformed through material objects and locales (Sorenson, 2000). Shoes are gendered, body adornments are gendered, public restrooms are gendered, weapons are gendered, ships are gendered, wrapping paper is gendered, and deodorants are gendered. The list is endless. The point is that these locales and objects are transformed into a medium for gender to operate within (Sorenson, 2000). They make gender seem "real," and they give it material consequences (Sorenson, 2000, p. 82).

In short, social scientific research underscores the complexity of the prism of gender and demonstrates how gender is constructed at multiple, interacting levels of society. The first reading by Risman is a detailed examination of the ways in which our gender structure is embedded in the individual, interactional, and institutional dimensions of our society, emphasizing that gender cannot be reduced to one level or dimension: individual, interactional, or institutional. We are literally and figuratively immersed in a gendered world—a world in which difference, opposition, and inequality are the culturally defined themes. And yet, that world is kaleidoscopic in nature. The lesson of the kaleidoscope is that "nothing in life is immune to change" (Baker, 1999, p. 29). Reality is in flux; you never know what's coming next. The metaphor of the kaleidoscope reminds us to keep seeking the shifting meanings as well as the recurring patterns of gender (Baker, 1999).

We live in an interesting time of kaleidoscopic change. Old patterns of gender difference and inequality keep reappearing, often in new guises, while new patterns of convergence, equality, and self-realization have emerged. Social science research is vital in helping us to stay focused on understanding the prism of gender as changeable, and responding to its context—as a social dialogue about societal membership and conventions—and "as the outcome of how individuals are made to understand their differences and similarities" (Sorenson, 2000, p. 203–204). With that focus in mind, we can more clearly and critically explore our gendered society.

References

Abrams, D. C. (2002). Father nature: The making of a modern dad. *Psychology Today,* March/April, 38–42.

Baker, C. (1999). *Kaleidoscopes: Wonders of wonder.* Lafayette, CA: C&T Publishing.

Beale, B. (2001). The sexes: New insights into the X and Y chromosomes. *The Scientist, 15* (15), 18. Retrieved July 23, 2001 from http://www.the-scientist.com/yr2001/jul/research1_010723.html

Bem, S. L. (1993). *The lenses of gender.* New Haven, CT: Yale University Press.

Bose, C. E., & Whaley, R. B. (2001). Sex segregation in the U.S. labor force. In D. Vannoy (Ed.), *Gender mosaics* (pp. 228–239). Los Angeles: Roxbury Publishing.

Connell, R. W. (2000). *The men and the boys.* Berkeley: University of California Press.

Dowling, C. (2000). *The frailty myth.* New York: Random House.

Epstein, C. F. (1988). *Deceptive distinctions.* New Haven, CT: Yale University Press.

Estioko-Griffin, A., & Griffin, P. B. (2001). Woman the hunter: The Agta. In C. Brettell & C. Sargent (Eds.), *Gender in cross-cultural perspective* (3rd ed., pp. 238–239). Upper Saddle River, NJ: Prentice Hall.

Fausto-Sterling, A. (2000). *Sexing the body: Gender politics and the construction of sexuality.* New York: Basic Books.

Fujimora, J. H. (2006). Sex genes: A critical sociomaterial approach to the politics and molecular genetics of sex determination. *Signs, 32*(1), 49–81.

Glick, P., & Fiske, S. T. (1999). Gender, power dynamics, and social interaction. In M. Ferree, J. Lorber, & B. Hess (Eds.), *Revisioning gender* (pp. 365–398). Thousand Oaks, CA: Sage.

Hewlett, B. S. (2001). The cultural nexus of Aka father-infant bonding. In C. Brettell & C. Sargent (Eds.), *Gender in cross-cultural perspective* (3rd ed., pp. 45–46). Upper Saddle River, NJ: Prentice Hall.

Kimmel, M. S. (2000). *The gendered society.* New York: Oxford University Press.

Lont, C. M. (2001). The influence of the media on gender images. In D. Vannoy (Ed.), *Gender mosaics.* Los Angeles, CA: Roxbury Publishing.

Lorber, J, (1994). *Paradoxes of gender.* New Haven, CT: Yale University Press.

Lorber, J., & Moore, L. J. (2007). *Gendered bodies.* Los Angeles: Roxbury Publishing.

O'Brien, J. (1999). *Social prisms: Reflections on the everyday myths and paradoxes.* Thousand Oaks, CA: Pine Forge Press.

Ridgeway, C. L., & Correll, S. J. (2004). Unpacking the gender system: A theoretical perspective on gender beliefs and social relations. *Gender & Society, 18*(4), 510–531.

Sorenson, M. L. S. (2000). *Gender archaeology.* Cambridge, England: Polity Press.

Steinberg, R. J. (2001). How sex gets into your paycheck and how to get it out: The gender gap in pay and comparable worth. In D. Vannoy (Ed.), *Gender mosaics.* Los Angeles: Roxbury Publishing.

Tavris, C. (1992). *The mismeasure of woman.* New York: Simon & Schuster.

Wharton, A. S. (2005). *The Sociology of Gender.* Malden, MA: Blackwell.

Van Ausdale, D., & Feagin, J. R. (2001). *The first r: How children learn race and racism.* Lanham, MD: Rowman & Littlefield Publishers.

Yoder, J. D. (2003). *Women and gender: Transforming psychology.* Upper Saddle River, NJ: Prentice Hall.

Introduction to Reading 1

Barbara Risman is a sociologist who has made significant contributions to research and writing on gender in heterosexual American families. In this article she argues that we need to conceptualize gender as a social structure so we can better analyze the ways in which gender is embedded in the individual, interactional, and institutional dimensions of social life. Her article also addresses the value of intersectional analysis while pointing to the importance of examining different structures of inequality, such as race and gender, for potentially different mechanisms that produce those inequalities. Finally, Risman addresses how her theory of gender as social structure might help us to understand how inequality can be transformed to create a more just world.

1. Why does Risman include the individual dimension of social life in her theory of gender as a social structure?
2. How does her work on men and mothering illustrate the value of applying gender as social structure at multiple levels of analysis?
3. Why does Risman caution against research on gender that operates only within an intersectional framework?

Gender as a Social Structure

Theory Wrestling With Activism

Barbara J. Risman

In this article, I briefly summarize my . . . argument that gender should be conceptualized as a social structure (Risman 1998) and extend it with an attempt to classify the mechanisms that help produce gendered outcomes within each dimension of the social structure.

From Risman, B., 2004. "Gender as a social structure," *Gender & Society 18*(4), p. 429. Reprinted with permission of Sage Publications, Inc.

I then provide evidence from my own and others' research to support the usefulness of this theoretical schema. Finally, using gender structure as a starting point, I engage in conversation with ideas currently emerging about intersectionality and wrestle with how we might use theory in the service of social change.

Gender as Social Structure

With this theory of *gender as a social structure,* I offer a conceptual framework, a scheme to organize the confusing, almost limitless, ways in which gender has come to be defined in contemporary social science. Four distinct social scientific theoretical traditions have developed to explain gender. The first tradition focuses on how individual sex differences originate, whether biological (Udry 2000) or social in origin (Bem 1993). The second tradition . . .emerged as a reaction to the first and focuses on how the social structure (as opposed to biology or individual learning) creates gendered behavior. The third tradition, also a reaction to the individualist thinking of the first, emphasizes social interaction and accountability to others' expectations, with a focus on how "doing gender" creates and reproduces inequality (West and Zimmerman 1987). The sex-differences literature, the doing gender interactional analyses, and the structural perspectives have been portrayed as incompatible in my own early writings as well as in that of others (Epstein 1988; Kanter 1977; Ferree 1990; Risman 1987; Risman and Schwartz 1989). England and Browne (1992) argued persuasively that this incompatibility is an illusion: All structural theories must make assumptions about individuals, and individualist theories must make presumptions about external social control. While we do gender in every social interaction, it seems naive to ignore the gendered selves and cognitive schemas that children develop as they become cultural natives in a patriarchal world (Bem 1993). The more recent integrative approaches (Connell 2002; Lorber 1994; Ferree, Lorber, and Hess 1999; Risman 1998) treat gender as a socially constructed stratification system. This article fits squarely in the current integrative tradition.

Lorber (1994) argued that gender is an institution that is embedded in all the social processes of everyday life and social organizations. She further argued that gender difference is primarily a means to justify sexual stratification. Gender is so endemic because unless we see difference, we cannot justify inequality. I share this presumption that the creation of difference is the very foundation on which inequality rests.

I build on this notion of gender as an institution but find the institutional language distracting. The word "institution" is too commonly used to refer to particular aspects of society, for example, the family as an institution or corporations as institutions. My notion of gender structure meets the criteria offered by Martin (forthcoming). . . . While the language we use may differ, our goals are complementary, as we seek to situate gender as embedded not only in individuals but throughout social life (Patricia Martin, personal communication).

I prefer to define gender as a social structure because this brings gender to the same analytic plane as politics and economics, where the focus has long been on political and economic structures. While the language of structure suits my purposes, it is not ideal because despite ubiquitous usage in sociological discourse, no definition of the term "structure" is widely shared. Smelser (1988) suggested that all structuralists share the presumption that social structures exist outside individual desires or motives and that social structures at least partially explain human action. Beyond that, consensus dissipates. Blau (1977) focused solely on the constraint collective life imposes on the individual. Structure must be conceptualized, in his view, as a force opposing individual motivation. Structural concepts must be observable, external to the individual, and independent of individual motivation. This definition of "structure" imposes a clear dualism between structure and action, with structure as constraint and action as choice.

Constraint is, of course, an important function of structure, but to focus only on structure as constraint minimizes its importance. Not only are women and men coerced into differential social roles; they often choose their gendered paths. A social structural analysis must help us understand how and why actors choose one alternative over another. A structural theory of action (e.g., Burt 1982) suggests that actors compare themselves and their options to those in structurally similar positions. From this viewpoint, actors are purposive, rationally seeking to maximize their self-perceived well-being under social-structural constraints. As Burt (1982) suggested, one can assume that actors choose the best alternatives without presuming they have either enough information to do it well or the options available to make choices that effectively serve their own interests. For example, married women may choose to do considerably more than their equitable share of child care rather than have their children do without whatever "good enough" parenting means to them if they see no likely alternative that the children's father will pick up the slack.

While actions are a function of interests, the ability to choose is patterned by the social structure. Burt (1982) suggested that norms develop when actors occupy similar network positions in the social structure and evaluate their own options vis-à-vis the alternatives of similarly situated others. From such comparisons, both norms and feelings of relative deprivation or advantage evolve. The social structure as the context of daily life creates action indirectly by shaping actors' perceptions of their interests and directly by constraining choice. Notice the phrase "similarly situated others" above. As long as women and men see themselves as different kinds of people, then women will be unlikely to compare their life options to those of men. Therein lies the power of gender. In a world where sexual anatomy is used to dichotomize human beings into types, the differentiation itself diffuses both claims to and expectations for gender equality. The social structure is not experienced as oppressive if men and women do not see themselves as similarly situated.

While structural perspectives have been applied to gender in the past (Epstein 1988; Kanter 1977), there has been a fundamental flaw in these applications. Generic structural theories applied to gender presume that if women and men were to experience identical structural conditions and role expectations, empirically observable gender differences would disappear. But this ignores not only internalized gender at the individual level . . . but the cultural interactional expectations that remain attached to women and men because of their gender category. A structural perspective on gender is accurate only if we realize that gender itself is a structure deeply embedded in society.

Giddens's (1984) structuration theory adds considerably more depth to this analysis of gender as a social structure with his emphasis on the recursive relationship between social structure and individuals. That is, social structures shape individuals, but simultaneously, individuals shape the social structure. Giddens embraced the transformative power of human action. He insisted that any structural theory must be concerned with reflexivity and actors' interpretations of their own lives. Social structures not only act on people; people act on social structures. Indeed, social structures are created not by mysterious forces but by human action. When people act on structure, they do so for their own reasons. We must, therefore, be concerned with why actors choose their acts. Giddens insisted that concern with meaning must go beyond the verbal justification easily available from actors because so much of social life is routine and so taken for granted that actors will not articulate, or even consider, why they act.

This nonreflexive habituated action is what I refer to as the cultural component of the social structure: The taken for granted or cognitive image rules that belong to the situational context (not only or necessarily to the actor's personality). The cultural component of the social structure includes the interactional expectations that each of us meet in every social encounter. My aims are to bring women and men back into a structural theory where gender is the structure under analysis and to identify when behavior is habit (an enactment of taken for granted gendered cultural norms) and when we do gender consciously, with intent, rebellion, or even with irony. When are we doing gender and re-creating inequality without intent? And what happens to interactional dynamics and male-dominated institutions when we rebel? Can we refuse to do gender or is rebellion simply doing gender differently, forging alternative masculinities and femininities?

Connell (1987) applied Giddens's (1984) concern with social structure as both constraint and created by action in his treatise on gender and power (see particularly chapter 5). In his analysis, structure constrains action, yet "since human action involves free invention . . . and is reflexive, practice can be turned against what constrains it; so structure can deliberately be the object of practice" (Connell 1987, 95). Action may turn against structure but can never escape it.

A theory of gender as a social structure must integrate this notion of causality as recursive with attention to gender consequences at multiple levels of analysis. Gender is deeply embedded as a basis for stratification not just in our personalities, our cultural rules, or institutions but in all these, and in complicated ways. The gender structure differentiates opportunities and constraints based on sex category and thus has consequences on three dimensions: (1) At the individual level, for the development of gendered selves; (2) during interaction as men and women face different cultural expectations even when they fill the identical structural positions; and (3) in institutional domains where explicit regulations regarding resource distribution and material goods are gender specific.

Advantages to Gender Structure Theory

This schema advances our understanding of gender in several ways. First, this theoretical model imposes some order on the encyclopedic research findings that have developed to explain gender inequality. Thinking of each research question as one piece of a jigsaw puzzle, being able to identify how one set of findings coordinates with others even when the dependent variables or contexts of interest are distinct, furthers our ability to build a cumulative science. Gender as a social structure is enormously complex. Full attention to the web of interconnection between gendered selves, the cultural expectations that help explain interactional patterns, and institutional regulations allows each research tradition to explore the growth of their own trees while remaining cognizant of the forest.

A second contribution of this approach is that it leaves behind the modernist warfare version of science, wherein theories are pitted against one another, with a winner and a loser in every contest. In the past, much energy . . . was devoted to testing which theory best explained gender inequality and by implication to discounting every alternative possibility.[1] Theory building that depends on theory slaying presumes parsimony is always desirable, as if this complicated world of ours were best described with simplistic monocausal explanations. While parsimony and theory testing were the model for the twentieth-century science, a more postmodern science should attempt to find complicated and integrative theories (Collins 1998). The conceptualization of gender as a social structure is my contribution to complicating, but hopefully enriching, social theory about gender.

A third benefit to this multidimensional structural model is that it allows us to seriously investigate the direction and strength of causal relationships between gendered phenomena on each dimension. We can try to identify the site where change occurs and at which level of analysis the ability of agentic women and men seem able at this, historical moment, to effectively reject habitualized gender routines. For example, we can empirically investigate the relationship between gendered selves and doing gender without accepting simplistic unidirectional arguments for inequality presumed to be either about identities or cultural ideology. It is quite possible, indeed likely, that socialized femininity does help explain why we do gender, but doing gender to meet others' expectations, surely, over time, helps construct our gendered selves. Furthermore, gendered institutions depend on our willingness to do gender, and when we rebel, we can sometimes change the institutions themselves. I have used the language of dimensions interchangeably with the language of levels because when we think of gender as a social structure, we must move away from privileging any particular dimension as higher than another. How social change occurs is an empirical question, not an a priori theoretical assumption. It may be that individuals struggling to change their own identities (as in consciousness-raising groups of the early second-wave women's movement) eventually bring their new selves to social interaction and create new cultural expectations. For example, as women come to see themselves (or are socialized to see themselves) as sexual actors, the expectations that men must work to provide orgasms for their female partners becomes part of the cultural norm. But this is surely not the only way social change can happen. When social movement activists name as inequality what has heretofore been considered natural (e.g., women's segregation into low-paying jobs), they can create organizational changes such as career ladders between women's quasi-administrative jobs and actual management, opening up opportunities that otherwise would have remained closed, thus creating change on the institutional dimension. Girls raised in the next generation, who know opportunities exist in these workplaces, may have an altered sense of possibilities and therefore of themselves. We need, however, to also study change and equality when it occurs rather than only documenting inequality.

Perhaps the most important feature of this conceptual schema is its dynamism. No one dimension determines the other. Change is fluid and reverberates throughout the structure dynamically. Changes in individual identities and moral accountability may change interactional expectations, but the opposite is possible as well. Change cultural expectations, and individual identities are shaped differently. Institutional changes must result from individuals or group action, yet such change is difficult, as institutions exist across time and space. Once institutional changes occur, they reverberate at the level of cultural expectations and perhaps even on identities. And the cycle of change continues. No mechanistic predictions are possible because human beings sometimes reject the structure itself and, by doing so, change it.

Social Processes Located by Dimension in the Gender Structure

When we conceptualize gender as a social structure, we can begin to identify under what conditions and how gender inequality is being produced within each dimension. The "how" is important because without knowing the mechanisms, we cannot intervene. If indeed gender inequality in the division of household labor at this historical moment were primarily explained (and I do not suggest that it is) by gendered selves, then we would do well to consider the most effective socialization mechanisms to create fewer gender-schematic children and resocialization for adults. If, however, the gendered division of household labor is primarily constrained today by cultural expectations and moral accountability, it is those cultural images we must work to alter. But then again, if the reason many men do not equitably do their share of family labor is that men's jobs are organized so they cannot succeed at work and do their share at home, it is the contemporary American workplace that must change (Williams 2000). We may never find a universal theoretical explanation for the gendered division of household labor because universal social laws may be an illusion of twentieth-century empiricism. But in any given moment for any particular setting, the causal processes should be identifiable empirically. Gender complexity goes beyond historical specificity, as the particular causal processes that constrain men and women to do gender may be strong in one institutional setting (e.g., at home) and weaker in another (e.g., at work).

The forces that create gender traditionalism for men and women may vary across space as well as time. Conceptualizing gender as a social structure contributes to a more postmodern, contextually specific social science. We can use this schema to begin to organize thinking about the causal processes that are most likely to be effective on each dimension. When we are concerned with the means by which individuals come to have a preference to do gender, we should focus on how identities are constructed through early childhood development, explicit socialization, modeling, and adult experiences, paying close attention to the internalization of social mores. To the extent that women and men choose to do gender-typical behavior cross-situationally and over time, we must focus on such individual explanations. Indeed, much attention has already been given to gender socialization and the individualist presumptions for gender. The earliest and perhaps most commonly referred to explanations in popular culture depend on sex-role training, teaching boys and girls their culturally appropriate roles. But when trying to understand gender on the interactional/cultural dimension, the means by which status differences shape expectations and the ways in which in-group and out-group membership influences behavior need to be at the center of attention. Too little attention has been paid to how inequality is shaped by such cultural expectations during interaction. I return to this in the section below. On the institutional dimension, we look to law, organizational practices, and formal regulations that distinguish by sex category. Much progress has been made in the post-civil rights era with rewriting formal laws and organizational practices to ensure gender neutrality. Unfortunately, we have often found that despite changes in gender socialization and gender neutrality on the institutional dimension, gender stratification remains.

What I have attempted to do here is to offer a conceptual organizing scheme for the study of gender that can help us to understand gender in all its complexity and try to isolate the social processes that create gender in each dimension. Table 1.1 provides a schematic outline of this argument.[2]

Cultural Expectations During Interaction and the Stalled Revolution

In *Gender Vertigo* (Risman 1998), I suggested that at this moment in history, gender inequality between partners in American heterosexual couples could be attributed particularly to the interactional expectations at the cultural level: the differential expectations attached to being a mother and father, a husband and wife. Here, I extend this argument in two ways. First, I propose that the stalled gender revolution in other settings can similarly be traced to the interactional/cultural dimension of the social structure. Even when women and men with feminist identities work in organizations with formally gender-neutral rules, gender inequality is reproduced during everyday interaction. The cultural expectations attached to our sex category, simply being identified as a woman or man, has remained relatively impervious to the feminist forces that

Table 1.1 Dimensions of Gender Structure, by Illustrative Social Processes[a]

	Dimensions of the Gender Structure		
	Individual Level	*Interactional Cultural Expectations*	*Institutional Domain*
Social Processes	Socialization	Status expectations	Organizational practices
	Internalization	Cognitive bias	Legal regulations
	Identity work	Othering	Distribution of resources
	Construction of selves	Trading power for patronage	Ideology
		Altercasting	

a. These are examples of social processes that may help explain the gender structure on each dimension. They are meant to be illustrative and not a complete list of all possible social processes or causal mechanisms.

have problematized sexist socialization practices and legal discrimination. I discuss some of those processes that can help explain why social interaction continues to reproduce inequality, even in settings that seem ripe for social change.

Contemporary social psychological writings offer us a glimpse of possibilities for understanding how inequality is reconstituted in daily interaction. Ridgeway and her colleagues (Ridgeway 1991, 1997, 2001; Ridgeway and Correll 2000; Ridgeway and Smith-Lovin 1999) showed that the status expectations attached to gender and race categories are cross-situational. These expectations can be thought of as one of the engines that re-create inequality even in new settings where there is no other reason to expect male privilege to otherwise emerge. In a sexist and racist society, women and all persons of color are expected to have less to contribute to task performances than are white men, unless they have some other externally validated source of prestige. Status expectations create a cognitive bias toward privileging those of already high status. What produces status distinction, however, is culturally and historically variable. Thus, cognitive bias is one of the causal mechanisms that help to explain the reproduction of gender and race inequality in everyday life. It may also be an important explanation for the reproduction of class and heterosexist inequality in everyday life as well, but that is an empirical question.

Schwalbe and his colleagues (2000, 419) suggested that there are other "generic interactive processes through which inequalities are created and reproduced in everyday life." Some of these processes include othering, subordinate adaptation, boundary maintenance, and emotion management. Schwalbe and his colleagues suggested that subordinates' adaptation plays an essential role in their own disadvantage. Subordinate adaptation helps to explain women's strategy to adapt to the gender structure. Perhaps the most common adaptation of women to subordination is "trading power for patronage" (Schwalbe et al. 2000, 426). Women, as wives and daughters, often derive significant compensatory benefits from relationships with the men in their families. Stombler and Martin (1994) similarly showed how little sisters in a fraternity trade affiliation for secondary status. In yet another setting, elite country clubs, Sherwood (2004) showed how women accept subordinate status as "B" members of clubs, in exchange for men's approval, and how when a few wives challenge men's privilege, they are threatened with social ostracism, as are their husbands. Women often gain the economic benefits of patronage for themselves and their children in exchange for their subordinate status.

One can hardly analyze the cultural expectations and interactional processes that construct gender inequality without attention to the actions of members of the dominant group. We must pay close attention to what men do to preserve their power and privilege. Schwalbe et al. (2000) suggested that one process involved is when superordinate groups effectively "other" those who they want to define as subordinate, creating devalued statuses and expectations for them. Men effectively do this in subversive ways through "politeness" norms, which construct women as "others" in need of special favors, such as protection. By opening doors and walking closer to the dirty street, men construct women as an "other" category, different and less than independent autonomous men. The cultural significance attached to male bodies signifies the capacity to dominate, to control, and to elicit deference, and such expectations are perhaps at the core of what it means for men to do gender (Michael Schwalbe, personal communication).

These are only some of the processes that might be identified for understanding how we create gender inequality based on embodied cultural expectations. None are determinative causal predictors, but instead, these are possible leads to reasonable and testable hypotheses about the production of gender. In the next section of this article, I provide empirical illustrations of this conceptual scheme of gender as a social structure.

Empirical Illustrations

I begin with an example from my own work of how conceptualizing gender as a social structure helps to organize the findings and even push forward an understanding of the resistance toward an egalitarian division of family work among contemporary American heterosexual couples. This is an area of research that incorporates a concern with nurturing children, housework, and emotional labor. My own question, from as early as graduate school, was whether men could mother well enough that those who care about children's well-being would want them to do so.

To ask the question, Can men mother, presuming that gender itself is a social structure leads us to look at all the ways that gender constrains men's mothering and under what conditions those change. Indeed, one of my most surprising, and unanticipated, findings was that single fathers who were primary caretakers came to describe themselves more often than other men with adjectives such as "nurturant," "warm," and "child oriented," those adjectives we social scientists use to measure femininity. Single fathers' identities changed based on their experiences as primary parents. In my research, men whose wives worked full-time did not, apparently, do enough mothering to have such experiences influence their own sense of selves. Most married fathers hoard the opportunity for leisure that frees them from the responsibilities of parenting that might create such identity change. My questions became more complicated but more useful when I conceptualized gender as a social structure. When and under what conditions do gendered selves matter? When do interactional expectations have the power to overcome previous internalized predispositions? What must change at the institutional level to allow for expectations to change at the interactional level? Does enough change on the interactional dimension shift the moral accountability that then leads to collective action in social organizations? Could feminist parents organize and create a social movement that forces workplaces to presume that valuable workers also have family responsibilities?

These questions led me to try to identify the conditions that enable women and men to actually succeed in creating egalitarian relationships. My next research project was an in-depth interview and qualitative study of heterosexual couples raising children who equally shared the work of earning a living and the family labor of child care, homemaking, and emotion work. The first interesting piece of data was how hard it was to find such people in the end of the twentieth century, even when recruiting at daycare centers, parent-teacher associations, university venues, and feminist newsletters (all in the southeastern United States). Three out of four volunteer couples failed the quite generous criteria for inclusion: Working approximately the same number of hours in the labor force (within five hours per week), sharing the household labor and child care tasks within a 60/40 split, and both partners' describing the relationship as equitable. There are clearly fewer couples who live equal lives than those who wish fervently that they did so.

What I did find from intensive interviews and home observations with 20 such couples was that the conditions that enabled their success spread across each dimension of the gender structure. Although I would have predicted otherwise (having once been committed to a purely structural theory of human behavior), selves and personalities matter. The women in my sample were strong, directive women married to relatively laidback men. Given the overwhelming gendered expectations for men's privilege in heterosexual marriage, this should have been expected, but to someone with my theoretical background, it was not. Less surprising to me, the women in these couples also had at least the income and career status of their partners and often bettered them. But this is not usually enough to dent men's privilege, or we would have far more egalitarian marriages by now. In addition, these couples were ideologically committed to equality and to sharing. They often tried explicitly to create social relationships with others who held similar values, for example, by joining liberal churches to meet like-minded others. Atypical gendered selves and shared feminist-inspired cultural expectations were important conditions for equality, but they were not enough. Men's workplace flexibility mattered as well. Nearly every father in this sample was employed in a job with flexible working hours. Many women worked in jobs with flexibility as well, but not as uniformly as their male partners. These were privileged, educated workers for whom workplace flexibility was sometimes simply luck (e.g., a father who lost a corporate job and decided to sell real estate) but more often was a conscious choice (e.g., clinical psychologists choosing to teach at a small college to have more control over working hours despite decreased earning power). Thus, these couples experienced enabling contexts at the level of their individual selves, feminist ideology to help shape the cultural expectations in their most immediate environments (within the dyad and among at least some friends), and the privilege within the economy to have or find flexible jobs. By attending to each dimension of the gender structure, I amassed a more effective explanation for their ability to negotiate fair relationships than I could have without attention to selves, couple interaction, and their workplaces. The implications for feminist social change are direct: We cannot simply attend to socializing children differently, nor creating moral accountability for men to share family work, nor fighting for flexible, family-friendly workplaces. We must attend to all simultaneously.

Gender structures are even more complicated than my discussion suggests thus far because how gender identities are constructed on the individual and cultural dimensions vary tremendously over time and space. Even within contemporary American society, gender structures vary by community, social class, ethnicity, and race.

Gender Structure and Intersectionality

Perhaps the most important development in feminist thought in the last part of the twentieth century was the increasing concern with intersectionality (Andersen and Collins 1994; Baca Zinn and Thornton Dill 1994; Collins 1990). Women of color had been writing about intersectionality from nearly the start of the second wave of feminist scholarship. It was, however, not until several decades into the women's movement when they were heard and moved from margin closer to center (Myers et al. 1998). There is now considerable consensus growing that one must always take into consideration multiple axes of oppression; to do otherwise presumes the whiteness of women, the maleness of people of color, and the heterosexuality of everyone.

I concur with this consensus that gender must be understood within the context of the intersecting domains of inequality. The balkanization of research and theory into specializations of race or ethnicity or gender or stratification has undermined a sophisticated analysis of inequality (but see Reskin 2002; Schwalbe et al. 2000; Tilly 1999). I do not agree, however, with an operational strategy for scholarship that suggests the appropriate analytic solution is to only work within an intersectionality framework. While various axes of domination are always intersecting, the systems of inequality are not necessarily produced or re-created with identical social processes. The historical and current mechanisms that support gender inequality may or may not be those that are most significant for other kinds of oppression; whether this is the case is an empirical question. To focus all investigations into the complexity or subjective experience of interlocking oppressions would have us lose access to how the mechanisms for different kinds of inequality are produced. Feminist scholarship needs a both/and strategy (Collins 1998). We cannot study gender in isolation from other inequalities, nor can we only study inequalities' intersection and ignore the historical and contextual specificity that distinguishes the mechanisms that produce inequality by different categorical divisions, whether gender, race, ethnicity, nationality, sexuality, or class.

Calhoun (2000) exemplifies this both/and strategy in her argument that heterosexism cannot simply be understood as gender oppression and merged into feminist theory. She argued that we must study heterosexism as a separate system of oppression. While it is clearly the case that gender subordination and heterosexism support one another, and a gendered analysis of homophobia is critical, the two oppressions should not be conflated. Calhoun . . . suggested . . . that challenging men's dominance is a necessary condition of ending the subordination of lesbians and gay men but not a sufficient condition to end such oppression. It is important for analytic clarity, and therefore to the scholarly contribution to social change, to identify causal mechanisms for heterosexism and gender oppression distinctly.

My argument is that race, gender, and sexuality are as equally fundamental to human societies as the economy and the polity. Those inequalities that are fundamentally embedded throughout social life, at the level of individual identities, cultural expectations embedded into interaction, and institutional opportunities and constraints are best conceptualized as structures: The gender structure, the race structure, the class structure, and the sexuality structure. This does not imply that the social forces that produced, nor the causal mechanisms at work in the daily reproduction of inequality within each structure, are of similar strength or type at any given historical moment. For example, gender and race structures extend considerably further into everyday life in the contemporary American context, at home and at work, than does the political structure.[3] I propose this structural language as a tool to help disentangle the means by which inequalities are constructed, recreated, and—it is hoped—transformed or deconstructed. The model for how gender structure works, with consequences for individuals, interactions/cultural expectations, and institutions, can be generalized to the study of other equally embedded inequalities such as race and sexuality. Each structure of inequality exists on its own yet coexists with every other structure of inequality. The subjective experience of actual human beings is always of intersecting inequalities, but the historical construction and contemporary reproduction of inequality on each axis may be distinct. Oppressions can be loosely or tightly coupled, can have both common and distinct generative mechanisms.

Theory Wrestling With Activism

Within any structure of inequality, perhaps the most important question a critical scholar must ask is, What mechanisms are currently constructing inequality, and how can these be transformed to create a more just world? If as critical scholars, we forget to keep our eye on social transformation, we may slip without intention into the implicitly value-free role of social scientists who study gender merely to satisfy intellectual curiosity (Risman 2003). The central questions for feminists must include a focus on social transformation, reducing inequality, and improving the status of women. A concern with social change brings us to the thorny and as yet too little explored issue of agency. When do subordinate groups collectively organize to challenge their oppression? When do superordinate groups mobilize to resist? How do we know agency when we see it, and how can we support feminist versions of it?

Feminist scholarship must seek to understand how and why gender gets done, consciously or not, to help those who hope to stop doing it. I end by focusing our attention on what I see as the next frontier for feminist change agents: A focus on the processes that might spur change at the interaction or cultural dimension of the gender structure. We have begun to socialize our children differently, and while identities are hardly post-gender, the sexism inherent in gender socialization is now widely recognized. Similarly, the organizational rules and institutional laws have by now often been rewritten to be gender neutral, at least in some nations. While gender-neutral laws in a gender-stratified society may have short-term negative consequences (e.g., displaced homemakers who never imagined having to support themselves after marriage), we can hardly retreat from equity in the law or organizations. It is the interactional and cultural dimension of gender that have yet to be tackled with a social change agenda.

Cognitive bias is one of the mechanisms by which inequality is re-created in everyday life.

There are, however, documented mechanisms for decreasing the salience of such bias (Bielby 2000; Reskin 2000; Ridgeway and Correll 2000). When we consciously manipulate the status expectations attached to those in subordinate groups, by highlighting their legitimate expertise beyond the others in the immediate social setting, we can begin to challenge the nonconscious hierarchy that often goes unnoticed. Similarly, although many subordinates adapt to their situation by trading power for patronage, when they refuse to do so, interaction no longer flows smoothly, and change may result. Surely, when wives refuse to trade power for patronage, they can rock the boat as well as the cradle.

These are only a few examples of interactive processes that can help to explain the reproduction of inequality and to envision strategies for disrupting inequality. We need to understand when and how inequality is constructed and reproduced to deconstruct it. I have argued before (Risman 1998) that because the gender structure so defines the category woman as subordinate, the deconstruction of the category itself is the best, indeed the only sure way, to end gender subordination. There is no reason, except the transitional vertigo that will accompany the process to dismantle it, that a utopian vision of a just world involves any gender structure at all. Why should we need to elaborate on the biological distinction between the sexes? We must accommodate reproductive differences for the process of biological replacement, but there is no a priori reason we should accept any other role differentiation simply based on biological sex category.

Feminist scholarship always wrestles with the questions of how one can use the knowledge we create in the interest of social transformation. As feminist scholars, we must talk beyond our own borders. This kind of theoretical work becomes meaningful if we can eventually take it public. Feminist sociology must be public sociology (Burawoy forthcoming). We must eventually take what we have learned from our theories and research beyond professional journals to our students and to those activists who seek to disrupt and so transform gender relations. We must consider how the knowledge we create can help those who desire a more egalitarian social world to refuse to do gender at all, or to do it with rebellious reflexiveness to help transform the world around them. For those without a sociological perspective, social change through socialization and through legislation are the easiest to envision. We need to shine a spotlight on the dimension of cultural interactional expectations as it is here that work needs to begin.

In conclusion, I have made the argument that we need to conceptualize gender as a social structure, and by doing so, we can analyze the ways in which gender is embedded at the individual, interactional, and institutional dimensions of our society. This situates gender at the same level of significance as the economy and the polity. In addition, this framework helps us to disentangle the relative strength of a variety of causal mechanisms for explaining any given outcome without dismissing the possible relevance of other processes that are situated at different dimensions of analysis. Once we have a conceptual tool to organize the encyclopedic research on gender, we can systematically build on our knowledge and progress to understanding the strength and direction of causal processes within a complicated multidimensional recursive theory. I have also argued that our concern with intersectionality must continue to be paramount but that different structures of inequality have different infrastructure and perhaps different influential causal mechanisms at any given historical moment. Therefore, we need to follow a both/and strategy, to understand gender structure, race structure, and other structures of inequality as they currently operate, while also systematically paying attention to how these axes of domination intersect. Finally, I have suggested that we pay more attention to doing research and writing theory with explicit attention to how our work can come to be "fighting words" (Collins 1998) to help transform as well as inform society. If we can identify the mechanisms that create gender, perhaps we can offer alternatives to them and so use our scholarly work to contribute to envisioning a feminist utopia.

Notes

1. See Scott (1997) for a critique of feminists who adopt a strategy where theories have to be simplified, compared, and defeated. She too suggested a model where feminists build on the complexity of each others' ideas.

2. I thank my colleague Donald Tomaskovic-Devey for suggesting the visual representation of these ideas as well as his usual advice on my ideas as they develop.

3. One can certainly imagine a case where political structures extend far into everyday life, a nation in the midst of civil war or in the grips of a fascist state. One can also envision a case when race retreats to the personal dimension, as when the Irish became white in twentieth-century America.

References

Andersen, Margaret, and Patricia Hill Collins. 1994. *Race, class, and gender: An anthology.* Belmont CA: Wadsworth.

Baca Zinn, Maxine, and Bonnie Thornton Dill. 1994. *Women of color in U.S. society.* Philadelphia: Temple University Press.

Bem, Sandra. 1993. *The lenses of gender.* New Haven, CT: Yale University Press.

Bielby, William T. 2000. Minimizing workplace gender and racial bias. *Contemporary Sociology 29*(1): 120–29.

Blau, Peter. 1977. *Inequality and heterogeneity.* New York: Free Press.

Burawoy, Michael. Forthcoming. Public sociologies contradictions, dilemmas and possibilities. *Social Forces.*

Burt, Ronald S. 1982. *Toward a structural theory of action.* New York: Academic Press.

Calhoun, Cheshire. 2000. *Feminism, the family, and the politics of the closet: Lesbian and gay displacement.* New York: Oxford University Press.

Collins, Patricia Hill. 1990. *Black feminist thought: Knowledge, consciousness, and the politics of empowerment.* New York: Routledge.

———. 1998. *Fighting words: Black women and the search for justice.* Minneapolis: University of Minnesota Press.

———. 2004. *Black sexual politics: African Americans, gender, and the new racism.* New York: Routledge.

Connell, R. W. 1987. *Gender and power: Society, the person, and sexual politics.* Stanford, CA: Stanford University Press.

———. 2002. *Gender: Short introductions.* Malden, MA: Blackwell.

England, Paula, and Irene Browne. 1992. Internalization and constraint in women's subordination. *Current Perspectives in Social Theory* 12:97–123.

Epstein, Cynthia, Fuchs. 1988. *Deceptive distinctions: Sex, gender, and the social order.* New Haven, CT: Yale University Press.

Espiritu, Yen Le. 1997. *Asian American women and men: Labor, laws, and love.* Thousand Oaks, CA: Sage.

Ferree, Myra Marx. 1990. Beyond separate spheres: Feminism and family research. *Journal of Marriage and the Family 53*(4): 866–84.

Ferree, Myra Marx, Judith Lorber, and Beth Hess. 1999. *Revisioning gender.* Thousand Oaks, CA: Sage.

Giddens, Anthony. 1984. *The constitution of society: Outline of the theory of structuration.* Berkeley: University of California Press.

Kanter, Rosabeth. 1977. *Men and women of the corporation.* New York: Basic Books.

Lorber, Judith. 1994. *Paradoxes of gender.* New Haven, CT: Yale University Press.

Martin, Patricia. Forthcoming. Gender as a social institution. *Social Forces.*

Myers, Kristen A., Cynthia D. Anderson, and Barbara J. Risman, eds. 1998. *Feminist foundations: Toward transforming society.* Thousand Oaks, CA: Sage.

Reskin, Barbara. 2000. The proximate causes of employment discrimination. *Contemporary Sociology 29*(2): 319–28.

———. 2002. How did the poison get in Mr. Barlett's stomach? Motives and mechanisms in explaining inequality. Presidential address given at the 97th annual meetings of the American Sociological Association, Chicago, August.

Ridgeway, Cecilia L. 1991. The social construction of status value: Gender and other nominal characteristics. *Social Forces 70*(2): 367–86.

———. 1997. Interaction and the conservation of gender inequality: Considering employment. *American Sociological Review 62*(2): 218–35.

———. 2001. Gender, status, and leadership. *Journal of Social Issues 57*(4): 637–55.

Ridgeway, Cecilia L., and Shelley J. Correll. 2000. Limiting inequality through interaction: The end(s) of gender. *Contemporary Sociology* 29:110–20.

Ridgeway, Cecilia L., and Lynn Smith-Lovin. 1999. The gender system and interaction. *Annual Review of Sociology* 25:191–216.

Risman, Barbara J. 1983. Necessity and the invention of mothering. Ph.D. diss, University of Washington.

———. 1987. Intimate relationships from a microstructural perspective: Mothering men. *Gender & Society* 1: 6–32.

———. 1998. *Gender vertigo: American families in transition.* New Haven, CT: Yale University Press.

———. 2003. Valuing all flavors of feminist sociology. *Gender & Society* 17:659–63.

Risman, Barbara J., and Pepper Schwartz. 1989. *Gender in intimate relationships.* Belmont, CA: Wadsworth.

Schwalbe, Michael, Sandra Godwin, Daphne Holden, Douglas Schrock, Shealy Thompson, and Michele Wolkomir. 2000. Generic processes in the reproduction of inequality: An interactionist analysis. *Social Forces 79*(2): 419–52.

Scott, Joan Wallach. 1997. Comment on Hawkesworth's "Confounding Gender." *Signs: Journal of Women in Culture and Society 22*(3): 697–702.

Sherwood, Jessica. 2004. Talk about country clubs: Ideology and the reproduction of privilege. Ph.D. diss., North Carolina State University.

Smelser, Neil J. 1988. Social structure. In *Handbook of sociology,* edited by Neil J. Smelser. Beverly Hills, CA: Sage.

Staples, Robert. 1990. Social inequality and Black sexual pathology. The essential relationship. *Black Scholar 21*(3): 29–37.

Stombler, Mindy, and Patricia Yancey Martin. 1994. Bring women in, keeping women down: Fraternity "little sister" organizations. *Journal of Contemporary Ethnography* 23:150–84.

Tilly, Charles. 1999. *Durable inequality.* Berkeley: University of California Press.

Udry, J. Richard. 2000. Biological limits of gender construction. *American Sociological Review* 65:443–57.

West, Candace, and Don Zimmerman. 1987. Doing gender. *Gender & Society* 1:125–51.

Williams, Joan. 2000. *Unbending gender: Why family and work conflict and what to do about it.* New York: Oxford University Press.

Introduction to Reading 2

Sociologist Betsy Lucal describes the rigidity of the American binary gender system and the consequences for people who do not fit by analyzing the challenges she faces in the course of her daily experience of negotiating the boundaries of our gendered society. Since her physical appearance does not clearly define her as a woman, she must navigate a world in which some people interact with her as though she is a man. Through analysis of her own story, Lucal demonstrates how gender is something we do, rather than something we are.

1. Why does Lucal argue that we cannot escape "doing gender"?
2. How does Lucal negotiate "not fitting" into the American two-and-only-two gender structure?
3. Have you ever experienced a mismatch between your gender-identity and the gender that others perceive you to be? If so, how did you feel and respond?

What It Means to Be Gendered Me

Betsy Lucal

I understood the concept of "doing gender" (West and Zimmerman 1987) long before I became a sociologist. I have been living with the consequences of inappropriate "gender display" (Goffman 1976; West and Zimmerman 1987) for as long as I can remember. My daily experiences are a testament to the rigidity of gender in our society, to the real implications of "two and only two" when it comes to sex and gender categories (Garfinkel 1967; Kessler and McKenna 1978). Each day, I experience the consequences that our gender system has for my identity and interactions. I am a woman who has been called "Sir" so many times that I no longer even hesitate to assume that it is being directed at me. I am a woman whose use of public rest rooms regularly causes reactions ranging from confused stares to confrontations over what a man is doing in the women's room. I regularly enact a variety of practices either to minimize the need for others to know my gender or to deal with their misattributions.

I am the embodiment of Lorber's (1994) ostensibly paradoxical assertion that the "gender bending" I engage in actually might serve to preserve and perpetuate gender categories. As a feminist who sees gender rebellion as a significant part of her contribution to the dismantling of sexism, I find this possibility disheartening.

In this article, I, examine how my experiences both support and contradict Lorber's (1994) argument using my own experiences to illustrate and reflect on the social construction of gender. My analysis offers a discussion of the consequences of gender for people who do not follow the rules as well as an examination of the possible implications of the existence of people like me for the gender system itself. Ultimately, I show how life on the boundaries of gender affects me and how my life, and the lives of others who make similar decisions about their participation in the gender system, has the potential to subvert gender.

Because this article analyzes my experiences as a woman who often is mistaken for a man, my focus is on the social construction of gender for women. My assumption is that, given the gendered nature of the gendering process itself, men's experiences of this phenomenon might well be different from women's.

The Social Construction of Gender

It is now widely accepted that gender is a social construction, that sex and gender are distinct, and that gender is something all of us "do." This conceptualization of gender can be traced to Garfinkel's (1967) ethnomethodological study of "Agnes."[1] In this analysis, Garfinkel examined the issues facing a male who wished to pass as, and eventually become, a woman. Unlike individuals who perform gender in culturally expected ways, Agnes could not take her gender for granted and always was in danger of failing to pass as a woman (Zimmerman 1992).

This approach was extended by Kessler and McKenna (1978) and codified in the classic "Doing Gender" by West and Zimmerman (1987). The social constructionist approach has been developed most notably by Lorber (1994, 1996). Similar theoretical strains have developed outside of sociology, such as work by Butler (1990) and Weston (1996). . . .

Given our cultural rules for identifying gender (i.e., that there are only two and that masculinity

From Lucal, B., "What it means to be gendered me" *Gender & Society 13*(6), p. 781. Reprinted with permission of Sage Publications, Inc.

is assumed in the absence of evidence to the contrary), a person who does not do gender appropriately is placed not into a third category but rather into the one with which her or his gender display seems most closely to fit; that is, if a man appears to be a woman, then he will be categorized as "woman," not as something else. Even if a person does not want to do gender or would like to do a gender other than the two recognized by our society, other people will, in effect, do gender for that person by placing her or him in one and only one of the two available categories. We cannot escape doing gender or, more specifically, doing one of two genders. (There are exceptions in limited contexts such as people doing "drag" [Butler 1990; Lorber 1994].)

People who follow the norms of gender can take their genders for granted. Kessler and McKenna asserted, "Few people besides transsexuals think of their gender as anything other than 'naturally' obvious"; they believe that the risks of not being taken for the gender intended "are minimal for nontranssexuals" (1978, 126). However, such an assertion overlooks the experiences of people such as those women Devor (1989) calls "gender blenders" and those people Lorber (1994) refers to as "gender benders." As West and Zimmerman (1987) pointed out, we all are held accountable for, and might be called on to account for, our genders.

People who, for whatever reasons, do not adhere to the rules, risk gender misattribution and any interactional consequences that might result from this misidentification. What are the consequences of misattribution for social interaction? When must misattribution be minimized? What will one do to minimize such mistakes? In this article, I explore these and related questions using my biography.

For me, the social processes and structures of gender mean that, in the context of our culture, my appearance will be read as masculine. Given the common conflation of sex and gender, I will be assumed to be a male. Because of the two-and-only-two genders rule, I will be classified, perhaps more often than not, as a man—not as an atypical woman, not as a genderless person. I must be one gender or the other; I cannot be neither, nor can I be both. This norm has a variety of mundane and serious consequences for my everyday existence. Like Myhre (1995), I have found that the choice not to participate in femininity is not one made frivolously.

My experiences as a woman who does not do femininity illustrate a paradox of our two-and-only-two gender system. Lorber argued that "bending gender rules and passing between genders does not erode but rather preserves gender boundaries" (1994, 21). Although people who engage in these behaviors and appearances do "demonstrate the social constructedness of sex, sexuality, and gender" (Lorber 1994, 96), they do not actually disrupt gender. Devor made a similar point: "When gender blending females refused to mark themselves by publicly displaying sufficient femininity to be recognized as women, they were in no way challenging patriarchal gender assumptions" (1989, 142). As the following discussion shows, I have found that my own experiences both support and challenge this argument. Before detailing these experiences, I explain my use of my self as data.

My Self as Data

This analysis is based on my experiences as a person whose appearance and gender/sex are not, in the eyes of many people, congruent. How did my experiences become my data? I began my research "unwittingly" (Krieger 1991). This article is a product of "opportunistic research" in that I am using my "unique biography, life experiences, and/or situational familiarity to understand and explain social life" (Riemer 1988, 121; see also Riemer 1977). It is an analysis of "unplanned personal experience" that is, experiences that were not part of a research project but instead are part of my daily encounters (Reinharz 1992).

This work also is, at least to some extent, an example of Richardson's (1994) notion of writing as a method of inquiry. As a sociologist who specializes in gender, the more I learned, the more I realized that my life could serve as a case study. As I examined my experiences, I found out things—about my experiences and about theory—that I did not know when I started (Richardson 1994).

It also is useful, I think, to consider my analysis an application of Mills's (1959) "sociological imagination." Mills (1959) and Berger (1963) wrote about the importance of seeing the general in the particular. This means that general social patterns can be discerned in the behaviors of particular individuals. In this article, I am examining portions of my biography, situated in U.S. society during the 1990s, to understand the "personal troubles" my gender produces in the context of a two-and-only-two gender system. I am not attempting to generalize my experiences; rather, I am trying to use them to examine and reflect on the processes and structure of gender in our society.

Because my analysis is based on my memories and perceptions of events, it is limited by my ability to recall events and by my interpretation of those events. However, I am not claiming that my experiences provide the truth about gender and how it works. I am claiming that the biography of a person who lives on the margins of our gender system can provide theoretical insights into the processes and social structure of gender. Therefore, after describing my experiences, I examine how they illustrate and extend, as well as contradict, other work on the social construction of gender.

Gendered Me

Each day, I negotiate the boundaries of gender. Each day, I face the possibility that someone will attribute the "wrong" gender to me based on my physical appearance. I am six feet tall and large-boned. I have had short hair for most of my life. For the past several years, I have worn a crew cut or flat top. I do not shave or otherwise remove hair from my body (e.g., no eyebrow plucking). I do not wear dresses, skirts, high heels, or makeup. My only jewelry is a class ring, a "men's" watch (my wrists are too large for a "women's" watch), two small earrings (gold hoops, both in my left ear), and (occasionally) a necklace. I wear jeans or shorts, T-shirts, sweaters, polo/golf shirts, button-down collar shirts, and tennis shoes or boots. The jeans are "women's" (I do have hips) but do not look particularly "feminine." The rest of the outer garments are from men's departments. I prefer baggy clothes, so the fact that I have "womanly" breasts often is not obvious (I do not wear a bra).

Sometimes, I wear a baseball cap or some other type of hat. I also am white and relatively young (30 years old).[2] My gender display—what others interpret as my presented identity—regularly leads to the misattribution of my gender. An incongruity exists between my gender self-identity and the gender that others perceive. In my encounters with people I do not know, I sometimes conclude, based on our interactions, that they think I am a man. This does not mean that other people do not think I am a man, just that I have no way of knowing what they think without interacting with them.

Living With It

I have no illusions or delusions about my appearance. I know that my appearance is likely to be read as "masculine" (and male) and that how I see myself is socially irrelevant. Given our two-and-only-two gender structure, I must live with the consequences of my appearance. These consequences fall into two categories: issues of identity and issues of interaction.

My most common experience is being called "Sir" or being referred to by some other masculine linguistic marker (e.g., "he," "man"). This has happened for years, for as long as I can remember, when having encounters with people I do not know.[3] Once, in fact, the same worker at a fast-food restaurant called me "Ma'am" when she took my order and "Sir" when she gave it to me.

Using my credit cards sometimes is a challenge. Some clerks subtly indicate their disbelief, looking from the card to me and back at the card and checking my signature carefully. Others challenge my use of the card, asking whose it is or demanding identification. One cashier asked to see my driver's license and then asked me whether I was the son of the cardholder. Another clerk told me that my signature on the receipt "had better match" the one on the card. Presumably, this was her way of letting me know that she was not convinced it was my credit card.

My identity as a woman also is called into question when I try to use women-only spaces. Encounters in public rest rooms are an adventure. I have been told countless times that "This is the ladies' room." Other women say nothing to me, but their stares and conversations with others let me know what they think. I will hear them say, for example, "There was a man in there. "I also get stares when I enter a locker room. However, it seems that women are less concerned about my presence, there, perhaps because, given that it is a space for changing clothes, showering, and so forth, they will be able to make sure that I am really a woman. Dressing rooms in department stores also are problematic spaces. I remember shopping with my sister once and being offered a chair outside the room when I began to accompany her into the dressing room. Women who believe that I am a man do not want me in women-only spaces. For example, one woman would not enter the rest room until I came out, and others have told me that I am in the wrong place. They also might not want to encounter me while they are alone. For example, seeing me walking at night when they are alone might be scary.[4]

I, on the other hand, am not afraid to walk alone, day or night. I do not worry that I will be subjected to the public harassment that many women endure (Gardner 1995). I am not a clear target for a potential rapist. I rely on the fact that a potential attacker would not want to attack a big man by mistake. This is not to say that men never are attacked, just that they are not viewed, and often do not view themselves, as being vulnerable to attack.

Being perceived as a man has made me privy to male-male interactional styles of which most women are not aware. I found out, quite by accident, that many men greet, or acknowledge, people (mostly other men) who make eye contact with them with a single nod. For example, I found that when I walked down the halls of my brother's all-male dormitory making eye contact, men nodded their greetings at me. Oddly enough, these same men did not greet my brother,

I had to tell him about making eye contact and nodding as a greeting ritual. Apparently, in this case I was doing masculinity better than he was! I also believe that I am treated differently, for example, in auto parts stores (staffed almost exclusively by men in most cases) because of the assumption that I am a man. Workers there assume that I know what I need and that my questions are legitimate requests for information.

I suspect that I am treated more fairly than a feminine-appearing woman would be. I have not been able to test this proposition. However, Devor's participants did report "being treated more respectfully" (1989, 132) in such situations. There is, however, a negative side to being assumed to be a man by other men. Once, a friend and I were driving in her car when a man failed to stop at an intersection and nearly crashed into us. As we drove away, I mouthed "stop sign" to him. When we both stopped our cars at the next intersection, he got out of his car and came up to the passenger side of the car, where I was sitting. He yelled obscenities at us and pounded and spit on the car window. Luckily, the windows were closed. I do not think he would have done that if he thought I was a woman. This was the first time I realized that one of the implications of being seen as a man was that I might be called on to defend myself from physical aggression from other men who felt challenged by me. This was a sobering and somewhat frightening thought.

Recently, I was verbally accosted by an older man who did not like where I had parked my car. As I walked down the street to work, he shouted that I should park at the university rather than on a side street nearby. I responded that it was a public street and that I could park there if I chose. He continued to yell, but the only thing I caught was the last part of what he said: "Your tires are going to get cut!" Based on my appearance that day—I was dressed casually and carrying a backpack, and I had my hat on backward—I believe he thought that I was a young male student rather than a female professor. I do not think he would have yelled at a person he thought to be a woman—and perhaps especially not a woman professor.

Given the presumption of heterosexuality that is part of our system of gender, my interactions with women who assume that I am a man also can be viewed from that perspective. For example, once my brother and I were shopping when we were "hit on" by two young women. The encounter ended before I realized what had happened. It was only when we walked away that I told him that I was pretty certain that they had thought both of us were men. A more common experience is realizing that when I am seen in public with one of my women friends, we are likely to be read as a heterosexual dyad. It is likely that if I were to walk through a shopping mall holding hands with a woman, no one would look twice, not because of their open-mindedness toward lesbian couples but rather because of their assumption that I was the male half of a straight couple. Recently, when walking through a mall with a friend and her infant, my observations of others' responses to us led me to believe that many of them assumed that we were a family on an outing, that is, that I was her partner and the father of the child.

Dealing With It

Although I now accept that being mistaken for a man will be a part of my life so long as I choose not to participate in femininity, there have been times when I consciously have tried to appear more feminine. I did this for a while when I was an undergraduate and again recently when I was on the academic job market. The first time, I let my hair grow nearly down to my shoulders and had it permed. I also grew long fingernails and wore nail polish. Much to my chagrin, even then one of my professors, who did not know my name, insistently referred to me in his kinship examples as "the son." Perhaps my first act on the way to my current stance was to point out to this man, politely and after class, that I was a woman.

More recently, I again let my hair grow out for several months, although I did not alter other aspects of my appearance. Once my hair was about two and a half inches long (from its original quarter inch), I realized, based on my encounters with strangers, that I had more or less passed back into the category of "woman." Then, when I returned to wearing a flat top, people again responded to me as if I were a man.

Because of my appearance, much of my negotiation of interactions with strangers involves attempts to anticipate their reactions to me. I need to assess whether they will be likely to assume that I am a man and whether that actually matters in the context of our encounters. Many times, my gender really is irrelevant, and it is just annoying to be misidentified. Other times, particularly when my appearance is coupled with something that identifies me by name (e.g., a check or credit card) without a photo, I might need to do something to ensure that my identity is not questioned. As a result of my experiences, I have developed some techniques to deal with gender misattribution.

In general, in unfamiliar public places, I avoid using the rest room because I know that it is a place where there is a high likelihood of misattribution and where misattribution is socially important. If I must use a public rest room, I try to make myself look as nonthreatening as possible. I do not wear a hat, and I try to rearrange my clothing to make my breasts more obvious. Here, I am trying to use my secondary sex characteristics to make my gender more obvious rather than the usual use of gender to make sex obvious. While in the rest room, I never make eye contact, and I get in and out as quickly as possible. Going in with a woman friend also is helpful; her presence legitimizes my own. People are less likely to think I am entering a space where I do not belong when I am with someone who looks like she does belong.[5]

To those women who verbally challenge my presence in the rest room, I reply, "I know," usually in an annoyed tone. When they stare or talk about me to the women they are with, I simply get out as quickly as possible. In general, I do not wait for someone I am with because there is too much chance of an unpleasant encounter.

I stopped trying on clothes before purchasing them a few years ago because my presence in the changing areas was met with stares and whispers. Exceptions are stores where the dressing rooms are completely private, where there are individual stalls rather than a room with stalls separated by curtains, or where business is slow and no one else is trying on clothes. If I am trying on a garment clearly intended for a woman, then I usually can do so without hassle. I guess the attendants assume that I must be a woman if I have, for example, a women's bathing suit in my hand. But usually, I think it is easier for me to try the clothes on at home and return them, if necessary, rather than risk creating a scene. Similarly, when I am with another woman who is trying on clothes, I just wait outside.

My strategy with credit cards and checks is to anticipate wariness on a clerk's part. When I sense that there is some doubt or when they challenge me, I say, "It's my card." I generally respond courteously to requests for photo ID, realizing that these might be routine checks because of concerns about increasingly widespread fraud. But for the clerk who asked for ID and still did not think it was my card, I had a stronger reaction. When she said that she was sorry for embarrassing me, I told her that I was not embarrassed but that she should be. I also am particularly careful to make sure that my signature is consistent with the back of the card. Faced with such situations, I feel somewhat nervous about signing my name—which, of course, makes me worry that my signature will look different from how it should.

Another strategy I have been experimenting with is wearing nail polish in the dark bright colors currently fashionable. I try to do this when I travel by plane. Given more stringent travel regulations, one always must present a photo ID. But my experiences have shown that my driver's license is not necessarily convincing. Nail polish might be. I also flash my polished nails when I enter airport rest rooms, hoping that they will provide a clue that I am indeed in the right place.

There are other cases in which the issues are less those of identity than of all the norms of interaction that, in our society, are gendered. My most common response to misattribution actually is to appear to ignore it, that is, to go on with the interaction as if nothing out of the ordinary has happened. Unless I feel that there is a good reason to establish my correct gender, I assume the identity others impose on me for the sake of smooth interaction. For example, if someone is selling me a movie ticket, then there is no reason to make sure that the person has accurately discerned my gender. Similarly, if it is clear that the person using "Sir" is talking to me, then I simply respond as appropriate. I accept the designation because it is irrelevant to the situation. It takes enough effort to be alert for misattributions and to decide which of them matter; responding to each one would take more energy than it is worth.

Sometimes, if our interaction involves conversation, my first verbal response is enough to let the other person know that I am actually a woman and not a man. My voice apparently is "feminine" enough to shift people's attributions to the other category. I know when this has happened by the apologies that usually accompany the mistake. I usually respond to the apologies by saying something like "No problem" and/or "It happens all the time." Sometimes, a misattributor will offer an account for the mistake, for example, saying that it was my hair or that they were not being very observant.

These experiences with gender and misattribution provide some theoretical insights into contemporary Western understandings. of gender and into the social structure of gender in contemporary society. Although there are a number of ways in which my experiences confirm the work of others, there also are some ways in which my experiences suggest other interpretations and conclusions.

What Does It Mean?

Gender is pervasive in our society. I cannot choose not to participate in it. Even if I try not to do gender, other people will do it for me. That is, given our two-and-only-two rule, they must attribute one of two genders to me. Still, although I cannot choose not to participate in gender, I can choose not to participate in femininity (as I have), at least with respect to physical appearance. That is where the problems begin. Without the decorations of femininity, I do not look like a woman. That is, I do not look like what many people's commonsense understanding of gender tells them a woman looks like. How I see myself, even how I might wish others would see me, is socially irrelevant. It is the gender that I appear to be (my "perceived gender") that is most relevant to my social identity and interactions with others. The major consequence of this fact is that I must be continually aware of which gender I "give off " as well as which gender I "give" (Goffman 1959).

Because my gender self-identity is "not displayed obviously, immediately, and consistently" (Devor 1989, 58), I am somewhat of a failure in social terms with respect to gender. Causing people to be uncertain or wrong about one's gender is a violation of taken-for-granted rules that leads to embarrassment and discomfort; it means that something has gone wrong with the interaction (Garfinkel 1967; Kessler and McKenna 1978). This means that my nonresponse to misattribution is the more socially appropriate response; I am allowing others to maintain face (Goffman 1959, 1967). By not calling attention to their mistakes, I uphold their images of themselves as competent social actors. I also maintain my own image as competent by letting them assume that I am the gender I appear to them to be.

But I still have discreditable status; I carry a stigma (Goffman 1963). Because I have failed to participate appropriately in the creation of meaning with respect to gender (Devor 1989), I can be called on to account for my appearance. If discredited, I show myself to be an incompetent social actor. I am the one not following the rules, and I will pay the price for not providing people with the appropriate cues for placing me in the gender category to which I really belong.

I do think that it is, in many cases, safer to be read as a man than as some sort of deviant woman. "Man" is an acceptable category; it fits properly into people's gender worldview. Passing as a man often is "the path of least resistance" (Devor 1989; Johnson 1997). For example, in situations where gender does not matter, letting people take me as a man is easier than correcting them.

Conversely, as Butler noted, "We regularly punish those who fail to do their gender right" (1990, 140). Feinberg maintained, "Masculine girls and women face terrible condemnation and brutality—including sexual violence—for crossing the boundary of what is 'acceptable' female expression" (1996, 114). People are more likely to harass me when they perceive me to be a woman who looks like a man. For example, when a group of teenagers realized that I was not a man because one of their mothers identified me correctly, they began to make derogatory comments when I passed them. One asked, for example, "Does she have a penis?"

Because of the assumption that a "masculine" woman is a lesbian, there is the risk of homophobic reactions (Gardner 1995; Lucal 1997). Perhaps surprisingly, I find that I am much more likely to be taken for a man than for a lesbian, at least based on my interactions with people and their reactions to me. This might be because people are less likely to reveal that they have taken me for a lesbian because it is less relevant to an encounter or because they believe this would be unacceptable. But I think it is more likely a product of the strength of our two-and-only-two system. I give enough masculine cues that I am seen not as a deviant woman but rather as a man, at least in most cases. The problem seems not to be that people are uncertain about my gender, which might lead them to conclude that I was a lesbian once they realized I was a woman. Rather, I seem to fit easily into a gender category—just not the one with which I identify. In fact, because men represent the dominant gender in our society, being mistaken for a man can protect me from other types of gendered harassment. Because men can move around in public spaces safely (at least relative to women), a "masculine" woman also can enjoy this freedom (Devor 1989).

On the other hand, my use of particular spaces—those designated as for women only—may be challenged. Feinberg provided an intriguing analysis of the public rest room experience. She characterized women's reactions to a masculine person in a public rest room as "an example of genderphobia" (1996, 117), viewing such women as policing gender boundaries rather than believing that there really is a man in the women's rest room. She argued that women who truly believed that there was a man in their midst would react differently. Although this is an interesting perspective on her experiences, my experiences do not lead to the same conclusion.[6]

Enough people have said to me that "This is the ladies' room" or have said to their companions that "There was a man in there" that I take their reactions at face value. Still, if the two-and-only-two gender system is to be maintained, participants must be involved in policing the categories and their attendant identities and spaces. Even if policing boundaries is not explicitly intended, boundary maintenance is the effect of such responses to people's gender displays.

Boundaries and margins are an important component of both my experiences of gender and our theoretical understanding of gendering processes. I am, in effect both woman and not woman. As a woman who often is a social man but who also is a woman living in a patriarchal society, I am in a unique position to see and act.

I sometimes receive privileges usually limited to men, and I sometimes am oppressed by my status as a deviant woman. I am, in a sense, an outsider within (Collins 1991). Positioned on the boundaries of gender categories, I have developed a consciousness that I hope will prove transformative (Anzaldua 1987). In fact, one of the reasons why I decided to continue my nonparticipation in femininity was that my sociological training suggested that this could be one of my contributions to the eventual dismantling of patriarchal gender constructs. It would be my way of making the personal political. I accepted being taken for a man as the price I would pay to help subvert patriarchy. I believed that all of the inconveniences I was enduring meant that I actually was doing something to bring down the gender structures that entangled all of us.

Then, I read Lorber's (1994) *Paradoxes of Gender* and found out, much to my dismay, that I might not actually be challenging gender after all. Because of the way in which doing gender works in our two-and-only-two system, gender displays are simply read as evidence of one of the two categories. Therefore, gender bending, blending, and passing between the categories do not question the categories themselves. If one's social gender and personal (true) gender do not correspond, then this is irrelevant unless someone notices the lack of congruence.

This reality brings me to a paradox of my experiences. First, not only do others assume that I am one gender or the other, but I also insist that I *really am* a member of one of the two gender categories. That is, I am female; I self-identify as a woman. I do not claim to be some other gender or to have no gender at all. I simply place myself in the wrong category according to stereotypes and cultural standards; the gender I present, or that some people perceive me to be presenting, is inconsistent with the gender with which I identify myself as well as with the gender I could be "proven" to be. Socially, I display the wrong gender; personally, I identify as the proper gender.

Second, although I ultimately would like to see the destruction of our current gender structure, I am not to the point of personally abandoning gender. Right now, I do not want people to see me as genderless as much as I want them to see me as a woman. That is, I would like to expand the category of "woman" to include people like me. I, too, am deeply embedded in our gender system, even though I do not play by many of its rules. For me, as for most people in our society, gender is a substantial part of my personal identity (Howard and Hollander 1997). Socially, the problem is that I do not present a gender display that is consistently read as feminine. In fact, I consciously do not participate in the trappings of femininity. However, I do identify myself as a woman, not as a man or as someone outside of the two-and-only-two categories.

Yet, I do believe, as Lorber (1994) does, that the purpose of gender, as it currently is constructed, is to oppress women. Lorber analyzed gender as a "process of creating distinguishable social statuses for the assignment of rights and responsibilities" that ends up putting women in a devalued and oppressed position (1994, 32). As Martin put it, "Bodies that clearly delineate gender status facilitate the maintenance of the gender hierarchy" (1998, 495).

For society, gender means difference (Lorber 1994). The erosion of the boundaries would problematize that structure. Therefore, for gender to operate as it currently does, the category "woman" is expanded to include people like me. The maintenance of the gender structure is dependent on the creation of a few categories that are mutually exclusive, the members of which are as different as possible (Lorber 1994). It is the clarity of the boundaries between the categories that allows gender to be used to assign rights and responsibilities as well as resources and rewards.

It is that part of gender—what it is used for—that is most problematic. Indeed, is it not *patriarchal*—or, even more specifically, *heteropatriarchal*—constructions of gender that are actually the problem? It is not the differences between men and women, or the categories themselves, so much as the meanings ascribed to the categories and, even more important, the hierarchical nature of gender under patriarchy that is the problem (Johnson 1997). Therefore, I am rebelling not against my femaleness or even my womanhood; instead, I am protesting contemporary constructions of femininity and, at least indirectly, masculinity under patriarchy. We do not, in fact, know what gender would look like if it were not constructed around heterosexuality in the context of patriarchy. Although it is possible that the end of patriarchy would mean the end of gender, it is at least conceivable that something like what we now call gender could exist in a postpatriarchal future. The two-and-only-two categorization might well disappear, there being no hierarchy for it to justify. But I do not think that we should make the assumption that gender and patriarchy are synonymous.

Theoretically, this analysis points to some similarities and differences between the work of Lorber (1994) and the works of Butler (1990), Goffman (1976, 1977), and West and Zimmerman (1987). Lorber (1994) conceptualized gender as social structure, whereas the others focused more on the interactive and processual nature of gender. Butler (1990) and Goffman (1976, 1977) view gender as a performance, and West and Zimmerman (1987) examined it as something all of us do. One result of this difference in approach is that in Lorber's (1994) work, gender comes across as something that we are caught in something that, despite any attempts to the contrary, we cannot break out of. This conclusion is particularly apparent in Lorber's argument that gender rebellion, in the context of our two-and-only-two system, ends up supporting what it purports to subvert. Yet, my own experiences suggest an alternative possibility that is more in line with the view of gender offered by West and Zimmerman (1987): If gender is a product of interaction, and if it is produced in a particular context, then it can be changed if we change our performances. However, the effects of a performance linger, and gender ends up being institutionalized. It is institutionalized, in our society, in a way that perpetuates inequality, as Lorber's (1994) work shows. So, it seems that a combination of these two approaches is needed.

In fact, Lorber's (1994) work seems to suggest that effective gender rebellion requires a more blatant approach—bearded men in dresses, perhaps, or more active responses to misattribution. For example, if I corrected every person who called me "Sir," and if I insisted on my right to be addressed appropriately and granted access to women-only spaces, then perhaps I could start to break down gender norms. If I asserted my right to use public facilities without being harassed, and if I challenged each person who gave me "the look," then perhaps I would be contributing to the demise of gender as we know it. It seems that the key would be to provide visible evidence of the nonmutual exclusivity of the categories. Would *this* break down the patriarchal components of gender? Perhaps it would, but it also would be exhausting.

Perhaps there is another possibility. In a recent book, *The Gender Knot,* Johnson (1997) argued that when it comes to gender and patriarchy, most of us follow the paths of least resistance; we "go along to get along," allowing our actions to be shaped by the gender system. Collectively, our actions help patriarchy maintain and perpetuate a system of oppression and privilege. Thus, by withdrawing our support from this system by choosing paths of greater resistance, we can start to chip away at it. Many people participate in gender because they cannot imagine any alternatives. In my classroom, and in my interactions and encounters with strangers, my presence can make it difficult for people not to see that there *are* other paths. In other words, following from West and Zimmerman (1987), I can subvert gender by doing it differently.

For example, I think it is true that my existence does not have an effect on strangers who assume that I am a man and never learn otherwise. For them, I do uphold the two-and-only-two system. But there are other cases in which my existence can have an effect. For example, when people initially take me for a man but then find out that I actually am a woman, at least for that moment, the naturalness of gender may be called into question. In these cases, my presence can provoke a "category crisis" (Garber 1992, 16) because it challenges the sex/gender binary system.

The subversive potential of my gender might be strongest in my classrooms. When I teach about the sociology of gender, my students can see me as the embodiment of the social construction of gender. Not all of my students have transformative experiences as a result of taking a course with me; there is the chance that some of them see me as a "freak" or as an exception. Still, after listening to stories about my experiences with gender and reading literature on the subject, many students begin to see how and why gender is a social product. I can disentangle sex, gender, and sexuality in the contemporary United States for them. Students can begin to see the connection between biographical experiences and the structure of society. As one of my students noted, I clearly live the material I am teaching. If that helps me to get my point across, then perhaps I am subverting the binary gender system after all. Although my gendered presence and my way of doing gender might make others—and sometimes even me—uncomfortable, no one ever said that dismantling patriarchy was going to be easy.

Notes

1. Ethnomethodology has been described as "the study of commonsense practical reasoning" (Collins 1988, 274). It examines how people make sense of their everyday experiences. Ethnomethodology is particularly useful in studying gender because it helps to uncover the assumptions on which our understandings of sex and gender are based.

2. I obviously have left much out by not examining my gendered experiences in the context of race, age, class, sexuality, region, and so forth. Such a project clearly is more complex. As Weston pointed out gender presentations are complicated by other statuses of their presenters: "What it takes to kick a person over into another gendered category can differ with race, class, religion, and time" (1996, 168). Furthermore, I am well aware that my whiteness allows me to assume that my experiences are simply a product of gender (see, e.g., hooks 1981; Lucal 1996; Spelman 1988; West and Fenstermaker 1995). For now, suffice it to say that it is my privileged position on some of these axes and my more disadvantaged position on others that combine to delineate my overall experience.

3. In fact, such experiences are not always limited to encounters with strangers. My grandmother, who does not see me often, twice has mistaken me for either my brother-in-law or some unknown man.

4. My experiences in rest rooms and other public spaces might be very different if I were, say, African American rather than white. Given the stereotypes of African American men, I think that white women would react very differently to encountering me (see, e.g., Staples [1986] 1993).

5. I also have noticed that there are certain types of rest rooms in which I will not be verbally challenged; the higher the social status of the place, the less likely I will be harassed. For example, when I go to the theater, I might get stared at, but my presence never has been challenged.

6. An anonymous reviewer offered one possible explanation for this. Women see women's rest rooms as their space; they feel safe, and even empowered,

there. Instead of fearing men in such space, they might instead pose a threat to any man who might intrude. Their invulnerability in this situation is, of course, not physically based but rather socially constructed. I thank the reviewer for this suggestion.

References

Anzaldua, G. 1987. *Borderlands/La Frontera.* San Francisco: Aunt Lute Books.

Berger, P. 1963. *Invitation to sociology.* New York: Anchor.

Bordo, S. 1993. *Unbearable weight.* Berkeley: University of California Press.

Butler, J. 1990. *Gender trouble.* New York: Routledge.

Collins, P. H. 1991. *Black feminist thought.* New York: Routledge.

Collins, R. 1988. *Theoretical sociology.* San Diego: Harcourt Brace Jovanovich.

Devor, H. 1989. *Gender blending: Confronting the limits of duality.* Bloomington: Indiana University Press.

Feinberg, L. 1996. *Transgender warriors.* Boston: Beacon.

Garber, M. 1992. *Vested interests: Cross-dressing and cultural anxiety.* New York: HarperPerennial.

Gardner, C. B. 1995. *Passing by: Gender and public harassment.* Berkeley: University of California.

Garfinkel, H. 1967. *Studies in ethnomethodology.* Englewood Cliffs, NJ: Prentice Hall.

Goffman, E. 1959. *The presentation of self in everyday life.* Garden City, NY. Doubleday.

———. 1963. *Stigma.* Englewood Cliffs, NJ: Prentice Hall.

———. 1967. *Interaction ritual.* New York: Anchor/ Doubleday.

———. 1976. Gender display. *Studies in the Anthropology of Visual Communication* 3:69–77.

———.1977. The arrangement between the sexes. *Theory and Society* 4:301–31.

hooks, b. 1981. *Ain't I a woman: Black women and feminism.* Boston: South End Press.

Howard, J. A., and J. Hollander. 1997. *Gendered situations, gendered selves.* Thousand Oaks, CA: Sage.

Johnson, A. G. 1997. *The gender knot: Unraveling our patriarchal legacy.* Philadelphia: Temple University Press.

Kessler, S. J., and W. McKenna. 1978. *Gender: An ethnomethodological approach.* New York: John Wiley.

Krieger, S. 1991. *Social science and the self.* New Brunswick, NJ: Rutgers University Press.

Lorber, J. 1994. *Paradoxes of gender.* New Haven, CT. Yale University Press.

———. 1996. Beyond the binaries: Depolarizing the categories of sex, sexuality, and gender. *Sociological Inquiry* 66:143–59.

Lucal, B. 1996. Oppression and privilege: Toward a relational conceptualization of race. *Teaching Sociology* 24:245–55.

———. 1997. "Hey, this is the ladies' room!": Gender misattribution and public harassment. *Perspectives on Social Problems* 9:43–57.

Martin, K. A. 1998. Becoming a gendered body: Practices of preschools. *American Sociological Review* 63:494–511.

Mills, C. W. 1959. *The sociological imagination.* London: Oxford University Press.

Myhre, J. R. M. 1995. One bad hair day too many, or the hairstory of an androgynous young feminist. In *Listen up: Voices from the next feminist generation,* edited by B. Findlen. Seattle, WA: Seal Press.

Reinharz, S. 1992. *Feminist methods in social research.* New York: Oxford University Press.

Richardson, L. 1994. Writing: A method of inquiry. In *Handbook of Qualitative Research,* edited by N. K. Denzin and Y. S. Lincoln. Thousand Oaks, CA: Sage.

Riemer, J. W. 1977. Varieties of opportunistic research. *Urban Life* 5:467–77.

———. 1988.Work and self. In *Personal sociology,* edited by P. C. Higgins and J. M. Johnson. New York: Praeger.

Spelman, E. V. 1988. *Inessential woman: Problems of exclusion in feminist thought.* Boston: Beacon.

Staples, B. 1993. Just walk on by. In *Experiencing race, class, and gender in the United States,* edited by V. Cyrus. Mountain View, CA: Mayfield. (Originally published 1986)

West, C., and S. Fenstermaker. 1995. Doing difference. *Gender & Society* 9:8–37.

West, C., and D. H. Zimmerman. 1987. Doing gender. *Gender & Society* 1:125–51.

Zimmerman, D. H. 1992. They were all doing gender, but they weren't all passing: Comment on Rogers. *Gender & Society* 6:192–98.

Introduction to Reading 3

Sharon Preves, a sociologist, has done ground-breaking research on what it is like to live in contemporary America with an intersexed body that doesn't conform to "standard" medical definitions of male and female. She recruited subjects for her study by contacting established intersex support groups in the United States and Canada, yet finding willing interviewees was difficult because intersexuality is largely invisible and intersexed people generally don't self-identify as intersexed. Despite this challenge, Preves interviewed thirty-seven intersexed adults in a private, face-to-face format, largely in participants' homes. Her book, from which this excerpt is taken, is built upon her subjects' personal narratives.

1. Why does the author state that "bodies that are considered normal or abnormal are not inherently this way"?
2. Why is intersexuality a social, not medical, problem?
3. What themes are shared between this reading and the piece by Lucal?

Beyond Pink and Blue

Sharon E. Preves

Recently I participated in a cultural diversity fieldtrip with twenty-two second graders in St. Paul, Minnesota. When I arrived at their school, the kids were squirrelly with anticipation. They were a colorful and varied bunch—some were tall and thin, others short and stout. Moreover, they were from a variety of racial and ethnic backgrounds and spoke nearly a half dozen native languages. When it was time to begin our community walking tour, the teachers attempted to bring the busy group to order quickly. How did they go about doing so? They told the children to form two lines: one for girls and the other for boys. The children did so seamlessly because they had been asked to line up in this manner countless number of times before. Within moments, the children were quiet and attentive. I was struck then, as I had been many times before, by how often and in the most basic ways societies are organized by a distinction between sexes. Even with children of every shape and color, the gender divide worked as a sure way to bring order to chaos. "Girls in one line, boys in the other." But sometimes the choice between the lines—and sexes—isn't so easy.

Which line would you join? Think about it seriously for a minute. How do you know whether to line up with the girls or the boys? For that matter, what sex or gender *are* you and how did you *become* the gender you are? Moreover, how do you *know* what sex and gender you are? Who decides? These questions may seem ridiculous. You may be saying to yourself, "Of course I know what gender I am; forget this book." But really stop to think about how you know what sex you are and how you acquired your gender. Most of us have been taught that sex is anatomical and gender is social. What's more, many of us have never had the occasion to explore our gender or sexual identities, because neither has given us cause for reflection. Much like Caucasians who say they "have none" when asked to explore their racial identity, many women and men find it difficult to be reflective about how they know and "do" gender.[1]

This article explores what happens to people who, from the time of their birth or early adolescence, inhabit bodies whose very anatomy does not afford them an easy choice between the gender lines.[2] Every day babies are born with bodies that are deemed sexually ambiguous, and with regularity they are surgically altered to reflect the sexual anatomy associated with "standard" female or male sex assignment. There are numerous ways to respond to this plurality of physical type, including no response at all. Because sex and gender operate as inflexible and central organizing principles of daily existence in this culture, such indifference is rare if not nonexistent. Instead, interference with sex and gender norms are cast as a major disturbance to social order, and people go to remarkable lengths to eradicate threats to the norm, even though they occur with great regularity.

Recent estimates indicate that approximately one or two in every two thousand infants are born with anatomy that some people regard as sexually ambiguous. These frequency estimates vary widely and are, at best, inconclusive. Those I provide here are based on an exhaustive review of recent medical literature.[3] This review suggests that approximately one or two per two thousand children are born with bodies considered appropriate for genital reconstruction surgery because they do not conform to socially accepted norms of sexual anatomy. Moreover, nearly 2 percent are born with chromosome, gonad, genital, or hormone features that could be considered "intersexed"; that is, children born with ambiguous genitalia, sexual organs, or sex chromosomes. Additional estimates report the frequency of this sexual variance as comprising approximately 1 to 4 percent of all births.[4]

These estimates differ so much because definitions of sexual ambiguity vary tremendously.[5] This is largely because distinctions between female and male bodies are actually on more of a continuum rather than a dichotomy. The criteria for what counts as female or male, or sexually ambiguous for that matter, are human standards. That is, bodies that are considered normal or abnormal are not inherently that way. They are, rather, classified as aberrant or customary by social agreement.[6] We have, as humans, created categories for bodies that fit the norm and those that don't, as well as a systematic method of surgically attempting to correct or erase sexual variation. That we have done so is evidence of the regularity with which sexual variation occurs.

Melanie Blackless and colleagues suggest that the total frequency of nongenital sexual variation (cases of intersexed chromosomes or internal sexual organs) is much higher than one in two thousand.[7] They conclude that using a more inclusive definition of sexual ambiguity would yield frequency estimates closer to one or two per one hundred births, bringing us back to the 1 to 2 percent range.

To put these numbers in perspective, although its occurrence has only recently begun to be openly discussed, physical sexual ambiguity occurs about as often as the well-known conditions of cystic fibrosis and Down syndrome.[8] Since there are approximately four million babies born annually in the United States, a conservative estimate is that about two to four thousand babies are born per year in this country with features of their anatomy that vary from the physical characteristics typically associated with females and males.[9] Some are born with genitalia that are difficult to characterize as clearly female or male. Others have sex chromosomes that are neither XX nor XY, but some other combination, such as X, XXY, or chromosomes that vary throughout the cells of their bodies, changing from XX to XY from cell to cell. Still others experience unexpected physical changes at puberty, when their bodies exhibit secondary sex characteristics that are surprisingly "opposite" their sex of assignment. Some forms of sexual ambiguity are inherited genetically, while others are brought on by hormonal activity during gestation, or by prescription medication women take during pregnancy. Regardless of its particular manifestation or cause, most forms of physical sexual anatomy that vary from the norm are medically classified and treated as forms of intersexuality, or hermaphroditism.

Take Claire's experience as an example.[10] Claire is a middle-class white woman and mother of two teenage daughters who works as a writer and editor. She was forty-four years old when she conveyed the following story to me during a four-hour interview that took place in her home. Claire underwent a clitorectomy when she was six years old at her parents' insistence, after clinicians agreed that her clitoris was just "too large" and they had to intervene. The size of her clitoris seemed to cause problems not for young Claire, but for the adults around her. Indeed, there was nothing ambiguous about Claire's sex before the surgery. She has XX chromosomes, has functioning female reproductive organs, and later in life went through a physically uneventful female puberty. Claire's experience illustrates that having a large clitoris is perceived as a physical trait dangerous to existing notions of gender and sexuality, despite the sexual pleasure it could have given Claire and her future sexual partners. In fact, doctors classify a large clitoris as a medical condition referred to as "clitoral megaly" or "clitoral hypertrophy." Conversely, small penises for anatomical boys, are classified as a medical problem called "micropenis."

Reflecting on the reasons for the clitorectomy she underwent at the age of six, Claire said, "I don't feel that my sex was ambiguous at all. There was never that question. But I'm sure that [clitorectomies] have been done forever because parents just [don't] like big clitorises because they look too much like a penis." Even more alarming are the physical and emotional outcomes of genital surgery that might be experienced by the patient. About the after effects of her surgery, Claire said,

> They just took the clitoris out and then whip stitched the hood together, so it's sort of an odd-looking thing. I don't know what they were hoping to preserve, although I remember my father thinking that if someone saw me, it would look normal because there's just a little skin poking out between my lips so it wouldn't look strange. I remember I was in the hospital for five days. And then it just got better and everything was forgotten, until I finally asked about it when I was twelve. [There was] total and complete silence. You know, it was never, never mentioned. I know you know what that does. I was just in agony trying to figure out who I was. And, you know, why . . . what sex I was. And feeling like a freak, which is a very common story. And then when I was twelve, I asked my father what had been done to me. And his answer was, "Don't be so self-examining." And that was it. I never asked again [until I was thirty-five].

During the course of my research I spoke with many other adults across North America who had childhood experiences remarkably similar to Claire's. Their stories are laden with family and medical secrecy, shame, and social isolation, as well as perseverance and strength of spirit, and eventual pride in their unique bodies and perspectives.

Personal Narratives

Being labeled as a misfit, by peers, by family members, or by medical diagnosis and treatment, is no doubt a challenge to one's identity development and stability. This is especially true for children whose bodies render traditional gender classification ineffective, for there is seemingly no place to belong without being gendered, especially during childhood. Negotiating identity, one's basic sense of place and self, is a challenge for many of us, and is potentially far more challenging for people whose sex is called into question. The social expectation for gender stability and conformity is prevalent across social spheres. Nearly every aspect of social life is organized by one's sex assignment—from schooling and relationships, to employment and religion, sports and entertainment, medicine and law. Because North American cultures are structured by gender, successful participation in society's organizations and personal relationships requires gender categorization. Many of us negotiate questions of sex and gender with little effort. Others, however, do not have the luxury or ease of fitting neatly into a dual-gendered culture.

In an attempt to understand how intersexuals experience and cope with their marginality in a society that demands sexual conformity, I turned to them directly for answers. In the end, I interviewed thirty-seven individuals throughout North America whose bodies have been characterized by others as intersexed.

Intersex Is a Social, Not Medical, Problem

While being born with indeterminate sexual organs indeed problematizes a binary understanding of sex and gender, several studies show—and there seems to be general consensus (even among the doctors performing the "normalizing" operations)—that most children with ambiguous sexual anatomy do not require medical intervention for their physiological health.[19] Nevertheless, the majority of sexually ambiguous infants are medically assigned a definitive sex, often undergoing repeated genital surgeries and ongoing hormone treatments, to "correct" their variation from the norm.

It is my argument that medical treatments to create genitally *unambiguous* children are not performed entirely or even predominantly for the sake of preventing stigmatization and trauma to the child. Rather, these elaborate, expensive, and risky procedures are performed to maintain social order for the institutions and adults that surround that child. Newborns are completely oblivious to the rigid social conventions to which their families and caregivers adhere. Threats to the duality of sex and gender undermine inflexibly gendered occupational, education, and family structures, as well as heterosexuality itself. After all, if one's sex is in doubt, how would they identify their sexual orientation, given that heterosexuality, homosexuality, and even bisexuality are all based on a sexual binary? So, when adults encounter a healthy baby with a body that is not easily "sexed," they may understandably experience an inability to imagine a happy and successful future for that child. They may wonder how the child will fit in at school and with its peers, and how the child will negotiate dating and sexuality, as well as family and a career. But most parents don't find a real need to address these questions until years after a child's birth. Furthermore, it is my contention that parents and caregivers of intersexed children don't need to be so concerned about addressing the "personal troubles" of their children either. Rather, we should all turn our attention to the "public issues" and problems wrought by unwavering, merciless adherence to sex and gender binarism.[20]

That medical sex assignment procedures could be considered cosmetic raises several important questions, including the human influences in constructing definitions of health and pathology. Bodies are classified as healthy or pathological (or as normal and abnormal) through social expectations, human discourse, and human interaction. As a result, what is seen as normal or standard in one culture and time is seen as aberrant and strange in another. Indeed, deviance itself is created socially through human actions, beliefs, and judgments.[21] Consider Gilbert Herdt's anthropological research in Papua New Guinea. The Sambia males of Papua New Guinea believe that they become masculine by ingesting the semen of adult men. In order to become virile, therefore, Sambian boys perform oral sex on and ingest the semen of adult men.[22] According to Herdt, such activity is considered a standard rite of passage to manhood among the Sambia in Papua New Guinea, much like a bar or bat mitzvah is seen as a customary ritual of young adulthood among Jews.

I offer the above example to illustrate that definitions of normal and abnormal vary tremendously by culture. That said, there is no reason to consider intersex as necessarily problematic in itself. In fact, since physical sexual ambiguity has been shown to be a cause of health problems in only very rare cases, sexual ambiguity could be considered a *social* problem, rather than a physical problem. Rare physical problems do occur in cases where eliminating bodily waste, such as urine and feces, is difficult because of internal physiological complications or infrequent cases of salt-wasting congenital adrenal hyperplasia, which is a condition where children have hyperactive adrenal glands and hormone therapy is required to regulate the endocrine system.[23] Because Western cultures place such strong emphasis on sexual (and other forms of) categorization, intersex ambiguity causes major social disruption and discomfort. If there were less concern about gender, there would be less concern about gender variation. Because intersex is often identified in a medical setting by physicians during childbirth or during a pediatric appointment in later childhood, the social response to intersex "deviance" is largely medical.

A Brief History of Sexual Ambiguity in Medicine

According to one theory, sex distinctions were based on a continuum of heat, with males' bodies being internally hotter than females,' thus creating the impetus for external male reproductive organs and colder, internal female organs.[16] In Thomas Laqueur's analysis, this differential temperature theory actually provided the basis for a one-sex conceptual model, with females being seen as the inverse of males.[17] (That is, female and male anatomy was viewed as identical, although the vagina was viewed as an inverted penis, the uterus as an internalized scrotum, the fallopian tubes as seminal vesicles, the ovaries as internal testicles, and so on).

In their historical review of medical literature, Myra Hird and Jenz Germon demonstrate that based on this humoral theory, sixteenth-century philosophers and physicians regarded hermaphroditism as evidence of two sexes existing in one body.[18] With the invention of microscopy and surgery on *living* patients at the beginning of the twentieth century (rather than upon autopsy after death), the notion of anatomical hermaphroditism was transformed. Hird and Germon argue that the newer concept of one "true sex" being hidden by sexually ambiguous anatomy didn't enter the fray until Klebs's classification system. Klebs identified five categories of sexual classification: female, male, female pseudohermaphrodite, male pseudohermaphrodite, and true hermaphrodite. Klebs identified people as true hermaphrodites only if they had the very rare combination of ovarian and testicular tissue in the same body. All other cases of hermaphroditism, for example people who had ambiguous genitalia but no combination of gonadal tissue, were identified as pseudohermaphrodites. Klebs's classification system served to drastically decrease the number of people who were defined as hermaphrodites, and thus reinforced the newly popular thought that there were two and only two sexes: female and male, with a very rare and unusual exception in the case of true hermaphroditism.

Early surgical attempts to regulate the appearance of sexual anatomy, such as lowering abdominal testicles, appeared in the beginning of the nineteenth century.[19] A primary motive for the social insistence upon outward displays of gender clarity was an underlying fear of homosexuality, or "hermaphroditism of the soul," a threat that was present in the sexually ambiguous (or, quite literally, *bisexual*) body of the hermaphrodite. By appearing, outwardly, to be of the "other" sex, it was feared hermaphrodites would tempt heterosexual partners into "homosexual" relations.[20]

As Alice Dreger, Myra Hird and Jenz Germon, and others point out, medical methods of response to physical sexual ambiguity have changed yet again in more recent times.[21] Unlike eighteenth- and nineteenth-century medical attempts to reveal a "true sex" that is disguised by pseudohermaphroditic (or false) sexual ambiguity, the current medical model relies on making a sex assignment that is most appropriate for heterosexual capacity. Hird and Germon refer to this as the "best sex" mentality in medical sex assignment.

Unanswered Questions

Despite the recent mobilization and activism of thousands of intersexuals worldwide, the social focus on gender categorization and genital regularity continues to be strong, in both medical and nonmedical sectors. In order to improve the quality of life not just for those labeled intersexed, but for us all, we must remove or reduce the importance of gender categorization and the need for gender categories, including the category of intersex itself. A more realistic and tangible goal is to respond far less to sexual variance. That is, to focus on the health of children born with genital variation, not on their difference. An outcome of this philosophy would prevent physicians from cutting into the bodies of intersexed patients unless a clear physiological need presents its necessity. Moreover, this philosophy would ask us to hold off on rushing kids into a patient mentality and instead send them to speak with counselors or other kids with divergent sexual anatomy. We should refrain from identifying them as different in any way, unless a child demonstrates a need and a desire for such special attention. Pushing the label of sexual ambiguity, sexual difference, or "sexual problem" onto children through medicalization, or remedicalization via social services, leads to stigmatization of the self. What's more, there is no inherent need for children to have therapy unless a need presents itself. We shouldn't restigmatize intersexuals by assuming a need for therapy or "preventative treatment" on the basis of physical variation alone. While these support resources are invaluable for people who have already been adversely affected by negative socialization—and family members and clinicians should be prepared to call on support services if necessary—we shouldn't assume that genital variation itself creates a pathological need.

Will the category and identity "intersex" disappear altogether if doctors stop treating, studying, and classifying children in this way, or will this category continue to expand further into academic and social realms until its presence is solidified in a way similar to the presence of transgender, gay, or alcoholic identities? Focusing less on articulating the category "intersex" may indeed be a step in the right direction toward the ultimate goal of focusing less on gender and sexual categorization. After all, returning to the elementary school children and the "gender lines" to which they are socialized to adhere, the clear message from my research is to decrease sex and gender categorization rather than to create yet a third rigid sex or gender line for us all to ponder. Instead, we should focus on loving and accepting children as they are, not because of or in spite of their differences, but rather just because they are terrific kids in their own right with or without bodies that vary from some mythical standard.

While my research addresses many questions about the construction and negotiation of contested sex, several important issues related to assessing and revising the treatment of sexual ambiguity remain untouched. Future research in this area should include intensive research on two additional groups: parents of intersex children and clinicians who specialize in intersex management. This research should focus on parents' and clinicians' personal and professional experiences with intersexuality and their perspectives on genital variation and the management thereof. Additional research should include observation of family consultations with medical staff and follow-up visits regarding the "sexing" of intersexed infants, in order to offer further insight into the interactional aspect of the social construction of sex and gender. Timely research in this area may provide further empirical data to inform the current debate and potential shift in intersex clinical management.

Despite my systematic analysis of former patients' experiences with medical sex assignment, the void remains for a rigorous analysis of the experiences of those with variant sexual anatomy who are not associated with intersex support and advocacy organizations and those who did not undergo medical sex assignment. This research is critical because proponents of surgical sex assignment will most likely resist clinical reform until the experiences of these populations are better understood. Until then, critics of my research and that of others will continue to shrug off the overwhelming dissatisfaction and trauma reported by intersex activists as representative of a disgruntled but vocal minority.

Far more could also be learned about the impact of peer social support as a means of reframing sexual ambiguity in a more positive light. In order to develop effective recommendations for clinical practice as they relate to peer advocacy, further investigation is needed that focuses specifically on the role of social support in coping with difference. Clinical mandates should also be informed by further research on the impact of age-appropriate disclosure of information to intersexuals and their family members. Developing a longitudinal research program with this focus is possible now that some children deemed sexually ambiguous are being raised in a new era of social support and have access to complete and accurate diagnostic information.

Finally, the strategies and efforts of intersex activists warrant further attention, as does the overall concept of the intersex social movement. Due to their prolific appearances in print, film, and radio media, substantive content analysis of their activism efforts is possible. This analysis could inform a more nuanced understanding of models of coming-out and community empowerment rather than assimilation to an existing social norm. This research is now possible because there are already substantial archives related to this area, and because intersex patient, parent, and doctor communication, education, and support networks continue to emerge.

Notes

1. West and Zimmerman 1987.
2. While there is tremendous cross-cultural variation in responding to physical sexual variation, I explore how North American intersexuals experience and cope with being labeled sexually ambiguous in a culture that demands sexual conformity.
3. Blackless et al. 2000.
4. Money 1989; Edgerton 1964; Fiedler 1978.
5. Dreger 1998b; Kessler 1998.
6. Hird 2000.
7. Blackless et al. 2000.
8. See Dreger 1998b:43; Desai 1997; and Roberts et al. 1998.
9. National Center for Health Statistics (2001). Note that others project an annual birthrate of 1,500–2,000 intersexed children in the United States (Beh and Diamond 2000).
10. Most research participants chose pseudonyms for themselves to be used in the study. Perhaps as evidence of their desire to overcome the secrecy and shame they associated with attempts to erase their intersexuality, 27 percent of those I interviewed chose to use their real names. I do not distinguish here or elsewhere between those who chose pseudonyms and those who did not.
11. Diamond and Sigmundson 1997b; Dreger 1998a; Kessler 1998.
12. Mills 1959:8.
13. Newman 2002; Becker 1963.
14. Herdt 1994, 1998.
15. Wilson and Reiner 1998; Kessler 1998; Diamond and Sigmundson 1997b.
16. Note that Aristotle valued "male" heat over "female" cold and viewed females' lack of heat as a sign of inferiority and even deformity (Cadden 1993).
17. Laqueur 1990.
18. Hird and Germon 2001.
19. Pagliassotti 1993.
20. Hekma 1994; Dreger 1995a.
21. Dreger 1998b; Hird and Germon 2001.

References

Becker, Howard. 1963. *The Outsiders.* New York: Free Press.

Blackless, Melanie, Anthony Charuvastra, Amanda Derryck, Anne Fausto-Sterling, Karl Lauzanne, and Ellen Lee. 2000. "How Sexually Dimorphic Are We?" *American Journal of Human Biology* *12*(2):151–166.

Cadden, Joan. 1993. *Meanings of Sex Differences in the Middle Ages: Medicine, Science, and Culture.* New York: Cambridge University Press.

Desai, Sindoor S. 1997. "Down Syndrome: A Review of the Literature." *Oral Surgery, Oral Medicine, Oral Pathology, Oral Radiology, and Endodontics.* September. *84*(3):279–285.

———. 1997b. "Management of Intersexuality: Guidelines for Dealing with Persons with Ambiguous Genitalia." *Archives of Pediatric Adolescent Medicine* 151(October):1046–1050.

Dreger, Alice Domurat. 1995a *Doubtful Sex: Cases and Concepts of Hermaphroditism in France and Britain, 1868–1915.* Ph.D. dissertation, Indiana University.

———. 1995b. "Doubtful Sex: The Fate of the Hermaphrodite in Victorian Medicine." *Victorian Studies* (Spring):335–370.

———. 1998a. "Ambiguous Sex'—or Ambivalent Medicine? Ethical Issues in the Treatment of Intersexuality." *Hastings Center Report 28*(3): 24–36.

———. 1998b. *Hermaphrodites and the Medical Invention of Sex.* Cambridge, Mass.: Harvard University Press.

Hekma, Gert. 1994. "A Female Soul in a Male Body: Sexual Inversion as Gender Inversion in Nineteenth-Century Sexology." In *Third Sex, Third Gender: Beyond Sexual Dimorphism in Culture and History,* ed. Gilbert Herdt, 213–239. New York: Zone Books.

Herdt, Gilbert. 1994. "Mistaken Sex: Culture, Biology and the Third Sex in New Guinea." In *Third Sex, Third Gender: Beyond Sexual Dimorphism in Culture and History,* ed. Gilbert Herdt, 419–445. New York: Zone Books.

———. 1998. *Same Sex, Different Cultures: Exploring Gay and Lesbian Lives.* Boulder, Colo. : Westview Press.

Hird, Myra J. 2000. "Gender's Nature: Intersexuality, Transsexualism and the 'Sex'/'Gender' Binary." *Feminist Theory 1*(3): 347–364.

Hird, Myra J., and Jenz Germon. 2001. "The Intersexual Body and the Medical Regulation of Gender." In *Constructing Gendered Bodies,* ed. Kathryn Backett-Milburn and Linda McKie, 162–178. New York: Palgrave.

———. 1998. *Lessons from the Intersexed.* New Brunswick, N.J.: Rutgers University Press.

Laqueur, Thomas. 1990. *Making Sex: Body and Gender from the Greeks to Freud.* Cambridge, Mass.: Harvard University Press.

Mills, C. Wright. 1959. *The Sociological Imagination.* New York: Oxford University Press.

———. 1989. *The Geraldo Rivera Show.* "Hermaphrodites: The Sexually Unfinished."

National Center for Health Statistics. 2001. "Births: Preliminary Data for 2000." NVSR 49, No. 5. 20 pp. (PHS) 2001–1120.

Nederman, Cary J., and Jacqui True. 1996. "The Third Sex: The Idea of the Hermaphrodite in Twelfth-Century Europe." *Journal of the History of Sexuality* 6(4):497–517.

Newman, David. 2002. *Sociology: Exploring the Architecture of Everyday Life.* 4th ed. Thousand Oaks, Calif.: Pine Forge Press.

Pagliassotti, Druann. 1993. "On the Discursive Construction of Sex and Gender." *Communication Research 20*(3):472–493.

Roberts, Helen E., Janet D. Cragan, Joanne Cono, Muin J. Khoury, Mark R. Weatherly, and Cynthia A. Moore. 1998. "Increased Frequency of Cystic Fibrosis among Infants with Jejunoileal Atresia." *American Journal of Medical Genetics* 78: 446–449.

West, Candace, and Don H. Zimmerman. 1987. "Doing Gender." *Gender & Society 1*(2): 125–151.

Wilson, Bruce E., and William G. Reiner. 1998. "Management of Intersex: A Shifting Paradigm." *Journal of Clinical Ethics* 9(4):360–369.

Introduction to Reading 4

Robert Sapolsky is professor of biology and neurology at Stanford University and a research associate with the Institute of Primate Research, National Museum of Kenya. He is the author of *The Trouble with Testosterone, Why Zebras Don't Get Ulcers* and, most recently, *A Primate's Memoir*. Sapolsky has lived as a member of a baboon troop in Kenya, conducting cutting-edge research on these beautiful and complex primates. In this article, he uses his keen wit and scientific understanding to debunk the widely held myth that testosterone causes aggression in males.

1. Why do many Americans want to believe that biological factors, such as hormones, are the basis of gender differences and inequalities?
2. Sapolsky says that hormones have a "permissive effect." What does "permissive effect" mean in terms of the relationship between testosterone and aggression?
3. How does research on testosterone, male monkeys, and spotted hyenas help one to grasp the role of social factors and environment in behavioral biology?

The Trouble with Testosterone

Robert Sapolsky

Face it, we all do it. We all believe in certain stereotypes about certain minorities. The stereotypes are typically pejorative and usually false. But every now and then, they are true. I write apologetically as a member of a minority about which the stereotypes are indeed true. I am male. We males account for less than 50 percent of the population, yet we generate an incredibly disproportionate percentage of the violence. Whether it is something as primal as having an ax fight in an Amazonian clearing or as detached as using computer-guided aircraft to strafe a village, something as condemned as assaulting a cripple or as glorified as killing someone wearing the wrong uniform, if it is violent, males excel at it. Why should that be? We all think we know the answer. A dozen millennia ago or so, an adventurous soul managed to lop off a surly bull's testicles and thus invented behavioral endocrinology. It is unclear from the historical records whether this individual received either a grant or tenure as a result of this experiment, but it certainly generated an influential finding something or other comes out of the testes that helps to make males such aggressive pains in the ass.

That something or other is testosterone.[1] The hormone binds to specialized receptors in muscles and causes those cells to enlarge. It binds to similar receptors in laryngeal cells and gives rise to operatic basses. It causes other secondary sexual characteristics, makes for relatively unhealthy blood vessels, alters biochemical events in the liver too dizzying to even contemplate, has a profound impact, no doubt, on the workings of cells in big toes. And it seeps into the brain, where it binds to those same "androgen" receptors and influences behavior in a way highly relevant to understanding aggression.

What evidence links testosterone with aggression? Some pretty obvious stuff. Males tend to have higher testosterone levels in their circulation than do females (one wild exception to that will be discussed later) and to be more aggressive. Times of life when males are swimming in testosterone (for example, after reaching puberty) correspond to when aggression peaks. Among numerous species, testes are mothballed most of the year, kicking into action and pouring out testosterone only during a very circumscribed mating season—precisely the time when male–male aggression soars.

Impressive, but these are only correlative data, testosterone repeatedly being on the scene with no alibi when some aggression has occurred. The proof comes with the knife, the performance of what is euphemistically known as a "subtraction" experiment. Remove the source of testosterone in species after species and levels of aggression typically plummet. Reinstate normal testosterone levels afterward with injections of synthetic testosterone, and aggression returns.

To an endocrinologist, the subtraction and replacement paradigm represents pretty damning proof: this hormone is involved. "Normal testosterone levels appear to be a prerequisite for normative levels of aggressive behavior" is the sort of catchy, hummable phrase that the textbooks would use. That probably explains why you shouldn't mess with a bull moose during rutting season. But that's not why a lot of people want to understand this sliver of science. Does the action of this hormone tell us anything about individual differences in levels of aggression, anything about why some males, some human males, are exceptionally violent? Among an array of males—human or otherwise—are the highest testosterone levels found in the most aggressive individuals?

Generate some extreme differences and that is precisely what you see. Castrate some of the well-paid study subjects, inject others with enough testosterone to quadruple the normal human levels, and the high-testosterone males are overwhelmingly likely to be the more aggressive ones. However, that doesn't tell us much about the real world. Now do something more subtle by studying the normative variability in testosterone—in other words, don't manipulate anything, just see what everyone's natural levels are like—and high levels of testosterone and high levels of aggression still tend to go together. This would seem to seal the case—interindividual differences in levels of aggression among normal individuals are probably driven by differences in levels of testosterone. But this turns out to be wrong.

Okay, suppose you note a correlation between levels of aggression and levels of testosterone among these normal males. This could be because (a) testosterone elevates aggression; (b) aggression elevates testosterone secretion; (c) neither causes the other. There's a huge bias to assume option a, while b is the answer. Study after study has shown that when you examine testosterone levels when males are first placed together in the social group, testosterone levels predict nothing about who is going to be aggressive. The subsequent behavioral differences drive the hormonal changes, rather than the other way around.

Because of a strong bias among certain scientists, it has taken forever to convince them of this point. Behavioral endocrinologists study what behavior and hormones have to do with each other. How do you study behavior? You get yourself a notebook and a stopwatch and a pair of binoculars. How do you measure the hormones? You need a gazillion-dollar machine, you muck around with radiation and chemicals, wear a lab coat, maybe even goggles—the whole nine yards. Which toys would you rather get for Christmas? Which facet of science are you going to believe in more? Because the endocrine aspects of the business are more high-tech, more reductive, there is the bias to think that it is somehow more scientific, more powerful. This is a classic case of what is often called physics envy, the disease among scientists where the behavioral biologists fear their discipline lacks the rigor of physiology, the physiologists wish for the techniques of the biochemists, the biochemists covet the clarity of the answers revealed by the molecular biologists, all the way down until you get to the physicists, who confer only with God.[2] Hormones seem to many to be more real, more substantive, than the ephemera of behavior, so when a correlation occurs, it must be because hormones regulate behavior, not the other way around.

As I said, it takes a lot of work to cure people of that physics envy, and to see that interindividual differences in testosterone levels don't predict subsequent differences in aggressive behavior among individuals. Similarly, fluctuations in testosterone levels within one individual over time do not predict subsequent changes in the levels of aggression in that one individual—get a hiccup in testosterone secretion one afternoon and that's not when the guy goes postal.

Look at our confusing state: normal levels of testosterone are a prerequisite for normal levels of aggression, yet changing the amount of testosterone in someone's bloodstream within the normal range doesn't alter his subsequent levels of aggressive behavior. This is where, like clockwork, the students suddenly start coming to office hours in a panic, asking whether they missed something in their lecture notes.

Yes, it's going to be on the final, and it's one of the more subtle points in endocrinology—what is referred to as a hormone having a "permissive effect." Remove someone's testes and, as noted, the frequency of aggressive behavior is likely to plummet. Reinstate precastration levels of testosterone by injecting that hormone, and precastration levels of aggression typically return. Fair enough. Now this time, castrate an individual and restore testosterone levels to only 20 percent of normal and . . . amazingly, normal precastration levels of aggression come back. Castrate and now generate twice the testosterone levels from before castration—and the same level of aggressive behavior returns. You need some testosterone around for normal aggressive behavior—zero levels after castration, and down it usually goes; quadruple it (the sort of range generated in weight lifters abusing anabolic steroids), and aggression typically increases. But anywhere from roughly 20 percent of normal to twice normal and it's all the same; the brain can't distinguish among this wide range of basically normal values.

We seem to have figured out a couple of things by now. First, knowing the differences in the levels of testosterone in the circulation of a bunch of males will not help you much in figuring out who is going to be aggressive. Second, the subtraction and reinstatement data seem to indicate that, nevertheless, in a broad sort of way, testosterone causes aggressive behavior. But that turns out not to be true either, and the implications of this are lost on most people the first thirty times you tell them about it. Which is why you'd better tell them about it thirty-one times, because it is the most important point of this piece.

Round up some male monkeys. Put them in a group together, and give them plenty of time to sort out where they stand with each other—affiliative friendships, grudges and dislikes. Give them enough time to form a dominance hierarchy, a linear ranking system of numbers 1 through 5. This is the hierarchical sort of system where number 3, for example, can pass his day throwing around his weight with numbers 4 and 5, ripping off their monkey chow, forcing them to relinquish the best spots to sit in, but, at the same time, remembering to deal with numbers 1 and 2 with shit-eating obsequiousness.

Hierarchy in place, it's time to do your experiment. Take that third-ranking monkey and give him some testosterone. None of this within-the-normal-range stuff. Inject a ton of it into him, way higher than what you normally see in a rhesus monkey; give him enough testosterone to grow antlers and a beard on every neuron in his brain. And, no surprise, when you then check the behavioral data, it turns out that he will probably be participating in more aggressive interactions than before. So even though small fluctuations in the levels of the hormone don't seem to matter much, testosterone still causes aggression. But that would be wrong. Check out number 3 more closely. Is he now raining aggressive terror on any and all in the group, frothing in an androgenic glaze of indiscriminate violence? Not at all. He's still judiciously kowtowing to numbers 1 and 2, but has simply become a total bastard to numbers 4 and 5. This is critical: testosterone isn't *causing* aggression, it's *exaggerating* the aggression that's already there.

Another example just to show we're serious. There's a part of your brain that probably has lots to do with aggression, a region called the amygdala.[3] Sitting right near it is the Grand Central Station of emotion-related activity in your brain, the hypothalamus. The amygdala communicates with the hypothalamus by way of a cable of neuronal connections called the stria terminalis. No more jargon, I promise. The amygdala has its influence on aggression via that pathway, with bursts of electrical excitation called action potentials that ripple down the stria terminals, putting the hypothalamus in a pissy mood.

Once again, do your hormonal intervention; flood the area with testosterone. You can do that by injecting the hormone into the bloodstream, where it eventually makes its way to this part of the brain. Or you can be elegant and surgically microinject the stuff directly into this brain region. Six of one, half a dozen of the other. The key thing is what doesn't happen next. Does testosterone now cause there to be action potentials surging down the stria terminalis? Does it turn on that pathway? Not at all. If and only if the amygdala is sending an aggression-provoking volley of action potentials down the stria terminalis, testosterone increases the rate of such action potentials by shortening the resting time between them. It's not turning on the pathway, it's increasing the volume of signaling if it is already turned on. It's not causing aggression, it's exaggerating the preexisting pattern of it, exaggerating the response to environmental triggers of aggression.

This transcends issues of testosterone and aggression. In every generation, it is the duty of behavioral biologists to try to teach this critical point, one that seems a maddening cliché once you get it. You take that hoary old dichotomy between nature and nurture, between biological influences and environmental influences, between intrinsic factors and extrinsic ones, and, the vast majority of the time, regardless of which behavior you are thinking about and what underlying biology you are studying, the dichotomy is a sham. No biology. No environment. Just the interaction between the two.

Do you want to know how important environment and experience are in understanding testosterone and aggression? Look back at how the effects of castration were discussed earlier. There were statements like "Remove the source of testosterone in species after species and levels of aggression typically plummet." Not "Remove the source . . . and aggression always goes to zero." On the average it declines, but rarely to zero, and not at all in some individuals. And the more social experience an individual had being aggressive prior to castration, the more likely that behavior persists sans cojones. Social conditioning can more than make up for the hormone.

Another example, one from one of the stranger corners of the animal kingdom: If you want your assumptions about the nature of boy beasts and girl beasts challenged, check out the spotted hyena. These animals are fast becoming the darlings of endocrinologists, sociobiologists, gynecologists, and tabloid writers. Why? Because they have a wild sex-reversal system—females are more muscular and more aggressive than males and are socially dominant over them, rare traits in the mammalian world. And get this: females secrete more of certain testosterone-related hormones than the males do, producing the muscles, the aggression (and, as a reason for much of the gawking interest in these animals, wildly masculinized private parts that make it supremely difficult to tell the sex of a hyena). So this appears to be a strong vote for the causative powers of high androgen levels in aggression and social dominance. But that's not the whole answer. High up in the hills above the University of California at Berkeley is the world's largest colony of spotted hyenas, massive bone-crunching beasts who fight with each other for the chance to have their ears scratched by Laurence Frank, the zoologist who brought them over as infants from Kenya. Various scientists are studying their sex-reversal system. The female hyenas are bigger and more muscular than the males and have the same weirdo genitals and elevated androgen levels that their female cousins do back in the savannah. Everything is in place except . . . the social system is completely different from that in the wild. Despite being stoked on androgens, there is a very significant delay in the time it takes for the females to begin socially dominating the males—they're growing up without the established social system to learn from.

When people first grasp the extent to which biology has something to do with behavior, even subtle, complex, human behavior, there is often an initial evangelical enthusiasm of the convert, a massive placing of faith in the biological components of the story. And this enthusiasm is typically of a fairly reductive type—because of physics envy, because reductionism is so impressive, because it would be so nice if there were a single gene or hormone or neurotransmitter or part of the brain that was it, the cause, the explanation of everything. And the trouble with testosterone is that people tend to think this way in an arena that really matters.

This is no mere academic concern. We are a fine species with some potential. Yet we are racked by sickening amounts of violence. Unless we are hermits, we feel the threat of it, often as a daily shadow. And regardless of where we hide, should our leaders push the button, we will all be lost in a final global violence. But as we try to understand and wrestle with this feature of our sociality, it is critical to remember the limits of the biology. Testosterone is never going to tell us much about the suburban teenager who, in his after-school chess club, has developed a particularly aggressive style with his bishops. And it certainly isn't going to tell us much about the teenager in some inner-city hellhole who has taken to mugging people. "Testosterone equals aggression" is inadequate for those who would offer a simple solution to the violent male—just decrease levels of those pesky steroids. And "testosterone equals aggression" is certainly inadequate for those who would offer a simple excuse: Boys will be boys and certain things in nature are inevitable. Violence is more complex than a single hormone. This is endocrinology for the bleeding heart liberal—our behavioral biology is usually meaningless outside the context of the social factors and environment in which it occurs.

Notes

1. Testosterone is one of a family of related hormones, collectively known as "androgens" or "anabolic steroids." They all are secreted from the testes or are the result of a modification of testosterone, they all have a similar chemical structure, and they all do roughly similar things. Nonetheless, androgen mavens spend entire careers studying the important differences in the actions of different androgens. I am going to throw that subtlety to the wind and, for the sake of simplification that will horrify many, will refer throughout to all of these related hormones as "testosterone."

2. An example of physics envy in action. Recently, a zoologist friend had obtained blood samples from the carnivores that he studies and wanted some hormones in the sample assays in my lab. Although inexperienced with the technique, he offered to help in any way possible. I felt hesitant asking him to do anything tedious but, so long as he had offered, tentatively said, "Well, if you don't mind some unspeakable drudgery, you could number about a thousand assay vials." And this scientist, whose superb work has graced the most prestigious science journals in the world, cheerfully answered, "That's okay, how often do I get to do *real* science, working with test tubes?"

3. And no one has shown that differences in the size or shape of the amygdala, or differences in the numbers of neurons in it, can begin to predict differences in normal levels of aggression. Same punch line as with testosterone.

Further Reading

For a good general review of the subject, see E. Monaghan and S. Glickman,"Hormones and Aggressive Behavior," in J. Becker, M. Breedlove, and D. Crews, eds., *Behavioral Endocrinology* (Cambridge, Mass.: MIT Press, 1992), 261. This also has an overview of the hyena social system, as Glickman heads the study of the Berkeley hyenas.

For technical papers on the acquisition of the female dominance in hyenas, see S. Jenks, M. Weldele, L. Frank, and S. Glickman, "Acquisition of Matrilineal Rank in Captive Spotted Hyenas: Emergence of a Natural Social System in, Peer-Reared Animals and Their Offspring," *Animal Behavior* 50 (1995): 893; and L. Frank, S. Glickman, and C. Zabel, "Ontogeny of Female Dominance in the Spotted Hyaena: Perspectives from Nature and Captivity," in P. Jewell and G. Maloiy, eds., "The Biology of Large African Mammals in Their Environment," *Symposium of the Zoological Society of London* 61 (1989): 127.

I have emphasized that while testosterone levels in the normal range do not have much to do with aggression, a massive elevation of exposure, as would be seen in anabolic steroid abusers, does usually increase aggression. For a recent study in which even elevating into that range (approximately five times normal level) still had no effect on mood or behavior, see S. Bhasin, T. Storer, N. Berman, and colleagues, "The Effects of Supraphysiologic Doses of Testosterone on Muscle Size and Strength in Normal Men," *New England Journal of Medicine* 335 (1996): 1.

The study showing that raising testosterone levels in the middle-ranking monkey exaggerates preexisting patterns of aggression can be found in A. Dixson and J. Herbert, "Testosterone, Aggressive Behavior and Dominance Rank in Captive Adult Male Talapoin Monkeys *(Miopithecus talapoin)," Physiology and Behavior* 18 (1977): 539.

For the demonstration that testosterone shortens the resting period between action potentials in neurons, see K. Kendrick and R. Drewert, "Testosterone Reduces Refractory Period of Stria Terminalis Neurons in the Rat Brain," *Science* 204 (1979): 877.

Introduction to Reading 5

The anthropologist, Serena Nanda, is widely known for her ethnography of India's Hijaras, titled *Neither Man nor Woman*. The article included here is from her more recent book on multiple sex/gender systems around the world. Nanda's analysis of multiple genders among Native North Americans is rich and detailed. As you read this piece, consider the long-term consequences of the failure of European colonists and early anthropologists to get beyond their ethnocentric assumptions so that they could understand and respect the gender diversity of North American Indian cultures.

1. Why does Serena Nanda use the term *gender variants* instead of *two-spirit* and *berdache?*
2. What was the relationship between sexual orientation and gender status among American Indians whose cultures included more than two sex/gender categories? How about hermaphroditism and gender status?
3. Why was there often an association between spiritual power and gender variance in Native American cultures?

Multiple Genders Among North American Indians

Serena Nanda

The early encounters between Europeans and Indian societies in the New World, in the fifteenth through the seventeenth centuries, brought together cultures with very different sex/gender systems. The Spanish explorers, coming from a society where sodomy was a heinous crime, were filled with contempt and outrage when they recorded the presence of men in American Indian societies who performed the work of women, dressed like women, and had sexual relations with men (Lang 1996; Roscoe 1995).

Europeans labeled these men "berdache," a term originally derived from an Arabic word meaning male prostitute. As such, this term is inappropriate and insulting, and I use it here only to indicate the history of European (mis)understanding of American Indian sex/gender diversity. The term berdache focused attention on the sexuality associated with mixed gender roles, which the Europeans identified, incorrectly, with the "unnatural" and sinful practice of sodomy in their own societies. In their ethnocentrism, the early European explorers and colonists were unable to see beyond their own sex/gender systems and thus did not understand the multiple sex/gender systems they encountered in the Americas. They also largely overlooked the specialized and spiritual functions of many of these alternative sex/gender roles and the positive value attached to them in many American Indian societies.

By the late-nineteenth and early-twentieth centuries, some anthropologists included accounts of North American Indian sex/gender diversity in their ethnographies. They attempted to explain the berdache from various functional perspectives, that is, in terms of the contributions these sex/gender roles made to social structure or culture. These accounts, though less contemptuous than earlier ones, nevertheless largely retained the emphasis on berdache sexuality. The berdache was defined as a form of "institutionalized homosexuality," which served as a social niche for individuals whose personality and sexual orientation did not match the definition of masculinity in their societies, or as a "way out" of the masculine or warrior role for "cowardly" or "failed" men (see Callender and Kochems 1983).

Anthropological accounts increasingly paid more attention to the association of the berdache with shamanism and spiritual powers and also noted that mixed gender roles were often central and highly valued in American Indian cultures, rather than marginal and deviant. These accounts were, nevertheless, also ethnocentric in misidentifying indigenous gender diversity with European concepts of homosexuality, transvestism, or hermaphroditism, which continued to distort their indigenous meanings.

In American Indian societies, the European homosexual/heterosexual dichotomy was not culturally relevant and the European labeling of the berdache as homosexuals resulted from their own cultural emphasis on sexuality as a central, even defining, aspect of gender and on sodomy as an abnormal practice and/or a sin. While berdache in many American Indian societies did engage in sexual relations and even married persons of the same sex, this was not central to their alternative gender role. Another overemphasis resulting from European ethnocentrism was the identification of berdache as *transvestites.* Although berdache often cross-dressed, transvestism was not consistent within or across societies. European descriptions of berdache as *hermaphrodites* were also inaccurate. Considering the variation in alternative sex/gender roles in native North America, a working definition may be useful: the berdache in the anthropological literature refers to people who partly or completely take on aspects of the culturally defined role of the other sex and who are classified neither as women nor men, but as genders of their own (see Callender and Kochems 1983:443). It is important to note here that berdache thus refers to variant gender roles, rather than a complete crossing over to an opposite gender role.

In the past twenty-five years there have been important shifts in perspectives on sex/gender diversity among American Indians and anthropologists, both Indian and non-Indian (Jacobs, Thomas, and Lang 1997:Introduction). Most current research rejects institutionalized homosexuality as an adequate explanation of American Indian gender diversity, emphasizing the importance of occupation rather than sexuality as its central feature. Contemporary ethnography views multiple sex/gender roles as a normative part of American Indian sex/gender systems, rather than as a marginal or deviant part (Albers 1989:134; Jacobs et al. 1997; Lang 1998). A new emphasis on the variety of alternative sex/gender roles in North America undercuts the earlier treatment of the berdache as a unitary phenomenon across North (and South) America (Callender and Kochems 1983; Jacobs et al. 1997; Lang 1998; Roscoe and Murray 1998). Current research also emphasizes the integrated and often highly valued position of gender variant persons and the association of sex/gender diversity with spiritual power (Roscoe 1996; Williams 1992).

A change in terminology has also taken place. Berdache generally has been rejected, but there is no unanimous agreement on what should replace it. One widely accepted suggestion is the term *two-spirit* (Jacobs et al. 1997; Lang 1998), a term coined in 1990 by urban American Indian gays and lesbians. Two-spirit has the advantage of conveying the spiritual nature of gender variance as viewed by gay, lesbian, and transgendered American Indians and also the spirituality associated with traditional American Indian gender variance, but the cultural continuity suggested by two-spirit is in fact a subject of debate. Another problem is that two-spirit emphasizes the Euro-American gender construction of only two genders. Thus, I use the more culturally neutral term, variant genders (or gender variants) and specific indigenous terms wherever possible.

Distribution and Characteristics of Variant Sex/Gender Roles

Multiple sex/gender systems were found in many, though not all, American Indian societies. Male gender variant roles (variant gender roles assumed by biological males) are documented for 110 to 150 societies. These roles occurred most frequently in the region extending from California to the Mississippi Valley and upper-Great Lakes, the Plains and the Prairies, the Southwest, and to a lesser extent along the Northwest Coast tribes. With few exceptions, gender variance is not historically documented for eastern North America, though it may have existed prior to European invasion and disappeared before it could be recorded historically (Callender and Kochems 1983; Fulton and Anderson 1992).

There were many variations in North American Indian gender diversity. American Indian cultures included three or four genders: men, women, male variants, and female variants (biological females who by engaging in male activities were reclassified as to gender). Gender variant roles differed in the criteria by which they were defined; the degree of their integration into the society; the norms governing their behavior; the way the role was acknowledged publicly or sanctioned; how others were expected to behave toward gender variant persons; the degree to which a gender changer was expected to adopt the role of the opposite sex or was limited in doing so; the power, sacred or secular, that was attributed to them; and the path to recruitment.

In spite of this variety, however, there were also some common or widespread features: transvestism, cross-gender occupation, same sex (but different gender) sexuality, some culturally normative and acknowledged process for recruitment to the role, special language and ritual roles, and associations with spiritual power.

Transvestism

The degree to which male and female gender variants were permitted to wear the clothing of the other sex varied. Transvestism was often associated with gender variance but was not equally important in all societies. Male gender variants frequently adopted women's dress and hairstyles partially or completely, and female gender variants partially adopted the clothing of men; sometimes, however, transvestism was prohibited. The choice of clothing was sometimes an individual matter and gender variants might mix their clothing and their accoutrements. For example, a female gender variant might wear a woman's dress but carry (male) weapons. Dress was also sometimes situationally determined: a male gender variant would have to wear men's clothing while engaging in warfare but might wear women's clothes at other times. Similarly, female gender variants might wear women's clothing when gathering (women's work), but male clothing when hunting (men's work) (Callender and Kochems 1983:447). Among the Navajo, a male gender variant, nádleeh, would adopt almost all aspects of a woman's dress, work, language and behavior; the Mohave male gender variant, called alyha, was at the extreme end of the cross-gender continuum in imitating female physiology as well as transvestism. . . . Repression of visible forms of gender diversity, and ultimately the almost total decline of transvestism, were a direct result of American prohibitions against it.

Occupation

Contemporary analysis emphasizes occupational aspects of American Indian gender variance as a central feature. Most frequently a boy's interest in the implements and activities of women and a girl's interest in the tools of male occupations signaled an individual's wish to undertake a gender variant role (Callender and Kochems 1983:447; Whitehead 1981). In hunting societies, for example, female gender variance was signaled by a girl rejecting the domestic activities associated with women and participating in playing and hunting with boys. In the arctic and subarctic, particularly, this was sometimes encouraged by a girl's parents if there were not enough boys to provide the family with food (Lang 1998). Male gender variants were frequently considered especially skilled and industrious in women's crafts and domestic work (though not in agriculture, where this was a man's task) (Roscoe 1991; 1996). Female gender crossers sometimes won reputations as superior hunters and warriors.

Male gender variants' households were often more prosperous than others, sometimes because they were hired by whites. In their own societies the excellence of male gender variants' craftwork was sometimes ascribed to a supernatural sanction for their gender transformation (Callender and Kochems 1983:448). Female gender variants opted out of motherhood, so were not encumbered by caring for children, which may explain their success as hunters or warriors. In some societies, gender variants could engage in both men's and women's work, and this, too, accounted for their increased wealth. Another source of income was payment for the special social activities of gender variants due to their intermediate gender status, such as acting as go-betweens in marriage. Through their diverse occupations, then, gender variants were often central rather than marginal in their societies.

Early anthropological explanations of male gender variant roles as a niche for a "failed" or cowardly man who wished to avoid warfare or other aspects of the masculine role are no longer widely accepted. To begin with, masculinity was not associated with warrior status in all American Indian cultures. In some societies, male gender variants were warriors and in many others, men who rejected the warrior role did not become gender variants. Sometimes male gender variants did not go to war because of cultural prohibitions against their using symbols of maleness, for example, the prohibition against their using the bow among the Illinois. Where male gender variants did not fight, they sometimes had other important roles in warfare, like treating the wounded, carrying supplies for the war party, or directing postbattle ceremonials (Callender and Kochems 1983:449). In a few societies male gender variants become outstanding warriors, such as Finds Them and Kills Them, a Crow Indian who performed daring feats of bravery while fighting with the United States Army against the Crow's traditional enemies, the Lakota Sioux (Roscoe and Murray 1998:23).

Gender Variance and Sexuality

Generally, sexuality was not central in defining gender status among American Indians. But in any case, the assumption by European observers that gender variants were homosexuals meant they did not take much trouble to investigate or record information on this topic. In some American Indian societies same-sex sexual desire/practice did figure significantly in the definition of gender variant roles; in others it did not (Callender and Kochems 1983:449). Some early reports noted specifically that male gender variants lived with and/or had sexual relations with women as well as men; in other societies they were reported as having sexual relations only with men, and in still other societies, of having no sexual relationships at all (Lang 1998:189–95).

The bisexual orientation of some gender variant persons may have been a culturally accepted expression of their gender variance. It may have resulted from an individual's life experiences, such as the age at which he or she entered the gender variant role, and/or it may have been one aspect of the general freedom of sexual expression in many American Indian societies. While male and female gender variants most frequently had sexual relations with, or married, persons of the same biological sex as themselves, these relationships were not considered homosexual in the contemporary Western understanding of that term. In a multiple gender system the partners would be of the same sex but different genders, and homogender, rather than homosexual, practices bore the brunt of negative cultural sanctions. The sexual partners of gender variants were never considered gender variants themselves.

The Navajo are a good example (Thomas 1997). The Navajo have four genders; in addition to man and woman there are two gender variants: masculine female-bodied nádleeh and feminine male-bodied nádleeh. A sexual relationship between a female nádleeh and a woman or a sexual relationship between a male-bodied nádleeh and a man were not stigmatized because these persons were of different genders, although of the same biological sex. However, a sexual relationship between two women, two men, two female-bodied nádleeh or two male-bodied nádleeh, was considered homosexual, and even incestual, and was strongly disapproved of.

The relation of sexuality to variant sex/gender roles across North America suggests that sexual relations between gender variants and persons of the same biological sex were a result rather than a cause of gender variance. Sexual relationships between a man and a male gender variant were accepted in most American Indian societies, though not in all, and appear to have been negatively sanctioned only when it interfered with child-producing heterosexual marriages. Gender variants' sexual relationships varied from casual and wide-ranging (Europeans used the term promiscuous), to stable, and sometimes even involved life-long marriages. In some societies, however, male gender variants were not permitted to engage in long-term relationships with men, either in or out of wedlock. In many cases, gender variants were reported as living alone.

There are some practical reasons why a man might desire sexual relations with a (male) gender variant: in some societies taboos on sexual relations with menstruating or pregnant women restricted opportunities for sexual intercourse; in other societies, sexual relations with a gender variant person were exempt from punishment for extramarital affairs; in still other societies, for example, among the Navajo, some gender variants were considered especially lucky and a man might hope to vicariously partake of this quality by having sexual relations with them (Lang 1998:349).

Biological Sex and Gender

Transformations

European observers often confused gender variants with hermaphrodites. Some American Indian societies explicitly distinguished hermaphrodites from gender variants and treated them differently; others assigned gender variant persons and hermaphrodites to the same alternative gender status. With the exception of the Navajo, in most American Indian societies biological sex (or the intersexedness of the hermaphrodite) was not the criterion for a gender variant role, nor were the individuals who occupied gender variant roles anatomically abnormal. The Navajo distinguished between the intersexed and the alternatively gendered, but treated them similarly, though not exactly the same (Thomas 1997; Hill 1935).

And even as the traditional Navajo sex/gender system had biological sex as its starting point, it was only a starting point, and Navajo nádleeh were distinguished by sex-linked behaviors, such as body language, clothing, ceremonial roles, speech style, and work. Feminine, male-bodied nádleeh might engage in women's activities such as cooking, weaving, household tasks, and making pottery. Masculine, female-bodied nádleeh, unlike other female-bodied persons, avoided childbirth; today they are associated with male occupational roles such as construction or firefighting (although ordinary women also sometimes engage in these occupations). Traditionally, female-bodied nádleeh had specific roles in Navajo ceremonials.

Thus, even where hermaphrodites occupied a special gender variant role, American Indian gender variance was defined more by cultural than biological criteria. In one recorded case of an interview with and physical examination of a gender variant male, the previously mentioned Finds Them and Kills Them, his genitals were found to be completely normal (Roscoe and Murray 1998).

If American Indian gender variants were not generally hermaphrodites, or conceptualized as such, neither were they conceptualized as transsexuals. Gender transformations among gender variants were recognized as only a partial transformation, and the gender variant was not thought of as having become a person of the opposite sex/gender. Rather, gender variant roles were autonomous gender roles that combined the characteristics of men and women and had some unique features of their own. This was sometimes symbolically recognized: among the Zuni a male gender variant was buried in women's dress but men's trousers on the men's side of the graveyard (Parsons quoted in Callender and Kochems 1983:454; Roscoe 1991:124, 145). Male gender variants were neither men—by virtue of their chosen occupations, dress, demeanor, and possibly sexuality—nor women, because of their anatomy and their inability to bear children. Only among the Mohave do we find the extreme imitation of women's physiological processes related to reproduction and the claims to have female sexual organs—both of which were ridiculed within Mohave society. But even here, where informants reported that female gender variants did not menstruate, this did not make them culturally men. Rather it was the mixed quality of gender variant status that was culturally elaborated in native North America, and this was the source of supernatural powers sometimes attributed to them.

Sacred Power

The association between the spiritual power and gender variance occurred in most, if not all, Native American societies. Even where, as previously noted, recruitment to the role was occasioned by a child's interest in occupational activities of the opposite sex, supernatural sanction, frequently appearing in visions or dreams, was also involved. Where this occurred, as it did mainly in the Prairie and Plains societies, the visions involved female supernatural figures, often the moon. Among the Omaha, for example, the moon appeared in a dream holding a burden strap—a symbol of female work—in one hand, and a bow—a symbol of male work—in the other. When the male dreamer reached for the bow, the moon forced him to take the burden strap (Whitehead 1981). Among the Mohave, a child's choice of male or female implements heralding gender variant status was sometimes prefigured by a dream that was believed to come to an embryo in the womb (Devereux 1937).

Sometimes, by virtue of the power associated with their gender ambiguity, gender variants were ritual adepts and curers, or had special ritual functions (Callender and Kochems 1983:453, Lang 1998). Gender variants did not always have important sacred roles in native North America, however. Where feminine qualities were associated with these roles, male gender variants might become spiritual leaders or healers, but where these roles were associated with male qualities they were not entered into by male gender variants. Among the Plains Indians, with their emphasis on the vision as a source of supernatural power, male gender variants were regarded as holy persons, but in California Indian societies, this was not the case and in some American Indian societies gender variants were specifically excluded from religious roles (Lang 1998:167). Sometimes it was the individual personality of the gender variant rather than his/her gender variance itself, that resulted in occupying sacred roles (see Commentary following Callender and Kochems 1983). Nevertheless, the importance of sacred power was so widely associated with sex/gender diversity in native North America that it is generally agreed to be an important explanation of the frequency of gender diversity in this region of the world.

In spite of cultural differences, some significant similarities among American Indian societies are particularly consistent with multigender systems and the positive value placed on sex/gender diversity (Lang 1996). One of these similarities is a cosmology (system of religious beliefs and way of seeing the world) in which transformation and ambiguity are recurring themes. Thus a person who contains both masculine and feminine qualities or one who is transformed from the sex/gender assigned at birth into a different gender in later life manifests some of the many kinds of transformations and ambiguities that are possible, not only for humans, but for animals and objects in the natural environment. Indeed, in many American Indian cultures, sex/gender ambiguity, lack of sexual differentiation, and sex/gender transformations play an important part in the story of creation. American Indian cosmology may not be "the cause" of sex/gender diversity but it certainly (as in India) provides a hospitable context for it (Lang 1996:187). . . .

As a result of Euro-American repression and the growing assimilation of Euro-American sex/gender ideologies, both female and male gender variant roles among American Indians largely disappeared by the 1930s, as the reservation system was well under way. And yet, its echoes may remain. The current academic interest in American Indian multigender roles, and particularly the testimony of contemporary two-spirits, remind us that alternatives are possible and that understanding American Indian sex/gender diversity in the past and present makes a significant contribution to understandings of sex/gender diversity in the larger society.

References

Albers, Patricia C. 1989. "From Illusion to Illumination: Anthropological Studies of American Indian Women." In *Gender and Anthropology: Critical Reviews for Research and Teaching,* edited by Sandra Morgen. Washington, DC: American Anthropological Association.

Callender, Charles, and Lee M. Kochems. 1983. "The North American Berdache." *Current Anthropology* 24(4): 443–56 (Commentary, pp. 456–70).

Devereux, George. 1937. "Institutionalized Homosexual of the Mojave Indians." in *Human Biology,* 9: 498–597.

Fulton, Robert, and Steven W. Anderson. 1992. "The Amerindian 'Man-Woman': Gender, Liminality, and Cultural Continuity." *Current Anthropology 33*(5): 603–10.

Hill, Willard W. 1935. "The Status of the Hermaphrodite and Transvestite in Navaho Culture." *American Anthropologist* 37:273–79.

Jacobs, Sue-Ellen, Wesley Thomas, and Sabine Lang, eds. 1997. *Two-Spirit People: Native American Gender Identity, Sexuality, and Spirituality.* Urbana and Chicago: University of Illinois.

Lang, Sabine. 1996. "There Is More than Just Men and Women: Gender Variance in North America." In *Gender Reversals and Gender Culture,* edited by Sabrina Petra Ramet, pp. 183–96. London and New York: Routledge

———. 1998. *Men as Women, Women as Men: Changing Gender in Native American Cultures.* Trans. from the German by John L. Vantine. Austin: University of Texas Press.

Roscoe, Will. 1991. *The Zuni Man-Woman.* Albuquerque: University of New Mexico Press.

_____. 1995. "Cultural Anesthesia and Lesbian and Gay Studies." *American Anthropologist,* 97: 448–52.

______. 1996. "How to Become a Berdache: Toward a Unified Analysis of Gender Diversity." In *Third Sex, Third Gender: Beyond Sexual Dimorphism in Culture and History,* edited by Gilbert Herdt, pp. 329–72. New York: Zone (MIT).

Roscoe, Will, and Stephen O. Murray, eds. 1998. *Boy-Wives and Female-Husbands: Studies in African Homosexualities.* New York: St. Martins.

Thomas, Wesley. 1997. "Navajo Cultural Constructions of Gender and Sexuality." In *Two-Spirit People: Native American Gender Identity, Sexuality, and Spirituality,* edited by Sue-Ellen Jacobs, Wesley Thomas, and Sabine Lang, pp. 156–73. Urbana and Chicago: University of Illinois.

Whitehead, Harriet. 1981. "The Bow and the Burden Strap: A New Look at Institutionalized Homosexuality in Native North America." In *Sexual Meanings: The Cultural Construction of Gender and Sexuality,* edited by Sherry B. Ortner and Harriet Whitehead, pp. 80–111. Cambridge: Cambridge University Press.

Williams, Walter. 1992. *The Spirit and the Flesh: Sexual Diversity in American Indian Culture.* Boston: Beacon.

⚜ Topics for Further Examination ⚜

- Visit the Web site of the Intersex Society of North America (http://www.isna.org) and read about the organization's mission statement and tips for parents. The blog posts are also interesting. In addition, you might find Web sites on male lactation and men breastfeeding to be helpful in expanding your understanding of biological overlap between males and females.
- Locate research on gender bending in the arts (e.g., performance art, literature, music videos). For example, visit Diane Torr's Web site (http://www.dianetorr.com).
- Google "doing gender" and explore the many Web sites that discuss the concept and its application.

2

The Interaction of Gender With Other Socially Constructed Prisms

After considering what gender is and isn't, we are going to complicate things a bit by looking at how a variety of other socially constructed categories of difference and inequality such as race, ethnicity, social class, religion, age, and sexuality shape gender. As is the case with prisms in a kaleidoscope, the interaction of gender with other social prisms creates complex patterns of identity and relationships for people across groups and situations. Because there are so many different social prisms that interact with gender in daily life, we can only discuss a few in this chapter; however, other social categories are explored throughout this book. The articles we have selected for this chapter illustrate three key arguments. First, gender is a complex and multifaceted array of experiences and meanings that cannot be understood without considering the social context within which they are situated. Second, variations in the meaning and display of gender are related to different levels of prestige, privilege, and power associated with membership in other socially constructed categories of difference and inequality. Third, gender intersects with other socially constructed categories of difference and inequality at all levels discussed in the introduction to this book—individual, interactional, and institutional.

Privilege

In our daily lives, there usually isn't enough time or opportunity to consider how the interaction of multiple social categories to which we belong affects beliefs, behaviors, and life chances. In particular, we are discouraged from critically examining our culture, as will be discussed in Chapter 3. People who occupy positions of privilege often do not notice how their privileged social positions influence them. In the United States, privilege is associated with white skin color, masculinity, wealth, heterosexuality, youth, able-bodiedness, and so on. It seems normal to those of us who occupy privileged positions and those we interact with, that our positions of privilege be deferred to, allowing us to move more freely in society. Peggy McIntosh is a pioneer in examining these hidden and unearned benefits of privilege. She argues that there are implicit benefits of privilege and that persons who have "unearned advantages" often do not understand how their privilege is a function of the disempowerment of others. For example, male privilege seems "normal" and White privilege seems "natural" to those who are male and/or White.

The struggle for women's rights also has seen the effects of privilege, with White women historically dominating this struggle. The privilege of race and social class created a view of woman as a universal category, and while many women of color stood up for women's rights, they did so in response to a universal definition of womanhood derived from White privilege (hooks, 1981). This pattern of White dominance in the women's movement is not a recent phenomenon. For example, in 1867 former slave Sojourner Truth responded to a White man who felt women were more delicate than men, told of the exertion by describing the toil of her work as a slave and asked "ain't I a woman" (Guy-Scheftall, 1995). One hundred years later, women of color including Audre Lorde (1982), bell hooks (1981), Angela Davis (1981), Gloria Anzaldúa (1987), and others continued to speak out against White privilege within the women's movement. These women, recognizing that the issues facing women of color were *not* always the same as those of White women, carried on a battle to make African American women visible in the second wave of feminism. For example, while White feminists were fighting for the right to abortion, African American women were fighting other laws and sterilization practices that denied them the right to control their own fertility. Women of color, including Patricia Hill Collins (this chapter) and others, some of who are listed in this introduction, continue to challenge the White-dominated definitions of gender and fight to include in the analysis of gender an understanding of the experience of domination and privilege of all women.

Understanding the Interaction of Gender With Other Categories of Difference and Inequality

Throughout this chapter introduction, you will read about social scientists and social activists who attempt to understand the interactions between "interlocking oppressions." Social scientists develop theories, with their primary focus on explanation, whereas social activists explore the topic of interlocking oppressions from the perspective of initiating social change (see the reading by Barbara Ryan in Chapter 10 for a discussion of the latter). Although the goals of explanation and change are rarely separated in feminist research, they reflect different emphases (Hill Collins, 1990). As such, these two different agendas shape attempts to understand the interaction of gender with multiple social prisms of difference and inequality. Much of this research on intersectionality is written by women, with a focus on women, because it came out of conflicts and challenging issues within the women's movement. As you read through this chapter and book, it is important to remember that the two socially defined gender categories in Western culture, masculinity and femininity, intersect with other prisms of difference and inequality.

The effort to include the perspectives and experiences of *all* women in understanding gender is complicated. Previous theories had to be expanded to include the interaction of gender with other social categories of difference and privilege. We group these efforts into three different approaches. The earliest approach is to treat each social category of difference and inequality as if it was separate and not overlapping. A second, more recent approach is to add up the different social categories that an individual belongs to and summarize the effects of the social categories of privilege and power. The third and newest approach attempts to understand the simultaneous interaction of gender with all other categories of difference and inequality. These three approaches are described in more detail in the following paragraphs.

Separate and Different Approach

Deborah King (1988) describes the earliest approach as the "race-sex analogy." She characterizes this approach as one in which oppressions related to race are compared to those related to gender, but each is seen as a separate influence. King quotes Elizabeth Cady Stanton, who in 1860 stated that "Prejudice against color, of which we hear so much, is no stronger than that against sex" (p. 43). This approach, the "race-sex analogy," continued well into the late twentieth century and can get in the way of a deeper understanding of the complexity of gender. For example, in the race-sex analogy, gender is assumed to have the same effect for African American and White women, while race is defined as the same experience for African American men and women. However, the reading by Karen Pyke and Denise Johnson in this chapter, which explores Asian American women's definitions of "femininity," illustrates how we cannot assume that gender will mean the same thing to different women from the same society. Although these Vietnamese American and Korean American women were born and raised in the United States, they live with two different cultural definitions of femininity, an idea we will consider in more depth in the next chapter. Although they may try, the reality is that Asian American women cannot draw a line down their bodies separating gender from ethnicity and culture. As these women's words suggest, the effects of prisms of difference on daily experience are inextricably intertwined. For example, African American females cannot always be certain that the discrimination they face is due to race or gender, or both. As individuals, we are complex combinations of multiple social identities. Separating the effects of multiple social prisms theoretically does not always make sense and it is almost impossible to do on an individual level.

Looking at these individual challenges, King (1988) argues that attempting to determine which "ism" is most oppressive and most important to overcome (e.g., sexism, racism, or classism) does not address the real life situations. This approach pits the interests of each subordinated group against the others and asks individuals to choose one group identity over others. For example, must poor, African American women decide which group will best address their situations in society: groups fighting racial inequality, gender inequality, or class inequality? The situations of poor, African American women are more complex than this single-issue approach can address.

The race-sex analogy of treating one "ism" at a time has been criticized because, although some needs and experiences of oppressed people are included, others are ignored. For instance, Patricia Hill Collins (1990) describes the position of African American women as that of "outsiders within" the feminist movement and in relations with women in general. She and others have criticized the women's movement for leaving out of its agenda and awareness the experiences and needs of African American women (e.g., King, 1988). This focus on one "ism" or another—the formation of social action groups around one category of difference and inequality—is called identity politics and will be discussed in detail in the article by Ryan in Chapter 10.

Additive Approach

The second approach used by theorists to understand how multiple social prisms interact at the level of individual life examines the effects of multiple social categories in an additive model. In this approach, the effects of race, ethnicity, class, and other social prisms are added together as static, equal parts of a whole (King, 1988). Returning to the earlier example of poor, African American women, the strategy is one of adding up the effects of racism, sexism, and classism to equal what is termed *triple jeopardy*. If that same woman was also a lesbian, according to the additive model, her situation would be that of *quadruple jeopardy*.

Although this approach takes into account multiple social identities in understanding oppression, King and others reject it as too simplistic. We cannot simply add up the complex inequalities across social categories of difference and inequality because the weight of each social category varies based upon individual situations. McIntosh, in her reading in this chapter, argues that privileges associated with membership in particular social categories interact to create "interlocking oppressions" whose implications and meanings shift across time and situations. For example, for African American women in some situations, their gender will be more salient, while in other situations their race will be more salient.

Interaction Approach

The third approach to understanding the social and personal consequences of membership in multiple socially constructed categories is called multiracial theory. Various terms are used by multiracial theorists to describe "interlocking oppressions," including: intersectional analysis (Baca Zinn & Thornton Dill, 1996), interrelated (Weber, 2001), simultaneous (Hill Collins, 1990; Weber, 2001), multiplicative or multiple jeopardy (King, 1988), matrix of domination (Hill Collins, 1990), and relational (Baca Zinn, Hondagneu-Sotelo, & Messner, 2001: Baca Zinn & Thornton Dill, 1996). For the purposes of this book, we describe this approach as consisting of "prismatic" or intersectional interactions, which occur when socially constructed categories of difference and inequality interact with other categories in the patterns of individuals' lives. A brief discussion of some of these different models for explaining "interlocking oppressions" is useful in deepening our understanding of the complex interactions of membership in multiple social prisms.

King (1988) brings these interactions to light, discussing the concept of multiple jeopardy, in which she refers not only "to several, simultaneous oppressions, but to the multiplicative relationships among them as well" (p. 47). As a result, socially constructed categories of difference and inequality fold into individual identities; not in an additive way, but in a way in which the total construction of an individual's identity incorporates the relationship of the identities to each other. King's model includes both multiple social identities and situational factors to understand individual differences. For example, being a submissive woman might matter more in certain religious groups where women have more restricted roles, while race or class may be less salient in that situation because the latter social prisms are likely to be similar across the religious group.

Baca Zinn, Hondagneu-Sotelo, and Messner (2001) emphasize that gender is relational. Focusing on gender as a process (Connell, 1987), they argue that "the meaning of *woman* is defined by the existence of women of different races and classes" (Baca Zinn, Hondagneu-Sotelo, & Messner 2001, p. 174). As Pyke and Johnson's reading illustrates, the fact that the Asian American women they interviewed were confused over whether to accept Asian standards of gender or American standards "normalizes" and makes White femininity dominant.

Patricia Hill Collins (1990), on the other hand, conceptualizes oppressions as existing in a "matrix of domination" in which individuals not only experience but also resist multiple inequalities. Collins argues that domination and resistance can be found at three levels: personal, cultural, and institutional. Individuals with the most privilege and power—White, upper-class men—control dominant definitions of gender in this model. Hill Collins discusses how "White skin privilege" has limited White feminists' understandings of gender oppression to their own experiences and created considerable tension in the women's movement. Tensions occur on all three levels—personal, within and between groups, and at the level of institutions such as the women's movement itself—all of which maintain power differentials. In her article in this chapter, Collins seeks a deeper understanding of gender oppression by considering how oppressions related to sexism, racism, and classism operate simultaneously. The reading by Williams in this chapter illustrates Hill's "matrix of domination" by explaining how toy store workers within two very different (social class) toy stores deal with both domination and resistance across gender, race, and class.

As you can see, this third approach does not treat interlocking oppressions as strictly additive. The unequal power of groups created by systems of inequality affects interpersonal power in all relationships, both within and across gender categories. The articles in this chapter illustrate how multiple socially constructed prisms interact to shape both the identities and opportunities of individuals. They also show how interpersonal relationships are intricately tied to the larger structures of society and how gender is maintained across groups in society, as the readings by Williams and Ward illustrate.

These efforts to understand gender through the lens of multiple social prisms of difference and inequality can be problematic. One concern is that gender could be reduced to what has been described as a continually changing quilt of life experiences (Baca Zinn, Hondagneu-Sotelo, & Messner, 2001; Connell, 1992). That is, if the third approach is taken to the extreme, gender is seen as a series of individual experiences and the approach could no longer be used as a tool for explaining patterns across groups, which would be meaningless in generating social action. Thus, the current challenge for researchers and theorists is to forge an explanation of the interaction of gender with other socially constructed prisms that both recognizes and reflects the experiences of individuals, while at the same time highlighting the patterns that occur across groups of individuals.

Since we live in a world that includes many socially constructed categories of difference and inequality, understanding the ways social prisms come together is critical for understanding gender. How we visualize the effects of multiple social prisms depends upon whether we seek social justice, theoretical understanding, or both. If multiple social categories are linked, then what are the mechanisms by which difference is created, supported, and changed? Are multiple identities multiplied, as King suggests; added, as others suggest; or combined into a matrix, as Hill Collins suggests? We raise these questions not to confuse you, but rather to challenge you to try to understand the complexity of gender relations.

Prismatic Interactions

We return now to the metaphor of the kaleidoscope to help us sort out this question of how to deal with multiple social identities in explaining gender. Understanding the interaction of several socially constructed identities can be compared to the ray of light passing through the prisms of a kaleidoscope. Socially constructed categories serve as prisms that create life experiences. Although the kaleidoscope produces a flowing and constantly changing array of patterns, so too do we find individual life experiences that are unique and flowing. However, similar colors and patterns often reoccur in slightly different forms. Sometimes, when we look through a real kaleidoscope, we find a beautiful image in which blue is dominant.

Although we may not be able to replicate the specific image, it would not be unusual for us to see another blue-dominant pattern. Gender differences emerge in a similar form; not as a single, fixed pattern, but as a dominant, broad pattern that encompasses many unique but similar patterns.

The prism metaphor offers an avenue for systematically envisioning the complexity of gender relations. We would argue that to fully understand this interaction of social influences, one must focus on power. The distribution of privilege and oppression is a function of power relations (Baca Zinn, Hondagneu-Sotelo, & Messner, 2001). All of the articles in this chapter examine differences defined by power relations, and power operates at every level of life, from the intimate and familial (see reading by Pyke & Johnson in this chapter) to the institutional (also see readings by Williams & Ward in this chapter). It is difficult to explain the combined effects of multiple social prisms without focusing on power. We argue that the power one accrues from a combination of socially constructed categories explains the patterns created by these categories. However, one cannot add up the effects from each category one belongs to, as in the additive approach described earlier. Instead, one must understand that, like the prisms in the kaleidoscope, the power of any single socially constructed identity is related to all other categories. The final patterns that appear are based upon the combinations of power that shape the patterns. Individuals' life experiences, then, take unique forms as race, class, ethnicity, religion, age, ability/disability, body type, and other socially constructed characteristics are combined to create patterns that emerge across contexts and daily life experiences.

Consider your own social identities and the social categories to which you belong. How do they mold you at this time and how did they, or might they, mold your experience of gender at other times and under different circumstances? Consider other social prisms such as age, ability/disability, religion, and national identity. If you were to build your own kaleidoscope of gender, what prisms would you include? What prisms interact with gender to shape your life? Do these prisms create privilege or disadvantage for you? Think about how these socially constructed categories combine to create your life experiences and how they are supported by the social structure in which you live. Keep your answers in mind as you read these articles to gain a better understanding of the role prisms play in shaping gender and affecting your life.

References

Anzaldúa, G. (1987). *Borderlands lafrontera: The new mestiza.* San Francisco: Aunt Lute Book Company.

Baca Zinn, M., Hondagneu-Sotelo, P., & Messner, M. A. (2001). Gender through the prism of difference. In M. L. Andersen & P. Hill Collins (Eds.), *Race, class, and gender: An anthology* (4th ed., pp. 168–176). Belmont, CA: Wadsworth.

Baca Zinn, M., & Thornton Dill, B. (1996). Theorizing difference from multiracial feminism. *Feminist Studies, 22*(2), 321–327.

Connell, R. W. (1987). *Gender and power: Society, the person, and sexual politics.* Stanford, CA: Stanford University Press.

Connell, R. W. (1992). A very straight gay: Masculinity, homosexual experience, and the dynamics of gender. *American Sociological Review, 57,* 735–751.

Davis, A. Y. (1981). *Women, race, and class.* New York: Random House.

Guy-Scheftall, B. (Ed.). (1995). *Words of fire: An anthology of African-American feminist thought.* New York: The New Press.

Hill Collins, P. (1990). *Black feminist thought: Knowledge, consciousness, and the politics of empowerment.* New York: Routledge.

hooks, b. (1981). *Ain't I a woman: Black women and feminism.* Boston: South End Press.

King, D. (1988). Multiple jeopardy: The context of a Black feminist ideology. *Signs, 14*(1), 42–72.

Lorde, A (1982). *Zami, A new spelling of my name.* Trumansberg, NY: The Crossing Press.

Lorde, A. (1984). *Sister outsider.* Trumansberg, NY: The Crossing Press.

McIntosh, P. (1998). *White privilege and male privilege: Unpacking the invisible knapsack.* Wellsley, MA: Wellesley College Center for Research on Women.

Weber, L. (2001). *Understanding race, class, gender, and sexuality: A conceptual framework.* Boston: McGraw-Hill.

Introduction to Reading 6

This excerpt from Christine L. Williams's book *Inside Toyland: Working, Shopping, and Social Inequality* builds upon Patricia Hill Collins's idea of a "matrix of domination" in which race, class, and gender intersect to create different life experiences for women. The data and findings in her book, and in this chapter, are based upon an ethnographic analysis of two toy stores.

Williams was a sales worker at Toy Warehouse, where staff was required to entertain the middle-class "guests," targeted by the corporate headquarters in advertising, and she also worked at Diamond Toys, a store that focused on upper-class, "discriminating" shoppers, where staff "served" customers' every need, much as butlers do in elite households. This reading presents an interesting matrix of domination as class, race, and gender of customers and staff form prismatic patterns across these two different toy store settings.

1. After reading this article, return to the definition for "matrix of domination" in the introduction to this chapter and apply it to Williams's reading.
2. How does the gender, race, and class of customers alter the experiences of staff in these two stores?
3. How do the experiences of staff in these toy stores relate to the three levels of analysis discussed in the Introduction to this book: individual, interactional/situational, and institutional/societal?

Inequality on the Shopping Floor

Christine L. Williams

In this Chapter, I explore the consequences of job sorting for interactions between clerks and customers. I argue that race and gender segregation of jobs perpetuates stereotypes that shape the meanings and consequences of routinized shopping encounters. When different kinds of people fill each role in a ritualized service encounter, the result is the reproduction of social inequality.

To illustrate this, imagine a service industry where all servers would be women and all customers men. This was the scenario encountered by Arlie Hochschild (1983) in her study of flight attendants in the 1970s. She argued that this stratification reinforced assumptions linking femininity with servility and masculinity with economic and social power. As Goffman (1977) observes, virtually any stereotype will come to seem natural and inevitable if ritualized interactions that reinforce it are repeated often enough (see also West and Zimmerman 1987). Because the sorting of people into jobs on the basis of not only gender but also race is very common in our economy, service work has become a major source of group stereotypes and prejudices.

Of course, in actual practice, service rituals are often modified. Despite their goal of precisely scripting the service encounter, corporations cannot control all interactions between clerks and customers . . . [T]here is always room for creativity and individual expression within the constraints of the role. Corporate ideals have to be translated into local contexts. In a study of Burger King and McDonald's restaurants in New York City, Jennifer Parker Talwar (2002) found that the scripts mandated by these companies had to be changed according to the ethnic makeup of the neighborhoods where the restaurants were located. For instance, the requirement that workers always smile upset many customers in a predominately Chinese neighborhood, who interpreted the behavior as overly pushy. In one franchise, greeters were hired to welcome customers respectfully at the front door and accompany them to the order counter to assist them in making their selections.

Not only local context but the actual mix of customer and clerk will shape the service encounter. In this chapter, I describe how white service workers encounter a different set of customer expectations than black service workers, resulting in different modifications of the scripted role. Likewise, white customers encounter a different set of service worker expectations than black customers.

Patricia Hill Collins (2000) uses the concept of a "matrix of domination" to analyze these shifting configurations of race and gender and their consequences for social interaction. In contrast to those who might assume that all women face similar forms of domination and oppression, she argues that sexism comes in a variety of forms that are shaped by the contexts of race, social class, and sexuality. Discrimination against black women is different from discrimination against white women because these groups are located at different places on what she calls the matrix of domination. Black women are typically subject to domination on the basis of gender, race, and class, while white women suffer from gender domination but are privileged by their race and often by their class as well. This matrix operates at all levels of society and culture. Its workings are evident in the scarcity of black women and the total absence of black men at the higher-paying Diamond Toys compared to their over-representation at the Toy Warehouse. It is also apparent in the division of tasks within the stores, where white men monopolize the director positions, white or light-skinned women are concentrated in cashiering jobs, and darker-skinned African American women and men work as stockers and gofers.

In this chapter I explore how the matrix of domination shapes, but does not fully determine, customer-worker interactions in the toy store. I focus on the interaction rules that govern the shopping floor and how they reproduce stereotypes about different groups. There are both formal rules, developed by corporations, and informal rules, developed by workers to protect their dignity and self-respect. I also discuss what happens when these rules are not followed and interactions break down into conflict. Whether the interaction can be repaired will depend on the matrix because different groups have different resources to draw on to assert their will in the toy store. It will also depend on the creativity and personalities of the individuals involved. The meanings of workplace rituals are not fixed and self-evident but change depending on the mix of individuals engaged in the interaction. Only through a process of symbolic interaction among active, creative, knowledgeable participants do the meanings and consequences of these rituals emerge (Blumer 1969). . . .

Shop-Floor Culture

Interactions between costumers and workers in toy stores are governed by informal rules that are shaped but not determined by the corporate cultures. These rules are sometimes known as "the ropes." New hires pick up the ropes from observing experienced workers.

One of the first lessons I learned on the shopping floor was that middle-class white women shoppers got whatever they wanted. I suppose that as a middle-class white woman I should have found this empowering. Instead I came to understand it as a result of race and class privilege.

Most of the customers at both stores were women. At the Toy Warehouse we were told that women made 90 percent of the purchase decisions, so we were to treat women deferentially. Olive told me that the store abided by the "$19,800 rule." If a customer wanted to return merchandise and it was questionable whether we should take the return (because it had been broken or worn out by the customer, or because the customer had lost the receipt), we should err on the side of the customer. A $200 loss today might please the customer so much that she would return to the store and spend the rest of the $20,000 on each of her children.

But in my experience only the white women got this kind of treatment. Not surprisingly, many developed a sense of entitlement and threw fits when they were not accommodated. . . .

Middle-class white women had a reputation at the store of being especially demanding and abusive toward salesclerks. Susan, a thirty-five-year-old Latina, agreed with my observation that rich white women were the most demanding customers; she said they always demanded to see the manager and always got appeased. Susan, one of the people who taught me the ropes at the service desk, said that Latinas/os never demanded to see the manager and never threatened to stop shopping at the Toy Warehouse. I asked if she had ever seen a black person do this; at first she said no, and then she said yes. She remembered that sometimes at the end of the month poor blacks got upset at the service desk and demanded to see a manager. She explained that at the end of the month if they ran out of money they might try to return the merchandise that they had purchased earlier in the month but that had already been used. If the workers at the service desk turned down the request for a refund, they demanded to see the manager. But, she said, they were not going to get satisfaction.

* * *

Susan's remarks about demanding and complaining shoppers illustrate some of the elaborate stereotypes that service desk workers used in the course of their daily transactions. Immediate assumptions were made about customers based on their race, gender, and apparent social class; workers responded to customers using these cues.

Because middle-class white women were the most coveted customers at the Toy Warehouse, many developed a sense of entitlement. On one occasion, a very pregnant white woman came up to the service desk to return a teddy bear mobile for over the baby crib. The mechanism that played music and moved the mobile wasn't working, so she wanted to exchange it for one that was working. We sent her into the store to find another one for an even exchange. She came back after quite a while with another one in a box that she had opened. (Customers weren't supposed to open the factory-sealed boxes, but we didn't say anything.) She said that one package that she had opened also had a broken mechanism but that she had found a mobile with a different motif (clowns instead of teddy bears) made by the same company that had a mechanism in it that worked. She asked if she could replace that for the broken one in her original box. She said, "So will you do this and make a customer happy? If you do it I will come back to shop here more, and if you don't I won't ever set foot in the store. So what's it going to be? Do you want a happy customer or not?" I couldn't believe the attitude but I kept my mouth shut. I thought about the floor workers who would get in trouble for all those open boxes. My supervisor Vannie said she could make the switch. The woman never said thank you, but then customers rarely did.

To me, one of the most eye-opening examples of white women's sense of entitlement that I witnessed in the Toy Warehouse was their refusal to check their bags at the counter. Since stealing was such a big problem in the store, customers were required to leave all large bags and backpacks at the service desk. A large sign indicating this policy was posted on the store's entrance. The vast majority of customers carrying bags immediately approached the desk to comply with the rule. The exception was white women, who almost universally ignored the sign. When challenged, they would argue—"But my wallet is in there!" or "I need my bag with me!"—and we would have to insist so as not to appear unfair to the other people. I guiltily recognized myself in their behavior. Since then, I have always turned over my bag.

White women developed a sense of entitlement because in most instances they got what they wanted. Members of other groups who wanted to return used merchandise, or who needed special consideration, were rarely granted their request. . . .

It has been well documented that African Americans suffer discrimination in public places, including stores (Feagin and Sikes 1995); this phenomenon is sometimes referred to jokingly as "shopping while black" (Williams 2004). They report being followed by security, treated harshly by attendants, and flatly refused service. The flip side of this discrimination is the privilege experienced by middle-class whites. This privilege is not recognized precisely because it is so customary. Whites expect first-rate service; when it is not forthcoming, some feel victimized, even discriminated against. This was especially apparent in the Toy Warehouse, where most of the salespeople were black. I noticed that when white women customers were subjected to long waits in line, or if they received what they perceived as uncaring attention, they would often sigh loudly, roll their eyes, and try to make eye contact with other whites, looking for a sign of recognition that the service they were receiving was inferior and unfair.

* * *

At Diamond Toys most customers didn't mind waiting for their turn to consult with me. When the lines were long they didn't make rude huffing noises or try to make eye contact with their fellow sufferers. The two stores were staffed and structured quite differently, and that certainly helps to explain some differences in the experiences (and satisfaction) of the customers. We were so understaffed at the Toy Warehouse that I felt as if I were running the whole time I was at work. However, I couldn't help thinking that customers—who were mostly white—cut us more slack at Diamond Toys because most of us were white. We were presumed to be professional, caring, and knowledgeable even when we weren't. My African American coworkers at the Toy Warehouse, in contrast, were assumed to be incompetent and uncaring. Like the people who employed domestic workers studied by Julia Wrigley (1995), white customers seemed less respectful of racial/ethnic minority service workers than white workers; they were willing to pay more and wait more for the services of whites because they assumed that whites were more refined and intelligent.

* * *

White customers at the Toy Warehouse seemed frequently unnerved by their interactions with the clerks in the store. One time at the Toy Warehouse a white woman customer stopped my coworker Gail to ask for a gift suggestion for a ten-year-old boy. Gail, who was swiftly walking across the floor to deal with another customer's request, practically shouted at her, "Don't ask me about no boys; I got girls, not sons!" and then took off. The white woman looked startled at the response and maybe even a bit mortified. What she didn't know was that Gail found a coworker who had sons to answer the woman's question. The caring, efficiency, and sense of humor of my coworkers at the Toy Warehouse often went unnoticed by white customers.

Realizing that white customers in particular treated them with disrespect and even disdain, my African American coworkers developed interactional skills to minimize their involvement with them. I noticed at the service desk that the black women who worked there didn't smile or act concerned when customers came up for complaints or returns. They did their work well and efficiently (or at least as efficiently as possible given the myriad demands on their energies when they were at the service desk), but they did not exude a sense that they really cared. Rather, they looked suspicious, or bored, or resigned, or maybe a little miffed. That might be a defense mechanism. If they looked solicitous, then the customers would walk all over them. Over time I learned that this attitude of ennui or suspicion was cultivated as a way to garner respect for their work. It was saying, "This isn't my problem, it's your problem, but I will see what can be done to fix it." If workers were more gung-ho and it turned out that the problem couldn't be fixed, then they would look incompetent, which was the assumption that too many white customers were willing to make of black women. So they made it look as if the problem were insurmountable, and then when they did resolve it (which was most of the time) they garnered a little bit of respect. But it was at the cost of appearing unfriendly, so the store got hit with bad customer service evaluations.

Manipulating customers through self-presentation constitutes an informal "feeling rule" (Hochschild

1983). These techniques for displaying affect were developed by workers to manage and minimize difficult customer interactions. I call them informal rules because management would prefer that workers always convey serious concern and solicitude. This was the case at both stores. With few exceptions, workers at Diamond Toys accepted this directive, but many workers at the Toy Warehouse resisted it, recognizing that, if they were African Americans, adopting an attitude of servility would reinforce racism among shoppers. The informal rules were sensitive to race and gender dynamics in a way that management rules could not be. Different groups had to use different means to do their jobs.

Moreover, the stratification of the jobs at the store meant that we all had different levels of formal and informal power to resolve situations. White men had the most power in the stores. The store directors in both places where I worked were white men who could and did trump any decision made by managers or sales associates. Sales associates had virtually no options for resolving disputes. Measures intended to keep us from stealing also prevented us from dealing with customer complaints or special requests. We couldn't offer small discounts or make up a price when the tag was missing, for example, because doing so would make the store vulnerable to employee theft. Thus maintaining control was dependent on the race, gender, and organizational authority of both workers and customers.

Interaction Breakdown: Social Control in the Toy Store

It may seem anathema to talk about power and control in the context of toy shopping. But customers frequently misbehaved in the stores where I worked. They didn't just throw fits at the service desk; every day I witnessed customers ripping open packages, hiding garbage, spilling Cokes, and generally making a mess of the store. One of my coworkers at Diamond Toys even found a dirty diaper on a shelf. There is a sort of gleeful abandon that some shoppers experience in a store. Barbara Ehrenreich (2001) describes how shoppers at Wal-Mart could tear down in minutes a clothing display that had taken her all day to arrange. She speculates that after full days of picking up after family members at home, some women shoppers like to experience a turn at being the messy ones.

As an older white woman, I could exercise some control over the extremes of this bad behavior. I could stand nearby, for example, and the customer might notice me and guiltily try to stuff the toy back into the box or replace the dozen toys she had pulled off the shelf. We weren't allowed to confront customers, even if we suspected them of stealing or destroying the merchandise, but we were expected to develop subtle ways to control them. I couldn't do this as well as my male coworkers could, but I was definitely more respected (and feared) than my women coworkers who were African American, Asian, and Latina.

One example of this involved Thelma, who was about fifty years old and was one of the two African American salesclerks at Diamond Toys. In the break room we talked about how rich the people were who shopped at the store. She said she easily got them to spend a lot of money, and she gave an example of how recently she had talked people into spending over $200. She said she was really good with the kids; she got them to behave by threatening to take away their toys. Their mothers were so grateful they sometimes asked her to come home with them and work for them. While she was proud of her selling ability, at the same time this particular reaction upset her. The women's offers clearly drew on cultural conventions linking black women to domestic service and to the "mammy" stereotype in particular. Thelma said that she objected to being treated like a servant. She said that the day before a customer had called her a "bitch" to one of the managers. She had asked the customer not to sit on the display tables because they were not sturdy and we had all been instructed to keep people off them. She asked her to not sit there in a polite way, but the woman got mad and complained to the manager, "That bitch told me to move. What if I had been disabled?" Thelma saw this treatment as racist and sexist. She said, "I am not her bitch," meaning her servant.

The customers that I had the hardest time controlling were men. Men were outnumbered by women in both stores. At the Toy Warehouse, I saw them mostly on the weekend, which seemed to be the most popular time for fathers to come in with their children. (I often wondered if they were divorced fathers.) At Diamond Toys, I observed men tourists shopping with their families, businessmen buying small gifts for their children back home, and, during the Christmas season, men buying high-end toys for their wives. In general men seemed annoyed to be in the stores, and they sometimes acted annoyed with me, especially when they were unaccompanied by women. One white man at the Toy Warehouse tossed his shopping list at me when I was working at the service desk. He expected me to assemble the items for him or get someone else to do it. One apparently very wealthy white man at Diamond Toys tossed his credit card at me while pointing to merchandise kept in the glass cases. I had to scurry after him to keep up with his numerous demands. Some men were just mean. On two occasions at Diamond Toys, men demanded to use my telephone, which the store had strict rules against. I said they weren't allowed to use it, and they just reached over the counter and did it anyway. I was terrified that [my supervisor] would walk by and yell at me, maybe even fire me. One time during the Christmas rush a white man complained to me about the wait at my register, an uncommon occurrence at Diamond Toys. I explained that two of the workers hadn't shown up that day. He told me I should fire them. (Why he thought I had this power I couldn't guess.) On another occasion at Diamond Toys, a business professor in town for a professional convention was upset because a Barney sippy cup he wanted to buy was missing its price tag and I couldn't find it listed in the store inventory. He made me call over the store director and subjected both of us to a critique of store operations, which he threatened to write up and submit for publication to a business journal unless we sold him the sippy cup.

The sense of entitlement I experienced with these men customers was different from that which I encountered with white women customers. Perhaps the expectation that shopping was "women's work" made these men feel entitled to make me to do their shopping for them or reorganize the store to make it more convenient for them. To assert masculinity while engaging in this otherwise feminine activity seemed to require them to disrupt the routinized clerk-customer relationship.

* * *

Some customers were considered more demanding and harder to control than others, in particular, black men and white women. But no one ever threatened to call the police on an angry white woman. In general, only white women could count on being appeased; for them, acting out seemed to get the results they desired.

The fact that police were called to control only black men customers reveals underlying cultural assumptions about gender and race. Toy stores catered to white mothers, who were believed to be behaving virtuously–if not civilly–on behalf of their children. Black men, in contrast, were assumed to be violent by nature; their anger was seen as evidence of an underlying animal nature that had to be controlled (Ferguson 2000).

Thus white women's demands were satisfied because, from a corporate perspective, their anger was assumed to serve a legitimate end. They also were appeased because it was assumed that they had economic resources ($20,000 for each child!), an assumption that was almost never invoked in the case of black men. White women were immediately seen as potential spenders; black men, as potential thieves. In the overall retail environment, these racialized assumptions might extend to white and black people in general (Chin 2001), but in a toy store that catered to women in particular, we might predict this polarized race/gender dynamic to emerge. Thus a shouting white woman got whatever she wanted and a shouting black man got threatened with arrest. (Other groups of customers, in my experience, didn't shout, with the notable exception of children.)

Some have argued that stealing and scamming in stores can be understood as a form of resistance to racism toward black consumers. This has certainly been the case in the past, when organized looting by African Americans expressed their collective anger at being excluded, mistreated, and/or overcharged by white merchants (Cohen 2003). Some of the customers caught stealing may have been motivated by such social and economic protest, but I don't have any way to know this. What I do know is that African Americans were typically assumed to be scamming whenever they complained, returned merchandise, or made special requests. Whites were given the benefit of the doubt.

Conclusion

As a nation of consumers, we spend a great deal of time in stores interacting with sales workers. In this chapter I have tried to make a case for paying attention to these interactions as sites for the reproduction of social inequality. I argue that where we shop and how we shop are shaped by and bolster race, class, and gender inequalities.

Corporations script the customer-server interaction in ways that are designed to appeal to a particular kind of customer. The fun, child-centered Toy Warehouse aimed to attract middle-class mothers who were looking for a wide assortment of toys at discount prices, while the sophisticated Diamond Toys catered to a more discriminating, upper-class clientele. These corporate driven agendas influenced who entered the stores as customers and also shaped the labor practices of the two stores. The Toy Warehouse attracted a dazzling mix of customers who were served by a mostly African American staff. At that store, my job was to swiftly and cheerfully process customer requests and check them out at the register. At Diamond Toys, in contrast, the service encounter was considered a more central part of my job. I was expected to provide "professional" advice to the well-to-do, mostly white clientele. It was not a coincidence that the majority of salesclerks at that store were white. Stores like Diamond Toys that marketed their "expert" staff might prefer to hire whites instead of African Americans because of the (presumed) racism of wealthy shoppers and a culture that associates professionalism with whiteness.

When interactions broke down, the ability to repair them depended on the characteristics of the customer and worker. As a white woman, I had a different repertoire of control strategies than my African American women coworkers. They had to reckon with racist as well as sexist assumptions from irate white customers, while most of my difficulties were due to customer sexism. White men had more power in the stores, but they seemed to have difficulty managing and controlling black customers. Control was an achievement that had to be negotiated anew with each service interaction.

I realize that most of my examples of conflict are from the Toy Warehouse. I believe that conflict was more common there than at Diamond Toys. The unusually diverse mix of people in the Toy Warehouse often provoked misunderstandings, as when Gail shouted at the customer that she didn't have sons. Diamond Toys protected itself from conflict by catering to an upper-class clientele, thereby functioning much like a gated community. This is not to say that diversity always results in conflict, but it did at the Toy Warehouse because race, class, and gender differences were overlain by power differences within the store. Interactions between clerks and customers took place within a context where these differences had been used to shape marketing agendas, hiring practices, and labor policies—all of which benefited some groups (especially middle-class white women and men) over others.

From a sociological perspective, there is nothing inevitable about these categories or their meanings. Race, class, and gender are not inherent properties of individuals; they are socially constructed categories that derive their significance through human interactions. On rare instances, I experienced interactions with customers that transcended their ritualized forms and momentarily broke through the matrix of domination . . . .

References

Blumer, Herbert. 1969. *Symbolic Interactionism.* Berkeley: University of California Press.

Chin, Elizabeth. 2001. *Purchasing Power: Black Kids and American Consumer Culture.* Minneapolis: University of Minnesota Press.

Cohen, Lizabeth. 2003. *A Consumer's Republic: The Politics of Mass Consumption in Postwar America.* New York: Knopf.

Collins, Patricia Hill. 2000. *Black Feminist Thought.* New York: Routledge.

Ehrenreich, Barbara. 2001. *Nickel and Dimed: On (Not) Getting By in America.* New York: Metropolitan Books.

Feagin, Joe, and Melvin Sikes. 1995. *Living with Racism: The Black Middle Class Experience.* New York: Beacon.

Ferguson, Ann Arnet. 2000. *Bad Boys: Public Schools in the Making of Black Masculinity.* Ann Arbor: University of Michigan Press.

Goffman, Erving. 1977. *Interaction Ritual: Essays on Face to Face Behavior.* Garden City, NY: Anchor Books.

Hochschild, Arlie. 1983. *The Managed Heart: Commercialization of Human Feeling.* Berkeley: University of California Press.

Talwar, Jennifer Parker. 2002. *Fast Food, Fast Track: Immigrants, Big Business, and the American Dream.* Boulder, CO: Westview Press.

West, Candace, and Don Zimmerman. 1987. "Doing Gender." *Gender & Society* 1: 125–51.

Williams, Jerome D. 2004. "Perspectives from the Classroom." Paper presented at the Consumer Racial Profiling Conference, University of Texas, Austin, March 30.

Wrigley, Julia. 1995. *Other People's Children.* New York: Basic Books.

Introduction to Reading 7

Patricia Hill Collins is a sociologist and a leader in the efforts to explain and understand the interaction of gender with race and social class. She has worked toward an understanding of gender that moves beyond theories that narrowly apply to White, middle- and upper-class women. This paper was an address to faculty attending a workshop on integrating race and social class into courses on gender. Although it is not intended for a student audience, it provides many ways that we can eliminate racism and sexism in our daily lives. She also challenges us to think more critically about the meaning of oppression and how we might apply it to social categories of difference and inequality.

1. What is wrong with an additive analysis of oppression?
2. What does Hill Collins mean by institutional oppression?
3. How does Hill Collins's discussion of privilege relate to Williams's discussion of women in toy stores in the previous article?

Toward a New Vision

Race, Class, and Gender as Categories of Analysis and Connection

Patricia Hill Collins

The true focus of revolutionary change is never merely the oppressive situations which we seek to escape, but that piece of the oppressor which is planted deep within each of us.

—Audre Lorde, *Sister Outsider,* 123

Audre Lorde's statement raises a troublesome issue for scholars and activists working for social change. While many of us have little difficulty assessing our own victimization within some major system of oppression, whether it be by race, social class, religion, sexual orientation, ethnicity, age or gender, we typically fail to see how our thoughts and actions uphold someone else's subordination. Thus, white feminists routinely point with confidence to their oppression as women but resist seeing how much their white skin privileges them. African-Americans who possess eloquent analyses of racism often persist in viewing poor White women as symbols of White power. The radical left fares little better. "If only people of color and women could see their true class interests," they argue, "class solidarity would eliminate racism and sexism." In essence, each group identifies the type of oppression with which it feels most comfortable as being fundamental and classifies all other types as being of lesser importance.

Oppression is full of such contradictions. Errors in political judgment that we make concerning how we teach our courses, what we tell our children, and which organizations are worthy of our time, talents and financial support flow smoothly from errors in theoretical analysis about the nature of oppression and activism. Once we realize that there are few pure victims or oppressors, and that each one of us derives varying amounts of penalty and privilege from the multiple systems of oppression that frame our lives, then we will be in a position to see the need for new ways of thought and action.

To get at that "piece of the oppressor which is planted deep within each of us," we need at least two things. First, we need new visions of what oppression is, new categories of analysis that are inclusive of race, class, and gender as distinctive yet interlocking structures of oppression. Adhering to a stance of comparing and ranking oppressions—the proverbial, "I'm more oppressed than you"—locks us all into a dangerous dance of competing for attention, resources, and theoretical supremacy. Instead, I suggest that we examine our different experiences within the more fundamental relationship of damnation and subordination. To focus on the particular arrangements that race or class or gender take in our time and place without seeing these structures as sometimes parallel and sometimes interlocking dimensions of the more fundamental relationship of domination and subordination may temporarily ease our consciences. But while such thinking may lead to short term social reforms, it is simply inadequate for the task of bringing about long term social transformation.

While race, class and gender as categories of analysis are essential in helping us understand

From Patricia Hill Collins, "Toward a new vision: Race, class and gender as categories of analysis and connection" in *Race, Class, and Gender,* the structural bases of domination and subordination, new ways of thinking that are not accompanied by new ways of acting offer incomplete prospects for change. To get at that "piece of the oppressor which is planted deep within each of us," we also need to change our daily behavior. Currently, we are all enmeshed in a complex web of problematic relationships that grant our mirror images full human subjectivity while stereotyping and objectifying those most different than ourselves. We often assume that the people we work with, teach, send our children to school with, and sit next to, will act and feel in prescribed ways because they belong to given race, social class or gender categories. These judgments by category must be replaced with fully human relationships that transcend the legitimate differences created by race, class and gender as categories of analysis. We require new categories of connection, new visions of what our relationships with one another can be.

Our task is immense. We must first recognize race, class and gender as interlocking categories of analysis that together cultivate profound differences in our personal biographies. But then we must transcend those very differences by reconceptualizing race, class and gender in order to create new categories of connection.

[This paper] addresses this need for new patterns of thought and action. I focus on two basic questions. First, how can we reconceptualize race, class and gender as categories of analysis? Second, how can we transcend the barriers created by our experiences with race, class and gender oppression in order to build the types of coalitions essential for social exchange? To address these questions I contend that we must acquire both new theories of how race, class and gender have shaped the experiences not just of women of color, but of all groups. Moreover, we must see the connections between these categories of analysis and the personal issues in our everyday lives. . . . As Audre Lorde points out, change starts with self, and relationships that we have with those around us must always be the primary site for social change.

How Can We Reconceptualize Race, Class and Gender as Categories of Analysis?

To me, we must shift our discourse away from additive analyses of oppression (Spelman 1982; Collins 1989). Such approaches are typically based on two key premises. First, they depend on either/or, dichotomous thinking. Persons, things and ideas are conceptualized in terms of their opposites. For example, Black/White, man/woman, thought/feeling, and fact/opinion are defined in oppositional terms. Thought and feeling are not seen as two different and interconnected ways of approaching truth that can coexist in scholarship and teaching. Instead, feeling is defined as antithetical to reason, as its opposite. In spite of the fact that we all have "both/and" identities (I am both a college professor and a mother—I don't stop being a mother when I drop my child off at school, or forget everything I learned while scrubbing the toilet), we persist in trying to classify each other in either/or categories. I live each day as an African-American woman—a race/gender specific experience. And I am not alone. Everyone has a race/gender/class specific identity. Either/or, dichotomous thinking is especially troublesome when applied to theories of oppression because every individual must be classified as being either oppressed or not oppressed. The both/and position of simultaneously being oppressed and oppressor becomes conceptually impossible.

A second premise of additive analyses of oppression is that these dichotomous differences must be ranked. One side of the dichotomy is typically labeled dominant and the other subordinate. Thus, Whites rule Blacks, men are deemed superior to women, and reason is seen as being preferable to emotion. Applying this premise to discussions of oppression leads to the assumption that oppression can be quantified, and that some groups are oppressed more than others. I am frequently asked, "Which has been most oppressive to you, your status as a Black person or your status as a woman?" What I am really being asked to do is divide myself into little boxes and rank my various statuses. If I experience oppression as a both/and phenomenon, why should I analyze it any differently?

Additive analyses of oppression rest squarely on the twin pillars of either/or thinking and the necessity to quantify and rank all relationships in order to know where one stands. Such approaches typically see African-American women as being more oppressed than everyone else because the majority of Black women experience the negative effects of race, class and gender oppression simultaneously. In essence, if you add together separate oppressions, you are left with a grand oppression greater than the sum of its parts.

I am not denying that specific groups experience oppression more harshly than others—lynching is certainly objectively worse than being held up as a sex object. But we must be careful not to confuse this issue of the saliency of one type of oppression in people's lives with a theoretical stance positing the interlocking nature of oppression. Race, class and gender may all structure a situation but may not be equally visible and/or important in people's self-definitions. . . . This recognition that one category may have salience over another for a given time and place does not minimize the theoretical importance of assuming that race, class and gender as categories of analysis structure all relationships.

In order to move toward new visions of what oppression is, I think that we need to ask new questions. How are relationships of domination and subordination structured and maintained in the American political economy? How do race, class and gender function as parallel and interlocking systems that shape this basic relationship of domination and subordination? Questions such as these promise to move us away from futile theoretical struggles concerned with ranking oppressions and towards analyses that assume race, class and gender are all present in any given setting, even if one appears more visible and salient than the others. Our task becomes redefined as one of reconceptualizing oppression by uncovering the connections among race, class and gender as categories of analysis.

1. Institutional Dimension of Oppression

Sandra Harding's contention that gender oppression is structured along three main dimensions—the institutional, the symbolic, and the individual—offers a useful model for a more comprehensive analysis encompassing race, class and gender oppression (Harding 1986). Systemic relationships of domination and subordination structured through social institutions such as schools, businesses, hospitals, the work place, and government agencies represent the institutional dimension of oppression. Racism, sexism and elitism all have concrete institutional locations. Even though the workings of the institutional dimension of oppression are often obscured with ideologies claiming equality of opportunity, in actuality, race, class and gender place Asian-American women, Native American men, White men, African-American women, and other groups in distinct institutional niches with varying degrees of penalty and privilege.

Even though I realize that many . . . would not share this assumption, let us assume that the institutions of American society discriminate, whether by design or by accident. While many of us are familiar with how race, gender and class operate separately to structure inequality, I want to focus on how these three systems interlock in structuring the institutional dimension of oppression. To get at the interlocking nature of race, class and gender, I want you to think about the antebellum plantation as a guiding metaphor for a variety of American social institutions. Even though slavery is typically analyzed as a racist institution, and occasionally as a class institution, I suggest that slavery was a race, class, gender specific institution. Removing any one piece from our analysis diminishes our understanding of the true nature of relations of domination and subordination under slavery.

Slavery was a profoundly patriarchal institution. It rested on the dual tenets of White male authority and White male property, a joining of the political and the economic within the institution of the family. Heterosexism was assumed and all Whites were expected to marry. Control over affluent White women's sexuality remained key to slavery's survival because property was to be passed on to the legitimate heirs of the slave owner. Ensuring affluent White women's virginity and chastity was deeply intertwined with maintenance of property relations.

Under slavery, we see varying levels of institutional protection given to affluent White women, working-class and poor White women, and enslaved African women. Poor White women enjoyed few of the protections held out to their upper class sisters. Moreover, the devalued status of Black women was key in keeping all White women in their assigned places. Controlling Black women's fertility was also key to the continuation of slavery, for children born to slave mothers themselves were slaves.

African-American women shared the devalued status of chattel with their husbands, fathers and sons. Racism stripped Blacks as a group of legal rights, education, and control over their own persons. African-Americans could be whipped, branded, sold, or killed, not because they were poor, or because they were women, but because they were Black. Racism ensured that Blacks would continue to serve Whites and suffer economic exploitation at the hands of all Whites.

So we have a very interesting chain of command on the plantation—the affluent White master as the reigning patriarch, his White wife helpmate to serve him, help him manage his property and bring up his heirs, his faithful servants whose production and reproduction were tied to the requirements of the capitalist political economy, and largely property-less, working class White men and women watching from afar. In essence, the foundations for the contemporary roles of elite White women, poor Black women, working class White men, and a series of other groups can be seen in stark relief in this fundamental American social institution. While Blacks experienced the most harsh treatment under slavery, and thus made slavery clearly visible as a racist institution, race, class and gender interlocked in structuring slavery's systemic organization of domination and subordination.

Even today, the plantation remains a compelling metaphor for institutional oppression. Certainly the actual conditions of oppression are not as severe now as they were then. To argue, as some do, that things have not changed all that much denigrates the achievements of those who struggled for social change before us. But the basic relationships among Black men, Black women, elite White women, elite White men, working class White men and working class White women as groups remain essentially intact.

A brief analysis of key American social institutions most controlled by elite White men should convince us of the interlocking nature of race, class and gender in structuring the institutional dimension of oppression. For example, if you are from an American college or university, is your campus a modern plantation? Who controls your university's political economy? Are elite White men over represented among the upper administrators and trustees controlling your university's finances and policies? Are elite White men being joined by growing numbers of elite White women helpmates? What kinds of people are in your classrooms grooming the next generation who will occupy these and other decision-making positions? Who are the support staff that produce the mass mailings, order the supplies, fix the leaky pipes? Do African-Americans, Hispanics or other people of color form the majority of the invisible workers who feed you, wash your dishes, and clean up your offices and libraries after everyone else has gone home? . . .

2. The Symbolic Dimension of Oppression

Widespread, societally-sanctioned ideologies used to justify relations of domination and subordination comprise the symbolic dimension of oppression. Central to this process is the use of stereotypical or controlling images of diverse race, class and gender groups. In order to assess the power of this dimension of oppression, I want you to make a list, either on paper or in your head, of "masculine" and "feminine" characteristics. If your list is anything like that compiled by most people, it reflects some variation of the following:

masculine	*feminine*
aggressive	passive
leader	follower
rational	emotional
strong	weak
intellectual	physical

Not only does this list reflect either/or dichotomous thinking and the need to rank both sides of the dichotomy, but ask yourself exactly which men and women you had in mind when compiling these characteristics. This list applies almost exclusively to middle-class White men and women. The allegedly "masculine" qualities that you probably listed are only acceptable when exhibited by elite White men, or when used by Black and Hispanic men against each other or against women of color. Aggressive Black and Hispanic men are seen as dangerous, not powerful, and are often penalized when they exhibit any of the allegedly "masculine" characteristics. Working-class and poor White men fare slightly better and are also denied the allegedly "masculine" symbols of leadership, intellectul competence, and human rationality. Women of color and working class and poor White women are also not represented on this list, for they have never had the luxury of being "ladies." What appear to be universal categories representing all men and women instead are unmasked as being applicable to only a small group.

It is important to see how the symbolic images applied to different race, class and gender groups interact in maintaining systems of domination and subordination. If I were to ask you to repeat the same assignment, only this time, by making separate lists for Black men, Black women, Hispanic women and Hispanic men, I suspect that your gender symbolism would be quite different. In comparing all of the lists, you might begin to see the interdependence of symbols applied to all groups. For example, the elevated images of White womanhood need devalued images of Black womanhood in order to maintain credibility.

While the above exercise reveals the interlocking nature of race, class and gender in structuring the symbolic dimension of oppression, part of its importance lies in demonstrating how race, class and gender pervade a wide range of what appears to be universal language. Attending to diversity . . . in our daily lives provides a new angle of vision on interpretations of reality thought to be natural, normal and "true." Moreover, viewing images of masculinity and femininity as universal gender symbolism, rather than as symbolic images that are race, class and gender specific, renders the experiences of people of color and of non-privileged White women and men invisible. One way to dehumanize an individual or a group is to deny the reality of their experiences. So when we refuse to deal with race or class because they do not appear to be directly relevant to gender, we are actually becoming part of some one else's problem.

Assuming that everyone is affected differently by the same interlocking set of symbolic images allows us to move forward toward new analyses. Women of color and White women have different relationships to White male authority and this difference explains the distinct gender symbolism applied to both groups. Black women encounter controlling images such as the mammy, the matriarch, the mule and the whore, that encourage others to reject us as fully human people. Ironically, the negative nature of these images simultaneously encourages us to reject them. In contrast, White women are offered seductive images, those that promise to reward them for supporting the status quo. And yet seductive images can be equally controlling. Consider, for example, the views of Nancy White, a 73-year-old Black woman, concerning images of rejection and seduction:

> My mother used to say that the black woman is the white man's mule and the white woman is his dog. Now, she said that to say this: we do the heavy work and get beat whether we do it well or not. But the white woman is closer to the master and he pats them on the head and lets them sleep in the house, but he ain't gon' treat neither one like he was dealing with a person. (Gwalatney 1980, 148)

Both sets of images stimulate particular political stances. By broadening the analysis beyond the confines of race, we can see the varying levels of rejection and seduction available to each of us due to our race, class and gender identity. Each of us lives with an allotted portion of institutional privilege and penalty, and with varying levels of rejection and seduction inherent in the symbolic images applied to us. This is the context in which we make our choices. Taken together, the institutional and symbolic dimensions of oppression create a structural backdrop against which all of us live our lives.

3. The Individual Dimension of Oppression

Whether we benefit or not, we all live within institutions that reproduce race, class and gender oppression. Even if we never have any contact with members of other race, class and gender groups, we all encounter images of these groups and are exposed to the symbolic meanings attached to those images. On this dimension of oppression, our individual biographies vary tremendously. As a result of our institutional and symbolic statuses, all of our choices become political acts.

Each of us must come to terms with the multiple ways in which race, class and gender as categories of analysis frame our individual biographies. I have lived my entire life as an African-American woman from a working-class family and this basic fact has had a profound impact on my personal biography. Imagine how different your life might be if you had been born Black, or White, or poor, or of a different race/class/gender group than the one with which you are most familiar. The institutional treatment you would have received and the symbolic meanings attached to your very existence might differ dramatically from what you now consider to be natural, normal and part of everyday life. You might be the same, but your personal biography might have been quite different.

I believe that each of us carries around the cumulative effect of our lives within multiple structures of oppression. If you want to see how much you have been affected by this whole thing, I ask you one simple question—who are your close friends? Who are the people with whom you can share your hopes, dreams, vulnerabilities, fears and victories? Do they look like you? If they are all the same, circumstance may be the cause. For the first seven years of my life I saw only low-income Black people. My friends from those years reflected the composition of my community. But now that I am an adult, can the defense of circumstance explain the patterns of people that I trust as my friends and colleagues? When given other alternatives, if my friends and colleagues reflect the homogeneity of one race, class and gender group, then these categories of analysis have indeed become barriers to connection.

I am not suggesting that people are doomed to follow the paths laid out for them by race, class and gender as categories of analysis. While these three structures certainly frame my opportunity structure, I as an individual always have the choice of accepting things as they are, or trying to change them. As Nikki Giovanni points out, "we've got to live in the real world. If we don't like the world we're living in, change it. And if we can't change it, we change ourselves. We can do something" (Tate 1983, 68). While a piece of the oppressor may be planted deep within each of us, we each have the choice of accepting that piece or challenging it as part of the "true focus of revolutionary change." . . .

Since I opened with the words of Audre Lorde, it seems appropriate to close with another of her ideas . . . :

> Each of us is called upon to take a stand. So in these days ahead, as we examine ourselves and each other, our works, our fears, our differences, our sisterhood and survivals, I urge you to tackle what is most difficult for us all, self-scrutiny of our complacencies, the idea that since each of us believes she is on the side of right, she need not examine her position. (1985)

I urge you to examine your position.

References

Collins, Patricia Hill. 1989. "The Social Construction of Black Feminist Thought." *Signs.* Summer 1989.

Gwalatney, John Langston. 1980. *Drylongso: A Self-Portrait of Black America.* New York:Vintage.

Harding, Sandra. 1986. *The Question in Feminism.* Ithaca, New York: Cornell University Press.

Lorde, Audre. 1984. *Sister Outsider.* Trumansberg, New York: The Crossing Press.

———. 1985. "Sisterhood and Survival." Keynote address, Conference on the Black Woman Writer and the Diaspora, Michigan State University.

Spelman, Elizabeth. 1982. "Theories of Race and Gender: The Erasure of Black Women." *Quest* 5: 36–32.

Tate, Claudia, ed. 1983. *Black Women Writers at Work.* New York: Continuum.

Introduction to Reading 8

In this article, Karen Pyke and Denise Johnson use both the social construction of gender and intersectional analysis to examine the experiences of second-generation Asian American women. They interviewed 100 daughters of Korean American (KA) and Vietnamese American (VA) immigrants to better understand how gender, ethnicity, and culture influenced the meaning respondents gave to their experiences. By living in two worlds, the Asian American women were acutely aware of the social construction of gender within culture, as they had to move between two cultural constructions of femininity. Thus, ethnicity and gender interact in ways that made the women conscious of their decision to "do gender" based upon the culturally defined, situational expectations for femininity.

1. Using this article as an example, explain what it means to "do gender."
2. Why don't these women just be "who they are" across situations?
3. How do these women's struggles between cultural definitions of femininity reinforce, and make dominant, White femininity?

Asian American Women and Racialized Femininities

"Doing" Gender Across Cultural Worlds

Karen D. Pyke and Denise L. Johnson

The study of gender in recent years has been largely guided by two orienting approaches: (1) a social constructionist emphasis on the day-to-day production or doing of gender (Coltrane 1989; West and Zimmerman 1987), and (2) attention to the interlocking systems of race, class, and gender (Espiritu 1997; Hill Collins 2000). Despite the prominence of these approaches, little empirical work has been done that integrates the doing of gender with the study of race. A contributing factor is the more expansive incorporation of social constructionism in the study of gender than in race scholarship where biological markers are still given importance despite widespread acknowledgment that racial oppression is rooted in social arrangements and not biology (Glenn 1999). In addition, attempts to theoretically integrate the doing of gender, race, and class around the concept of "doing difference" (West and Fenstermaker 1995) tended to downplay historical macrostructures of power and domination and to privilege gender over race and class (Hill Collins et al. 1995). Work is still needed that integrates systems of oppression in a social constructionist framework without granting primacy to any one form of inequality or ignoring larger structures of domination.

The integration of gender and race within a social constructionist approach directs attention to issues that have been overlooked. Little research has examined how racially and ethnically subordinated women, especially Asian American women, mediate cross-pressures in the production of femininity as they move between mainstream and ethnic arenas, such as family, work, and school, and whether distinct and even contradictory gender displays and strategies are enacted across different arenas. Many, if not most, individuals move in social worlds that do not require dramatic inversions of their gender performances, thereby enabling them to maintain stable and seemingly unified gender strategies. However, members of communities that are racially and ethnically marginalized and who regularly traverse interactional arenas with conflicting gender expectations might engage different gender performances depending on the local context in which they are interacting. Examining the ways that such individuals mediate conflicting expectations would address several unanswered questions. Do marginalized women shift their gender performances across mainstream and subcultural settings in response to different gender norms? If so, how do they experience and negotiate such transitions? What meaning do they assign to the different forms of femininities that they engage across settings? Do racially subordinated women experience their production of femininity as inferior to those forms engaged by privileged white women and glorified in the dominant culture?

We address these issues by examining how second-generation Asian American women experience and think about the shifting dynamics involved in the doing of femininity in Asian ethnic and mainstream cultural worlds. We look specifically at their assumptions about gender dynamics in the Euro-centric mainstream and Asian ethnic social settings, the way they think about their gendered selves, and their strategies in doing gender. Our analysis draws on and elaborates the theoretical literature concerning the construction of femininities across race, paying particular attention to how controlling images and ideologies shape the subjective experiences of women of color. This is the first study to our knowledge that examines how intersecting racial and gender hierarchies affect the everyday construction of gender among Asian American women.

Constructing Femininities

Current theorizing emphasizes gender as a socially constructed phenomenon rather than an innate and stable attribute (Lorber 1994; Lucal 1999; West and Zimmerman 1987). Informed by symbolic interactionism and ethnomethodology, gender is regarded as something people do in social interaction. Gender is manufactured out of the fabric of culture and social structure and has little, if any, causal relationship to biology (Kessler and McKenna 1978; Lorber 1994). Gender displays are "culturally established sets of behaviors, appearances, mannerisms, and other cues that we have learned to associate with members of a particular gender" (Lucal 1999, 784). These displays "cast particular pursuits as expressions of masculine and feminine 'natures'" (West and Zimmerman 1987, 126). The doing of gender involves its display as a seemingly innate component of an individual.

The social construction of gender provides a theoretical backdrop for notions of multiple masculinities put forth in the masculinities literature (Coltrane 1994; Connell 1987, 1995; Pyke 1996). We draw on this notion in conceptualizing a plurality of femininities in the social production of women. According to this work, gender is not a unitary process. Rather, it is splintered by overlapping layers of inequality into multiple forms of masculinities (and femininities) that are both internally and externally relational and hierarchical. The concepts of hegemonic and subordinated masculinities are a major contribution of this literature. . . .

The concept of femininities has served mostly as a placeholder in the theory of masculinities where it remains undertheorized and unexamined. Connell (1987, 1995) has written extensively about hegemonic masculinity but offers only a fleeting discussion of the role of femininities. He suggested that the traits of femininity in a patriarchal society are tremendously diverse, with no one form emerging as hegemonic. Hegemonic masculinity is centered on men's global domination of women, and because there is no configuration of femininity organized around women's domination of men, Connell (1987, 183) suggested the notion of a hegemonic femininity is inappropriate. He further argued that women have few opportunities for institutionalized power relations over other women. However, this discounts how other axes of domination, such as race, class, sexuality, and age, mold a hegemonic femininity that is venerated and extolled in the dominant culture, and that emphasizes the superiority of some women over others, thereby privileging white upper-class women. To conceptualize forms of femininities that are subordinated as "problematic" and "abnormal," it is necessary to refer to an oppositional category of femininity that is dominant, ascendant, and "normal" (Glenn 1999, 10). We use the notion of hegemonic and subordinated femininities in framing our analysis.

Ideas of hegemonic and subordinated femininities resonate in the work of feminist scholars of color who emphasize the multiplicity of women's experiences. Much of this research has focused on racial and class variations in the material and (re)productive conditions of women's lives. More recently, scholarship that draws on cultural studies, race and ethnic studies, and women's studies centers the cultural as well as material processes by which gender and race are constructed, although this work has been mostly theoretical (Espiritu 1997; Hill Collins 2000; St. Jean and Feagin 1998). Hill Collins (2000) discussed "controlling images" that denigrate and objectify women of color and justify their racial and gender subordination. Controlling images are part of the process of "othering," whereby a dominant group defines into existence a subordinate group through the creation of categories and ideas that mark the group as inferior (Schwalbe et al. 2000, 422). Controlling images reaffirm whiteness as normal and privilege white women by casting them as superior.

White society uses the image of the Black matriarch to objectify Black women as overly aggressive, domineering, and unfeminine. This imagery serves to blame Black women for the emasculation of Black men, low marriage rates, and poverty and to control their social behavior by undermining their assertiveness (Hill Collins 2000). While Black women are masculinized as aggressive and overpowering, Asian women are rendered hyperfeminine: passive, weak, quiet, excessively submissive, slavishly dutiful, sexually exotic, and available for white men (Espiritu 1997; Tajima 1989). This Lotus Blossom imagery obscures the internal variation of Asian American femininity and sexuality, making it difficult, for example, for others to "see" Asian lesbians and bisexuals (Lee 1996). Controlling images of Asian women also make them especially vulnerable to mistreatment from men who view them as easy targets. By casting Black women as not feminine enough and Asian women as too feminine, white forms of gender are racialized as normal and superior. In this way, white women are accorded racial privilege.

The dominant culture's dissemination of controlling imagery that derogates nonwhite forms of femininity (and masculinity) is part of a complex ideological system of "psychosocial dominance" (Baker 1983, 37) that imposes elite definitions of subordinates, denying them the power of self-identification. In this way, subordinates internalize "commonsense" notions of their inferiority to whites (Espiritu 1997; Hill Collins 2000). Once internalized, controlling images provide the template by which subordinates make meaning of their everyday lives (Pyke 2000), develop a sense of self, form racial and gender identities, and organize social relations (Osajima 1993; Pyke and Dang in press). For example, Chen (1998) found that Asian American women who joined predominately white sororities often did so to distance themselves from images of Asian femininity. In contrast, those who joined Asian sororities were often surprised to find their ideas of Asian women as passive and childlike challenged by the assertive, independent women they met. By internalizing the racial and gendered myth making that circumscribes their social existence, subordinates do not pose a threat to the dominant order. As Audre Lorde (1984, 123) described, "the true focus of revolutionary change is never merely the oppressive situations which we seek to escape, but that piece of the oppressor which is planted deep within us."

Hegemonies are rarely without sites of resistance (Espiritu 2001; Gramsci 1971; Hill Collins 2000). Espiritu (1997) described Asian American writers and filmmakers whose portraits of Asians defy the gender caricatures disseminated in the white-dominated society. However, such images are often forged around the contours of the one-dimensional stereotypes against which the struggle emerges. Thus, controlling images penetrate all aspects of the experience of subordinates, whether in a relationship of compliance or in one of resistance (Osajima 1993; Pyke and Dang in press).

The work concerning the effects of controlling images and the relational construction of subordinated and hegemonic femininities has mostly been theoretical. The little research that has examined how Asian American women do gender in the context of racialized images and ideologies that construct their gender as "naturally" inferior to white femininity provides only a brief look at these issues (Chen 1998; Lee 1996). Many of the Asian American women whom we study here do not construct their gender in one cultural field but are constantly moving between sites that are guided by ethnic immigrant cultural norms and those of the Euro-centric mainstream. A comparison of how gender is enacted and understood across such sites brings the construction of racialized gender and the dynamics of hegemonic and subordinated femininities into bold relief. We examine how respondents employ cultural symbols, controlling images, and gender and racial ideologies in giving meanings to their experiences.

Gender in Ethnic and Mainstream Cultural Worlds

We study Korean and Vietnamese Americans, who form two of the largest Asian ethnic groups in southern California, the site of this research. We focus on the daughters of immigrants as they are more involved in both ethnic and mainstream cultures than are members of the first generation. . . . The second generation, who are still mostly children and young adults, must juggle the cross-pressures of ethnic and mainstream cultures without the groundwork that a long-standing ethnic enclave might provide. This is not easy. Disparities between ethnic and mainstream worlds can generate substantial conflict for children of immigrants, including conflict around issues of gender (Kibria 1993; Zhou and Bankston 1998).

Respondents dichotomized the interactional settings they occupy as ethnic, involving their immigrant family and other coethnics, and mainstream, involving non-Asian Americans in peer groups and at work and school. They grew up juggling different cultural expectations as they moved from home to school and often felt a pressure to behave differently when among Asian Americans and non-Asian Americans. Although there is no set of monolithic, stable norms in either setting, there are certain pressures, expectations, and structural arrangements that can affect different gender displays (Lee 1996). Definitions of gender and the constraints that patriarchy imposes on women's gender production can vary from culture to culture. The Confucian moral code, which accords male superiority, authority, and power over women in family and social relations, has influenced the patriarchal systems of Korea and Vietnam (Kibria 1993; Min 1998). Women are granted little decision-making power and are not accorded an individual identity apart from their family role, which emphasizes their service to male members. A woman who violates her role brings shame to herself and her family. Despite Western observers' tendency to regard Asian families as uniformly and rigidly patriarchal, variations exist (Ishii-Kuntz 2000). Women's resistance strategies, like the exchange of information in informal social groups, provide pockets of power (Kibria 1990). Women's growing educational and economic opportunities and the rise of women's rights groups in Korea and Vietnam challenge gender inequality (Palley 1994). Thus, actual gender dynamics are not in strict compliance with the prescribed moral code.

As they immigrate to the United States, Koreans and Vietnamese experience a shift in gender arrangements centering on men's loss of economic power and increased dependency on their wives' wages (Kibria 1993; Lim 1997; Min 1998). Immigrant women find their labor in demand by employers who regard them as a cheap labor source. With their employment, immigrant women experience more decision-making power, autonomy, and assistance with domestic chores from their husbands. However, such shifts are not total, and male dominance remains a common feature of family life (Kibria 1993; Min 1998). Furthermore, immigrant women tend to stay committed to the ethnic patriarchal structure as it provides resources for maintaining their parental authority and resisting the economic insecurities, racism, and cultural impositions of the new society (Kibria 1990, 1993; Lim 1997). The gender hierarchy is evident in parenting practices. Daughters are typically required to be home and performing household chores when not in school, while sons are given greater freedom.

Native-born American women, on the other hand, are perceived as having more equality, power, and independence than women in Asian societies, reflecting differences in gender attitudes. A recent study of Korean and American women found that 82 percent of Korean women agreed that "women should have only a family-oriented life, devoted to bringing up the children and looking after the husband," compared to 19 percent of U.S. women (Kim 1994). However, the fit between egalitarian gender attitudes and actual behavior in the United States is rather poor. Patriarchal arrangements that accord higher status to men at home and work are still the norm, with women experiencing lower job status and pay, greater responsibility for family work even when employed, and high rates of male violence. Indeed, the belief that gender equality is the norm in U.S. society obscures the day-to-day materiality of American patriarchy. Despite cultural differences in the ideological justification of patriarchy, gender inequality is the reality in both Asian and mainstream cultural worlds.

* * *

Gender Across Cultural Terrains: "I'm Like a Chameleon. I Change My Personality"

The 44 respondents who were aware of modifying their gender displays or being treated differently across cultural settings framed their accounts in terms of an oppressive ethnic world and an egalitarian mainstream. They reaffirmed the ideological constructions of the white-dominated society by casting ethnic and mainstream worlds as monolithic opposites, with internal variations largely ignored. Controlling images that denigrate Asian femininity and glorify white femininity were reiterated in many of the narratives. Women's behavior in ethnic realms was described as submissive and controlled, and that in white-dominated settings as freer and more self-expressive.

Some respondents suggested they made complete personality reversals as they moved across realms. They used the behavior of the mainstream as the standard by which they judged their behavior in ethnic settings. As Elizabeth (19, VA) said,

> I feel like when I'm amongst other Asians . . . I'm much more reserved and I hold back what I think. . . . But when I'm among other people like at school, I'm much more outspoken. I'll say whatever's on my mind. It's like a diametric character altogether. . . . I feel like when I'm with other Asians that I'm the *typical* passive [Asian] person and I feel like that's what's expected of me and if I do say something and if I'm the *normal* person that I am, I'd stick out like a sore thumb. So I just blend in with the situation. (emphasis added)

Elizabeth juxtaposes the "typical passive [Asian] person" and the "normal," outspoken person of the mainstream culture, whom she claims to be. In so doing, she reaffirms the stereotypical image of Asians as passive while glorifying Americanized behavior, such as verbal expressiveness, as "normal." This implies that Asian ethnic behavior is aberrant and inferior compared to white behavior, which is rendered normal. This juxtaposition was a recurring theme in these data (Pyke 2000). It contributed to respondents' attempts to distance themselves from racialized notions of the typical Asian woman who is hyperfeminine and submissive by claiming to possess those traits associated with white femininity, such as assertiveness, self-possession, confidence, and independence. Respondents often described a pressure to blend in and conform with the form of gender that they felt was expected in ethnic settings and that conflicted with the white standard of femininity. Thus, they often described such behavior with disgust and self-loathing. For example, Min-Jung (24, KA) said she feels "like an idiot" when talking with Korean adults:

> With Korean adults, I act more shy and more timid. I don't talk until spoken to and just act shy. I kind of speak in a higher tone of voice than I usually do. But then when I'm with white people and white adults, I joke around, I laugh, I talk, and I communicate about how I feel. And then my voice gets stronger. But then when I'm with Korean adults, my voice gets really high. . . . I just sound like an idiot and sometimes when I catch myself I'm like, "Why can't you just make conversation like you normally do?"

Many respondents distanced themselves from the compliant femininity associated with their Asianness by casting their behavior in ethnic realms as a mere act not reflective of their true nature. Repeatedly, they said they cannot be who they really are in ethnic settings and the enactment of an authentic self takes place only in mainstream settings. . . .

Wilma (21, VA) states, "Like some Asian guys expect me to be passive and let them decide on everything. Non-Asians don't expect anything from me. They just expect me *to be me*" (emphasis added). Gendered behavior engaged in Asian ethnic settings was largely described as performative, fake, and unnatural, while that in white-dominated settings was cast as a reflection of one's true self. The femininity of the white mainstream is glorified as authentic, natural, and normal, and Asian ethnic femininity is denigrated as coerced, contrived, and artificial. The "white is right" mantra is reiterated in this view of white femininity as the right way of doing gender.

The glorification of white femininity and controlling images of Asian women can lead Asian American women to believe that freedom and equity can be acquired only in the white-dominated world. For not only is white behavior glorified as superior and more authentic, but gender relations among whites are constructed as more egalitarian. . . .

Controlling images of Asian men as hypermasculine further feed presumptions that whites are more egalitarian. Asian males were often cast as uniformly domineering in these accounts. Racialized images and the construction of hegemonic (white) and subordinated (Asian) forms of gender set up a situation where Asian American women feel they must choose between white worlds of gender equity and Asian worlds of gender oppression. Such images encourage them to reject their ethnic culture and Asian men and embrace the white world and white men so as to enhance their power (Espiritu 1997). . . .

In these accounts, we can see the construction of ethnic and mainstream cultural worlds—and Asians and whites—as diametrically opposed. The perception that whites are more egalitarian than Asian-origin individuals and thus preferred partners in social interaction further reinforces anti-Asian racism and white superiority. The cultural dominance of whiteness is reaffirmed through the co-construction of race and gender in these narratives. The perception that the production of gender in the mainstream is more authentic and superior to that in Asian ethnic arenas further reinforces the racialized categories of gender that define white forms of femininity as ascendant. In the next section, we describe variations in gender performances within ethnic and mainstream settings that respondents typically overlooked or discounted as atypical.

Gender Variations Within Cultural Worlds

Several respondents described variations in gender dynamics within mainstream and ethnic settings that challenge notions of Asian and American worlds as monolithic opposites. Some talked of mothers who make all the decisions or fathers who do the cooking. These accounts were framed as exceptions to Asian male dominance. For example, after Vietnamese women were described in a group interview as confined to domesticity, Ngâ (22, VA), who immigrated at 14 and spoke in Vietnamese-accented English, defined her family as gender egalitarian. She related,

> I guess I grow[sic] up in a *different* family. All my sisters don't have to cook, her husbands[sic] cooking all the time. Even my oldest sister. Even my mom—my dad is cooking. . . . My sisters and brothers are all very strong. (emphasis added)

Ngâ does not try to challenge stereotypical notions of Vietnamese families but rather reinforces such notions by suggesting that her family is different. Similarly, Heidi (21, KA) said, "Our family was kind of *different* because . . . my dad cooks and cleans and does dishes. He cleans house" (emphasis added). Respondents often framed accounts of gender egalitarianism in their families by stating they do not belong to the typical Asian family, with "typical" understood to mean male dominated. This variation in gender dynamics within the ethnic community was largely unconsidered in these accounts.

Other respondents described how they enacted widely disparate forms of gender across sites within ethnic realms, suggesting that gender behavior is more variable than generally framed. Take, for example, the case of Gin (29, KA), a law student married to a Korean American from a more traditional family than her own. When she is with her husband's kin, Gin assumes the traditional obligations of a daughter-in-law and does all the cooking, cleaning, and serving. The role exhausts her and she resents having to perform it. When Gin and her husband return home, the gender hierarchy is reversed. . . .

Controlling images of Asian men as hyper-domineering in their relations with women obscures how they can be called on to compensate for the subservience exacted from their female partners in some settings. Although respondents typically offered such stories as evidence of the patriarchy of ethnic arenas, these examples reveal that ethnic worlds are far more variable than generally described. Viewing Asian ethnic worlds through a lens of racialized gender stereotypes renders such variation invisible or, when acknowledged, atypical.

Gender expectations in the white-dominated mainstream also varied, with respondents sometimes expected to assume a subservient stance as Asian women. These examples reveal that the mainstream is not a site of unwavering gender equality as often depicted in these accounts and made less so for Asian American women by racial images that construct them as compliant. Many respondents described encounters with non-Asians, usually whites, who expected them to be passive, quiet, and yielding. Several described non-Asian (mostly white) men who brought such expectations to their dating relationships. Indeed, the servile Lotus Blossom image bolsters white men's preference for Asian women (Espiritu 1997). As Thanh (22, VA) recounted,

> Like the white guy that I dated, he expected me to be the submissive one—the one that was dependent on the guy. Kind of like the "Asian persuasion," that's what he'd call it when he was dating me. And when he found out that I had a spirit, kind of a wild side to me, he didn't like it at all. Period. And when I spoke up—my opinions—he got kind of scared.

So racialized images can cause Asian American women to believe they will find greater gender equality with white men and can cause white men to believe they will find greater subservience with Asian women. This dynamic promotes Asian American women's availability to white men and makes them particularly vulnerable to mistreatment.

There were other sites in the mainstream, besides dating relationships, where Asian American women encountered racialized gender expectations. Several described white employers and coworkers who expected them to be more passive and deferential than other employees and were surprised when they spoke up and resisted unfair treatment. Some described similar assumptions among non-Asian teachers and professors. Diane (26, KA) related,

> At first one of my teachers told me it was okay if I didn't want to talk in front of the class. I think she thought I was quiet or shy because I'm Asian. . . . [Laughing.] I am very outspoken, but that semester I just kept my mouth shut. I figured she won't make me talk anyway, so why try. I kind of went along with her.

Diane's example illustrates how racialized expectations can exert a pressure to display stereotyped behavior in mainstream interactions. Such expectations can subtly coerce behavioral displays that confirm the stereotypes, suggesting a kind of self-fulfilling prophecy. Furthermore, as submissiveness and passivity are denigrated traits in the mainstream, and often judged to be indicators of incompetence, compliance with such expectations can deny Asian American women personal opportunities and success. Not only is passivity unrewarded in the mainstream; it is also subordinated. The association of extreme passivity with Asian women serves to emphasize their otherness. Some respondents resist this subordination by enacting a more assertive femininity associated with whiteness. Lisa (18, KA) described being quiet with her relatives out of respect, but in mainstream scenes, she consciously resists the stereotype of Asian women as passive by adjusting her behavior. . . .

To act Asian by being reserved and quiet would be to "stand out in a negative way" and to be regarded as "not cool." It means one will be denigrated and cast aside. Katie consciously engages loud and gregarious behavior to prove she is not the typical Asian and to be welcomed by white friends. Whereas many respondents describe their behavior in mainstream settings as an authentic reflection of their personality, these examples suggest otherwise. Racial expectations exert pressure on these women's gender performances among whites. Some go to great lengths to defy racial assumptions and be accepted into white-dominated social groups by engaging a white standard of femininity. As they are forced to work against racial stereotypes, they must exert extra effort at being outspoken and socially gregarious. Contrary to the claim of respondents, gender production in the mainstream is also coerced and contrived. The failure of some respondents to recognize variations in gender behavior within mainstream and ethnic settings probably has much to do with the essentialization of gender and race. That is, as we discuss next, the racialization of gender renders variations in behavior within racial groups invisible.

The Racialization of Gender: Believing Is Seeing

In this section, we discuss how respondents differentiate femininity by race rather than shifting situational contexts, even when they were consciously aware of altering their own gender performance to conform with shifting expectations. Racialized gender was discursively constructed as natural and essential. Gender and race were essentialized as interrelated biological facts that determine social behavior.

Among our 100 respondents, there was a tendency to rely on binary categories of American (code for white) and Asian femininity in describing a wide range of topics, including gender identities, personality traits, and orientations toward domesticity or career. Racialized gender categories were deployed as an interpretive template in giving meaning to experiences and organizing a worldview. Internal variation was again ignored, downplayed, or regarded as exceptional. White femininity, which was glorified in accounts of gender behavior across cultural settings, was also accorded superiority in the more general discussions of gender.

Respondents' narratives were structured by assumptions about Asian women as submissive, quiet, and diffident and of American women as independent, self-assured, outspoken, and powerful. That is, specific behaviors and traits were racialized. As Ha (19, VA) explained, "sometimes I'm quiet and passive and shy. That's a Vietnamese part of me." Similarly, domesticity was linked with Asian femininity and domestic incompetence or disinterest, along with success in the work world, with American femininity. Several women framed their internal struggles between career and domesticity in racialized terms. Min-Jung said,

> I kind of think my Korean side wants to stay home and do the cooking and cleaning and take care of the kids whereas my American side would want to go out and make a difference and become a strong woman and become head of companies and stuff like that.

This racialized dichotomy was central to respondents' self-identities. Amy (21, VA) said, "I'm not Vietnamese in the way I act. I'm American because I'm not a good cook and I'm not totally ladylike." In fact, one's ethnic identity could be challenged if one did not comply with notions of racialized gender. In a group interview, Kimberly (21, VA) described "joking around" with coethnic dates who asked if she cooked by responding that she did not. . . .

Similarly, coethnic friends tell Hien (21, VA), "You should be able to cook, you are Vietnamese, you are a girl." To be submissive and oriented toward family and domesticity marks Asian ethnicity. Conformity to stereotypes of Asian femininity serves to symbolically construct and affirm an Asian ethnic identity. Herein lies the pressure that some respondents feel to comply with racialized expectations in ethnic settings, as Lisa (18, KA) illustrates in explaining why she would feel uncomfortable speaking up in a class that has a lot of Asians:

> I think they would think that I'm not really Asian. Like I'm whitewashed . . . like I'm forgetting my race. I'm going against my roots and adapting to the American way. And I'm just neglecting my race.

American (white) women and Asian American women are constructed as diametric opposites. Although many respondents were aware that they contradicted racialized notions of gender in their day-to-day lives, they nonetheless view gender as an essential component of race. Variation is ignored or recategorized so that an Asian American woman who does not comply is no longer Asian. This was also evident among respondents who regard themselves as egalitarian or engage the behavioral traits associated with white femininity. There was the presumption that one cannot be Asian and have gender-egalitarian attitudes. Asian American women can engage those traits associated with ascendant femininity to enhance their status in the mainstream, but this requires a rejection of their racial/ethnic identity. This is evident by the use of words such as "American," "whitewashed," or "white"—but not Asian—to describe such women. Star (22, KA) explained, "I look Korean but I don't act Korean. I'm whitewashed. [Interviewer asks, 'How do you mean you don't act Korean?'] I'm loud. I'm not quiet and reserved."

As a result, struggles about gender identity and women's work/family trajectories become superimposed over racial/ethnic identity. The question is not simply whether Asian American women like Min-Jung want to be outspoken and career oriented or quiet and family oriented but whether they want to be American (whitewashed) or Asian. Those who do not conform to racialized expectations risk challenges to their racial identity and charges that they are not really Asian, as occurs with Lisa when she interacts with her non-Asian peers. She said,

> They think I'm really different from other Asian girls because I'm so outgoing. They feel that Asian girls have to be the shy type who is very passive and sometimes I'm not like that so they think, "Lisa, are you Asian?"

These data illustrate how the line drawn in the struggle for gender equality is superimposed over the cultural and racial boundaries dividing whites and Asians. At play is the presumption that the only path to gender equality and assertive womanhood is via assimilation to the white mainstream. This assumption was shared by Asian American research assistants who referred to respondents' gender egalitarian viewpoints as evidence of assimilation. The assumption is that Asian American women can be advocates of gender equality or strong and assertive in their interactions only as a result of assimilation, evident by the display of traits associated with hegemonic femininity, and a rejection of their ethnic culture and identity. This construction obscures gender inequality in mainstream U.S. society and constructs that sphere as the only place where Asian American women can be free. Hence, the diversity of gender arrangements practiced among those of Asian origin, as well as the potential for social change within Asian cultures, is ignored. Indeed, there were no references in these accounts to the rise in recent years of women's movements in Korea and Vietnam. Rather, Asian ethnic worlds are regarded as unchanging sites of male dominance and female submissiveness.

Discussion and Summary

Our analysis reveals dynamics of internalized oppression and the reproduction of inequality that revolve around the relational construction of hegemonic and subordinated femininities. Respondents' descriptions of gender performances in ethnic settings were marked by self-disgust and referred to as a mere act not reflective of one's true gendered nature. In mainstream settings, on the other hand, respondents often felt a pressure to comply with caricatured notions of Asian femininity or, conversely, to distance one's self from derogatory images of Asian femininity to be accepted. In both cases, the subordination of Asian femininity is reproduced.

In general, respondents depicted women of Asian descent as uniformly engaged in subordinated femininity marked by submissiveness and white women as universally assertive and gender egalitarian. Race, rather than culture, situational dynamics, or individual personalities, emerged as the primary basis by which respondents gave meaning to variations in femininity. That is, despite their own situational variation in doing gender, they treat gender as a racialized feature of bodies rather than a sociocultural product. Specific gender displays, such as a submissive demeanor, are required to confirm an Asian identity. Several respondents face challenges to their ethnic identity when they behave in ways that do not conform with racialized images. Indeed, some claimed that because they are assertive or career oriented, they are not really Asian. That is, because they do not conform to the racialized stereotypes of Asian women but identify with a hegemonic femininity that is the white standard, they are different from other women of Asian origin. In this way, they manipulate the racialized categories of gender in attempting to craft identities that are empowering. However, this is accomplished by denying their ethnicity and connections to other Asian American women and through the adoption and replication of controlling images of Asian women.

Respondents who claim that they are not really Asian because they do not conform with essentialized notions of Asian femininity suggest similarities to transgendered individuals who feel that underneath, they really belong to the gender category that is opposite from the one to which they are assigned. The notion that deep down they are really white implies a kind of transracialized gender identity. In claiming that they are not innately Asian, they reaffirm racialized categories of gender just as transgendered individuals reaffirm the gender dichotomy (Kessler and McKenna 1978; Lorber 1994). However, there are limitations to notions of a transracialized identity as racial barriers do not permit these women to socially pass into the white world, even though they might feel themselves to be more white than Asian. Due to such barriers, they use terms that are suggestive of a racial crossover, such as "whitewashed" or "American" rather than "white" in describing themselves. Such terms are frequently used among Asian Americans to describe those who are regarded as assimilated to the white world and no longer ethnic, further underscoring how racial categories are essentialized (Pyke and Dang in press). Blocked from a white identity, these terms capture a marginalized space that is neither truly white nor Asian. As racial categories are dynamic, it remains to be seen whether these marginalized identities are the site for new identities marked by hybridity (Lowe 1991) or whether Asian Americans will eventually be incorporated into whiteness. This process may be hastened by outmarriage to whites and high rates of biracial Asian Americans who can more easily pass into the white world, thereby leading the way for other Asian Americans. While we cannot ascertain the direction of such changes, our data highlight the contradictions that strain the existing racial and gender order as it applies to second-generation Asian American women.

While respondents construct a world in which Asian American women can experience a kind of transracial gender identity, they do not consider the same possibility for women of other races. A white woman who is submissive does not become Asian. In fact, there was no reference in these accounts to submissive white women who are rendered invisible by racialized categories of gender. Instead, white women are constructed as monolithically self-confident, independent, assertive, and successful—characteristics of white hegemonic femininity. That these are the same ruling traits associated with hegemonic masculinity, albeit in a less exaggerated, feminine form, underscores the imitative structure of hegemonic femininity. That is, the supremacy of white femininity over Asian femininity mimics hegemonic masculinity. We are not arguing that hegemonic femininity and masculinity are equivalent structures. They are not. Whereas hegemonic masculinity is a superstructure of domination, hegemonic femininity is confined to power relations among women. However, the two structures are interrelated with hegemonic femininity constructed to serve hegemonic masculinity, from which it is granted legitimacy.

Our findings illustrate the powerful interplay of controlling images and hegemonic femininity in promoting internalized oppression. Respondents draw on racial images and assumptions in their narrative construction of Asian cultures as innately oppressive of women and fully resistant to change against which the white-dominated mainstream is framed as a paradigm of gender equality. This serves a proassimilation function by suggesting that Asian American women will find gender equality in exchange for rejecting their ethnicity and adopting white standards of gender. The construction of a hegemonic femininity not only (re)creates a hierarchy that privileges white women over Asian American women but also makes Asian American women available for white men. In this way, hegemonic femininity serves as a handmaiden to hegemonic masculinity.

By constructing ethnic culture as impervious to social change and as a site where resistance to gender oppression is impossible, our respondents accommodate and reinforce rather than resist the gender hierarchal arrangements of such locales. This can contribute to a self-fulfilling prophecy as Asian American women who hold gender egalitarian views feel compelled to retreat from interactions in ethnic settings, thus (re)creating Asian ethnic cultures as strongholds of patriarchy and reinforcing the maintenance of a rigid gender hierarchy as a primary mechanism by which ethnicity and ethnic identity are constructed. This marking of ethnic culture as a symbolic repository of patriarchy obscures variations in ethnic gender practices as well as the gender inequality in the mainstream. Thus, compliance with the dominant order is secured.

Our study attempts to bring a racialized examination of gender to a constructionist framework without decentering either race or gender. By examining the racialized meaning systems that inform the construction of gender, our findings illustrate how the resistance of gender oppression among our respondents draws ideologically on the denigration and rejection of ethnic Asian culture, thereby reinforcing white dominance. Conversely, we found that mechanisms used to construct ethnic identity in resistance to the proassimilation forces of the white-dominated mainstream rest on narrow definitions of Asian women that emphasize gender subordination. These findings underscore the crosscutting ways that gender and racial oppression operates such that strategies and ideologies focused on the resistance of one form of domination can reproduce another form. A social constructionist approach that examines the simultaneous production of gender and race within the matrix of oppression, and considers the relational construction of hegemonic and subordinated femininities, holds much promise in uncovering the micro-level structures and complicated features of oppression, including the processes by which oppression infiltrates the meanings individuals give to their experiences.

References

Baker, Donald G. 1983. *Race, ethnicity and power.* Boston: Routledge Kegan Paul.

Chang, Edward T. 1999. The post-Los Angeles riot Korean American community: Challenges and prospects. *Korean American Studies Bulletin* 10:6–26.

Chen, Edith Wen-Chu. 1998. The continuing significance of race: A case study of Asian American women in white, Asian American, and African American sororities. Ph.D. diss., University of California, Los Angeles.

Coltrane, Scott. 1989. Household labor and the routine production of gender. *Social Problems* 36:473–90.

———. 1994. Theorizing masculinities in contemporary social science. In *Theorizing masculinities,* edited by Harry Brod and Michael Kaufman. Thousand Oaks, CA: Sage.

Connell, R. W. 1987. *Gender and power.* Stanford, CA: Stanford University Press.

———. 1995. *Masculinities.* Los Angeles: University of California Press.

Espiritu, Yen L. 1997. *Asian American women and men.* Thousand Oaks, CA: Sage.

———. 2001. "We don't sleep around like white girls do": Family, culture, and gender in Filipina American life. *Signs: Journal of Women in Culture and Society* 26:415–40.

Glaser, Barney G., and Anselm L. Straus. 1967. *The discovery of grounded theory.* New York: Aldine.

Glenn, Evelyn Nakano. 1999. The social construction and institutionalization of gender and race. In *Revisioning gender,* edited by Myra Marx Ferree, Judith Lorber, and Beth B. Hess. Thousand Oaks, CA: Sage.

Gramsci, Antonio. 1971. *Selections from the prison notebooks of Antonio Gramsci,* edited and translated by Quintin Hoare and Geoffrey Nowell Smith. New York: International.

Hill Collins, Patricia. 2000. *Black feminist thought.* New York: Routledge.

Hill Collins, Patricia, Lionel A. Maldonado, Dana Y. Takagi, Barrie Thorne, Lynn Weber, and Howard Winant. 1995. Symposium: On West and Fenstermaker's "Doing difference." *Gender & Society* 9:491–513.

Ishii-Kuntz, Masako. 2000. Diversity within Asian American families. In *Handbook of family diversity,* edited by David H. Demo, Katherine Allen, and Mark A. Fine. New York: Oxford University Press.

Kessler, Suzanne, and Wendy McKenna. 1978. *Gender: An ethnomethodological approach.* Chicago: University of Chicago Press.

Kibria, Nazli. 1990. Power, patriarchy, and gender conflict in the Vietnamese immigrant community. *Gender & Society* 4:9–24.

———. 1993. *Family tightrope: The changing lives of Vietnamese Americans.* Princeton, NJ: Princeton University Press.

———. 1997. The construction of "Asian American": Reflections on intermarriage and ethnic identity among second generation Chinese and Korean Americans. *Ethnic and Racial Studies* 20:523–44.

Kim, Byong-suh. 1994. Value orientations and sex-gender role attitudes on the comparability of Koreans and Americans. In *Gender division of labor in Korea,* edited by Hyong Cho and Oil-wha Chang. Seoul, Korea: Ewha Women's University Press.

Lee, Jee Yeun. 1996. Why Suzie Wong is not a lesbian: Asian and Asian American lesbian and bisexual women and femme/butch/gender identities. In *Queer studies,* edited by Brett Beemyn and Mickey Eliason. New York: New York University Press.

Lim, In-Sook. 1997. Korean immigrant women's challenge to gender inequality at home: The interplay of economic resources, gender, and family. *Gender & Society* 11:31–51.

Lorber, Judith. 1994. *Paradoxes of gender:* New Haven, CT: Yale University Press.

Lorde, Audre. 1984. *Sister outsider:* Trumansberg, NY: Crossing Press.

Lowe, Lisa. 1991. Heterogeneity, hybridity, multiplicity: Marking Asian American differences. *Diaspora* 1:24–44.

Lucal, Betsy. 1999. What it means to be gendered me: Life on the boundaries of a dichotomous gender system. *Gender & Society* 13:781–97.

Min, Pyong Gap. 1998. *Changes and conflicts.* Boston: Allyn & Bacon.

Oropesa, R. S., and Nancy Landale. 1995. *Immigrant legacies: The socioecnomic circumstances of children by ethnicity and generation in the United States.* Working paper 95–01R. State College: Population Research Institute, Pennsylvania State University.

Osajima, Keith. 1993. The hidden injuries of race. In *Bearing dreams, shaping visions: Asian Pacific American perspectives,* edited by Linda Revilla, Gail Nomura, Shawn Wong, and Shirley Hune. Pullman: Washington State University Press.

Palley, Marian Lief. 1994. Feminism in a Confucian society: The women's movement in Korea. In *Women of Japan and Korea,* edited by Joyce Gelb and Marian Lieff. Philadelphia: Temple University Press.

Pyke, Karen. 1996. Class-based masculinities: The interdependence of gender, class, and interpersonal power. *Gender & Society* 10:527–49.

———. 2000. "The normal American family" as an interpretive structure of family life among grown children of Korean and Vietnamese immigrants. *Journal of Marriage and the Family* 62:240–55.

Pyke, Karen, and Tran Dang. In press. "FOB" and "whitewashed": Intra-ethnic identities and internalized oppression among second generation Asian Americans. *Qualitative Sociology.*

St. Jean, Yanick, and Joe R. Feagin. 1998. *Double burden: Black women and everyday racism.* Armonk, NY: M. E. Sharpe.

Schwalbe, Michael, Sandra Godwin, Daphne Holden, Douglas Schrock, Shealy Thompson, and Michele Wolkomir. 2000. Generic processes in the reproduction of inequality: An interactionist analysis. *Social Forces* 79:419–52.

Tajima, Renee E. 1989. Lotus blossoms don't bleed: Images of Asian women. In *Making waves,* edited by Asian Women United of California. Boston: Beacon.

West, Candace, and Sarah Fenstermaker. 1995. Doing difference. *Gender & Society* 9:8–37.

West, Candace, and Don H. Zimmerman. 1987. Doing gender. *Gender & Society* 1:125–51.

Zhou, Min. 1999. Coming of age: The current situation of Asian American children. *Amerasia Journal* 25:1–27.

Zhou, Min, and Carl L. Bankston III. 1998. *Growing up American.* New York: Russell Sage.

Introduction to Reading 9

Jane Ward's study of gendered relationships in Bienestar, a Latino health organization, provides an intersectional analysis from an organizational or institutional perspective, as well as an intersectional analysis of gender, ethnicity, and sexualities. Bienestar serves the lesbian, bisexual, gay, and transgendered (LBGT) Latino community in Los Angeles, including programs targeted to gay and lesbian adults and youth. A focus of care provision is on HIV/AIDS education and prevention, primarily because this is a major funding stream into the organization. Ward gained entry to the organization through her partner/girlfriend and conducted interviews with eight women employees, two men administrators, and a woman who served on the organization's governing board. Both the composition of the leadership—most workers at Bienestar are gay men—and the mission—HIV/AIDS prevention and education—contribute to why Ward defines the organization as gendered.

1. What does it mean to define an organization as gendered?
2. What are the expectations for and treatment of women employees in Bienestar's organization?
3. How does the intersection of sexuality and gender affect women in a predominately LBGT organization?

"Not All Differences Are Created Equal"

Multiple Jeopardy in a Gendered Organization

Jane Ward

The triad of race, class, and gender oppression, or "triple jeopardy," continues to be the normative framework used by many scholars to discuss the matrix of oppression. In the essay in which she proposes the more apt term "multiple jeopardy," Deborah King (1990) asserted that identities are not additive because oppression is about the quality, and not the quantity, of one's experiences. In addition to clarifying the relationship between race, class, and gender inequality, King's multiple jeopardy theory has allowed for an expanded understanding of the "matrix of domination," including, for example, a recognition of the multiplicative effects of heterosexism and homophobia for lesbians and gay men already affected by other forms of structural inequality (Anzaldua 1990; Trujillo 1991). This theory of the intersectional nature of oppressions, commonly referred to as intersectional feminist theory, developed in the 1980s largely in response to racism in the feminist movement of the 1970s and sexism in the civil rights movement of the 1950s and 1960s. Thus, a critique of "single-identity" social movements that combat one form of oppression while supporting others has also been central to the feminist intersectional approach (Combahee River Collective 1983; hooks 1981; Robnett 1996).

More recently, social movement scholars have applied feminist intersectional theory to demonstrate that single-identity movements are inevitably ineffective because they exclude vast numbers of constituents, support the matrix of domination by addressing one form of inequality and not others, and fail to recognize the complex mechanisms behind the very inequality they intend to address (Kurtz 2002; Stockdill 2001). Yet the application of feminist intersectional theory to social movement work is difficult in part due to unanswered questions within intersectional theory itself about how to recognize and respond to multiple oppressions simultaneously, particularly within organizational contexts in which one or more forms of structural inequality emerge as more salient obstacles than others. An extensive body of intersectional feminist theory has explored how the multiplicative nature of structural inequalities produces "multiple consciousness," or the consciousness that develops from being at the center of intersecting and mutually reliant systems of oppression (Baca Zinn and Thorton Dill 1996; Hill Collins 1990; King 1990). Multiple consciousness, paired with coalition politics that temporarily unite differently located groups working toward a common goal, has been posited as the corrective to single-identity movements and the problematic practice of counting or ranking oppressions (Kurtz 2002; Stockdill 2001). Multiple consciousness, it has been argued, allows individuals and groups to recognize that no system of domination is more primary than another.

Yet many feminists are still counting and/or ranking oppressions, and the notion that "not all differences are created equal" has developed as a particularly common theme within writing by queer women of color (Takagi 1996, 24). For example, Dana Takagi argued that while some white lesbian and gay activists "think of themselves as patterned on the 'ethnic model'" of identity, there are important ideological, discursive, and historical differences between ethnicity

From Ward, J., "'Not all differences are created equal': Multiple jeopardy in a gendered organization" *Gender & Society, 18*(1): 82–102.

and sexuality, particularly with respect to the way that race/ethnicity is "more obviously written on the body" than one's sexual identity (1996, 23–25). According to Takagi, oppression claims by lesbians and gay men raise many questions about whether homosexuality is a form of marginalization that people "choose" and is therefore less urgent than racism given the "quality of volunteerism in being gay/lesbian" (p. 25). Takagi's assertions imply that counting oppressions is problematic for the very reason that it treats oppressions as "equal," or unrankable, by obscuring important contextual differences between them. In contrast, Barbara Smith argued that lesbians of color often have been the most astute in their opposition to the "'easy way out' of choosing a 'primary oppression'" (1993, 100). Yet Smith positioned homophobia as particularly complicated and urgent by explaining that "homophobia is usually the last oppression to be mentioned, the last to be taken seriously, and the last to go" and identifying the "numerous reasons for otherwise sensitive people's reluctance to confront homophobia," such as the misconception that "lesbian and gay oppression is not as serious as other oppressions" (pp. 100–101). Even within queer multicultural feminist theory, the tension between efforts not to create a hierarchy of oppressions and the highly contextual sense of urgency that individuals often feel with respect to one or more forms of oppression (and not others) is evident and difficult to avoid.

This article demonstrates that there is a disjuncture between multiple jeopardy theory and the experiences of individuals moving in and out of organizations with different political ideologies, structures, and leadership. These meso-level factors may result in the perception that one form of inequality is more oppressive than others. The dictate to not count oppressions is therefore difficult to reconcile with the experience of many lesbians of color that not all oppressions are created equal inside social movement organizations. I focus on the experiences of lesbian staff and clients at Bienestar, a large, Latino health organization in Los Angeles focusing primarily on HIV/AIDS prevention and education. This study shows that oppression can be experienced as layered, or additive. Counting oppressions may remain a common practice, and an important political strategy, in the context of some social movement organizations. Emphasizing gender oppression, or ranking it as primary, may be an important strategy for women in social movement organizations in which combating other, and multiple, forms of oppression—racism, homophobia, and AIDS phobia—are central to organizational ideology. . . .

One way to explore context is to examine the ways in which organizations are influenced by other organizations and bureaucracies, social movements, and a vast array of exogenous forces that may have implications for organizational gender ideology. External forces or events may compromise organizational gender ideology and practices or may provide legitimacy and external support for their institutionalization. They can lead an organization to adapt its goals to ensure survival. In nonprofit or social movement organizations, exogenous forces can lead to a willingness of members to allow organizational failure rather than compromise ideology (Ward 2000). This article examines the impact of one exogenous force, the AIDS epidemic, on the evolving gender ideology and practices of an organization founded as a lesbian and gay activist group and transformed by the epidemic.

Institutionalized Sexism and the AIDS Infrastructure

In the late 1980s, gay and lesbian activists began to take notice of the drain of funds from gay-liberation causes to AIDS-related projects perceived by many funding agencies to be more urgent. Many activists became concerned that the political agenda of the gay and lesbian movement was being "de-gayed" and "swallowed by AIDS" (Rofes 1990; Vaid 1995). By the late 1980s, government grants for gay men and lesbians appeared to be earmarked almost exclusively for

HIV/AIDS, and many gay and lesbian organizations shifted their attention to new HIV/AIDS-related components to survive or grow (Vaid 1995). The phenomenal need for HIV-related services, combined with the need to distribute public and private funds to the organizations delivering these services, resulted in the development of a national AIDS infrastructure, including a network of AIDS service organizations (ASOs), AIDS-related government bureaucracies, and a community of natural and social scientists studying the epidemic and the communities it affects (Altman 1994).

Yet the extent to which the AIDS infrastructure has helped or hindered queer organizing and service delivery cannot be generalized, as it has differently affected lesbians and gay men, queer whites and queer people of color. On one hand, the availability of government funds for HIV/AIDS services has been a catalyst for the growth of organizations established by and for queer communities of color. Prior to the epidemic, white lesbian and gay organizations were funded largely by wealthy white gay men, while queer communities of color had fewer wealthy community members from whom to solicit private funding (Clendinen and Nagourney 1999; Vaid 1995). Thus, public HIV/AIDS grants have allowed for the development of organizations that aim to provide culturally sensitive services to queer people of color in general, and particularly gay men of color who are believed to be at high risk for contracting HIV. The now widely accepted epidemiological contention that all gay men are at high risk for HIV infection translates into more and better social services for all gay men, both HIV positive and HIV negative. Most urban ASOs and lesbian, gay, bisexual, transgendered (LGBT) community centers offer a host of publicly funded HIV-prevention programs designed to encourage HIV-negative gay men, and particularly gay men of color who are now understood to be at even higher risk than white gay men, to practice safer sex (Gay and Lesbian Medical Association 2000). Based on the principles of community and individual empowerment, these prevention programs address factors such as culture, communication, and self-esteem through fun, community-building activities that offer alternatives to high-risk behavior.

On the other hand, the AIDS infrastructure has had a different impact on lesbians, lesbians of color specifically. While the link between HIV-prevention and empowerment programs for gay men of color is relatively easy to draw by pointing to data on infection rates, the use of HIV/AIDS money for lesbian empowerment programs, in light of lesbians' reportedly low risk for HIV infection, requires more complicated rationales. Rates of infection among lesbians demonstrate that the threat of HIV/AIDS in the lesbian community has been, and continues to be, considerably less urgent than the threat to the gay male community. Yet lesbians remain at high risk for a number of psychosocial dangers, such as depression, suicide, and sexual violence—risk categories that are presumably addressed by mental health services in lesbian and gay organizations and funded by non-AIDS-related funding sources (Gay and Lesbian Medical Association 2000).

There are limited comfortable community spaces and service providers for lesbians of color. Few organizations provide an environment that is both queer and culturally familiar. LGBT community centers, generally run by queer whites, are often unknown, too far, too uncomfortable, or too racist to be a realistic alternative for lesbians of color (Stockdill 2001). Instead, lesbians of color may meet, socialize, organize, and receive services and support within their racial/ethnic communities even in AIDS organizations not structured to receive lesbians and run by and for gay men (Schwartz 1993). In organizations such as Bienestar, lesbians of color may find cultural solidarity in a queer environment yet at the cost of an internal struggle with sexism. According to Stockdill (2001, 208), "AIDS-related inequality . . . emanates not only from dominant social institutions, but from within oppressed communities themselves." This is an important finding not only because racism, homophobia, and sexism "are major obstacles to effective AIDS prevention and intervention" but also because these inequalities structure the opportunities of even those groups presumably least at risk, such as lesbians.

This article is concerned with how the nexus of sexism and racism leads lesbians of color to seek community and identity in AIDS organizations and, second, how the financial and ideological support given to racial, cultural, and queer pride in ASOs run by gay men of color leads lesbians of color to experience gender oppression as a primary obstacle in their organizational lives. Yet it is important to recognize that lesbians of color are not the only group of women affected by the focus on men, or heterosexual focus in the case of notions about women's risk, that has been central to the very structure of local and national AIDS organizations. For example, some research demonstrates that middle-class and male-centered notions about HIV risk have contributed to the failure of health educators to identify and address risky behaviors among young Black women living in poverty (White 1999). Other scholars and activists have suggested that lesbians, across racial groups, engage in AIDS activism even while they perceive it as a distraction from their own health concerns (Schwartz 1993; Stoller 1997). While some historical accounts of the lesbian and gay movement explain this trend by emphasizing the important work done by lesbians to take care of gay men living with AIDS and heal the political divisions between them (Schneider and Stoller 1995), essays such as Schwartz's (1993) and empirical research such as this focus on the limited structural opportunities available to lesbians, particularly lesbians of color, outside of AIDS-funded organizations.

* * *

Why Bienestar? Counting Oppressions and Making Organizational Choices

In the Summer of 2000, the women's program at Bienestar had only 1 employee and no funding and was suffering by comparison to a thriving gay men's component that was staffed by approximately 10 gay men across seven locations. In light of these disparities and the organization's primary identification as an HIV/AIDS organization, it is important to examine the intersection of factors that bring Latina lesbians to Bienestar, as opposed to organizations that have funding for lesbian-specific programs, such as the Los Angeles Gay and Lesbian Center. In Los Angeles, a city in which there are dozens of Latino community organizations, several white queer community organizations, and a handful of queer Latino organizations (all ASOs), the Latina lesbians interviewed for this project explained that three layers of inequality prevented them from finding an organization intended for their specific community. For Latina lesbians in Los Angeles, without organizations of their own, triple jeopardy has a practical and rotating meaning–race is the oppressive category in queer white organizations, sexual identity is the oppressive category in Latina organizations, and gender is the oppressive category in gay Latino organizations. It is the experience of being othered in each of these contexts that can make counting one's oppression a relevant practice in the lives of lesbians of color.

According to the organization's male leaders, Bienestar's women's component provides an exclusively Latina environment that fosters *jota* (or queer Latina) pride, community building, and romantic relationships among Chicana/Latina lesbians. Martin, the executive director, explains that queer Latinas "choose" Bienestar not only because it produces events and offers programs that are culturally familiar but also because of the racism that Latinas experience in large gay and lesbian organizations with white leadership. Gay and lesbian resource centers often appear to have achieved racial/ethnic/cultural integration but maintain a white leadership and a European American cultural hegemony that is manifested in the look and feel of the organization and its events (Vaid 1995). Martin argues that Bienestar is needed because staff in gay and lesbian organizations with predominantly white leadership will claim that they "open the door and welcome [Latinos]," but despite this ideological commitment, "the music, the food, the everything, is not really welcome to Latinos." In response to my question about whether a lesbian component is appropriate in an AIDS organization that is managed by gay men and has no funds for lesbians, David, the director of QUE PASA, explains,

> It's appropriate [because] it's not just about HIV and AIDS issues; it's about cultural relevancy. I have seen and heard stories from women that go to the agencies that do have lesbian-specific money, and because the women are only able to identify along a sexuality basis, they still have to deal with issues of racism at those agencies and may not feel comfortable. It's amazing to me, even taking into consideration how small the women's program is here, the overwhelming response we get from the community is that it's needed.

Both Martin and David assert that cultural sensitivity and an environment free of racism is often more important to Bienestar's lesbian clients than having a large and well-supported queer women's component. For the gay Latino men in leadership at Bienestar, it was therefore not unexpected that Latina lesbians would seek services at their organization, regardless of whether the organization has made an explicit commitment to hiring queer women or including service to lesbians in its mission.

In contrast, the women staff at Bienestar offered a more nuanced explanation for the presence of young Latina lesbians in the organization. According to lesbian staff, it is the lack of alternative, more woman-centered, political and social organizations for queer Latinas in Los Angeles that brings young Latinas to Bienestar. While some Latino gay and lesbian organizations, such as GLLU, were growing and active in the late 1970s and 1980s, the undeniable impact of the AIDS epidemic on gay Latino men, as well as the availability of AIDS-related funds, led many of these organizations to turn their attention to HIV/AIDS or to disband altogether. Elena, a previous manager of the women's program, suggests that she might have become involved in an organization that was both culturally sensitive and gender sensitive had such an organization existed:

> If I wanted to work for something that was more specific to being a jota and being queer identified and female, I couldn't find that. . . . There was Lesbianas Unidas. They kind of have been defunct for a while. . . . I think there has been like a backlash as far as identity politics go . . . anything that [comes] to queer communities of color [is] in the form of AIDS monies.

Similarly, according to Tina, a transgender Latina staff member,

> You have to earn your respect there, because if we start first with society in general, men are always in the position of power. . . . You're a woman, so that's not good because you're not in a position of power. Apart from that, you're a lesbian. And the last part is you're Latina. So you're triple-fucked.

While Martin and David's comments imply that many of Bienestar's lesbian clients prioritize racial/ethnic unity over the greater stability and centrality of lesbian programming offered in non-Latino organizations, Elena and Tina emphasize that it is the limited choices for women of color in general and within the queer Latino community that lead women to Bienestar. If an ideal organization for Latina lesbians is one that offers cultural empowerment and lesbian leadership and programming, the attention of David and Martin to the failings of white organizations distracted from, or overlooked, work that might be done at Bienestar to produce such an environment.

Yet according to Martin, the supportive presence of Latinas in an organization run predominantly by and for Latino men is also a reflection of a "comfortable" and "natural" gender division of labor in Latino culture. For example, lesbian board members, he argues, are invested in the success of Bienestar in part because of the cultural role of Latinas as caretakers:

> I think that in all of our family units, it is our mothers, our sisters, that are primarily our caretakers, the ones who give us that support, and even in an environment of machismo, it is always the

> women. And I think they feel comfortable just keeping on that role in coming every month to the board meetings and doing the work because they are taking care of their brothers. Without really stopping and analyzing, I think it's just a very natural progression of our culture, why they are in that capacity.

While many of the women staff and clients at Bienestar were dissatisfied with the marginal role of the women's program, ultimately leading them to organize against sexism in the organization as described below, it is also important to recognize the partial truth in Martin's assessment that the Latinas at Bienestar were willing to manage their coworkers' machismo in exchange for being accepted as queer in a distinctly Latino/Chicano space. When I asked one lesbian staff member why she did not "just quit" after being treated so poorly by male management, she explained,

> Bienestar is sexist, but if I leave, I'm afraid I'll never get to work with my community again. . . . I may be *pocha* and all that, but here there are other Mexican women who need me and respect me. And that's not going to be easy to find for a butch Mexican woman.

A gay-male-dominated organization. Lesbian staff members at Bienestar describe examples of inequalities that are well documented in the literature on women working in male-dominated environments and particularly emphasized male coworkers' ignorance of male privilege and their entitlement to conversational and physical space (Lorber 1994; West and Zimmerman 1987). Lesbian staff perceived that they and their clients had been relegated to the least desirable physical space in the centers in which they worked. Fatima, a staff member who was holding a lesbian support group in a makeshift area of one of Bienestar's offices, recalls her supervisor's response to her request that he share the "nicer" meeting room where the gay men's group was being held:

> He was telling me, ". . . I've been facilitating this group with the same men for about two years, and we worked hard to get this . . . nice group discussion room. And I feel, for lack of a better word, *offended* that you are even asking me to share the room." And I was like, ". . . It's not like I am taking something away from you, I'm asking you to *share* what we have together within the agency." And he had to *think* about it!

Fatima adds that the discussion that ensued around sharing the meeting room signaled to her that her supervisor lacked the ability to recognize male privilege or draw connections between systems of oppression. . . .

Women staff explained that this kind of experience led them to believe that male coworkers could understand oppression only from their position of disadvantage (i.e., racism) but not from the position of privilege (i.e., sexism).

Fatima and other lesbian staff members assert that they had more or less expected to struggle with sexism of this kind at Bienestar based on its reputation as a gay-male-dominated organization. Yet they found that gender inequality at Bienestar was complicated by the centrality of gay men to the organization's HIV-focused mission, the external legitimacy given to Bienestar's programmatic priorities, the urgency of the AIDS epidemic, and the availability of funding for gay men's programs. If lesbian programs had no funding, did their support groups deserve to use the better meeting room? If the mission of the organization is HIV prevention and most clients are gay men, on what grounds could lesbian staff demand an environment more sensitive to their needs?

Lesbian Programs: Appropriate but Not Necessary

The psychocultural barriers model. Understanding the intersection of clients' racial/ethnic, sexual, and cultural identities is central to Bienestar's HIV-prevention model. . . .

In theory, Bienestar's funding structure necessitated that the queer women's component prioritize the provision of HIV-prevention services to lesbian clients. Given that lesbian clients come to the organization to come out and meet other women, this project would likely have failed immediately had staff prioritized HIV prevention in practice or relied on the prevention models David refers to above (i.e., talking about biological risk, handing out condoms, etc.). Thus, the psychocultural barriers model, which David generalizes from Diaz's [1997] study of gay Latino men to include the gender-inclusive term "communities of color," justifies the delivery of HIV-prevention services to lesbians because the method de-emphasizes HIV and instead emphasizes issues that are of interest to young Latina lesbians (e.g., family, communication, dating).

If lesbians are being served at Bienestar, then why should it matter who funds these services and with what intention? While Bienestar boasts that it is the only organization in Los Angeles that provides culturally relevant services to Latina lesbians, it does so without an ideological commitment to lesbian programs, lesbian staff, or lesbian clients. Although the use of the psychocultural model validates the importance of issues that might be important to young Latina lesbians, it is based on accountability to funders and not lesbian clients themselves or the Latina lesbian community. Women staff and clients were therefore simultaneously grateful for the opportunity to be part of the only queer Latino organization that provided lesbian services but also aware that their role in the organization was fragile and auxiliary. Bienestar's financial rationale for queer women's marginalization (i.e., "there just isn't funding for your program") made it difficult for lesbian staff to determine whether lack of support for their programs, and for themselves as employees, stemmed from a funding obstacle that the male leadership could do nothing about or a more personal conflict between gay men and lesbians in the organization. Monica, a queer staff member, points out that limited funding is an obstacle for the queer women's program but also emphasizes that gay male staff felt threatened by the program and the possibility that women could be "taking over." . . .

Such comments highlight the awareness of lesbian staff that difficulty finding funds to support lesbian youth programs is a real obstacle, likely faced by organizations other than Bienestar. Yet the constant attention to HIV/AIDS funding as a means to explain all disparities between gay men and lesbians in the organization begged other questions for lesbian staff, questions that focused on inequalities between gay and lesbian employees that had little or nothing to do with external funding for lesbian programs.

"We are the population we serve": Hiring and promotions at Bienestar. Lesbian staff suggested that a second philosophy adopted at Bienestar supported gender discrimination in hiring and promotions. Committed to the notion that staff members should represent the population they serve, Bienestar's leaders emphasized the importance of peer-led programs in which the identities of employees closely matched those of their clients. According to Elena, "[David] was always proud to say that, 'We are the population that we serve; we are the ones who have been here before.'" Yet Elena points out that this same philosophy was used to justify the absence of lesbians in leadership positions and in the organization in general.

While the psychocultural model could be used to explain why a lesbian program was appropriate at Bienestar, the number of AIDS cases could be used to explain why developing the program, and concomitantly hiring and promoting more lesbian staff, was not necessary. Together these logics led lesbian staff to believe that their presence at Bienestar was a privilege or luxury, in contrast with the feeling of entitlement that they perceived in gay men. At Bienestar, the good news of being theoretically at low risk for HIV infection became bad news for lesbians who hoped for job advancement in an organization in which the staff was expected to reflect AIDS demographics.

* * *

Analysis and Conclusions

The data presented here indicate that the external sociopolitical environment and ideological currents within social movements can assist in forming an institutional context in which gender inequality is mystified, legitimized, and/or naturalized. In the case of Bienestar, which began as a subcommittee of a lesbian and gay organization (GLLU), the urgency of the AIDS epidemic resulted in a shift in organizational identity and goals such that the organization's constituency became HIV positive gay men. While GLLU women struggled with sexism, they were nonetheless central to GLLU's early mission and functioning. As GLLU became Bienestar, it was the new emphasis on HIV, paired with external logics about HIV risk and prevention methods, that institutionalized the marginal position of lesbian staff and lesbian programs in the organization. In conjunction with other research findings (Schwartz 1993; Stockdill 2001; White 1999), the data presented here suggest that public health discourse, the AIDS infrastructure, gendered norms regarding women's caretaking of gay men affected by HIV, and the prioritization of racial/cultural pride over gender equality in the AIDS movement have converged to provide both scientific and political justifications for the marginal position of women in AIDS-funded organizations.

While sexism has divided the gay liberation movement since its inception (D'Emilio 2000; Faderman 1991), lesbian marginalization has witnessed a new level as the centrality of gay men in LGBT organizations has been rendered less visible by a new set of tools and logics made available by the AIDS epidemic. On the surface, the case of Bienestar tells a familiar story of women struggling with sexism in a male-dominated work organization. Yet what is sociologically significant in their accounts is the way in which lesbian staff at Bienestar were compelled to believe in the legitimacy of their own marginalization, based on external logics about risk associated with HIV/AIDS. Conflicts regarding sexism at Bienestar were complicated by the interconnectedness of, on one hand, sexist attitudes and practices that male coworkers could conceivably change and, on the other hand, external obstacles, such as a shortage of funding for women's programs, about which male managers could do little. The interplay of internal organizational dynamics and external forces, micro-level interactions between men and women coworkers and macro-level trends in the national response to an epidemic, resulted in additional challenges for women attempting to make sense of, and respond to, their position within a gendered organization.

These effects are particularly consequential for lesbians of color, whose experience of triple jeopardy is both cause and effect of their limited organizational options. This article has also demonstrated that in the context of multicultural social movement organizations, it is difficult for organizational actors not to count, emphasize, and prioritize particular oppressions for the very reason that the histories and meanings of racism, classism, sexism, homophobia, and other forms of oppression are personally, politically, and organizationally distinct. Several women interviewed for this research referred to their experience of triple jeopardy, or the primary yet rotating significance of three forms of oppression in their lives: sexism, racism, and homophobia. In the context of Bienestar, in which both racism and homophobia received significant ideological attention from male staff, sexism emerged as the most salient form of oppression for women staff. Although racism and homophobia bring queer Latinas to Bienestar to begin with, sexism became the form of oppression with the greatest material consequences given the centrality of Bienestar to their economic security and sense of safety and community. The data presented here support the assertion that oppressions are multiplicative and qualitatively experienced. Yet the case of Bienestar simultaneously, and perhaps paradoxically, suggests the need for an expansion of intersectional feminist theory, multiple jeopardy theory specifically, to include an analysis of organizational context and the rotating significance of experientially countable and distinct forms of oppressions.

Britton (2000) has stressed the importance of examining context as a means of discovering how gendered characteristics of organizations change and develop over time, how members of an organization understand and interact with these characteristics, and how organizations might become less oppressively gendered. The organized response of lesbian and gay communities to the AIDS epidemic serves as one example of a sociopolitical context that has produced change in the gendered characteristics of LGBT organizations. . . .

References

Altman, Dennis. 1994. *Power and community: Organizational and cultural responses to AIDS.* London: Taylor & Francis.

Anzaldua, Gloria. 1990. *Bridge,* drawbridge, sandbar or island: Lesbians-of-color hacienda alianzas. In *Bridges of power: Women's multicultural alliances,* edited by Lisa Albrecht and Rose M. Brewer. Philadelphia: New Society.

Baca Zinn, Maxine, and Bonnie Thorton Dill. 1996. Theorizing difference from multiracial feminism. *Feminist Studies* 22:321–31.

Britton, Dana. 2000. The epistemology of the gendered organization. *Gender & Society* 14:418–34.

Clendinen, Dudley, and Adam Nagourney. 1999. *Out for good: The struggle to build a gay rights movement in America.* New York: Simon & Schuster.

Combahee River Collective. 1983. The Combahee River Collective statement. In *Home girls: A Black feminist anthology,* edited by Barbara Smith. New York: Kitchen Table Women of Color Press.

D' Emilio, John. 2000. Cycles of change, questions of strategy: The gay and lesbian movement after fifty years. In *Politics of gay rights,* edited by C. Rimmerman, K. Wald, and C. Wilcox. Chicago: University of Chicago Press.

Diaz, Rafael Jorge. 1997. Latino gay men and psychocultural barriers to AIDS prevention. In *In changing times: Gay men and lesbians encounter HIV and AIDS,* edited by M. Levine, P. Nardi, and J. Gagnon. Chicago: University of Chicago Press.

Faderman, Lillian. 1991. *Odd girls and twilight lovers: A history of lesbian life in twentieth-century America.* New York: Penguin.

Gay and Lesbian Medical Association. 2000. *Healthy people 2010: A companion document for lesbian, gay, bisexual and transgender health.* San Francisco: Gay and Lesbian Medical Association.

Hill Collins, Patricia. 1990. *Black feminist thought: Knowledge, consciousness, and the politics of empowerment.* Boston: Unwin Hyman.

hooks, bell. 1981. *Ain't I a woman: Black women and feminism.* Boston: South End.

King, Deborah. 1990. Multiple jeopardy, multiple consciousness: The context of a Black feminist ideology. In *Black women in America: Social science perspectives,* edited by M. Malson, Elisabeth Mudimbe-Boyi, Jean F. O'Barr, and Mary Wyer. Chicago: University of Chicago Press.

Kurtz, Sharon. 2002. *Workplace justice: Organizing multi-identity movements.* Minneapolis: University of Minnesota Press.

Lorber, Judith. 1994. *Paradoxes of gender.* New Haven, CT: Yale University Press.

Reskin, Barbara. 1988. Bringing the men back in: Sex differentiation and the devaluation of women's work. *Gender & Society* 2: 58–81.

Robnett, Belinda. 1996. African-American women in the civil rights movement, 1954–1965: Gender, leadership, and micromobilization. *American Journal of Sociology* 101 (6): 1661–93.

Rofes, Eric. 1990. Gay lib vs. AIDS: Averting civil war in the 1990s. *OUT/LOOK* 2 (4): 8–17.

Schneider, Beth, and Nancy Stoller. 1995. *Women resisting AIDS: Feminist strategies of empowerment.* Philadelphia: Temple University Press.

Schwartz, Ruth. 1993. New alliances, strange bedfellows: Lesbians, gay men, AIDS. In *Sisters, sexperts, queers: Beyond the lesbian nation,* edited by A. Stein. New York: Penguin.

Smith, Barbara. 1993. Homophobia: Why bring it up? In *The lesbian and gay studies reader,* edited by Henry Abelove, Michile Aina Barale, and David Halperin. New York: Routledge.

Stockdill, Brett. 2001. Forging a multidimensional oppositional consciousness: Lessons from community-based AIDS activism. In *Oppositional consciousness: The subjective roots of social protest,* edited by J. Mansbridge and A. Morris. Chicago: University of Chicago Press.

Stoller, Nancy. 1997. From feminism to polymorphous activism: Lesbians in AIDS organizations. *In changing times: Gay men and lesbians encounter HIV/AIDS,* edited by M. Levine, P. Nardi, and S. Gagnon. Chicago: University of Chicago Press.

Takagi, Dana. 1996. Maiden voyage: Excursion into sexuality and identity politics in Asian America. In *Asian American sexualities: Dimensions of the gay & lesbian experience,* edited by Russell Leong. New York: Routledge.

Trujillo, Carla, ed. 1991. *Chicana lesbians: The girls our mothers warned us about.* Berkeley, CA: Third Woman Press.

Vaid, Urvashi. 1995. *Virtual equality: The mainstreaming of gay & lesbian liberation.* New York: Anchor Books.

Ward, Jane. 2000. A new kind of AIDS: Adapting to the success of protease inhibitors in an AIDS care organization. *Qualitative Sociology* 23: 247–66.

West, Candace, and Don Zimmerman. 1987. Doing gender. *Gender & Society* 1: 125–51.

White, Renee. 1999. *Putting risk in perspective: Black teenage lives in the era of AIDS.* Lanham, MD: Rowman and Littlefield.

✤ Topics for Further Examination ✤

- Using an academic database, look up the work of Patricia Hill Collins, Bonnie Thornton Dill, Maxinne Baca Zinn, Michael Messner, or others mentioned in the Introduction to Chapter 2 to find out what is currently being done on intersectionality. (Use parentheses around their names and ask for referred journals only.)
- Do a Web search using "feminist theory" and another category of difference and inequality (i.e., "feminist theory" and "race.")
- Using the Web, locate information on those cited in the Introduction to Chapter 2: Audre Lorde, bell hooks, Angela Davis, Gloria Anzaldua, or Sojourner Truth. When doing so, try to find the names of others who challenged the whiteness of the women's movement.

3

Gender and the Prism of Culture

Now that we have introduced you to the ways in which the American gender system interacts with, and is modified by, a complex set of categories of difference and inequality, we turn to an exploration of the ways in which the prism of culture interacts with gender definitions and arrangements. Generations of researchers in the social sciences, especially anthropology and sociology, have opened our eyes to the array of "genderscapes" around the globe. When we look through our kaleidoscope at the interaction between the prisms of gender and culture, we see different patterns that blur, blend, and are cast into a variety of culturally gendered configurations (Baker, 1999).

What Is Culture?

Culture consists of the beliefs, practices, and material objects that are created and shared within a group of people, thus constituting their way of life (Stone & McKee, 1999). The anthropological and sociological view of culture makes it clear that, without culture, human experience would have little shape or meaning (Schwalbe, 2005). A group's culture provides members with the assumptions and expectations on which their social interaction is built and in which their identities are forged. It also makes groups distinct from each other.

Indeed, the members of different cultural groups create very different realities. For example in some cultures, such as the Sambia of New Guinea, people do not perceive or categorize people as homosexual, and yet members of such cultures may regularly engage in same-gender sex (Herdt, 1997). Not only do different groups of people produce different cultures, the cultures they produce are dynamic. That is, people continually generate and alter culture both as individuals and as members of particular groups; as a result, all cultures undergo change as their members evaluate, resist, and challenge beliefs and practices (Stone & McKee, 1999). To illustrate, in the United States today, racist beliefs about intellectual and other differences between so-called racial groups are eroding, in part because scientists and social scientists have demonstrated that race is a social fiction rather than a biological fact and in part because the ongoing civil rights movement has been successful in dismantling much of the racist infrastructure of American society.

The prisms of gender and culture are inextricably intertwined. That is, people construct specific gender beliefs and practices in relation to particular cultural traditions and societal conditions. Cultures are gendered in distinctive ways and gender systems and, in turn, shape both material and symbolic cultural products (see, for example, Chapter 5). As you will discover, the crosscultural analyses of gender presented in this chapter provide critical support for the social constructionist argument that gender is a situated, negotiated, contested, and changing set of practices and understandings.

Let's begin with a set of observations about gender in different cultures. Do you know that there are cultures in which individuals can move from one gender category to another without being stigmatized? If you traveled from country to country around the world, you would find cultures in which men are gentle, soft-spoken, and modest; and cultures in which women are viewed as strong and take on roles that are labeled masculine in the United States. Although we hear news about extreme forms of oppression of women in some places in the world (e.g., bride-burning in parts of India), there are other places where women and men live in relative harmony and equality. Also, there are cultural groups in which the social prisms of difference and inequality that operate in the United States (e.g., social class, race, sexual orientation) are minimal, inconsequential, or nonexistent (see chapter readings by Christine Helliwell and Maria Lepowsky).

The Problem of Ethnocentrism in Crosscultural Research

If you find any of these observations unsettling or even shocking, then you have probably tapped into the problem of bias in crosscultural studies. One of the great challenges of crosscultural research is learning to transcend one's own cultural assumptions to be able to value and understand another culture in and of itself. It takes practice, conscious commitment, and self-awareness to get outside of one's own cultural box. After all, seeing what our culture wants us to see is precisely what socialization is about. Not only do cultural blinders make it difficult for us to see what gender is and how it is configured and reconfigured by various social prisms within our own culture, they can make it even more challenging to grasp the profoundly different ways in which people in other cultures think about and organize human relations.

We tend to "see what we believe," which means that we are likely to deny gender patterns that vary from our own cultural experience and/or to misinterpret patterns that are different from our own. For example, the Europeans who first explored and colonized Africa were horrified by the ways in which African forms of gender and sexuality diverged from their own. They had no framework in which to understand warrior women, such as Nzinga of the Ndongo kingdom of the Mbundu (Murray & Roscoe, 1998). Nzinga was king of her people, dressed as man, and was surrounded by a harem of young men who dressed as women and were her "wives" (Murray & Roscoe, 1998). However, her behavior made sense in the context of her culture, one in which people perceived gender as situational and symbolic, thus allowing for alternative genders (Murray & Roscoe, 1998).

It is a challenge to resist the tendency toward ethnocentrism (i.e., the belief that the ideas and practices of one's own group are the standard and that divergent cultures are substandard or inferior). However the rewards for bracketing the ethnocentric attitude are extremely valuable because one is then able to understand how and why gender operates across cultures. Thanks to the wide-ranging research of sociologists and anthropologists, we are increasingly able to grasp the peculiarities of our gender system and understand more deeply lifeways, including genderscapes, in other places in the world. With this capacity for escaping enthnocentrism in mind, how about testing your capacity to transcend ethnocentrism? Read the fifth endnote to the

Newland article, in this chapter, on female circumcision. Was your initial reaction ethnocentric? If so, how might you adjust your thinking and feeling to overcome that reaction?

The readings in this chapter will introduce you to some of the variety in gender beliefs and practices across cultures and illustrate three of the most important findings of crosscultural research on gender: (1) there is no universal definition or experience of gender; indeed gender is not constituted as oppositional and binary in all societies; (2) gender inequality, specifically the dominance of men over women, is not the rule everywhere in the world; and (3) gender arrangements, whatever they may be, are socially constructed and, thus, ever-evolving.

There Is No Universal Definition or Experience of Gender

Although people in many contemporary cultures perceive at least some differences between women and men, and assign different tasks and responsibilities to people based on gender categories, these differences vary both from culture to culture and within cultures. There is no unified ideal or definition of masculinity or femininity across cultures. In some cultures, such as the Ju/'hoansi of Namibia and Botswana, women and men alike can become powerful and respected healers, while in others, such as the United States today, powerful healing roles are dominated by men (Bonvillain, 2001). Among the seminomadic, pastoral Tuareg of the Sahara and the Sahel, women have considerable economic independence as livestock owners, herders, gardeners, and leathersmiths, while in other cultural groups, such as the Taliban of Afghanistan, women are restricted to household labor and economic dependence on men (Rasmussen, 2001).

The readings in this chapter highlight some of the extraordinary crosscultural differences in beliefs about men and women and in the tasks and rights assigned to them. They offer insights into how gender is shaped across cultures by a number of factors including ideology, participation in economic production, and control over sexuality and reproduction. For example, in the chapter reading with the titillating title, "It's Only a Penis," Christine Helliwell provides an account of the Gerai of Borneo, a cultural group in which rape does not occur. Helliwell argues that the Gerai belief in the biological sameness of women and men is a key to understanding their rape-free society. Her research offers an important account of how assumptions about human biology, in this case femaleness and maleness, are culturally shaped. In a similar vein, the reading on woman-woman marriage among the Gikũyũ of Kenya illustrates the capacity of humans to invent and insitutionalize marital relationships that are not based on the assumption of heterosexual attraction.

In addition, the two-sex (male or female), two-gender (feminine or masculine) and two sexual orientation (homosexual or heterosexual) system of Western culture is not a universal mode of categorization and organization. As you know from reading Serena Nanda's article on gender variants in Native North America (Chapter 1), the two-spirit role was widespread and accepted in many American Indian tribes. Gilbert Herdt (1997), an expert on the anthropology of sexual orientation and gender, points out that the two-spirit role reached a high point in its cultural elaboration among the Mojave Indians who "sanctioned both male (alyha) and female (hwame) two-spirit roles, each of which had its own distinctive social positions and worldviews" (92).

Gender Inequality Is Not the Rule Everywhere

Gender and power go together but not in only one way. The relationship of power to gender in human groups varies from extreme male dominance to relative equality between women and men. Most societies are organized so that men, in general, have greater access to and control over valued resources such as wealth, authority, and prestige. At the extreme are intensely patriarchal societies, such as traditional China and India, in which women were dominated by men in multiple contexts and relationships. In traditional China, for example, sons were preferred, female infanticide was common, divorce could only be initiated by husbands, restrictions on girls and women were embodied in the mutilating practice of foot binding, and the suicide rate among young wives—who typically endured extreme isolation and hardship—was higher than any other age and gender category (Bonvillain, 2001).

The United States also has a history of gender relations in which White men as a group have had power over women as a group (and over men of color). For many decades, men's power was overt and legal. For example, in the nineteenth century, husbands were legally empowered to beat wives, women did not have voting rights, and women were legally excluded from many occupations (Stone & McKee, 1999). Today gender inequality takes more covert and subtle forms. For example, women earn less, on average, than do men of equal educational and occupational level; women are far more likely to be sexually objectified; and women are more likely to shoulder the burden of a double day of work inside and outside home (Coltrane & Adams, 2001; Chapters 7 and 8).

Understanding the relationship between power and gender requires us to use our sharpest sociological radar. To start, it is important to understand that power does not reside in individuals per se. For example, neither presidents nor bosses have power in a vacuum. They require the support of personnel and special resources such as media, weapons, and money. Power is a group phenomenon, and it exists only so long as a powerful group, its ruling principles, and its control over resources are sustained (Kimmel, 2000).

In addition, not all members of an empowered group have the same amount of power. In the United States and similar societies, male power benefits some men more than others. In fact, many individual men do not occupy formal positions of power, and many do not feel powerful in their everyday lives (Johnson, 2001; Kimmel, 2000; see the reading by Pyke in Chapter 8). Yet, major institutions and organizations (e.g., government, big business, the military, the mass media) in the United States are gendered masculine, with controlling positions in those arenas dominated by men, but not just any men (Johnson, 2001). Controlling positions are overwhelmingly held by White men who typically come from privileged backgrounds and whose lives appear to conform to a straight and narrow path (Johnson, 2001; Kimmel, 2000). As we learned in Chapter 2, the relationship of gender to power in a nation such as the United States is complicated by interactions among structures of domination and subordination such as race, social class, and sexual orientation.

Not all societies are as highly and intricately stratified by gender, race, social class, and other social categories of privilege and power as is the United States. Many cultural groups organize relationships in ways that give most or all adults access to similar rights, prestige, decision making authority, and autonomy in their households and communities. Traditional Agta, Ju/'hoansi, and Iroquois societies are good examples. In other cultural groups, such as the precontact Tlingit and Haida of the Canadian Pacific coastal region, relations among people were based on their position in economic and status hierarchies, yet egalitarian valuation of women and men prevailed (Bonvillain, 2001).

The point is that humans do not inevitably create inequalities out of perceived differences. Thus, even though there is at least some degree of division of labor by gender in every cultural group today, differences in men and women's work do not inexorably lead to patriarchal relations in which men monopolize high-status positions in important institutions and women are relegated to a restricted world of low-status activities and tasks. To help illustrate, Maria Lepowsky's ethnography of Vanatinai social relations provides us with a model of a society in which the principles of personal autonomy and freedom of choice prevail. The gender ideology of the Vanatinai is egalitarian, and their belief in equality manifests itself in daily life. For example, women as well as men own and inherit land and other valuables. Women choose their own marriage partners and lovers, and they divorce at will. Any individual on Vanatinai may try to become a leader by demonstrating superior knowledge and skill.

Gender inequality is not the rule everywhere. Male dominance, patriarchy, gender inequality—whatever term one uses—is not the inevitable state of human relations. Additionally, patriarchy itself is not unitary. Patriarchy does not assume a particular shape, and it does not mean that women have no control or influence in their communities. Even in the midst of patriarchy, women and men may create identities and relationships that allow for autonomy and independence. See, for example, Annie George's discussion of the emergence of more egalitarian relations between working-class Indian women and men.

Gender Arrangements Are Ever-Evolving

The crosscultural story of gender takes us back to the metaphor of the kaleidoscope. Life is an ongoing process of change from one pattern to another. We can never go back to "the way things were" at some earlier moment in time, nor can we predict exactly how the future will unfold. This is, of course, the story of gender around the world. For example, the chapter reading by Annie George explores the links between changes in the meaning and practice of male honor and the rise of greater autonomy among married women in a working-class neighborhood of Mumbai, India.

Two of the major sources of change in gender meanings and practices across cultures are culture contact and diffusion of beliefs and practices across the globe (Sorenson, 2000; Ritzer, 2004). Among the most well-documented accounts of such change have been those that demonstrate how Western gender systems were imposed on people whose gender beliefs and arrangements varied from Western assumptions and practices. For example, Native American multiple gender systems were actively, and sometimes violently, discouraged by European colonists (Herdt, 1997; see Nanda in Chapter 1). Today, globalization—a complex process of worldwide diffusion of practices, images, and ideas (Ritzer, 2004)—raises the problem of the development of a world order, including a gender order, that may be increasingly dominated by Western cultural values and patterns (Held, McGrew, Goldbatt, & Perraton, 1999; see Chapter 10 for further discussion). Newland's analysis of female circumcision (and male circumcision) as practiced by Muslim Sundanese people in West Java draws careful attention to the dangers of new forms of Western colonialism that cast "third world" women as victims always in need of being saved by so-called "first world" women and men.

Culture contact and diffusion via globalization are by no means the only source of changing gender arrangements (see Chapter 10 for detailed discussion of gender change). The forces of change are many and complex, and they have resulted in a mix of tendencies toward rigid, hierarchical gender relations and toward gender flexibility and equality, depending upon the specific cultural context and forces of change experienced by particular groups of people. In all of this, there is one fact: people are not bound by any set of gender beliefs and practices. Culture change is inevitable and thus is change in the genderscape.

References

Baker, C. (1999). *Kaleidoscopes: Wonders of wonder.* Lafayette, CA: C&T Publishing.

Bonvillain, N. (2001). *Women and men: Cultural constructs of gender* (3rd ed.). Upper Saddle River, NJ: Prentice Hall.

Coltrane, S., & Adams, M. (2001). Men, women, and housework. In D. Vannoy (Ed.), *Gender mosaics* (pp. 145–154). Los Angeles: Roxbury Publishing.

Held, D., McGrew, A., Goldblatt, D., & Perraton, J. (1999) *Global transformations.* Stanford, CA: Stanford University Press.

Herdt, G. (1997). *Same sex, different cultures.* Boulder, CO: Westview Press.

Johnson, A. G. (2001). *Privilege, power, and difference.* Mountain View, CA: Mayfield Publishing.

Kimmel, M. (2000). *The gendered society.* New York: Oxford University Press.

Murray, S. O., & Roscoe, W. (1998). *Boy-wives and female-husbands: Studies of African homosexualities.* New York: St. Martin's Press.

O'Brien, J. (1999). *Social prisms.* Thousand Oaks, CA: Pine Forge Press.

Ortner, S. B. (1996). *Making gender: The politics and erotics of culture.* Boston, MA: Beacon Press.

Rasmussen, S. (2001). Pastoral nomadism and gender. In C. B. Brettell & C. F. Sargent (Eds.), *Gender in cross-cultural perspective* (pp. 280–293). Upper Saddle River, NJ: Prentice Hall.

Ritzer, G. (2004). *The globalization of nothing.* Thousand Oaks, CA: Pine Forge Press.

Schwalbe, M. (2005). *The sociologically examined life.* New York: McGraw-Hill.

Sorenson, M. L. Stig. (2000). *Gender archeology.* Cambridge, England: Polity Press.

Stone, L. & McKee, N. P. (1999). *Gender and culture in America.* Upper Saddle River, NJ: Prentice Hall.

Introduction to Reading 10

Anthropologist Christine Helliwell provides a challenging account of a cultural group, the Gerai of Indonesia, in which rape does not occur. She links the freedom from rape among the Gerai people to peculiar aspects of the relatively egalitarian nature of their gender relations. Helliwell's research questions many gender beliefs held by members of Western cultures today.

1. How are men's and women's sexual organs conceptualized among the Gerai, and what are the consequences for Gerai understandings of sexual intercourse?
2. Genitalia do not determine identity in Gerai. What does?
3. What does Helliwell mean when she states that "rape imposes difference as much as it is produced by difference" (p. 812)?

"It's Only a Penis"

Rape, Feminism, and Difference

Christine Helliwell

In 1985 and 1986 I carried out anthropological fieldwork in the Dayak community of Gerai in Indonesian Borneo. One night in September 1985, a man of the village climbed through a window into the freestanding house where a widow lived with her elderly mother, younger (unmarried) sister, and young children. The widow awoke, in darkness, to feel the man

---

inside her mosquito net, gripping her shoulder while he climbed under the blanket that covered her and her youngest child as they slept (her older children slept on mattresses nearby). He was whispering, "be quiet, be quiet!" She responded by sitting up in bed and pushing him violently, so that he stumbled backward, became entangled with her mosquito net, and then, finally free, moved across the floor toward the window. In the meantime, the woman climbed from her bed and pursued him, shouting his name several times as she did so. His hurried exit through the window, with his clothes now in considerable disarray, was accompanied by a stream of abuse from the woman and by excited interrogations from wakened neighbors in adjoining houses.

I awoke the following morning to raucous laughter on the longhouse verandah outside my apartment where a group of elderly women gathered regularly to thresh, winnow, and pound rice. They were recounting this tale loudly, and with enormous enjoyment, to all in the immediate vicinity. As I came out of my door, one was engaged in mimicking the man climbing out the window, sarong falling down, genitals askew. Those others working or lounging near her on the verandah—both men and women—shrieked with laughter.

When told the story, I was shocked and appalled. An unknown man had tried to climb into the bed of a woman in the dead, dark of night? I knew what this was called: attempted rape. The woman had seen the man and recognized him (so had others in the village, wakened by her shouting). I knew what he deserved: the full weight of the law. My own fears about being a single woman alone in a strange place, sleeping in a dwelling that could not be secured at night, bubbled to the surface. My feminist sentiments poured out. "How can you laugh?" I asked my women friends; "this is a very bad thing that he has tried to do." But my outage simply served to fuel the hilarity. "No, not bad," said one of the old women (a particular friend of mine), "simply stupid."

I felt vindicated in my response when, two hours later, the woman herself came onto the verandah to share betel nut and tobacco and to broadcast the story. Her anger was palpable, and she shouted for all to hear her determination to exact a compensation payment from the man. Thinking to obtain information about local women's responses to rape, I began to question her. Had she been frightened? I asked. Of course she had—Wouldn't I feel frightened if I awoke in the dark to find an unknown person inside my mosquito net? Wouldn't I be angry? Why then, I asked, hadn't she taken the opportunity, while he was entangled in her mosquito net, to kick him hard or to hit him with one of the many wooden implements near at hand? She looked shocked. Why would she do that? she asked—after all, he hadn't hurt her. No, but he had wanted to, I replied. She looked at me with puzzlement. Not able to find a local word for rape in my vocabulary, I scrabbled to explain myself: "He was trying to have sex with you." I said, "although you didn't want to. He was trying to hurt you." She looked at me, more with pity than with puzzlement now, although both were mixed in her expression. "Tin [Christine], it's only a penis" she said. "How can a penis hurt anyone?"

Rape, Feminism, and Difference

A central feature of many feminist writings about rape in the past twenty years is their concern to eschew the view of rape as a natural function of male biology and to stress instead its bases in society and culture. It is curious, then, that so much of this work talks of rape in terms that suggest—either implicitly or explicitly—that it is a universal practice. To take only several examples: Pauline Bart and Patricia O'Brien tell us that "every female from nine months to ninety years is at risk" (1985, 1); Anna Clark argues that "all women know the paralyzing fear of walking down a dark street at night. . . . It seems to be a fact of life that the fear of rape imposes a curfew on our movements" (1987, 1); Catharine MacKinnon claims that "sexuality is central to women's definition and forced sex is central to sexuality," so "rape is indigenous, not exceptional, to women's social condition" (1989b, 172) and "all women live all the time under the shadow of the threat of sexual abuse" (1989a, 340); Lee Madigan and Nancy Gamble write of "the global terrorism of rape" (1991, 21–2); and Susan Brison asserts that "the fact that all women's lives are restricted by sexual violence is indisputable" (1993, 17). . . . This is particularly puzzling given that Peggy Reeves Sanday, for one, long ago demonstrated that while rape occurs widely throughout the world, it is by no means a human universal: some societies can indeed be classified as rape free (1981).

There are two general reasons for this universalization of rape among Western feminists. The first of these has to do with the understanding of the practice as horrific by most women in Western societies. In these settings, rape is seen as "a fate worse than, or tantamount to, death" (S. Marcus 1992, 387): a shattering of identity that, for instance, left one North American survivor feeling "not quite sure whether I had died and the world went on without me, or whether I was alive in a totally alien world" (Brison 1993, 10). . . .

A second, equally deep-seated reason for the feminist tendency to universalize rape stems from Western feminism's emphasis on difference between men and women and from its consequent linking of rape and difference. Two types of difference are involved here. The first of these is difference in social status and power; thus rape is linked quite explicitly, in contemporary feminist accounts, to patriarchal social forms. Indeed, this focus on rape as stemming from difference in social position is what distinguishes feminist from other kinds of accounts of rape (see Ellis 1989, 10). In this view, inequality between men and women is linked to men's desire to possess, subjugate, and control women, with rape constituting a central means by which the freedom of women is limited and their continued submission to men ensured. Since many feminists continue to believe that patriarchy is universal—or, at the very least, to feel deeply ambivalent on this point—there is a tendency among us to believe that rape, too, is universa1.[1]

However, the view of women as everywhere oppressed by men has been extensively critiqued within the anthropological literature. A number of anthropologists have argued that in some societies, while men and women may perform different roles and occupy different spaces, they are nevertheless equal in value, status, and power.[2] . . .

But there is a second type of difference between men and women that also, albeit largely implicitly, underlies the assumption that rape is universal, and it is the linkage between this type of difference and the treatment of rape in feminist accounts with which I am largely concerned in this article. I refer to the assumption by most Western feminists writing on rape that men and women have different bodies and, more specifically, different genitalia: that they are, in other words differently sexed. Furthermore, it is taken for granted in most feminist accounts that these differences render the former biologically, or "naturally," capable of penetrating and therefore brutalizing the latter and render the latter "naturally" able to be brutalized. . . . Rape of women by men is thus assumed to be universal because the same "biological" bodily differences between men and women are believed to exist everywhere.

Unfortunately, the assumption that preexisting bodily difference between men and women underlies rape has blinded feminists writing on the subject to the ways the practice of rape itself creates and inscribes such difference. This seems particularly true in contemporary Western societies where the relationship between rape and bodily/genital dimorphism appears to be an extremely intimate one. Judith Butler (1990, 1993) has argued (following Foucault 1978) that the Western emphasis on sexual difference is a product of the heterosexualization of desire within Western societies over the past few centuries, which "requires and institutes the production of discrete and asymmetrical oppositions between 'feminine' and 'masculine' where these are understood as expressive attributes of 'male' and 'female'" (1990, 17).[3] The practice of rape in Western contexts can only properly be understood with reference to this heterosexual matrix, to the division of humankind into two distinct—and in many respects opposed—types of body (and hence types of person).[4] While it is certainly the case that rape is linked in contemporary

Western societies to disparities of power and status between men and women, it is the particular discursive form that those disparities take—their elaboration in terms of the discourse of sex—that gives rape its particular meaning and power in these contexts.

Sharon Marcus has already argued convincingly that the act of rape "feminizes" women in Western settings, so that "the entire female body comes to be symbolized by the vagina, itself conceived of as a delicate, perhaps inevitably damaged and pained inner space" (1992, 398). I would argue further that the practice of rape in these settings—both its possibility and its actualization—not only feminizes women but masculinizes men as well.[5] This masculinizing character of rape is very clear in, for instance, Sanday's ethnography of fraternity gang rape in North American universities (1990b) and, in particular, in material on rape among male prison inmates. In the eyes of these rapists the act of rape marks them as "real men" and marks their victims as not men, that is, as feminine.[6] In this iconography, the "masculine" body (along with the "masculine" psyche), is viewed as hard, penetrative, and aggressive, in contrast to the soft, vulnerable, and violable "feminine" sexuality and psyche. Rape both reproduces and marks the pronounced sexual polarity found in these societies.

Western understandings of gender difference have almost invariably started from the presumption of a presocial bodily difference between men and women ("male" and "female") that is then somehow acted on by society to produce gender. In particular, the possession of either male genitals or female genitals is understood by most Westerners to be not only the primary marker of gender identity but, indeed, the underlying cause of that identity. . . .

I seek to do two things in this article. First, in providing an account of a community in which rape does not occur, I aim to give the lie to the widespread assumption that rape is universal and thus to invite Western feminists to interrogate the basis of our own tendency to take its universality for granted.[7] The fundamental question is this: Why does a woman of Gerai see a penis as lacking the power to harm her, while I, a white Australian/New Zealand woman, am so ready to see it as having the capacity to defile, to humiliate, to subjugate and, ultimately, to destroy me?

Second, by exploring understandings of sex and gender in a community that stresses identity, rather than difference, between men and women (including men's and women's bodies), I aim to demonstrate that Western beliefs in the "sexed" character of bodies are not "natural" in basis but, rather, are a component of specifically Western gendering and sexual regimes. And since the practice of rape in Western societies is profoundly linked to these beliefs, I will suggest that it is an inseparable part of such regimes. This is not to say that the practice of rape is always linked to the kind of heterosexual regime found in the West; even the most cursory glance at any list of societies in which the practice occurs indicates that this is not so.[8] But it is to point out that we will be able to understand rape only ever in a purely localized sense, in the context of the local discourses and practices that are both constitutive of and constituted by it. In drawing out the implications of the Gerai stress on identity between men and women for Gerai gender and sexual relations, I hope to point out some of the possible implications of the Western emphasis on gender difference for Western gender and sexual relations—including the practice of rape.

Gender, Sex, and Procreation in Gerai

Gerai is a Dayak community of some seven hundred people in the Indonesian province of Kalimantan Barat (West Borneo).[9] In the twenty months I spent in the community, I heard of no cases of either sexual assault or attempted sexual assault (and since this is a community in which privacy as we understand it in the West is almost nonexistent—in which surveillance by neighbors is at a very high level [see Helliwell 1996]—I would certainly have heard of any such cases had they occurred). In addition, when I questioned men and women about sexual assault, responses ranged from puzzlement to outright incredulity to horror.

While relations between men and women in Gerai can be classified as relatively egalitarian in many respects, both men and women nevertheless say that men are "higher" than women (Helliwell1995, 364). This greater status and authority does not, however, find expression in the practice of rape, as many feminist writings on the subject seem to suggest that it should. This is because the Gerai view of men as "higher" than women, although equated with certain kinds of increased potency vis-à-vis the world at large, does not translate into a conception of that potency as attached to and manifest through the penis—of men's genitals as able to brutalize women's genitals.

Shelly Errington has pointed out that a feature of many of the societies of insular Southeast Asia is a stress on sameness, even identity, between men and women (1990, 35, 39), in contrast to the Western stress on difference between the passive "feminine" object and the active, aggressive "masculine" subject.[10] Gerai understandings of gender fit Errington's model very well. In Gerai, men and women are not understood as fundamentally different types of persons: there is no sense of a dichotomized masculinity and femininity. Rather, men and women are seen to have the same kinds of capacities and proclivities, but with respect to some, men are seen as "more so" and with respect to others, women are seen as "more so." Men are said to be braver and more knowledgeable about local law (adat), while women are said to be more persistent and more enduring. All of these qualities are valued. Crucially, in terms of the central quality of nurturance (perhaps the most valued quality in Gerai), which is very strongly marked as feminine among Westerners, Gerai people see no difference between men and women. As one (female) member of the community put it to me: "We all must nurture because we all need."[11] The capacity both to nurture and to need, particularly as expressed through the cultivation of rice as a member of a rice group, is central to Gerai conceptions of personhood: rice is the source of life, and its (shared) production humanizes and socializes individuals (Helliwell, forthcoming). Women and men have identical claims to personhood based on their equal contributions to rice production (there is no notion that women are somehow diminished as persons even though they may be seen as less "high"). As in Strathern's account of Hagen (1988), the perceived mutuality of rice-field work in Gerai renders inoperable any notion of either men or women as autonomous individual subjects.

It is also important to note that while men's bravery is linked to a notion of their greater physical strength, it is not equated with aggression—aggression is not valued in most Gerai contexts.[12] As a Gerai man put it to me, the wise man is the one "who fights when he has to, and runs away when he can"; such avoidance of violence does not mark a man as lacking in bravery. . . . While it is recognized that a man will sometimes need to fight—and skill and courage in fighting are valued—aggression and hotheadedness are ridiculed as the hallmarks of a lazy and incompetent man. In fact, physical violence between adults is uncommon in Gerai, and all of the cases that I did witness or hear about were extremely mild.[13] Doubtless the absence of rape in the community is linked to this devaluing of aggression in general. However, unlike a range of other forms of violence (slapping, beating with a fist, beating with an implement, knifing, premeditated killing, etc.), rape is not named as an offense and accorded a set punishment under traditional Gerai law. In addition, unlike these other forms of violence, rape is something that people in the community find almost impossible to comprehend ("How would he be able to do such a thing?" one woman asked when I struggled to explain the concept of a man attempting to put his penis into her against her will). Clearly, then, more is involved in the absence of rape in Gerai than a simple absence of violence in general.

Central to all of the narratives that Gerai people tell about themselves and their community is the notion of a "comfortable life": the achievement of this kind of life marks the person and the household as being of value and constitutes the norm to which all Gerai people aspire. Significantly, the content of such a life is seen as identical for both men and women: it is marked by the production of bountiful rice harvests each year and the successful raising of a number of healthy children to maturity. The core values and aspirations of men and women are thus identical; of the many life histories that I collected while in the community—all of which are organized around this central image—it is virtually impossible to tell those of men from those of women. Two points are significant in this respect. First, a "comfortable life" is predicated on the notion of a partnership between a man and a woman (a conjugal pair). This is because while men and women are seen to have the same basic skills and capacities, men are seen to be "better" at certain kinds of work and women to be 'better" at other kinds. Second, and closely related to this, the Gerai notion of men's and women's work does not constitute a rigid division of labor: both men and women say that theoretically women can perform all of the work routinely carried out by men, and men can perform all of the work routinely carried out by women. However, men are much better at men's work, and women are much better at women's work. Again, what we have here is a stress on identity between men and women at the expense of radical difference.

This stress on identity extends into Gerai bodily and sexual discourses. A number of people (both men and women) assured me that men sometimes menstruate; in addition, menstrual blood is not understood to be polluting, in contrast to how it is seen in many societies that stress more strongly the difference between men and women. While pregnancy and childbirth are spoken of as "women's work," many Gerai people claim that under certain circumstances men are also able to carry out this work—but, they say, women are "better" at it and so normally undertake it. In line with this claim, I collected a Gerai myth concerning a lazy woman who was reluctant to take on the work of pregnancy and childbirth. Her husband instead made for himself a lidded container out of bark, wood, and rattan ("like a betel nut container"), which he attached around his waist beneath his loincloth and in which he carried the growing fetus until it was ready to be born. On one occasion when I was watching a group of Gerai men cut up a boar, one, remembering an earlier conversation about the capacity of men to give birth, pointed to a growth in the boar's body cavity and said with much disapproving shaking of the head: "Look at this. He wants to carry his child. He's stupid." In addition, several times I saw fathers push their nipples into the mouths of young children to quiet them; while none of these fathers claimed to be able to produce milk, people nevertheless claimed that some men in the community were able to lactate, a phenomenon also attested to in myth. Men and women are thought to produce the same genital fluid, and this is linked in complex ways to the capacity of both to menstruate. All of these examples demonstrate the community's stress on bodily identity between men and women.

Furthermore, in Gerai, men's and women's sexual organs are explicitly conceptualized as the same. This sexual identity became particularly clear when I asked several people who had been to school (and hence were used to putting pencil to paper) to draw men's and women's respective organs for me: in all cases, the basic structure and form of each were the same. One informant, endeavoring to convince me of this sameness, likened both to wooden and bark containers for holding valuables (these vary in size but have the same basic conical shape, narrower at the base and wider at the top). In all of these discussions, it was reiterated that the major difference between men's and women's organs is their location: inside the body (women) and outside the body (men).[14] In fact, when I pressed people on this point, they invariably explained that it makes no sense to distinguish between men's and women's genitalia themselves; rather, it is location that distinguishes between penis and vulva."[15]

Heterosexuality constitutes the normative sexual activity in the community and, indeed, I was unable to obtain any information about homosexual practices during my time there. In line with the stress on sameness, sexual intercourse between a man and a woman in Gerai is understood as an equal coming together of fluids, pleasures, and life forces. The same stress also underlies beliefs about conception. Gerai people believe that repeated acts of intercourse between the same two people are necessary for conception, since this "prepares" the womb for pregnancy. The fetus is deemed to be created through the mingling of equal quantities of fluids and forces from both partners. Again, what is seen as important here is not the fusion of two different types of bodies (male and female) as in Western understandings; rather, Gerai people say, it is the similarity of the two bodies that allows procreation to occur. As someone put it to me bluntly: "If they were not the same, how could the fluids blend? It's like coconut oil and water: they can't mix!"

What needs to be stressed here is that both sexual intercourse and conception are viewed as involving a mingling of similar bodily fluids, forces, and so on, rather than as the penetration of one body by another with a parallel propulsion of substances from one (male) body only into the other, very different (female) one. What Gerai accounts of both sexual intercourse and conception stress are tropes of identity, mingling, balance, and reciprocity. In this context it is worth noting that many Gerai people were puzzled by the idea of gender-specific "medicine to prevent contraception—such as the injectable or oral contraceptives promoted by state-run health clinics in the area. Many believed that, because both partners play the same role in conception, it should not matter whether husband or wife received such medicine (and indeed, I knew of cases where husbands had taken oral contraceptives meant for their wives). This suggests that such contraceptive regimes also serve (like the practice of rape) to reinscribe sex difference between men and women (see also Tsing 1993, 104–20). . . .

While Gerai people stress sameness over difference between men and women, they do, nevertheless, see them as being different in one important respect: their life forces are, they say, oriented differently ("they face different ways," it was explained to me). This different orientation means that women are "better" at certain kinds of work and men are "better" at other kinds of work—particularly with respect to rice-field work. Gerai people conceive of the work of clearing large trees for a new rice field as the definitive man's work and regard the work of selecting and storing the rice seed for the following year's planting—which is correlated in fundamental ways with the process of giving birth—as the definitive woman's work. Because women are perceived to lack appropriate skills with respect to the first, and men are perceived to lack appropriate skills with respect to the second, Gerai people say that to be viable a household must contain both adult males and adult females. And since a "comfortable life" is marked by success in production not only of rice but also of children, the truly viable household must contain at least one conjugal pair. The work of both husband and wife is seen as necessary for the adequate nurturance of the child and successful rearing to adulthood (both of which depend on the successful cultivation of rice). Two women or two men would not be able to produce adequately for a child since they would not be able to produce consistently successful rice harvests; while such a household might be able to select seed, clear a rice field, and so grow rice in some rudimentary fashion, its lack of expertise at one of these tasks would render it perennially poor and its children perennially unhealthy, Gerai people say. . . .

Gender difference in Gerai, then, is not predicated on the character of one's body, and especially of one's genitalia as in many Western contexts. Rather, it is understood as constituted in the differential capacity to perform certain kinds of work, a capacity assigned long before one's bodily being takes shape.[16] In this respect it is important to note that Gerai ontology rests on a belief in predestination, in things being as they should (see Helliwell 1995). In this understanding, any individual's semongan is linked in multifarious and unknowable ways to the cosmic order, to the "life" of the universe as a whole. Thus the new fetus is predestined to become someone "fitted" to carry out either men's work or women's work as part of the maintenance of a universal balance. Bodies with the appropriate characteristics—internal or external genitalia, presence or absence of breasts, and so on—then develop in line with this prior destiny. At first sight this may not seem enormously different from Western conceptions of gender, but the difference is in fact profound. While, for Westerners, genitalia, as significant of one's role in the procreative process, are absolutely fundamental in determining ones identity, in Gerai the work that one performs is seen as fundamental, and genitalia along with other bodily characteristics, are relegated to a kind of secondary, derivative function.

Gerai understandings of gender were made quite clear through circumstances surrounding my own gender classification while in the community. Gerai people remained very uncertain about my gender for some time after I arrived in the community because (as they later told me) "I did not . . . walk like a woman, with arms held out from the body and hips slightly swaying; I was 'brave' trekking from village to village through the jungle on my own; I had bony kneecaps; I did not know how to tie a sarong in the appropriate way for women; I could not distinguish different varieties of rice from one another; I did not wear earrings; I had short hair; I was tall" (Helliwell 1993, 260). This was despite the fact that people in the community knew from my first few days with them both that I had breasts (this was obvious when the sarong that I wore clung to my body while I bathed in the river) and that I had a vulva rather than a penis and testicles (this was obvious from my trips to defecate or urinate in the small stream used for that purpose, when literally dozens of people would line the banks to observe whether I performed these functions differently from them). As someone said to me at a later point, "Yes, I saw that you had a vulva, but I thought that Western men might be different." My eventual, more definitive classification as a woman occurred. . . . (a)s I learned to distinguish types of rice and their uses, I became more and more of a woman (as I realized later), since this knowledge—including the magic that goes with it—is understood by Gerai people as foundational to femininity.

. . . Gerai people talk of two kinds of work as defining a woman: the selection and storage of rice seed and the bearing of children.[17] But the first of these is viewed as prior, logically as well as chronologically. People are quite clear that in the womb either "someone who can cut down the large trees for a ricefield is made, or someone who can select and store rice." When I asked if it was not more important whether or not someone could bear a child, it was pointed out to me that many women do not bear children (there is a high rate of infertility in the community), but all women have the knowledge to select and store rice seed. In fact, at the level of the rice group the two activities of "growing" rice and "growing" children are inseparable: a rice group produces rice in order to raise healthy children, and it produces children so that they can in turn produce the rice that will sustain the group once their parents are old and frail (Helliwell, forthcoming). For this reason, any Gerai couple unable to give birth to a child of their own will adopt one, usually from a group related by kinship. The two activities of growing rice and growing children are constantly talked about together, and the same imagery is used to describe the development of a woman's pregnancy and the development of rice grains on the plant. . . .

Gerai, then, lacks the stress on bodily—and especially genital—dimorphism that most feminist accounts of rape assume. Indeed, the reproductive organs themselves are not seen as "sexed." In a sense it is problematic even to use the English categories woman and man when writing of this community, since these terms are saturated with assumptions concerning the priority of biological (read, bodily) difference. In the Gerai context, it would be more accurate to deal with the categories of, on the one hand, "those responsible for rice selection and storage" and, on the other, "those responsible for cutting down the large trees to make a ricefield." There is no discursive space in Gerai for the distinction between an active, aggressive, penetrating male sexual organ (and sexuality) and a passive, vulnerable, female one. Indeed, sexual intercourse in Gerai is understood by both men and women to stem from mutual "need" on the part of the two partners; without such need, people say, sexual intercourse cannot occur, because the requisite balance is lacking. . . . the sexual act is understood as preeminently mutual in its character, including in its initiation. The idea of having sex with someone who does not need you to have sex with them—and so the idea of coercing someone into sex—is thus almost unthinkable to Gerai people. In addition, informants asserted that any such action would destroy the individual's spiritual balance and that of his or her rice group and bring calamity to the group as a whole.[18]

In this context, a Gerai man's astonished and horrified question "How can a penis be taken into a vagina if a woman doesn't want it?" has a meaning very different from that of the same statement uttered by a man in the West. In the West, notions of radical difference between men and women—incorporating representations of normative male sexuality as active and aggressive, normative female sexuality as passive and vulnerable, and human relationships (including acts of sexual intercourse) as occurring between independent, potentially hostile, agents—would render such a statement at best naive, at worst misogynist. In Gerai, however, the stress on identity between men and women and on the sexual act as predicated on mutuality validates such a statement as one of straightforward incomprehension (and it should be noted that I heard similar statements from women). In the Gerai context, the penis, or male genitalia in general, is not admired, feared, or envied. . . . In fact, Gerai people see men's sexual organs as more vulnerable than women's for the simple reason that they are outside the body, while women's are inside. This reflects Gerai understandings of "inside" as representing safety and belonging, while "outside" is a place of strangers and danger, and it is linked to the notion of men as braver than women.[19] In addition, Gerai people say, because the penis is "taken into" another body, it is theoretically at greater risk during the sexual act than the vagina. This contrasts, again, quite markedly with Western understandings, where women's sexual organs are constantly depicted as more vulnerable during the sexual act—as liable to be hurt, despoiled, and so on (some men's anxieties about vagina dentata not withstanding). In Gerai a penis is "only a penis": neither a marker of dimorphism between men and women in general nor, in its essence, any different from a vagina.

Conclusions

With this background, I return now to the case with which I began this article—and, particularly, to the great differences between my response to this case and that of the Gerai woman concerned. On the basis of my own cultural assumptions concerning the differences—and particularly the different sexual characters—of men and women, I am inclined (as this case showed me) to read any attempt by a man to climb into a woman's bed in the night without her explicit consent as necessarily carrying the threat of sexual coercion and brutalization. The Gerai woman, in contrast, has no fear of coerced sexual intercourse when awakened in the dark by a man. She has no such fear because in the Gerai context . . . women's sexuality and bodies are no less aggressive and no more vulnerable than men's.

In fact, in the case in question, the intruding man did expect to have intercourse with the woman.[20] He claimed that the woman had already agreed to this through her acceptance of his initiatory gifts of soap.[21] The woman, however, while privately agreeing that she had accepted such gifts, claimed that no formal agreement had yet been reached. Her anger, then, did not stem from any belief that the man had attempted to sexually coerce her ("How would he be able to do such a thing?"). Because the term "to be quiet" is often used as a euphemism for sexual intercourse in Gerai, she saw the man's exhortation that she "be quiet" as simply an invitation to engage in sex with him, rather than the implicit threat that I read it to be.[22] Instead, her anger stemmed from her conviction that the correct protocols had not been followed, that the man ought to have spoken with her rather than taking her acceptance of the soap as an unequivocal expression of assent. She was, as she put it, letting him know that "you have sexual relations together when you talk together. Sexual relations cannot be quiet."[23]

Yet, this should not be taken to mean that the practice of rape is simply a product of discourse: that brutality toward women is restricted to societies containing particular, dimorphic representations of male and female sexuality and that we simply need to change the discourse in order to eradicate such practices.[24] Nor is it to suggest that a society in which rape is unthinkable is for that reason to be preferred to Western societies. To adopt such a position would be still to view the entire world through a sexualized Western lens.

In order to understand the practice of rape in countries like Australia and the United States, then—and so to work effectively for its eradication there—feminists in these countries must begin to relinquish some of our most ingrained presumptions concerning difference between men and women and, particularly, concerning men's genitalia and sexuality as inherently brutalizing and penetrative and women's genitalia and sexuality as inherently vulnerable and subject to brutalization. Instead, we must begin to explore the ways rape itself produces such experiences of masculinity and femininity and so inscribes sexual difference onto our bodies.

Notes

1. Among "radical" feminists such as Andrea Dworkin and Catharine MacKinnon this belief reaches its most extreme version, in which all sexual intercourse between a man and a woman is viewed as akin to rape (Dworkin 1987; MacKinnon 1989a, 1989b).

2. Leacock 1978 and Bell 1983 are well-known examples. Sanday 1990a and Marcus 1992 are more recent examples, on Minangkabau and Turkish society, respectively.

3. See Laqueur 1990 for a historical account of this process.

4. On the equation of body and person within Western (especially feminist) thought, see Moore 1994.

5. See Plaza 1980: "[Rape] is very sexual in the sense that [it] is frequently a sexual activity, but especially in the sense that it opposes men and women: it is social sexing which is latent in rape. . . . Rape is sexual essentially because it rests on the very social difference between the sexes" (31).

6. The material on male prison inmates is particularly revealing in this respect. As an article by Stephen Donaldson, a former prisoner and the president of the U.S. advocacy group Stop Prisoner Rape, makes clear, "hooking up" with another prisoner is the best way for a prisoner to avoid sexual assaults, particularly gang rapes. Hooking up involves entering a sexual liaison with a senior partner ("jocker," "man," "pitcher," "daddy") in exchange for protection. In this arrangement, the rules are clear: the junior partner gives up his autonomy and comes under the authority of the senior partner; he is often expected by the senior partner to be as feminine in appearance and behavior as possible," including shaving his legs, growing long hair, using a feminine nickname, and performing work perceived as feminine (laundry, cell cleaning, giving backrubs. etc.) (Donaldson 1996, 17, 20). See also the extract from Jack Abbott's prison letters in Halperin 1993 (424–25).

7. While I am primarily concerned here with the feminist literature (believing that it contains by far the most useful and insightful work on rape), it needs to be noted that many other (nonfeminist) writers also believe rape to be universal. See, e.g., Ellis 1989; Palmer 1989.

8. For listings of "rape-prone" societies, see Minturn, Grosse, and Haider 1969; Sanday 1981.

9. I carried out anthropological fieldwork in Gerai from March 1985 to February 1986 and from June 1986 to January 1987. The fieldwork was funded by an Australian National University Ph.D. scholarship and carried out under the sponsorship of Lembaga Ilmu Pengetahuan Indonesia. At the time that I was conducting my research a number of phenomena were beginning to have an impact on the community—these had the potential to effect massive changes in the areas of life discussed in this article. These phenomena included the arrival of a Malaysian timber company in the Gerai region and the increasing frequency of visits by Malay, Bugis, Chinese, and Batak timber workers

to the community; the arrival of two American fundamentalist Protestant missionary families to live and proselytize in the community; and the establishment of a Catholic primary school in Gerai, resulting in a growing tendency among parents to send their children (both male and female) to attend Catholic secondary school in a large coastal town several days' journey away.

10. The Wana, as described by Jane Atkinson (1990), provide an excellent example of a society that emphasizes sameness. Emily Martin points out that the explicit Western opposition between the "natures" of men and women is assumed to occur even at the level of the cell, with biologists commonly speaking of the egg as passive and immobile and the sperm as active and aggressive even though recent research indicates that these descriptions are erroneous and that they have led biologists to misunderstand the fertilization process (1991). See also Lloyd 1984 for an excellent account of how (often latent) conceptions of men and women as having opposed characteristics are entrenched in the history of Western philosophical thought.

11. The nurture-need dynamic (that I elsewhere refer to as the "need-share dynamic") is central to Gerai sociality. Need for others is expressed through nurturing them; such expression is the primary mark of a "good" as opposed to a "bad" person. See Helliwell (forthcoming) for a detailed discussion.

12. In this respect, Gerai is very different from, e.g., Australia or the United States, where, as Michelle Rosaldo has pointed out, aggression is linked to success, and women's constitution as lacking aggression is thus an important element of their subordination (1980b, 416; see also Myers 1988, 600).

13. See Helliwell 1996, 142–43, for an example of a "violent" altercation between husband and wife.

14. I have noted elsewhere that the inside-outside distinction is a central one within this culture (Helliwell 1996).

15. While the Gerai stress on the sameness of men's and women's sexual organs seems, on the face of it, to be very similar to the situation in Renaissance Europe as described by Laqueur 1990, it is profoundly different in at least one respect: in Gerai, women's organs are not seen as emasculated versions of men's—"female penises"—as they were in Renaissance Europe. This is clearly linked to the fact that, in Gerai, as we have already seen, people is not synonymous with men, and women are not relegated to positions of emasculation or abjection, as was the case in Renaissance Europe.

16. In this respect Gerai is similar to a number of other peoples in this region (e.g., Wana, Ilongot), for whom difference between men and women is also seen as primarily a matter of the different kinds of work that each performs.

17. In Gerai, pregnancy and birth are seen not as semimystical "natural" processes, as they are for many Westerners, but simply as forms of work, linked very closely to the work of rice production.

18. Sanday 1986 makes a similar point about the absence of rape among the Minangkabau. See Helliwell (forthcoming) for a discussion of the different kinds of bad fate that can afflict a group through the actions of its individual members.

19. In Gerai, as in nearby Minangkabau (Sanday 1986), vulnerability is respected and valued rather than despised.

20. The man left the community on the night that this event occurred and went to stay for several months at a nearby timber camp. Community consensus—including the view of the woman concerned—was that he left because he was ashamed and distressed, not only as a result of having been sexually rejected by someone with whom he thought he had established a relationship but also because his adulterous behavior had become public, and he wished to avoid an airing of the details in a community moot. Consequently, I was unable to speak to him about the case. However, I did speak to several of his close male kin (including his married son), who put his point of view to me.

21. The woman in this particular case was considerably younger than the man (in fact, a member of the next generation). In such cases of considerable age disparity between sexual partners, the older partner (whether male or female) is expected to pay a fine in the form of small gifts to the younger partner, both to initiate the liaison and to enable its continuance. Such a fine rectifies any spiritual imbalance that may result from the age imbalance and hence makes it safe for the relationship to proceed. Contrary to standard Western assumptions, older women appear to pay such fines to younger men as often as older men pay them to younger women (although it was very difficult to obtain reliable data on this question, since most such liaisons are adulterous and therefore highly secretive).

While not significant in terms of value (women usually receive such things as soap and shampoo, while men receive tobacco or cigarettes), these gifts are crucial in their role of "rebalancing" the relationship. It would be entirely erroneous to subsume this practice under the rubric of "prostitution."

22. Because Gerai adults usually sleep surrounded by their children, and with other adults less than a meter or two away (although the latter are usually inside different mosquito nets), sexual intercourse is almost always carried out very quietly.

23. In claiming that "sexual relations cannot be quiet," the woman was playing on the expression "be quiet" (meaning to have sexual intercourse) to make the point that while adulterous sex may need to be even "quieter" than legitimate sex, it should not be so "quiet" as to preclude dialogue between the two partners. Implicit here is the notion that in the absence of such dialogue, sex will lack the requisite mutuality.

24. Foucault, e.g., once suggested (in a debate in French reprinted in La Folie Encerclee [see Plaza 1980]) that an effective way to deal with rape would be to decriminalize it in order to "desexualize" it. For feminist critiques of his suggestion, see Plaza 1980; de Lauretis 1987; Woodhull 1988.

References

Atkinson, Jane Monnig. 1990. "How Gender Makes a Difference in Wana Society" In *Power and Difference: Gender in Island Southeast Asia,* ed. Jane Monnig Atkinson and Shelly Errington, 59–93. Stanford, Calif.: Stanford University Press.

Barry, Kathleen. 1995. *The Prostitution of Sexuality.* New York and London: New York University Press.

Bart, Pauline B., and Patricia H. O'Brien. 1985. *Stopping Rape: Successful Survival Strategies.* New York: Pergamon.

Bell, Diane. 1983. *Daughters of the Dreaming.* Melbourne: McPhee Gribble.

Bourque, Linda B. 1989. *Defining Rape.* Durham, N.C., and London: Duke University Press.

Brison, Susan J. 1993. "Surviving Sexual Violence: A Philosophical Perspective." *Journal of Social Philosophy 24*(1): 5–22.

Butler, Judith. 1990. *Gender Trouble: Feminism and the Subversion of Identity.* New York and London: Routledge.

———. 1993. *Bodies That Matter: On the Discursive Limits of "Sex."* New York and London: Routledge.

Clark, Anna. 1987. *Women's Silence, Men's Violence: Sexual Assault in England, 1770–845.* London and New York: Pandora.

de Lauretis, Teresa. 1987. "The Violence of Rhetoric: Considerations on Representation and Gender." In her *Technologies of Gender: Essays on Theory, Film and Fiction,* 31–50. Bloomington and Indianapolis: Indiana University Press.

Dentan, Robert Knox. 1968. *The Semai: A Nonviolent People of Malaya.* New York: Holt, Rinehart & Winston.

———. 1978. "Notes on Childhood in a Nonviolent Context: The Semai Case (Malaysia)." In *Learning Non-Aggression: The Experience of Non-Literate Societies,* ed. Ashley Montagu, 94–143. New York: Oxford University Press.

Donaldson, Stephen. 1996. "The Deal behind Bars" *Harper's* (August): 17–20.

Dubinsky, Karen. 1993. *Improper Advances: Rape and Heterosexual Conflict in Ontario, 1880–1929.* Chicago and London: University of Chicago Press.

Dworkin, Andrea. 1987. *Intercourse.* London: Secker & Warburg.

Ellis, Lee. 1989. *Theories of Rape: Inquiries into the Causes of Sexual Aggression.* New York: Hemisphere.

Errington, Shelly. 1990. "Recasting Sex, Gender, and Power: A Theoretical and Regional Overview" In *Power and Difference: Gender in Island Southeast Asia,* ed. Jane Monnig Atkinson and Shelly Errington, 1–58. Stanford, Calif.: Stanford University Press.

Foucault, Michel. 1978. *The History of Sexuality.* Vol. 1, *An Introduction.* Harmondsworth: Penguin.

Gatens, Moira. 1996. "Sex, Contract, and Genealogy? *Journal of Political Philosophy 4*(1): 29–44.

Geddes, W. R. 1957. *Nine Dayak Nights.* Melbourne and New York: Oxford University Press.

Gibson, Thomas. 1986. *Sacrifice and Sharing in the Philippine Highlands: Religion and Society among the Buid of Mindoro.* London and Dover: Athlone.

Gilman, Sander L. 1985. "Black Bodies, White Bodies: Toward an Iconography of Female Sexuality in Late Nineteenth-Century Art, Medicine, and Literature." In *"Race," Writing, and Difference,* ed. Henry Louis Gates, Jr., 223–40. Chicago and London: University of Chicago Press.

Gordon, Margaret T., and Stephanie Riger. 1989. *The Female Fear.* New York: Free Press.

Gregor, Thomas. 1990. "Male Dominance and Sexual Coercion." In *Cultural Psychology: Essays on Comparative Human Development,* ed. James W. Stigler, Richard A. Shweder, and Gilbert Herdt, 477–95. Cambridge: Cambridge University Press.

Griffin, Susan. 1986. *Rape: The Politics of Consciousness.* San Francisco: Harper & Row.

Halperin, David M. 1993. "Is There a History of Sexuality?" In *The Lesbian and Gay Studies Reader,* ed. Henry Abelove, Michele Barale, and David M. Halperin, 416–31. New York and London: Routledge.

Helliwell, Christine 1993. "Women in Asia: Anthropology and the Study of Women." In *Asia's Culture Mosaic,* ed. Grant Evans, 260–86. Singapore: Prentice Hall.

———. 1995. "Autonomy as Natural Equality: Inequality in 'Egalitarian' Societies." *Journal of the Royal Anthropological Institute* 1(2) : 359–75.

———. 1996. "Space and Sociality in a Dayak Longhouse." In *Things as They Are: New Directions in Phenomenological Anthropology,* ed. Michael Jackson, 128–48. Bloomington and Indianapolis: Indiana University Press.

———. Forthcoming. *"Never Stand Alone": A Study of Borneo Sociality.* Williamsburg: Borneo Research Council.

Howell, Signe. 1989. *Society and Cosmos: Chewong of Peninsular Malaysia.* Chicago and London: University of Chicago Press.

Kilpatrick, Dean G., Benjamin E. Saunders, Lois J. Veronen, Connie L. Best, and Judith M. Von. 1987. "Criminal Victimization: Lifetime Prevalence, Reporting to Police, and Psychological Impact." *Crime and Delinquency 33*(4) : 479–89.

Koss, Mary P., and Mary R. Harvey. 1991. *The Rape Victim: Clinical and Community Interventions.* 2d ed. Newbury Park, Calif.: Sage.

Laqueur, Thomas. 1990. *Making Sex: Body and Gender from the Greeks to Freud.* Cambridge, Mass., and London: Harvard University Press.

Leacock, Eleanor. 1978. "Women's Status in Egalitarian Society: Implications for Social Evolution." *Current Anthropology 19*(2): 247–75.

Lloyd, Genevieve. 1984. *The Man of Reason: "Male" and "Female" in Western Philosophy.* London: Methuen.

MacKinnon, Catharine A. 1989a. "Sexuality, Pornography, and Method: 'Pleasure under Patriarchy.'" *Ethics* 99: 314–46.

———. 1989b. *Toward a Feminist Theory of the State.* Cambridge, Mass., and London: Harvard University Press.

Madigan, Lee, and Nancy C. Gamble. 1991. *The Second Rape: Society's Continued Betrayal of the Victim.* New York: Lexington.

Marcus, Julie. 1992. *A World of Difference: Islam and Gender Hierarchy in Turkey.* Sydney: Allen & Unwin.

Marcus, Sharon. 1992. "Fighting Bodies, Fighting Words: A Theory and Politics of Rape Prevention." In *Feminists Theorize the Political,* ed. Judith Butler and Joan W. Scott, 385–403. New York and London: Routledge.

Martin, Emily 1991. "The Egg and the Sperm: How Science Has Constructed a Romance Based on Stereotypical Male-Female Roles." *Signs: Journal of Women in Culture and Society 16*(3): 485–501.

McColgan, Aileen. 1996. *The Case for Taking the Date Out of Rape.* London: Pandora.

Minturn, Leigh, Martin Grosse, and Santoah Haider. 1969. "Cultural Patterning of Sexual Beliefs and Behaviour." *Ethnology 8*(3): 301–18.

Mohanty, Chandra Talpade. 1991. "Under Western Eyes: Feminist Scholarship and Colonial Discourses." In *Third World Women and the Politics of Feminism,* ed. Chandra Talpade Mohanty, Ann Russo, and Lourdes Torres, 51–80. Bloomington and Indianapolis: Indiana University Press.

Moore, Henrietta L. 1994. *A Passion for Difference: Essays in Anthropology and Gender.* Cambridge and Oxford: Polity.

Myers, Fred R. 1988. "The Logic and Meaning of Anger among Pintupi Aborigines." *Man 23*(4): 589–610.

Palmer, Craig. 1989. "Is Rape a Cultural Universal? A Re-Examination of the Ethnographic Data." *Ethnology 28*(1): 1–16.

Plaza, Monique. 1980. "Our Costs and Their Benefits." *m/f* 4: 28–39.

Rosaldo, Michelle Z. 1980a. *Knowledge and Passion: Ilongot Notions of Self and Social Life.* Cambridge: Cambridge University Press.

———. 1980b. "The Use and Abuse of Anthropology: Reflections on Feminism and Cross-cultural Understanding." *Signs* 5(3): 389–417.

Russell, Diana E. H. 1984. *Sexual Exploitation: Rape, Child Abuse, and Workplace Harassment.* Beverly Hills, Calif.: Sage.

Sanday, Peggy Reeves. 1981. "The Socio-Cultural Context of Rape: A Cross-Cultural Study." *Journal of Social Issues* 37(4): 5–27.

———. 1986. "Rape and the Silencing of the Feminine." In *Rape,* ed. Sylvana Tomaselli and Roy Porter, 84–101. Oxford: Blackwell.

———. 1990a. "Androcentric and Matrifocal Gender Representations in Minangkabau Ideology." In *Beyond the Second Sex: New Directions in the Anthropology of Gender,* ed. Peggy Reeves Sanday and Ruth Gallagher Goodenough, 141–68. Philadelphia: University of Pennsylvania Press.

———. 1990b. *Fraternity Gang Rape: Sex, Brotherhood, and Privilege on Campus.* New York and London: New York University Press.

Strathern, Marilyn. 1988. *The Gender of the Gift: Problems With Women and Problems With Society in Melanesia.* Berkeley: University of California Press.

Trumbach, Randolph. 1989. "Gender and the Homosexual Role in Modern Western Culture: The Eighteenth and Nineteenth Centuries Compared." In *Homosexuality, Which Homosexuality?* ed. Dennis Altman, 149–69. Amsterdam: An Dekker/Schorer; London: GMP.

———. 1993. "London's Sapphists: From Three Sexes to Four Genders in the Making of Modern Culture." In *Third Sex, Third Gender: Beyond Sexual Dimorphism in Culture and History,* ed. Gilbert Herdt, 111–36. New York: Zone.

Tsing, Anna Lowenhaupt. 1993. *In the Realm of the Diamond Queen: Marginality in an Out-of-the-Way Place.* Princeton, N.J.: Princeton University Press.

van der Meer, Theo. 1993. "Sodomy and the Pursuit of a Third Sex in the Early Modern Period." In *Third Sex, Third Gender: Beyond Sexual Dimorphism in Culture and History,* ed. Gilbert Herdt, 137–212. New York: Zone.

Woodhull, Winifred. 1988. "Sexuality, Power, and the Question of Rape." In *Feminism and Foucault: Reflections on Resistance,* ed. Irene Diamond and Lee Quinby, 167–76. Boston: Northeastern University Press.

Young, Iris Marion. 1990. "Throwing like a Girl: A Phenomenology of Feminine Body Comportment, Motility, and Spatiality." In her *Throwing like a Girl and Other Essays in Feminist Philosophy and Social Theory,* 141–59. Bloomington and Indianapolis: Indiana University Press.

Introduction to Reading 11

Lynda Newland, a sociologist at the University of the South Pacific, Fiji, conducted participant observation among the Sundanese people of West Java, Indonesia. Here, the Sundanese view Islam as a primary marker of their identity. Female and male circumcision is among a number of rituals that are defined as essential to transforming a child into a socially acceptable member of the Sundanese community. This article explores the nature and context of female circumcision in particular. Newland argues that Western condemnation of all forms of female circumcision is problematic because practices vary widely and include types that pose no significant disbenefits for girls and women.

1. Discuss the gender differences and similarities between the Sundanese rituals that transform a child into a member of the community.
2. What is the zero tolerance campaign and why is Newland critical of it?
3. What is colonialism and how has colonial history affected the Sundanese view of Westerners?

Female Circumcision

Muslim Identities and Zero Tolerance Policies in Rural West Java

Lynda Newland

It was another early morning of participant-observation fieldwork among the Sundanese in the tea plantations of West Java, Indonesia. I accompanied my reluctant midwife-informant to a tiny bamboo house on a round of 7-month pregnancy rituals. The usual elements, including the seven flowers in a bowl of water, a bowl of uncooked rice with Rp 1000 money and an egg on top of it, were set out in front of us (Newland, 2001). As we sat on the floor waiting for relevant parties to arrive, the midwife apologetically said something to me in Sundanese that I did not understand. I nodded and waited for the pregnant woman to enter. Instead, a woman brought out a tiny swaddled baby. While praying in Arabic, the midwife drew a length of cotton thread through turmeric to color it gold and then loosened the baby's swaddling to reveal the little girl's genitals. Shocked, I realized I was not at a 7-month pregnancy ritual but a baby girl's circumcision.

In this ritual as conducted by this midwife, no knife was used—just a needle. There was no significant bloodletting. The midwife simply used the needle to scrape the clitoral area. When the baby started to cry, she stopped, pushed the needle through a banana leaf and put it in the flower water. Still praying, she tied gold thread around the baby's neck and then tied the yellowed cotton around the baby's left wrist, the mother's right wrist, the baby's right leg, and finally the right leg of a chicken. Then she held the chicken above the baby's head and drew three circles in the air. Finally, she gave the chicken to the grandmother with a handful of rice, and the grandmother left the room. This, I was to learn, was the *gnahuripan* or ritual blessing that tied souls into and between mother, child, and chicken, but it was the circumcision that impressed itself most upon me. Having had no idea that female circumcision was practised in West Java, it took me some time to accept what I had witnessed.

At the same time as my fieldwork experiences in the mid-1990s, the move to eradicate female circumcision at the international level had escalated to become a major platform for improving the rights of women and girls. At the national level, the Indonesian government had already ratified the Convention on the Rights of the Child in 1990 (Budiharsana, Amaliah, Utomo, & Erwinia, 2003; Moore & Rompies, 2004). Although this Convention did not specifically isolate practices of circumcision as oppositional to the rights of the child, UNICEF has since counted the criminalisation of female genital mutilation in sub-Saharan Africa as one of its successes in terms of targeting child survival and development (UNICEF, n.d.).

In 1995, the United Nations declared female circumcision a violation of women's rights on the strength of the Beijing Declaration and Platform for Action (UNESCO, 2005). The United States Congress later supported this by refusing financial assistance to countries where the government has not taken steps to eradicate female circumcision (Budiharsana et al., 2003). By 1998, a United Nations campaign was launched to eradicate female circumcision worldwide, which many United Nations organisations actively supported (Estabrooks, 2000; UNFPA, 1996; UNHCHR 1994, 1995, 2003). As a result, female circumcision became a highly politicised practice associated with violence against women and

From Newland, L. "Female circumcision: Muslim identities and zero tolerance policies in rural West Java" *Women's Studies International Forum 29,* copyright © 2006. Reprinted with permission of Elsevier.

girls (sometimes even with torture) or, at the very least, was described as a harmful practice (see, for instance, UNESCO, 1995).

Although many UN documents do not differentiate between the different types of circumcision practised, the WHO has distinguished four 'types' and two extra categories in the following way:

- Type I–excision of the prepuce, with or without excision of part or all of the clitoris;
- Type II–excision of the clitoris with partial or total excision of the labia minora;
- Type III–excision of part or all of the external genitalia and stitching/narrowing of the vaginal opening (infibulation);
- Type IV–pricking, piercing or incising of the clitoris and/or labia; stretching of the clitoris and/or labia; cauterization by burning of the clitoris and surrounding tissue;
- scraping of tissue surrounding the vaginal orifice (angurya cuts) or cutting of the vagina (gishiri cuts);
- introduction of corrosive substances or herbs into the vagina to cause bleeding or for the purpose of tightening or narrowing it; and any other procedure that falls under the definition given above (WHO, 2000).

While the practices vary widely, the term 'female genital mutilation' is often used as a gloss for all of them. Strategically, the term provides an emotional force behind a policy of zero tolerance towards any cutting that may occur in the female genital area, other than genital surgery (whether for medical reasons or for sexual reassignment). The label also carries the assumptions that all female circumcision practices intentionally limit female sexual pleasure in order to control women, have a detrimental effect on women's health, and, because they harm women and girls, must be eradicated regardless of the extent of the procedure or the context in which they occur. Moreover, the policy of zero tolerance framed by the UN and advocated for adoption by member states suggests that the educated first world (from which this term emerged) has the responsibility to 'save' third world women—without recourse to their very different histories and 'manifestations of differently structured desires' (Abu Lughod, 2002:783; Walley, 1997; Wood, 2001).

Yet, although many of these documents assume all female genital mutilation results in health issues and therefore harm to women's sexuality, studies conducted in Indonesia suggest there are no significant health risks related to female circumcision and that infection and bleeding seem to be more likely to result from male circumcision than from female circumcision (Darwin, Faturochman, Ptranti, Purwatiningsih, & Octaviatie, 2002), indicating that these practices are probably less harmful to women than they are to men. If the issue of harm is questionable, the stereotype that female circumcision impairs sexual enjoyment is also undermined by an idea that the practice enhances sexual enjoyment, which seems to be widespread throughout Java (Berninghausen & Kerstan, 1992; Darwin et al., 2002). While the impact of Indonesian forms of circumcision on sexual pleasure cannot be measured, such a response calls dominant and broadly Western notions of sexual embodiment into question.

Although many of the assumptions regarding sexual embodiment and the 'saving' of third world women that are implicit in this term have since been deconstructed (e.g., Bell, 2005; Parker, 1999; Walley, 1997), the term, 'female genital mutilation,' has retained emotive force and currency at United Nations levels, serving to justify the universalisation of zero tolerance policies. Despite ethnographic work in this area (e.g., Abusharaf, 2001; Boddy, 1989), one of the major aspects that tends to be overlooked is the variable contexts in which female circumcision occurs. In this article, I explore some of these issues in relation to my own fieldwork experiences to argue against the notion that female circumcision is universally harmful to women and girls because it does not recognise the diversity of circumcision practises and does not allow for the fact that female circumcision in regions such as West Java is often part of an elaborated process of socialisation in which a child is embedded within a community. In this regard, I argue that such processes are not necessarily oppressive to women and girls in and of themselves. Nor are they necessarily violent. Rather, they are a representation of parental responsibility towards the child.

The Making of Persons

Children are highly valued in Sundanese society, not only as agricultural labour in a peasant society (as theorised in the 'value of children' literature in the 1970s and early 1980s, e.g., Nag, White, & Peet, 1978; Hull, 1977), but in a rich cosmology where children link preceding generations with those yet to come. In the simplest sense, having children bestows the status of adulthood on the parents in the titles of *Ibu* and *Bapak* (literally 'mother' and 'father' with usage similar to 'Mrs.' and 'Mr.'). Children are also closely associated with both material and non-material forms of wealth and well-being as evident in the Indonesia-wide saying, '*banyak anak, banyak rezeki*' or 'many children, much fortune.'

With the high value of children comes the responsibility of giving newborn children a defined position in the religious community that deeply roots them into their social world, away from the world of spirits (cf. Geertz, 1961; Jaspan & Hill, 1987).[1] A child's entry into this world is conceptualised as the beginning of a process of affirmation of a destined personhood, although this personhood will not be complete until after death at Judgement Day. . . .

Thus, the rituals around birth are replete with ideas about how to imbue the child with the values of the preceding generation that will maintain the quality of the family vertically through the generations as well as enhancing horizontal communal relationships.

To this end, female circumcision is performed in one of a series of birth rituals which include the 7-month pregnancy ritual (*tingkeban*), the burial of the placenta, the process of naming, a Muslim *marhaba* when the umbilical stub falls off the child, piercing the ears of girls, the animal sacrifice (*aqiqah*), the tying of the mother's and child's spirits *(ngahuripan),* and the haircutting ritual *(cukuran).* [2] The parents participate in the separation of the baby from its companion spirits in the womb, in shaping the baby's character towards desirable gender values and a fortunate future, and in introducing the baby to the values of the Muslim community. While male circumcisions are performed, when the boy is around 7 years old and are therefore considered part of the *rite-de-pasage* to adulthood, female circumcision has become part of the rituals around birth. In this section, then, I explore some of these rituals to show the way in which circumcision at birth or during childhood is embedded in ideas about the creation of a moral person, which, in this community, means to be socialised into the values of a local version of Islam.

The responsibility of the parent in appropriately socialising the child begins from the 7-month pregnancy ritual *(tingkeban).* Although primarily intended to aid the woman in experiencing a smooth or fluid birth, the *tingkeban* is also an attempt to counter some of the physical and moral faults of the parents that may be later transmitted to the baby. During pregnancy, for instance, *lahir* (outside) and *batin* (inside) mirror each other in such a way that, if a pregnant woman sees a deformed person, the child will also be deformed. . . .

In an acknowledgement of gender difference, girl's ears are pierced while they are babies. Although there is no religious function attributed to piercing, several informants noted that it was *haram* (forbidden) or in the very least *jelek* (ugly, awful) if a boy's ears are pieced. One village midwife also made a brief and intriguing association between the piercing of ears and female circumcision, where the first makes holes in the ears with sharpened studs and the second opens the flesh around the vagina. While both effectively reinscribe the baby's sexual destiny onto her body, her first earrings of tiny gold studs predict a lifetime of receiving gold jewellery, from sleepers (looped earrings) to necklaces and rings. Gold jewellery is significant in showing the financial status of the household and gold is also associated with fortune and fertility.

The village midwives also say that through female circumcision and the other rituals around birth the child can inherit desirable characteristics from ritual objects. Rice, eggs, needles, thread, flowered water, and oil are often used. While these elements together attract certain spirits, repel others, and, as discussed above, tie the child's soul into his or her body, to the mother's soul, and to the home (see Newland, 1999, 2001, 2002b), they may also represent desirable qualities that need to be transmitted to the child. For example, in explaining the symbolism of the *tingkeban* (or pregnancy ritual), one midwife claimed that the roundness of the egg is associated with the roundness of thought (so the child would not be easily influenced by malevolent forces); the needle's sharpness is associated with sharpness of memory and cleverness, and the thread represents its association with wisdom and judgement. . . . The fact that this transference of qualities is called an inheritance points to the way each child is born into a highly structured position with mutual obligations within the vertical relationships of the generations.

Already bound to women (and particularly to the lineage of its mother) through blood and milk,[3] the child must also be bound to the family and then to the wider community. This is done with the three main *selamatan* or ritual feasts: the *marhaba* after the child's umbilical stub drop off, in the goat sacrifice *(aqiqah),* and the hair-cutting ritual *(cukuran)*. . . .

The last of the *selamatan* is performed at 40 days after the birth as this is the time both mother and baby are considered out of danger. This *selamatan* is called the *cukuran* or 'the cutting' because several of the baby's locks of hair are cut. This ritual is meant formally to introduce the baby to the male community for the first time. The men pray and sing while the father carries the baby to the centre. Three different men then cut the child's hair and, between each cut, they dip the scissors in a bowl of water, containing flowers of seven colours. The symbolic resonances between cutting hair, prayers, and circumcision are significant. Cutting the hair is a symbol of submission to the Covenant and a sacrifice of individuality in order to conform to the community's values. In the same way, circumcision and its cutting of flesh, the cutting and covering of hair, the sacrifice of chicken and goats, blood from humans, chickens and goats all refer to the same set of ideas about submission to Allah and the Muslim community.

These *selamatan* place both male and female circumcision in a context where babies must be made socially acceptable, by undergoing both a physical and ritual separation from the mother's body and spirit in order to be incorporated into the broader Muslim community. Clearly, none of these rituals are intended to violate the child. In this context, female circumcision is perceived as roughly equal to ear piercing in its physical effects, but absolutely essential to the embodiment of Islam.

Female Circumcision

When I asked the Sundanese villagers why women should be circumcised, many were quite shocked that women should go uncircumcised. For example, Ibu Acih replied, 'If you're Muslim you must be circumcised, because circumcision is the direction given by Islam. If you're outside Islam, it's not necessary to be circumcised. If children aren't circumcised, they're not permitted to participate in the prayers at the mosque.' Likewise Imas responded, 'Circumcision is the identifying feature of Islam. Boys must be circumcised, girls also. The old people said circumcision is for throwing away what's dirty. I often heard that at *pengajian* [meetings for recitation of the *Qur'an*].' For them as for the *ulama* (religious leaders), circumcision positions the child in the Muslim community by physically tracing Islamic identity on to the body. It is also an expression of the idea that both men and women are considered equal before Allah.

For boys, circumcision occurs when they reach the age of seven. A specialist removes the foreskin, and then the boy sits on a chair greeting guests. The mother and female relatives cook a feast that includes fish, chicken, and cakes.

While boys have a public circumcision, the timing of a girl's circumcision occurs shortly after childbirth and therefore is a quiet affair, entailing the presence of a midwife, mother and child. The exact timing varies according to the midwife: one midwife performed circumcisions on babies between 7 and 15 days old and a second circumcised girls from between 40 days and 2 months.[4] The extent of the practice also seems to depend on the midwife. Ma Eha, for instance, merely scratches the clitoral area with a needle. . . . By contrast, Ma Juju scoffed at this practice:

> *A girl is circumcised by knife. If it's by a needle, she will get sick. With a knife, you pry off a little. For Muslims, if they're not circumcised, they're not allowed to go into a mosque.*

The amount of flesh cut is described as a *mata holang,* the size a grain of rice and white, unlike the surrounding flesh.

Yet another midwife, Ayah Enjum, one of the rare males who served the poorer villages, espoused another view, saying that: 'Women can be circumcised, but they don't have to be. But if they're not circumcised they must have a ritual ablution when they reach adulthood.' He also interpreted circumcision fairly broadly, saying: 'If you cut the umbilical cord while reciting the confession of faith, it's the same as being circumcised with the confession of faith. Girls whose birth I assist shouldn't be circumcised because when I cut their umbilical cord, the baby becomes a Muslim straight away.' Here, circumcision is displaced through a visual metaphor where the cutting of the umbilical cord is a conscious analogy for the cutting of the clitoral area. What matters for Ayah Enjum is that the confession of faith has been and an organ has been cut, not the excision of a specific organ or the removal of any flesh.

Clearly, all the village midwives see some sort of actual or symbolic female circumcision as necessary, although the extent of the practice varies significantly. Further, throughout their explanations, female circumcision is explained in relation to Islam. Despite arguments that female circumcision is cultural rather than religious, this is clearly not how Muslim Sundanese experience the embodiment of their religion. Circumcision is considered to inscribe the major distinction between Muslim and heathen: as indicated above, only the circumcised may enter a mosque to pray.

Circumcision for boys and girls is also one of several measures strongly associated with the notion of cleanliness or purification. The *ulama,* Bapak Syamsudin, described circumcision to me this way:

> *Circumcision is included in the five types of cleanliness.*
>
> *Circumcision originated from the word* ***khotana****, that is, throwing away the bud, so that the defiling filth that sits in the bud is thrown away. So circumcision is cleansing . . .* [The second type of cleanliness is] *shaving the armpits. The third type is cutting finger and toenails, the fourth is shaving the genitals, and the fifth is shaving the moustache. With shaving the genitals,* [cleanliness] *is assured for 40 days.*

The notion that circumcision was the most important in a range of cleansing acts to remove dirt or filth is commonly emphasized. Indeed, both men and women are barred from the mosque when considered polluted by semen or menstrual blood which are considered dirty and which cancel out the effect of prayers. From this perspective, Islam treats both men and women equally in that both must rid the 'defiling filth' from their bodies in order to come before Allah.

. . . (C)ircumcision does not limit women more than men nor is it used to mark gender difference but rather it reinforces the idea that both sexes can attain equal purity. Nor is circumcision in this context intended to control female sexuality. On the contrary Bapak Symasudin, an *ulama* for the Muslim organisation, Persis, explained:

> *Rasulullah* [Mohammed] *governed his wives should be circumcised. Circumcision for women is the getting rid of 'the eye of* ***holang****' or a little bit inside the vagina. Circumcision for women is connected with* [their] *sexual enjoyment. So for circumcision, there is sexual enjoyment, cleanliness, and performance of acts of devotion.*

As previously mentioned, the notion that female circumcision enhances sexual enjoyment seems to be widespread throughout Java (Berninghausen & Kerstan, 1992; Darwin et al., 2002). Certainly, women do not perceive female circumcision as harming their sexual desire. In my own fieldwork, Ma Eha who had married twice noted that:

> *All women must marry. I know the reason for marriage. I wanted it because Allah gave me passion. The religious leader says that there are many more passionate women than men. Women's desire is nine while men's is one, so women's is bigger.*

Another study also noted that many of their informants in various locations throughout Indonesia associated female circumcision with the *hadith* or saying where Mohammed tells the circumciser to 'Cut off only the foreskin . . . and do not cut off deeply . . . for this is brighter for the face (of the girl) and more favourable with the husband' (cited in Budiharsana et al., 2003, p. 8). Even if such ideas play into Muslim notions that the wife should be sexually available to her husband, they show that this availability does not entail the lack of sexual pleasure on her part.

In this way, female circumcision is not perceived as a physically or emotionally harmful act against women and girls. While some argue that girls are unable to give or refuse their consent or even that it is a form of child abuse (e.g., Family Law Council, 1994), in this region circumcision is considered a parental duty through which to position their child in the *ummat* or Muslim community. To refuse to do this would amount to neglect of parental responsibilities.

In this light, suggestions that the practice may have increased because of Islamic revivalism (Hull & Budiharsana, 2001) are difficult to substantiate, given that in West Java the practice had remained unquestioned and unproblematic during my fieldwork in the mid-1990s. From my own informants, it seemed that female circumcision had been practised throughout the aging midwives' careers and probably for many generations.

Zero Tolerance

Despite the fact that the United Nations had launched a campaign to eradicate female circumcision worldwide in 1998, a report notes that in the year 2000: 'public awareness is low. The subject is not addressed in schools and rarely in the media' (United States Department of State, 2001). In Indonesia, various arrangements were made to commit the Ministry of Women's Empowerment and the Ministry of Religion to study religious teachings that obstruct women's rights and the agreement of the National Ulama's Council had been given to eliminate female circumcision in stages (United States Department of State, 2001). Yet, little more seems to have been done until the government campaign of zero tolerance towards female circumcision was finally launched in January 2003, which was also met with a relaxed attitude (Kompas, 2004). It seems that there was little inclination among the citizenry to support such campaigns. If this is the case, then the concerns about female circumcision are perceived at state level to be foreign and inappropriate for the vast Muslim population in Indonesia.

However, one of the effects of the campaign of zero tolerance towards female circumcision seems to have resulted in the medicalisation of the practice (despite the WHO recommendation against it in 1982), which has ironically led to more invasive procedures being performed (Budiharsana et al., 2003). Although the hospitals were considered more hygienic places in which to conduct the procedure, the medical practice involved the use of scissors to cut away more of the genital tissue than the village midwives ever removed using needles and penknives (Budiharsana et al., 2003; Moore & Rompies, 2004). Moreover, the hospital clinic midwives were not trained by the Health Department in such procedures but relied on information given by senior staff or village midwives (Budiharsana et al., 2003). Thus, in 2004, female circumcision was being offered as part of a package of surgical procedures performed in hospitals for just-born girls. Girls could be vaccinated, have their ears pierced and be circumcised at the same time for prices between Rp 15,000 and Rp 95,000 depending on the hospital (Moore & Rompies, 2004). The Indonesian health authorities announced a ban on medics (presumably meaning the clinic midwives) performing female circumcisions by mid-2005 in an effort to prevent hospitals from continuing the practice (News Unleashed, 2005), but the effect of this ban is not yet known.

Overall, the immediate response to the zero tolerance campaign continues to be muted. A report in the Indonesian newspaper, *Kompas,* in 2004 raised the concern that, if the 'calm' attitude of the government continued, Indonesia would be grouped with countries such as Somalia, Ethiopia, Yemen, and Malaysia, many of which practise more intrusive forms of female circumcision. To be grouped with such countries would give Indonesia a questionable reputation, but, on the other hand, enforcing zero tolerance would entail legislating against parents who take their girl children out of the country to have them circumcised (Kompas, 2004). For many Indonesians, such a ruling would come as a shock, because taking their moral responsibilities as a parent seriously could lead to a confrontation with the law. Yet, for the Sundanese in West Java, there is no correspondence between such arguments for zero tolerance and the values in which female circumcision is locally embedded. Instead, such legislation criminalises parents if they proceed with any type of female circumcision, and, if enacted, could cause harm greater to the child than the practice of female circumcision through removing and imprisoning parents.[5]

During the 1990s when I conducted fieldwork, the Sundanese had an ambivalent relationship with the state as institutionalised and led by the Suharto regime. In its early years, the regime had depoliticised Islamic parties and distanced itself from Muslim influence in its pursuit of development. Moreover, state programs such as the family planning campaign, which dictated that *'Dua anak cukup'* ('Two children are enough'), were widely perceived to be in contradiction with rural Sundanese values (Newland, 1999, 2002a). By 2003, there had been successive changes in government and successive problems with unrest in the provinces (e.g., East Timor, Aceh, Ambon, Kalimantan, and Irian Jaya). Given such upheavals, it would seem that any Indonesian government would be wary of antagonising Muslims by overtly supporting and legislating zero tolerance policies that might be perceived as anti-Islam in a country that hosts the largest number of Muslims anywhere in the world. Indeed, any policy that is perceived as a foreign imposition helps contribute to local stereotypes of the Western world as rich and corrupt. Because of Sundanese colonial history, the rural Sundanese in my fieldwork area continue to view Westerners with suspicion since, in their eyes, Westerners have the power to perpetuate new forms of colonialism. International law and international campaigns such as that regarding genital mutilation exacerbate such images, especially when they fail to understand the variety of practices that are glossed by labels such as female circumcision or female genital mutilation and the very different contexts in which they occur. Universalising prohibitions on female circumcision disregards the more immediate needs of women and their kin groups and thus may be interpreted as simply another imposition from outside.

While my fieldwork experiences shocked me at first, it quickly became clear that female circumcision was not performed with any intention of violence, abuse, or even harm towards girl children and did not seem to have any measurable effect on their lives. Instead, parents were fulfilling their obligations by circumcising boys and girls to conform to a moral order deeply identified with Islam and to position them appropriately in the Muslim community. In this context, zero tolerance policies towards female circumcision seem out of touch with the realities experienced at the grass-roots level.

Notes

1. In local terms, if certain ritual procedures that separate the child from the 'sibling' spirits are not properly done, the child will suffer from deformities. Although never voiced, the suggestion seems to be

that, having come from the spirit world (a notion which is itself under contestation from differing Muslim interpretations) the child must be grounded in the actual world to aid in the avoidance of infant death, which entails the child's return to the spirit world.

2. Due to space considerations, not all of these rituals are discussed here, but an elaboration of them is available in Newland (1999).

3. While being born to a woman is a primary factor in kinship, kinship ties are further set by breastfeeding, as made evident by old ideas about cross-sex twins. Cross-sex twins are considered to be married (dijodohkan) in the womb. If the twins are separated at birth and breastfed by different mothers, they will be encouraged to marry in adulthood as they are considered a destined and ideal match (Newland, 1999).

4. Age at circumcision seems to vary throughout Indonesia. Two different research teams have since noted that the girl can be anywhere from roughly 7 days to 9 years according to the district they live in (Budiharsana et al., 2003; Darwin et al., 2002).

5. In other countries, parents have already faced such rulings. A case that came before the courts in France led to a Gambian mother being gaoled for having her daughters circumcised, despite having no other carer for her children (Pitt, 1993, p. 29).

References

Abu Lughod, Lila (2002), Do Muslim women really need saving? Anthropological reflections on cultural relativism and its others. *American Anthropologist, 104,* 783–790.

Abusharaf, Rogaia Mustafa (2001). Virtuous cuts: Female genital circumcision in an African ontology. *Differences: A Journal of Feminist Cultural Studies, Vol. 12* (pp. 65–140).

Bell, Kerstan (2005). Genital cutting and Western discourses on sexuality. *Medical Anthropology Quarterly, 19,* 125–148.

Beminghausen, Jutta, & Kerstan, Birgit (1992). *Forging new paths: Feminist social methodology and rural women in West Java.* London: Zed Books.

Boddy, Janice (1989). *Wombs and alien spirits: Women, men, and the Zar cult in northern Sudan.* London: Zed Books.

Budiharsana, Meiwita, Amaliah, Lila, Utomo, Budi, & Erwinia (2003). *Female circumcision in Indonesia: Extent, implications and possible interventions to uphold women's health rights, Jakarta: Population Council.* http://www.synergyaids.com/resources .asp?id=5049 Accessed 2/11/05.

Darwin, Muhadjir, Faturochman, Ptranti, Basilica Dyah, Purwatiningsih, Sri, & Octaviatie, Isaac Tri (2002). *Male and female genital cutting among Yogyakartans and Madurans.* http:// demography.anu.edu.au/G&SH/projects.html Accessed 21/7/04.

Estabrooks, Elizabeth A. (2000). *Female genital mutilation.* http://www.munfw.org/archive/50th/wh02 .html Accessed 1/10/03.

Family Law Council (1994). *Female genital mutilation: A report to the Attorney-General prepared by the Family Law Council, June 1994.* Commonwealth of Australia. http://law.gov.au/flca/Female% 20Genital%20Mutilation.htm Accessed 15/6/04.

Geertz, Hildren (1961). *The Javanese family: A study of kinship and socialization.* New York: Free Press of Glencoe Inc.

Hull, Terence (1977). The influence of social class on the need and effective demand for children in a Javanese village. In Lado T. Ruzicka (Ed.), *The economic and social supports for high fertility: Proceedings of the conference held in Canberra, 16–18 November 1976.* Canberra: Demography Dept., ANU.

Hull, Terence, & Budiharsana, Meiwita (2001). Putting men in the picture: Problems of male reproductive health in Southeast Asia. *IUSSP XXIV Congress, Salvador; Brazil.* http://www.iussp .org/Brazil2001/s20/S22_02_Hull.pdf Accessed 30/11/05.

Jaspan, Helen, & Hill, Lewis (1987). *The child in the Family: A study of childbirth and child-rearing in rural Central Java in the late 1950s.* Hull: The University of Hull.

Kompas (2004). *Sunat Perempuan? Jangan deh . . .* 04.03.04. http://www.kompas.com/kesehatan/news/ 0403/04/115051.htm Accessed 26/10/04.

Moore, Matthew Rompies, Karuni (2004). In the cut. *Sydney Morning Herald* 13/1/2004. http://www .smh.com.au/articles/2004/01/12/107387 7760147.html Accessed 16/6/2004.

Nag, Moni, White, Benjamin N. F., & Peet, R. Creighton (1978). An anthropological approach to the study of the economic value of children in Java and Nepal. *Current Anthropology, 19,* 293–306.

Newland, Lynda (1999). *Cosmologies in conflict: The politics of reproduction in West Java.* Unpublished PhD thesis. Sydney: Macquarie University.

Newland, Lynda (2001). Syncretism and the politics of the Tingkeban in West Java. *The Australian Journal of' Anthropology, 12,* 312–326.

Newland, Lynda (2002a). The deployment of the prosperous family: Family planning in West Java. *National Women's Studies Association Journal, 13,* 22–48.

Newland, Lynda (2002b). Of Paraji and Bidan: Hierarchies of knowledge among Sundanese midwives. In Santi Rozario & Geoffrey Samuel (Eds.), *Daughters of Hariti: Childbirth and female healers in South and Southeast Asia* (pp. 256–278). London: Routledge.

News Unleashed (2005). Indonesia to forbid medics from conducting female circumcisions. 02.06.05. http://newsunleashed.com/indonesia-to-forbid-medics-from-conducting-female-circumcision Accessed 28/11/05.

Parker, Melissa (1999). Female circumcision and cultures of sexuality. In Tracey Skelton & Tim Allen (Eds.), *Culture and Global Change* (pp. 201–211). London: Routledge.

United Nations Educational, Scientific, and Cultural Organization (UNESCO). (1995). Beijing declaration and platform for action http://www.unesco.org/education/information/nfsunesco/pdf/BEIJIN_E.PDF

United Nations Educational, Scientific, and Cultural Organization (UNESCO). (2005). *Beijing declaration and platform for action. Fourth World Conference on Women, 1995.* http://www.unesco.org/education/information/nfsunesco/pdf/BEIJIN_E.PDF

United Nations Children's Fund (UNICEF). (n.d.), *Convention on the rights of the child.* http://www.unicef.org/crc/crc.htm

United Nations High Commission on Human Rights (UNHCHR). (1994). Review of further developments in fields with which the sub-commission has been concerned: Plan of action for the elimination of harmful traditional practices affecting the health of women and children. *Report from the Economic and Social Council.* http://www.unhchr.ch/huridocda/huridoca.nsf/0/ Accessed 2/10/03.

United Nations High Commission on Human Rights (UNHCHR). (1995). *Concluding observations of the Committee on the Elimination of Discrimination Against Women: Uganda.* http://www.unhchr.ch/tbs/doc.nsf/0/ Accessed 2/10/03.

United Nations High Commission on Human Rights (UNHCHR). (2003). Traditional practices affecting the health of women and girls: A human rights issue. *Women's rights are human rights.* http://www.unhchr.ch/women/focus-tradpract.html Accessed 2/10/03.

United Nations Family Planning Association (UNFPA). (1996). Uganda—FGM breakthrough. *Dispatches—News from UNFPA, No. 6, March.* http://www.un.org/popin/unfpa/dispatches/mar96.html Accessed 2/10/03.

United States Department of State (2001). Indonesia: Report on female genital mutilation (FGM) or female genital cutting (FGC). http://www.state.gov/g/wi/rls/rep/crfgm/10102.htm Accessed 21/ 7/04.

Walley, Christine (1997). Searching for 'voices': Feminism, anthropology, and the global debate over female genital operations. *Cultural Anthropology, 12,* 405–437.

WHO. (2000). Female genital mutilation. *World Health Organisation. Fact Sheet No. 241.* http://www.who.int/health_topics/female_genital_mutilation/en/ Accessed 1/10/03.

Wood, Cynthia (2001). Authorising gender and development: 'Third World Women,' native informants, and speaking nearby. *Neplanta: Views from South, vol. 2* (pp. 429–447).

Introduction to Reading 12

Wairimũ Ngarũiya Njambi, a Kenyan-born sociologist, and William E. O'Brien offer an incisive account of the institution of woman-woman marriage as practiced in Kenya today. Based on in-depth interviews with Gĩkũyũ women married to other women, the authors argue for an extensive revision of scholarship on woman-woman marriage that reflects the situated, complex, and empowering nature of this tradition.

1. What are the major shortcomings of past studies of woman-woman marriage?
2. How are woman-woman marriages regarded in the Gĩkũyũ community today?
3. Why do the authors view woman-woman marriage as a mode of empowerment?

Revisiting "Woman-Woman Marriage"

Notes on Gĩkũyũ Women

Wairimũ Ngarũiya Njambi and William E. O'Brien

Introduction

The practice of women marrying women is somewhat common in certain societies in West Africa, Southern Africa, East Africa, and the Sudan (O'Brien 1977). Yet, besides a total lack of discussion in the popular media, what is typically called woman-woman marriage is the subject of a very small body of academic literature.[1] Early scholarship is limited to the margins of several colonial-era ethnographies such as those of Evans-Pritchard, Herskovits, and Leakey. . . . More recent work remains equally marginal. Precious few writings address woman-woman marriage practices exclusively (e.g., Amadiume 1987; Burton 1979; Krige 1974; Oboler 1980); within others the subject remains little more than a footnote (e.g., Davis and Whitten 1987; Mackenzie 1990; Okonjo 1992). Since O'Brien's (1977) call for field research into woman-woman marriages more than two decades ago, there has been no study of Gũkũyũ woman-woman marriages, and few studies anywhere else. Our study attempts to revive this dormant discourse in relation to the Gĩkũyũ .

Based on interviews with members of households containing woman-woman marriages, we attempt to provide images of this institution as practiced in central Kenya. Relying upon these women's voices, we present these Gĩkũyũ woman-woman marriages in relation to major themes in the literature.[2] . . . Our attention is on the ambiguities and flexibility inherent in women's decision to marry women. In addition,

From Njambi, Wairimũ Ngarũiya and William E. O'Brien. 2000. "Revisiting 'woman-woman marriage'" *NWSA Journal 12*(1), pp. 1–23. Reprinted with permission of Indiana University Press.

we point to the strong emotional bonds to one another expressed by these women, shedding critical light on the omissions of purely functionalist perceptions of woman-woman marriage relationships. We also challenge the generalized conceptualizations of women who initiate such marriages as "female husbands." That term, used by Leakey and virtually all other authors on the topic, regardless of cultural context, imposes a "male" characterization upon a situation where none necessarily exists. Emphasizing a term such as "female husband" prompts sex-role presumptions that do not fit these Gĩkũyũ women, who bristle at the implied male-identification regarding their roles.

This study is based on interviews with women in eight households in a small village in Murangá District in central Kenya. This case study approach does not attempt to portray a generalized picture of woman-woman marriages, but relies upon the women's situated words to explain why they have married women, allowing them to present their own illuminating perspectives (see Smith 1987). . . .

The Gĩkũyũ are the largest ethnic group in Kenya, generally occupying the administrative unit of Central Province. "Kikuyuland," as it is commonly called, is bounded by Nairobi to the south and Mt. Kirinyaga (Mt. Kenya) to the north, the Rift Valley and Nyandarua Range (Aberdares) to the west, and the Mbeere Plain to the east. . . .

Most of the woman-woman marriage households in the study engaged in peasant farming for a living, dividing their agricultural production between cash crops and subsistence crops, a pattern typical of this rural setting. However, some of the women were engaged in other occupations including shop ownership, market trading of small commodities, and, in one case, *matatu* (mini bus) driving. The initiators of these relationships, who are called *ahikania,* were all landowners, and the households all had modest living standards similar to most others in the locality. Though the interviews took place in a rural setting, two of the subjects were residents of Nairobi, while another lived and worked in a nearby small urban center.

The majority of the *ahikania* were middle-aged at the time of marriage, and two were in their early 30s. All of the *ahiki,* the women who accepted the marriage offer, were between the ages of 20 and 30 when they were married. Education patterns of the subjects show that most of the initiators of the marriages were educated through the traditional Gĩkũyũ educational system of *githomo gia ugikuyu:* one had a high school education, one primary school. Almost all of the women who accepted the marriage offer had at least a primary school education. The wide range of age and education suggests to us that woman-woman marriage continues to be a relevant potential life-option for Gĩkũyũ women.

Kuhikania, the process of getting married, and *uhiki,* the marriage ceremony, takes place in the same manner for woman-woman marriages as with woman-man marriages. In fact, there is no separate term to differentiate a woman-woman marriage from a woman-man marriage. Even the term which describes the marriage initiator, *muhikania,* is used to describe a woman or a man.[3] As woman-woman marriages are not sanctioned by the various Christian churches in the region, *kuhikania* and *uhiki* continue to be performed through customary guidelines. The woman seeking a marriage partner, the *muhikania,* announces, either through a *kiama* (a customary civic organization) or through her own effort, her desire to find a marriage partner, or *muhiki.* Once the word is out, interested women go to visit, and once a suitable partner is found the *muhikania's* friends and family bring *ruracio* (gifts associated with *uhiki*) to those of the future wife and vice-versa. *Uhiki* takes place after this gift exchange and is performed with ceremonial blessings, termed *irathimo,* by elders of both families as the new wife moves into the *muhikania's* house.

Woman-Woman: Marriages and Family Definitions

While woman-woman marriage may be familiar to most anthropologists, at least in passing, the topic remains relatively obscure to most people outside Africa. In family studies discourse, the topic is pushed to the extreme margins by an historical fixation on western nuclear families as a universal ideal. This normative presumption of nuclearity makes it very difficult for particular non-western family forms, such as the woman-woman marriages in this study, to be evaluated as anything but bizarre novelties. As Skolnick and Skolnick argue:

> The assumption of universality has usually defined what is normal and natural both for research and therapy and has subtly influenced our thinking to regard deviations from the nuclear family as sick or perverse or immoral. (1989, 7)

. . . The Gĩkũyũ woman-woman marriages we studied challenge this thinking on all counts. Not only are the adults involved in these marriages of the same sex, but also there may be more than two, and the form of the family is not necessarily permanent once a union is made, but may change periodically. Furthermore, men are often absent from such relationships, though they may be involved in married relationships as spouses of women who initiate woman-woman marriages.

One example of such a relationship in our study is Kũhĩ's household. In this complex case, Kũhĩ (a woman) and Huta (a man) were originally married to each other. Later, they decided together that Huta would marry a second woman, Kara, creating a polygymous marriage.[4] Later still, Kũhĩ entered into a woman-woman marriage with a woman named Wamba. Wamba came to that family as Kũhĩ's marriage partner, and to assist in raising the children of that household. In this particular case, Wamba could have a sexual relationship with Huta (whom she also informally regarded as a husband), and was not restricted from having sexual relationships with other men outside their household. Later in her life, while still married to Kũhĩ, Wamba married a woman named Wambũi. The result is that this single household contains four marriages: two woman-man marriages and two woman-woman marriages. Such complex relationships do not break any "rules," expectations, or ideals of woman-woman marriages, but are an accepted aspect of such relationships in Gĩkũyũ contexts. . . .

The idea of same sex relationships has spurred discussion of the sexuality of women in such marriages. A few texts imply that there may be sexual involvement in these marriages. Herskovits, for example, suggested that Dahomey woman-woman marriages sometimes involved sexual relations between the women (1937). Davis and Whitten go so far as to state that the main issue in explaining these relationships generally is over whether reasons for such partnerships are in fact "homoerotic" or strictly socio-economic (1987, 87). While sexuality was not directly discussed in our interviews, we can glean from the experience that this dichotomy makes little sense.[5]

In our Gĩkũyũ locale, women in these relationships did not talk about sexual involvement with one another, although some did indicate sharing the same bed at night. . . . Given the ambiguity in this Gĩkũyũ context, one might borrow Obbo's assertion regarding the Kamba of Kenya that while there may be no clear indication of sexual relations among women in these marriages, we simply cannot dismiss the possibility (1976). We agree with Carrier that this possibility has been too quickly dismissed by some authors, and suggest that the subject deserves more careful investigation (1980).

On the other side of the dichotomy, to suggest that such relationships are based solely on socio-economic factors like access to land and other resources or lineage ignores the close emotional ties experienced by these women. Such functionalist views have strongly influenced historical, and still-held stereotypes of African marriages generally. African family relations, compared to the privileged, western nuclear family form, are often portrayed as relatively primitive since they are presumed to be based on practical considerations alone, such as access to resources, as opposed to having a significant emotional aspect (e.g., Albert 1971; Ainsworth 1967; Beeson 1990; Kilbride and Kilbride 1990; Le Vine 1970). The women interviewed help undermine such rigid notions, demonstrating clear emotional commitment to the women they marry. For example, one participant, Nduta, proclaims her feelings for her muka wakwa, or co-wife, Cirũ:[6]

> No one dare to disturb my co-wife in any way, and especially knowing what I would do to them. No one dares point a finger at her. I tell her to proudly proclaim her belongingness to me, and I to her. . . . What I hate most is when people come to gossip to me about my co-wife's whereabouts or whom they have seen her with. I don't care as long as she is here for me now and even after I am gone. . . . Regardless of what she does, she is here because of me. Then why should I tell her what to do and what not to do. She is a free woman. And that is what I want her to be. So, when they come here to gossip, I tell them to leave her alone. She is mine and she is here on my property, not yours. . . . She who sincerely loved me and I loved back, let her stay mine. It is she who shall enshrine and take over this household when my time comes. (in interview)

In addition to expressing love (*wendo*) for Cirũ, Nduta also alludes to the fact that Cirũ is not restricted from having sexual encounters with men outside the woman-woman marriage relationship. Such liaisons, however, in no way undermine Cirũ's reciprocated love and appreciation for Nduta. In a separate interview, Cirũ, who has been married to Nduta for over 25 years, presents her deep feelings for her marriage partner:

> I know that some people do talk negatively about our marriage. Although honestly I have never caught anybody personally. But I ask myself, "What is it that women who are married to men have that I don't have? Is it land?" I have land. "Is it children?" I have children. I don't have a man, but I have a woman who cares for me. I belong to her and she belongs to me. And I tell you, I don't have to worry about a man telling me what to do. Here, I make all the decisions for myself. Nduta likes women who are able to stand on their own, like herself. I do what I want and the same goes for Nduta. Now I'm so used to being independent, and I like that a lot. I married Nduta because I knew we could live together well. She is a very wonderful woman with a kind heart. (in interview)[7]

While functionalist interpretations perceive African family relationships in terms of the purposes they serve in the functioning of a society, our interviewees highlight the complex and intertwined aspects of relationships that one would expect to find in a discussion of any committed, caring marriage partnership, undermining prevailing notions of the non-emotional African "Other."

One other point in the ideology of the nuclear family that remains strong, even among scholars, but is challenged by the woman-woman marriage data, is the alleged need for a father figure to maintain "functionality" (Cheal 1991). . . . The presence of a father is apparently not so important in many woman-woman marriages. During interviews, some women downplayed the importance of men in their households. Of the eight households in our study, six did not include permanent relationships with male partners. Among these six households, it seemed clear based on our interviews that male involvement with children, beyond procreation, was restricted, even identities of designated male genitors could not be revealed. Cirũ's comments support the view that males are viewed principally as friends and/or sex partners with no claim on children or property. What does she desire from men? Not much, apparently, except perhaps sex, and she can get that when she wants on her own terms:

> I have freedom to have sex with any man that I desire, for pleasure and for conceiving babies. And none of these men can ever settle here at our home or claim the children. They can't. They are not supposed to, and they know that very well. They come and go. (in interview)

Nduta's comments present the same lack of interest in having a man around as the ideal situation, expressing the independence provided by keeping men out of the household:

> We have no interest with a man who wants to stay in our home. We only want the *arume a mahutini* [men met in "the bush," a term for "male genitors"]—meaning those who are met only for temporary

needs. *The meaning for this is for a woman to be independent enough so that she can make her own homestead shrine.* Cirũ sees also that I myself do not keep a man here. What for? To make me miserable? If I kept a man here who will then start asking me for money to buy alcohol, where would I find such money? (in interview, emphasis added)

Another case that downplays the importance of a male presence is that of Mbura, who had been married to a man, though he had died over 40 years ago. She was more recently married to a woman, Nimũ, who subsequently left after a couple of years. Mbura was later married to a woman named Kabura on the last day of this fieldwork. Mbura responds as well to the question of the place of men in the woman-woman marriage household, adding that, to her, men are not trustworthy, though she still appreciates their temporary presence:

> Men, even the good friends, know that they are not welcome here. They are here just for a visit and to leave. Whatever they come here to do, they must leave. They cannot be trusted. That is not good. One is given respect and that's all. (in interview)

Despite the fact that the other two households in the study *did* have men present as partners of one of the *ahikania,* or marriage initiators, the need for a "father figure," an ideal of most heterosexual nuclear families is clearly not a universal reality for all family situations.

Beyond Common Explanations

An overview of the literature on woman-woman marriages in African societies might tempt a reader to make three intertwined cross-cultural generalizations. The first generalization regards access to children. Sudarkasa suggests that the basis for woman-woman marriage, as with African marriages generally, is the desire "to acquire rights over a woman's childbearing capacity" ([1986] 1989, 155). That is, the woman who initiates a marriage seeks access to children that she herself does not have. Rights over childbearing capacity are often linked to a second general theme: that children are desired by such women as a means of transferring property through inheritance. . . . Connected to both general circumstances is the third common assertion that women's "barrenness" is a fundamental factor prompting woman-woman marriages. In fact, one of the most widely held general assumptions, as Burton points out, is that woman-woman marriages must involve women who cannot themselves have children (1979). . . .

Gĩkũyũ women in our relatively small study sample, living within a very proscribed spatial setting, expressed multiple and heterogeneous reasons for marrying women. . . . The women initiating these marriages pursued various objectives: companionship to appease loneliness, to be remembered after death, to have children to increase the vibrancy of the household, to fulfill social obligations in accordance with indigenous spiritual beliefs, and not least to avoid direct domination by male partners in a strongly patriarchal society, including men's control of both the women's behavior and household finances.

Our study does not deny the inability to bear children, inheritance, or lineage as partial explanations for some, or even many, Gĩkũyũ woman-woman marriages. Expressed reasons for marrying women in our study *did* often include the desire for the *muhikania* to have a child to inherit property and/or to perpetuate her family lineage. However, such explanations are never offered as the exclusive reasons, nor are they offered by all women. Such women appear to have much greater latitude in choosing how and why they participate in woman-woman marriages. For example, situations that defy Leakey's account include those in which women who are already married to men (who are still alive) and have their own children then initiate *uhiki,* or marriage, with a woman, as in the above described case of Kũhĩ (married to Wamba).

Mbura's explanation for *kuhikia,* or marrying a woman superficially resembles Leakey's account, since she expresses a desire for children that she herself cannot bear, as indicated in the following statement:

> I married Nimũ because I could never have children myself. I did not even give birth to children who later died, nor did I experience any miscarriage. I remained the way I came out of my mother's womb. And now I'm getting old and there is no way I can sit, think and decide to have a baby because my time is over, unless Ngai's [God's] miracle happens to me [she laughs].[8] I think a lot about how my husband left me and how I can't have a baby. . . . I ask *Ngai wenda mdathima na mutumia ungi* [God, please bless me with another woman]. . . . "Won't you please send that woman here to my home." Who knows, that woman might . . . give me a child. . . . Don't you see when I die I will be satisfied that I have left somebody in that home, who shall continue and revive that home? (in interview)

While she seems to portray a conventional account—marrying a woman to have a child to continue a lineage—Mbura's explanation is more complicated, indicating a desire for children beyond their role as inheritors of land and name. This is not to suggest, however, that lineage is not important in Mbura's decision to marry a woman. But the lineage she seeks to perpetuate is not necessarily her husband's, as Leakey and others would argue. Rather, Mbura is most interested in being remembered herself, as she indicates in the following statement:

> If I were to die even as we speak, that would be the end of it. I would be completely forgotten. No one would ever mention my name. That is simply because there would be no one to carry on my name. Since my husband died he is still remembered by many. But the key reason why he is still remembered is because of me. Someone may pass through here and demand to know "Whose home is that?" Then turn around and ask, "What about the next one?" One would reply, "Did you know so and so? This is his wife's home." Now do you see that the reason he is being remembered is because of me? Because I can be seen. But if I were to die, who will make me be remembered? . . . That is why the idea of marrying another woman came to me. Even now as we speak, if Ngai would bless me with another woman I would appreciate her.

Mbura continues, suggesting that companionship to appease loneliness is another strong motivation for marrying a woman:

> Let me tell you, I'm not the only one or the first one to marry a woman. And certainly, there are many others out there like me. I'm all alone just like that. No husband, no child. Just poor me. No one is here to keep me company or even to ask me "Did you sleep well?," except for occasional visits by some people like those you met here the other day. (in interview)

While Leakey's explanation may partly account for Mbura's case, Nduta's case clearly has emerged under a set of circumstances not fully considered by Leakey. First of all, Nduta's decision is the result of women's collaboration, namely between Nduta and her mother-in-law. Nduta married a man named Ndũngũ with whom she had three sons and a daughter. However, early in their marriage, her husband and their three sons were poisoned to death by some people in her husband's clan who wanted their land. After their deaths, Nduta's mother-in-law advised her to marry a woman as a way of protecting their family and land from male relatives who were trying to take her land, a sign of the tenuous hold that women have over land in Gĩkũyũ society (Mackenzie 1990). Rather than being victimized by men within their family, Nduta's case shows how women collaborate to look out for one another to protect women's interests:

> When a woman is left alone, she should not be frightened, but must be brave. You must make yourself a queen, otherwise, be a coward and everything you stand for will be taken away from you by those who are hungry for what you have. . . . If you were a woman, and you had properties, you will be the first one to be stolen from by the men

who thought they were more important than women. So, she must act. . . . I had a lot of properties and if it were not for *karamu* [the "pen"] that cheated me out of many of them, I would still have a lot.[9] I lost many of them because I was a woman and I had no sons. So, my mother-in-law advised me to marry my own woman because all my people had been finished [i.e., killed] except for my daughter. And that is the piece of advice that I myself chose to follow. So I married her. When I married her [Cirũ], she said "It is better to live with a woman. I'm tired of men." I responded, "Is that so?! I love that." We became good friends and partners and thereafter I gave *ruracio* to her family. (in interview) . . .

Nduta's case is similar to Mackenzie's and Leakey's images of woman-woman marriage presented by those authors in that she had been married to a man who died and she had no sons (they died as well). However, upon marrying a woman after her husband's death, she asserts that she could have passed her land to her daughter, Ceke. Indeed, Ceke was given half of Nduta's land. While Nduta explained that she could have left all of her land to Ceke, she decided against doing so because she did not want to constrain her daughter with the social expectations that "staying at home" entails:

. . . I didn't want my daughter, Ceke, to stay here. I gave her freedom to fly and land wherever she wanted. That is the same freedom that brought me here. So why would I want to hold her here? Women like to go far. They don't like to be held down at their birth home. (in interview)

While the issue of inheritance is important in Nduta's case, related to her difficult struggle as a woman to maintain control over land resources, Nduta adds an important dimension drawn from Gĩkũyũ mythology. This reason becomes clear when we hear Nduta, who is about 90 years old, speak of her dead sons who, she says, visit her in her sleep to thank her for marrying a woman:

Roho wa anake akwa makwrire [the spirits of my dead sons] come to visit me to show appreciation for what I have done for them. One time they came and told me, "Thank you, mother for marrying Cirũ for us. We are very grateful for bringing us dead people back home again. We are grateful indeed. For that we will always be watching over you. Nothing will ever harm you. We will take care of you." And then I would say, "If I didn't marry Cirũ for them, who else would I have married her for?" Then the other day they came to tell me that I have got only five years to live; that I'm going to die soon [she laughs hard]. I said, "Is that so? Thanks a lot and may *Ngai* be praised!" That is fine for me. I need rest. (in interview)

Nduta's sons died long ago, very young, and had not been able to accomplish much in their lives. Some Gĩkũyũ still believe that if someone dies suddenly, his or her life activities can be carried out as if they are still alive so that their opportunities would not be denied. Thus, when their mother married Cirũ, she married her in the same way her sons would have married had they lived. In this sense, even though these sons were already dead, they feel quite at home because of Cirũ's presence.

While Nduta's and Mbura's cases push the limits of Leakey's narrow inheritance-focused account of woman-woman marriage, the case of Nduta's daughter, Ceke, falls largely outside the scope of his scenario. Ceke's decision to marry a woman appears to be heavily influenced by the example set by her mother, who acted as a role model. However, unlike her mother she was at the same time still married and living with her husband, Ngigĩ, together with her daughter, Wahu, along with Wahu's six children. Having grown attached to Wahu's children, Ceke was insecure about whether Wahu would move away with them, leaving Ceke in a household without children. Ceke's marriage to a woman (Ngware) was thus viewed as a way Ceke could have more children. Ceke's intention was that her wife, Ngware, would have children with her husband, Ngigĩ. After having a child, however, Ngware left the household. Ceke and Wahu (her daughter) then reached an agreement that the children would be welcome to remain with Ceke even if Wahu decides to leave:

> Although my daughter was living with me at the time, and had all these children that you see here, I did not know what to expect from her. I did not know whether one day I will wake up and find her gone with all her children that I personally have raised and who actually call me *maitu* [mother], or whether she had already made up her mind that she will never leave. I made that move of wanting to find out when my wife [Ngware] left us. After that, my husband and I made an agreement with Wahu that she will live with us permanently and that if she will ever feel like leaving, her children that we have raised as our own will be welcome to remain with us where they are already guaranteed good care as well as land settlement when they grow up. In any case, this is her land too, you know. Since we have got no other children, everything we have belongs to her and her children and to my other son borne by my wife before she left. (in interview)

While this example supports the general claim that women marry women to acquire rights over childbearing capacity (Sudarkasa [1986] 1989), Ceke's decision is not linked to property inheritance, "barrenness," or widowhood: the three essential criteria for a Gĩkũyũ woman-woman marriage, according to Leakey. Like Mbura, Ceke's strong desire for children was an important factor in her decision. The option of woman-woman marriage as a means to fulfill this desire was immediately apparent, given the influence and example of her mother, Nduta.

Finally, we have already alluded to the more overtly political motivations for marrying women expressed by some of our interviewees. The relative freedom from male control, which appears to be built into Gĩkũyũ woman-woman marriages, is expressed most forcefully by Cirũ and Nduta in previous quotations. Recall, for example, Nduta's conversation with her then wife-to-be, Cirũ, who commented, "I'm tired of men," to which Nduta responded, "I love that." And Nduta's comment about why she doesn't live with a man, stating "What for? To make me miserable?" Recall also these women's comments regarding the sexual freedom they find in these relationships. And finally, recall the opening quote in which Cirũ states that her woman-woman marriage allows her to avoid having "a man telling me what to do."

These examples demonstrate that flexibility, heterogeneity, and ambiguity appear as guiding principles in explaining such marriages, rather than being governed by somewhat rigid social rules, as the literature so often implies. However, contributions to the woman-woman marriage literature have continually, since the early-twentieth century, presented these relationships in functionalist terms. Cheal suggests that functionalist explanations continue to be perceived as having a "subterranean" influence on the study of families, describing such relationships in terms of "the ways in which they meet society's needs for the continuous replacement of its members" (1991, 4). Our alternative has been to present the institution of woman-woman marriage, at least in the Gĩkũyũ context, as a flexible option available to women within which they may pursue any number of interests: political, social, economic, and personal.

What's in a Name? Rethinking the "Female Husband"

Another area of concern for us in the literature is the unquestioned use of the term "female husband," the general term used to describe women who initiate woman-woman marriages. . . . Not surprisingly, the major debate regarding the term "female husband" is over the male social traits often attributed to such women. Some have criticized the emphasis placed on gendered assumptions regarding sex-roles. For example, Krige suggests that one cannot assume that female husbands generally are taking on male roles (1974). Rather, one must carefully study sex-roles in particular societies. For the Lovedu, Krige points out that numerous roles involve both males and females. Oyewumi, writing about the Yoruba, argues that local terms for both "husband" and "wife" are not gender-specific since both males and females can be husbands or wives (1994). As a result, as Burton (1979, 69) contends, the assumption that "husband" and "male" are automatically connected "confounds roles with people" since "husband" is a role that can be carried out by women as well as by men. Amadiume (1987), Burton (1979), Krige (1974), Oyewumi (1994), and Sudarkasa (1986) all suggest that in many societies, "masculinity" and "femininity" are not as clearly defined categories as they are in the West; presuming that "husband" automatically connotes "male" and that "wife" connotes "female" imposes western sex-role presumptions on other societies, ignoring local ambiguity regarding these roles (Sudarkasa [1986] 1989).

While our study supports views that women initiating marriages are not characterized as "male," we question the continued use of the term "female husband" to describe such women. Burton (1979), Krige (1974), and Sudarkasa (1986), while criticizing those who confuse social roles with genders, implicitly suggest that the term "female husband" is adequate and that the only task is to transform its connotative meaning. We argue that the term "female husband" should be reconsidered on the grounds that the male connotation of "husband" cannot be so easily disposed of; just as the term "wife" conjures an association with "female," so does "husband" with "male." Especially in contexts where gender roles are ambiguous, this implicit association will easily mislead readers to impose western presumptions upon woman-woman marriages. Thus, in our view, efforts to theoretically disassociate gender from such role-centered terms—like "husband" and "wife" in this instance—imposed originally by western researchers in colonial contexts, will in a practical way continue to impose a male/female dichotomy. . . .

We acknowledge that there is nothing essential about the term "husband" that necessitates domination and control. But we also acknowledge, as does Oyewumi, that historically the term "husband" in most western contexts is normally associated with the role of "breadwinner," "decision-maker," and "head of household" (1994). We feel that the use of the term "female husband" serves to mask the relatively egalitarian woman-woman marriage relationships we encountered. . . .

The relative absence of domination, for example, is evident in the terms the women used to describe one another. The women interviewed never used the Gĩkũyũ term for "husband" (*muthuri*) to describe their partners. Instead, they consistently referred to each other using the terms *mutumia wakwa* and *muka wakwa,* which when used by these women translates as "co-wife," or *muiru wakwa,* which translates as "partner in marriage," indicating the mutual respect and relative equality between them. While most women in our study who initiate the marriages tended to be women with social influence and/or relatively greater material wealth, within the marriages both women interpreted their relationship as semiotically and materially equal.

Furthermore, women in our study rejected any male-association with their position of initiator of the marriage. None of the women interviewed indicated that they aspired to be like "males." As a tentative alternative to "female husband" we have been using the phrase "marriage initiator" to describe women in that position. However, we acknowledge that such description can be problematic, especially if it is used to focus more attention on the "initiator" at the expense of the agency of the one "initiated" into the marriage. We also acknowledge that descriptions of such concepts will differ from one culture to another.

Notes

1. Other terms include "woman-marriage" and "woman-to-woman marriage."

2. The names of interviewees have been changed to protect their identities.

3. Note that multiple Gĩkũyũ terms seem to describe the same concept. Choice of term depends upon the context in which the concept is employed. For example, while "marriage initiator" in one context is expressed as muhikania, the plural form of the concept is *ahikania.*

4. It is important to acknowledge that in most cases, polygamous marriages among the Gĩkũyũ

come as a result of negotiation between the first wives and husbands.

5. While the sexuality of the women involved in woman-woman marriages is clearly one of the most interesting unresolved issues on the topic, the Human Subjects Review Board reviewing the research proposal decided that the topic was too sensitive, and therefore declared such questions off limits.

6. Interviews were conducted in the Gĩkũyũ language and were translated by the primary author.

7. *Miario miuru,* or "negative talk," that is mentioned by Cirũ in this quotation points to fundamental changes that have occurred in Gĩkũyũ society over the course of the twentieth century with colonialist religious and educational training. These changes are reflected in complex local attitudes toward indigenous practices and are discussed briefly in the last section of this paper.

8. *Ngai* commonly translates as "God," although the Gĩkũyũ term carries no gendered connotation.

9. *Karamu,* or "the pen," refers to the use of title deeds (by those who could read and write—mainly men) that conferred private ownership of property since the 1960s. This private ownership was started under colonial rule and undermined (though it did not eliminate completely) more customary land tenure rules (Mackenzie 1990).

References

Ainsworth, Mary D. Salter. 1967. *Infancy in Uganda: Infant Care and the Growth of Love.* Baltimore, MD: Johns Hopkins University.

Albert, Ethel M. 1971. "Women of Burundi: A Study of Social Values." In *Women of Tropical Africa,* ed. D. Paulme. Los Angeles: University of California Press.

Amadiume, Ifi. 1987. *Male Daughters and Female Husbands: Gender and Sex in an African Society.* London: Zed Press.

Beeson, R. W. 1990. "The Clinical Distribution of Family Systems." *International Journal of Contemporary Sociology* 27:89–127.

Burton, Clare. 1979. "Woman-Marriage in Africa: A Critical Study for Sex-Role Theory?" *Australian and New Zealand Journal of Sociology 15*(2):65–71.

Carrier, Joe. 1980. "Some Comments on Woman-Woman Marriage in Africa." *Anthropological Research Group on Homosexuality Newsletter* 2:2–4.

Cheal, David. 1991. *Family and the State of Theory.* Toronto, Canada: University of Toronto Press.

Davis, D. L., and R. G. Whitten. 1987. "The Cross-Cultural Study of Human Sexuality." *Annual Review of Anthropology* 16:69–98.

Haraway, Donna. 1997. *Modest Witness@Second Millennium. FemaleMan Meets OncoMouse: Feminism and Technoscience.* New York: Routledge.

Herskovits, Melville D., 1937. "A Note on Woman Marriage in Dahomey." *Africa* 2(3): 335–41.

Kilbride, Philip Levey, and Janet Capriotti Kilbride. 1990. *Changing Family Life in East Africa: Women and Children at Risk.* University Park: The Pennsylvania State University Press.

Krige, Eileen Jensen. 1974. "Woman-Marriage, With Special Reference to the Lovedu: Its Significance for the Definition of Marriage." *Africa 44*:11–37.

Le Vine, R. 1970. "Personality and Change." In *The African Experience, Vol 1,* eds. J. N. Paden and E. W. Soja. Evanston, IL: Northwestern University Press.

Mackenzie, Fiona. 1990. "Gender and Land Rights in Murang'a District, Kenya." *The Journal of Peasant Studies* 17:609–43.

Muriuki, Godfrey. 1974. *A History of the Kikuyu 1500–1900.* New York: Oxford University Press.

Obbo, Christine. 1976. "Dominant Male Ideology and Female Options: Three East African Case Studies." *Africa* 46(4):371–89.

Oboler, Regina Smith. 1980. "Is the Female Husband a Man? Woman/Woman Marriage Among the Nandi of Kenya." *Ethnology* 19:69–88.

O'Brien, Denise. 1977. "Female Husbands in Southern Bantu Societies." In *Sexual Stratification: A Cross-Cultural View,* ed. A. Schlegel, 109–26. New York: Columbia University Press.

Okonjo, Kamene. 1992. "Aspects of Continuity and Change in Mate Selection Among the Igbo West of the Niger River." *Journal of Comparative Family Studies* 23:339–60.

Oyewumi, Oyeronke. 1994. "Inventing Gender: Questioning Gender in Precolonial Yorubaland." In *Problems in African History: The Precolonial*

Centuries, eds. R. Collins et al., 244–50. New York: Marcus Wiener Publishing, Inc.

Skolnick, Arlene S., and Jerome H. Skolnick. 1989. "Introduction: Family in Transition." In *Families in Transition,* 6th ed., eds. A. S. Skolnick and J. H. Skolnick, 1–18. Boston: Scott, Foresman and Company.

Smith, Dorothy E. 1987. *The Everyday World as Problematic: A Feminist Sociology.* Boston: Northeastern University Press.

Sudarkasa, Niara. (1986) 1989. "'The Status of Women' in Indigenous African Societies." In *Feminist Frontiers II: Rethinking Sex, Gender, and Society,* eds. L. Richardson and V. Taylor, 152–58. New York: McGraw-Hill, Inc.

Talbot, Percy A. (1926) 1969. *The Peoples of Southern Nigeria.* London: Cass.

Introduction to Reading 13

This article offers a wonderful analysis of the relational nature of masculinities and femininities in the context of the changing lives of working-class married people in a suburban community in Mumbai, India. Sociologist Annie George discusses the ways in which women's "discourses" on men's actions provoke men to assume a new honorable masculinity that is softer and more accepting of women's autonomy.

1. Describe the research approach employed by Annie George for this study.
2. Discuss the role of sexual self-control in the idea of honorable masculinity.
3. How did women's "discourses" contribute to the reshaping of definitions of (honorable) masculinity?

Reinventing Honorable Masculinity

Discourses from a Working-Class Indian Community

Annie George

Maintenance of personal and family honor and avoidance of shame are central concerns of Indian communities. Honor in Indian contexts is typically viewed as engendered and embodied. Yet notions of honor and gendered identities—masculinities and femininities—have ambiguous and contradictory meanings that shift in relation to contexts and histories over time. In this article, I provide an interpretive account of honor and masculinities that operates among a group of working-class men and women of Mumbai, India,[1] through analyses of two related discourses: men's accounts of sexual self-control and women's accounts of "understanding" men.

For a section of working-class men of Mumbai, sexual self-control forms a central constituent of a gendered, embodied notion of honorable masculinity. For working-class women, however, economic provisioning and absence of physical and sexual violence are critical markers of honorable masculinity. The linkage of these conditions is surprising because Indian discourses on honor hold that women's bodies and actions, and not those of men, are the primary markers of personal and family honor (Dube 1986; Gold 1994; Jefferey, Jeffrey, and Lyon 1989; Kumar 1994; Ram 1991; Sharma and Vanjare 1993).

Men's claims about sexual control within marriage—when a popular interpretation of marriage is men's social license to have relatively unregulated sexual access, the so-called male right in marriage—destabilize received ideas of both honor and masculinities in Indian literature and are my starting point to explore emerging forms of honorable masculinities in particular Indian communities. I make two related arguments about emerging masculinities and honor. First, men's actions and women's discourses on men's actions critically shape men's honor. Second, the masculinities men seek must be considered contemporaneously with the femininities[2] that are emerging around them. Traditional avenues for men to gain honor were that of providing adequately for their families and exercising control over their wives and children. Men controlled their wives' sexuality, movements outside the house, and access to and control over productive resources. However, consequent to social and economic changes in their everyday lives and increased participation of women in economic activities outside the house, men and women are redefining hegemonic ideas about gendered identities and creating new, even contradictory, discourses on gendered honorability.

Study Locale and Methods

This study was conducted for a period of twelve months from November 1995 in Kaamgar Nagar (a pseudonym), a community of approximately three hundred thousand people in suburban Mumbai. Kaamgar Nagar comprised a government-recognized shantytown surrounded by people living "illegally" on land that was not designated for slum settlements. Some residents were born in Kaamgar Nagar; others migrated to the city from all over India. We gained access to potential participants through nongovernment organizations' that worked in Kaamgar Nagar introducing us to the people of the area.[3] We contacted thirty-seven currently married men, of whom twenty-three agreed to participate in the study.[4] All men had middle school education, were married more than five years, and had at least one live child.

Although all men participants were employed, they had a hard time supporting their families, which typically had three adults and three children. Ten men were self-employed in skilled and unskilled trades as plumber, electrician, flower seller, newspaper vendor, and so on. Seven had permanent jobs in the private and public sector. Of the remaining, three were contract labor, and the others were daily wage earners. The range of monthly income in this group varied from one thousand to five thousand rupees.[5] Only one man owned the 10 × 12 feet tenement that was their home.

We also conducted repeated in-depth interviews with sixty-five married women residents of Kaamgar Nagar, similar in social class to the men participants. We gathered information from separate focus-group discussions with married men and women and observations and interviews with other residents and regulars of the study site. But data were primarily obtained through repeated in-depth interviews because this method has been used successfully to elicit sensitive information (Balmer et al. 1995; Hammersley and Atkinson 1995; Helitzer-Allen, Makhambera, and Wangel 1994). Each participant was interviewed two to three times, and interviews lasted between half an hour and two hours. They were interviewed at their homes; in local teashops; their work sites, if self-employed and eventually, because of the lack of privacy and frequent interruptions, in a local school classroom after school hours. Areas of discussion included sexual experiences, negotiation and decision making within marriage, sexual experiences outside marriage, perceptions of sexual risk, and the use of risk-reduction practices. Discussions, conducted in Hindi and Marathi, were audiotaped, transcribed, translated into English, sorted, and coded by categories that were generated through the data themselves. An in-depth content analysis was then conducted on the data in relation to the category of *sexual control* to provide insights into men's and women's ideas of marital sex and ideal gendered masculinity.

Sexual Control in Marriage

Men used the English word *control* to refer to three distinct sets of activities: periodic sexual abstinence with one's wife, avoidance of sexual relations outside marriage, and the practice of withdrawal as a form of contraception.[6] In this article, I focus on the first two uses of this concept.

Men's narratives indicated that they sought to observe normative expectations for sex within marriage. These rules of action related to frequency of sex at various stages in life, times and places for sex, and permissible sex acts and partners. During the early years of marriage, couples were expected to have frequent sex for physical pleasure and procreation. After fathering a number of children, however, men were expected to focus attentions as well on their families and the world around and gradually to decrease their interest in sex. Men considered their sexuality to be unruly by nature and in need of control. They believed that, in contrast to women's sexual interests and needs, which declined sharply after having several children, men's sexual urges did not decrease even after many years of marriage and being sexually active. Thus, most men said they were unable to practice the expectation of decreasing sexual activity even when they had achieved their desired family size. They desired sex more often than their wives did, leading them to conclude that their desire was "in excess." Men reported practicing sexual control by not having *jyada* (excessive) relations with their wives. Sambhaji,[7] thirty-three-years-old and married thirteen years, discussed this dynamic of sexual control as follows:

> After two children usually a woman's urge decreases whereas in the case of a man it remains the same even till fifty years of age. Even if I try to have sexual intercourse with my wife she does not get involved in it. I want it every day whereas she does not. At this age what is the use of a strong sex urge? We have three children. In young age we were too enthusiastic about sex, we enjoyed as much as we could. Now we have to "control."

Here, Sambhaji's notions about appropriate activities for a man of his age and marital status guided his practice of sexual self-control. Men who reported being in a position similar to Sambhaji's made similar remarks about having to struggle with their sexuality. These views suggest that men considered honor to be embodied through self-control. Married men had social permission to have sexual access to their wives; yet some married men reportedly chose self-control.

Women's narratives corroborated these views, and most women appreciated monogamous, understanding[8] men who moderated their demands for marital sex. Gauri, married twelve years, described her husband as follows:

> He is understanding *(samajdhar).* He does not want [sex] everyday, just once in fifteen days, once in a month. . . . If I tell my husband that today I don't have the mood he says, "That's okay, let's forget it." People in our family and even other people say your husband is very understanding.

Here, the woman, her family, and people in her daily life evaluate as honorable her husband's refusal to engage in coercive marital sex. In contrast, dishonorable husbands were reported to coerce wives to have sexual relations. Many women we interviewed reported regular experience of coercive sex, and some mentioned that their husbands also engaged in sex outside marriage.[9] One such woman was Zeenat, who, when seeking treatment from a local doctor for symptoms of a sexually transmitted infection she contracted from her husband, noted that the doctor evaluated her husband negatively. When the doctor learned that Zeenat's husband was a taxi driver she said, "[Sexually transmitted infections] are a specialty among wives of drivers. Drivers are greedy for sex. One woman is just not enough for them. They go to outside women and come to their wives for sex without a care that this [sexual ill health] happens."

Through her narrative Zeenat sought to convey that it was not only she but also others such as medical doctors who perceived her husband as dishonorable because of his actions: first of having sex outside marriage, and then of coercing his wife to have sexual relations and face the consequences of his lack of sexual control.

Honor at Home, Dishonor Outside

Sex outside marriage was the foil against which married men spoke of marital sex, sexual self-control, and, by extension, honorable masculinity. Since marriage is the only culturally sanctioned means by which a person could have sexual relations, men talked about their wives as being "women of the house" and "proper women." In contrast, they described sex outside marriage as "wrong, bad work" *(galat kaam)* that men did with "wrong, bad women" *(galat aurat).* These cultural rules on marital sexuality contextualize Khalil's answer to a question about whether he ever had a sexual relationship outside marriage. He said, "When we have everything in the house, why should we walk on the wrong path?" This quotation, which is typical of other men's views, foregrounds the notion that sex outside marriage with an "outside woman" outside the house was a loss of sexual control. Men knew that sexual illnesses could be prevented through monogamy,[10] and sexual control was a means of preventing sexual illness—the bodily marker of dishonor and the absence of self-control.[11]

A study (Verma, Rangaiyan, and Narkhede n.d.) conducted in Kaamgar Nagar of practitioners providing treatment for male sexual problems supports this interpretation. Verma et al. found that medical practitioners causally linked sexual illness to absence of sexual control. The practitioners attributed perceived causes of various male sexual problems like boils and sores in the genital area, white discharge, thinning of semen, wet dreams and lack of erection to an excess of unsatisfied sexual desire, masturbation, and sexual intercourse. Oral and anal sex and sex with "cheap" women are also considered important reasons for *garmi,* a serious illness. Participants in our study shared these beliefs. Thus, Vinayak's answer that his "nature was clean" in response to a question on whether he thought he was susceptible to a sexually transmitted disease shows the connections he made between sexual control and its trade-offs, such as the avoidance of stigmatizing illness and the consolidation of personal honor.

Our discussions with men did not specifically include the type of sexual act they practiced with their wives. We learned about cultural expectations of permissible sex acts when men talked about its transgression. Many men described their current married sex life as "boring." Only Bhimrao said that he had extramarital sex because he was "fed up of doing the same thing again and again" with his wife. He went to sex workers after marriage because they did things that "one could not do with the wife." From local constructions of honorable masculinity, this man was doubly out of control in this instance: he had sex outside marriage, and engaged in sexual acts that were outside the realm of marital sex.

Vasant, who had watched pornographic films after marriage, spoke elliptically about sexual acts permitted in marriage. He opined that any man would feel like doing some of the things he saw being done in these films, but a man "will not do such a thing with his own wife, maybe with an outside woman." Because he considered his desire as reasonable, he claimed that control was even harder. Few Indian studies on sexual behavior list the type of sex acts performed by sex workers. One study of Calcutta sex workers (Jana and Chakroborty 1994) reported that while the "usual mode" of sexual intercourse was vaginal, the majority of sex workers also practiced oral sex, and one-quarter practiced group sex.[12] Sex outside marriage could provide men with alternative sexual practices, which they considered more interesting than marital sex. Yet this analysis of men's narratives suggests that men sought honor through claims of resisting nonnormative sex.

Sex Outside Marriage: Loss of Control and Honor

Male respondents frequently mentioned controlling one's desire to have sex with a woman other than one's wife as a form of sexual control. I did not get the impression that all male respondents, if given social sanction, would have sex outside marriage. Rather, men spoke of the nonavailability of extramarital sex as a kind of loss. Societal double standards ensured that men who had extramarital sexual relations were censured less than their female partners were; yet the existence of censures helped men maintain control. Several men spoke of wanting to avoid the dishonor of having wives and kin discover their extramarital relationships. Gopal said that men would not admit having sexual relations outside marriage because of "shame."

> Articulating a widespread local belief that regular provisioning was a fundamental mark of a responsible man, Ramesh said he did not want to "lower his head with the shame of being known as one who did not provide for his family." Some men juxtaposed sex outside marriage and economic provisioning. Men reportedly avoided sex outside marriage, even when sex at home was boring, because they were uneasy paying for transient pleasures with scarce money that could be spent on the family. Even a one-time extramarital episode was expensive, and men weighed the benefit against other ways the family could use the money.

. . . (M)en viewed the desire for sex outside marriage as their unreasonable craving for sexual pleasure which, putatively, was already available to them through marriage. Also related was the notion of the issues to which responsible men should give primacy, namely, family provision and stability and sexual control.

Competing Masculinities, Competing Claims

Control was seen as a necessary practice to further a family's social advancement. Men who practiced sexual control made statements similar to the following, which was Bashir's rationale for sexual control:

> One does feel sexual desire, but we have to keep control over ourselves because we want to move ahead in life. So, even if we have the desire in the mind, the body should avoid it.

Honorable men were those who, from the perspective of their families, neighbors, and communities, had moved ahead in their lives. In the congested living environment of slum characterized by one-room houses with porous walls, the boundaries between private and public life were slippery. Family life spilled onto the streets; neighbors saw, heard, and talked about the goings-on in each other's houses. A man's wife and the community around empathized with men who tried to advance in life through the performance of honorable masculinity and were critical of men who did not. Men who did not adhere to local notions of honorable masculinity, particularly economic provisioning, were labeled "weak" or a "mouth to feed." Mahmood expressed this belief as follows: "My wife doesn't work. When I am working, where is the need for her to work?" In reality, women in Kaamgar Nagar worked to support their families and reportedly persuaded husbands to put aside concerns about male honor in favor of economic stability through women's wage work. Lakshmi's narrative elaborates this point.

> The children are growing up and expenses are increasing. I wanted to work, but he would say,

> "No, don't work. The woman of the house should not go out." Now that I am on this job, he feels ashamed. He says that in our family, no woman before me went out to earn money. People talk. They say, "Is this man so weak that he is sending his wife to work?" That's why he told me that if I am going to work, it is on my own responsibility. Nobody should accuse him that he sent his wife to work, that he is weak. There should be no talk like this.

Men's concerns about personal honor constrained customary ideas about gendered divisions of labor and marital power relations. Like Lakshmi, two of every three women we interviewed worked to complement the man's earnings, and one in three women was the main wage earner of the family. In contrast, all the men we interviewed claimed to be regular providers with stay-at-home wives, have no addictions, and control their wives without excessive use of force.[13] Ten male respondents went further and claimed to practice sexual control. In sharp contrast to men's claims–and some women's claims for their men–we learned of dishonorable masculinities from female respondents. In such narratives, men were portrayed as being unreliable or nonproviders who were addicted to gambling or alcohol, having extramarital sexual relations, and being "excessively" physically violent toward the wife and or children.[14] I present segments of three women's narratives, each of which highlights women's perspectives on (dis)honor and masculinity.

> Farida highlighted the reversal of traditional gendered division of labor when she described her husband as follows: "Today he is not feeding me. I am feeding him." Farida's husband worked irregularly, although he was a skilled mechanic who, she claimed, could find work easily. She worked as a construction laborer to make ends meet. As their financial situation deteriorated, Farida removed her two older children from school and put them to work with her. Additionally, she herself started working night shifts at an export factory.

However, after some time, he stopped working and at the time we interviewed her, Farida was working her day job and night shifts at the export factory, and she reported that she and her husband were . . . objects of neighborhood talk. This vignette reveals the engendered nature of honorability where the man's "dishonorable" actions were as severely censured by his neighbors as the necessary wage-earning, yet allegedly honor-damaging, actions of his wife.

Girija recounted the story of her neighbors' reactions to her husband's extramarital activities.

> My husband was going behind one woman in our area. That woman's brother saw him with me. He asked, "Who is she, what is she to you?" My husband said, "My wife," Then that woman's brother said, "Despite having a wife, why are you behind my sister?" He slapped my husband a few times. Now [that man] meets me in the bazaar sometimes and asks me whether my husband behaves properly.

This vignette indicates that neighbors differentiated among actions of the individuals in a marital dyad and awarded respect accordingly. It also indicates that cultural expectations and public surveillance constrain men to control their actions. Similar to the case of Farida above and that of Sheila to follow, Girija's narrative shows the workings of public evaluation of individual actions of men and women to accord personal and family honor.

Sheila noted that word had spread in their community that her husband's family was "not good." She observed,

> Marriages do not take place in their family. They are unreliable providers, alcoholics, wife beaters. Only [her husband] is married. His older brother is single. The one older than him, he married a woman but would not work and feed the family. She left him and married somebody else. And there is yet another one older than him. He tricked a girl to marry him saying that things are good at our place. But when the girl came here and saw the situation, she left within three months. When he went to bring her back, her parents beat him and sent him away. Now, marriages do not take place in their family.

This narrative shows that, contrary to received ideas about women's actions shaping family honor, it was the cumulative effect of actions of individual men of a family and public talk about these actions that damaged their personal and family honor.

Women's narratives, like the ones just presented, indicate that individual (in)actions were consequential for personal and family honor and that men lost honor through public knowledge of and women's discourses on men's actions. Instead of being tied to women's actions alone, personal and family honor in this framework are also defined by men's actions. Women's discourses suggest that urban residents differentiate between family and individual honor and the personal honor of individuals in marital dyads. This explains why women can claim personal honor while being married to men who dishonor families through their actions.

Emerging Honorable Masculinities

The question still remains, however, what do men gain through claims to practice sexual control, a private action, when they can gain honor through public actions such as providing for their families and being moderate in other activities? This analysis of honorable and dishonorable masculinities in a working-class community of Mumbai suggests that masculinities and men's personal honor are co-constituted through men's own actions, private and public,[15] sexual and otherwise, and women's discourses about men's actions. Contemporaneously, emerging masculinities are shaped by, and in turn shape, emerging femininities.

Mature, responsible men in the *grihastha*—householder—stage were expected to maintain a balance among public actions of pursuing economic advancement and social and religious commitments and private actions of pursuing sexual pleasure. Working-class men of Mumbai found it hard to provide regularly and/or adequately for their families; often, they resorted to borrowing large sums of money to meet family needs. When men did not make regular economic provisions for the family, their wives had to seek wage work. Women became de facto heads of families and acted in ways that could be perceived by their husbands and communities as being out of the bounds of male control, although women and families also acknowledged the circumstances and the necessity of women taking on these responsibilities. Women worked outside the house for wages and made independent decisions relating to family concerns, children, and expenditure of money; some assertive women resisted coercive marital sex. Thus, autonomous wives challenged men's efforts to forge traditional, honorable masculinities.

In situations of fluid gender roles and authority, men sought honor to command influence and renegotiate power relations between themselves and their wives. . . . Working-class men in contemporary Mumbai sought to reestablish a dominant position in their families through the accrual of personal honor, at least in the sexual arena, which then would legitimate their claim to power and position as head of family. Alter's (1994, 55) point. . . . that men in positions of power, when confronted by what they perceive to be an almost apocalyptic transformation of society, are forced to see the extreme contingency and the fictional basis of their gendered position. As a result, they critically deconstruct elements of that ideology in search of a more primary, natural truth about themselves and claim to practice celibacy to embody their gender. By locating a "truth" of their gendered identities in sexual self-control, some contemporary working-class Mumbai men translate an ideology of domination into an insidious form of biological determinism and as a means to claim moral piety and personal honor.

Women and men worked to create positive evaluations for their responses to changing social realities such as autonomous wives and men who "allowed" the existence of such wives. Discursively, men constructed themselves as honorable by presenting themselves as moderate and in control. The concepts of *moderation* and *excess* show the linkages between sexual control and honor. The shared belief of the local people was that men who practiced moderation in observable activities such as earning money, consuming alcohol, and controlling their wives were also assumed to be moderate in their private actions. Men's primary gain through sexual control is the honor of being perceived by their wives, families, and neighbors as honorable men. Some men reportedly sought honor through the practice of moderation in their sexual and familial lives, while others indulged in excesses and continued to be unreliable providers. Therefore, men who claimed to practice sexual control, despite their difficult economic situation, increasingly autonomous wives, and social sanction to enjoy conjugal rights, could claim personal honor. This is a source of their view of themselves as honorable, in sharp contrast to men in similar circumstances who lost control of themselves and resorted to "unmanly" behavior.

Women, in contrast to men, used the concept of *understanding* husbands to seek positive valuation and honor for an emerging form of masculinity that allowed women to work for wages, assume greater responsibility and decision-making power in the home, and assume greater control of marital sexual experiences and that did not readily resort to violence to exert authority over women. Normalizing this new masculinity serves to normalize a newly emerging autonomous femininity. Formerly, gendered identities and honor were based on men's ideas of women's actions, thereby allowing men to define idealized gendered identities and relations of domination and resistance between men and women. By reversing this pattern and constructing honor also through women's ideas of men's actions, urban working-class women are recasting and stretching the contours of honorable masculinities and femininities and gender relations. When women discursively construct male honor as related to men's actions in the area of provisioning, (absence of) violence against women, and sexual control, they provoke men to assume responsibility for their personal and family honor. . . . Women's discourses on men's actions also facilitate the emergence and acceptability of a "softer" masculinity in contrast to other masculinities that are characterized by consumption (Anandhi, Jeyaranjan, and Krishnan 2002; Osella and Osella 2000), violence (Butalia 2000; Menon and Bhasin 1998) or sexual dominance of women (George 1998; Khan et al. 1996). Through discourse, women seek to normalize an assertive, autonomous femininity that exists alongside this emerging masculinity.

Notes

1. Formerly known as Bombay, Mumbai lies on India's west coast and is India's commercial capital.

2. And other emerging social realities. For an account of emerging Dalit (formerly untouchable) masculinity in the context of changing caste relations in a rural Indian context, see Anandhi, Jeyaranjan, and Krishnan (2002).

3. The author conceptualized the study of which these data form one part. She interviewed female respondents and participated in focus-group discussions and observations. A male researcher assistant, under the supervision of the author, conducted interviews with male respondents.

4. The men who refused to participate probably differed from those who did; they were, however, demographically similar to the participants. The participants came from Hindu, Muslim, and Buddhist communities, mainly from Maharashtra State (of which Mumbai is the capital), with a few from other Indian states. However, there were no major differences along community lines with respect to data described in this article.

5. In 1995, the exchange rate for the U.S. dollar was approximately thirty-eight rupees for one dollar.

6. It is interesting to note that none of the men mentioned masturbation in their discussion of sexual control. However, for a discussion of masturbation as a sexual illness that resulted from "excess" and, hence, implicitly, from lack of self-control, see Verma, Rangaiyan, and Narkhede (n.d.). I thank Adele Clarke for bringing this point to my attention.

7. All names used are pseudonyms.

8. For a discussion of urban middle-class Indian women's constructions of "understanding" husbands, see Puri (1999, 141–43).

9. Thirty-one of sixty-five female respondents reported the experience of sexual coercion by husbands, and fifteen reported that husbands had extramarital relationships.

10. Condom use was rare among this group of men. Less than 1 percent of them reported regular use of condoms.

11. The deep sense of shame involved when a man contracts a sexually transmitted infection has been reported by HIV-positive men; see Bharat (1997).

12. Recent unpublished studies of male sexual activities from various rural and urban locations in India indicate that men who practice extramarital sex reported a variety of sexual acts: vaginal, oral, and anal sex practiced individually with a woman partner and also in groups of men with one or more women (Bert Pelto, pers. comm., March 9, 2004).

13. This is no doubt an artifact of self-selection, as we only formally interviewed those men and women who volunteered to do so.

14. Of the sixty-five we interviewed formally, thirty-four claimed they could not rely on their husbands for regular economic support, thirty-two reported experiences of physical violence from the husband, thirty-five reported to have husbands who abused alcohol, fifteen reported that husbands had extramarital relations and/or "second wives," and thirty-one reported to experience of sexual coercion by the husband. These statistics also are an artifact of self-selection. It appears that the majority of men and women who participated considered themselves to embody local ideas of honorable gender identity.

15. Indeed, characterization of actions as private or public are erroneous, as such distinctions blur in real-life situations.

References

Alter, J. S. 1994. Celibacy, sexuality, and the transformation of gender into nationalism in north India. *Journal of Asian Studies* 53:45–66.

Anandhi, S., J. Jeyaranjan, and R. Krishnan. 2002. Work, caste and competing masculinities: Notes from a Tamil village. *Economic and Political Weekly* 26 (October): 1–16.

Balmer, D. H., E. Gikundi, M. Kanyotu, and R. Waithaka. 1995. The negotiating strategies determining coitus in stable heterosexual relationships. *Health Transition Review* 5:85–95.

Bharat. S. 1997. Household and community responses to HIV/AIDS: Executive summary of a study in Mumbai. *Indian Journal of Social Work* 58:90–98.

Butalia, U. 2000. *The other side of silence: Voices from the partition of India.* Durham, NC: Duke University Press.

Dube, L. 1986. On the construction of gender. Hindu girls in patriarchal India. *Economic and Political Weekly* 30 (April): WS11–19.

Edwards, J. 1983. Semen anxiety in South Asian cultures: Cultural and transcultural significance. *Medical Anthropology Quarterly* 7:51–63.

George, A. 1998. Differential perspectives of men and women in Mumbai, India, on sexual relations and negotiations within marriage. *Reproductive Health Matters* 6:87–96.

Gold, A. G. 1994. Gender, violence and power: Rajasthani stories of shakti. In *Women as subjects: South Asian histories,* edited by N. Kumar, 68–90. New Delhi, India: Stree.

Hammersley, M., and P. Atkinson. 1995. *Ethnography principles in practice.* New York: Routledge.

Helitzer-Allen, D., M. Makhambera, and A. M Wangel. 1994. Obtaining sensitive information: The need for more than focus groups. *Reproductive Health Matters* 3:75–81

Jana, S., and A. K. Chakroborty. 1994. Community based survey of STD/HIV infection among commercial sex workers in Calcutta, India. Part II. Sexual behavior, knowledge and attitude towards STD. *Journal of Communicable Diseases* 26:168–71.

Jefferey, P., R. Jeffrey, and A. Lyon. 1989. *Labour pains and labour powers: Women and childbearing in India.* London: Zed Books.

Khan, M. E., R. Townsend, R. Sinha, and S. Lakhanpal. 1996. Sexual violence within marriage. *Seminar,* no. 447:32–35.

Kumar, N. 1994. *Women as subjects: South Asian histories.* New Delhi, India: Stree.

Menon, R., and K. Bhasin. 1998. *Borders & boundaries: Women in India's partition.* New Brunswick, NJ: Rutgers University Press.

Osella, F., and C. Osella. 2000. Migration, money and masculinity in Kerala. *Journal of the Royal Anthropological Institute* 6: 117–33.

Puri, J. 1999. *Women, body, desire in post-colonial India: Narratives of gender and sexuality.* New York: Routledge.

Ram, K. 1991. *Mukkuvar women: Gender, hegemony and capitalist transformation in a south Indian fishing community.* London: Zed Press.

Sharma, M., and U. Vanjare. 1993. The political economy of reproductive activities in a Rajasthan village. In *Explorations of South Asian systems,* edited by A. W. Clarke, 24–65. New Delhi, India: Oxford University Press.

Verma, R. K., G. Rangaiyan, and S. Narkhede. n.d. Cultural perceptions and categorization of male sexual health problems by practitioners and men in a Mumbai slum population. Unpublished report.

Introduction to Reading 14

Maria Lepowsky is an anthropologist who lived among the Melanesian people of Vanatinai, a small, remote island near New Guinea, from 1977 to 1979, for two months in 1981, and again for three months in 1987. She chose Vanatinai, which literally means "motherland," because she wanted to do research in a place where "the status of women" is high. The egalitarianism of the Vanatinai challenges the Western belief in the universality of male dominance and female subordination.

1. What is the foundation of women's high status and gender equality among the people of Vanatinai?
2. What does gender equality mean on Vanatinai? Does it mean that women and men split everything fifty-fifty? Are men and women interchangeable?
3. What are the similarities and differences between the egalitarianism of the Gerai people (depicted in Helliwell's article in this chapter) and that of the people of Vanatinai?

Gender and Power

Maria Alexandra Lepowsky

Vanatinai customs are generally egalitarian in both philosophy and practice. Women and men have equivalent rights to and control of the means of production, the products of their own labor, and the products of others. Both sexes have access to the symbolic capital of prestige, most visibly through participation in ceremonial exchange and mortuary ritual. Ideologies of male superiority or right of authority over women are notably absent, and ideologies of gender equivalence are clearly articulated. Multiple levels of gender ideologies are largely, but not entirely, congruent. Ideologies in turn are largely congruent with practice and individual actions in expressing gender equivalence, complementarity, and overlap.

There are nevertheless significant differences in social influence and prestige among persons. These are mutable, and they fluctuate over the lifetime of the individual. But Vanatinai social relations are egalitarian overall, and sexually egalitarian in particular, in that at each stage in the life cycle all persons, female and male, have equivalent autonomy and control over their own actions, opportunity to achieve both publicly and privately acknowledged influence and power over the actions of others, and access to valued goods, wealth, and prestige. The quality of generosity, highly valued in both sexes, is explicitly modeled after parental nurture. Women are not viewed as polluting or dangerous to themselves or others in their persons, bodily fluids, or sexuality.

Vanatinai sociality is organized around the principle of personal autonomy. There are no chiefs, and nobody has the right to tell another adult what to do. This philosophy also results in some extremely permissive childrearing and a strong degree of tolerance for the idiosyncrasies of other people's behavior. While working together, sharing, and generosity are admirable, they are strictly voluntary. The selfish and antisocial person might be ostracized, and others will not give to him or her. If kinfolk, in-laws, or neighbors disagree, even with a powerful and influential big man or big woman, they have the option, frequently taken, of moving to another hamlet where they have ties and can expect access to land for gardening and foraging. Land is communally held by matrilineages, but each person has multiple rights to request and be given space to make a garden on land held by others, such as the mother's father's matrilineage. Respect and tolerance for the will and idiosyncrasies of individuals is reinforced by fear of their potential knowledge of witchcraft or sorcery.

Anthropological discussions of women, men, and society over the last one hundred years have been framed largely in terms of "the status of women," presumably unvarying and shared by all women in all social situations. Male dominance and female subordination have thus until recently been perceived as easily identified and often as human universals. If women are indeed universally subordinate, this implies a universal primary cause: hence the search for a single underlying reason for male dominance and female subordination, either material or ideological.

More recent writings in feminist anthropology have stressed multiple and contested gender statuses and ideologies and the impacts of historical forces, variable and changing social contexts, and conflicting gender ideologies. Ambiguity and contradiction, both within and between levels of ideology and social practice, give both women and men room to assert their value and exercise power. Unlike in many cultures where men stress women's innate inferiority, gender relations on Vanatinai are not contested, or antagonistic: there are no male versus female ideologies which vary markedly or directly contradict each other. Vanatinai mythological motifs, beliefs about supernatural power, cultural ideals of the sexual division of labor and of the qualities inherent to men and women, and the customary freedoms and restrictions upon each sex at different points in the life course all provide ideological underpinnings of sexual equality.

Since the 1970s writings on the anthropology of women, in evaluating degrees of female power and influence, have frequently focused on the disparity between the "ideal" sex role pattern of a culture, often based on an ideology of male dominance, publicly proclaimed or enacted by men, and often by women as well, and the "real" one, manifested by the actual behavior of individuals. This approach seeks to uncover female social participation, overt or covert, official or unofficial, in key events and decisions and to learn how women negotiate their social positions. The focus on social and individual "action" or "practice" is prominent more generally in cultural anthropological theory of recent years. Feminist analyses of contradictions between gender ideologies of female inferiority and the realities of women's and men's daily lives—the actual balance of power in household and community—have helped to make this focus on the actual behavior of individuals a wider theoretical concern.[1]

In the Vanatinai case gender ideologies in their multiple levels and contexts emphasize the value of women and provide a mythological charter for the degree of personal autonomy and freedom of choice manifested in real women's lives. Gender ideologies are remarkably similar (though not completely, as I discuss below) as they are manifested situationally, in philosophical statements by women and men, in the ideal pattern of the sexual division of labor, in taboos and proscriptions, myth, cosmology, magic, ritual, the supernatural balance of power, and in the codifications of custom. Women are not characterized as weak or inferior. Women and men are valorized for the same qualities of strength, wisdom, and generosity. If possessed of these qualities an individual woman or man will act in ways which bring prestige not only to the actor but to the kin and residence groups to which she or he belongs.

Nevertheless, there is no single relationship between the sexes on Vanatinai. Power relations and relative influence vary with the individuals, sets of roles, situations, and historical moments involved. Gender ideologies embodied in myths, beliefs, prescriptions for role-appropriate behavior, and personal statements sometimes contradict each other or are contradicted by the behavior of individuals.

* * *

Material and Ideological Bases of Equality

Does equality or inequality, including between men and women, result from material or ideological causes? We cannot say whether an idea preceded or followed specific economic and social circumstances. Does the idea give rise to the act, or does the act generate an ideology that justifies it or mystifies it?

If they are congruent ideology and practice reinforce one another. And if multiple levels of ideology are in accord social forms are more likely to remain unchallenged and fundamentally unchanged. Where levels of ideology, or ideology and practice, are at odds, the circumstances of social life are more likely to be challenged by those who seek a reordering of social privileges justified according to an alternative interpretation of ideology. When social life embodies these kinds of contradictions, the categories of people in power—aristocrats, the rich, men—spend a great deal of energy maintaining their power. They protect their material resources, subdue the disenfranchised with public or private violence, coercion, and repression, and try to control public and private expressions of ideologies of political and religious power.

On Vanatinai, where there is no ideology of male dominance, the material conditions for gender equality are present. Women—and their brothers—control the means of production. Women own land, and they inherit land, pigs, and valuables from their mothers, their mothers' brothers, and sometimes from their fathers equally with men. They have the ultimate decision-making power over the distribution of staple foods that belong jointly to their kinsmen and that their kinsmen or husbands have helped labor to grow. They are integrated into the prestige economy, the ritualized exchanges of ceremonial valuables. Ideological expressions, such as the common saying that the woman is the owner of the garden, or the well-known myth of the first exchange between two female beings, validate material conditions.

I do not believe it would be possible to have a gender egalitarian society, where prevailing expressions of gender ideology were egalitarian or valorized both sexes to the same degree, without material control by women of land, means of subsistence, or wealth equivalent to that of men. This control would encompass anything from foraging rights, skills, tools, and practical and sacred knowledge to access to high-paying, prestigious jobs and the knowledge and connections it takes to get them. Equal control of the means of production, then, is one necessary precondition of gender equality. Vanatinai women's major disadvantage is their lack of access to a key tool instrumental in gaining power and prestige, the spear. Control of the means of production is potentially greater in a matrilineal society.

* * *

Gender Ideologies and Practice in Daily Life

In Melanesian societies the power of knowing is privately owned and transmitted, often through ties of kinship, to heirs or younger supporters. It comes not simply from acquiring skills or the experience and the wisdom of mature years but is fundamentally a spiritual power that derives from ancestors and other spirit forces.

In gender-segregated societies, such as those that characterize most of Melanesia, this spiritual knowledge power is segregated as well into a male domain through male initiations or the institutions of men's houses or male religious cults. Most esoteric knowledge—and the power over others that derives from it—is available to Vanatinai women if they can find a kinsperson or someone else willing to teach it to them. There are neither exclusively male nor female collectivities on Vanatinai nor characteristically male versus female domains or patterns of sociality (cf. Strathern 1987:76).

Decisions taken collectively by Vanatinai women and men within one household, hamlet, or lineage are political ones that reverberate well beyond the local group, sometimes literally hundreds of miles beyond. A hundred years ago they included decisions of war and peace. Today they include the ritualized work of kinship, more particularly of the matrilineage, in mortuary ritual. Mortuary feasts, and the interisland and interhamlet exchanges of ceremonial valuables that support them, memorialize the marriages that tied three matrilineages together, that of the deceased, the deceased's father, and the widowed spouse. Honoring these ties of alliance, contracted by individuals but supported by their kin, and threatened by the dissolution of death, is the major work of island politics. . . .

The small scale, fluidity (cf. Collier and Rosaldo 1981), and mobility of social life on Vanatinai, especially in combination with matriliny, are conducive of egalitarian social relations between men and women and old and young. They promote an ethic of respect for the individual, which must be integrated with the ethic of cooperation essential for survival in a subsistence economy. People must work out conflict through face to face negotiation, or existing social ties will be broken by migration, divorce, or death through sorcery or witchcraft.

Women on Vanatinai are physically mobile, traveling with their families to live with their own kin and then the kin of their spouse, making journeys in quest of valuables, and attending mortuary feasts. They are said to have traveled for these reasons even in precolonial times when the threat of attack was a constant danger. The generally greater physical mobility of men in human societies is a significant factor in sexual asymmetries of power, as it is men who generally negotiate and regulate relationships with outside groups (cf. Ardener 1975:6).

Vanatinai women's mobility is not restricted by ideology or by taboo, and women build their own far-ranging personal networks of social relationships. Links in these networks may be activated as needed by the woman to the benefit of her kin or hamlet group. Women are confined little by taboos or community pressures. They travel, choose their own marriage partners or lovers, divorce at will, or develop reputations as wealthy and generous individuals active in exchange.

Big Men, Big Women, and Chiefs

Vanatinai giagia, male and female, match Sahlins's (1989) classic description of the Melanesian big man, except that the role of gia is gender-blind. There has been renewed interest among anthropologists in recent years in the big man form of political authority.[2] The Vanatinai case of the female and male giagia offers an intriguing perspective. . . .

Any individual on Vanatinai, male or female, may try to become known as a gia by choosing to exert the extra effort to go beyond the minimum contributions to the mortuary feasts expected of every adult. He or she accumulates ceremonial valuables and other goods both in order to give them away in acts of public generosity and to honor obligations to exchange partners from the local area as well as distant islands. There may be more than one gia in a particular hamlet, or even household, or there may be none. A woman may have considerably more prestige and influence than her husband because of her reputation for acquiring and redistributing valuables. While there are more men than women who are extremely active in exchange, there are some women who are far more active than the majority of men.

Giagia of either sex are only leaders in temporary circumstances and if others wish to follow, as when they host a feast, lead an exchange expedition, or organize the planting of a communal yam garden. Decisions are made by consensus, and the giagia of both sexes influence others through their powers of persuasion, their reputations for ability, and their knowledge, both of beneficial magic and ritual and of sorcery or witchcraft. . . .

On Vanatinai power and influence over the actions of others are gained by achievement and demonstrated superior knowledge and skill, whether in the realm of gardening, exchange, healing, or sorcery. Those who accumulate a surplus of resources are expected to be generous and share with their neighbors or face the threat of the sorcery or witchcraft of the envious. Both women and men are free to build their careers through exchange. On the other hand both women and men are free not to strive toward renown as giagia but to work for their own families or simply to mind their own business. They can also achieve the respect of their peers, if they seek it at all, as loving parents, responsible and hard-working lineage mates and affines, good gardeners, hunters, or fishers, or skilled healers, carvers, or weavers.

Mead (1935) observes that societies vary in the degree to which "temperament types" or "approved social personalities" considered suitable for each sex or a particular age category differ from each other. On Vanatinai there is wide variation in temperament and behavior among islanders of the same sex and age. The large amount of overlap between the roles of men and women on Vanatinai leads to a great deal of role flexibility, allowing both individual men and women the freedom to specialize in the activities they personally enjoy, value, are good at performing, or feel like doing at a particular time. There is considerable freedom of choice in shaping individual lifestyles.

An ethic of personal autonomy, one not restricted to the powerful, is a key precondition of social equality. Every individual on Vanatinai from the smallest child to an aged man or woman possesses a large degree of autonomy. Idiosyncrasies of personality and character are generally tolerated and respected. When you ask why someone does or does not do something, your friends will say, emphatically and expressively, "We [inclusive we: you and I both] don't know," "It is something of theirs" [their way], or, "She doesn't want to."

Islanders say that it is not possible to know why a person behaves a certain way or what thoughts generate an action. Persisting in a demand to "know" publicly the thoughts of others is dangerous, threatening, and invasive. Vanatinai people share, in part, the perspectives identified with postmodern discussions of the limits of ethnographic representation: it is impossible to know another person's thoughts or feelings. If you try they are likely to deceive you to protect their own privacy or their own interests. Your knowing is unique to you. It is your private property that you transmit only at your own volition, as when you teach magical spells to a daughter or sister's son.[3]

The prevailing social sanction is also individualistic: the threat of somebody else's sorcery or witchcraft if you do not do what they want or if you arouse envy or jealousy. But Vanatinai cultural ideologies stress the strength of individual will in the face of the coercive pressures of custom, threat of sorcery, and demands to share. This leads to a Melanesian paradox: the ethic of personal autonomy is in direct conflict to the ethic of giving and sharing so highly valued on Vanatinai, as in most Melanesian cultures. Nobody can make you share, short of stealing from you or killing you if you refuse them. You have to want to give: your nurture, your labor, your valuables, and your person. This is where persuasion comes in. It comes from the pressure of other people, the force of shame, and magical seduction made potent by supernatural agency. Vanatinai custom supplies a final, persuasive argument to resolve this paradox: by giving, you not only strengthen your lineage and build its good name, you make yourself richer and more powerful by placing others in your debt.

What can people in other parts of the world learn from the principles of sexual equality in Vanatinai custom and philosophy? Small scale facilitates Vanatinai people's emphasis on face-to-face negotiations of interpersonal conflicts without the delegation of political authority to a small group of middle-aged male elites. It also leaves room for an ethic of respect for the will of the individual regardless of age or sex. A culture that is egalitarian and nonhierarchical overall is more likely to have egalitarian relations between men and women.

Males and females on Vanatinai have equivalent autonomy at each life cycle stage. As adults they have similar opportunities to influence the actions of others. There is a large amount of overlap between the roles and activities of women and men, with women occupying public, prestige-generating roles. Women share control of the production and the distribution of valued goods, and they inherit property. Women as well as men participate in the exchange of valuables, they organize feasts, they officiate at important rituals such as those for yam planting or healing, they counsel their kinfolk, they speak out and are listened to in public meetings, they possess valuable magical knowledge, and they work side by side in most subsistence activities. Women's role as nurturing parent is highly valued and is the dominant metaphor for the generous men and women who gain renown and influence over others by accumulating and then giving away valuable goods.

But these same characteristics of respect for individual autonomy, role overlap, and public participation of women in key subsistence and prestige domains of social life are also possible in large-scale industrial and agricultural societies. The Vanatinai example suggests that sexual equality is facilitated by an overall ethic of respect for and equal treatment of all categories of individuals, the decentralization of political power, and inclusion of all categories of persons (for example, women and ethnic minorities) in public positions of authority and influence. It requires greater role overlap through increased integration of the workforce, increased control by women and minorities of valued goods—property, income, and educational credentials—and increased recognition of the social value of parental care. The example of Vanatinai shows that the subjugation of women by men is not a human universal, and it is not inevitable. Sex role patterns and gender ideologies are closely related to overall social systems of power and prestige. Where these systems stress personal autonomy and egalitarian social relations among all adults, minimizing the formal authority of one person over another, gender equality is possible.

Notes

1. See, for example, Rogers (1975) and Collier and Rosaldo (1981) on ideal versus real gender relations. Ortner (1984) summarizes approaches to practice; cf. Bourdieu (1977).

2. The appropriateness of using the big man institution to define Melanesia versus a Polynesia characterized by chiefdoms, the relationship of big men to social equality, rank, and stratification, and the interactions of this form of leadership with colonialism and modernization are central issues in recent anthropological writings on big men (e.g., Brown 1987, Godelier 1986, Sahlins 1989, Strathern 1987, Thomas 1989, Lederman 1991). I discuss the implications of the Vanatinai case of the giagia at greater length in Lepowsky (1990).

3. See, for example, Clifford (1983), Clifford and Marcus (1986), and Marcus and Fischer (1986) on representations. In this book I have followed my own cultural premises and not those of Vanatinai by publicly attributing thoughts, motives, and feelings to others and by trying to find the shapes in a mass of chaotic and sometimes contradictory statements and actions. But my Vanatinai friends say, characteristically, that my writing is "something of mine"—my business.

REFERENCES

Ardener, Edwin. 1975. "Belief and the Problem of Women." In Shirley Ardener, ed., *Perceiving Women.* London: Malaby.

Bourdieu, Pierre. 1977. *Outline of a Theory of Practice.* T. R. Nice. Cambridge: Cambridge University Press.

Brown, Paula. 1987. "New Men and Big Men: Emerging Social Stratification in the Third World, A Case Study from the New Guinea Highlands." *Ethnology* 26:87–106.

Clifford, James. 1983. "On Ethnographic Authority." *Representations* 1:118–146.

Clifford, James, and George Marcus, eds. 1986. *Writing Culture:The Poetics and Politics of Ethnography.* Berkeley: University of California Press.

Collier, Jane, and Michelle Rosaldo. 1981. "Politics and Gender in Simple Societies." In Sherry Ortner and Harriet Whitehead, eds., *Sexual Meanings: The Cultural Construction of Gender and Sexuality.* Cambridge: Cambridge University Press.

Gailey, Christine. 1980. "Putting Down Sisters and Wives: Tongan Women and Colonization." In Mona Etienne and Eleanor Leacock, eds., *Women and Colonization.* New York: Bergin/Praeger.

Godelier, Maurice. 1986. *The Making of Great Men: Male Domination and Power Among the New Guinea Baruya.* Cambridge: Cambridge University Press.

Kan, Sergei. 1989. *Symbolic Immortality:The Tlingit Potlatch of the Nineteenth Century.* Washington, D.C.: Smithsonian Institution Press.

Lederman, Rena. 1991. "'Interests' in Exchange: Increment, Equivalence, and the Limits of Bigmanship." In Maurice Godelier and Marilyn Strathern, eds., *Big Men and Great Men: Personifications of Power in Melanesia.* Cambridge: Cambridge University Press.

Lepowsky, Maria. 1990. "Big Men, Big Women, and Cultural Autonomy." *Ethnology* 29(10):35–50.

Linnekin, Jocelyn. 1990. *Sacred Queens and Women of Consequence: Rank, Gender, and Colonialism in the Hawaiian Islands.* Ann Arbor: University of Michigan Press.

Marcus, George, and Michael Fischer, eds. 1986. *Anthropology as Cultural Critique: An Experimental Moment in the Human Sciences.* Chicago: University of Chicago Press.

Mead, Margaret. 1935. *Sex and the Temperament in Three Primitive Societies.* New York: William Morrow.

Ortner, Sherry. 1984. "Theory in Anthropology Since the Sixties." *Comparative Studies in Society and History* 26(1):126–166.

Rogers, Susan Carol. 1975. "Female Forms of Power and the Myth of Male Dominance: A Model of Female/Male Interaction in Peasant Society." *American Ethnologist* 2:727–756.

Sahlins, Marshall. 1989. "Comment: The Force of Ethnology: Origins and Significance of the Melanesia/Polynesia Division." *Current Anthropology* 30:36–37.

Silverblatt, Irene. 1987. *Moon, Sun, and Witches: Gender Ideologies and Class in Inca and Colonial Peru.* Princeton: Princeton University Press.

Strathern, Marilyn. 1987. "Introduction." In Marilyn Strathern, ed., *Dealing with Inequality: Analysing Gender Relations in Melanesia and Beyond.* Cambridge: Cambridge University Press.

Thomas, Nicholas. 1989. "The Force of Ethnology: Origins and Significance of the Melanesia/Polynesia Division." *Current Anthropology* 30:27–34.

✤ Topics for Further Examination ✤

- Locate and read scholarly research on the Hijras of India, the Faáfafines of Samoa, and the Kathoey of Thailand.
- Find research articles and Web sites on female and male circumcision and its meanings and consequences in different societies today.
- Look up scholarly studies that discuss the egalitarian gender system of the Iroquois Confederacy.

PART II

PATTERNS

4

LEARNING AND DOING GENDER

We began this book by discussing the shaping of gender in Western and non-Western cultures. Part II expands upon the idea of prisms by examining the patterns of gendered experiences that emerge from the interaction of gender with other socially constructed prisms. As multiple patterns are created by the refraction of light as it travels through a kaleidoscope containing prisms, so too are the patterns of individuals' life experiences influenced by gender and other social prisms discussed in Part I.

GENDERED PATTERNS

Social patterns are the center of social scientists' work. Schwalbe (1998), a sociologist, defines social patterns as "a regularity in the way the world works" (101). For example, driving down the "right" side of the street is a regularity American people appreciate. You will read about different gendered patterns in Part II, many of which are regularities you will find problematic because they deny the individuality of women and men. Clearly, there are exceptions to social patterns; however, these exceptions are in the details, not in the regularity of social behavior itself (Schwalbe, 1998).

A deeper understanding of how and why particular social patterns exist helps us to interpret our own behavior and the world around us. Gender, as we discussed in Part I, is not a singular pattern of masculinity or femininity that carries from one situation to another. Instead it is complex, multifaceted, and ever-changing depending upon the social context, whom we are with, and where we are. Our behavior in almost all situations is framed within our knowledge of ideal gender—hegemonic masculinity and emphasized femininity—as discussed in the Introduction and Chapter 1.

Keep the concepts of hegemonic masculinity and emphasized femininity in mind as we examine social patterns of gender. To illustrate this, let's return to the stereotype discussed in the Introduction—that women talk more than men. We know from research that the real social pattern in mixed-gender groups is that men talk more, interrupt more, and change the topic more often than women (Anderson & Leaper, 1998; Wood, 1999). The stereotype, while trivializing women's talk and ignoring the dominance of men in mixed-gender groups, maintains the patterns of dominance and subordination associated with hegemonic masculinity and emphasized femininity, influencing women's as well as men's behaviors. Girls are encouraged to use a nice voice and not talk too much. Later, as they grow older and join mixed-gender groups at work or play, women's voices are often ignored and women are subordinated as they monitor what they say and how often they talk, and check to make sure they are not dominating the conversation. By examining how these idealized versions of masculinity and femininity pattern daily practices, we can understand better the patterns and meanings of our behavior and the behaviors of others.

Gendered patterns of belief and behavior influence us throughout our lives, in almost every activity in which we engage. Readings in Part II examine the process and consequences of learning to do gender and then describe gendered patterns in work (Chapter 7) and in daily intimate relationships with family and friends (Chapter 8). We also explore how gendered patterns affect our bodies, sexualities, and emotions (Chapter 6), and how patterns of dominance, control, and violence enforce gender patterns (Chapter 9). This chapter examines the process and consequences of learning to do gender.

The patterns that emerge from the gender kaleidoscope are not unique experiences in individual lives; they are regularities that occur in many people's lives. Institutions and groups enforce certain types of gendered relationships in the home, workplace, and daily life as described in the readings throughout Part II. These patterns overlap and reinforce gender differences and inequalities. For example, gender discrimination in wages affects families' decisions about parenting roles and relationships. Since most men still earn more than most women, the choices of families who wish to break away from traditional gender patterns are limited (see Chapters 7 and 8).

Learning Gender

We begin this part of the book by examining the processes by which we acquire self-perceptions and behaviors that fit our culture's patterns of masculinities and femininities (Chapters 4 and 5). The readings in this chapter emphasize that, regardless of our inability or unwillingness to attain idealized femininity and masculinity, almost everyone in a culture learns what idealized gender is and organizes their lives around those expectations. The term sociologists use to describe how we learn gender is "gender socialization," and sociologists approach it from a variety of different perspectives (Coltrane, 1998).

Socialization is the process of teaching members of a society the values, expectations, and practices of the larger culture. Socialization takes place in all interactions and situations, with families and schools typically having formal responsibility for socializing new members in Western societies. Early attempts to explain gender socialization gave little attention to the response of individuals to agents of socialization such as parents, peers, and teachers. There was an underlying assumption in this early perspective that individuals were blank tablets (tabulae rasae) upon which the cultural definitions of gender and other appropriate behaviors were written. This perspective assumed that, as individuals developed, they took on a gender identity appropriate to their biological sex category (Howard & Alamilla, 2001).

Social scientists now realize that individuals are not blank tablets; gender socialization is not just something that is "done" to us. Theorists now describe socialization into gender as a series of complex and dynamic processes. Individuals create, as well as respond to, social stimuli (Carlton-Ford & Houston, 2001; Howard & Alamilla, 2001). For example, the Urla and Swedlund reading in Chapter 5 discusses children's reactions to and interpretations of gender-specific toys such as Barbie. Moreover, socialization doesn't simply end after childhood. Socialization is a process that lasts across one's lifetime, from birth to death (Bush & Simmons, 1981). Throughout our lives, we assess cues around us and behave as situations dictate. Gender is a key factor in determining what is appropriate.

Socializing Children

There are many explanations for why children gravitate toward gender-appropriate behavior. It is not just family members who teach children to behave as "good boys" or "good girls." Almost every person a child comes into contact with, and virtually all aspects of a child's material world (e.g., toys, books, clothing) reinforce gender. Even stories repeated across many generations portray gender ideals as they are told today in movies and books, as described in the reading by the Lori Baker-Sperry and Liz Grauerholz in this chapter. It is not long, then, before most children come to understand that they are "boys" or "girls" and segregate themselves accordingly. Adults play a major role in teaching gender, as the reading by Emily Kane in this chapter finds. It is not just parents but also teachers who teach gender. When teachers separate children into gender-segregated spaces in lunch lines or playground areas, they reinforce gender differences (Sadker & Sadker, 1994; Thorne, 1993).

Most children quickly understand the gender-appropriate message directed toward them and behave accordingly. Although not all boys are dominant and not all girls are subordinate, studies in a variety of areas find that most White boys tend toward active and aggressive behaviors, while most White girls tend to be quieter and focus on relationships. The patterning for African American boys and girls is similar, as suggested in the reading in this chapter by Prudence Carter, where boys who do not act in gender-appropriate ways are seen as "soft" or feminine. These patterns have been documented in schools and in play (e.g., Sadker & Sadker, 1994; Thorne, 1993; McGuffey & Rich, in this chapter).

The consequences for gender-appropriate behavior are considerable. Gender-appropriate behavior is related to lower self-confidence and self-esteem for girls (e.g., Eder, 1995; Orenstein, 1994; Spade, 2001) whereas boys are taught to "mask" their feelings and compete with everyone for control, thus isolating themselves and ignoring their own feelings (e.g., Connell, 2000; Messner, 1992; Pollack, 2000).

The dominant pattern of gender expectations, the "pink and blue syndrome" described in Chapter 1, begins at birth. Once external genital identification takes place, immediate expectations for masculine and feminine behavior follow. Exclamations of "he's going to be a great baseball (or football or soccer) player" and "she's so cute" are accompanied by gifts of little sleepers in pink or blue with gender-appropriate decorations. Try as we might, it is very difficult to find gender-neutral clothing for children (see Nelson in Chapter 5). These expectations, and the way we treat young children, reinforce idealized gender constructions of dominance and subordination.

Schools and Socialization

Schools reinforce separate spheres for boys and girls (Orenstein, 1994; Sadker & Sadker, 1994; Thorne, 1993). Considerable research by the American Association of University Women (1992, 1998, 1999) documents how schools "shortchange" girls. Schools are social institutions that maintain patterns of power and dominance. Indeed, we teach dominance in schools in patterns of teacher/student interactions such as respecting the responses of boys while encouraging girls to be helpers in the classroom (Sadker & Sadker, 1994).

In effect, the structure of society pronounces boys and girls as different and teaches them how to behave accordingly. In Chapter 5, you will read more about how capitalist societies reinforce and maintain gender for children and adults. Television, music, books, clothing, and toys differentiate and prescribe appropriate behavior for girls and boys. For example, studies of children's books find some distinctive patterns that reinforce idealized forms of gender. One study of children's readers from the early 1970s found that while Jane looked on, Dick did exciting and interesting things. Stories about boys outnumbered those about girls by five to two in that study, with boys engaged in adventures. Any adventures or discoveries on the part of girls were attributed to an accident or luck (Women on Words and Images, 1974). This research was the beginning of careful analyses of how the books American children read socialize children into gender patterns. A recent study of award-winning books from 1995 to 1999 found that although boys and girls were equally represented as main characters, portrayals of male characters were likely to be dominant, while female characters were likely to be subordinate. Although other evidence also supported the depiction of traditional gender patterns, on the plus side the researchers found that girls and women are more likely to be portrayed in gender-atypical roles in many recent children's books (Gooden & Gooden, 2001). Unfortunately, boys and men continue to be depicted in gender-typical roles in these recent, award-winning children's books (Gooden & Gooden, 2001). Providing further evidence of perpetuating gendered patterns, the reading by Lori Baker-Sperry and Liz Grauerholz in this chapter finds that our most enduring stories—fairy tales—emphasize the "feminine beauty ideal," further reinforcing gender. All of those little girls in Cinderella costumes dancing around our neighborhoods are clearly doing gender.

However, not all boys and men are allowed to be dominant across settings (Eder, 1995). Ann Arnett Ferguson (2000) describes how schools discourage African American boys from claiming their blackness and masculinity. Although White boys may be allowed to be "rambunctious" and disrespectful, African American boys are punished more severely than White peers when they "act out." Girls also exist within a hierarchy of relationships (Eder, 1995). Girls from racial, ethnic, economically disadvantaged, or other subordinated groups must fight even harder to succeed under multiple systems of domination and inequality in schools. To help illustrate this, Bettie (2002) compared the paths to success for upwardly mobile White and Mexican high school girls and found some similarities in gender experiences at home and school that facilitated mobility, such as participation in sports. There were also differences in their experiences because race was always salient for the Mexican American girls. However, Bettie (2002) believes that achieving upward mobility may be easier for these Mexican girls than their brothers because it is easier for them to transgress gender boundaries. Their brothers, on the other hand, feel pressure to "engage in the rituals of proving masculinity" (Bettie, 2002, p. 419).

Bettie's (2002) study emphasizes the fact that multiple social prisms of difference and inequality create an array of patterns, which would not be possible if gender socialization practices were universal. Individuals' lives are constructed around many factors, including gender. Cultural values and expectations influence, and frequently contradict, the maintenance of hegemonic masculinity and emphasized femininity in Western societies. The readings by Karen Pyke and Denise Johnson in Chapter 2 and those of Prudence Carter and Scott Coltrane, Ross Parke and Michele Adams in this chapter illustrate how the practice of gender is strongly influenced by culture. The process of gender socialization is rooted in the principle that people are not equal and that the socially constructed categories of difference and inequality (gender, race, ethnicity, class, religion, age, culture, etc.) are legitimate.

Sports and Socialization

Sports, particularly organized sports, provide other examples of how institutionalized activities reinforce the gender identities that children learn. Boys learn the meaning of competition and success, including the idea that winning is everything (e.g., Messner, 1992). Girls, on the other hand, often are found on the edge of the playing field, or on the sides of the playgrounds, watching the boys (Thorne, 1993). The reading by C. Shawn McGuffey and B. Lindsay Rich in this chapter illustrates the complex gendering processes involved in children's play activities. Yet not all children play in the same ways. Goodwin (1990) finds that children from urban, lower-class, high-density neighborhoods, where households are closer together, are more likely to play in mixed-gender and mixed-age groups. In suburban middle-class households, which are farther apart than urban households, parents are more likely to drive their children to sporting activities or houses to play with same-gender, same-age peers. The consequences of social class and place of residence are that lower-class children are more comfortable with their sexuality as they enter preadolescence and are less likely to gender segregate in schools (Goodwin, 1990).

Gender Transgressions

Children learn to display gender-appropriate behavior; however, there are times when they step out of gender-appropriate zones as described by McGuffey and Rich in this chapter. The patterns they found are similar to the trend found in children's books of the late 1990s—girls and women are more likely to transgress and do masculine things than boys and men are to participate in feminine activities. McGuffey and Rich's reading helps us to understand that hegemonic masculinity is complex and constantly negotiated, even among children at play. They find that girls who transgress into the "boys' zone" may eventually be respected by their male playmates if they are good at conventionally male activities such as playing baseball. Boys, however, are harassed and teased when they try to participate in any activity associated with girls. By denying boys access to girls' activity, the dominance of hegemonic masculinity is maintained, even when boys are ridiculed because they "throw like a girl."

As you can see, learning gender is complicated. Clearly, gender is something that we "do" as well as learn, and in doing gender we are responding to structured expectations from institutions in society. Every time we enter a new social situation, we look around for cues and guides to determine how to behave in a gender-appropriate manner. In some situations, we might interpret gender cues as calling for a high degree of gender conformity, while in other situations, the clues allow us to be more flexible. We create gender as well as respond to expectations for it.

Doing Masculinity and Femininity Throughout Our Lives

Most men have learned to "do" the behaviors that maintain hegemonic masculinity, while at the same time suppress feelings and behaviors that might make them seem feminine (Connell, 1987). Elizabeth Gilbert, a White woman journalist, helps us to understand how adult Americans define and enact masculinity and femininity. In her article in this chapter, she tells how she learned to see the world through the eyes of men by becoming a "drag king." She became a man in appearance, physical posturing, and cognitive and emotional approaches to others. In doing so, she discovered how hegemonic masculinity limits one's identity, including how she felt about herself and interacted with others. The ease with which she was able to "become a man" points to the relational nature of gender. That is, by learning how to be a woman she also learned how the "opposite" gender, men, must act.

As you can see, hegemonic masculinity is maintained in a hierarchy that is realized by only a few men, with everyone else subordinated to them—women, poor White men, men of color, gay men, and men from devalued ethnic and religious groups. Furthermore, this domination is not always one-on-one, but can be institutionalized in the structure of the situation. As you read these and other articles in this chapter, you will see that gender is not something that we learn once in one setting, such as an inoculation or shot for rabies. Instead, we learn to do gender over time in virtually everything we undertake.

Moreover, learning to do gender is complicated by the other prisms that interact in our lives. Recall the lessons from Chapter 2 and remember that gender does not stand alone, but rather is reflected in other social identities. The last reading in this chapter, by Leora Tanenbaum, illustrates the intersection of gender with sexuality. Just as race was used to reinforce status hierarchies in the reading by Carter in this chapter on African American boys and girls, sexuality is also used to reinforce and control individuals' behaviors. By determining what is "right" and what is not, sexuality keeps most individuals neatly subordinated within the framework of hegemonic masculinity and emphasized femininity.

It is not easy to separate the learning and doing of gender from other patterns. As you read selections in other chapters in this part of the book, you will be able to see the influence of social processes and institutions on how we learn and do gender across all aspects of our lives. Before you start to read, ask yourself how you learned gender and how well you do it. Not succeeding at doing gender is normal. That is, if we all felt comfortable with ourselves, no one would be striving for idealized forms of gender—hegemonic masculinity or emphasized femininity. Imagine a world in which we all felt comfortable with who we are! As you read through the rest of this book, ask yourself why that world doesn't exist.

References

American Association of University Women. (1992). *How schools shortchange girls.* Washington, DC: American Association of University Women Educational Foundation.

American Association of University Women. (1998). *Gender gaps: Where schools still fail our children.* Washington, DC: American Association of University Women Educational Foundation.

American Association of University Women. (1999). *Voices of a generation: Teenage girls on sex, school, and self.* Washington, DC: American Association of University Women Educational Foundation.

Anderson, K. J., & Leaper, C. (1998). Meta-analysis of gender effects on conversational interruption: Who, what, when, where, and how. *Sex Roles, 39*, (3–4), p. 225–252.

Barajas, L. H., & Pierce, J. L. (2001). The significance of race and gender in school success among Latinas and Latinos in college. *Gender & Society, 15* (5), p. 859–878.

Bettie, J. (2002). Exceptions to the rule: Upwardly mobile White and Mexican American high school girls. *Gender & Society, 16*(3), p. 403–422.

Bush, D. M., & Simmons, R. G. (1981). Socialization processes over the life course. In M. Rosenberg & R. H. Turner (Eds.), *Social psychology: Sociological perspectives* (pp. 133–164). New York: Basic Books.

Carlton-Ford, S., & Houston, P. V. (2001). Children's experience of gender: Habitus and field. In D. Vannoy (Ed.), *Gender mosaics: Societal perspectives* (pp. 65–74). Los Angeles: Roxbury Publishing Company.

Coltrane, S. (1998). *Gender and families.* Thousand Oaks, CA: Pine Forge Press.

Connell, R. W. (1987). Gender and power: Society, the person, and sexual politics. Stanford, CA: Stanford University Press.

Connell, R. W. (2000). *The men and the boys.* Berkeley: University of California Press.

Eder, D. (1995). *School talk: Gender and adolescent culture.* New Brunswick, NJ: Rutgers University Press.

Ferguson, A. A. (2000). *Bad boys: Public schools in the making of Black masculinity.* Ann Arbor: University of Michigan Press.

Gooden, A. M., & Gooden, M. A. (2001). Gender representation in notable children's picture books: 1995–1999. *Sex Roles, 45*(1/2), 89–101.

Goodwin, M. H. (1990). *He-said-she-said: Talk as social organization among Black children.* Bloomington: Indiana University Press.

Howard, J. A., & Alamilla, R. M. (2001). Gender and identity. In D. Vannoy (Ed.), *Gender mosaics: Societal perspectives* (pp. 54–64). Los Angeles: Roxbury Publishing Company.

Messner, M. A. (1992). *Power at play: Sports and masculinity.* Boston: Beacon Press.

Orenstein, P. (1994). *School girls: Young women, self-esteem, and the confidence gap*. New York: Anchor Books.

Pollack, W. S. (2000). *Real boys' voices.* New York: Penguin Putnam.

Sadker, D., & Sadker, M., (1994). *Failing at fairness: How our schools cheat girls.* New York: Simon & Schuster.

Schwalbe, M. (1998). *The sociologically examined life: Pieces of the conversation.* Mountain View, CA: Mayfield Publishing Company.

Spade, J. Z. (2001). Gender and education in the United States. In D. Vannoy (Ed.), *Gender mosaics: Societal perspectives* (pp. 85–93). Los Angeles: Roxbury Publishing Company.

Thorne, B. (1993). *Gender play: Girls and boys in school.* New Brunswick, NJ: Rutgers University Press.

Women on Words and Images. (1974). Look Jane look. See sex stereotypes. In J. Stacey, S. Bereaud, & J. Daniels (Eds.), *And Jill came tumbling after: Sexism in American education* (pp. 159–177). New York: Dell Publishing Co., Inc.

Wood, J. T. (1999). *Gendered lives: Communication, gender, and culture* (3rd ed.). Belmont, CA: Wadsworth.

Introduction to Reading 15

C. Shawn McGuffey and B. Lindsay Rich use their sociological radar to examine the ways five- to twelve-year-olds from various racial and ethnic backgrounds create and maintain gender patterns in a summer day camp. The first author was a counselor at this camp and spent over nine weeks during the summer of 1996 observing children's play. He kept daily logs and met weekly with the second author to discuss his observations. He also interviewed twenty-two children from the camp and six parents to get a deeper understanding of the ways children constructed gendered meanings and activities. Their conceptualization of the gender transgression zone helps us to understand how gendered boundaries are maintained and violated by young children.

1. What do the authors mean by the "organization of homosocial status systems"?
2. How easy is it for boys and girls across racial and class backgrounds to enter the gender transgression zone?
3. How does hegemonic masculinity combine with boys' racial and class backgrounds to affect their ability to transgress gender?

Playing in the Gender Transgression Zone

Race, Class, and Hegemonic Masculinity in Middle Childhood

C. Shawn McGuffey and B. Lindsay Rich

By now, R. W. Connell's concept of "hegemonic masculinity" has wide currency among students of gender.[1] The concept implies that there is a predominant way of doing gender relations (typically by men and boys, but not necessarily limited to men and boys) that enforces the gender order status quo: It elevates the general social status of masculine over feminine qualities and privileges some masculine qualities over others. The notion that "masculinities" and

C. Shawn McGuffey and B. Lindsay Rich, "Playing in the gender transgression zone" *Gender & Society 13*(5), pp. 608–627.

"femininities" exist and can be interrogated as negotiated realities allows us to further our understanding about the larger gender order in which they are embedded.

We want to caution, however, against the temptation to overgeneralize the concept of hegemonic masculinity. To do so runs the risk of glossing the modalities, both historical and social-spatial (in terms of class, ethnoracial, sexual, and age variations), in which hegemonic masculinity emerges. We believe that hegemonic masculinity, while having general qualities as a form of social power, may take on many valences and nuances, depending on the social setting and the social actors involved. Connell (1987, 1995, 36–37) is himself careful to make the sorts of qualifications we make here while similarly claiming the general analytic utility of the concept. We agree with advocates of the concept that it indeed gives us great theoretical leverage and explanatory power toward clarifying and refining how and why men's dominance works at higher levels of social organization, perhaps even at the global level (Connell 1995; Hawkesworth 1997). In this article, we provide evidence of how hegemonic masculinity is manifest in middle childhood play and used to re-create a gender order among children wherein the larger social relations of men's dominance are learned, employed, reinforced, and potentially changed. Specifically, we present and discuss the results of a preliminary participant observation study of microlevel processes of gender boundary negotiation in middle childhood (ages 5–12).

Providing empirical evidence about the ways in which boys and girls negotiate gender relations within specific social contexts can further understanding about why gender relations take the forms they do in childhood. Using the concept of hegemonic masculinity as a heuristic tool, we decided to focus on how gender relations—specifically, the enactment of masculine hegemony within these relations—were "done" (West and Zimmerman 1987). We eschew the notion that men and women (or boys and girls) merely enact "sex roles" as handed-down scripts. Rather, while acknowledging structural gender socialization implied by the concept of role, we focus on the ways in which the relations between girls and boys are negotiated (Connell 1987, 1995; Messner 1998).

* * *

[W]hen we refer to specific children in this study, we designate ethnoracial, gender, and age distinctions as follows: W = White, B = Black, A = Asian, B = Boy, G = Girl, and a number representing the age of the child. For example, an African American girl who is seven years old will be represented as (BG7). If a child is of another racial category other than White, Black, or Asian, his or her specific classification will be marked accordingly.

Organization of Homosocial Status Systems

When examined as two separate social groups, boys and girls organize themselves differently based on distinct systems of valuing. We must reiterate that masculinity and femininity are not bipolar or opposites but are rather "separate and relatively independent dimensions" (Absi-Semaan, Crombie, and Freeman 1993, 188). Gender is a social construction that is constantly being modified as individuals mature. What may be gender appropriate at one stage in life may be gender inappropriate at a later stage. Boys, for example, are free to touch each other affectionately in early middle childhood, but this is subsequently stigmatized, with a few exceptions (such as victory celebrations). As "independent dimensions," one can develop a clearer view of masculinity and femininity by studying how they differ in context to intragender (homosocial) relations and then how they interact in intergender (heterosocial) relations. Homosocial relationships—nonsexual attractions held by members of the same sex—define how heterosocial relationships are maintained. Thus, it is essential to understand how boys and girls organize themselves within each homosocial group to understand how they negotiate boundaries between the two (Bird 1996).[2]

Structural Formation of Boys in Middle Childhood

Boys in middle childhood organize themselves in a definite hierarchical structure in which the high-status boys decide what is acceptable and valued—that which is hegemonically masculine—and what is not. A boy's rank in the hierarchy is chiefly determined by his athletic ability. Researchers have identified sports as a central focus in boys' development (Fine 1992; Messner 1992, 1994). Boys in this context were observed using words such as captain, leader, and various other ranking references, even when they were not playing sports. Messner (1994, 209) explains the attraction of sports in hegemonic masculinity as a result of young males finding the "rulebound structure of games and sports to be a psychologically 'safe' place in which [they] can get (non-intimate) connection with others within a context that maintains clear boundaries, distance, and separation from others." Sharon Bird (1996) identifies three characteristics in maintaining hegemonic masculinity: emotional detachment, competitiveness, and the sexual objectification of women, in which masculinity is thought of as different from and better than femininity. As another essential feature of hegemonic masculinity, we want to add to these characteristics the ability to draw attention to one's self. Because hegemony is sustained publicly, being able to attract positive attention to one's self is vital. The recognition a boy receives from his public performance of masculinity allows him to maintain his high status and/or increase his rank in the hierarchy.

Conflicts and disagreements in the boys' hierarchy are resolved by name-calling and teasing, physical aggression, and exclusion from the group. These forms of aggression structure and maintain the hierarchy by subordinating alternate propositions and identities that threaten hegemonic masculinity. Although direct and physical aggression are the most physically damaging, the fear of being exiled from the group is the most devastating since the hierarchy confirms masculinity and self-worth for many young boys. According to Kaufman (1995, 16), the basis for a hegemonic masculinity is "unconsciously rooted before the age of six" and "is reinforced as the child develops." Lower-status boys adhere to the hegemonic rules as established by the top boys even if they do not receive any direct benefits from the hierarchy within the homosocial context. The overwhelming majority of boys support hegemonic masculinity in relation to subordinated masculinities and femininities because it not only gives boys power over an entire sex (i.e., girls), but it also gives them the opportunity to acquire power over members of their own sex. This helps maintain the hierarchical frame by always giving boys—even low-status boys—status and power over others. Connell (1987, 183) states that hegemonic masculinity "is always constructed in relation to various subordinated masculinities as well as in relation to women." Connell (1995, 79) describes this panmasculine privilege over girls and women as the "patriarchal dividend." Hegemonic masculinity is publicly used to sustain the power of high-status boys over subordinate boys and boys over girls.

Emotional detachment, competitiveness, and attention arousal could be witnessed in any game of basketball. High-status boys in our study generally performed the best and always distinguished themselves after scoring points. Three high-status boys demonstrate this particularly well. After scoring, Adam[3] (WB11) usually jumped in the air, fist in hand, and shouted either, "In your face!" or "You can't handle this!" Brian's (BB11) style consisted of a little dance followed by, "It's all good and it's all me!" Darrel (BB11) also had a shuffle he performed and ended his routine with, "You can't handle my flow!" or "Pay attention and take notes on how a real 'G' [man] does it." These three are also the most aggressive, often times running over their own teammates. By constantly displaying their athletic superiority, these high-status boys are validating their position and maintaining separation from lower-status boys. Most boys usually did some "attention getting" as well when they scored.

The sexual objectification of women can easily be seen in boys' homosocial interactions. Sexually degrading remarks by boys about women and girls at the pool were common; harassment by young boys occasionally occurred. In one instance at a nearby swimming pool, an adolescent girl, approximately 16 years old, was on her stomach sunbathing with the top portion of her bikini unfastened. Adam (WB11)—the highest-ranked boy—walked over to the young lady and asked if he could put some tanning lotion on her back. After she refused his offer, Adam—with a group of boys urging him on—poured cold water on her back, causing her to instinctually raise up and reveal her breasts. While he was being disciplined, the other boys cheered him on, and Adam smiled with pride. In "The Dirty Play of Little Boys," Gary Fine (1992, 137) argues that "given the reality that many talkers have not reached puberty, we can assume that their sexual interests are more social than physiological. Boys wish to convince their peers that they are sexually mature, active, and knowledgeable" and, we might add, definitely heterosexual.

Despite the fact that there were definite racial and class differences in the boys' hierarchy, these factors had surprisingly little consequence for rankings in the power structure. Black and/or economically disadvantaged boys were just as likely to hold high positions of authority as their white and/or middle-class counterparts. Though a white middle-class boy (Adam) was the highest-ranked youth in the boys' hierarchy, two poor Black youths (Brian and Darrel) held the second and third positions in the hierarchy. Furthermore, when Adam went on a two-week vacation with his family, Darrel surpassed Brian and assumed the alpha position in the boys' social order. Nonetheless, upon his return, Adam reasserted his dominance in the group.

Structural Formation of Girls in Middle Childhood

Girls' homosocial organizational forms are distinct from boys.' The tendency toward a single hierarchy, for example, is quite rare. Social aggression (e.g., isolating a member of the group) is used to mark boundaries of femininity. These boundaries do not seem to involve a singular notion of hegemonic femininity with which to subordinate other forms or to heighten public notice of a higher-status femininity. Girls' boundaries are less defined than boys.' The girls in our study generally organized themselves in small groups ranging from two to four individuals. These groups, nonetheless, usually had one girl who was of higher status than the other girls in the clique. The highest-status girl was generally the one considered the most sociable and the most admired by others in the immediate clique as well as others in the camp. Much as Luria and Thorne (1994, 52) observed, the girls were connected by shifting alliances. Girls deal with personal conflicts by way of exclusion from the group and social manipulation. Social manipulation includes gossiping, friendship bartering, and indirectly turning the group against an individual. Contrary to the findings of many sociologists and anthropologists who only characterize aggression in physical aspects, we—like Kaj Bjorkqvist (1994) in "Sex Differences in Physical, Verbal, and Indirect Aggression"—found that girls display just as much aggression as boys but in different ways. When Elaine (WG8), for example, would not share her candy with her best friends Brandi (WG8) and Darlene (WG7), Brandi and Darlene proclaimed that Elaine could no longer be their friend. Elaine then joined another group of girls. This is quite representative of what happens when one girl is excluded from a clique. To get back at Brandi and Darlene, Elaine told her new "best friends" that Brandi liked Kevin (biracial B11) and that Darlene urinated on herself earlier that day. This soon spread throughout the camp, and Darlene and Brandi were teased for the rest of the day, causing them to cry. As Bjorkqvist (1994, 180) suggests, there is no reason to believe that girls are any less aggressive than boys. In fact, social manipulation may be more damaging than physical aggression because though physical wounds heal, gossip and group exclusion can persist eternally (or at least until the end of summer).

Unlike the boys who perform or comply with a predominant form of masculinity, no such form of hegemonic femininity was observed. Connell (1987) explains the lack of a hegemonic form of femininity as the result of the collective subordination of women to the men's homosocial hierarchy. According to Connell, since power rests in the men's (boys') sphere, there is no reason to form power relations over other women (girls). Hence, "no pressure is set up to negate or subordinate other forms of femininity in the way hegemonic masculinity must negate other masculinities" (Connell 1987, 187). Girls were inclined to gather in different groups, or cliques, reflective of various ways to define femininity; they gathered with those who defined their girlhood on the same terms. Just as Ann Beutel and Margaret Marini (1995, 436) discovered in their work, "Gender and Values," girls in our study also formed girl cliques "characterized by greater emotional intimacy, self-disclosure, and supportiveness." Intimacy helps ensure faithfulness to the group. All the girl groups—regardless of racial makeup, socioeconomic background, or age differences—had an idea of being "nice," which enhanced clique solidarity. Various girls were asked to give a definition of what it meant to be nice: "Nice just means, you know, helping each other out" (BG12); "Nice just means doing the right thing" (BG7); "Nice means . . . getting along" (WG11). Despite this notion of being nice, however, being nice in one group may be seen as being mean in another. In some groups, for example, it was considered nice for one girl to ask another if the former could have some of the other's chips at lunch. In others, though, this was considered rude; the nice, or proper, conduct was to wait until one was offered some chips. Nice was relative to the particular group. Being nice among the girls observed in this study generally entailed sharing, the aversion of physical and direct aggression, and the avoidance of selfish acts.

The organization of African American girls was somewhat unique. As mentioned, campers were divided into four age groups. In each age group, there were no more than four or five Black girls. Within these age groups, African American girls had the same structural patterns as Caucasians—small cliques of two to four in which a person may drift from clique to clique at a given time. There were no problems with the Black girls mixing with the white girls in age groups or organized activities.

During times when the campers were not restricted to specific groups (e.g., snack time, most field trips, at the pool, group games, and free time), however, the preponderance of African American girls gravitated to each other, despite age differences. This differs from previous research that notes that children in middle childhood associate with near-aged members (Absi-Semaan, Crombie, and Freeman 1993; Andersen 1993; Beutel and Marini 1995; Block 1984; Curran and Renzetti 1992; Luria and Thorne 1994). The first author also visited another camp with similar demographics and observed a similar lack of age segregation among African American girls. African American girls formed larger groups and occupied more space than white girls.

In general, Black girls were more assertive and therefore less likely to be bothered by boys. A loose hierarchy formed in which the older girls made most of the decisions for the younger ones in the group. This hierarchy was by no means hegemonic as in the boys' hierarchy. Rather, this hierarchy used a communal approach to decision making, with the older girls working to facilitate activities for the group. This process was illustrated every day as this group of girls decided which activity they would participate in at free time. The oldest girls—Brittany (BG11), Alexia (BG12), and Melanie (BG11)—would give options such as arts and crafts, checkers, basketball, and jump roping for the group to choose from. After considering all the options—taking into account what they had played the day before, the time left to participate in the activity, and the consensus of the group—the older girls indirectly shifted the focus to a particular activity that seldom received objection from the younger girls in the clique. . . .

Moreover, whereas the boys displayed little class segregation, the girls were clearly marked by class affiliations. Girls usually formed groups with other girls from their neighborhood. Most of the girls in a clique knew each other as neighbors or schoolmates. Even when girls switched groups or bartered for friendship, they often did so along class lines. This was especially evident in unstructured activities in which children could freely choose to associate with whomever they wanted (e.g., snack time and free time). The data here suggest that class and racial distinctions are more salient for girls than boys in middle childhood.

The Gender Transgression Zone

How do boys and girls negotiate boundaries in the GTZ? This area of activity—where boys and girls conduct heterosocial relations in hopes of either expanding or maintaining current gender boundaries in child culture—is where gender transgression takes place. A boy playing hand-clapping games (e.g., patty cake) or a girl completing an obstacle course that is designed to determine one's "manliness" are instances of transgression that occur in this zone. . . .

Hegemonic Masculinity in the Gender Transgression Zone

Boys spend the majority of their time trying to maintain current gender boundaries. It is through the enforcement of gender boundaries that boys construct their social status. High-status boys are especially concerned with gender maintenance because they have the most to lose. By maintaining gender boundaries, top boys secure resources for themselves—such as playing area, social prestige/status, and power. The social prestige procured by high-status boys causes lower-status boys and girls to grant deference to high-ranked boys. If a high-ranked boy insults a lower-status boy or interrupts girls' activities, he is much less likely to be socially sanctioned by boys or girls. The position of lower-status boys in the hierarchy prevents them from challenging the higher-ranked boy's authority, while the collective subordination of girls to boys inhibits much dissension from girls. Connell (1987, 187) would likely suggest that girls' deference to high-status boys is an adaptive strategy to "the global dominance of heterosexual men."

To young children, "masculinity is power" (Kaufman 1995, 16). As a social construction, then, masculinity is maintained through a hegemonic process that excludes femininity and alternate masculinities. Hence, in the GTZ, boys seldom accept deviant boys or girls. Just as boys actively participate in the maintenance of the hegemonic hierarchy by using name-calling, physical aggression, and exclusion to handle personal conflicts, these same tactics are used to handle gender transgressors. The GTZ, then, is where hegemonic masculinity flexes its social muscle.

Boys Patrolling Boys in the GTZ

High-status boys maximize the influence of hegemonic masculinity and minimize gender transgressors by identifying social deviants and labeling them as outcasts. A continuous process occurs of homosocial patrolling and stigmatizing anomalies. Boys who deviate are routinely chastised for their aberrant behavior. Two examples of this process are particularly obvious. Joseph (WB7) is a seven-year-old who is recognized as a "cry baby." He is not very coordinated and gets along better with girls than boys. Because Joseph is so young, he is not directly affected by the full scrutiny of the solidified form of hegemonic masculinity. His age still allows him the luxury of displaying certain behaviors (e.g., crying) that are discredited in subsequent stages of middle childhood. Although the first author did not observe any kids in Joseph's own age group calling Joseph names, many older boys figure that he will "probably be gay when he grows up," as stated by Daniel (WB10). Fewer of the older boys associate with Joseph during free time, and he is not allowed around the older boys as are some of the other more "hegemonically correct" younger boys.

Phillip (WB10) was rejected by all the boys, which, in turn, aided in the maintenance of hegemonic masculinity. Phillip acted rather feminine and looked feminine as well. He lacked coordination, was small in stature, and had shoulder-length hair. Phillip often played with girls and preferred stereotypically feminine activities (e.g., jump rope). It was not uncommon to hear him being referred to as a faggot, fag, or gay. He was the ultimate pariah in the boys' sphere. He was constantly rejected from all circles of boys but got along quite fine with girls. His untouchable status was exemplified clearly in two instances. First, during a game of trains and tunnels—which requires partners linking arms—all players voluntarily paired up with same-sex companions except Phillip. As parents came to pick up their children, however, cross-gendered pairs began to form. This caused little disruption. However, there came a point when a hegemonically masculine boy, Sean (WB9), should have paired up with Phillip. Upon finding out who his new partner would be, Sean violently rejected Phillip. Sean was told that if he did not accept Phillip as his partner, he would have to sit out the rest of the game. Sean screamed, "I don't care if I have to sit out the whole summer 'cause I'm not going to let that faggot touch me!" In another situation during an arts and crafts activity, Phillip finished early. When kids finished early, the staff usually asked them to help an individual who was having problems. Usually everyone accepted help. However, when Phillip attempted to assist Markus (WB9), Markus rejected him harshly. Nonetheless, Markus did accept help from Karen (WG10). Phillip threatened a boy's masculinity because Phillip had been labeled homosexual; receiving help from a girl in this particular area is nonthreatening. If Joseph's behavior continues, we expect that he will experience the same harsh rejections that Phillip received. By stigmatizing Joseph and rejecting Phillip, homophobia emerges as a cautionary tale in the GTZ that deters other boys from deviating from the norm out of fear of rejection.

The boys in our case study used Joseph and Phillip to represent what would happen to other boys who transgressed the bounds of hegemonic masculinity. If a boy started slipping from gender appropriate activities, then other boys would simply associate him with one of the two pariahs, Joseph or Phillip, or call him a fag to get him back in the hegemonic group. The boys devalue homosexuality; the threat of being labeled gay is used as a control mechanism to keep boys conforming to the norms of hegemonic masculinity. Gregory Lehne (1992, 389) says that the fear of being labeled gay "is a threat used by societies and individuals to enforce social conformity in the male role, and maintain social control . . . used in many ways to encourage certain types of male behavior and to define the limits of 'acceptable' masculinity." Talk of faggots and gays is also used to help define a boy's own masculinity.[4] By negatively talking about gays and excluding members who are presumed homosexual, individual boys are defining their own heterosexuality, while collectively they are endorsing hegemonic masculinity. Because most of these boys are not sexually mature or knowledgeable, many do not have an accurate conception of homosexuality (or, for that matter, heterosexuality) at this age. Gaybashing is another way boys can separate themselves from gender-deviant behavior.

Boys Patrolling Girls in the GTZ

Just as it is important for boys to patrol their own sex, it is equally important for boys to monitor the activities of girls and to keep them out of the boys' domain. If girls entered the boys' sphere in substantial numbers, the hegemonic hierarchy would be jeopardized. Girls who enter the boys' realm, therefore, are made to feel inadequate by the boys. The few girls who do succeed in the boys' sphere, nevertheless, are either marginalized or adopted into boys' middle-childhood culture (masculinized). Marginalization or masculinization depends on the girl's overall athletic prowess and emotional detachment while in the boys' sphere. This is illustrated by the following incident.

During one of the camp field trips, campers went to a university athletic training center. During the tennis rotation, Adrianne (WG10) was put with three boys. Adrianne, who took tennis lessons, was ignored by the boys. While the boys were arguing over the proper way to hit a backhand, Adrianne sat quietly on the sideline. When one boy finally asked her if she knew how to hit a backhand, she shook her head no. The first author knew this was incorrect because Adrianne had explained to him the proper way to hit a backhand earlier that day. Therefore, the first author asked Adrianne why she responded no. She replied, "When you're with boys, sometimes it's better to pretend like you don't know stuff because they're going to ignore you or tell you you're wrong."

Marginalization also occurs when girls meet some, but not all, of the requirements of hegemonic masculinity. The group of African American girls, for example, was marginalized. Many were just as assertive, and two were more athletic than some boys of high status. Yet, these girls remained marginal, retaining too many feminine characteristics, such as expressive acts of emotion when comforting teammates when they performed poorly in an activity. When a group of boys was asked why they did not associate with these girls who were more athletic than many of the boys in the hierarchy, Adam replied, "They're just different. I don't know about them. That whole group of them are just different. They're all weird."

Girl masculinization occurs when boys dissociate a girl from her feminine gender. The best example of a girl being adopted into a hegemonic masculine identity is Patricia (WG11). She is very athletic and can outplay many boys in basketball, the game that seemed to most signify one's masculinity at this site.[5] She also remained emotionally detached while interacting with boys. One time at the playground, Adam (WB11) created an obstacle course that he contended proved whether or not one was a "man." Some of the "manhood" tests were very dangerous—such as balancing on the rails of a high overhang—and had to be stopped. Each boy who completed a task successfully received applause and high fives. Those who did not complete successfully were laughed at because, according to the other boys, they were not "men." There was one catch to this test of masculinity—Patricia. She completed the numerous tasks faster and better than many of the boys. She did not get the screams of jubilation and high fives as did the other boys at first. As she proved her "manhood," however, she began to be accepted by the boys. By the end of the tests, Patricia was proclaimed a "man." About eight weeks later, when the first author asked a group of boys why Patricia was accepted as a member of their group, Adam, the apparent spokesman for the hierarchy, said, "Well, Patricia is not really a girl. Technically she is, but not really. I mean, come on, she acts like a boy most of the time. She even passed the 'manhood' test, remember?" Though this reveals Patricia's acceptance into the boys' hierarchy, she had to forfeit her feminine gender. As Thorne (1994) recognizes, girls who successfully transgress into the boys' activities under boys' terms do not challenge stereotypical gender norms. Hence, Patricia's participation in boys' activities "does little to challenge existing arrangements" (Thorne 1994, 133).

Hegemonic masculinity in middle childhood maintains itself in regards to girls in the GTZ. Girls are not welcome into the boys' sphere, which occupies more space. If girls partially meet standards, they are marginalized and thought of as "weird." In a way, they are almost degendered. Girls who fit all hegemonic requirements (tomboys) are conceptualized as masculine, or a boy/man. This reasoning is especially disturbing because masculinity is not only maintaining and defining itself, but it is also defining femininity.

Femininity in the Gender Transgression Zone

As previously stated, girls find various forms of femininity acceptable, despite how different the form may be from their own. With this in mind, one can understand that while some girls do not challenge gender boundaries, those who do are not stigmatized by other girls. To test this observation, a group of girls (who were stereotypically "gender appropriate") were asked during lunch one day their views about the behavior of various girls who transgressed into the boys' sphere. Speaking of Patricia (WG11)—the girl who was proclaimed a "man"—Melissa (WG11) said, "She's pretty nice," and Lucia (WG9) added, "Yeah, she's pretty cool. . . . She just likes to do different stuff. There's nothing wrong with that." They were then asked about the various members of the Black girl clique, and Melissa responded, "They're nice to [us]." When the girls were specifically asked if there was anything wrong with the way these gender transgressors behaved, Melissa and Lucia simply said no. Even Robin (WG10), who was not completely comfortable with the actions of these transgressors, replied, "I guess not. They just have their own way of acting. I just don't think it's very lady-like acting." As one can see, gender transgression is virtually accepted among even gender-traditional girls. Nevertheless, girls deal with clique deviants—those who are not "nice" relative to the clique's definition—just as they handle personal conflicts: exclusion from a particular group and social manipulation.

Girls Patrolling Girls in the GTZ

As girls get older, they recognize the higher value that society puts on masculine traits as well as the resources accumulated in the boys' sphere. Girls also see masculinity as power (Connell 1987). With increasing encouragement from the larger society (parents, teachers, and other pro-feminist role models), many girls attempt to access these resources as they mature. It should be noted, however, that high-status girls in these small groups also have social power in their cliques. The highest-ranked girl largely dictated who was gossiped about and who would be banished from the group. Yet, girls' resources were limited in comparison to their masculine-gendered playmates' because their resources did not extend much further than their small clique. Interestingly, girls who dare to participate in the boys' realm not only avoid stigmatization from most girls but are often praised by other girls if they succeed in the boys' sphere. For the most part, girls only receive restrictions from the prime agents of hegemonic masculinity at play—boys. Though girls' relations generally consist of small, intimate groupings when dealing with each other, large group affiliation and support seem to be the gender strategy when girls transgress onto traditional boys' turf. This was observed frequently throughout the summer.

Whenever a girl beat a boy in an athletic event, girls, as a collectivity, cheered them on despite age differences. During a Connect Four contest, Travis (WB9), the champion, was bragging about winning—especially when he beat girls. He would say, "It only takes me two minutes to beat girls," and "Girls aren't a challenge." This changed, however, when Corisa (BG6) started to play. Corisa beat Travis four times in a row. Girls of all ages rallied behind Corisa. For the duration of the day, girls praised Corisa, and some even introduced their parents to Corisa in admiration. One introduction went as follows: "Mommy, this is Corisa. She beats boys in Connect Four."

The best example of group solidarity in resistance to boys' dominance was provided one day when leaving the swimming pool. Molly (WG9)—whose eyes were irritated by chlorine and was basically walking to the locker room with her eyes closed—accidentally entered the boys' locker room while the campers were changing. Many of the boys laughed at her and ridiculed Molly for her mistake. Brian (BB11) said, "She just wanted to look at our private stuff," and Thomas (WB12) called her a "slut." Molly started to cry. Girls, however, came to Molly's rescue. While Molly's immediate clique comforted her, the other girls scared off boys who attempted to harass Molly for the rest of the day. Crysta (BG9) and Brittany (BG11) were the most effective protectors. This was surprising because even though Molly and Crysta were in the same age group, they did not get along, and Brittany—who is in the oldest group—to the best of our knowledge, had never even talked to Molly. As a gender strategy, girls—regardless of age, class, or racial differences—united together to combat the dominance of boys.

Girls Patrolling Boys in the GTZ

Without a uniform or constant form of femininity, girls were more lenient to both girls and boys when either ventured into the GTZ. Girls accepted Joseph and Phillip, both gender-deviant boys, into all their activities without a problem. These boys, nevertheless, had to adhere to the same principles of "niceness" as did the girls. If the boys did not, they were punished in the same manner as girls—exclusion and social manipulation. When Joseph (WB7) did not share his "Now and Later" candy during lunch one day with the group of girls he was eating with, he soon found himself eating alone and the subject of much gossip in the girls' sphere.

"Alphas Rule! Others Drool!" or How High-Status Boys Direct Change in the Gtz

The top-ranked boys in the hierarchy direct the actions of all the boys who aspire to hegemonic masculinity (or are, at least, complicit with it). High-status boys are primarily concerned with maintaining gender boundaries to retain status and all the luxuries that are a result of being hegemonically masculine. Dominant boys make decisions for the group and can manipulate the other boys to sustain high status and its privileges. Examples of status privileges include being picked first for teams, getting first dibs on other people's lunches, being allowed to cut in line, and being freed by other males during prison ball—a game similar to dodge ball—with no reciprocal obligation to free low- or middle-status boys.

High-status boys have the unique power of negotiating gender boundaries by accepting, denying, or altering gender codes. The power of high-status boys to alter gender boundaries was strikingly borne out by a series of events that, for weeks, redefined a feminine gender-stereotyped activity, hand-clapping games, into a hegemonically masculine one. . . .

Here is how the defeminization occurred. One day, right before the closing of the camp, Adam—the highest-ranked boy—was the only boy left waiting for his mother to pick him up. Four girls remained as well and were performing the *Rockin' Robin* hand-clapping routine. When one of the girls left, one of the three remaining girls asked Adam if he would like to learn the routine. He angrily replied, "No, that's girly stuff." Having been a camp counselor for three years, the first author knows every clapping routine from *Bo Bo See Aut In Totin* to *Miss Susie's Steamboat.* He, therefore, volunteered. The girls were amazed that he knew so many of what they referred to as "their" games. After a while, only two girls remained and *Rockin' Robin* requires four participants. Surprisingly, Adam asked to learn. Before he left, Adam had learned the sequence and was having a good time.

We believe that Adam transgressed for three reasons. One, all the other boys were gone, so there were no relevant or important (to him) witnesses to his transgression. Thorne (1994, 54) repeatedly states that witnesses hinder gender deviance: "Teasing makes cross-gender interaction risky, increases social distance between girls and boys, and has the effect of making and policing gender boundaries." Second, Adam saw the first author participate freely in an activity that was previously reserved for girls. Third, as the highest-ranked boy, Adam has a certain degree of freedom that allows him to transgress with little stigmatization. Thorne asserts that the highest-status boy in a hierarchy has "extensive social leeway" (p. 123) since his masculinity is rarely questioned. The next day, Adam was seen perfecting the routine he learned the day before. Many of the boys looked curiously and questioned why Adam was partaking in such an activity. Soon after, other boys started playing, and boys and girls were interacting heterosocially in what was formerly defined as a "girls-only" activity. Cross-gendered hand-clapping games continued for the rest of the summer and remained an area in which both girls and boys could come together. Defeminization occurred because Adam—the highest-status boy—set the standard and affirmed this type of entertainment as acceptable for boys. This incident supports our view that high-status boys control gender negotiations by showing that gender boundaries can be modified if someone of high status changes the standard of hegemonic masculinity.

To make hand-clapping more masculine, nonetheless, the first author documented boys changing the verses of the most popular hand-clapping game, *Rockin' Robin,* to further defeminize the activity. One of the original verses is "All the little birdies on J-Bird Street like to hear the robin go tweet, tweet, tweet." The boys changed this to "All the little birdies on J-Bird Street like to hear robin say eat my meat!" About a month later, the first author discovered another altered verse from the boys. They changed "Brother's in jail waiting for bail" to "Brother's in jail raising hell!" Since these verses were not condoned at the camp—though we are sure the children used them out of the hearing distance of counselors—girls cleverly modified one of the profane verses by singing, "Brother's in jail raising H-E- double hockey sticks!" This, too, the boys picked up and started applying as their own. Hand-clapping games moved from the girls' sphere to the GTZ. Defeminization of hand clapping exposes the constant fluctuation and restructuring of gender norms in childhood play.

Boys in middle childhood organize themselves in a definite hierarchy that is run by high-status boys in accordance with the hegemonic form of masculinity that they embody and police in the GTZ. Boys are not accepting of deviant boys or girls. Gender deviants are handled by teasing and name-calling, marginalization and exclusion from the group, and physical aggression. High-status boys, though, have the unique power to negotiate gender boundaries by either accepting, denying, or altering gender codes. Girls who enter the boys' realm are made to feel inadequate by the boys. Those girls who do succeed in the boys' sphere, nevertheless, are either marginalized from or masculinized into boys' middle-childhood culture. They are forced to leave their femininity behind if they want to cross the border fully. Therefore, no feminization of hegemonic masculinity is allowed. As can be seen in the hand-clapping phenomenon, the redefinition entails defeminization.

Notes

1. To the best of our knowledge, this concept was first set out in his book *Gender and Power* (Connell 1987, 183–88).

2. Bird (1996) explains how homosocial interactions maintain gender boundaries among adult men. Beutel and Marini (1995) discuss the contrasting value systems of males and females.

3. The names of children in this study are pseudonyms.

4. This may be part and parcel of what McCreary (1994) refers to as the universal avoidance of femininity: Homophobia may be a rejection of the "abnormality" of being attracted to boys (i.e., being "girlish").

5. At the other site that the first author visited, football was the most masculinizing athletic activity.

References

Absi-Semaan, N., G. Crombie, and C. Freeman. 1993. Masculinity and femininity in middle childhood: Development and factor analyses. *Sex Roles 28*(3/4): 187–206.

Andersen, Margaret L. 1993. *Thinking about women.* New York: Macmillan.

Beutel, Ann M., and Margaret M. Marini. 1995. Gender and values. *American Sociological Review 60*(3): 436–38.

Bird, Sharon R. 1996. Welcome to the men's club: Homosociality and the maintenance of hegemonic masculinity. *Gender & Society 10*(2): 120–32.

Bjorkqvist, Kaj. 1994. Sex differences in physical, verbal, and indirect aggression: A review of recent research. *Sex Roles* 30 (3/4): 177–88.

Connell, R. W. 1987. *Gender & power.* Stanford, CA: Stanford University Press.

———. 1995. Masculinities. Berkeley: University of California Press.

Curran, Daniel J., and Claire M. Renzetti. 1992. *Women, men, and society,* 2d ed. Needham Heights, MA: Allyn & Bacon.

Fine, Gary Alan. 1992. The dirty play of little boys. In *Men's lives,* edited by Michael S. Kimmel and Michael A. Messner. New York: Macmillan.

Hawkesworth, Mary. 1997. Confounding gender. *Signs: Journal of Women in Culture and Society 22*(3): 649–86.

Kaufman, Michael. 1995. The construction of masculinity and the triad of men's violence. In *Men's lives,* edited by Michael Kimmel and Michael Messner. New York: Macmillan.

Lehne, Gregory K. 1992. Homophobia among men: Supporting and defining the male role. In *Men's lives,* edited by Michael S. Kimmel and Michael Messner. New York: Macmillan.

Luria, Zella, and Barrie Thorne. 1994. Sexuality and gender in children's daily worlds. In *Sociology: Windows on society,* edited by John W. Heeren and Marylee Mason. Los Angeles: Roxbury.

McCreary, Donald R. 1994. The male role and avoiding femininity, *Sex Roles 31*(9): 517–32.

Messner, Michael A. 1992. *Power at play: Sports and the problem of masculinity.* Boston: Beacon.

———. 1994. The meaning of success: The athletic experience and the development of male identity. In *Sociology: Windows on Society,* edited by John W. Heeren and Marylee Mason. Los Angeles: Roxbury.

———. 1998. The limits of "the male sex role": An analysis of the men's liberation and men's rights movements discourse. *Gender & Society 12*(3): 255–76.

Thorne, Barrie. 1994. *Gender play: Girls and boys in school.* New Brunswick, NJ: Rutgers University Press.

West, Candace, and Don H. Zimmerman. 1987. Doing gender. *Gender & Society 1*(2): 125–51.

Introduction to Reading 16

The following reading by sociologist Emily Kane helps us to understand the role of parents in enforcing gender, particularly when children prefer not to conform to normative gender patterns. She describes her findings from interviews with forty-two New England parents, primarily from southern and central Maine. These twenty-four mothers and eighteen fathers reflect a diverse group across race, social class, and sexual orientations. Although the parents typically had more than one child, the interviews focused on preschool children—twenty-two sons and twenty daughters. Hearing how parents respond to gender-atypical behavior provides deeper meaning to the process of gender socialization.

1. Why are girls not challenged more by parents when they do not conform to gender expectations?
2. At what point do parents stop preschoolers who are not conforming to gender, particularly boys?
3. How do mothers and fathers see their roles differently in the gender socialization process?

"No Way My Boys Are Going to Be Like That!"

Parents' Responses to Children's Gender Nonconformity

Emily W. Kane

Parents begin gendering their children from their very first awareness of those children, whether in pregnancy or while awaiting adoption. Children themselves become active participants in this gendering process by the time they are conscious of the social relevance of gender, typically before the age of two. I address one aspect of this process of parents doing gender, both for and with their children, by exploring how parents respond to gender nonconformity among preschool-aged children. As West and Zimmerman (1987, 136) note, "to 'do' gender is not always to live up to normative conceptions of femininity or masculinity; it is to engage in behavior *at the risk of gender assessment.*" I argue that many parents make efforts to stray from and thus expand normative conceptions of gender. But for their sons in particular, they balance this effort with conscious attention to producing a masculinity approximating hegemonic ideals. This balancing act is evident across many parents I interviewed regardless of gender, race/ethnicity, social class, sexual orientation, and partnership status. But I also argue that within that broader pattern are notable variations. Heterosexual fathers play a particularly central role in accomplishing their sons' masculinity and, in the process, reinforce their own as well. Their expressed motivations for that accomplishment work often involve personal endorsement of hegemonic masculinity. Heterosexual mothers and gay parents, on the other hand, are more likely to report motivations that invoke accountability to others for crafting their sons' masculinity in accordance with hegemonic ideals.

* * *

Responses to Gender Nonconformity

Mothers and fathers, across a variety of social locations, often celebrated what they perceived as gender nonconformity on the part of their young daughters. They reported enjoying dressing their daughters in sports-themed clothing, as well as buying them toy cars, trucks, trains, and building toys. Some described their efforts to encourage, and pleased reactions to, what they considered traditionally male activities such as t-ball, football, fishing, and learning to use tools. Several noted that they make an effort to encourage their young daughters to aspire to traditionally male occupations and commented favorably on their daughters as "tomboyish," "rough and tumble," and "competitive athletically." These positive responses were combined with very little in the way of any negative response. The coding of each interviewee for the combination of positive/neutral and negative responses summarizes this pattern clearly: Among parents commenting about daughter(s), the typical combination was to express only positive responses. For example, a white, middle-class, heterosexual mother noted approvingly that her five-year-old daughter "does a lot of things that a boy would do, and we encourage that," while a white, upper-middle-class, lesbian mother reported that she and her partner intentionally "do [a lot] of stuff that's not stereotypically female" with their daughter. Similarly, a white, upper-middle-class, heterosexual father indicated with relief that his daughter is turning out to be somewhat "boyish": "I never wanted a girl who was a little princess, who was so fragile. . . . I want her to take on more masculine characteristics." An African American, working-class, heterosexual father also noted this kind of preference: "I don't want her just to color and play with dolls, I want her to be athletic."

A few parents combined these positive responses with vague and general negative responses. But these were rare and expressed with little sense of concern, as in the case of an African American, low-income, heterosexual mother who offered positive responses but also noted limits regarding her daughter: "I wouldn't want her to be too boyish, because she's a girl." In addition, no parents expressed only negative responses. These various patterns suggest that parents made little effort to accomplish their daughters' gender in accordance with any particular conception of femininity, nor did they express any notable sense of accountability to such a conception. Instead, parental responses may suggest a different kind of gendered phenomenon closely linked to the pattern evident in responses toward sons: a devaluing of traditionally feminine pursuits and qualities. Although many parents of daughters reported positive responses to what they consider typical interests and behaviors for a girl, most also celebrated the addition of atypical pursuits to their daughters' lives, and very few noted any negative response to such additions.

It is clear in the literature that there are substantial gendered constraints placed on young girls, and any devaluation of the feminine is potentially such a constraint. But the particular constraint of negative responses by parents to perceived gender nonconformity was not evident in my interview results. It is possible that negative response from parents to perceived departures from traditional femininity would be more notable as girls reach adolescence. Pipher (1998, 286) argues that parents of young girls resist gender stereotypes for their daughters but that "the time to really worry is early adolescence. That's when the gender roles get set in cement, and that's when girls need tremendous support in resisting cultural definitions of femininity." Thorne (1994, 170) invokes a similar possibility, claiming that girls are given more gender leeway than boys in earlier childhood, "but the leeway begins to tighten as girls approach adolescence and move into the heterosexualized gender system of teens and adults." The question of whether negative parental responses might be less gender differentiated in adolescence cannot be addressed with my interview data and remains instead an intriguing question for future research.

In stark contrast to the lack of negative response for daughters, 23 of 31 parents of sons expressed at least some negative responses, and 6 of these offered only negative responses regarding what they perceived as gender nonconformity. Of 31 parents, 25 did indicate positive responses as well, but unlike references to their daughters, they tended to balance those positive feelings and actions about sons with negative ones as well. The most common combination was to indicate both positive and negative responses.

Domestic Skills, Nurturance, and Empathy

Parents accepted, and often even celebrated, their sons' acquisition of domestic abilities and an orientation toward nurturance and empathy. Of the 25 parents of sons who offered positive/neutral responses, 21 did so in reference to domestic skills, nurturance, and/or empathy. For example, they reported allowing or encouraging traditionally girl toys such as dolls, doll houses, kitchen centers, and tea sets, with that response often revolving around a desire to encourage domestic competence, nurturance, emotional openness, empathy, and nonviolence as attributes they considered nontraditional but positive for boys. These parents were reporting actions and sentiments oriented toward accomplishing gender in what they considered a less conventional manner. One white, low-income, heterosexual mother taught her son to cook, asserting that "I want my son to know how to do more than boil water, I want him to know how to take care of himself." Another mother, this one a white, working-class, heterosexual parent, noted that she makes a point of talking to her sons about emotions: "I try to instill a sense of empathy in my sons and try to get them to see how other people would feel." And a white, middle-class, heterosexual father emphasized domestic competence when he noted that it does not bother him for his son to play with dolls at his cousin's house: "How then are they going to learn to take care of their children if they don't?" This positive response to domestic activities is consistent with recent literature on parental coding of toys as masculine, feminine, or neutral, which indicates that parents are increasingly coding kitchens and in some cases dolls as neutral rather than exclusively feminine (Wood, Desmarais, and Gugula 2002).

In my study, mothers and fathers expressed these kinds of efforts to accomplish gender differently for their sons with similar frequency, but mothers tended to express them with greater certainty, while fathers were less enthusiastic and more likely to include caveats. For example, this mother described her purchase of a variety of domestic toys for her three-year-old son without ambivalence: "One of the first big toys [I got him] was the kitchen center. . . . We cook, he has an apron he wears. . . . He's got his dirt devil vacuum and he's got his baby [doll]. And he's got all the stuff to feed her and a highchair" (white, low-income, heterosexual mother).

Some mothers reported allowing domestic toys but with less enthusiasm, such as a white, low-income, heterosexual mother who said, regarding her three-year-old son, "He had been curious about dolls and I just said, you know, usually girls play with dolls, but it's okay for you to do it too." But this kind of caution or lack of enthusiasm, even in a response coded as positive or neutral due to its allowance of gender-atypical behavior, was more evident among fathers, as the following quote illustrates: "Occasionally, if he's not doing something, I'll encourage him to maybe play with his tea cups, you know, occasionally. But I like playing with his blocks better anyway" (white, middle-class, heterosexual father).

Thus, evident among both mothers and fathers, but with greater conviction for mothers, was widespread support among parents for working to "undo" gender at the level of some of their sons' skills and values. However, this acceptance was tempered for many parents by negative responses to any interest in what I will refer to as iconic feminine items, attributes, or activities, as well as parental concern about homosexuality.

Icons of Femininity

A range of activities and attributes considered atypical for boys were met with negative responses, and for a few parents (3 of 31 parents of sons) this even included the kind of domestic toys and nurturance noted above. But more common were negative responses to items, activities, or attributes that could be considered icons of femininity. This was strikingly consistent with Kimmel's (1994, 119) previously noted claim that the "notion of anti-femininity lies at the heart of contemporary and historical constructions of manhood," and it bears highlighting that this was evident among parents of very young children. Parents of sons reported negative responses to their sons' wearing pink or frilly clothing; wearing skirts, dresses, or tights; and playing dress up in any kind of feminine attire. Nail polish elicited concern from a number of parents too, as they reported young sons wanting to have their fingernails or toenails polished. Dance, especially ballet, and Barbie dolls were also among the traditionally female activities often noted negatively by parents of sons. Of the 31 parents of sons, 23 mentioned negative reactions to at least one of these icons.

Playing with nail polish and makeup, although tolerated by some parents, more often evoked negative responses like this one, from a white, upper-middle-class, gay father, speaking about his four-year-old son's use of nail polish: "He put nail polish on himself one time, and I said 'No, you can't do that, little girls put nail polish on, little boys don't.' "

Barbie dolls are an especially interesting example in that many parents reported positive responses to baby dolls, viewing these as encouraging nurturance and helping to prepare sons for fatherhood. Barbie, on the other hand, an icon of femininity, struck many parents of sons as more problematic. Barbie was often mentioned when parents were asked whether their child had ever requested an item or activity more commonly associated with the other gender. Four parents–three mothers and one father–indicated that they had purchased a Barbie at their son's request, but more often parents of sons noted that they would avoid letting their son have or play with Barbie dolls. Sometimes this negative response was categorical, as in the quote above in which a mother of a three-year-old son noted that "there's not many toys I wouldn't get him, except Barbie." A father offers a similar negative reaction to Barbie in relation to his two young sons: "If they asked for a Barbie doll, I would probably say no, you don't want [that], girls play with [that], boys play with trucks" (white, middle-class, heterosexual father).

Along with material markers of femininity, many parents expressed concern about excessive emotionality (especially frequent crying) and passivity in their sons. For example, a white, upper-middle-class, heterosexual father, concerned about public crying, said about his five-year-old son, "I don't want him to be a sissy. . . . I want to see him strong, proud, not crying like a sissy." Another father expressed his frustration with his four-year-old son's crying over what the father views as minor injuries and indicated action to discourage those tears: "Sometimes I get so annoyed, you know, he comes [crying], and I say, 'you're not hurt, you don't even know what hurt is yet,' and I'm like 'geez, sometimes you are such a little wean,' you know?" (white, middle-class, heterosexual father).

Passivity was also raised as a concern, primarily by fathers. For example, one white, middle-class, heterosexual father of a five-year-old noted that he has told his son to "stop crying like a girl," and also reported encouraging that son to fight for what he wants: "You just go in the corner and cry like a baby. I don't want that. If you decide you want [some] thing, you are going to fight for it, not crying and acting like a baby and hoping that they're going to feel guilty and give it to you."

A mother who commented negatively about passivity even more directly connected her concern to how her son might be treated: "I do have concerns. . . . He's passive, not aggressive. . . . He's not the rough and tumble kid, and I do worry about him being an easy target" (white, working-class, heterosexual mother).

Taken together, these various examples indicate clearly the work many parents are doing to accomplish gender with and for their sons in a manner that distances those sons from any association with femininity. This work was not evident among all parents of sons. But for most parents, across racial, class, and sexual orientation categories, it was indeed evident.

Homosexuality

Along with these icons of feminine gender performance, and arguably directly linked to them, is the other clear theme evident among some parents' negative responses to perceived gender nonconformity on the part of their sons: fear that a son either would be or would be perceived as gay. Spontaneous connections of gender nonconformity and sexual orientation were not evident in parents' comments about daughters, nor among gay and lesbian parents, but arose for 7 of the 27 heterosexual parents who were discussing sons.

The fact that the connection between gender performance and sexual orientation was not raised for daughters, and that fear of homosexuality was not spontaneously mentioned by parents of daughters whether in connection to gender performance or not, suggests how closely gender conformity and heterosexuality are linked within hegemonic constructions of masculinity. Such connections might arise more by adolescence in relation to daughters, as I noted previously regarding other aspects of parental responses to gender nonconformity. But for sons, even among parents of very young children, heteronormativity appears to play a role in shaping parental responses to gender nonconformity, a connection that literature on older children and adults indicates is made more for males than females (Antill 1987; Hill 1999; Kite and Deaux 1987; Sandnabba and Ahlberg 1999). Martin's (2005) recent analysis also documents the importance of heteronormativity in the advice offered to parents by experts. She concludes that expert authors of child-rearing books and Web sites are increasingly supportive of gender-neutral child rearing. But especially for sons, that expert support is limited by implicit and even explicit invocations of homosexuality as a risk to be managed. As McCreary (1994, 526) argues on the basis of experimental work on responses to older children and adults, "the asymmetry in people's responses to male and female gender role deviations is motivated, in part, by the implicit assumption that male transgressions are symptomatic of a homosexual orientation." This implicit assumption appears to motivate at least some parental gender performance management among heterosexual parents, even for children as young as preschool age. Given the connections between male heterosexuality and the rejection of femininity noted previously as evident in theories of hegemonic masculinity, the tendency for parents to associate gender performance and sexual orientation for sons more than daughters may also reflect a more general devaluation of femininity.

Mothers Versus Fathers in the Accomplishment of Masulinity

. . . Although both mothers and fathers were equally likely to express a combination of positive and negative responses to their sons' perceived gender nonconformity, with domestic skills and empathy accepted and icons of femininity rejected, the acceptance was more pointed for mothers, and the rejection was more pointed for fathers. More fathers (11 of 14) than mothers (12 of 17) of sons indicated negative reactions to at least one of the icons discussed. Fathers also indicated more categorically negative responses: 7 of the 14 fathers but only 2 of the 17 mothers reported simply saying "no" to requests for things such as Barbie dolls, tea sets, nail polish, or ballet lessons, whether actual requests or hypothetical ones. Although fewer parents referred to excessive emotionality and passivity as concerns, the 6 parents of sons who did so included 4 fathers and 2 mothers, and here too, the quotes indicate a more categorical rejection by fathers.

Another indication of more careful policing of icons of femininity by fathers is evident in comments that placed age limitations on the acceptability of such icons. Four fathers (but no mothers) commented with acceptance on activities or interests that they consider atypical for boys but went on to note that these would bother them if they continued well past the preschool age range. The following quote from a father is typical of these responses. After noting that his four-year-old son sometimes asks for toys he thinks of as "girl toys," he went on to say, "I don't think it will ruin his life at this age but . . . if he was 12 and asking for it, you know, My Little Pony or Barbies, then I think I'd really worry" (white, middle-class, heterosexual father). While comments like this one were not coded as negative responses, since they involved acceptance, I mention them here as they are consistent with the tendency for fathers to express particular concern about their sons' involvement with icons of femininity.

Three of 15 heterosexual mothers and 4 of 12 heterosexual fathers of sons responded negatively to the possibility of their son's being, or being perceived as, gay. These numbers are too small to make conclusive claims comparing mothers and fathers. But this pattern is suggestive of another arena in which fathers—especially heterosexual fathers—may stand out, especially taken together with another pattern. Implicit in the quotes offered above related to homosexuality is a suggestion that heterosexual fathers may feel particularly responsible for crafting their sons' heterosexual orientation. In addition, in comparison to mothers, their comments are less likely to refer to fears for how their son might be treated by others if he were gay and more likely to refer to the personal disappointment they anticipate in this hypothetical scenario. I return to consideration of these patterns in my discussion of accountability below.

Parental Motivations for the Accomplishment of Masculinity

The analysis I have offered thus far documents that parents are aware of their role in accomplishing gender with and for their sons. Although some parents did speak of their sons as entirely "boyish" and "born that way," many reported efforts to craft a hegemonic masculinity. Most parents expressed a very conscious awareness of normative conceptions of masculinity (whether explicitly or implicitly). Many, especially heterosexual mothers and gay parents, expressed a sense that they felt accountable to others in terms of whether their sons live up to those conceptions. In numerous ways, these parents indicated their awareness that their sons' behavior was at risk of gender assessment, an awareness rarely noted with regard to daughters. Parents varied in terms of their expressed motivations for crafting their sons' masculinity, ranging from a sense of measuring their sons against their own preferences for normative masculinity (more common among heterosexual fathers) to concerns about accountability to gender assessment by peers, other adults, and society in general (more common among heterosexual mothers and gay parents, whether mothers or fathers).

* * *

Conclusion

The interviews analyzed here, with New England parents of preschool-aged children from a diverse array of backgrounds, indicate a considerable endorsement by parents of what they perceive as gender nonconformity among both their sons and their daughters. This pattern at first appears encouraging in terms of the prospects for a world less constrained by gendered expectations for children. Many parents respond positively to the idea of their children's experiencing a greater range of opportunities, emotions, and interests than those narrowly defined by gendered stereotypes, with mothers especially likely to do so. However, for sons, this positive response is primarily limited to a few attributes and abilities, namely, domestic skills, nurturance, and empathy. And it is constrained by a clear recognition of normative conceptions of masculinity (Connell 1987, 1995). Most parents made efforts to accomplish, and either endorsed or felt accountable to, an ideal of masculinity that was defined by limited emotionality, activity rather than passivity, and rejection of material markers of femininity. Work to accomplish this type of masculinity was reported especially often by heterosexual fathers; accountability to approximate hegemonic masculinity was reported especially often by heterosexual mothers, lesbian mothers, and gay fathers. Some heterosexual parents also invoked sexual orientation as part of this conception of masculinity, commenting with concern on the possibility that their son might be gay or might be perceived as such. No similar pattern of well-defined normative expectations or accountability animated responses regarding daughters, although positive responses to pursuits parents viewed as more typically masculine may well reflect the same underlying devaluation of femininity evident in negative responses to gender nonconformity among sons.

In the broader study from which this particular analysis was drawn, many parents invoked biology in explaining their children's gendered tendencies. Clearly, the role of biological explanations in parents' thinking about gender merits additional investigation. But one of the things that was most striking to me in the analyses presented here is how frequently parents indicated that they took action to craft an appropriate gender performance with and for their preschool-aged sons, viewing masculinity as something they needed to work on to accomplish. These tendencies are in contrast to what Messner (2000) summarizes eloquently in his essay on a gender-segregated preschool sports program. He observes a highly gender-differentiated performance offered by the boys' and girls' teams during the opening ceremony of the new soccer season, with one of the girls' teams dubbing themselves the Barbie Girls, while one of the boys' teams called themselves the Sea Monsters. He notes that parents tended to view the starkly different approaches taken by the boys and girls as evidence of natural gender differences. "The parents do not seem to read the children's performances of gender as social constructions of gender. Instead, they interpret them as the inevitable unfolding of natural, internal differences between the sexes" (Messner 2000, 770).

I agree with Messner (2000) that this tendency is evident among parents, and I heard it articulated in some parts of the broader project from which the present analysis is drawn. I began this project expecting that parents accept with little question ideologies that naturalize gender difference. Instead, the results I have presented here demonstrate that parents are often consciously aware of gender as something that they must shape and construct, at least for their sons. This argument extends the literature on the routine accomplishment of gender in childhood by introducing evidence of conscious effort and awareness by parents as part of that accomplishment. This awareness also has implications for efforts to reduce gendered constraints on children. Recognition that parents are sometimes consciously crafting their children's gender suggests the possibility that they could be encouraged to shift that conscious effort in less gendered directions.

In addition to documenting this parental awareness, I am also able to extend the literature by documenting the content toward which parents' accomplishment work is oriented. The version of hegemonic masculinity I have argued underlies parents' responses is one that includes both change and stability. Parental openness to domestic skills, nurturance, and empathy as desirable qualities in their sons likely represents social change, and the kind of agency in the accomplishment of gender to which Fenstermaker and West (2002) refer. As Connell (1995) notes, hegemonic masculinity is historically variable in its specific content, and the evidence presented in this article suggests that some broadening of that content is occurring. But the clear limits evident within that broadening suggest the stability and power of hegemonic conceptions of masculinity. The parental boundary maintenance work evident for sons represents a crucial obstacle limiting boys' options, separating boys from girls, devaluing activities marked as feminine for both boys and girls, and thus bolstering gender inequality and heteronormativity.

Finally, along with documenting conscious awareness by parents and the content toward which their accomplishment work is oriented, my analysis also contributes to the literature by illuminating the process motivating parental gender accomplishment. The heterosexual world in general, and heterosexual fathers in particular, play a central role in that process. This is evident in the direct endorsement of hegemonic masculinity many heterosexual fathers expressed and in the accountability to others (presumably heterosexual others) many heterosexual mothers, lesbian1nothers, and gay fathers expressed. Scholarly investigations of the routine production of gender in childhood, therefore, need to pay careful attention to the role of heterosexual fathers as enforcers of gender boundaries and to the role of accountability in the process of accomplishing gender. At the same time, practical efforts to loosen gendered constraints on young children by expanding their parents' normative conceptions of gender need to be aimed at parents in general and especially need to reach heterosexual fathers in particular. The concern and even fear many parents—especially heterosexual mothers, lesbian mothers, and gay fathers—expressed about how their young sons might be treated if they fail to live up to hegemonic conceptions of masculinity represent a motivation for the traditional accomplishment of gender. But those reactions could also serve as a motivation to broaden normative conceptions of masculinity and challenge the devaluation of femininity, an effort that will require participation by heterosexual fathers to succeed.

References

Antill, John K. 1987. Parents' beliefs and values about sex roles, sex differences, and sexuality. *Review of Personality and Social Psychology* 7:294–328.

Connell, R. W. 1987. *Gender and power.* Stanford, CA: Stanford University Press.

———. 1995. *Masculinities.* Berkeley: University of California Press.

Fenstermaker, Sarah, and Candace West, eds. 2002. *Doing gender, doing difference.* New York: Routledge.

Golombock, Susan, and Fiona Tasker. 1994. Children in lesbian and gay families: Theories and evidence. *Annual Review of Sex Research* 5:73–100.

Gottman, Julie Schwartz. 1990. Children of gay and lesbian parents. *Marriage and Family Review* 14:177–96.

Hill, Shirley A. 1999. *African American children.* Thousand Oaks, CA: Sage.

Kimmel, Michael S. 1994. Masculinity as homophobia. In *Theorizing masculinities,* edited by Harry Brod. Thousand Oaks, CA: Sage.

Kite, Mary E., and Kay Deaux. 1987. Gender belief systems: Homosexuality and the implicit inversion theory. *Psychology of Women Quarterly* 11:83–96.

Martin, Karin A. 2005. William wants a doll, can he have one? Feminists, child care advisors, and gender-neutral child rearing. *Gender & Society* 20:1–24.

McCreary, Donald R. 1994. The male role and avoiding femininity. *Sex Roles* 31:517–31.

Messner, Michael. 2000. Barbie girls versus sea monsters: Children constructing gender. *Gender & Society* 14:765–84.

Patterson, Charlotte J. 1992. Children of lesbian and gay parents. *Child Development* 63:1025–42.

Pipher, Mary. 1998. *Reviving Ophelia.* New York: Ballantine Books.

Sandnabba, N. Kenneth, and Christian Ahlberg. 1999. Parents' attitudes and expectations about children's cross-gender behavior. *Sex Roles* 40:249–63.

Stacey, Judith, and Timothy J. Bibliarz. 2001. (How) does the sexual orientation of parents matter? *American Sociological Review* 66:159–83.

Thorne, Barrie. 1994. *Gender play.* New Brunswick, NJ: Rutgers University Press.

West, Candace, and Don Zimmerman. 1987. Doing gender. *Gender & Society* 1:124–51.

Wood, Eileen, Serge Desmarais, and Sara Gugula. 2002. The impact of parenting experience on gender stereotyped toy play of children. *Sex Roles* 47:39–49.

Introduction to Reading 17

Children's books are an important element of learning, but as Lori Baker-Sperry and Liz Grauerholz find, they also reinforce gender ideals for girls by emphasizing feminine beauty. This analysis coded feminine beauty for tales with human characters and compares Grimm's Brothers fairy tales from 1857 (Grimm & Grimm, 1857) with those published in 1992. Baker-Sperry and Grauerholz compared those fairy tales that have endured over time to those that have not. Unfortunately, the message for children in these stories that have endured time is that of reinforcing gender-normative behavior.

1. How does the description of feminine beauty differ from masculine beauty in these fairy tales?
2. What is the relationship between fairy tales most likely to emphasize feminine beauty and their likelihood of being reproduced in popular culture today?
3. Compare these findings to those of Kane and consider how socialization into gender shapes the behaviors of boys and girls.

The Pervasiveness and Persistence of the Feminine Beauty Ideal in Children's Fairy Tales

Lori Baker-Sperry and Liz Grauerholz

The institution of gender relies in part on what Lorber (1994, 30–31) referred to as gender imagery—"the cultural representations of gender and embodiment of gender in symbolic language and artistic productions that reproduce and legitimate gender statuses." Children's fairy tales, which emphasize such things as women's passivity and beauty, are indeed gendered scripts and serve to legitimatize and support the dominant gender system.

The present study focuses on one prominent message that is represented in many children's fairy tales: the feminine beauty ideal. The feminine beauty ideal—the socially constructed notion that physical attractiveness is one of women's most important assets, and something all women should strive to achieve and maintain—is of particular interest to feminist scholars. While the feminine beauty ideal is viewed largely as an oppressive, patriarchal practice that objectifies, devalues, and subordinates women (e.g., Bartky 1990; Bordo 1993; Freedman 1986; Wolf 1991), it is acknowledged that many women willingly engage in "beauty rituals" and perceive being (or becoming) beautiful as empowering, not oppressive (Dellinger and Williams 1997). A further paradox of the feminine beauty ideal is that in a patriarchal system, those women who seek or gain power through their attractiveness are often those who are most dependent on men's resources.

This study investigates the extent to which the feminine beauty ideal has persisted over nearly 150 years by examining its pervasiveness, and tracing its survival, in children's fairy tales. We begin by investigating the pervasiveness of feminine beauty in the Grimms' fairy tales. We then analyze tales according to whether they survived into the twentieth century and explore the extent to which women's beauty predominates in these surviving tales. This study of beauty's significance in children's fairy tales can provide insight into the dynamic relationship between gender, power, and culture, as well as the cultural and social significance of beauty to women's lives.

* * *

The Importance of Children's Literature

Cultural products embody societal values and provide a means to observe shifts in such values (Schudson 1989). One of the most useful sets of cultural products for investigating cultural motifs and values is children's stories, which according to Bettelheim (1962) are a major means by which children assimilate culture. According to Pescosolido, Grauerholz, and Milkie (1997, 444), "the intended clarity and moral certainty with which adults provide children with tales of their world offer a fortuitous opportunity to examine social relations and belief systems." Children's literature is especially useful for studying value constructs such as the beauty ideal. Fox (1977, 807) suggested that where normative restriction prevails, one is likely to "find an elaboration of socialization structures that conduce toward the internalization" of such values. Thus, we would expect to find these values expressed in media, especially those marketed toward children.

Research since the early 1970s has shown that children's literature contains explicit and implicit messages about dominant power structures in

---

Baker-Sperry, L. & Grauerholz, L., "The pervasiveness and persistence of the feminine beauty ideal in children's fairy tales," *Gender & Society, 15*(5), pp. 711–726.

society, especially those concerning gender (Clark, Lennon, and Morris 1993; Crabb and Bielawski 1994; Kortenhaus and Demarest 1993; Weitzman et al. 1972). Fairy tales written during the eighteenth and nineteenth centuries were intended to teach girls and young women how to become domesticated, respectable, and attractive to a marriage partner and to teach boys and girls appropriate gendered values and attitudes (Zipes 1988a, 1988b).

But these messages are not static. Children's media have been found to be powerfully responsive to social change and not simply in a way that mirrors society. Research by Pescosolido, Grauerholz, and Milkie (1997) found that during periods of intense racial conflict and significant political gains by African Americans, Black characters virtually disappeared from children's books. They suggested that children's media both reflect and are shaped by shifting social and power relations among groups. As such, it is possible to study children's literature for insight into important political and social struggles over time.

In the present study, we investigate the gendered messages concerning feminine beauty as contained in children's media. We chose a classic set of children's literature—fairy tales written by the Grimm brothers in the nineteenth century—to investigate the extent to which the pervasiveness of the feminine beauty ideal has shifted over time. These tales were originally used as primers for relatively affluent European children and served to impart moral lessons to them (Zipes 1988a). Today, these tales, at least those that survived into the twentieth century, are read by children across various social class and racial groups (Zipes 1997), while continuing to contain symbolic imagery that legitimates existing race, class, and gender systems.

We first document the prominence of a feminine beauty ideal and the ways in which beauty is presented in these tales. Our main concern, however, is not whether these fairy tales contain stereotypic images (they do) but rather whether women's beauty appears to play a more important role in fairy tales during certain time periods, possibly serving as a means of normative social control. Thus, we document which tales have survived (i.e., were reproduced in books and films) into the twentieth century and whether those that survived placed greater emphasis on women's beauty than those that did not survive. Furthermore, we examine the time periods when tales were reproduced. If normative social control is more critical during times when many women have gained greater social power, we would expect a large increase in reproductions of tales that focus on women's beauty during the latter half of the twentieth century.

Of course, we would expect some variation in the number of reproductions given changes in the children's book publishing industry during the twentieth century. When publishing houses first established separate children's sections around 1920, there was an increase in the production of children's books (Tebbel 1978). The 1930s and 1940s saw some decline in sales, although it was during this time period (1932) that Western Printing and Lithographing Company—the largest lithographic company and publisher of children's books in the world—entered into an exclusive contract with Walt Disney Inc. to produce its books (Gottlieb 1978; Tebbel 1978). Children's book publishing increased significantly during the 1950s and 1960s, as the baby boom market increased sales and interest in children's reading, and federal aid was made available for library materials (Gottlieb 1978; Turow 1978). Finally, by the 1970s, the growth in children's book publishing subsided (Gottlieb 1978).

In the present study, we are not interested in whether reproductions of tales follow these general patterns but in whether those tales that highlight a feminine beauty ideal actually had increased reproductions during periods when normative control would be more necessary, such as since the 1970s. This study represents one of the few attempts to analyze long-term changes in children's literature and the only one to offer a historical analysis of the reproduction of a beauty ideal in fairy tales. As such, it provides critical insight into ways in which children's literature has been shaped by political and social forces over time and yet continues to provide traditional gendered prescriptions for children.

* * *

Findings

Physical Appearance and Beauty in Fairy Tales

There is frequent mention of characters' physical appearances (their looks, physiques, clothing, etc.) in these fairy tales, and this is true regardless of their gender or age. . . . For instance, 94 percent of the tales make some mention of physical appearance, and the average number of times per story is 13.6 (among those stories that have at least one mention, the average is 14.5). There is no substantive gender difference in the number of times physical appearance is mentioned (the average number of times that physical appearance is mentioned in reference to men is 6.0 and for women is 7.6), but there is a notable difference in the range of references for men and women. The number of references to men's physical appearance ranges from 0 to 35 per story, whereas the range for women is 0 to 114.

More detailed examination of physical beauty/handsomeness by gender and age reveals some interesting patterns. . . . [W]omen's beauty is highlighted more than men's attractiveness and that beauty plays a more dominant role for younger women than for older ones. Overall, there are approximately five times more references to women's beauty per tale than to men's handsomeness (the average number of references to women is 1.25 and 0.21 for references to men's handsomeness). The average number of references to younger women's beauty in all tales (1.17) outnumbers those of younger men (0.20), older women (0.08), and older men (0.02) combined. Although the actual number of references to younger women's beauty is not all that great, what is striking is the way in which women's beauty is mentioned. For instance, in *The Pink Flower* a maiden is described as "so beautiful that no painter could ever have made her look more beautiful" (Grimm and Grimm 1992, 286), and in *The Goose Girl at the Spring* a young woman is said to be "so beautiful that the entire world considered her a miracle" (Grimm and Grimm 1992, 566).

Of the tales that contain younger women, 57 percent described them as "pretty," "beautiful," or "the fairest," and on average there are 1.74 references to their beauty. By contrast, only 5.2 percent of tales that contain older women make reference to their beauty, with the average number of references to older women's beauty being 0.14. For male characters, 18.3 percent of the tales that contain younger men describe them as "handsome" (average number of references was 0.25). Only 1.7 percent of the tales with older men characters describe them as handsome (average number of references is 0.02).

Discourse analyses reveal several themes in relationship to beauty. Often there is a clear link between beauty and goodness, most often in reference to younger women, and between ugliness and evil (31 percent of all stories associate beauty with goodness, and 17 percent associate ugliness with evil). *Mother Holle* incorporates both of these themes. The story begins, "A widow had two daughters, one who was beautiful and industrious, the other ugly and lazy" (Grimm and Grimm 1992, 96). As the tale unfolds, both daughters have the opportunity to work for Mother Holle. While staying with Mother Holle, the beautiful and industrious daughter admitted that she was homesick:

> "I'm pleased that you want to return home," Mother Holle responded. . . . She took the maiden by the hand and led her to a large door. When it was opened and the maiden was standing right beneath the doorway, an enormous shower of gold came pouring down, and all the gold stuck to her so that she became completely covered with it.
>
> "I want you to have this because you have been so industrious," said Mother Holle. (Grimm and Grimm 1992, 97)

When the ugly, lazy daughter began her work for Mother Holle, all did not go as well:

> On the first day she made an effort to work hard and obey Mother Holle when the old woman told her what to do, for the thought of gold was on her mind. On the second day she started loafing, and on the third day she loafed even more. . . . Soon Mother Holle became tired of this and dismissed the maiden from her service. The lazy maiden was quite happy to go and expected that now the shower of gold would come. Mother Holle led her to the door, but as the maiden was standing beneath the doorway, a big kettle of pitch came pouring down on her head instead of gold. . . . The pitch did not come off the maiden and remained on her as long as she lived. (Grimm and Grimm 1992, 99)

Thus, while beauty is often rewarded, lack of beauty is punished.

Another theme identified through the discourse analysis, as evidenced by the example of *Mother Holle,* is that beauty is sometimes linked to race and class. The "lazy" daughter in *Mother Holle* is covered in (black) pitch. In *The White Bride and the Black Bride,* the mother and daughter are "cursed" with blackness and ugliness. Many tales connote goodness with industriousness, and both with beauty, and characters are "rewarded" for their hard work (*Cinderella* is another classic example). In this way, beauty becomes associated not only with goodness but also with whiteness and economic privilege.

Although beauty is often rewarded in Grimms' tales, it is also a source of danger. Of the tales in which danger or harm is associated with physical attractiveness (28 percent of all tales), 89 percent involve harm to women. Forty percent of these acts of victimization are the direct result of the character's physical appearance. For instance, there are examples of women who must flee or disguise themselves for protection because they are so beautiful. Such was the case for the princess in *All Fur* who was "so beautiful that her equal could not be found anywhere on earth" (Grimm and Grimm 1992, 258). She was forced to run away from the castle because her father "fell passionately in love with her and said to his councillors, 'I'm going to marry my daughter'" (Grimm and Grimm 1992, 260).

Finally, in 17 percent of the stories there are links between beauty and jealousy. These issues almost exclusively concern female characters. *Snow White* offers strong messages concerning competition among women and the importance of beauty for women: "When a year had passed, the king married another woman, who was beautiful but proud and haughty, and she could not tolerate anyone else who might rival her beauty" (Grimm and Grimm 1992, 196). The murderous actions taken by the stepmother remind readers of the symbolic lengths some women go to maintain or acquire beauty.

In sum, messages concerning feminine beauty pervade these fairy tales. Although the tales are not devoid of references to men's beauty, or handsomeness, it is women's beauty that is emphasized in terms of the number of references to beauty, the ways it is portrayed, and the role feminine beauty plays in moving the story along.

Social Reproduction of the Feminine Beauty Ideal

Of the 168 tales analyzed, 43 (25.6 percent) have been reproduced in children's books or movies. The most frequently reproduced tale is *Cinderella,* for which 332 reproductions were recorded. In fact, just 5 fairy tales—*Cinderella, Snow White, Briar Rose* (also known as *Sleeping Beauty), Little Red Cap* (also known as *Little Red Riding Hood*), and *Hansel and Gretel*—constitute more than two-thirds (72.7 percent) of all reproductions.

There are many more references to women's physical appearances in reproduced versus nonreproduced tales (11.3 vs. 6.15), and this is somewhat true for references to men's physical appearance (8.0 vs. 5.2) (see Table 4.1). In terms of beauty, the average number of references to women's beauty in those tales that have been reproduced is 2.11 for women, which is more than twice the number in nonreproduced tales (0.93) and much higher than the average number of references to men's handsomeness in reproduced tales (0.37) and nonreproduced tales (0.15).

Table 4.1

Reference	*All Tales Ever Reproduced* (n = 43)	*Tales Reproduced 101 or More Times* (n = 5)	*Tales Reproduced Between 1 and 100 Times* (n = 38)	*Tales Never Reproduced* (n = 125)
Women's appearance	11.30	33.80	8.60	6.15
Men's appearance	8.00	2.60	8.70	5.20
Women's beauty	2.11	7.20	1.50	0.93
Men's handsomeness	0.37	0	0.41	0.15

Table 4.1 groups tales according to the number of times they have been reproduced. Interestingly, of the top five most reproduced tales—those that have been reproduced more than 100 times—there are two exceptions to the "beauty rule": *Little Red Cap* or *Little Red Riding Hood,* for which 227 reproductions were documented, and *Hansel and Gretel,* which trails the other tales at 143 reproductions. There are no references to women's or men's beauty in *Hansel and Gretel* and *Little Red Cap.* In fact, in *Hansel and Gretel* there are more references to men's appearance (8) than women's appearance (5). When analyses are conducted on just the top three most reproduced tales, which eliminates *Little Red Cap* and *Hansel and Gretel,* the references to women's beauty and women's appearance are much higher (12 references to beauty for the top three vs. 7.2 for the top five; 41.7 references to appearance for the top three vs. 33.8 for the top five) and those references for men's appearance decline (0.67 for top three vs. 2.6 for top five). Note that there are no references to men's handsomeness in any of the top five tales.

Because references to men's handsomeness and older women's beauty are so low (e.g., 98.8 percent of all tales have no mention of older men's handsomeness), we combined older and younger men, and older and younger women, to perform the regression analyses. Preliminary analyses suggested that it was appropriate to do so since there is no interaction effect between age and gender with respect to appearance or beauty.

Even after controlling for length of the tale, references to women's beauty are associated with the likelihood that a tale has been reproduced many times, as is the number of references to women's physical appearance. For men, physical handsomeness and appearance are not significantly related to a tale's reproduction, nor is length of a tale.

We explored alternative factors that may help account for tales' reproduction, such as themes of romantic love or victimization. We found that even after controlling the regression analysis for tales that have a romantic theme, the number of times women's beauty is mentioned in a tale remains strongly related to the number of times it has been reproduced, as does women's physical appearance. Furthermore, there is a moderate bivariate correlation between women's victimization and number of reproductions (.202), but women's victimization becomes nonsignificant when number of reproductions is regressed on women's beauty, women's victimization, and pages (and women's beauty remains significant). The general presence of violence or men's victimization was not linked to reproductions.

Examination of reproductions over time reveals an interesting pattern. The vast majority of tales were reproduced in the latter part of the twentieth century. For instance, the average number of reproductions before 1900 was 4.07 (*SD* = 10.32) versus 24.79 between 1981 and 2000 (*SD* = 51.72). This is particularly true for the most reproduced tales. For instance, there were 46 reproductions of *Cinderella* before 1900, 5 or 6 for each of the time periods between 1901 and 1960, 42 between 1961 and 1980, and 227 between 1981 and 2000. When we correlated time period in which tales were most often reproduced (which ranges from 0 to 6) with mentions of beauty, handsomeness, and physical appearance, we found no significant correlation between physical appearance and time, for men or women. However, the number of mentions of women's beauty is significantly correlated with a larger number of reproductions in the latest time period (r = .159), and the same is true for mentions of men's handsomeness (r=.203). In fact, all but one tale that mention a man being handsome were reproduced most often in the latter period.

Discussion and Conclusion

Not surprisingly, among the many messages contained in fairy tales, those concerning the importance off feminine beauty, especially for younger women, are paramount. Young women are more often described as "beautiful," "pretty," or "fair" than are older women or than men of any age are described as handsome, Furthermore, beauty is often associated with being white, economically privileged, and virtuous. Fairy tales, like other media (Currie 1997), convey messages about the importance of feminine beauty not only by making "beauties" prominent in stories but also in demonstrating how beauty gets its rewards. So ingrained is the image of women's beauty in fairy tales that it is difficult to imagine any that do not highlight and glorify it. Recent Disney films and even contemporary feminist retellings of popular fairy tales often involve women who differ from their earlier counterparts in ingenuity, activity, and independence but not physical attractiveness.

Several of the tales have been reproduced in books and movies since their original publication. Our findings suggest that those that have been reproduced the most (*Cinderella* and *Snow White*) are precisely the ones that promote a feminine beauty ideal. Tales that make frequent reference to physical appearance and beauty for women are likely to have been reproduced, Even after controlling for length of a tale, references to feminine beauty and women's physical appearance are related to the number of times a tale is reproduced, However, the same is not true for men.

Our findings further suggest that attention to attractiveness may have become increasingly prevalent over the past century. Tales that were reproduced mostly in the latter part of the twentieth century tend to make more mentions of women's beauty and men's handsomeness, which is consistent with earlier studies that have found an increased emphasis on physical attractiveness in the late twentieth century for men (Berger, Wallis, and Watson 1995). In fact, of the 11 tales that have been reproduced and mention men's handsomeness, 10 were reproduced most often in the last time period. This finding suggests that both men and women are being increasingly manipulated by media messages concerning attractiveness, a trend that is undoubtedly linked to efforts to boost consumerism. This trend does not necessarily contradict a social control perspective that suggests such messages should be directed more toward women than men. We found that messages concerning women's beauty are far more dominant than those for men. Only 2 of the reproduced tales that mention men's handsomeness are fairly popular (*Rapunzel* and *Puss and Boots*), and each makes only one mention of men's handsomeness. Passing mentions of men's handsomeness in these 11 tales simply do not compare to the tales in which women's beauty is glorified and in which beauty, for beauty's sake, plays a major role in the story, as in *Cinderella* or *Snow White.* Thus, while there does appear to be an increased emphasis on men's handsomeness along with women's beauty in the late twentieth century, there remains a profound difference in the prevalence and persistence of messages concerning attractiveness for men and for women, which is consistent with a social control perspective.

Clearly, beauty is not the only reason certain tales have survived. Some tales become popular during particular historical periods because they resonant so deeply with individuals' and societies' economic, social, or political struggles during these times (Zipes 1988b). Certainly, much of the success of certain tales can be attributed to the work of Walt Disney. For instance, the three top tales had all been made into Disney movies before 1960 and have enjoyed continued popularity. We were not able, however, to determine any other clear links between survival of a tale and themes. Mentions of women's beauty are far more likely to be linked to reproductions than are other popular cultural motifs such as victimization or romance.

We suggest that this emphasis on a feminine beauty ideal may operate as a normative social control for girls and women. The fact that women's beauty is particularly salient in tales in the latter part of the twentieth century suggests that normative social controls (such as internalization of a feminine beauty ideal) may have become increasingly important over the course of the twentieth century as external constraints on women's lives diminished. We do not propose that there is a direct relationship between cultural values concerning feminine beauty and women's behavior and identities, but the feminine beauty ideal may operate indirectly as a means of social control insofar as women's concern with physical appearance (beauty) absorbs resources (money, energy, time) that could otherwise be spent enhancing their social status. Women may "voluntarily" withdraw from or never pursue activities or occupations they fear will make them appear "unattractive" (e.g., "hard labor," competitive sports). The competition women may feel toward other women over physical appearance may limit their ability to mobilize as a group. In these ways, the focus on and glorification of feminine beauty in children's fairy tales may represent a means by which gender inequality is reproduced via cultural products.

Of course, the effect of media on behavior is not clear (Currie 1997). As with other literature, children's media should not be viewed simply as gender scripts. Children (or their parents, through their readings of the texts to children) have the ability to use these texts to challenge or "rewrite" these scripts (stories). Zipes (1988b, 191) suggested that by

> introducing unusual elements into the fairy tale . . . the child is compelled to shatter a certain uniform reception of fairy tales, to re-examine the elements of the classical tales, and to reconsider their function and meaning and whether it might not be better to alter them.

The recent film Shrek, whose main woman character is ultimately transformed into an ogre rather than the beautiful maiden she was believed to be, may begin to challenge the value and meaning of women's beauty. But such retellings of fairy tales are rare, and the cumulative effect of the more traditional tales, in conjunction with the unidirectional nature of media, makes such agency difficult. Indeed, the "beauty" of messages that may serve as normative controls is that so few question or challenge their legitimacy.

References

Bartky, Sandra. 1990. *Femininity and domination.* New York: Routledge.

Berger, Maurice, Brian Wallis, and Simon Watson. 1995. *Constructing masculinity.* London: Routledge.

Bettelheim, Bruno. 1962. *Uses of enchantment.* New York: Collier.

Bianchi, Susan M. 1999. Feminization and juvenilization of poverty: Trends, relative risks, causes, and consequences. *Annual Review of Sociology* 25:307–33.

Bordo, Susan. 1993. *Unbearable weight.* Berkley: University of California Press.

Clark, Roger, Rachel Lennon, and Leanna Morris. 1993. Of Caldecotts and kings: Gendered images in recent American children's books by Black and non-Black illustrators. *Gender & Society* 7:227–45.

Crabb, Peter, and Dawn Bielawski. 1994. The social representation of material culture and gender in children's books. *Sex Roles* 30:69–79.

Currie, Dawn. 1997. Decoding femininity: Advertisements and their teenage readers. *Gender & Society* 11:453–77.

Dellinger, Kirsten, and Christine L. Williams. 1997. Makeup at work: Negotiating appearance rules in the workplace. *Gender & Society* 11:151–77.

Flexner, Eleanor, and Ellen Fitzpatrick. 1996. *Century of struggle: The woman's rights movement in the United States.* Cambridge, MA: Harvard University Press.

Fox, Greer Litton. 1977. "Nice girl": Social control of women through a value construct. *Signs: Journal of Women in Culture and Society* 2:805–17.

Freedman, Rita. 1986. *Beauty bound.* Lexington, MA: Lexington Books.

Gottlieb, Robin. 1978. *Publishing children's books in America, 1919–1976.* New York: Children's Book Council.

Grimm, Jacob, and Wilhelm Grimm. 1857. *Children's and household tales.* Berlin: R. Dummler.

———. 1992. *The complete fairy tales of the Brothers Grimm.* Translated by Jack Zipes. New York: Bantam.

Kortenhaus, Carole, and Jack Demarest. 1993. Gender role stereotyping in children's literature: An update. *Sex Roles* 28:219–32.

Lorber, Judith. 1994. *Paradoxes of gender.* New Haven, CT: Yale University Press.

Pescosolido. Bernice, Elizabeth Grauerholz, and Melissa Milkie. 1997. Culture and conflict: The portrayal of Blacks in U.S. children's picture books through the mid- and late-twentieth century. *American Sociological Review* 62:443–64.

Russo, Ann. 2001. *Taking back our lives.* New York: Routledge.

Schudson, Michael. 1989. How culture works. *Theory and Society* 18:153–80.

Sorensen, Elaine. 1991. *Gender and racial pay gaps in the 1980s: Accounting for different trends.* Washington, DC: Urban Institute.

Tebbel, John. 1978. *A history of book publishing in the United States: Volume III.* New York: Bowker.

Turow, Joseph. 1978. *Getting books to children: An exploration of publisher-market relations.* Chicago: American Library Association.

Weitzman, Lenore J., Deborah Eifler, Elizabeth Hokada, and Catherine Ross. 1972. Sex role socialization in picture books for preschool children. *American Journal of Sociology* 77:1125–49.

Williams, Joan. 2000. *Unbending gender.* Oxford, UK: Oxford University Press.

Wolf, Naomi. 1991. *The beauty myth.* New York: Doubleday.

Zipes, Jack. 1988a. *The brothers Grimm.* New York: Routledge.

———. 1988b. Fairy tales and the art of subversion. New York: Methuen.

———. 1997. *Happily ever after.* New York: Routledge.

Introduction to Reading 18

Prudence Carter, an anthropologist, conducted a series of interviews with sixty-eight Latino and African American teenagers from Yonkers, New York. This chapter from her book, *Keeping It Real: School Success Beyond Black and White*, describes how these youth, from low-income households, define gender in daily life, including their relationships to school. Although the teens in this study define gender behavior as "soft" and "hard," you will see many corollaries with gender socialization throughout this chapter, albeit with a slightly different twist. Carter illustrates the social construction of gender through many instances in which negative peer pressure maintains the "hard" and "soft" personas of gender.

1. How does gender influence these teenagers' relationships to school and their plans for their futures?
2. What mechanisms do boys and girls use to establish and maintain their masculinity and femininity and which would be more problematic for performance in school?
3. Why can't girls be too "soft" and how is this different from concerns about boys being too "soft"?

BETWEEN A "SOFT" AND A "HARD" PLACE

GENDER, ETHNICITY, AND CULTURE IN THE SCHOOL AND AT HOME

Prudence L. Carter

Girls are outperforming boys in the classroom (Jacobs 1996; Mickelson 1989). Whether they are actually smarter remains to be seen; there are no proven differences in intelligence between the sexes. Still, research shows that throughout their scholastic careers, girls perform better in subjects that require verbal competence. Although studies show that boys do better than girls in math and science (Catsambis 1994; Maccoby and Jacklin 1974; Nowell and Hedges 1998; Stevenson and Newman 1986), overall, girls have higher grade point averages and college attainment rates (Jacobs 1996; Mickelson 1989).

Among Blacks and Latinos, however, the gender gap is starker. A *Newsweek* cover story announces that Black women have outpaced Black men in professional and managerial careers, with almost a quarter of Black women holding such jobs, compared to 17 percent of Black men (Cose and Samuels 2003). Weekly e-zines from the *Journal of Blacks in Higher Education* reveal that women outnumber men in professional schools by significant ratios. Twice as many African American females as African American males earned college degrees in the 1990s (Special Report, *Journal of Blacks in Higher Education* 1999). Black female students at many elite colleges and universities personify this statistic (Massey et al. 2003) as they lament the fact that they outnumber Black males almost two to one, which signifies a shallow dating pool, especially if in-group, heterosexual dating is preferred. Although various studies show that the gender gap is not as wide for Latinos as it is for African Americans (Community College Week 2000), Latinas also exhibit stronger relationships to schooling and outnumber Latinos in high school graduation and college enrollment rates (Valenzuela 1999; Lopez 2003). Even among the students in this study, females outpaced males, earning significantly higher mean and median grade point averages—80.12 and 82.25, respectively, compared to 74.61 and 75 for males. Furthermore, females were more than twice as likely to report being B students, while males were more likely to earn Cs.

What is happening in urban minority communities in the United States to create such gender differences in achievement and attainment? And why are Black and Latino males now more predisposed to seek "smarts" on the streets of Atlanta, Chicago, Los Angeles, New York, Philadelphia, and other urban centers than in the lecture halls of Howard, Hampton, Brown and New York universities, to name a few? Trapped in the "codes of the streets," many ultimately head toward the university of the pen[itentiary] instead of the University of Penn[sylvania]. Since there are no innate or essential intelligence differences in our society, the explanations must be social and cultural. More specifically, they must be tied to how gender, as a social process, is lived, experienced, prescribed, and enforced in our society (Lopez 2003; Collins 1991; Frankenberg 1993; Crenshaw 1992). . . . Street smarts include knowing how to look someone in the eye, to avert one's gaze at the right moment, to avoid life-threatening fights and encounters, to defend oneself by fighting, and to navigate through dangerous gang territories (Anderson 1980, 1990). Street smarts, says Maxwell, are a male domain where men stand up

---

From pp. 77–105 of *Keepin' It Real: School Success Beyond Black and White* (2005) by Carter, Prudence. By permission of Oxford University Press, Inc.

for themselves and avoid submitting to the control of others ("You ain't gon' let somebody play you, play you and put their hands on you"), Girls, according to Maxwell, did not possess the same interpersonal negotiation skills that boys possessed. Instead, he believed that girls fought over petty matters. "They fight over stupid stuff," he claimed. "'She got my sneakers on!'" Maxwell said mockingly, imitating what he thought girls did.

Sylvestre agreed with Maxwell's assessment of the differences between males and females. "Girls and women are a little bit smarter than us," he declared. "It's true. I'm serious. I see it. They think smarter than guys." When I asked why he believed that girls were smarter, he said, "I know that they got a little more push, a little more drive." . . .

Girls believed that they were "book" smarter, too. Though fifteen-year-old Tiara Mitchell's reasons differed from Sylvestre's, her conclusions were similar: "Well, boys, I know, are not into books. Most girls I know love to read, Whites, Black, all, in the library. Boys just love gym. Boys that I know just love to go to gym. But girls are not. I'm not interested in gym, but I play gym." Tiara further explained her beliefs as I probed: "Most boys just come to school to play. Boys always are in gym; that's one class that boys do not miss. I always think about that. But I'm saying, most girls, I mean, they're just in class. Most boys are always cutting" . . .

Both Sylvestre and Tiara outlined a script that many boys appear in this study to follow. They gambled and risked reprimands and suspensions by being truant or by hanging out with the guys. . . . 57 percent of the males reported getting a kick out of engaging in risky and dangerous behaviors compared to only 37 percent of the females. Also, nearly four times as many boys (more than one out of three) as girls (fewer than one out of ten) reported cutting school. What are the reasons for these differences? Sylvestre believed that males and females have different developmental trajectories, that girls mature faster than boys, and girls figure out earlier the importance of school. Tiara stayed away from the psychological and instead offered a sociological explanation. She said that the threat of her mother's sanctions motivated her to attend classes and to complete her schoolwork. Results are mixed on Sylvestre's claim about gender and development between boys and girls, though some research shows that females experience several stages of social development more quickly than males (Streitmatter 1993). Although my data would not allow me to explore Sylvestre's claims about social psychological and developmental differences, I did follow up on Tiara's explanation. I found that males' and females' parental monitoring were significantly different. Mothers often gave sons more social latitude than daughters. Further, while conveying messages of masculinity and femininity, parents often reproduced the differential levels of social control that males and females experience in our society.

Gender comprises a set of constructed acts and performances that are both individually and socially meaningful, and sociohistorical scripts about maleness and femaleness are observed, watched, guarded, protected, and reinforced. If men and women cross gender boundaries, their identities can be questioned, and they can face social sanctions (Butler 1993; Guy-Sheftall 1996; Fenstermaker and West 2002). These acts, like ethno-specific acts, are neither biological nor intrinsic traits; rather, identification as male or female is best understood as yet another classification with political, economic, and social effects.

Gender roles enforced by parents, community, and society influence students' personal identities and their approaches to school (Adler, Kless, and Adler 1992); and these roles create disparate schooling patterns for males and females (Epstein, Hey, and Maw 1998). As I investigated how the Yonkers youths' ideas about masculinity and femininity shape how they approached school and how their gender ideology influenced their perspectives on socioeconomic mobility, three striking patterns emerged. First, all of the youths participated in our system of gender relations and reproduced common ideas about what it means to be male and female.

These ideas determined how they moved through the world. Metaphors of hardness and softness emerged to characterize male and female ways of interacting in school and among their peers. Males experienced social pressure to be hard. Hardness is akin to the behaviors that urban sociologists have discovered for decades in poor communities of color where males develop "codes of the street," putting on tough coats of armor to protect themselves against bleak social and economic conditions (Anderson 1990; Liebow 1967; Hannerz 1969). During one group discussion, thirteen-year-old Marcus Smith accused thirteen-year-old Michael Jones of being "soft" (like a girl) because he refused to fight. Marcus's actions matched Maxwell's beliefs. Maxwell had exclaimed when we talked one-on-one, "What kind of man would not stand up and fight for himself?" And he believed that guys weren't supposed to let anyone "push up" (physically challenge) on them.

The constructions of masculinity among the males in the study raised an interesting paradox: most aspired to white-collar professions, yet what challenged their aspirations were the conflicts and tensions between their ideas about masculinity and many of the tenets of mobility, especially behaviors in which acculturation is required. I found that ethno-specific cultural behaviors embraced by these youths intersected with their gender-specific cultural behaviors to create different educational, social, and economic experiences for males and females. One noticeable outcome of their blending gender role expectations with ethnic and cultural expectations was the construction of many of the behaviors associated with "acting white" as feminine. . . .

Samurai suggests that "acting white" is just what it sounds like—acting, a social performance in which (Black) girls use affected speech to impress or to be like "certain people," to move up to a more respectable status. He also implicitly associates whiteness with respectability when he describes how girls would tell a peer that they have to be home by 8 PM as opposed to 10 PM and, therefore, suggests that White parents maintain tighter social control over their children. Then he emasculates whiteness by suggesting that tight, social control is what females face, and thus, emphatically he declared that he knew no Black males who "act white," only girls.

Samurai also conflated the ideas of Standard English, slang, and the dialects of white youth subcultures. As he compared "white talk" to "black talk," he made slippery distinctions between English as a language system and slang as an informal language practice. White youth slang, like Black youth urban vernacular, is not Standard English. But Samurai deemed "white talk" to be proper talk and he claimed his (female) peers "talked white," even if it were a slang or dialect. . . .

In Adrienne Ingrams' case, however, "white talk" was equated with Standard English. "I don't see a lot of boys who talk like me," Samurai's peer and neighbor told me. After some musing, she continued, "Yeah, my friend Daniel . . . he says I 'act white,' too. But his friends pick on him because he talks soft and stuff . . . like they call him gay or something or they say he 'acts white,' too." With the exception of two references to allegedly gay males, all of the principal characters in the fifty-one references to acting white that I heard were female.[1] . . .

In a society where masculinity has long been associated with heterosexuality, a person with both X and Y chromosomes who does not conform to constructed norms and cultural expectations is liable to be considered a gender and sexual deviant (Connell 1995). According to Adrienne, Daniel's sexuality was questioned because he talked too "soft," not forcefully or sufficiently deep-voiced. Anthropologist Signithia Fordham, whose early work heightened the debate about the impact of "burden of acting white," argues that African American and Latino females may not reject schooling to the same extent as their male co-ethnics because they do not experience the same level of estrangement from the normalized American female roles as males do from the normalized male roles. . . .

The idea of assimilation, in itself, threatens the already tenuous power that a racial and ethnic minority male holds in the larger society

(Connell 1995). In other words, aspects of cultural assimilation—or closeness to whiteness by means of body norms, language, clothing styles, and interactions—directly challenge many of these males' perceptions of masculinity. As Ann Ferguson writes: "Transgressive behavior is that which constitutes masculinity. Consequently, African American males (like the ones whom I interviewed) in the very act of identification, of signifying masculinity, are likely to be breaking rules," even rules that White men follow (Ferguson 2000, 170). Therefore, it should not be surprising that more males than females in my study fell into the noncompliant believer group or that girls significantly outnumbered males among the cultural straddlers and cultural mainstreamers. Specifically, two-thirds of the males, or rather twenty out of thirty, fell into the noncompliant believer group, compared to less than half of the females, eighteen out of thirty-eight. Moreover, twenty of the thirty students deemed as either cultural straddlers or cultural mainstreamers believers were female. . . .Noncompliant believers are significantly less likely to perform as well in school as cultural straddlers and cultural mainstreamers. Without sufficient social and economic resources, the consequences are that poor African American and Latino males, in their quests to assert manhood in a patriarchal and male-dominant society, are likely to collaborate in their own academic and socioeconomic marginalization (Noguera 1996).

"Hard" And "Soft": Constructions of Masculinity

The soft/hard and feminine/masculine dichotomies continue to reproduce the boundaries between boys and girls that maintain differing results between the two groups. Although male and female students in the study shared cultural meanings of blackness and "Spanishness," girls could not be too "hard," a masculine form, and boys could not be too "soft," a feminine expression. Hardness surfaced as a coat of armor developed by males in the low-income communities where the Yonkers youths live. Though many of the boys bemoaned the negative stereotypes of Black and Latino men, often the men in their neighborhoods conformed to those stereotypes, to images of street corner men (Liebow 1967; Hannerz 1969). In my afternoon visits to their housing complexes, I observed men hanging out in the parking lots and street corners. And the Yonkers youths would describe them either as unemployed or as "street pharmacists." The rate to which of these men have had experiences with the U.S. criminal justice system, its courts and prisons, is disproportionate to their overall demographic representation in the nation (Noguera, 2003).

As I moved through the students' neighborhoods, I came to understand how hardness provided marginalized men protection against severe social and economic oppressions, from racial insubordination to the fines of deindustrialization and the loss of jobs in the central cities. The contemporary meanings of hardness stemmed from lives of inner-city poverty where an underground economy "thrives," where turf wars of gang members endanger the lives of its residents, where individuals must generate creative and often illicit plans to survive, and where a different communicative system and styles of interaction materialize (Liebow 1967; Anderson 1990; Kelley 1994; Wilson 1987; Massey and Denton 1993; Dance 2002). In contrast to hardness, softness connotes the feminine forms perceived as compliant and nurturing. Sitting still and paying attention in class are imbued with feminine meanings and therefore might be avoided by students with more masculine identities. Yet the more "feminine" the gender identity of boys and girls, the better their performance when measured by classroom grade point averages (Burke 1989). Meanwhile, hard boys might excel on the "masculine" turfs, especially on the athletic fields and courts, often the channels through which many poor, racial and ethnic minority males become engaged in school (Solomon 1991; Davis 2001). . . .

All boys are not lower achievers than girls, however, and some of them disagreed with the beliefs that girls do better in school than boys.

For example, John Jamison, a thirteen-year-old cultural straddler and a high achiever, argued that the idea that girls are smarter is a cliché. "It's the stereotype right now that girls gotta do this and that . . . meaning do better in school. But it's not true. But that's the stereotype right now." Yet John was only one of five males in the study who could be categorized as a high achiever, and who was enrolled in advanced courses and maintained more than a B+ average. Others included fourteen-year-old Ramon Diaz who had his sights on earning a degree at the Massachusetts Institute of Technology in a field like computer engineering. There was also Jeremy James, a ninth grader perceived by his teachers to have the ability to enter the International Baccalaureate program, but who was squabbling with his mother about whether or not to enroll in it.

For most of these boys and their male peers, team sports and rap music were major preoccupations. One out of two of them participated in a team sport, compared to a little more than one in five of the girls. Some thinkers have suggested that a fixation with excelling in sports has become an alternative to classroom success for minority—particularly Black males (Solomon 1991; Hoberman 1997). The love of sports does not mean that these young men did not acknowledge the value of a high-school diploma. But it does suggest that these young men, influenced by normative cultural values about work and manhood, sought a high-school credential as a path to financial success. Unfortunately, most desired to obtain this success through one of the few avenues that an exclusionary limited opportunity structure opens to many men of color—professional sports and the hip-hop entertainment industry—even though less than one percent of college athletes make the cut into professional sports, and few break into fame in the music world. Yet, Black professional athletes and rappers, in particular, have become exemplars of manhood in America's poor, urban communities (Collins 2004), although their fame and wealth have not given much social and political authority to men of color as a group. Still, African American and Latino boys gravitate toward those niches occupied by men whose backgrounds and experiences resemble theirs. These career niches do not require high academic achievement. And the fierce competition for jobs in professional sports and in the recording industry prevents most athletically and musically inclined boys from realizing their aspirations to become professional athletes and entertainers. At the same time, the extremely low representation of Black and Latino men in professional fields, such as medicine, corporate management, engineering, and computer science lends itself to a limited consideration of these career paths by boys like the Yonkers males.

Balancing gender roles and identity are not easy. The social expectations of males and females are inscribed throughout U.S. society; they are inescapable. . . . the African American and Latino males in this study subscribed to conventional beliefs about the connection between manhood and job success. Sixty percent believed that respect for men came with the acquisition of a good job. Almost an equal percentage of females maintained a similar belief. However, the males were more likely than the females to cling to traditional beliefs about a "man's role" and a "woman's place." More than one out of three of them believed that it was better that a man be the provider and a woman the homemaker, compared to about one out of ten females. The Latino males adhered to these beliefs more so than did the African American males, which accounts for much of the significant difference between males and females. . . .

These boys understood what it means conventionally to be a "man" in U.S. society. Yet a strong contradiction exists in their lives. Massive unemployment and urban poverty, along with institutional racism, have powerfully interacted to reshape the notions of masculinity among Black and Latino men (Liebow 1967; Mirande 1997; Staples 1982).

Soft Work for the Hard Man

The African American and Latino males whom I interviewed desired the comforts and security of middle-class America; they wanted white-collar jobs and big houses and cars with lots of money in the bank. "I just want to make a decent amount of money, support my family, buy a house, and be able to pay my mortgage, and I want to be able to have a little money in my pocket," Sylvestre stated. He also hauled furniture part-time, but blue-collar work reeked heavily of toil and subordination, and Jermaine, one of Sylvestre's neighbors who had taken a different path from Sylvestre and enrolled in college agreed. . . .

While they aspire to the white-collar job world, the dilemma is that these jobs require that these young men embrace "soft" skills, in addition to acquiring a certain level of academic training. Employers look for "soft" or noncognitive skills such as a certain type of interactional and communicative style and demeanor (Kirschenman and Neckerman 1991); and according to Philip Moss and Chris Tilly (1996), soft skills are, in part, culturally defined. Studying employers' behaviors toward potential employees, Moss and Tilly found that employers rated African American men poorly on various subjective, noncognitive measures of evaluation. Employers often interpreted the "hard" persona as aggressive and threatening, consequently denying these young, urban, low-income males job opportunities.

Many of the male youths were potentially susceptible to similar treatment by employers. As they constructed their masculine selves through language and dress styles, they risked losing the opportunities they desired. Both males and females in this study had come to believe that employers preferred to hire African American females and Latinas because they were more likely than their male counterparts to possess soft skills. When I asked these youths which gender was more likely to succeed, the consensus was that African American women and Latinas were more likely because they are perceived as less threatening and as more capable of conforming to the "right" (as opposed to "ghetto") dress styles and codes. . . .

Seventeen-year-old Sylvia Escuela, another of Alberto's neighbors, explained why she believed managers prefer females. "I think that it's easier for a woman to get a job now because of the way they dress, the way they present themselves. You look good, you have a job if it's a guy boss. But men, they don't really get hired. I don't know why. That's how I see it because they don't really present themselves. They are very slow." "How do they tend to present themselves?" I asked. Sylvia replied, "Ghetto! There's a lot of ghetto people out here now. A lot of teenagers . . . they ghetto. They like to dress with their pants down to here. They like to have their hair wild, with the braids. You can't really get a job like that."

As young adults eager for money of their own, Alberto, DeAndre, and Sylvia had all looked for work in the service industry, an economic sector sought after by low-skilled young workers. Their encounters with potential employers shaped their perceptions that the soft skills of females make them more employable. . . . Economist Harry Holzer (1996) found that among Blacks, the hiring of females instead of males increased with the number of tasks performed and credentials required, although both groups' hiring percentages were significantly and strikingly lower than their White counterparts. In addition, the probability of being hired was greater for Black females than Black males when the job required customer contact. Some of these differences might result from the actual feminization of certain job sectors; there are more females than males in clerical and sales jobs. These patterns might also occur because of the differences in occupational choices by gender (ibid.).

. . . Gender norms influence the presentation of self that Anderson (1980) described from his ethnographic research almost two decades ago and that characterize the behaviors of the young men whom I interviewed recently. These young males do not wholeheartedly embrace various social behaviors because they perceive these behaviors as soft. Yet given that minority men who do not conform to the dominant cultural behaviors often fail to obtain jobs because of innate organizational and cultural biases (Gould 1999), these males' chances for academic, social, and economic mobility remain in jeopardy.

Girls Can't Be Too Soft

Although girls approach schooling and mobility in ways more in accordance with school authorities' expectations, they, nonetheless, struggle and often do not perform up to par. Using their noncompliant status as an indicator, I found that at least half of the girls in this study underperformed in school (compared to two-thirds of the boys). Low achievement was a problem for some, and as I explored why, I discovered that males do not hold exclusive rights to a demeanor of "hardness." Frequently, girls projected "tough" images so that they could avoid being picked on by others. Protecting their egos and self-images, they fought much among themselves, a fact that I heard about during a gathering with six young women over pepperoni pizza, chocolate chip cookies, and orange and grape soda in the social room of their housing complex. Beforehand, in one-on-one interviews, the topic of fighting among girls had arisen, so I inquired about the reasons for this antagonism.

All six of the girls had been in fights with other girls (and occasionally with boys) at school. Bickering continually over limited resources and social goods—for example, respect and boys' attention—girls were often described as "fighters" by girls and boys alike. Respect was inextricably linked to a "tough" persona, meaning that the more a girl's demeanor and behavior announced that she could not be reckoned with, the greater her formidability. When pushed further to explain why they fought, these girls often ridiculed and criticized their reasons for fighting. Yet they still fought, and one essential motive was to avoid being considered too soft.

"Soft" has several meanings. It connotes not only the feminine but also the weak. Often, the girls verbalized their understandings of prevalent beliefs about "female" roles. Yet to be female in poor urban America does not mean one is weak. If a girl cannot defend herself, then she's considered weak, which is synonymous with "soft." Still conventional feminine values held in their communities. "Your parents is on you, 'Be a lady. Cross your legs. Don't curse,'" Rayisha told me. "[They say], 'Don't use certain language. Don't scream too loud. Be a lady. Put on skirts and . . . [don't go] punching [others] in the face and fight. My mother's like, 'Don't fight. Be a lady.' " Recognizing the double standard, nineteen-year-old Hannah Cummings, agreeing with Rayisha, also criticized the expectations held for females. "Yeah because a Black woman is supposed to . . . is expected to be nice and conservative. But then the guys, there are no expectations for guys, because guys are going to be guys regardless. And if a girl doesn't, if she doesn't act the way that they feel a woman should act, then they call her all kinds of names and stuff like that."

Hannah was exasperated about the double standards, and she expressed her frustration that female agency is limited compared to that of males, since women are penalized by society for asserting sexual freedom. Although Hannah and other female informants understood that their femininity was linked to how "nice," "conservative," or "ladylike" they could be, they also engaged in behaviors that defied these feminine ideals. Competition, conflict, and tension surfaced in many of their relationships. Tension among these girls, often flamed through hearsay, would escalate to a breaking point, eventually resulting in fights. Consequently, several were suspended from school. At one point during that same group discussion, several commented, "Girls around here will turn they[sic] back on you. They tell your business. You don't want to beat them up, but you have to." Their comments confirmed psychologist Niobe Way's argument that as adolescent girls mature and become involved in romantic relationships, trust among same-sex peers diminishes and their sense of betrayal augments as they experience others "stealing" their romantic partners. "These heightened feelings of self-protection and vulnerability may cause adolescents to become increasingly cautious about whom they trust among their peers, who are also physically and sexually maturing and, perhaps, becoming more wary and self-protected" (Way 1996, 187).

Much of the literature on gender stresses different cultures between males and females, but these generalizations about girls' and boys' cultures come primarily from research done on those who are class-privileged and White. Some researchers argue that boys stress position and hierarchy, whereas girls emphasize intimacy and connection (Gilligan 1982). In contrast, studies of racial and ethnic minority and working-class girls have shown that these patterns do not necessarily hold in these communities (Browne 1999). To some extent, physical aggression and verbal assertiveness are common modes of expression among both African American males and females (Stevens 1997; Goodwin 1991). "Playing the dozens"—a colloquial term for brassy exchanges and name-calling defined as ritualized forms of insult to express power and aggression (Gates 1988; Major 1994)—characterizes some social interactions among African American females (Goodwin 1991). The girls channeled a great deal of energy into not being manipulated, not coming across as too soft, and not being made fun of or taken advantage of. Instead, they distinguished themselves through boisterous and assertive language.

Juggling femininity while not appearing too soft also means being able to take care of oneself. Frequently, social science research ignores the interactions of race, ethnicity, gender, and class, and it misses those patterns that do not conform to career choices and career development patterns of White, middle-class youth. In environments where men have difficulties securing jobs, the prospects of both marriage and a dual-income family have dimmed (Wilson 1987). Unlike their White counterparts, many females of color are socialized to view marriage separately from economic security, since, historically, it has not been expected that marriage will remove them from the labor market (Furstenburg, Herschberg, and Modell 1975; Jones 1985; Fuller 1980; Holland and Eisenhart 1990; Fernández Kelly 1995), Thus, some African American and Latino mothers appeared to be raising their daughters to be not too "soft" but rather to be self-sufficient and financially independent of male partners. To understand the family and community context in which these young women were being raised is important. Their family realities resembled those in many poor, urban areas: daily in-home parental contact came from their mothers, who as unmarried heads of households were raising their children alone with whatever financial means they could muster. Only about 15 percent of my research informants lived in a two-parent household, yet almost three-quarters (71 percent) of the youths' mothers worked outside the home.

The high incidence of mother-headed families among these youth raises questions about how children who live in these contexts form expectations about their economic futures and family life courses. . . . Based on parents' responses to questions about their children in the previous year, 89 percent of the females' mothers expected their daughters to have a high chance of a better life than they, compared with only 59 percent of the males' mothers. Eighty-three percent of the females' mothers believed that their daughters had a high chance of attending college, compared with only 56 percent of the males' mothers. Eighty percent of the females' mothers, versus 59 percent of the males' mothers, believed that their daughters would some day have a well-paying job. Still, roughly equal percentages of females' mothers and males' mothers maintained high educational aspirations for their children; 91 percent and 89 percent, respectively, of mothers desired that their children attend at least junior college or part of a four-year college program.

Unlike studies that show girls tending to avoid high-prestige occupations, and boys trained toward high achievement (Taylor, Gilligan, and Sullivan 1995), I found that these girls aspired to more high-status jobs than did the boys. Eighty-four percent of the African American girls aspired to professional, specialty jobs; their modal job preference was lawyer. Seventy-five percent of the Latina girls aspired to similar professional, specialty occupations; their modal job preferences were physician and fashion designer. In contrast, only 40 percent of Latino males aspired to professional, specialty occupations (excluding professional sports); auto mechanic was the modal job preference. Overall, the Latino males had blue-collar, entry-level aspirations;

this may be linked to the overarching social processes of their families' immigrant status, social position, and labor conditions (Waldinger 1996). Excluding professional sports—the modal job preference for African American males—the analysis shows that only 44 percent of the Black males aspired to professional, specialty jobs. When professional sports are included, the percentage of males aspiring to professional, specialty occupations increases to 50 percent and 75 percent for Latino and African American males, respectively. These findings are striking, considering the current body of literature on gender role socialization and career development, and they mirror labor statistics, which reveal that higher percentages of women of color are represented in professional occupations than are their male counterparts (Jaynes and Williams 1989).

Several of the respondents speculated that different socialization patterns and parental expectations help explain gender differences in schooling behaviors and career aspirations. Daughters asserted that they were raised and encouraged by their mothers differently from their brothers and male friends. And certainly these girls valued intelligence for themselves much more than males valued it for themselves. For example, a significantly higher percentage of girls than boys felt what it was more important for women to be intelligent than attractive—90 percent versus 77 percent, respectively. Several girls felt that their mothers and fathers pushed girls harder to do better and not to be dependent on males for income and resources. Consequently, parental monitoring of schooling behaviors appeared to be different for sons and daughters. Fourteen-year-old Rakeisha Shaw, like Tiara, shared the almost unanimous belief among my informants that girls did better in school because they feared their parents' retribution more than the boys did.

Teresa, Tameka, Rayisha, and the others affirm that even within impoverished communities with limited access to good jobs—places where the underground drug economy thrives—pervasive gender dynamics hold: males maintain a higher level of status and privilege within society. Referring to status symbols like shoes and to romantic relationships—matters of primacy in adolescent minds—the girls noted three social spaces where conspicuous gender differences exist: the school, the bedroom, and the home. More minority boys than girls spend time in trouble and in detention for school disruption and delinquency (Ferguson 2000); and parental controls do little to deter the boys from getting in trouble. Meanwhile, their sisters and girlfriends perceive that the boys have more personal and sexual freedom (boys don't get called "bitches and hos," Hannah had told me), in addition to more solicitude from mothers who buy sons the more costly Nike shoes that basketball great Michael Jordan made popular, unlike the Reebok, which has developed a strong marketing campaign targeting females. Boys get more attention, they exclaimed, while girls fear their parents' reprisals. However, rather than allow this imbalance to immobilize them, the girls claimed that it made them better prepare themselves for self-sufficiency.

Parents' reports about their treatment of their children may also support claims made by Teresa and the other young women that they "get less attention." When comparing across ethnicity, I found that African American mothers, rather than the Latina mothers, were more likely to report that they "baby and protect" their sons than their daughters. The African American males' mothers were more than five times more likely to report pampering their sons than their daughters. In comparison, 67 percent of Latinos' mothers reported pampering their sons, compared to 45 percent of the Latinas' mothers. In response to questions about understanding their children's concerns and worries, being emotionally cold, or allowing their children freedom, the youths' mothers reported that they were likely to treat sons and daughters similarly. Still, while mothers gave more emotional protection or support to their sons, their expectations for success in life were greater for their daughters. These findings resonate with an adage that African American mothers raise their daughters and love their sons (Collins 1991; Guy-Sheftall 1996). . . .

These data highlight the importance of examining career choices and interventions from a sociocultural perspective that stresses the crucial role of structural influences on the behaviors of ethnic minorities (Cauce et al. 1996). I would speculate that the boys' mothers had become more pessimistic about their sons' economic futures based on labor market outcomes for minority men. Both my personal observations and interviews indicated that within their neighborhoods, there were numerous unemployed men hanging out in the parking lots, rejected by an increasingly technology-based service economy that has little use for the unskilled, manual laborer. Thus, if their daughters become self-sufficient heads of households in the future, they would need to be prepared, perhaps better prepared than their mothers.

Another explanation for differing expectations might be that parents themselves were complicit in the reproduction of the gender hierarchy and encouraged distinctions between "hardness" (masculinity) and "softness" (femininity) that might influence school performance and behavior. The fact remains that differential schooling patterns, career aspirations, and parental expectations existed for the males and females in this study. The analyses in this chapter suggest that gender, race, ethnicity, and economic conditions intersect and likely influence the different achievement patterns in schools and in the labor market. Although their academic performances would not be described as remarkable, the relatively better performance of minority girls in schools could place them at a modest advantage in a limited opportunity structure. In addition, because of the association of "softness" with girls (and, by extension, their higher proclivity for "soft" skills), minority females would likely fare better in the labor market. These patterns have critical educational, social, and economic implications for many African American and Latino youths, particularly since their class and racial statuses already preclude the attainment of certain social and political resources in U.S. society.

Note

1. The African American and Latino girls in the study also adhered to styles and tastes that are distinctive from those of their White counterparts. Not only did they speak "slang," which Samurai associated and conflated with "black talk," but these girls also embraced conspicuous differences in interactional and other styles, such as nail and hair care. Many argue that African American, Latino, and White females will experience different social and cultural responses because of the varied impact of gender, race, and class dynamics in their lives (Collins 1991; Anzaldua 1990; hooks 1984; King 1996; Rollins 1985). Parker et al. (1995) have found that African American adolescent females and White adolescent females differ significantly in their concepts of beauty, body image, and weight concerns. African American females were found to be more flexible than their White counterparts and spoke more about "making what you've got work for you." In contrast, many White females expressed dissatisfaction with their body shape and were found to be rigid in their concepts of beauty.

References

Adler, Patricia A., Steven J. Kless, and Peter Adler. 1992. "Socialization to Gender Roles: Popularity among Elementary School Boys and Girls." *Sociology of Education* 65 (3): 169–87.

Anderson, Elijah. 1980. "Some Observations on Black Youth Employment." In *Youth Employment and Public Policy,* edited by B. E. Anderson and I. V. Sawbell. Englewood Cliffs, N.J.: Prentice–Hall.

———. 1990. *Streetwise: Race, Class and Change in an Urban Community*. Chicago: University of Chicago Press.

Anzaldua, Gloria, ed. 1990. *Making Face, Making Soul: Haciendo Caras*. San Francisco: Aunt Lute Foundation Books.

Beale, Frances. [1979] 1995. "Double Jeopardy: To Be Black and Female." In *Words of Fire: An*

Anthology of African-American Feminist Thought, edited by B. Guy-Sheftall. New York: New Press.

Browne, Irene, ed. 1999. *Latinas and African American Women at Work: Race, Gender and Economic Inequality*. New York: Russell Sage Foundation.

Burke, Peter J. 1989. "Gender Identity, Sex, and School Performance." *Social Psychology Quarterly 52* (2): 159–69.

Butler, Judith. 1993. *Bodies That Matter: On the Discursive Limits of "Sex."* New York: Routledge.

Catsambis, Sophia. 1994. "The Path to Math: Gender and Racial-Ethnic Differences in Mathematics Participation from Middle School to High School." *Sociology of Education 67*(3): 199–215.

Cauce, Ana M., Yumi Hirage, Diane Graves, Nancy Gonzales, Kimberly Ryan-Finn, and Kwai Grove. 1996. "African American Mothers and Their Adolescent Daughters: Closeness, Conflict, and Control." In *Urban Girls: Resisting Stereotypes, Creating Identities,* edited by B. J. R. Leadbeater and N. Way. New York: New York University Press.

Collins, Patricia Hill. 1991. *Black Feminist Thought.* New York: Routledge.

———. 2004. *Black Sexual Politics: African Americans, Gender and the New Racism.* New York: Routledge.

Community College Week. 2000. "ACE Study Explodes Widespread Gender 'Crisis' in College Enrollment." December 11, 2000, vol. *13*(9): 3.

Connell, Robert W. 1995. *Masculinities.* Berkeley and Los Angeles: University of California Press.

Cose, Ellis, and Allison Samuels. 2003. "The Black Gender Gap." *Newsweek,* March 3, 2003, 46.

Crenshaw, Kimberlé. 1992. "Whose Story Is It Anyway? Feminist and Anti-Racist Appropriations of Anita Hill." In *Race-ing Justice, En-gendering Power,* edited by T. Morrison. New York: Pantheon.

Dance, L. Janelle. 2002. *Tough Fronts: The Impact of Street Culture on Schooling.* New York: Routledge Falmer.

Davis, James E. 2001. "Transgressing the Masculine: African American Boys and the Failure of Schools." In *What about the Boys? Issues of Masculinity in Schools,* edited by W. Martino and B. Meyenn. Buckingham: Open University Press.

de Beauvoir, Simone. 1989. *The Second Sex.* New York: Vintage.

Epstein, Debbie, V. Hey, and J. Maw, eds. 1998. *Failing Boys: Issues in Gender and Achievement.* Buckingham: Open University Press.

Fenstermaker, Sarah, and Candace West., eds. 2002. *Doing Gender, Doing Difference: Inequality, Power, and Institutional Change.* New York: Routledge.

Ferguson, Ann Arnett. 2000. *Bad Boys: Public Schools in the Making of Black Masculinity.* Ann Arbor: University of Michigan Press.

Fernández Kelly, M. Patricia. 1995. "Social and Cultural Capital in the Urban Ghetto: Implications for the Economic Sociology of Immigration." In *The Economic Sociology of Immigration: Essays on Networks, Ethnicity and Entrepreneurship,* edited by A. Portes. New York: Russell Sage Foundation.

Frankenberg, Ruth. 1993. *White Women, Race Matters: The Social Construction of Whiteness.* Minneapolis: University of Minnesota Press.

Fuller, Mary. 1980. "Black Girls in a London Comprehensive School." In *Schooling for Women's Work,* edited by R. Deem. London: Routledge and Kegan Paul.

Furstenburg, Frank, Theodore Herschberg, and John Modell. 1975. "The Origins of the Black Female-Headed Family: The Impact of Urban Experience." *Journal of Interdisciplinary History 7*(2): 211–33.

Gates, Harry L. 1988. *The Signifying Monkey.* New York: Oxford University Press.

———. 1991. "Ethnicity, Gender and Social Class: The School Adaptation Patterns of West Indian Youths." In *Minority Status and Schooling: A Comparative Study of Immigrant and Involuntary Minorities,* edited by M. A. Gibson and J. U. Ogbu. New York: Garland.

Gilligan, Carol. 1982. In a Different Voice: Psychological Theory and Women's Development. Cambridge, Mass.: Harvard University Press.

Goodwin, Marjorie. 1991. *He-Said-She-Said: Talk as Social Organization among Black Children.* Bloomington: Indiana University Press.

Gould, Mark. 1999. "Race and Theory: Culture, Poverty and Adaptation to Discrimination in Wilson and Ogbu." *Sociological Theory 17*(2): 171–200.

Gramsci, Antonio. 1994. *Letters from Prison.* Edited by F. Rosengarten. New York: Columbia University Press.

Guy-Sheftall, Beverly, ed. 1996. *Words of Fire: An Anthology of African-American Feminist Thought*. New York: New Press.

Hannerz, Ulf. 1969. *Soulside: Inquiries into Ghetto Culture and Community.* New York: Columbia University Press.

Hoberman, John. 1997. *Darwin's Athletes: How Sports Has Damaged Black America and Preserved the Myth of Race*. New York: Houghton Mifflin.

Hochschild, Arlie Russell. 1973. "A Review of Sex Role Research." *The American Journal of Sociology 78*(4): 1011–29.

Holland, Dorothy C., and Margaret A. Eisenhart. 1990. *Educated in Romance: Women, Achievement and College Culture.* Chicago: University of Chicago Press.

Holtzer, Harry J. 1996. "Employer Skill Needs and Labor Market Outcomes by Race and Gender Institute for Research on Poverty." Discussion Paper # 1087–96. Institute for Research on Poverty: University of Wisconsin-Madison.

hooks, bell. 1984. *From Margin to Center.* Boston: South End Press.

Jacobs, Jerry A. 1996. "Gender Inequality and Higher Education." *Annual Review of Sociology* 22: 153–85.

Jaynes, Gerald, and R. M. Williams. 1989. *A Common Destiny: Blacks and American Society.* Washington, D.C.: National Academy Press.

Jones, Jacqueline. 1985. *Labor of Love, Labor of Sorrow: Black Women, Work and the Family from Slavery to Present*. New York: Basic.

Journal of Blacks in Higher Education. 1999. "Special Report: College Degree Awards: The Ominous Gender Gap in African American Higher Education." 23 (Spring): 6–9.

Kelley, Robin. 1994. *Race Rebels: Culture, Politics and the Black Working Class*. New York: Free Press.

King, Deborah K. 1996. "Multiple Jeopardy, Multiple Consciousness: The Context of Black Feminist Ideology." In *Words of Fire: An Anthology of African-American Feminist Thought,* edited by B. Guy-Sheftall. New York: New Press.

King, Mary C. 1993. "Black Women's Breakthrough into Clerical Work: An Occupational Tipping Model." *Journal of Economic Issues 27*(4): 1097–1126.

Kirschenman, Joleen, and Kathryn M. Neckerman. 1991. "'We'd Love to Hire Them But': The Meaning of Race for Employers." In *The Urban Underclass,* edited by C. Jencks and P. E. Peterson. Washington, D.C.: The Brookings Institution.

Liebow, Elliot. 1967. *Tally's Corner: A Study of Negro Streetcorner Men*. Boston: Little Brown.

Lopez, Nancy. 2003. *Hopeful Girls, Troubled Boys.* New York: Routledge.

Maccoby, Eleanor E., and Carol N. Jacklin. 1974. *Psychology of Sex Differences.* Stanford, Calif.: Stanford University Press.

Major, Clarence, ed. 1994. *Juba to Jive: A Dictionary of African-American Slang*. New York: Penguin.

Massey, Douglas, Camille Charles, Garvey F. Lundy, and Mary Fischer. 2003. *The Source of the River: The Social Origins of Freshmen at America's Selective Colleges and Universities.* Princeton, N.J.: Princeton University Press.

Massey, Douglas, and Nancy Denton. 1993. *American Apartheid.* Cambridge, Mass.: Harvard University Press.

Mickelson, A. 1989. "Why Does Jane Read and Write So Well? The Anomaly of Women's Achievement." *Sociology of Education 62*(1): 47–63.

Mirande, Alfredo. 1997. *Hombres y Machos: Masculinity and Latino Culture*. Boulder, Colo.: Westview.

Moss, Philip, and Chris Tilly. 1996. "'Soft' Skills and Race: An Investigation of Black Men's Employment Problems." *Work and Occupations* 23:252–76.

Noguera, Pedro A. 1996. "Responding to the Crisis Confronting California's Black Male Youth: Providing Support without Furthering Marginalization." *Journal of Negro Education* 65 (2): 219–36.

———. 2003. "The Trouble with Black Boys: The Impact of Social and Cultural Forces on the

Academic Achievement of African American Males." *Urban Education* 38 (4): 431–50.

Nowell, Amy, and Larry V. Hedges. 1998. "Trends in Gender Differences in Academic Achievement from 1960 to 1994: An Analysis of Differences in Mean Variance, and Extreme Scores." *Sex Roles* 39 (1/2): 21–43.

Parker, Sheila, Mimi Nichter, Mark Nichter, Nancy Vuckovic, Colette Sims, and Cheryl Ritenbaugh. 1995. "Body Image and Weight Concerns among African American and White Adolescent Females: Differences That Make a Difference." *Human Organization* 54 (2): 103–14.

Rollins, Judith. 1985. *Between Women: Domestics and Their Employers.* Philadelphia: Temple University Press.

Solomon, R. Patrick. 1991. *Black Resistance in High School, Frontiers in Education Series.* Albany: State University of New York Press.

Staples, Robert. 1982. *Black Masculinity: The Black Male's Role in American Society.* San Francisco, Calif.: Black Scholar Press.

Stevens, Joyce W. 1997. "African American Female Adolescent Identity Development: A Three-Dimensional Perspective." *Child Welfare* 76 (1): 145–72.

Stevenson, Harold W., and Richard S. Newman. 1986. "Long-Term Prediction of Achievement and Attitudes in Mathematics and Reading." *Child Development* 57:646–59.

Streitmatter, Janice. 1993. "Gender Differences in Identity Development: An Examination of Longitudinal Data." *Adolescence* 28 (109): 55–66.

Taylor, Jill McLean, Carol Gilligan, and Amy M. Sullivan, eds. 1995. *Between Voices and Silence: Women and Girls, Race and Relationship.* Cambridge, Mass.: Harvard University Press.

Valenzuela, Angela. 1999. *Subtractive Schooling: Issues of Caring in Education of U.S.-Mexican Youth.* Albany: State University of New York Press.

Waldinger, Roger. 1996. *Still the Promised City: African Americans and New Immigrants in Postindustrial New York.* Cambridge, Mass.: Harvard University Press.

Way, Niobe. 1996. "Between Experiences of Betrayal and Desire: Close Friendships among Urban Adolescents." In *Urban Girls: Resisting Stereotypes,* edited by B. J. Leadbeater and N. Way. New York: NYU Press.

Wilson, William J. 1987. *The Truly Disadvantaged: The Inner City, the Underclass, and Public Policy.* Chicago: University of Chicago Press.

Introduction to Reading 19

Journalist Elizabeth Gilbert wrote this article for *GQ* in 2001. She became a "man" with the help of Diane Torr, a performance artist who runs workshops designed to help women experience masculinity by becoming men. Gilbert learned that it takes more to become a man than simply bandaging breasts and wearing a birdseed penis. This article helps us to understand the complex emotional and interactional basis upon which we all "do gender," as well as the close relationship between doing femininity and doing masculinity.

1. How does Gilbert become a "man"?
2. What effect does this transformation have on the way she interacts with and looks at women?
3. What effect does this transformation have on Gilbert's long-term relationship with her husband and others she interacts with?

My Life as a Man

Elizabeth Gilbert

The first time I was ever mistaken for a boy, I was 6 years old. I was at the county fair with my beautiful older sister, who had the long blond tresses one typically associates with storybook princesses. I had short messy hair, and I had scabs all over my body from falling out of trees. My beautiful sister ordered a snow cone. The lady at the booth asked, "Doesn't your little brother want one, too?"

I was mortified. I cried all day.

The last time I was mistaken for a boy was only a few weeks ago. I was eating in a Denny's with my husband, and the waitress said, "You fellas want some more coffee?"

This time I didn't cry. It didn't even bother me, because I've grown accustomed to people making the mistake. Frankly, I can understand why they do. I'm afraid I'm not the most feminine creature on the planet. I don't exactly wish to hint that Janet Reno and I were separated at birth, but I do wear my hair short, I am tall, I have broad shoulders and a strong jaw, and I have never really understood the principles of cosmetics. In many cultures, this would make me a man already. In some very primitive cultures, this would actually make me a king.

But sometime after the Denny's incident, I decided, *Ah, to hell with it. If you can't beat 'em, join 'em.* What would it take, I began to wonder, for me to actually transform into a man? To live that way for an entire week? To try to fool everyone? . . .

Fortunately, I have plenty of male friends who rally to my assistance, all eager to see me become the best man I can possibly be. And they all have wise counsel to offer about exactly How to Be a Guy:

> "Interrupt people with impunity from now on," says Reggie. "Curse recklessly. And never apologize."
>
> "Never talk about your feelings," says Scott. "Only talk about your accomplishments."
>
> "The minute the conversation turns from something that directly involves you," says Bill, "let your mind wander and start looking around the room to see if there's anything nearby you can have sex with."
>
> "If you need to win an argument," says David, "just repeat the last thing the guy you're fighting with said to you, but say it much louder."

So I'm thinking about all this, and I'm realizing that I already do all this stuff. I always win arguments, I'm shamefully slow to apologize, I can't imagine how I could possibly curse any more than I already goddamn do, I've spent the better part of my life looking around to see what's available to have sex with, I can't shut up about my accomplishments, and I'm probably interrupting you right this moment.

Another one of my friends warns, "You do this story, people are gonna talk. People might think you're gay." Aside from honestly not caring what people think, I'm not worried about this possibility at all. I'm worried about something else entirely: that this transformation thing might be too easy for me to pull off.

What I'm afraid I'll learn is that I'm *already* a man.

My real coach in this endeavor, though, is a woman. Her name is Diane Torr. Diane is a performance artist who has made her life's work the exploration of gender transformation. As a famous drag king, she has been turning herself

---

into a man for twenty years. She is also known for running workshops wherein groups of women gather and become men for a day.

I call Diane and explain my goal, which is not merely to dress up in some silly costume but to genuinely pass as male and to stay in character for a week.

"That's a tough goal," Diane says, sounding dubious. "It's one thing to play with gender for the afternoon, but really putting yourself out there in the world as a man takes a lot of balls, so to speak. . . ."

Diane agrees to give me a private workshop on Monday. She tells me to spend the weekend preparing for my male life and buying new clothes. Before hanging up, I ask Diane a question I never thought I would ever have to ask anybody:

"What should I bring in terms of genitalia?"

This is when she informs me of the ingredients for my penis.

"Of course," I say calmly.

I write *birdseed* on my hand, underline it twice and make a mental note to stay away from the aviary next week.

I Spend the Weekend Inventing My Character.

One thing is immediately clear: I will have to be younger. I'm 31 years old, and I look it, but with my smooth skin, I will look boyish as a man. So I decide I will be 21 years old for the first time in a decade.

As for my character, I decide to keep it simple and become Luke Gilbert—a midwestern kid new to the city, whose entire background is cribbed from my husband, whose life I know as well as my own.

Luke is bright but a slacker. He really doesn't give a damn about his clothes, for instance. Believe me, I know—I'm the one who shopped for Luke all weekend. By Sunday night, Luke owns several pairs of boring Dockers in various shades of khaki, which he wears baggy. He has Adidas sneakers. He has some boxy short-sleeve buttondown shirts in brown plaids. He has a corduroy jacket, a bike messenger's bag, a few baseball caps and clean underwear. He also has, I'm sorry to report, a really skinny neck.

I haven't even met Luke yet, but I'm beginning to get the feeling he's a real friggin' geek.

The Transformation Begins Painlessly Enough.

It starts with my hair. Rayya, my regular hairdresser, spends the morning undoing all her work of the past months—darkening out my brightest blond highlights, making me drab, brownish, inconsequential; chopping off my sassy Dixie Chick pixie locks and leaving me with a blunt cut.

"Don't wash it all week," Rayya advises. "Get good and greasy; you'll look more like a guy."

Once the hair is done, Diane Torr gets to work on me. She moves like a pro, quick and competent. Together we stuff my condom ("This is the arts-and-crafts portion of the workshop!"), and Diane helps me insert it into my Calvins. She asks if I want my penis to favor the left or right side. Being a traditionalist, I select the right. Diane adjusts me and backs away; I look down and there it is—my semierect penis, bulging slightly against my briefs. I cannot stop staring at it and don't mind saying that it freaks me out to no end. Then she tries to hide my breasts. To be perfectly honest, my breasts are embarrassingly easy to make disappear. Diane expertly binds them down with wide Ace bandages. Breathing isn't easy, but my chest looks pretty flat now—in fact, with a men's undershirt on, I almost look as if I have well-developed pectoral muscles.

But my ass? Ah, here we encounter a more troublesome situation. I don't want to boast, but I have a big, fat, round ass. You could lop off huge chunks of my ass, make a nice osso buco out of it, serve it up to a family of four and still eat the leftovers for a week. This is a woman's ass, unmistakably. But once I'm fully in costume, I turn around before the mirror and see that I'm going to be OK. The baggy, low-slung pants are good ass camouflage, and the boxy plaid shirt completely eliminates any sign of my waist, so

I don't have that girlie hourglass thing happening. I'm a little pear-shaped, perhaps, but let us not kid ourselves, people. There are pear-shaped men out there, walking among us every day.

Then Diane starts on my facial transformation. She has brought crepe hair—thin ropes of artificial hair in various colors, which she trims down to a pile of golden brown stubble. I elect, in homage to Tom Waits, to go with just a small soul patch, a minigoatee, right under my bottom lip. Diane dabs my face with spirit gum—a kind of skin-friendly rubber cement—and presses the hair onto me. It makes for a shockingly good effect. I suggest sideburns, too, and we apply these, making me look like every 21-year-old male art student I've ever seen. Then we muss up and darken my eyebrows. A light shadow of brown under my nose gives me a hint of a mustache. When I look in the mirror, I can't stop laughing. *I am a goddamn man, man!*

Well, more or less.

Diane looks me over critically. "Your jaw is good. Your height is good. But you should stop laughing. It makes you look too friendly, too accessible, too feminine." I stop laughing. She stares at me. "Let's see your walk."

I head across the floor, hands in my pockets. "Not bad," Diane says, impressed.

Well, I've been practicing. I'm borrowing my walk from Tim Goodwin, a guy I went to high school with. Tim was short and slight but an amazing basketball player (we all called him "Tim Godwin"), and he had an athletic, knee-knocking strut that was very cool. There's also a slouch involved in this walk. But it's—and this is hard to explain—a *stiff* slouch. Years of yoga have made me really limber, but as Luke, I need to drop that ease of motion with my body, because men are not nearly as physically free as women. Watch the way a man turns his head: His whole upper torso turns with it. Unless he's a dancer or a baseball pitcher, he's probably operating his entire body on a ramrod, unyielding axis. On the other hand, watch the way a woman drinks from a bottle. She'll probably tilt her whole head back to accommodate the object, whereas a man would probably hold his neck stiff, tilting the bottle at a sharp angle, making the bottle accommodate *him.* Being a man, it seems, is sometimes just about not budging.

Diane goes on to coach my voice, telling me to lower the timbre and narrow the range. She warns me against making statements that come out as questions, which women do constantly (such as when you ask a woman where she grew up and she replies, "Just outside Cleveland?"). But I don't do that begging-for-approval voice anyway, so this is no problem. As I'd suspected, in fact, all this turning-male stuff is coming too easily to me.

But then Diane says, "Your eyes are going to be the real problem. They're too animated, too bright. When you look at people, you're still too engaged and interested. You need to lose that sparkle, because it's giving you away."

The rest of the afternoon, she's on me about my eyes. She says I'm too flirtatious with my eyes, too encouraging, too appreciative, too attentive, too *available.* I need to intercept all those behaviors, Diane says, and erase them. Because all that stuff is "shorthand for girl." Girls typically flirt and engage and appreciate and attend; men typically don't. It's too generous for men to give themselves away in such a manner. Too dangerous, even. Granted, there are men in this world who are engaging, attentive and sparkly eyed, but Luke Gilbert cannot be one of them. Luke Gilbert's looks are so on the border of being feminine already that I can't afford to express any behavior that is "shorthand for girl," or my cover is blown. I can only emit the most stereotypical masculine code, not wanting to offer people even the faintest hint that I'm anything but a man.

Which means that gradually throughout Monday afternoon, I find myself shutting down my entire personality, one degree at a time. It's very similar to the way I had to shut down my range of physical expression, pulling in my gestures and stiffening up my body. Similarly, I must not budge emotionally. I feel as if I'm closing down a factory, silencing all the humming machines of my character, pulling shut the gates, sending home the workers. All my most animated and familiar facial expressions have to go, and with them go all my most animated and familiar emotions. Ultimately, I am left with only two options for expression—boredom and aggression. Only with boredom and aggression do I truly feel male. It's not a feeling I like at all, by the way. In fact, I am amazed by how much I don't like it. We've been laughing and joking and relating all morning, but slowly now, as I turn into Luke, I feel the whole room chill.

Toward the end of the afternoon, Diane gives me her best and most disturbing piece of advice.

"Don't look at the world from the surface of your eyeballs," she says. "All your feminine availability emanates from there. Set your gaze back in your head. Try to get the feeling that your gaze originates from two inches behind the surface of your eyeballs, from where your optic nerves begin in your brain. Keep it right there."

Immediately, I get what she's saying. I pull my gaze back. I don't know how I appear from the outside, but the internal effect is appalling. I feel—for the first time in my life—a dense barrier rise before my vision, keeping me at a palpable distance from the world, roping me off from the people in the room. I feel dead eyed. I feel like a reptile. I feel my whole face change, settling into a hard mask.

Everyone in the room steps back. Rayya, my hairdresser, whistles under her breath and says, "Whoa . . . you got the guy vibe happenin' now, Luke."

Slouching and bored, I mutter a stony thanks.

Diane finally takes me outside, and we stroll down the street together. She has dressed in drag, too. She's now Danny King—a pompous little man who works in a Pittsburgh department store. She seems perfectly at ease on the street, but I feel cagey and nervous out here in the broad daylight, certain that everyone in the world can see that my face is covered with fake hair and rubber cement and discomfort. The only thing that helps me feel even remotely relaxed is the basketball I'm loosely carrying under my arm—a prop so familiar to me in real life that it helps put me at ease in disguise. We head to a nearby basketball court. We have a small crowd following us—my hairdresser, the makeup artist, a photographer. Diane and I pose for photos under the hoop. I set my basketball down, and almost immediately, a young and muscular black guy comes over and scoops it off the pavement.

> "Hey," he says to the crowd. "Whose basketball is this?"

Now, if you want to learn how to define your personal space as a man, you could do worse than take lessons from this guy. His every motion is offense and aggression. He leads with his chest and chin, and he's got a hard and cold set of eyes.

> "I said, whose basketball is this?" he repeats, warning with his tone that he doesn't want to have to ask again.
>
> "It's hers," says my hairdresser, pointing at me. "Hers?" The young man looks at me and snorts in disgust. "What are you talkin' about, hers? That ain't no *her*. That's a *guy*." My first gender victory!

But there's no time to celebrate this moment, because this aggressive and intimidating person needs to be dealt with. Now, here's the thing. Everyone on the court is intimidated by this guy, but I am not. In this tense moment, mind you, I have stopped thinking like Luke Gilbert; I'm back to thinking like Liz Gilbert. And Liz Gilbert always thinks she can manage men. I don't know if it's from years of tending bar, or if it's from living in lunatic-filled New York City, or if it's just a ridiculous (and dangerously naive) sense of personal safety, but I have always believed in my heart that I can disarm any man's aggression. I do it by paying close attention to the aggressive man's face and finding the right blend of flirtation, friendliness and confidence to put on my face to set him at ease, to remind him: You don't wanna hurt me, you wanna like me. I've done this a million times before. Which is why I'm looking at this scary guy and I'm thinking, Give me thirty seconds with him and he'll be on my side.

> I step forward. I open up my whole face in a big smile and say teasingly, "Yeah, that's my basketball, man. Why, you wanna play? You think you can take me?"

> "You don't know nothin' about this game," he says.
>
> In my flirtiest possible voice, I say, "Oh, I know a *little* somethin' about this game. . . ."
>
> The guy takes a menacing step forward, narrows his eyes and growls, "You don't know *shit* about this game."

This is when I snap to attention. This is when I realize I'm on the verge of getting my face punched. What the hell am I doing? This guy honestly thinks I'm a man! Therefore, my whole cute, tomboyish, I'm-just-one-of-the-guys act is not working. One-of-the-guys doesn't work when you actually are one of the guys. I have forgotten that I am Luke Gilbert—a little white loser on a basketball court who has just challenged and pissed off and flirted with an already volatile large black man. I have made a very bad choice here. I've only been on the job as a male for a few minutes, but it appears as though I'm about to earn myself a good old-fashioned New York City ass-kicking.

> He takes another step forward and repeats, "You don't know shit about nothin.'"
>
> "You're right, man," I say. I drop my eyes from his. I lower my voice, collapse my posture, show my submission. I am a stray dog, backing away from a fight, head down, tail tucked. "Sorry, man. I was just kidding. I don't know anything about basketball."
>
> "Yeah, that's right," says the guy, satisfied now that he has dominated me. "You don't know shit."

He drops the ball and walks away. My heart is slamming. I'm angry at my own carelessness and frightened by my newfound helplessness. Luke didn't know how to handle that guy on the court, and Luke almost got thrown a beating as a result (and would have deserved it, too—the moron). Realizing this makes me feel suddenly vulnerable, suddenly aware of how small I've become.

My hands, for instance, which have always seemed big and capable to me, suddenly appear rather dainty when I think of them as a man's hands. My arms, so sturdy only hours before, are now the thin arms of a weenie-boy. I've lost this comfortable feeling I've always carried through the world of being strong and brave. A five-foot-nine-inch, 140-pound woman can be a pretty tough character, after all. But a five-foot-nine-inch, 140-pound man? Kinda small, kinda wussy. . . .

* * *

My world-famously tolerant husband seems to have no trouble with my transformation at first. He unwinds my breast bandages every night before bed and listens with patience to my complaints about my itching beard. In the mornings before work, he binds up my breasts again and lends me his spice-scented deodorant so I can smell more masculine. We vie for mirror space in the bathroom as he shaves off his daily stubble and I apply mine. We eat our cereal together, I take my birth control pills, I pack my penis back into my slacks. . . . It's all very domestic.

Still, by Wednesday morning, my husband confesses that he doesn't want to hang around with me in public anymore. Not as long as I'm Luke. It's not that he's grossed out by my physical transformation, or threatened by the sexual politics at play, or embarrassed by the possibility of exposure. It's simply this: He is deeply, emotionally unsettled by my new personality.

"I miss you," he says. "It's seriously depressing for me to be around you this way."

What's upsetting to Michael is that as a man, I can't give him what he has become accustomed to getting from me as a woman. And I'm not talking about sex. Sex can always be arranged, even this week. (Although I do make a point now of falling asleep immediately after it's over, just to stay in character.) What Michael hates is that I don't engage him anymore. As Luke, I don't laugh at my husband's jokes or ask him about his day. Hell, as Luke, I don't even have a husband—just another drinking buddy whose jokes and workday concerns I don't really care about. Michael, still seeing his wife under her goatee, keeps thinking I'm mad at him, or—worse—bored by him. But I can't attend to him on this, can't reassure him, or I risk coming across like a girl.

The thing is, I don't like Luke's personality any more than Michael does. As Luke, I feel completely and totally bound—and not just because of the tight bandage wrapped around my chest. I keep thinking back to my drag-king workshop, when Diane Torr talked about "intercepting learned feminine habits." She spoke of those learned feminine habits in slightly disparaging terms. Women, she said, are too attentive, too concerned about the feelings of others, too *available*. This idea of women as lost in empathy is certainly a standard tenet of feminism (Oprah calls it the Disease to Please), and, yes, there are many women who drown in their own overavailability. But I've never personally felt that attentiveness and engagement are liabilities. As a writer—indeed, as a *human being*—I think the most exciting way you can interact with this fantastic and capricious world is by being completely available to it. Peel me wide open; availability is my power.

I would so much rather be vulnerable and experience existence than be strong and defend myself from it. And if that makes me a girlie-girl, then so be it—I'll be a goddamn girlie-girl.

Only, this week I'm not a girl at all. I'm Luke Gilbert. And poor Luke, I must say, is completely cut off from the human experience. The guy is looking at the world from a place two inches behind his eyeballs. No wonder my husband hates being around him. I'm not crazy about him myself.

* * *

[Wednesday], I'm walking home alone. Just ahead of me, a blond woman steps out of a bar, alone. She's screamingly sexy. She's got all the props—the long hair, the tiny skirt, the skimpy top, the wobbly stiletto heels, the eternal legs. I walk right behind this woman for several blocks and observe the tsunami she causes on 23rd Street in every man she passes—everyone has to react to her somehow. What amazes me, though, is how many of the men end up interacting with *me* after passing *her*. What happens is this: She saunters by, the guy stares at her in astonishment and then makes a comment about her to me because I'm the next man on the scene. So we have a little moment together, the guy and me, in which we share an experience. We get to bond. It's an ice-breaker for us.

The best is the older construction worker who checks out the babe, then raises his eyebrows at me and declares: "Fandango!"

"You said it!" I say, but when I walk on by, he seems a little disappointed that I haven't stuck around to talk more about it with him.

This kind of interaction happens more than a dozen times within three blocks. Until I start wondering whether this is actually the game. Until I start suspecting that these guys maybe don't want to talk to the girl at all, that maybe they just desperately want to talk to one another.

Suddenly, I see this sexy woman in front of me as being just like sports; she's an excuse for men to try to talk to one another. She's like the Knicks, only prettier—a connection for people who otherwise cannot connect at all. It's a very big job, but I don't know if she even realizes she's doing it.

* * *

[Friday] night, taking a friend's advice, I go out drinking in the East Village, where seven out of ten young men look just like Luke Gilbert. I end up at a bar that is crawling with really cute pierced-nosed girls. I'm wondering whom I should try to pick up when an opportunity falls into my lap. A pretty red-haired girl in a black camisole walks into the bar alone. She has cool tattoos all over her arms. The bouncer says to her, "Hey, Darcy, where's your crowd tonight?"

> "Everyone copped out," Darcy says. "I'm flying solo."
>
> "So lemme buy you a drink," I call over from the bar.
>
> "Rum and Coke," she says, and comes over to sit next to me.

Fandango!

We get to talking. Darcy's funny, friendly, from Tennessee. She tells me all about her roommate problems. She asks me about myself, but I don't share—Luke Gilbert is not available for sharing. Instead, I compliment Darcy on her pretty starfish necklace, which Darcy tells me was a gift from a childhood neighbor who was like a grandmother to her. I ask Darcy about her job, and she tells me she works for a publishing house that prints obscure journals with titles like *Catfish Enthusiast Monthly.*

"Damn, and here I just let my subscription to *Catfish Enthusiast Monthly* run out," I say, and she laughs. Darcy actually does that flirty thing girls do sometimes where they laugh and touch your arm and move closer toward you all at the same time. I know this move. I've been doing this move my whole life. And it is with this move and this touch and this laugh that I lose my desire to play this game anymore, because Darcy, I can tell, actually likes Luke Gilbert. Which is incredible, considering that Luke is a sullen, detached, stiff guy who can't make eye contact with the world. But she still likes him. This should feel like a victory, but all I feel like is a complete shitheel. Darcy is nice. And here I'm lying to her already. Now I really *am* a guy.

> "You know what, Darcy?" I say. "I have to go. I'm supposed to hook up with some friends for dinner."
>
> She looks a little hurt. But not as hurt as she would look if, say, we dated for a month and then she found out the truth about me.
>
> I give her a little kiss good-bye on the cheek. "You're great," I tell her. And then I'm done.

Undoing It All Takes a Few Days.

Rubbing alcohol gets the last of the spirit gum and fake hair off my face. I pluck my eyebrows and put on my softest bra (my skin has become chafed from days of binding and taping). I scatter my penis across the sidewalk for the pigeons. I make an appointment to get my hair lightened again. I go to yoga class and reawaken the idea of movement in my body. I cannot wait to get rid of this gender, which I have not enjoyed. But it's a tricky process, because I'm still walking like Luke, still standing like Luke, still thinking like Luke.

In fact, I don't really get my inner Liz back until the next weekend. It's not until the next Saturday night, when I am sitting at a bar on my own big fat ass, wearing my own girlie jeans, talking to an off-duty New York City fireman, that I really come back into myself. The fireman and I are both out with big groups, but somehow we peel off into our own private conversation. Which quickly gets serious. I ask him to tell me about the crucifix around his neck, and he says he's been leaning on God pretty hard this year. I want to know why. The fireman starts telling me about how his beloved father died this winter, and then his fiancée left him, and now the pressures of his work are starting to kill him, and there are times when he just wishes he could cry but he doesn't want people to see him like that. My guy friends are all playing darts in the corner, but I'm the one sitting here listening to this fireman tell me about how he never cries because his dad was such a hard-ass Irish cop, don'tcha know, because he was raised to hang so tough.

I'm looking right into this guy. I'm not touching him at all, but I'm giving him my entire self. He needs me right now, to tell all this to. He can have me. I've got my eyes locked on him, and I can feel how bad he wants to cry, and with my entire face I am telling this man: *Tell me everything.*

He says, "Maybe I was hard on her, maybe that's why she left me, but I was so worried about my father. . . ."

The fireman digs at his eye with a fist. I hand him a bar napkin. He blows his nose. He keeps talking. I keep listening. He can talk to me all night because I am unbound and I am wide-open. I'm open around the clock, open twenty-four hours a day; I never close. I'm really concerned for this guy, but I'm smiling while he spills his story because it feels so good to catch it. It feels so good to be myself again, to be open for business again—open once more for the rewarding and honest human business of complete *availability.*

Introduction to Reading 20

Leora Tanenbaum uses her own life experiences to help us understand the complex and powerful link between gender and sexuality. In a world permeated by sexual images (as will be described in Chapter 6) and increased sexual activity, she argues that being sexually active is still stigmatized for girls and women. In this selection from her book of the same title, she also addresses responses to her book that she received from various sources. Her responses to these comments help us to understand what the situation is for young girls who are thought to be sexually active and how we might change it.

1. What does Tanenbaum mean by "slut-bashing"? Have you seen instances of it?
2. What does power have to do with the social construction of femininity and sexuality?
3. How does "slut-bashing" maintain traditional gender patterns and pit women against women?

Slut!

Growing Up Female With a Bad Reputation

Leora Tanenbaum

Women living in the United States are fortunate indeed. Unlike women living in Muslim countries, who are beaten and murdered for the appearance of sexual impropriety, we enjoy enormous sexual freedom.[1] Yet even we are routinely evaluated and punished for our sexuality. In 1991, Karen Carter, a twenty-eight-year-old single mother, lost custody of her two-year-old daughter in a chain of events that began when she called a social service hot line to ask if it's normal to feel sexual arousal while breast feeding. Carter was charged with sexual abuse in the first degree, even though her daughter showed no signs of abuse; when she revealed in court that she had had a lifetime total of eight (adult male) lovers, her own lawyer referred to her "sexual promiscuity."[2] In 1993, when New Mexico reporter Tamar Stieber filed a sex discrimination lawsuit against the newspaper where she worked because she was earning substantially less than men in similar positions, defense attorneys deposed her former lover to ask him how often they'd had sex.[3] In the 1997 sexual-harassment lawsuits against Mitsubishi Motor Manufacturing, a company lawyer asked for the gynecological records of twenty-nine women employees charging harassment, and wanted the right to distribute them to company executives.[4] And in 1997 a North Carolina woman sued her husband's secretary for breaking up their nineteen-year-marriage and was awarded $1 million in damages by a jury. During the seven-day trial the secretary was described as a "matronly" woman who deliberately began wearing heavy makeup and short skirts in order to entice the husband into an affair.[5] . . .

In the realm of sexual choices we are light-years beyond the 1950s. Today a teenage girl can explore her sexuality without getting married, and most do. By age eighteen over half of all girls and nearly three quarters of all boys have had intercourse at least once.[6] Yet at the same time, a fifties-era attitude lingers: Teens today are fairly conservative about sex. A 1998 *New York Times*/CBS News poll of a thousand teens found that 53 percent of girls believe that sex before marriage is "always wrong," while 41 percent of boys agree.[7] Teens may be having sex, but they also look down on others, especially girls, who are sexually active. Despite the sexual revolution, despite three decades of feminism, despite the Pill, and despite legalized abortion, teenage girls today continue to be defined by their sexuality. The sexual double standard—and the division between "good" girls and "bad" or "slutty" ones—is alive and well. Some of the rules have changed, but the playing field is startlingly similar to that of the 1950s. . . .

In 1988, educators Janie Victoria Ward and Jill McLean Taylor surveyed Massachusetts teenagers across six different ethnic groups—black, white, Hispanic, Haitian, Vietnamese, and Portuguese—and found that the different groups upheld different sexual values. But one thing was universal: The sexual double standard. Regardless of race or ethnicity, "boys were generally allowed more freedom and were assumed to be more sexually active than girls." Ward and Taylor found that "sexual activity for adolescent males usually met cultural expectations and was generally accepted by adults and peers as part of normal male adolescence. . . . In general, women are often seen in terms of their sexual reputation rather than in terms of their personal characteristics."[8]

The double standard, we know, does not vaporize after high school. Sociologist Lillian Rubin surveyed six hundred students in eight colleges around the country in the late 1980s and found that 40 percent of the sexually active women said that they routinely understate their sexual experience because "my boyfriend wouldn't like it if he knew," "people wouldn't understand," and "I don't want him to think I'm a slut." Indeed, these women had reason to be concerned. When Rubin queried the men about what they expected of the women they might marry, over half said that they would not want to marry a woman who had been "around the block too many times," that they were looking for someone who didn't "sleep around," and that a woman who did was a "slut."[9]

Similarly when sex researcher Shere Hite surveyed over 2,500 college men and women, 92 percent of the men claimed that the double standard was unfair. Yet overwhelmingly they themselves upheld it. When asked, "If you met a woman you liked and wanted to date, but then found out she had had sex with ten to twenty men during the preceding year, would you still like her and take her seriously?," 65 percent of the men admitted that they would not take her seriously. At the same time only 5 percent said they would lose respect if a male friend had had sex with ten to twenty women in one year.[10]

Teenage girls who are called sluts today experience slut-bashing at its worst. Caught between the conflicting pressures to have sex and maintain a "good" reputation, they are damned when they do and damned when they don't. Boys and girls both are encouraged to have sex in the teen years—by their friends, magazines, and rock and rap lyrics—yet boys alone can get away with it. "There's no way that anyone who talks to girls thinks that there's a new sexual revolution out there for teenagers," sums up Deborah Tolman, a developmental psychologist at the Wellesley College Center for Research on Women. "It's the old system very much in place." It is the old system, but with a twist: Today's teenage girls have grown up after the feminist movement of the late 1960s and 1970s. They have been told their whole lives that they can, and should, do anything that boys do. But soon enough they discover that sexual equality has not arrived. Certain things continue to be the privilege of boys alone.

With this power imbalance, it's no wonder high school girls report feeling less comfortable with their sexual experiences than their male counterparts do. While 81 percent of adolescent boys say that "sex is a pleasurable experience," only 59 percent of girls feel the same way.[11] The statistical difference speaks volumes. Boys and girls both succumb to early sex due to peer and media pressures, but boys still get away with it while girls don't.

* * *

Most people who meet me for the first time are surprised by two things: that I am the type of person who would ever write a book with a title like Slut!, and that I was once known as a "slut." I have been described in print as "demure" and as "a petite brunette with wire-rimmed glasses"—code words for nice, shy, bookish. On the Oprah Winfrey show, I was presented as a nice, middle-class woman married to a nice, middle-class man. Over and over I am told, "But you're so clean-cut—you don't seem like a slut at all."

My point exactly: Any girl or woman can be labeled a "slut." Looks and attitude often have nothing to do with it. Yet the word continues to evoke for most people an image of someone trampy and pathetic—the kind of girl or woman who wears short, tight, cleavage-enhancing clothes, always makes a beeline for the guy who enters the room, and can't string two sentences together without making a non sequitur. In short, she deserves to be called a "slut." . . .

"Slut" is, of course, a disturbing insult. But it is part of the vocabulary of adolescents—and adults—and a key word in the vocabulary of the sexual double standard. The severity of the word might offend some people, such as a racial or ethnic insult would, but refraining from using it in serious discussion serves only to reinforce its power. After all, "nigger" is a profoundly disturbing word, but can we have an honest conversation about racism without using it? I don't think so. Likewise, we must use the word "slut" and openly discuss its ramifications in order to eliminate the sexual double standard.

Below are some of the comments I've received from men and women in bookstores and radio call-in shows, and from television and radio show hosts. It's clear that most people are far more concerned with the sexuality of girls than with that of boys. My responses point out that females as well as males should be entitled to express their sexual desires. Hardly a radical concept, but it can stir up a lot of hostility.

Slut-bashing is a terrible thing, but let's face it: it affects only a small number of girls. Why write a whole book about it?

A reputation acquired in adolescence can damage a young woman's self-perception for years. She may become a target for other forms of harassment and even rape, since her peers see her as "easy" and therefore not entitled to say "no." She may become sexually active with a large number of partners (even if she had not been sexually active before her reputation). Or she may shut down her sexual side completely, wearing baggy clothes and being unable to allow a boyfriend to even kiss her.

It's true that most girls escape adolescence unscathed by slut-bashing. Nevertheless, just about every girl is affected by it. Every girl internalizes the message that sex is bad—because it can earn you a reputation. The result is that even years later, when she is safely out of adolescence, a woman may suffer from a serious hangup about sex and intimacy—even if she was not herself called a "slut." Second, the fear of being called a "slut" makes many girls unlikely to carry or use contraceptives, leading of course to the risk of pregnancy or disease.

Slut-bashing also affects boys. It fosters a culture of sexual entitlement that says that "easy" girls are expendable while only "good" girls deserve to be treated well. And that means that only some girls are treated with the respect that they all deserve.

You make it seem as if we're living in the 1950s. But this is the twenty-first century. Lots of girls are having sex; most of them are not called "sluts."

Of course, a girl today has many freedoms that her 1950s counterpart did not possess, including the license to sexually experiment before marriage. But even today, the prevailing attitude is that there is something wrong with the girl who behaves just as a boy does. Compare, for example, the fate of two recent movies about teenagers and the pursuit of sex, one involving girls, the other involving boys. *Coming Soon,* a witty comedy set in the world of wealthy teen Manhattanites and boasting a star-filled cast (Mia Farrow, Spalding Gray, Ryan O'Neal, Gaby Hoffman), follows a female high school senior who has never had an orgasm and wonders why her boyfriend leaves her sexually unfulfilled. She worries that there is something wrong with her; after all, her girlfriends report that they feel completely fulfilled. (It turns out they're lying.) The movie is actually far from raunchy. There is no nudity, and the raciest scene involves only the protagonist and a Jacuzzi. It garnered positive reviews from Variety, The Hollywood Reporter, and many prestigious film festivals.

Yet the Motion Picture Association of America gave Coming Soon the dreaded NC-17 rating, effectively barring it from theaters—until director Colette Burson agreed to cut several scenes that the MPAA deemed "lurid." Now it has been granted a "respectable" R rating (and has been released by a small distributor in a few theaters), but at the expense of a serious exploration of female sexuality. Burson explains that the MPAA "really didn't like the idea of girls and orgasm."

While Burson was busy tranquilizing *Coming Soon,* kids lined up at theaters across the country to see the vulgar antics of four teenaged boys desperate to lose their virginity in *American Pie.* In a weak nod to the notion that one's partner should enjoy sex too, one of the buddy-boy characters works hard to give his girlfriend an orgasm—not because he cares about her satisfaction, but because that will induce her to go "all the way" with him. The movie is far more sexually explicit than *Coming Soon* ever was and utterly contemptuous of girls (the tag line is "There's something about your first piece"), yet it merited an R rating. *American Pie* has grossed over a hundred million dollars in United States box-office receipts alone.

Put side by side, these two movies demonstrate that the idea of females exploring sex is taboo, while the idea of males exploring sex is an opportunity for slapstick and knowing guffaws. With this double standard in place, it's no wonder that any girl who asserts her sexual desire (or is presumed to) is treated like a freak. Her behavior is considered so deviant that it can't even be represented in the same theaters that screen bloody, ultra-violent films like *Reservoir Dogs* or incest-themed films like *Spanking the Monkey.* . . .

The sexual double standard was also in full force in the summer of 1999 when a dozen Virginia junior high school girls were discovered to have engaged in oral sex throughout the school year during parties and at local parks. The Washington Post broke the story, which, was subsequently picked up by the Associated Press and reprinted in newspapers across the country. Parents, health educators, and guidance counselors weighed in with a loud chorus of condemnation. Certainly it was disturbing that kids so young were engaging in meaningless sexual encounters. But much more disturbing was that, first, the girls had been nothing more than sexual servicers to the boys; and second, that all of the censure was directed to the girls. It turns out that the school principal had called the parents of the girls to a special meeting to discuss the matter—but none of the boys or their parents was approached. Boys will be boys—but girls will be "sluts."

Girls today dress so provocatively, even to school, in skimpy outfits that expose a lot of flesh. They practically invite people to call them "sluts" and other names.

I have to admit that I am often appalled by some of the outfits I see young girls wearing these days: It's one thing for an adult woman to showcase her sexual appeal and a different thing entirely for an eighth grader to do likewise. But I don't blame the girls. On the contrary, I am sympathetic to them. These girls believe that if they attract a boyfriend and fall in love, their lives would be better and they would be happier. Sadly, many of these girls believe that their sexuality is the only power or appeal they have, and so they play it up to the hilt. They also feel competitive with other girls in a battle for the most desirable guys, so they feel the need to outdress their peers. Dressing in sexy outfits, then, is both a strategy to obtain romance and a competition with other girls. But just because a girl dresses in a sexually provocative way doesn't mean that she is sexually promiscuous. In reality, she may not be any more sexually active than the prissy girl in tailored pants, loafers, and sweater set.

Some of your interviewees were called "sluts" even though they weren't sexually active at all. They were innocent victims. But the girls who were sexually promiscuous are a different story: they deserved what they got.

I don't believe that there should be a distinction between those who deserve a bad reputation and those who don't. Because frankly, I don't think that any girl deserves to be called a "slut." After all, boys who are sexually active are congratulated as studs.

Dividing "sluts" into the innocent and the guilty merely reinforces the sexual double standard. This is why, when I have been asked about my own sexual history—believe it or not, radio show hosts, aping Howard Stern, have felt perfectly comfortable quizzing me about the details of my sex life—I have refused to respond. Besides the fact that the answers are no one's business, they would serve only to buttonhole me as either "innocent" or "guilty," and I reject both categories.

* * *

If females practiced an ethic of sexual modesty, males would be more likely to treat them with respect.

Ideally, I think sex should be harnessed within a romantic relationship, but that ideal isn't possible or desirable for everybody. There are young women who perhaps would like to wait and initiate their first sexual encounter in a loving relationship, but for whatever reason, they have desire and want to act on it before they've met the "right" person—or they may never meet the "right" person. I worry that these young women are going to feel guilty and ashamed of their own sexual desire. I'm also concerned that they are going to make bad choices about who their mate is going to be, perhaps marrying too young. A sex drive is a natural appetite for males and females. If you say that females are innately modest, then you're also saying that a girl or woman who isn't modest is doing something unfemale and wrong.

In her book *A Return to Modesty,* Wendy Shalit argues that girls have to be "good" in order for boys to behave properly—to stop sexually harassing them. But I worry about the implications of being a "good" girl. Once you start characterizing some females as "good," you inevitably label others as "bad." And once you start thinking of some girls as "bad," in essence you are saying that those girls don't deserve to be treated with respect. The irony is that so many girls who are regarded as slutty aren't even particularly sexually active, and they are rarely more sexually active than their peers are. So the whole good girl/bad girl thing is a sham. Its purpose is to elevate some girls and to degrade others, and in the long run it hurts everyone. Boys will treat girls with respect, and loveless, casual sexual encounters will decrease, when we have one standard for both sexes—that is, when we have sexual equality. . . .

School sexual harassment lawsuits are getting out of hand. How can a school be monetarily responsible for sexual harassment? These lawsuits hurt everyone, since they take money away from education.

The Supreme Court ruled in May 1999 (*Davis v. Monroe County Board of Education*) that school districts receiving federal money can be liable for monetary damages if they fail to prevent severe, persistent sexual harassment among students. The ruling was in response to a case brought by the mother of a fifth-grade girl in rural Georgia. The girl, LaShonda Davis, was harassed by a male classmate who made repeated unwanted sexual advances over the course of five months. At least two teachers, as well as the principal, were aware of the incidents, but no disciplinary action was taken against the boy. Meanwhile, Davis's grades dropped and her father discovered that she had written a suicide note.

The ruling is important because it sends the message that schools must be vigilant in halting sexual harassment—which includes slut-bashing, a verbal form of sexual harassment. I agree that a school should be liable if the sexual harassment is severe and persistent and if the school is aware of the behavior but does not take steps to halt it. If, on the other hand, the school makes a good-faith effort to stop the behavior, then I don't believe it should be liable.

It's unfortunate that a ruling against a school results in a monetary loss, but it's also unfortunate that the threat of monetary payment is the most effective wake-up call to school administrators. As for the argument that these payments take money away from education, sexual harassment also impedes the ability of teachers to effectively educate and the ability of students to effectively learn.

* * *

Why are girls often worse than boys when it comes to slut-bashing?

All of us yearn for one arena in which we can wield power. For girls, this desire is often thwarted. After all, girls may get better grades—but boys, especially athletes, by and large receive more attention and congratulatory pats on the back from school administrators and teachers. Boys call out more in class and get away with it. They rule the playground. Many feel a sense of entitlement to grope girls' bodies. With these depressing realities, it's no wonder that many girls develop a sense of self-hatred. Sensing that femininity is devalued, they may feel, at some level, uncomfortable with being a girl, and therefore are reluctant to bond with other girls. Instead, they latch on to one small sphere of power they can call their own: the power to make or break reputations. Slut-bashing is a cheap and easy way to feel powerful. If you feel insecure or ashamed about your own sexual desires, all you have to do is call a girl a "slut" and suddenly you're the one who is "good" and on top of the social pecking order.

What can we do to stop slut-bashing?

Teachers must recognize that slut-bashing is a serious problem. Too often, they dismiss it as part of the normal fabric of adolescent life. But slut-bashing is a form of sexual harassment, and it is illegal under Title IX, which entitles students to a harassment-free education. If a teacher witnesses slut-bashing, she must make sure that it stops. She must confront the ringleader and other name callers. Of course, teachers and school administrators shouldn't wait for slut-bashing to occur. They must create and publicize awareness through sexual harassment policies for their schools.

Parents should be open about sexuality with their kids—and that means being open about female sexuality as well as male sexuality. They should teach their daughters and sons that girls as well as boys have sexual feelings, and that sexual feelings are entirely normal. That way they won't have to pin their sexual anxieties on a scapegoat and then distance themselves from her.

But the most important thing that all of us need to work on is this: to stop calling or thinking of women as "sluts." Face it: At one time or another, many of us have called a woman a "slut." We see a woman who's getting away with something we wish we could get away with. What do we call her? A "slut." We see a woman who dresses provocatively, and maybe we wish we had the guts to dress that way ourselves. What do we call her? A "slut."

If we think of a woman or girl as a "slut," it's like she's not one of us. She's one of *them.* She is other. "Slut," like any other derogatory label, is a shorthand for one who is different, strange—and not worth knowing or caring about. Unlike other insults, however, it carries a unique sting: the stigma of the out-of-control, trampy female. Most of us recognize that this stigma is unjust and unwarranted. Yet we have used the "slut" insult anyway: Our social conditioning runs too deep. We must will ourselves to be aware of the sexual double standard and of how we lapse into slut-bashing on an everyday level. If we become aware of our behavior, then we have the power to stop.

And never again be slut-bashers or self-bashers.

Notes

1. In Jordan in 1993 a sixteen-year-old girl who had been raped by her older brother was killed by her family because, it was said, she had seduced him into sleeping with her. Kristen Golden, "Rana Husseini: A Voice for Justice," *Ms.*, July/August 1998, p. 36; Tali Edut, "Global Woman: Rana Husseini," *HUES,* Summer 1998, p. 41. In Afghanistan, where women must remain covered from head to toe in shrouds called *burqas,* the General Department for the Preservation of Virtue and Prevention of Vice beats women for wearing white socks or plastic sandals with no socks, attire that is said to provoke "impure thoughts" in men. John F. Burns, "Sex and the Afghan Woman: Islam's Straitjacket," *The New York Times,* August 29, 1997, p. A4. And in Turkey in 1998 five girls attempted suicide by eating rat poison and jumping into a water tank to avoid a forced virginity examination. An unmarried woman discovered not to be a virgin risks being beaten or killed. The virginity tests were carried out as the girls recovered in their hospital beds; when one girl did succeed in killing herself, her father had the exam performed on her corpse. Kelly Couturier, "Suicide Attempts Fuel Virginity Test Debate," *The Washington Post,* January 27, 1998, p. A18.

2. Lauri Umansky, "Breastfeeding in the 1990s: The Karen Carter Case and the Politics of Maternal Sexuality" in Molly Ladd-Taylor and Lauri Umansky, eds., *"Bad" Mothers: The Politics of Blame in Twentieth-Century America* (New York: New York University Press, 1998), pp. 299–309. Karen Carter is a pseudonym.

3. Tamar Stieber, "Viewpoint," *Glamour* August 1996, p. 138.

4. Stieber, p. 138.

5. Jon Jeter, "Woman Who Sued Ex-Husband's Mistress Is Awarded $1 Million," *The Washington Post,* August 7, 1997, p. A3.

6. *Sex and America's Teenagers* (New York and Washington: The Alan Guttmacher Institute), p. 20.

7. Laurie Goodstein with Marjorie Connelly, "Teen-Age Poll Finds a Turn to the Traditional," *The New York Times,* April 30, 1998, p. A20. The poll, of 1,048 teenagers ages thirteen to seventeen, was conducted by telephone in April 1998. The poll also found that only 18 percent of thirteen- to fifteen-year-olds said they had ever had sex, as against 38 percent of sixteen- and seventeen-year-olds.

8. Janie Victoria Ward and Jill McLean Taylor, "Sexuality Education for Immigrant and Minority Students: Developing a Culturally Appropriate Curriculum," in Janice M. Irvine, *Sexual Cultures and the Construction of Adolescent Identities* (Philadelphia: Temple University Press, 1994), p. 63.

9. Lillian B. Rubin, *Erotic Wars: What Happened to the Sexual Revolution?* (New York: HarperPerennial, 1991), p. 119.

10. Shere Hite, *Women and Love* (New York: St. Martin's Press, 1987), p. 205.

11. Tamar Lewin, "Boys Are More Comfortable With Sex Than Girls Are, Survey Finds," *The New York Times,* May 18, 1994.

⚜ Topics for Further Examination ⚜

- Go to the main Web sites for Girl Scouts (http://www.gsusa.org) and Boy Scouts (http://www.scouting.org) and compare the two organizations. Are their programs similar or different? What similarities and differences do you observe in the Web sites themselves? What effect do these organizations have on gender socialization?
- Go to the U.S. Department of Education Web site (http://www.dol.gov/) and search for differences in men and women in higher education.

5

Buying and Selling Gender

In the video called *Adventures in the Gender Trade,* Kate Bornstein, a transgender performance artist and activist, looks into the camera and says, "Once you buy gender, you'll buy anything to keep it." Her observation goes to the heart of deep connections between economic processes and institutionalized patterns of gender difference, opposition, and inequality in contemporary society. Readings in this chapter examine the ways in which modern marketplace forces such as commercialization, commodification, and consumerism exploit and construct gender. However, before we explore the buying and selling of gender, we want to review briefly the major elements of contemporary American economic life—elements that embody corporate capitalism—which form the framework for the packaging and delivery of gender to consumers.

Defining Corporate Capitalism

Corporate capitalism is an economic system in which large, national and transnational corporations are the dominant forces. The basic goal of corporate capitalism is the same as it was when social scientists such as Karl Marx studied early capitalist economies: converting money into more money (Johnson, 2001). Corporate capitalists invest money in the production of all sorts of goods and services for the purpose of selling at a profit. Capitalism, as Gitlin (2001) observes, requires a consumerist way of life.

In today's society, corporate capitalism affects virtually every aspect of life—most Americans work for a corporate employer, whether a fast food chain or a bank, and virtually everyone buys the products and services of capitalist production (Johnson, 2001; Ritzer, 1999). Those goods and services include things we must have in order to live (e.g., food and shelter) and, most important for contemporary capitalism's survival and growth, things we have learned to want or desire (e. g., microwave ovens, televisions, cruises, fitness fashions, cosmetic surgery), even though we do not need them in order to live (Ritzer, 1999).

From an economic viewpoint, we are a nation of consumers, people who buy and use a dizzying array of objects and services conceived, designed, and sold to us by corporations. George Ritzer (1999), a leading analyst of consumerism, observes that consumption plays such as big role in the lives of contemporary Americans that it has, in many respects, come to define our society.

In fact, as Ritzer notes, Americans spend most of their available resources on consumer goods and services. Corporate, consumer capitalism depends on luring people into what he calls the "cathedrals of consumption," such as book superstores, shopping malls, theme parks, fast food restaurants, and casinos, where we will spend money to buy an array of goods and services.

Our consumption-driven economy counts on customers whose spending habits are relatively unrestrained and who view shopping as pleasurable. Indeed, Americans spend much more today than they did just forty years ago (Ritzer, 1999). Most of our available resources go to purchasing and consuming "stuff." Americans consume more of everything and more varieties of things than people in other nations. We are also more likely to go into debt than Americans of earlier generations and people in other nations today. Some social scientists (e.g, Schor, 1998, p. 2004) use the term *hyperconsumption* to describe what seems to be a growing American passion for and obsession with consumption.

Marketing Gender

Gender is a fundamental element of the modern machinery of marketing. It is an obvious resource from which the creators and distributors of goods and services can draw ideas, images, and messages. The imagery of consumer culture thrives on gender difference and asymmetry. For example, consumer emblems of hyperfemininity and hypermasculinity, such as Barbie and GI Joe, stand in stark physical contrast to each other (Schiebinger, 2000). This is not happenstance. Barbie and GI Joe intentionally reinforce beliefs in essential differences between women and men. The exaggerated, gendered appearances of Barbie and GI Joe can be purchased by adult consumers who have the financial resources to pay for new cosmetic surgeries, such as breast and calf implants, that literally inscribe beliefs about physical differences between women and men into their flesh (Sullivan, 2001). As Walters observes (2001), turning difference into "an object of barter is perhaps the quintessentially American experience" (p. 289). Indeed, virtually every product and service, including the most functional, can be designed and consumed as masculine or feminine (e.g., deodorants, bicycles, greeting cards, wallpaper, cars, and hair styles).

Gender-coding of products and services is a common strategy employed by capitalist organizations to sell their wares. It is also integral to the processes by which gender is constructed, because it frames and structures gender practices. Let's look at the gender-coding of clothing to illustrate how consumer culture participates in the construction of gender through ordinary material forms. As the gender archeologist Sorenson (2000) observes, clothing is an ideal medium for the expression of a culture's gender beliefs because it is an extension of the body and an important element in identity and communication. No wonder corporate capitalists have cashed in on the business of fabricating gender through dress (Sorenson, 2000). Sorenson (2000) notes that simple observation of the clothing habits of people reveals a powerful pattern of "dressing gender" (p. 124). Throughout life, she argues, the gender-coding of colors, patterns, decorations, fabrics, fastenings, trimmings, and other aspects of dress create and maintain differences between boys and girls and men and women. Even when clothing designers and manufacturers create what appear to be "unisex" fashions (e.g., tuxedos for women), they incorporate just enough gendered elements (e.g., lacy trim or a revealing neckline) to insure that the culturally created gender categories—feminine and masculine—are not completely erased. Consider the lengths to which the fashion industry has gone to create dress that conveys a "serious yet feminine" business appearance for the increasing number of women in management and executive levels of the corporate world (Kimle & Damhorst, 1997). Contemplate the ferocity of the taboo against boys and men wearing skirts and dresses. Breaking the taboo (except on a few occasions such as Halloween) typically results in negative sanctions. The reading in this chapter by Adie Nelson examines the extent to which even fantasy dress for children ends up conforming to gender stereotypes.

Gender-coded clothing is one example of corporate exploitation of gender to sell all kinds of goods and services, including gender itself. Have we arrived at a moment in history when identities, including gender identity, are largely shaped within the dynamics of consumerism? Will we, as Bornstein observes, buy anything to keep up gender appearances? The readings in this chapter help us to answer these questions. They illuminate some of the key ways in which capitalist, consumer culture makes use of cultural definitions and stereotypes of gender to produce and sell goods and services.

In our "consumers' republic" (Cohen, 2003), the mass media (e.g., television and magazines) play a central role in delivering potential consumers to advertisers whose job it is to persuade us to buy particular products and services (Kilbourne, 1999; Ritzer, 1999). The advertising industry devotes itself to creating and keeping consumers in the marketplace, and it is very good at what it does. Today's advertisers use sophisticated strategies for hooking consumers. The strategies work because they link our deepest emotions and most beloved ideals to products and services by persuading us that identity and self-worth can be fashioned out of the things we buy (Featherstone, 1991; Zukin, 2004)). Advertisers transform gender into a commodity, and convince consumers that we can transform ourselves into more masculine men and more feminine women by buying particular products and services. Men are lured into buying cars that will make them feel like hypermasculine machines, and women are sold a wondrous array of cosmetic products and procedures that are supposed to turn them into drop-dead beauties.

Jacqueline Urla and Alan Swedlund's article explores the story that Barbie, a well advertised and wildly popular toy turned icon, tells about femininity in consumer culture. They note that although Barbie's long, thin body and big breasts are remarkably unnatural, she stands as an ideal that has played itself out in the real body trends of *Playboy* magazine centerfolds and Miss America contestants. The authors provide evidence that between 1959 and 1978, the average weight and hip size for women centerfolds and beauty contestants decreased steadily. A follow-up study for 1979–88 found the acceleration of this trend with "approximately 69 percent of Playboy centerfolds and 60 percent of Miss America contestants weighing in at 15 percent or more below their expected age and height category" (p. 298). One lesson we might glean from this story is that a toy (Barbie) and real women (centerfolds and beauty contestants) are converging in a culture in which the bonds of beauty norms are narrowing and tightening their grip on both products and persons (Sullivan, 2001). To illustrate the extent of media's influence even further, Kirsten Firminger's piece on representations of males in teenage girls' magazines demonstrates the power of print media to guide readers not only toward consumption of gendered products and services but also toward consumption of (stereo)types of people who are packaged much like other gendered products.

Any analysis of the marketing of femininity and masculinity has to take into account the ways in which the gendering of products and services is tightly linked to prisms of difference and inequality such as sexuality, race, age, and ability/disability. Consumer culture thrives, for example, on heterosexuality, whiteness, and youthfulness. Automobile advertisers market cars made for heterosexual romance and marriage. Liquor ads feature men and women in love (Kilbourne, 1999). Recent research on race and gender imagery in the most popular advertising medium, television, confirms the continuing dominance of images of White, affluent, young adults. "Virtually all forms of television marketing perpetuate images of White hegemonic masculinity and White feminine romantic fulfillment" (Coltrane & Messineo, 2000, p. 386). In spite of what is called niche marketing or marketing to special audiences such as Latinos, gay men, and older Americans, commercial television imagery continues to rely on stereotypes of race, gender, age, and the like (Coltrane & Messineo, 2000). Stereotypes sell.

Two readings in this chapter address intersections of prisms of difference and inequality in consumer culture. The first, by Toni Calasanti and Neal King, offers detailed insight into the mass-marketing of "successful aging" products, services, and activities to old men. They highlight the fact that marketing that targets old people plays upon the stigma of aging in American culture and, in the case of men, the often desperate attempts of aging men to hang onto youthful manliness. The second, by Minjeong Kim and Angie Chung, is a close analysis of multicultural advertising strategies that rely on racialized, sexualized, and gendered stereotypes of Asian American women as the "Other" not only to sell products but also to sell Orientalism itself.

Can You Buy In Without Selling Out?

The tension between creativity, resistance, and rebellion, on the one hand; and the lure and power of commercialization on the other, is a focus of much research on consumerism and consumer culture (Quart, 2003; Schor, 2004). Can we produce and consume the gendered products and services of corporate capitalism without wanting and trying to be just like Barbie or Madonna, the Marlboro Man or Brad Pitt? Does corporate, commercial culture consume everything and everyone in its path, including the creators of countercultural forms?

The latter question is important. Consider the fact that "grunge," which began as antiestablishment fashion, became a national trend when companies such as Diesel and Urban Outfitters coopted and commercialized it (O'Brien, 1999). Then contemplate how commercial culture has cleverly exploited the women's movement by associating serious social issues and problems with trivial or dangerous products. "New Freedom" is a maxipad. "ERA" is a laundry detergent. Cigarette ads often portray smoking as a symbol of women's liberation (Kilbourne, 1999). Commercial culture is quite successful in enticing artists of all sorts to "sell out." For example, Madonna began her career as a rebel who dared to display a rounded belly. But, over time, she has been "normalized," as reflected in the transformation of her body to better fit celebrity appearance norms (Bordo, 1997).

The culture of the commodity is also successful in mainstreaming the unconventional by turning nonconformity into obedience that answers to Madison Avenue (Harris, 2000). Analysts of the commodification of gayness have been especially sensitive to the potential problems posed by advertising's recent creation of a largely fictional identity of gay as "wealthy White man" with a lifestyle defined by hip fashion (Walters, 2001). What will happen if lesbian and gay male styles are increasingly drawn into mass-mediated, consumer culture? Will those modes of rebellion against the dominance of heterosexism lose their political clout? Will they become mere "symbolic forms of resistance, ineffectual strategies of rebellion" (Harris, 2000, p. xxiii)?

The Global Reach of American Gender Images and Ideals

The global reach of American culture is yet another concern of consumer culture researchers. Transnational corporations are selling American popular culture and consumerism as a way of life in countries around the world (Kilbourne, 1999; Ritzer, 1999). People across the globe are now regularly exposed to American images, icons, and ideals. For example, *Baywatch,* with its array of perfect (albeit cosmetically enhanced) male and female bodies, has been seen by more people in the world than any other television show (Kilbourne, 1999). American popular music and film celebrities dominate the world scene. Everyone knows Marilyn Monroe and James Dean, Tom Cruise and Julia Roberts.

You might ask, and quite legitimately, so what? The answer to that question is not a simple one, in part because cultural import-export relations are intricate. As Gitlin (2001) observes, "the cultural gates . . . swing both ways. For example, American rhythm and blues influenced Jamaican ska, which evolved into reggae, which in turn was imported to the United States via Britain" (p. 188). However, researchers have been able to document some troubling consequences of the global advantage of American commercial, consumer culture for the lifeways of people outside the United States. Thus, social scientists (e.g., Connell, 1999; Herdt, 1997) are tracing how American categories of sexual orientation are altering the modes of organization and perception of same-gender relations in some non-Western societies that have traditionally been more fluid and tolerant of sexual diversity than the United States.

Scientists are also documenting the impact of American mass media images of femininity and masculinity on consumers in far corners of the world. The island country of Fiji is one such place. Researchers have discovered that as the young women of Fiji consume American television on a regular basis, eating disorders such as anorexia nervosa are being recorded for the first time. The ultra-thin images of girls and women that populate U.S. TV shows and TV ads have become the measuring stick of femininity in a culture in which, previously, an ample, full body was the norm for women and men (Goode, 1999). The troubling consequences of the globalization of American consumer culture do not end with these examples. Consider the potential negative impact of idealized images of whiteness in a world in which most people are brown. Or how about the impact of America's negative images of older women and men on the people of cultures in which the elderly are revered?

Although corporate, capitalist economies provide many people with all the creature comforts they need and more, as well as making consumption entertaining and more accessible, there is a price to pay (Ritzer, 1999). This chapter explores one troubling aspect of corporate, consumer culture—the commodification and commercialization of gender.

A few final questions emerge from our analysis of patterns of gender in relationship to consumer capitalism. How can the individual develop an identity and self-worth that are *not* contingent upon and defined by a whirlwind of products and services? How do we avoid devolving into caricatures of stereotyped images of femininity and masculinity, whose needs and desires can only be met by gendered commodities? Is Kate Bornstein correct when she states that "Once you buy gender, you'll buy anything to keep it?" Or can we create and preserve alternative ways of life, even ways of life that undermine the oppression of dominant images and representations?

References

Bordo, S. (1997). Material girl: The effacements of postmodern culture. In R. Lancaster & M. di Leonardo (Eds.), *The gender/sexuality reader* (pp. 335—358). New York: Routledge.

Coltrane, S., & Messineo, M. (2000). The perpetuation of subtle prejudice: Race and gender imagery in 1990s television advertising. *Sex roles,* (42), 363–389.

Cohen, L. (2003). *A consumers' republic: The politics of mass consumption in postwar America.* New York: Vintage Books

Connell, R. W. (1999). Making gendered people: Bodies, identities, sexualities. In M. Ferree, J. Lorber & B. Hess (Eds.), *Revisioning gender* (pp. 449–471). Thousand Oaks, CA: Sage.

Featherstone, M. (1991). The body in consumer culture. In Featherstone, Hepworth, & Turner (Eds.), *The body: Social process and cultural theory* (pp. 170—196). London: Sage.

Gitlin, T. (2001). *Media unlimited: How the torrent of images and sounds overwhelms our lives.* New York: Henry Holt and Company.

Goode, E. (1999). Study finds TV alters Fiji girls' view of body. *New York Times,* May 20, p. A17.

Harris, D. (2000). *Cute, quaint, hungry and romantic: The aesthetics of consumerism.* Cambridge, MA: Da Capo Press.

Herdt, G. (1997). *Same sex, different cultures.* Boulder, CO: Westview.

Johnson, A. (2001). *Privilege, power, and difference.* Mountain View, CA: Mayfield.

Kilbourne, J. (1999). *Can't buy my love.* New York: Simon & Schuster.

Kimle, P. A., & Damhorst, M. L. (1997). A grounded theory model of the ideal business image for women. *Symbolic Interaction, 20* (1), 45–68.

Marenco, S., with Bornstein, K. (1993). *Adventures in the gender trade: A case for diversity.* Filmakers Library.

O'Brien, J. (1999). *Social prisms.* Thousand Oaks, CA: Pine Forge.

Quart, A. (2003). *Branded: The buying and selling of teenagers.* New York: Basic Books.

Ritzer, G. (1999). *Enchanting a disenchanted world.* Thousand Oaks, CA: Pine Forge.

Schiebinger, L. (2000). Introduction. In L. Schiebinger (Ed)., *Feminism and the body* (pp. 1–21). New York: Oxford University Press.

Schor, J. (1998). *The overspent American.* New York: Basic Books.

Schor, J. (2004). *Born to buy.* New York: Scribner.

Sorenson, M. L. Stig. (2000). *Gender archaeology.* Cambridge, England: Polity Press.

Sullivan, D. A. (2001). *Cosmetic surgery: The cutting edge of commercial medicine in America.* New Brunswick, NJ: Rutgers University Press.

Walters, S. D. (2001). *All the rage: The story of gay visibility in America.* Chicago: University of Chicago Press.

Zukin, S. (2004). *Point of purchase: How shopping changed American culture.* New York: Routledge.

Introduction to Reading 21

Adie Nelson's article offers a marvelously detailed analysis of one way in which the modern marketplace reinforces gender stereotypes—the gender coding of children's Halloween costumes. Nelson describes the research process she employed to label costumes as masculine, feminine, or neutral. She provides extensive information about how manufacturers and advertisers use gender markers to steer buyers, in this case parents, toward "gender-appropriate" costume choices for their children. Overall, Nelson's research indicates that gender-neutral costumes, whether they are ready-to-wear or sewing patterns, are a tiny minority of all the costumes on the market.

1. Many perceive Halloween costumes as encouraging children to engage in fantasy play. How does Nelson's research call this notion into question?
2. Describe some of the key strategies employed by manufacturers to "gender" children's costumes.
3. How do Halloween costumes help to reproduce an active-masculine/passive-feminine dichotomy?

The Pink Dragon Is Female

Halloween Costumes and Gender Markers

Adie Nelson

* * *

The celebration of Halloween has become, in contemporary times a socially orchestrated secular event that brings buyers and sellers into the marketplace for the sale and purchase of treats, ornaments, decorations, and fanciful costumes. Within this setting, the wearing of fancy dress costumes has such a prominent role that it is common, especially within large cities, for major department stores and large, specialty toy stores to begin displaying their selection of Halloween costumes by mid-August if not earlier. It is also evident that the range of masks and costumes available has broadened greatly beyond those identified by McNeill (1970), and that both

From Nelson, Adie. 2000. "The pink dragon is female: Halloween costumes and gender markers" *Psychology of Women Quarterly 24.*

children and adults may now select from a wide assortment of readymade costumes depicting, among other things, animals, objects, superheroes, villains, and celebrities. In addition, major suppliers of commercially available sewing patterns, such as Simplicity and McCall's, now routinely include an assortment of Halloween costumes in their fall catalogues. Within such catalogues, a variety of costumes designed for infants, toddlers, children, adults, and, not infrequently, pampered dogs are featured.

On the surface, the selection and purchase of Halloween costumes for use by children may simply appear to facilitate their participation in the world of fantasy play. At least in theory, asking children what they wish to wear or what they would like to be for Halloween may be seen to encourage them to use their imagination and to engage in the role-taking stage that Mead (1934) identified as play. Yet, it is clear that the commercial marketplace plays a major role in giving expression to children's imagination in their Halloween costuming. Moreover, although it might be facilely assumed that the occasion of Halloween provides a cultural "time out" in which women and men as well as girls and boys have tacit permission to transcend the gendered rules that mark the donning of apparel in everyday life, the androgyny of Halloween costumes may be more apparent than real. If, as our folk wisdom proclaims, "clothes make the man" (or woman), it would be presumptuous to suppose that commercially available children's Halloween costumes and sewing patterns do not reflect both the gendered nature of dress (Eicher & Roach-Higgens, 1992) and the symbolic world of heroes, villains, and fools (Klapp, 1962, 1964). Indeed, the donning of Halloween costumes may demonstrate a "gender display" (Goffman, 1966, p. 250) that is dependent on decisions made by brokering agents to the extent that it is the aftermath of a series of decisions made by commercial firms that market ready-made costumes and sewing patterns that, in turn, are purchased, rented, or sewn by parents or others. . . .

Building on Barnes and Eicher's (1992, p. 1) observation that "dress is one of the most significant markers of gender identity," an examination of children's Halloween costumes provides a unique opportunity to explore the extent to which gender markers are also evident within the fantasy costumes available for Halloween. To the best of my knowledge, no previous research has attempted to analyze these costumes nor to examine the ways in which the imaginary vistas explored in children's fantasy dress reproduce and reiterate more conventional messages about gender.

In undertaking this research, my expectations were based on certain assumptions about the perspectives of merchandisers of Halloween costumes for children. It was expected that commercially available costumes and costume patterns would reiterate and reinforce traditional gender stereotypes. Attempting to adopt the marketing perspective of merchandisers, it was anticipated that the target audience would be parents concerned with creating memorable childhood experiences for their children, envisioning them dressed up as archetypal fantasy characters. In the case of sewing patterns, it was expected that the target audience would be primarily mothers who possessed what manufacturers might imagine to be the sewing skills of the traditional homemaker. However, these assumptions about merchandisers are not the subject of the present inquiry. Rather, the present study offers an examination of the potential contribution of marketing to the maintenance of gender stereotypes. In this article, the focus is on the costumes available in the marketplace; elsewhere I examine the interactions between children and their parents in the selection, modification, and wearing of Halloween costumes (Nelson, 1999).

Method

The present research was based on a content analysis of 469 unique children's Halloween ready-made costumes and sewing patterns examined from August 1996 to November 1997 at craft stores, department stores, specialty toy stores, costume rental stores, and fabric stores containing catalogues of sewing patterns. Within retail stores, racks of children's Halloween costumes typically appeared in August and remained in evidence, albeit in dwindling numbers, until early November each year. In department stores, a subsection of the area generally devoted to toys featured such garments; in craft stores and/or toy stores, children's Halloween costumes were typically positioned on long racks in the center of a section devoted to the commercial paraphernalia now associated with the celebration of Halloween (e.g., cardboard witches, "Spook trees," plastic pumpkin containers). Costumes were not segregated by gender within the stores (i.e., there were no separate aisles or sections for boys' and girls' costumes); however, children's costumes were typically positioned separately from those designed for adults. . . .

All costumes were initially coded as (a) masculine, (b) feminine, or (c) neutral depending on whether boys, girls, or both were featured as the models on the packaging that accompanied a ready-to-wear costume or were used to illustrate the completed costume on the cover of a sewing pattern. . . . The pictures accompanying costumes may act as safekeeping devices, which discourage parents from buying "wrong"-sexed costumes. The process of labeling costumes as masculine, feminine, or neutral was facilitated by the fact that these public pictures (Goffman, 1979) commonly employed recognizable genderisms. For example, a full-body costume of a box of crayons could be identified as feminine by the long curled hair of the model and the black patent leather pumps with ribbons she wore. In like fashion, a photograph depicting the finished version of a sewing pattern for a teapot featured the puckish styling of the model in a variant of what Goffman (1979, p. 45) termed "the bashful knee bend" and augmented this subtle cue by having the model wear white pantyhose and Mary-Jane shoes with rosettes at the base of the toes. Although the sex of the model could have been rendered invisible, such feminine gender markers as pointy-toed footwear, party shoes of white and black patent leather, frilly socks. makeup and nail polish, jewelry, and elaborately curled (and typically long and blonde) hair adorned with bows/barrettes/ hairbands facilitated this initial stage of costume placement. By and large, female models used to illustrate Halloween costumes conformed to the ideal image of the "Little Miss" beauty pageant winner; they were almost overwhelmingly White, slim, delicate-boned blondes who did not wear glasses. Although male child models were also overwhelmingly White, they were more heterogeneous in height and weight and were more likely to wear glasses or to smile out from the photograph in a bucktooth grin. At the same time, however, masculine gender markers were apparent. Male models were almost uniformly shod in either well-worn running shoes or sturdy-looking brogues, while their hair showed little variation from the traditional little boy cut of short back and sides.

The use of gender-specific common and proper nouns to designate costumes (e.g., Medieval Maiden, Majorette, Prairie Girl) or gender-associated adjectives that formed part of the costume title (e.g., Tiny Tikes Beauty, Pretty Witch, Beautiful Babe, Pretty Pumpkin Pie) also served to identify feminine costumes. Similarly, the use of the terms "boy," "man," or "male" in the advertised name of the costume (e.g., Pirate Boy, Native American Boy, Dragon Boy) or the noted inclusion of advertising copy that announced "Cool dudes costumes are for boys in sizes" was used to identify masculine costumes. Costumes designated as neutral were those in which both boys and girls were featured in the illustration or photograph that accompanied the costume or sewing pattern or in which it was impossible to detect the sex of the wearer. By and large, illustrations for gender-neutral ads featured boys and girls identically clad and depicted as a twinned couple or, alternatively, showed a single child wearing a full-length animal costume complete with head and "paws," which, in the style of spats, effectively covered the shoes of the model. In addition, gender-neutral costumes were identified by an absence of gender-specific nouns and stereotypically gendered colors.

Following this initial division into three categories, the contents of each were further coded into a modified version of Klapp's (1964) schema of heroes, villains, and fools. In his work, Klapp suggested that this schema represents three dimensions of human behavior. That is, heroes are praised and set up as role models, whereas villains and fools are negative models, with the former representing evil to be feared and/or hated and the latter representing figures of absurdity inviting ridicule. However, although Klapp's categories were based on people in real life, I applied them to the realm of make-believe. For the purposes of this study, the labels refer to types of personas that engender or invite the following emotional responses, in a light-hearted way from audiences: heroes invite feelings of awe, admiration, and respect, whereas villains elicit feelings of fear and loathing, and fools evoke feelings of laughter and perceptions of cuteness. All of the feelings, however, are mock emotions based on feelings of amusement, which make my categories quite distinct from Klapp's. For example, although heroes invite awe, we do not truly expect somebody dressed as a hero to be held in awe. . . .

For the purposes of this secondary classification of costumes, the category of hero was broadened to include traditional male or female heroes (e.g., Cowboy, Robin Hood, Cinderella, Cleopatra), superheroes possessing supernatural powers (e.g., Superman, Robocop, Xena, the Warrior Princess) as well as characters with high occupational status (e.g., Emergency Room Doctor, Judge) and characters who are exemplars of prosocial conformity to traditional masculine and feminine roles (e.g., Team USA Cheerleader, Puritan Lady, Pioneer Boy). The category of villain was broadly defined to include symbolic representations of death (e.g., the Grim Reaper, Death, The Devil, Ghost), monsters (e.g., Wolfman, Frankenstein, The Mummy), and antiheroes (e.g., Convict, Pirate, The Wicked Witch of the West, Catwoman). Fool was a hybrid category, distinguished by costumes whose ostensible function was to amuse rather than to alarm. Within this category, two subcategories were distinguished. The first subcategory, figures of mirth, referred to costumes of clowns, court jesters, and harlequins. The second, nonhuman/inanimate objects, was composed of costumes representing foodstuffs (e.g., Peapod, Pepperoni Pizza, Chocolate Chip Cookie), animals and insects, and inanimate objects (e.g., Alarm Clock, Bar of Soap, Flower Pot). Where a costume appeared to straddle two categories, an attempt was made to assign it to a category based on the dominant emphasis of its pictorial representation. For example, a costume labeled Black Widow Spider could be classified as either an insect or a villain. If the accompanying illustration featured a broadly smiling child in a costume depicting a fuzzy body and multiple appendages, it was classified as an insect and included in the category of nonhuman/inanimate objects; if the costume featured an individual clad in a black gown, long black wig, ghoulish makeup, and a sinister mien, the costume was classified as a villain. Contents were subsequently reanalyzed in terms of their constituent parts and compared across masculine and feminine categories. In all cases, costumes were coded into the two coding schemes on the basis of a detailed written description of each costume. . . .

Results

The initial placement of the 469 children's Halloween costumes into masculine, feminine, or neutral categories yielded 195 masculine costumes, 233 feminine costumes, and 41 gender-neutral costumes. The scarcity of gender-neutral costumes was notable; costumes that featured both boys and girls in their ads or in which the gender of the anticipated wearer remained (deliberately or inadvertently) ambiguous accounted for only 8.7% of those examined. Gender-neutral costumes were more common in sewing patterns than in ready-to-wear costumes and were most common in costumes designed for newborns and very young infants. In this context, gender-neutral infant costumes largely featured a winsome assortment of baby animals (e.g., Li'l Bunny, Beanie the Pig) or foodstuffs (e.g., Littlest Peapod). By and large, few costumes for older children were presented as gender-neutral; the notable exceptions were costumes for scarecrows and emergency room doctors (with male/female models clad identically in olive-green "scrubs"), ready-made plastic costumes for Lost World/ Jurassic Park hunters, a single costume labeled Halfman/Halfwoman, and novel sewing patterns depicting such inanimate objects as a sugar cube, laundry hamper, or treasure chest.

Beginning most obviously with costumes designed for toddlers, gender dichotomization was promoted by gender-distinctive marketing devices employed by the manufacturers of both commercially made costumes and sewing patterns. In relation to sewing patterns for children's Halloween costumes, structurally identical costumes featured alterations through the addition or deletion of decorative trim (e.g., a skirt on a costume for an elephant) or the use of specific colors or costume names, which served to distinguish masculine from feminine costumes. For example, although the number and specific pattern pieces required to construct a particular pattern would not vary, View A featured a girl-modeled Egg or Tomato, whereas View B presented a boy-modeled Baseball or Pincushion. Structurally identical costumes modeled by both boys and girls would be distinguished through the use of distinct colors or patterns of material. Thus, for the peanut M & M costumes, the illustration featured girls clad in red or green and boys clad in blue, brown, or yellow. Similarly, female clowns wore costumes of soft pastel colors and dainty polka dots, but male clowns were garbed in bold primary colors and in material featuring large polka dots or stripes. Illustrations for ready-to-wear costumes were also likely to signal the sex of the intended wearer through the advertising copy: models for feminine costumes, for example, had long curled hair, were made up, and wore patent leather shoes. Only in such costumes as Wrinkly Old Woman, Grandma Hag, Killer Granny, and Nun did identifiably male children model female apparel. . . .

[A]lthough hero costumes constituted a large percentage of both masculine and feminine costumes, masculine costumes contained a higher percentage of villain costumes, and feminine costumes included substantially more fool costumes, particularly those of nonhuman/inanimate objects. It may be imagined that the greater total number of feminine costumes would provide young girls with a broader range of costumes to select from than exists for young boys, but in fact the obverse is true. . . . [W]hen finer distinctions were made within the three generic categories, hero costumes for girls were clustered in a narrow range of roles that, although distinguished by specific names, were functionally equivalent in the image they portray. It would seem that, for girls, glory is concentrated in the narrow realm of beauty queens, princesses, brides, or other exemplars of traditionally passive femininity. The ornate, typically pink, ball-gowned costume of the princess (with or without a synthetic jeweled tiara) was notable, whether the specific costume was labeled Colonial Belle, the Pumpkin Princess, Angel Beauty, Blushing Bride, Georgia Peach, Pretty Mermaid, or Beauty Contest Winner. In contrast, although hero costumes for boys emphasized the warrior theme of masculinity (Doyle, 1989; Rotundo, 1993), with costumes depicting characters associated with battling historical, contemporary, or supernatural Goliaths (e.g., Broncho Rider, Dick Tracy, Sir Lancelot, Hercules, Servo Samaurai, Robin the Boy Wonder), these costumes were less singular in the visual images they portrayed and were more likely to depict characters who possessed supernatural powers or skills.

Masculine costumes were also more likely than feminine costumes to depict a wide range of villainous characters (e.g., Captain Hook, Rasputin, Slash), monsters (e.g., Frankenstein, The Wolfman), and, in particular, agents or symbols of death (e.g., Dracula, Executioner, Devil boy, Grim Reaper). Moreover, costumes for male villains were more likely than those of female villains to be elaborate constructions that were visually repellant; to feature an assortment of scars, mutations, abrasions, and suggested amputations; and to present a wide array of ingenious, macabre, or disturbing visual images. For example, the male-modeled, ready-to-wear Mad Scientist's Experiment costume consisted of a full-body costume of a monkey replete with a half-head mask featuring a gaping incision from which rubber brains dangled. Similarly, costumes for such characters as Jack the Ripper, Serial Killer, Freddy Krueger, or The Midnight Stalker were adorned with the suggestion of bloodstains and embellished with such paraphernalia as plastic knives or slip-on claws.

In marked contrast, the costumes of female villains alternated between relatively simple costumes of witches in pointy hats and capes modeled by young girls, costumes of the few female arch villains drawn from the pages of comic books, and, for older girls, costumes that were variants of the garb donned by the popular TV character Elvira, Mistress of the Dark (i.e., costumes that consisted of a long black wig and a long flowing black gown cut in an empire-style, which, when decorated with gold brocade or other trim at the top of the ribcage, served to create the suggestion of a bosom). The names of costumes for the female villains appeared to emphasize the erotic side of their villainy (e.g., Enchantra, Midnite Madness, Sexy Devil, Bewitched) or to neutralize the malignancy of the character by employing adjectives that emphasized their winsome rather than wicked qualities (e.g., Cute Cuddley Bewitched, Little Skull Girl, Pretty Little Witch).

Within the category of fools, feminine costumes were more likely than masculine costumes to depict nonhuman/inanimate objects (33.1% of feminine costumes vs. 17.4% of masculine costumes). Feminine costumes were more likely than masculine costumes to feature a wide variety of small animals and insects (e.g., Pretty Butterfly, Baby Cricket, Dalmatian Puppy), as well as flowers, foodstuffs (BLT Sandwich, IceCream Cone, Lollipop), and dainty, fragile objects such as Tea Pot. For example, a costume for Vase of Flowers was illustrated with a picture of a young girl wearing a cardboard cylinder from her ribcage to her knees on which flowers were painted, while a profusion of pink, white, and yellow flowers emerged from the top of the vase to form a collar of blossoms around her face. Similarly, a costume for Pea Pod featured a young girl wearing a green cylinder to which four green balloons were attached; on the top of her head, the model wore a hat bedecked with green leaves and tendrils in a corkscrew shape. When costumed as animals, boys were likely to be shown modeling larger, more aggressive animals (e.g., Veliceraptor, Lion, T-Rex); masculine costumes were unlikely to be marketed with adjectives emphasizing their adorable, "li'l," cute, or cuddly qualities. In general, boys were rarely cast as objects, but when they were, they were overwhelmingly shown as items associated with masculine expertise. For example, a costume for Computer was modeled by a boy whose face was encased in the computer monitor and who wore, around his midtorso, a keyboard held up by suspenders. Another masculine costume depicted a young boy wearing a costume for Paint Can; the lid of the can was crafted in the style of a chef 's hat, and across the cylindrical can worn from midchest to midknee was written "Brand X Paint" and, in smaller letters, "Sea Blue." Although rarely depicted as edibles or consumable products, three masculine costumes featured young boys as, variously, Root Beer Mug, Pepperoni Pizza, and Grandma's Pickle Jar.

Discussion

Although the term "fantasy" implies a "play of the mind" or a "queer illusion" (Barnhart, 1967, p. 714), the marketing illustrations for children's Halloween costumes suggest a flight of imagination that remains largely anchored in traditional gender roles, images, and symbols. Indeed, the noninclusive language commonly found in the names of many children's Halloween costumes reverberates throughout many other dimensions of the gendered social life depicted in this fantastical world. For example, the importance of participation in the paid-work world and financial success for men and of physical attractiveness and marriage for women is reinforced through costume names that reference masculine costumes by occupational roles or titles but describe feminine costumes via appearance and/or relationships (e.g., "Policeman" vs. "Beautiful Bride"). Although no adjectives are deemed necessary to describe Policeman, the linguistic prompt contained in Beautiful Bride serves to remind observers that the major achievements for females are getting married and looking lovely. In addition to costume titles that employ such sex-linked common nouns as Flapper, Bobby Soxer, Ballerina, and Pirate Wench, sex-marked suffixes such as the *-ess* (e.g., Pretty Waitress, Stewardess, Gypsy Princess, Sorceress) and *-ette* (e.g., Majorette) also set apart male and female fantasy character costumes. Costumes for suffragettes or female-modeled police officers, astronauts, and fire fighters were conspicuous only by their absence.

Gender stereotyping in children's Halloween costumes also reiterates an active-masculine/passive-feminine dichotomization. The ornamental passivity of Beauty Queen stands in stark contrast to the reification of the masculine action figure, whether he is heroic or villainous. In relation to hero figures, the dearth of female superhero costumes in the sample would seem to reflect the comparative absence of such characters in comic books. Although male superheroes have sprung up almost "faster than a speeding bullet" since the 1933 introduction of *Superman,* the comic book life span of women superheroes has typically been abbreviated, "rarely lasting for more than three appearances" (Robbins, 1996, p. 2). Moreover, the applicability of the term "superhero" to describe these female characters seems at least somewhat dubious. Often their role has been that of the male hero's girlfriend or sidekick "whose purpose was to be rescued by the hero" (Robbins, 1996, p. 3).

In 1941 the creation of *Wonder Woman* (initially known as Amazon Princess Diana) represented a purposeful attempt by her creator, psychologist William Marston, to provide female readers with a same-sex superhero. . . . Nevertheless, over half a decade later, women comic book superheroes remain rare and, when they do appear, are likely to be voluptuous and scantily clad. If, as Robbins (1996, p. 166) argued, the overwhelmingly male comic book audience "expect, in fact demand that any new superheroines exist only as pinup material for their entertainment," it would seem that comic books and their televised versions are unlikely to galvanize the provision of flat-chested female superhero Halloween costumes for prepubescent females in the immediate future.

The relative paucity of feminine villains would also seem to reinforce an active/passive dichotomization on the basis of gender. Although costumes depict male villains as engaged in the commission of a wide assortment of antisocial acts, those for female villains appear more nebulous and are concentrated within the realm of erotic transgressions. Moreover, the depiction of a female villain as a sexual temptress or erotic queen suggests a type of "active passivity" (Salamon, 1983), whereby the act of commission is restricted to wielding her physical attractiveness over (presumably) weak-willed men. The veritable absence of feminine agents or symbols of death may reflect not only the stereotype of women (and girls) as life-giving and nurturing, but also the attendant assumption that femininity and lethal aggressiveness are mutually exclusive.

Building on the Sapir–Whorf hypothesis that the language we speak predisposes us to make particular interpretations of reality (Sapir, 1949; Whorf, 1956) and the assertion that language provides the basis for developing the gender schema identified by Bem (1983), the impact of language and other symbolic representations must be considered consequential. The symbolic representations of gender contained within

Halloween costumes may, along with specific costume titles, refurbish stereotypical notions of what women/girls and men/boys are capable of doing even within the realm of their imaginations. Nelson and Robinson (1995) noted that deprecatory terms in the English language often ally women with animals. Whether praised as a "chick," "fox," or "Mother Bear" or condemned as a "bitch," "sow," or an "old nag," the imagery is animal reductionist. They also noted that language likens women to food items (e.g., sugar, tomato, cupcake), with the attendant suggestion that they look "good enough to eat" and are "toothsome morsels." Complementing this, the present study suggests that feminine Halloween costumes also employ images that reduce females to commodities intended for amusement, consumption, and sustenance. A cherry pie, after all, has only a short shelf life before turning stale and unappealing. Although a computer may become obsolete, the image it conveys is that of rationality, of a repository of wisdom, and of scientifically minded wizardry.

In general, the relative absence of gender-neutral costumes is intriguing. Although it must remain speculative, it may be that the manufacturers of ready-to-wear and sewing pattern costumes subscribe to traditional ideas about gender and/or believe that costumes that depart from these ideas are unlikely to find widespread acceptance. Employing a supply–demand logic, it may be that marketing analysis of costume sales confirms their suspicions. Nevertheless, although commercial practices may reflect consumer preferences for gender-specific products rather than biases on the part of merchandisers themselves, packaging that clearly depicts boys or girls—but not both—effectively promotes gendered definitions of products beyond anything that might be culturally inherent in them. This study suggests that gender-aschematic Halloween costumes for children compose only a minority of both ready-to-wear costumes and sewing patterns. It is notable that, when male children were presented modeling female garments, the depicted character was effectively desexed by age (e.g., a wizened, hag-like "grandmother") or by calling (e.g., a nun).

The data for this study speak only to the gender practices of merchandisers marketing costumes and sewing patterns to parents who themselves may be responding to their children's wishes. Beyond this, the findings do not identify precisely whose tastes are represented when these costumes are purchased. It is always possible that, despite the gendered nature of Halloween costumes presented in the illustrations and advertising copy used to market them, parents and children themselves may engage in creative redefinitions of the boundary markers surrounding gender. A child or parent may express and act on a preference for dressing a male in a pink, ready-to-wear butterfly costume or a female as Fred Flinstone and, in so doing, actively defy the symbolic boundaries that gender the Halloween costume. Alternatively, as a strategy of symbolic negotiation, those parents who sew may creatively experiment with recognizable gender markers, deciding, for example, to construct a pink dragon costume for their daughter or a brown butterfly costume for their son. Such amalgams of gender-discordant images may, on the surface, allow both male and female children to experience a broader range of fantastical roles and images. However, like Persian carpets, deliberately flawed to forestall divine wrath, such unorthodox Halloween costumes, in their structure and design, may nevertheless incorporate fibers of traditional gendered images.

References

Barnes, R., & Eicher, J. B. (1992). *Dress and gender: Making and meaning in cultural contexts.* New York: Berg.

Barnhart, C. L. (1967). *The world book dictionary: A–K.* Chicago: Field Enterprises Educational Corporation.

Bem, S. L. (1983). Gender schema theory and its implications for child development: Raising gender-aschematic children in a gender-schematic society. *Signs: Journal of Women in Culture and Society, 8,* 598–616.

Doyle, J. A. (1989). *The male experience.* Dubuque, IA: Wm. C. Brown.

Eicher, J. B., & Roach-Higgins, M. E. (1992). Definition and classification of dress: Implications for analysis of gender roles. In R. Barnes & J. B. Eicher (Eds.), *Dress and gender. Making and meaning in cultural contexts* (pp. 8–28). New York: Berg.

Goffman, E. (1966). Gender display. *Philosophical Transactions of the Royal Society of London, 279,* 250.

Goffman, E. (1979). *Gender advertisements.* London: Macmillan.

Klapp, O. (1962). *Heroes, villains and fools.* Englewood Cliffs, NJ: Prentice-Hall.

Klapp, O. (1964). *Symbolic leaders.* Chicago: Aldine.

McNeill, F. M. (1970). *Hallowe'en: Its origins, rites and ceremonies in the Scottish tradition.* Edinburgh: Albyn Press.

Mead, G. H. (1934). *Mind, self and society.* Chicago: University of Chicago Press.

Nelson, E. D. (1999). *Dressing for Halloween, doing gender.* Unpublished manuscript.

Nelson, E. D., & Robinson, B. W. (1995). *Gigolos & Madame's bountiful: Illusions of gender, power and intimacy.* Toronto: University of Toronto Press.

Peretti, P. O., & Sydney, T. M. (1985). Parental toy stereotyping and its effect on child toy preference. *Social Behavior and Personality, 12,* 213–216.

Robbins, T. (1996). *The Great Women Super Heroes.* Northampton, MA: Kitchen Sink Press.

Rotundo, E. A. (1993). *American manhood: Transformations in masculinity from the revolution to the modern era.* New York: Basic Books.

Salamon, E. (1983). *Kept women: Mistress of the '80s.* London: Orbis.

Sapir, E. (1949). *Selected writings of Edward Sapir on language, culture and personality.* Berkeley: University of California Press.

Whorf, B. L. (1956). The relation of habitual thought and behavior to language. In J. B. Carroll (Ed.), *Language, thought, and reality* (pp. 134–159). Cambridge, MA: Technology Press of MIT.

Introduction to Reading 22

This reading by Jacqueline Urla and Alan Swedlund offers an interesting approach to understanding the relationship between the success of the Barbie doll and the everyday body ideals and practices of girls and women in North America today. The authors apply the science of measuring bodies, or anthropometry, to Barbie doll and her "friends," comparing the extreme deviation of Barbie's body to the anthropometry of real women. Urla and Swedlund point out that Barbie exemplifies the commodification of gender in modern, consumer culture. They argue that the success of Barbie points to the strong desire of consumers for fantasy and for products that will transform them. Finally, the authors discuss the multiple meanings of Barbie for the girls and women who are her fans.

1. Why is Barbie a "perfect icon" of late capitalist constructions of femininity?

2. How has anthropometry, the science of measuring bodies, altered how we think and feel about gendered bodies?

3. Discuss the link between hyperthin bodies and hyperconsumption.

The Anthropometry of Barbie

Unsettling Ideals of the Feminine Body in Popular Culture

Jacqueline Urla and Alan C. Swedlund

It is no secret that thousands of healthy women in the United States perceive their bodies as defective. The signs are everywhere: from potentially lethal cosmetic surgery and drugs to the more familiar routines of dieting, curling, crimping, and aerobicizing, women seek to take control over their unruly physical selves. Every year at least 150,000 women undergo breast implant surgery (Williams 1992), while Asian women have their noses rebuilt and their eyes widened to make themselves look "less dull" (Kaw 1993). Studies show that the obsession with body size and the sense of inadequacy start frighteningly early; as many as 80 percent of 9-year-old suburban girls are concerned about dieting and their weight (Bordo 1991: 125). Reports like these, together with the dramatic rise in eating disorders among young women, are just some of the more noticeable fallout from what Naomi Wolf calls "the beauty myth." Fueled by the hugely profitable cosmetic, weight-loss, and fashion industries, the beauty myth's glamorized notions of the ideal body reverberate back upon women as "a dark vein of self hatred, physical obsessions, terror of aging, and dread of lost control" (Wolf 1991: 10).

It is this conundrum of somatic femininity, that female bodies are never feminine enough, that they must be deliberately and oftentimes painfully remade to be what "nature" intended—a condition dramatically accentuated under consumer capitalism—that motivates us to focus our inquiry . . . on images of the feminine ideal. Neither universal nor changeless, idealized notions of both masculine and feminine bodies have a long history that shifts considerably across time, racial or ethnic group, class, and culture. Body ideals in twentieth-century North America are influenced and shaped by images from classical or "high" art, the discourses of science and medicine, and increasingly via a multitude of commercial interests, ranging from mundane life insurance standards to the more high-profile fashion, fitness, and entertainment industries.

Making her debut in 1959 as Mattel's new teenage fashion doll, Barbie rose quickly to become the top-selling toy in the United States. Thirty-four years and a woman's movement later, Barbie dolls remain Mattel's best-selling item, netting over one billion dollars in revenues worldwide (Adelson 1992), or roughly one Barbie sold every two seconds (Stevenson 1991). Mattel estimates that in the United States over 95 percent of girls between the ages of three and eleven own at least one Barbie, and that the average number of dolls per owner is seven (E. Shapiro 1992). Barbie is clearly a force to contend with, eliciting over the years a combination of critique, parody, and adoration. A legacy of the postwar era, she remains an incredibly resilient visual and tactile model of femininity for prepubescent girls headed straight for the twenty-first century.

Urla, J., & Swedlund A. C., 1995, "The anthropometry of Barbie: Unsettling ideals of the feminine body in popular culture" in *Deviant Bodies: Critical Perspectives on Difference in Science and Popular Culture*, ed. by Jennifer Terry and Jacqueline Urla. Reprinted with permission of Indiana University Press.

It is not our intention to settle the debate over whether Barbie is a good or bad role model for little girls or whether her unrealistic body wrecks havoc on girls' self-esteem. . . . We want to suggest that Barbie dolls, in fact, offer a much more complex and contradictory set of possible meanings that take shape and mutate in a period marked by the growth of consumer society, intense debate over gender and racial relations, and changing notions of the body. . . . We want to explore not only how it is that this popular doll has been able to survive such dramatic social changes, but also how she takes on new significance in relation to these changing contexts.

We begin by tracing Barbie's origins and some of the image makeovers she has undergone since her creation. From there we turn to an experiment in the anthropometry of Barbie to understand how she compares to standards for the "average American woman" that were emerging in the postwar period.[1] Not surprisingly, our measurements show Barbie's body to be thin—very thin—far from anything approaching the norm. Inundated as our society is with conflicting and exaggerated images of the feminine body, statistical measures can help us to see that exaggeration more clearly. But we cannot stop there. First, as our brief foray into the history of anthropometry shows, the measurement and creation of body averages have their own politically inflected and culturally biased histories. Standards for the "average" American body, male or female, have always been imbricated in histories of nationalism and race purity. Secondly, to say that Barbie is unrealistic seems to beg the issue. Barbie *is* fantasy: a fantasy whose relationship to the hyperspace of consumerist society is multiplex. What of the pleasures of Barbie bodies? What alternative meanings of power and self-fashioning might her thin body hold for women/girls? Our aim is not, then, to offer another rant against Barbie, but to clear a space where the range of her contradictory meanings and ironic uses can be contemplated: in short, to approach her body as a meaning system in itself, which, in tandem with her mutable fashion image, serves to crystallize some of the predicaments of femininity and feminine bodies in late-twentieth-century North America.

A Doll Is Born

. . . Making sense of Barbie requires that we look to the larger sociopolitical and cultural milieu that made her genesis both possible and meaningful. Based on a German prototype, the "Lili" doll, Barbie was from "birth" implicated in the ideologies of the Cold War and the research and technology exchanges of the military-industrial complex. Her finely crafted durable plastic mold was, in fact, designed by Jack Ryan, well known for his work in designing the Hawk and Sparrow missiles for the Raytheon Company. Conceived at the hands of a military-weapons-designer–turned-toy-inventor, Barbie dolls came onto the market the same year that the infamous Nixon-Khrushchev "kitchen debate" took place at the American National Exhibition in Moscow. Here, in front of the cameras of the world, the leaders of the capitalist and socialist worlds faced off, not over missile counts, but over "the relative merits of American and Soviet washing machines, televisions, and electric ranges" (May 1988:16). As Elaine Tyler May has noted in her study of the Cold War, this much-celebrated media event signaled the transformation of American-made commodities and the model suburban home into key symbols and safeguards of democracy and freedom. It was thus with fears of nuclear annihilation and sexually charged fantasies of the perfect bomb shelter running rampant in the American imaginary, that Barbie and her torpedo-like breasts emerged into popular culture as an emblem of the aspirations of prosperity, domestic containment, and rigid gender roles that were to characterize the burgeoning postwar consumer economy and its image of the American Dream.

Marketed as the first "teenage" fashion doll, Barbie's rise in popularity also coincided with, and no doubt contributed to, the postwar creation of a distinctive teenage lifestyle.[2] Teens, their tastes, and their behaviors were becoming the object of both sociologists and criminologists as well as market survey researchers intent on capturing their discretionary dollars. While J. Edgar Hoover was pronouncing "the juvenile jungle" a menace to American society, retailers, the music industry, and moviemakers declared the thirteen to nineteen-year-old age bracket "the seven golden years" (Doherty 1988:51–52).

Barbie dolls seemed to cleverly reconcile both of these concerns by personifying the good girl who was sexy, but didn't have sex, and was willing to spend, spend, spend. . . . Every former Barbie owner knows that to buy a Barbie is to lust after Barbie accessories. . . . As Paula Rabinowitz has noted, Barbie dolls, with their focus on frills and fashion, epitomize the way that teenage girls and girl culture in general have figured as accessories in the historiography of post-war culture; that is as both essential to the burgeoning commodity culture as consumers, but seemingly irrelevant to the central narrative defining cold war existence (Rabinowitz 1993). Over the years, Mattel has kept Barbie's love of shopping alive, creating a Suburban Shopper Outfit and her own personal Mall to shop in (Motz 1983:131). More recently, in an attempt to edge into the computer game market, we now have an electronic "Game Girl Barbie" in which (what else?) the object of the game is to take Barbie on a shopping spree. In "Game Girl Barbie," shopping takes skill, and Barbie plays to win.

Perhaps what makes Barbie such a perfect icon of late capitalist constructions of femininity is the way in which her persona pairs endless consumption with the achievement of femininity and the appearance of an appropriately gendered body. By buying for Barbie, girls practice how to be discriminating consumers knowledgeable about the cultural capital of different name brands, how to read packaging, and the overall importance of fashion and taste for social status (Motz 1983: 131–32). . . . In making this argument, we want to stress that we are drawing on more than just the doll. "Barbie" is also the packaging, spin-off products, cartoons, commercials, magazines, and fan club paraphernalia, all of which contribute to creating her persona. Clearly, as we will discuss below, children may engage more or less with those products, subverting or ignoring various aspects of Barbie's "official" presentation. However, to the extent that little girls *do* participate in the prepackaged world of Barbie, they come into contact with a number of beliefs central to femininity under consumer capitalism. Little girls learn, among other things, about the crucial importance of their appearance to their personal happiness and to their ability to gain favor with their friends. Barbie's social calendar is constantly full, and the stories in her fan magazines show her frequently engaged in preparation for the rituals of heterosexual teenage life: dates, proms, and weddings. . . .

Barbie exemplifies the way in which gender in the late twentieth century has become a commodity itself, "something we can buy into . . . the same way we buy into a style" (Willis 1991: 23). In her insightful analysis of the logics of consumer capitalism, cultural critic Susan Willis pays particular attention to the way in which children's toys like Barbie and the popular muscle-bound "He-Man" for boys link highly conservative and narrowed images of masculinity and femininity with commodity consumption (1991: 27). In the imaginary world of Barbie and teen advertising, observes Willis, being or becoming a teenager, having a "grown-up" body, is inextricably bound up with the acquisition of certain commodities, signaled by styles of clothing, cars, music, etc. . . .

Barbie Is a Survivor

. . . In the past three decades, this popular children's doll has undergone numerous changes in her fashion image and "occupations" and has acquired a panoply of ethnic "friends" and analogues that have allowed her to weather the dramatic social changes in gender and race relations that arose in the course of the sixties and seventies. . . .

[A] glance at Barbie's resumé, published in *Harper's* magazine in August 1990, while incomplete, shows Mattel's attempt to expand Barbie's career options beyond the original fashion model:

Positions Held

1959–present	Fashion model
1961–present	Ballerina
1961–64	Stewardess (American Airlines)
1964	Candy striper
1965	Teacher
1965	Fashion editor
1966	Stewardess (Pan Am)
1973–75	Flight attendant (American Airlines)
1973–present	Medical doctor
1976	Olympic athlete
1984	Aerobics instructor
1985	TV news reporter
1985	Fashion designer
1985	Corporate executive
1988	Perfume designer
1989–present	Animal rights volunteer

It is only fitting, given her origin, to note that Barbie has also had a career in the military and aeronautics space industry: she has been an astronaut, a marine, and, during the Gulf War, a Desert Storm trooper. Going from pink to green, Barbie has also acquired a social conscience, taking up the causes of UNICEF, animal rights, and environmental protection. . . .

For anyone tracking Barbiana, it is abundantly clear that Mattel's marketing strategies are sensitive to a changing social climate. Just as Mattel has sought to present Barbie as a career woman with more than air in her vinyl head, they have also tried to diversify her otherwise lily-white suburban world. . . . With the expansion of sales worldwide, Barbie has acquired multiple national guises (Spanish Barbie, Jamaican Barbie, Malaysian Barbie, etc.).[3] In addition, her cohort of "friends" has become increasingly ethnically diversified, as has Barbie advertising, which now regularly features Asian, Hispanic, and African American little girls playing with Barbie. . . . This diversification has not spelled an end to reigning Anglo beauty norms and body image. Quite the reverse. When we line the dolls up together, they look virtually identical. Cultural difference is reduced to surface variations in skin tone and costumes that can be exchanged at will. . . .

"The icons of twentieth-century mass culture," writes Susan Willis, "are all deeply infused with the desire for change," and Barbie is no exception (1991: 37). In looking over the course of Barbie's career, it is clear that part of her resilience, appeal, and profitability stems from the fact that her identity is constructed primarily through fantasy and is consequently open to change and reinterpretation. As a fashion model, Barbie continually creates her identity anew with every costume change. In that sense, we might want to call Barbie the prototype of the "transformer dolls" that cultural critics have come to see as emblematic of the restless desire for change that permeates postmodern capitalist society (Wilson 1985: 63). Not only can she renew her image with a change of clothes, Barbie also is seemingly able to clone herself effortlessly into new identities—Malibu Barbie; Totally Hair Barbie; Teen Talk Barbie; even Afrocentric Barbie, Shani—without somehow suggesting a serious personality disorder. . . . The multiplication of Barbie and her friends translates the challenge of gender inequality and racial diversity into an ever-expanding array of costumes, a new "look" that can be easily accommodated into a harmonious and illusory pluralism that never ends up rocking the boat of WASP beauty. What is striking, then, is that, while Barbie's identity may be mutable—one day she might be an astronaut, another a cheerleader—*her hyperslender, big-chested body has remained fundamentally unchanged over the years*—a remarkable fact in a society that fetishizes the new and improved. . . . We turn now from Barbie's "persona" to the conundrum of her body and to our class experiment in the anthropometry of feminine ideals. In so doing, our aim is deliberately subversive. We wish to use the tools of calibration and measurement—tools of normalization that have an unsavory history for women and racial or ethnic minorities—to destabilize the ideal. . . . We begin with a very brief historical overview of the anthropometry of women and the emergence of an "average" American female body in the postwar United States, before using our calipers on Barbie and her friends.

The Measured Body: Norms and Ideals

* * *

As the science of measuring human bodies, anthropometry belongs to a long line of techniques of the eighteenth and nineteenth centuries concerned with measuring, comparing, and interpreting variability in different zones of the human body: craniometry, phrenology, physiognomy, and comparative anatomy. Early anthropometry shared with these an understanding and expectation that the body was a window into a host of moral, temperamental, racial, or gender characteristics. It sought to distinguish itself from its predecessors, however, by adhering to rigorously standardized methods and quantifiable results that would, it was hoped, lead to the "complete elimination of personal bias" that anthropometrists believed had tainted earlier measurement techniques (Hrdlicka 1939: 12).[4]

Under the aegis of Earnest Hooton, Ales Hrdlicka, and Franz Boas, located respectively at Harvard University, the Smithsonian, and Columbia University, anthropometric studies within U.S. physical anthropology were utilized mainly in the pursuit of three general areas of interest: identifying racial and or national types; the measurement of adaptation and "degeneracy"; and a comparison of the sexes. Anthropometry was, in other words, believed to be a useful technique in resolving three critical border disputes: the boundaries between races or ethnic groups; the normal and the degenerate; and the border between the sexes.

As is well documented by now, women and non-Europeans did not fare well in these emerging sciences of the body (see the work of Blakey 1987; Gould 1981; Schiebinger 1989, 1993; Fee 1979; Russett 1989; also Horn and Fausto Sterling, this volume); measurements of women's bodies, their skulls in particular, tended to place them as inferior to or less intelligent than males. In the great chain of being, women as a class were believed to share certain atavistic characteristics with both children and so-called savages. Not everything about women was regarded negatively. In some cases it was argued that women possessed physical and moral qualities that were superior to those of males. Above all, woman's body was understood through the lens of her reproductive function; her physical characteristics, whether inferior or superior to those of males, were inexorably dictated by her capacity to bear children. . . . With males as the unspoken prototype, women's bodies were frequently described (subtly or not) as deviations from the norm: as subjects, the measurement of their bodies was occasionally risky to the male scientists,[5] and as bodies they were variations from the generic or ideal type (their body fat "excessive," their pelvises maladaptive to a bipedal [i.e., more evolved] posture, their musculature weak). Understood primarily in terms of their reproductive capacity, women's bodies, particularly their reproductive organs, genitalia, and secondary sex characteristics, were instead more carefully scrutinized and measured within "marital adjustment" studies and in the emerging science of gynecology, whose practitioners borrowed liberally from the techniques used by physical anthropologists. . . .

In the United States, an attempt to elaborate a scientifically sanctioned notion of a normative "American" female body, however, was taking place in the college studies of the late nineteenth and early twentieth centuries. By the 1860s, Harvard and other universities had begun to regularly collect anthropometric data on their male student populations, and in the 1890s comparable data began to be collected from the East Coast women's colleges as well. Conducted by departments of hygiene, physical education, and home economics, as well as physical anthropology, these large-scale studies gathered data on the elite, primarily WASP youth, in order to determine the dimensions of the "normal" American male and female. . . . Effectively excluded from these attempts to define the "normal" or average body, of course, were those "other" Americans—descendants of African slaves, North American Indians, and the many recent European immigrants from Ireland, southern Europe, and eastern Europe—whose bodies were the subject of racist, evolution-oriented studies concerned with "race crossing," degeneracy, and the effects of the "civilizing" process (see Blakey 1987). . . . Between the two wars, nationalist interests had fueled eugenic interests and provoked a deepening concern about the physical fitness of the American people. Did Americans constitute a distinctive physical "type"; were they puny and weak as some Europeans had alleged, or were they physically bigger and stronger than their European ancestors? Could they defend themselves in time of war? And who did this category of "Americans" include? Questions such as these fed into an already long-standing preoccupation with defining a specifically American national character and, in 1945, led to the creation of one of the most celebrated and widely publicized anthropometric models of the century: Norm and Norma, the average American male and female. Based on the composite measurements of thousands of young people, described only as "native white Americans," across the United States, the statues of Norm and Norma were the product of a collaboration between obstetrician-gynecologist Robert Latou Dickinson, well known for his studies of human reproductive anatomy, and Abram Belskie, the prize student of Malvina Hoffman, who had sculpted the Races of Mankind series.[6] . . .

Described in the press as the "ideal" young woman, Norma was said to be everything an American woman should be in a time of war: she was fit, strong-bodied, and at the peak of her reproductive potential. Commentators waxed eloquent about the model character traits—maturity, modesty, and virtuosity—that this perfectly average body suggested. . . . Norma and Norman were . . . more than statistical composites, they were ideals. It is striking how thoroughly racial and ethnic differences were erased from these scientific representations of the American male and female. Based on the measurements of white Americans, eighteen to twenty-five years old, Norm and Norma emerged carved out of white alabaster, with the facial features and appearance of Anglo-Saxon gods. Here, as in the college studies that preceded them, the "average American" of the postwar period was to be visualized only as a youthful white body. However, they were not the only ideal. The health reformers, educators, and doctors who approved and promoted Norma as an ideal for American women were well aware that her sensible, strong, thick-waisted body differed significantly from the tall slim-hipped bodies of fashion models in vogue at the time.[7] . . . As the postwar period advanced, Norma would continue to be trotted out in home economics and health education classes. But in the iconography of desirable female bodies, she would be overshadowed by the array of images of fashion models and pinup girls put out by advertisers; the entertainment industry, and a burgeoning consumer culture. These idealized images were becoming, as we will see below, increasingly thin in the sixties and seventies while the "average" woman's body was in fact getting heavier. With the thinning of the American feminine ideal, Norma and subsequent representations of the statistically average woman would become increasingly aberrant, as slenderness and sex appeal—not physical fitness—became the premier concern of postwar femininity.

The Anthropometry of Barbie: Turning the Tables

As the preceding discussion makes abundantly clear, the anthropometrically measured "normal" body has been anything but value-free. Formulated in the context of a race-, class-, and gender-stratified society, there is no doubt that quantitatively defined ideal types or standards have been both biased and oppressive. Incorporated into weight tables, put on display in museums and world's fairs, and reprinted in popular magazines, these scientifically endorsed standards produce what Foucault calls "normalizing effects," shaping, in not altogether healthy ways, how individuals understand themselves and their bodies. Nevertheless, in the contemporary cultural context, where an impossibly thin image of women's bodies has become the most popular children's toy ever sold, it strikes us that recourse to the "normal" body might just be the power tool we need for destabilizing a fashion fantasy spun out of control. It was with this in mind that we asked students in one of our social biology classes to measure Barbie to see how her body compared to the average measurements of young American women of the same period. Besides estimating Barbie's dimensions if she were life-sized, we see the experiment as an occasion to turn the anthropometric tables from disciplining the bodies of living women to measuring the ideals by which we have come to judge ourselves and others. We also see it as an opportunity for students who have grown up under the regimes of normalizing science—students who no doubt have been measured, weighed, and compared to standards since birth—to use those very tools to unsettle a highly popular cultural ideal. . . .

Since one objective of the course was to learn about human variation, our first task in understanding more about Barbie was to consider the fact that Barbie's friends and family do represent some variation, limited though it may be. Through colleagues and donations from students or (in one case) their children we assembled seventeen dolls for analysis. The sample included:

11 early '60s Barbie

4 mid-'70s-to-contemporary Barbies, including a Canadian Barbie

3 Kens

2 Skippers

1 Scooter

Assorted Barbie's friends, including Christie, Barbie's "black" friend

Assorted Ken's friends

To this sample we subsequently added the most current versions of Barbie and Ken (from the "Glitter Beach" collection) and also Jamal, Nichelle, and Shani, Barbie's more recent African American friends. As already noted, Mattel introduced these dolls (Shani, Asha, and Nichelle) as having a more authentic African American appearance, including a "rounder and more athletic" body. Noteworthy also are the skin color variations between the African American dolls, ranging from dark to light, whereas Barbie and her white friends tend to be uniformly pink or uniformly suntanned. . . .

Before beginning the actual measurements we discussed the kinds of data we thought would be most appropriate. Student interest centered on height and chest, waist, and hip circumference. Members of the class also pointed out the apparently small size of the feet and the general leanness of Barbie. As a result, we added a series of additional standardized measurements, including upper arm and thigh circumference, in order to obtain an estimate of body fat and general size. . . . In scaling Barbie to be life-sized, the students decided to translate her measurements using two standards: (a) if Barbie were a fashion model (5'10") and (b) if she were of average height for women in the United States (5' 4"). We also decided to measure Ken, using both an average male stature, which we designated as 5' 8" and the more "idealized" stature for men, 6'. We took measurements of dolls in the current Glitter Beach and Shani collection that were not available for our original classroom experiment, and all measurements were retaken to confirm estimates. We report here only the highlights of the measurements taken on the newer Barbie and newer Ken, Jamal, and Shani, scaled at their ideal fashion-model height. For purposes of comparison, we include data on average body measurements from the standardized published tables of the 1988 Anthropometric Survey of Army Personnel. We have dubbed these composites for the female and male recruits Army "Norma" and Army "Norm," respectively.

Barbie and Shani's measurements reveal interesting similarities and subtle differences.

First, considering that they are six inches taller than "Army Norma," . . . their measurements tend to be considerably less *at all points.* "Army Norma" is a composite of the fit woman soldier; Barbie and Shani, as high-fashion ideals, reflect the extreme thinness expected of the runway model. To dramatize this, had we scaled Barbie to 5' 4," her chest, waist, and hip measurements would have been 32"–17"–28," clinically anorectic to say the least. There are only subtle differences in size, which we presume intend to facilitate the exchange of costumes among the different dolls.

We were curious to see the degree to which Mattel had physically changed the Barbie mold in making Shani. Most of the differences we could find appeared to be in the face. The nose of Shani is broader and her lips are ever so slightly larger. However, our measurements also showed that Barbie's hip circumference is actually larger than Shani's, and so is her hip breadth. If anything, Shani might have thinner legs than Barbie, but her back is arched in such a way that it tilts her buttocks up. This makes them appear to protrude more posteriorly, even though the hip depth measurements of both dolls are virtually the same (7.1"). Hence, the tilting of the lumbar dorsal region and the extension of the sacral pelvic area produce the visual illusion of a higher, rounder butt. . . . This is, we presume, what Mattel was referring to in claiming that Shani has a realistic, or ethnically correct, body (Jones 1991).

One of our interests in the male dolls was to ascertain whether they represent a form closer to average male values than Barbie does to average female values. Ken and Jamal provide interesting contrasts to "Army Norm," but certainly not to each other. Their postcranial bodies are identical in all respects. They, in turn, represent a somewhat slimmer, trimmer male than the so-called fit soldier of today. Visually, the newer Ken and Jamal appear very tight and muscular and "bulked out" in impressive ways. The U.S. Army males tend to carry slightly more fat, judging from the photographs and data presented in the 1988 study.[8] Indeed, it would appear that Barbie and virtually all her friends characterize a somewhat extreme ideal of the human figure, but in Barbie and Shani, the female cases, the degree to which they vary from "normal" is much greater than in the male cases, bordering on the impossible. Barbie truly is the unobtainable representation of an imaginary femaleness. But she is certainly not unique in the realm of female ideals. Studies tracking the body measurements of Playboy magazine centerfolds and Miss America contestants show that between 1959 and 1978 the average weight and hip size for women in both of these groups have decreased steadily (Wiseman et al. 1992). Comparing their data to actuarial data for the same time period, researchers found that the thinning of feminine body ideals was occurring at the same time that the average weight of American women was actually increasing. A follow-up study for the years 1979–88 found this trend continuing into the eighties: approximately sixty-nine percent of Playboy centerfolds and sixty percent of Miss America contestants were weighing in at fifteen percent or more below their expected age and height category. In short, the majority of women presented to us in the media as having desirable feminine bodies were, like Barbie, well on their way to qualifying for anorexia nervosa.

Our Barbies, Our Selves

* * *

On the surface, at least, Barbie's strikingly thin body and the repression and self-discipline that it signifies would appear to contrast with her seemingly endless desire for consumption and self-transformation. And yet, as Susan Bordo has argued in regard to anorexia, these two phenomena—hyperthin bodies and hyperconsumption—are very much linked in advanced capitalist economies that depend upon commodity excess. Regulating desire under such circumstances is a constant, ongoing problem that plays itself out on the body. As Bordo argues:

> [In a society where we are] conditioned to lose control at the very sight of desirable products, we

> can only master our desires through a rigid defense against them. The slender body codes the tantalizing ideal of a well-managed self in which all is "in order" despite the contradictions of consumer culture. (1990:97)

The imperative to manage the body and "be all that you can be"—in fact, the idea that you can *choose* the body that you want to have—is a pervasive feature of consumer culture. Keeping control of one's body, not getting too fat or flabby—in other words, conforming to gendered norms of fitness and weight—are signs of an individual's social and moral worth. But, as feminists Bordo, Sandra Bartky, and others have been quick to point out, not all bodies are subject to the same degree of scrutiny or the same repercussions if they fail. It is in women's bodies and desires in particular where the structural contradictions—the simultaneous incitement to consume and social condemnation for overindulgence—appear to be most acutely manifested in bodily regimes of intense self-monitoring and discipline. . . . Just as it is women's appearance that is subject to greater social scrutiny, so it is that women's desires, hungers, and appetites are seen as most threatening and in need of control in a patriarchal society.

This cultural context is relevant to making sense of Barbie and the meaning her body holds in late consumer capitalism. In dressing and undressing Barbie, combing her hair, bathing her, turning and twisting her limbs in imaginary scenarios, children acquire a very tactile and intimate sense of Barbie's body. Barbie is presented in packaging and advertising as a role model, a best friend or older sister to little girls. Television jingles use the refrain, "I want to be just like you," while look-alike clothes and look-alike contests make it possible for girls to live out the fantasy of being Barbie. . . . In short, there is no reason to believe that girls (or adult women) separate Barbie's body shape from her popularity and glamour.[9] This is exactly what worries many feminists. As our measurements show, Barbie's body differs wildly from anything approximating "average" female body weight and proportions. Over the years her wasp-waisted body has evoked a steady stream of critique for having a negative impact on little girls' sense of self-esteem.[10] While her large breasts have always been a focus of commentary, it is interesting to note that, as eating disorders are on the rise, her weight has increasingly become the target of criticism. . . .

There is no doubt that Barbie's body contributes to what Kim Chernin (1981) has called "the tyranny of slenderness." But is repression all her hyperthin body conveys? Looking once again to Susan Bordo's work on anorexia, we find an alternative reading of the slender body—one that emerges from taking seriously the way anorectic women see themselves and make sense of their experience:

> For them, anorectics, [the slender ideal] may have a very different meaning; it may symbolize not so much the containment of female desire, as its liberation from a domestic, reproductive destiny. The fact that the slender female body can carry both these (seemingly contradictory) meanings is one reason, I would suggest, for its compelling attraction in periods of gender change. (Bordo 1990: 103)

. . . One could argue that, like the anorectic body she resembles, Barbie's body displays conformity to dominant cultural imperatives for a disciplined body and contained feminine desires. As a woman, however, her excessive slenderness also signifies a rebellious manifestation of willpower, a visual denial of the maternal ideal symbolized by pendulous breasts, rounded stomach and hips. Hers is a body of hard edges, distinct borders, self-control. It is literally impenetrable. Unlike the anorectic, whose self-denial renders her gradually more androgynous in appearance, in the realm of plastic fantasy Barbie is able to remain powerfully sexualized, with her large, gravity-defying breasts, even while she is distinctly nonreproductive. Like the "hard bodies" in fitness advertising, Barbie's body may signify for women the pleasure of control and mastery, both of which are highly valued traits in American society and predominantly associated with masculinity (Bordo 1990: 105). Putting these elements together with her apparent independent wealth can make for a very different reading of Barbie than the one we often find in the popular press. To paraphrase one Barbie-doll owner: she owns a Ferrari and doesn't have a husband—she must be doing something right![11] . . .

It is clear that a next step we would want to take in the cultural interpretation of Barbie is an ethnographic study of Barbie-doll owners.[12] In the meanwhile, we can know something about these alternative appropriations by looking to various forms of popular culture and the art world. Barbie has become a somewhat celebrated figure among avant-garde and pop artists, giving rise to a whole genre of Barbie satire, known as "Barbie Noire" (Kahn 1991). According to Peter Galassi, curator of *Pleasures and Terrors of Domestic Comfort,* an exhibit at the Museum of Modern Art, in New York "Barbie isn't just a doll. She suggests a type of behavior—something a lot of artists, especially women, have wanted to question" (quoted in Kahn 1991: 25). Perhaps the most notable sardonic use of Barbie dolls to date is the 1987 film *Superstar: The Karen Carpenter Story,* by Todd Haynes and Cynthia Schneider. In this deeply ironic exploration into the seventies, suburbia, and middle-class hypocrisy, Barbie and Ken dolls are used to tell the tragic story of Karen Carpenter's battle with anorexia and expose the perverse underbelly of the popular singing duo's candy-coated image of happy, apolitical teens. It is hard to imagine a better casting choice to tell this tale of femininity gone astray than the ever-thin, ever-plastic, ever-wholesome Barbie.

For Barbiana collectors it should come as no surprise that Barbie's excessive femininity also makes her a favorite persona of female impersonators, alongside Judy, Marilyn, Marlene, and Zsa Zsa. Appropriations of Barbie in gay camp culture have tended to favor the early, vampire Barbie look: with the arched eyebrows, heavy black eyeliner, and coy sideways look—the later superstar version of Barbie, according to BillyBoy, is just *too* pink. . . .

In the world of Barbie Noire, the hyper-rigid gender roles of the toy industry are targeted for inversion and subversion. While Barbie is transformed into a dominatrix drag queen, Ken, too, has had his share of spoofs and gender bending. Barbie's somewhat dull steady boyfriend has never been developed into much more than a reliable escort and proof of Barbie's appropriate sexual orientation and popularity. In contrast to that of Barbie, Ken's image has remained boringly constant over the years. He has had his "mod," "hippie" and Malibu-suntan days, and he has gotten significantly more muscular. But for the most part, his clothing line is less diversified, and he lacks an independent fan club or advertising campaign.[13] In a world where boys' toys are G.I. Joe-style action figures, bent on alternately saving or destroying the world, Ken is an anomaly. Few would doubt that his identity was primarily another one of Barbie's accessories. His secondary status vis-à-vis Barbie is translated into emasculation and/or a secret gay identity: cartoons and spoofs of Ken have him dressed in Barbie clothes, and rumors abound that Ken's seeming lack of sexual desire for Barbie is only a cover for his real love for his boyfriends, Alan, Steve, and Dave.

Inscrutable with her blank stare and unchanging smile, Barbie is thus available for any number of readings and appropriations. What we have done here is examine some of the ways she resonates with the complex and contradictory cultural meanings of femininity in postwar consumer society and a changing politics of the body. Barbie, as we, and many other critics, have observed, is an impossible ideal, but she is an ideal that has become curiously normalized. In a youth-obsessed society like our own, she is an ideal not just for young women, but for all women who feel that being beautiful means looking like a skinny, buxom, white twenty-year-old. It is this cultural imperative to remain ageless and lean that leads women to have skewed perceptions of their bodies, undergo painful surgeries, and punish themselves with outrageous diets. Barbie, in short, is an ideal that constructs women's bodies as hopelessly imperfect. It has been our intention to unsettle this ideal and, at the same time, to be sensitive to other possible readings, other ways in which this ideal body figures and reconfigures the female body. . . .

We have explored some of the battleground upon which the serious play of Barbie unfolds. If Barbie has taught us anything about gender, it is that femininity in consumer culture is a question of carefully performed display, of paradoxical fixity and malleability. One outfit, one occupation, one identity can be substituted for another, while Barbie's body has remained ageless, changeless untouched by the ravages of age or cellulite. She is always a perfect fit, always able to consume and be consumed. Mattel has skillfully managed to turn challenges of feminist protest, ethnic diversity, and a troubled multiculturalism to a new array of outfits and skin tones, annexing these to a singular anorectic body ideal. Cultural icon that she is, Barbie nevertheless cannot be permanently located in any singular cultural space. Her meaning is mobile as she is appropriated and relocated into different cultural contexts, some of which, as we have seen, make fun of many of the very notions of femininity and consumerism she personifies. As we consider Barbie's many meanings, we should remember that Barbie is not only a denizen of subcultures in the United States, she is also a world traveler. A product of the global assembly line, Barbie dolls owe their existence to the internationalization of the labor market and global flows of capital and commodities that today characterize the toy industry, as well as other industries in the postwar era. Designed in Los Angeles, manufactured in Taiwan or Malaysia, distributed worldwide, Barbie™ is American-made in name only. Speeding her way into an expanding global market, Barbie brings with her some of the North American cultural subtext we have outlined in this analysis. How this teenage survivor then gets interpolated into the cultural landscapes of Mayan villages, Bombay high-rises, and Malagasy towns is a rich topic that begs to be explored.

Notes

1. At the time of this writing, there was no definitive history of Barbie and the molds that have been created for her body. However, Barbie studies are booming and we expect new work in press, including M. G. Lord's *Forever Barbie: The Unauthorized Biography of a Real Doll* (1994), to provide greater insight into Barbie's history and the debates surrounding her body within Mattel and the press.

2. While the concept of adolescence as a distinct developmental stage between puberty and adulthood was not new to the fifties, Thomas Doherty (1988) notes that it wasn't until the end of World War II that the term "teenager" gained standard usage in the American language.

3. Recent work by Ann duCille promises to offer an incisive cultural critique of the "ethnification" of Barbie and its relationship to controversies in the United States over multiculturalism and political correctness (duCille 1995). More work, however, needs to be done on how Barbie dolls are adapted to appeal to various markets outside the U.S. For example, Barbie dolls manufactured in Japan for Japanese consumption have noticeably larger, rounder eyes than those marketed in the United States (see BillyBoy 1987). For some suggestive thoughts on the cultural implications of the transnational flow of toys like Barbie dolls, TransFormers, and He-Man, see Carol Breckenridge's (1990) brief but intriguing editorial comment to *Public Culture*.

4. Closely aligned with the emergence of statistics, it was Hrdlicka's hope that the two would be joined, and that one day the state would be "enlightened" enough to incorporate regular measurements of the population with the various other tabulations of the periodic census, in order to "ascertain whether and how its human stock is progressing or regressing" (1939: 12).

5. In *Practical Anthropometry* Hrdlicka goes to some trouble to instruct field-workers (presumably male) working among "uncivilized groups" about the steps they need to take not to offend, and thereby put themselves at risk, when measuring women (1939: 57–59).

6. Norma is described in the press reports as being based on the measurements of 15,000 "real American girls." Although we cannot be sure, it is likely this data come from the Bureau of Home Economics, which conducted extensive measurements of students "to provide more accurate dimensions and proportions for sizing women's ready-made garments"

(Shapiro 1945). For further information on the Dickinson collection and Dickinson's methods of observation, see Terry (1992).

7. Historians have noted a long-standing conflict between the physical culture movement, eugenicists and health reformers, on the one hand, and the fashion industry, on the other, that gave rise in American society to competing ideals of the fit and the fashionably fragile, woman (e.g., Banner 1983; Cogan 1989).

8. One aspect of the current undertaking that is clearly missing is the possible variation that exists *within* individual groups of dolls that would result from mold variation and casting processes. Determining this variation would require a much larger doll collection at our disposal. We are considering a grant proposal, but not seriously.

9. This process of identification becomes mimesis, not only in Barbie look-alike contests, but also in the recent Barbie workout video. In her fascinating analysis of the semiotics of workout videos, Margaret Morse (1987) has shown how these videos structure the gaze in such a way as to establish identification between the exercise leader's body and the participant-viewer. Surrounded by mirrors, the viewer is asked to exactly model her movements on those of the leader, literally mimicking the gestures and posture of the "star" body she wishes to become. In Barbie's video, producers use animation to make it possible for Barbie to occasionally appear on the screen as the exercise leader/cheerleader—the star whose body the little girls mimic.

10. In response to this anxiety, Cathy Meredig, an enterprising computer software designer, created the "Happy to Be Me" doll. Described as a healthy alternative for little girls, "Happy to Be Me" has a shorter neck, shorter legs, wider waist, larger feet, and a lot fewer clothes—designed to make her look more like the average woman ("She's No Barbie, nor Does She Care to Be." *New York Times,* August 15, 1991, C-11).

11. "Dolls in Playland." 1992. Colleen Toomey, producer. BBC.

12. While not exactly ethnographic, Hohmann's 1985 study offers a sociopsychological view of how children experiment with social relations during play with Barbies.

13. Signs of a Ken makeover, however, have begun to appear. In 1991, a Ken with "real" hair that can be styled was introduced and, most dramatically, in 1993, he had his hair streaked and acquired an earring in his left ear. This was presented as a "big breakthrough" by Mattel and was received by the media as a sign of a broader trend in the toy industry to break down rigid gender stereotyping in children's toys (see Lawson 1993). It doesn't appear, however, that Ken is any closer to getting a "realistic" body than Barbie. Ruth Handler notes that when Mattel was planning the Ken doll, she had wanted him to have genitals—or at least a bump, and claims the men in the marketing group vetoed her suggestion. Ken did later acquire his bump (see "Dolls in Playland," Colleen Toomey, producer. BBC. 1993).

References

Adelson, Andrea 1992 "And Now, Barbie Looks Like a Billion." *New York Times,* November 26, sec. D, p. 3.

Banner, Lois W. 1983 *American Beauty.* New York: Knopf.

Bartky, Sandra Lee 1990 "Foucault, Femininity, and the Modernization of Patriarchal Power." In *Femininity and Domination: Studies in the Phenomenology of Oppression,* pp. 63–82. New York: Routledge.

Billyboy. 1987 *Barbie, Her Life and Times, and the New Theater of Fashion.* New York: Crown.

Blakey, Michael L. 1987 "Skull Doctors: Intrinsic Social and Political Bias in the History of American Physical Anthropology" *Critique of Anthropology* 7(2): 7–35.

Bordo, Susan R. 1990 "Reading the Slender Body." In *Body/Politics: Women and the Discourses of Science.* Ed. Mary Jacobus, Evelyn Fox Keller, and Sally Shuttleworth, pp. 83–112. New York: Routledge.

Breckenridge, Carol A. 1990 "Editor's Comment: On Toying with Terror." *Public Culture* 2(2): i–iii.

Brumberg, Joan Jacobs 1988 *Fasting Girls: The History of Anorexia Nervosa.* Cambridge: Harvard University Press. Reprint, New York: New American Library.

Chernin, Kim 1981 *The Obsession: Reflections on the Tyranny of Slenderness.* New York: Harper and Row.

Cogan, Frances B. 1989 *All-American Girl: The Ideal of Real Woman-hood in Mid-Nineteenth-Century*

America. Athens and London: University of Georgia Press.

Davis, Kathy 1991 "Remaking the She-Devil: A Critical Look at Feminist Approaches to Beauty." *Hypatia 6*(2): 21–43.

Doherty, Thomas 1998 *Teenagers and Teen Pies: The Juvenilization of American Movies in the 1950s.* Boston: Unwin Hyman.

duCille, Ann 1995 "Toy Theory: Blackface Barbie and the Deep Play of Difference." In *The Skin Trade: Essays on Race, Gender, and the Merchandising of Difference.* Cambridge: Harvard University Press.

Fee, Elizabeth 1979 "Nineteenth-Century Craniology: The Study of the Female Skull." *Bulletin of the History of Medicine* 53:415–33.

France, Kim 1992 "Tits 'R' Us." *Village Voice,* March 17, p. 22.

Goldin, Nan 1993 *The Other Side.* New York: Scalo.

Gould, Stephen Jay 1981 *The Mismeasure of Man.* New York: Norton.

Halberstam, Judith 1994 "F2M: The Making of Female Masculinity." In *The Lesbian Postmodern,* ed. Laura Doan, pp. 210–28. New York: Columbia University Press.

Hohmann, Delf Maria 1985 "Jennifer and Her Barbies: A Contextual Analysis of a Child Playing Barbie Dolls." *Canadian Folklore Canadien 7*(1–2): 111–20.

Hrdlicka, Ales 1925 "Relation of the Size of the Head and Skull to Capacity in the Two Sexes." *American Journal of Physical Anthropology* 8:249–50.

Hrdlicka, Ales 1939 *Practical Anthropometry.* Philadelphia: Wistar Institute of Anatomy and Biology.

Jones, Lisa 1991 "Skin Trade: A Doll Is Born." *Village Voice,* March 26, p. 36

Kahn, Alice 1991 "A Onetime Bimbo Becomes a Muse." *New York Times,* September 29.

Kaw, Eugenia 1993 "Medicalization of Racial Features: Asian American Women and Cosmetic Surgery." *Medical Anthropology Quarterly 7*(1): 74–89.

Lawson, Carol 1993 "Toys Will Be Toys: The Stereotypes Unravel." *New York Times.* February 11, sec. C, pp. 1, 8.

Lord, M. G. 1994 *Forever Barbie: The Unauthorized Biography of a Real Doll.* New York: William Morrow.

"Material Girl: The Effacements of Postmodern Culture." In *The Female Body: Figures, Styles, Speculations.* Ed. Laurence Goldstein, pp. 106–30. Ann Arbor, Mich.: The University of Michigan Press.

May, Elaine Tyler 1988 *Homeward Bound: American Families in the Cold War Era.* New York: Basic Books.

Morse, Margaret 1987 "Artemis Aging: Exercise and the Female Body on Video." *Discourse 10*(1987/88): 20–53.

Motz, Marilyn Ferris 1983 "I Want to Be a Barbie Doll When I Grow Up: The Cultural Significance of the Barbie Doll." In *The Popular Culture Reader,* 3d ed. Ed Christopher D. Geist and Jack Nachbar, pp. 122–36. Bowling Green: Bowling Green University Popular Press.

Rabinowitz, Paula 1993 Accessorizing History: Girls and Popular Culture. Discussant Comments, Panel #150: Engendering Post-war Popular Culture in Britain and America. Ninth Berkshire Conference on the History of Women. Vassar College, June 11–13, 1993.

Russett, Cynthia Eagle 1989 *Sexual Science: The Victorian Construction of Womanhood.* Cambridge: Harvard University Press.

Schiebinger, Londa 1989 *The Mind Has No Sex?: Women in the Origins of Modern Science.* Cambridge: Harvard University Press.

Schiebinger, Londa 1993 Nature's Body: Gender in the Making of Modern Science. Boston: Beacon.

Schwartz, Hillel 1986 *Never Satisfied: A Cultural History of Diets, Fantasies and Fat.* New York: Free Press.

Shapiro, Eben 1992 "'Totally Hot, Totally Cool.' Long-Haired Barbie Is a Hit." *New York Times.* June 22, sec. D, p. 9.

Shapiro, Harry L. 1945 *Americans: Yesterday, Today, Tomorrow.* Man and Nature Publications. (Science Guide No. 126). New York: The American Museum of Natural History.

Spencer, Frank 1992 "Some Notes on the Attempt to Apply Photography to Anthropometry during the Second Half of the Nineteenth Century" In *Anthropology and Photography,* 1860–1920. Ed. Elizabeth Edwards, pp. 99–107. New Haven: Yale University Press.

Sprague Zones, Jane 1989 "The Dangers of Breast Augmentation." *The Network News* (July/August), pp. 1, 4, 6, 8. Washington, D.C.: National Women's Health Network.

Stevenson, Richard 1991 "Mattel Thrives as Barbie Grows." *New York Times.* December 2.

Terry, Jennifer C. 1992 Siting Homosexuality: A History of Surveillance and the Production of Deviant Subjects (1935–1950). Ph.D. diss., University of California at Santa Cruz.

Williams, Lena 1992 "Woman's Image in a Mirror: Who Defines What She Sees?" *New York Times,* February 6, sec. A, p. 1, sec. B, p. 7.

Willis, Susan 1991 *A Primer for Daily Life.* London and New York: Routledge.

Wilson, Elizabeth 1985 *Adorned in Dreams: Fashion and Modernity.* London: Virago.

Wiseman, C., J. Gray, J. Mosimann, and A. Ahrens 1992 "Cultural Expectations of Thinness in Women: An Update." *International Journal of Eating Disorders 11*(1): 85–89.

Wolf, Naomi 1991 *The Beauty Myth: How Images of Beauty Are Used against Women.* New York: William Morrow.

Introduction to Reading 23

This reading is a good example of the application of intersectional analysis, employing categories of gender, age, and social class. The authors studied a mass-marketed program of so-called "successful aging" that targets old men in an effort to persuade them to spend their money on products and activities that will supposedly make them look and feel youthfully and heterosexually virile and successful. Toni Calasanti and Neal King analyze the ageism of "successful aging" consumer campaigns and their implications for old men's "physical health, unequal access to wealth, heterosexual dominance, and fears of impotence" (from abstract).

1. How does ageism permeate "successful aging" consumer campaigns?
2. Why is it important to examine age relations and their intersections with other inequalities?
3. Discuss the "dirty"/"impotent" double bind and its link to the rise of "successful aging" consumer programs.

Firming the Floppy Penis

Age, Class, and Gender Relations in the Lives of Old Men

Toni Calasanti and Neal King

The rise of a consumer market that targets old people and their desire to remain young brings into sharp relief the problems that old age poses to manhood. This article proposes an expansion of research approaches to the lives of old men so that they may enrich our understandings of masculinities at a time when scientific breakthroughs and high-priced

Calasanti, T., & King, N. 2005. "Firming the floppy penis: Age, class, and gender relations in the lives of old men," *Men and Masculinities 8*(1), p. 3. Reprinted with permission of Sage Publications, Inc.

regimens sell visions of manhood renewed. We begin with a brief review of the (relative lack of) research on old men, continue with a look at the mass marketing of "successful aging," and conclude with an overview of the potential rewards that sustained scholarship on the old, and a theorizing of age relations as a dimension of inequality, can offer the studies of men and masculinities.

(Young) Men's Studies

Studies of old men are common in the gerontological literature, but those that theorize masculinity remain rare. As in many academic endeavors, men's experiences have formed the basis for much research, but this androcentric foundation goes largely unexplored because manhood has served as invisible norm rather than as explicit focus of theory. Men's lives have formed the standard for scholarship on retirement, for example, to such an extent that even the *Retirement History Study,* a longitudinal study conducted by the Social Security Administration, excluded married women as primary respondents (Calasanti 1993). In recent years, feminist gerontologists have urged that scholars examine not only women but gender relations as well, and a handful of scholars such as Woodward (1999), Cruikshank (2003), and Davidson (2001) have done so. Despite the proliferation of feminist theorizing, however, most mainstream gerontological studies of women still ignore gender (Hooyman 1999), and research on men lags further. Few studies examine old men *as men* or attend to masculinity as a research topic.

At the same time, profeminist studies of masculinity have studied neither old men nor the age relations that subordinate them. Ageism, often inadvertent, permeates this research, stemming from failures to study the lives of old men, to base questions on old men's accounts of their lives, or to theorize age the way we have theorized relations of gender, race, and class. Mentions of age inequality arise as afterthoughts, usually at the ends of lists of oppressions, but they remain unexamined. As a result, our understanding and concepts of manhood fall short because they assume, as standards of normalcy, men of middle age or younger. Aging scholars' inattention to old *men,* combined with men's studies' lack of concern with *old* men, not only renders old men virtually invisible but also reproduces our own present and future oppression. This article examines a range of popular representations of old men in the context of research about their lives to outline some ways in which the vital work on men and masculinity might benefit by taking age relations into account as a form of inequality that intersects with gender, race, sexuality, and class.

Denial of Aging

Our ageism—both our exclusion of the old and our ignorance of age relations as an inequality affecting us all—surfaces not only in our choices of what (not) to study but also in how we theorize men and masculinities. Listening to the old and theorizing the inequality that subordinates them require that we begin with elementary observations. People treat signs of old age as stigma and avoid notice of them in both personal and professional lives. For instance, we often write or say "older" rather than "old," usually in our attempts to avoid negative labels. But rather than accept this stigma attached to the old and help people to pass as younger than that, we should ask what seems so wrong with that stage of life. In a more aggregate version of this ageism, one theorizes old age as social construction and then suggests that people do not automatically become old at a particular age. One continues to treat "old age" as demeaning and merely seeks to eradicate recognition of it by granting reprieves from inclusion in the group. As well intended as such a theoretical move may be, it exacts a high price. It maintains the stigma rather than examining or removing it. As Andrews (1999) observed, all life cycle stages are social constructions, but "there is not much serious discussion about eliminating infancy, adolescence, or adulthood from the developmental landscape. It is only old age which comes under the scalpel" (302). Emphasis on the socially constructed status of this age category does nothing to eliminate its real-world consequences.

Old age has material dimensions, the consequences of actors both social and biological: bodies *do* age, even if at variable rates, just as groups categorize and apportion resources accordingly. Emphasizing their subjective nature makes age categories no less real. Bodies matter; and the old are not, in fact, just like the middle-aged but only older. They are different, even though cultures and people within them define the differences in divergent ways. We need to consider the social construction of old age in conjunction with the aging of bodies (which, in a vexing irony, we understand only through social constructions).

Successful Aging

A more refined form of ageism attempts to portray old age in a positive light but retains the use of middle age as an implicit standard of goodness and health, in contrast to which the old remain deviant. One may see this ageism in the popular notion that men should "age successfully." From this "anti-aging" perspective, some of the changes that occur with age might seem acceptable—gray hair and even, on occasion, wrinkles—but other age-related changes do not, such as losses of libido, income, or mobility. Aging successfully requires that the old maintain the activities popular among the middle-aged. Successful aging, in effect, requires well-funded resistance to culturally designated markers of old age, including relaxation. Within this paradigm, those signs of seniority remain thoroughly stigmatized.

To be sure, a research focus on men who have aged "successfully" flows from good intentions. Study of successful agers helps us negate stereotypes of the old as "useless," unhappy, and the like. Nevertheless, a theory of the age relations underlying this movement must recognize their interrelations with class, sexual, and racial inequalities. The relevant standards for health and happy lifestyles have been based on leisure activities accessible only to the more well-to-do and middle-aged: tennis, traveling, sipping wine in front of sunsets, and strolls on the beaches of tony resorts that appear in the advertising campaigns for such lifestyles.

The dictate to age successfully by remaining active is both ageist and ignorant of the lives of the working classes. Spurred by the new anti-aging industry, the promotional images of the "active elder" are bound by gender, race, class, and sexuality. The sort of consumption and lifestyles implicated in ads for posh retirement communities with their depiction of "'imagineered' landscapes of consumptions marked by 'compulsively tidy lawns' and populated by 'tanned golfers'" (McHugh 2000, 110) assumes a sort of "active" lifestyle available only to a select group: men whose race and class make them most likely to be able to afford it, and their spouses.

Regimens of successful aging also encourage consumers to define any old person in terms of "what she or he is no longer: a mature productive adult" (McHugh 2000, 104). One strives to remain active to show that one is not really old. In this sense, successful aging means not aging and not being old because our constructions of old age contain no positive content. Signs of old age continue to operate as stigma, even in this currently popular model with its many academic adherents. The successful aging movement disapproves implicitly of much about the lives of the old, pressuring those whose bodies are changing to work hard to preserve their "youth" so that they will not be seen as old. As a result, the old and their bodies have become subject to a kind of disciplinary *activity.* This emphasis on productive activity means that those who are chronically impaired, or who prefer to be contemplative, become "problem" old people, far too comfortable just being "old" (Katz 2000; Holstein 1999).

This underlying bias concerning successful aging and "agelessness" is analogous to what many white feminists have had to learn about race relations, or indeed many men have had to learn about gender relations. Many whites began with the notion that nonwhites were doing fine as long as they acted like whites (just as women in many workplaces were deemed OK to the extent that they acted like men). That actual diversity would benefit our society was news to many, its recognition hard-won by activists of color who championed an awareness of the structuring effects of race relations. Only when we can acknowledge and validate these constructed differences do we join the fights against racism and sexism. The same is true of age relations and the old. We must see the old as legitimately different from the middle-aged, separated by a systematic inequality—built on some set of biological factors—that affects all of our lives. To theorize this complex and ever-changing construction is to understand age relations.

The experience of ageism itself varies by gender and other social inequalities (just as the experience of manhood varies by age and the like). Others have already pointed to the double standard of aging whereby women are seen to be old sooner than men (Calasanti and Slevin 2001). But the experience of ageism varies among different social hierarchies. Women with the appropriate class background, for instance, can afford to use various technologies to "hide" signs of aging bodies (such as gray hair and wrinkles) that will postpone their experiences of ageism. Some women of color, such as African Americans, accept more readily the superficial bodily signs of aging that might bother middle-class white women. Within their communities, signs of aging may confer a status not affirmed in the wider culture (Slevin and Wingrove 1998). By failing to reflect on our own ageism and its sources, we have left age relations and its intersections with such other inequalities unquestioned and misunderstood. We have given lip service to age relations by placing it on a list of oppressions, but we have only begun to theorize them. And so we have left unexplored one of the most important systems shaping manhood.

Examining age relations and its intersections with other inequalities will allow us to address ageism in its deepest form and address the structural inequities that deny power to subgroups of the old. It involves breaking the ethical hold that successful, active aging has on our views of aging. Just as feminists have argued for women's emancipation from stigmatizing pressure to avoid the paths that they might like to take, so too must the old be free to choose ways to be old that suit them without having to feel like slackards or sick people. Old age should include acceptance of inactivity as well as activity, contemplation as well as exertion, and sexual assertiveness as well as a well-earned break. Old people will have achieved greater equality with the young when they feel free not to try to be young, when they need not be "exceptional," and when they can be frail, or flabby, or have "age spots" without feeling ugly. *Old* will have positive content and not be defined mainly by disease, mortality, or the absence of economic value.

Old Men in Popular Culture

The study of masculinity benefits from a look at mass-produced images of old men, because they suggest much about the changing definitions of their problems and the solutions offered. Viewed in context of the experiences of diverse old men as well as the structural constraints on various groups, these popular images illustrate the pressures to be masculine and ways in which men respond to accomplish old manhood. On one hand, the goal of consumer images is to convince others to buy products that will help them better their lives. What is instructive about such images is what they reveal about how people—in this case, aging men—should go about improving their lives (i.e., what it is that they should strive for). On the other hand, images of powerful older men—such as CEOs and politicians—periodically appear in the news media, demonstrating what old men should be striving for in the consumer ads: money, power, and the like. We use mass-produced images of old men, then, to explore the ways that men and masculinities intersect with other systems of inequality—including age relations—to influence various experiences of manhood.

Current Images: New Manhood in Old Age

The recent demographic shift toward an aged population has inspired consumer marketers to address the old with promises of "positive" or successful aging. A massive ad campaign sells anti-aging–the belief that one should deny or defy the signs and even the fact of aging, and treat the looks and recreation of middle-aged as the appropriate standards for beauty, health, and all around success. As Katz (2001–2002) recently put it, "The ideals of positive aging and anti-ageism have come to be used to promote a widespread anti-aging culture, one that translates their radical appeal into commercial capital" (27). These ads present a paradox for old men, whom ads depict as masculine but unable by virtue of infirmity and retirement to achieve the hegemonic ideals rooted in the lives of the young. Thus, old masculinity is always wanting, ever in need of strenuous affirmation. Even when blessed with the privileges of money and whiteness, old men lack two of hegemonic masculinity's fundamentals: hard-charging careers and robust physical strength. The most current ads promise successful aging with interesting implications for these forms of male privilege.

"Playing Hard"

The first image in this "new masculinity" shows men "playing hard," which differs from previous ads in important ways. It emphasizes activities modeled after the experiences of middle-aged, white, middle-class men. Men pursue leisure but not in terms of grandparenting, reading, or other familial and relaxing pastimes. Instead, they propel themselves into hard play as consumers of expensive sports and travel. Having maintained achievement orientations during their paid-work years, they now intensify their involvement in the expanding consumerist realm, trading production or administration for activity-based consumption. They compete not against other men for salaries and promotions but against their own and nature's incursions into their health as they defy old age to hobble them.

Katz (2001–2002) noted that many ads portray the "older person as an independent, healthy, flexi-retired 'citizen'; who bridges middle age and old age without suffering the time-related constraints of either. In this model . . . 'retirement is not old age'" (29). For instance, McHugh observed that the marketing of sunbelt retirement communities includes the admonition to seniors to busy themselves in the consumption of leisure, to "rush about as if their very lives depended upon it" (McHugh 2000, 112). Similarly, Aetna advertisements selling retirement financial planning show pictures of retired men in exotic places, engaging in such activities as surfing or communing with penguins. Captions offer such invitations as

> Who decided that at the age of 65 it was time to hit the brakes, start acting your age, and smile sweetly as the world spins by? . . . [W]hen you turn 65, the concept of retirement will be the only thing that's old and tired. (*Newsweek* January 5, 1998, 9)

This active consumer image reinforces a construction of old age that benefits elite men in two ways. First, it favors the young in that the old men pictured do nothing that would entitle them to pay. Instead, they purchase expensive forms of leisure. Readers can infer that old men neither need money nor deserve it. Retired, their roles center around spending their money (implicitly transferring it to the younger generations who do need and deserve it). Such ads affirm younger men's right to a cushion from competition with senior men for salaried positions, power, and status. Second, this active consumer image favors the monied classes by avoiding any mention of old men's financial struggles or (varied) dependence on the state. Indeed, age relations work to heighten economic inequalities, such that the greatest differences in income and wealth appear among the old (Calasanti and Slevin 2001). This polarization of income and wealth creates a demographic situation in which only the most privileged men—white, middle-class or better, and physically similar to middle-aged men—can engage in the recreation marketed.

There, we see an additional benefit to the young of such images of men—the emphasis on the physical abilities that the young are more likely to have. Featherstone and Hepworth (1995) noted that the consumer images of "positive aging" found in publications for those of retirement age or planning retirement ultimately have "serious shortcomings" because they do not counter the ageist meanings that adhere to "other" images of the old, that is, "decay and dependency." In other words, we look more kindly on those old persons engaged in "an extended plateau of active middle age typified in the imagery of positive aging as a period of youthfulness and active consumer lifestyles" (46). In this sense, the new, "positive," and consumer-based view of the old is one steeped in middle-aged, middle-class views and resources. The wide variety of retirement and other magazines—and, more recently, a large and expanding number of Web sites—convey the idea that the body can be "serviced and repaired, and . . . cultivate the hope that the period of active life can be extended and controlled" through the use of a wide range of advertised products (44). This image does not recognize or impute value to those more often viewed to be physically dependent, for example. As a result, those men who are able to achieve this masculine version of "successful aging" appear acceptable within this paradigm, but this new form of acceptance does not mitigate the ways in which we view the old. It denies the physical realities of aging and is thus doomed to failure. Not only are the majority of old men left out of this image of new masculinity for old men, but also the depiction is in itself illusory and transitory. Note the gender inequality in these depictions of aging denied through consumption. Most women participate in the lifestyles of the well-to-do as parts of married couples, dependent on men. Old men may lose status relative to younger men but still maintain privilege in relation to old women.

However hollow such promises of expensive recreation might be for most men, the study of men's physical aggression and self-care suggests that illusions drive many indeed and that men will often sacrifice health and even their lives to accomplish this exaggerated sense of physical superiority to women and resistance to the forces of nature. Researchers of health, violence, and manhood have already documented the harms that men do to themselves. Whether disenfranchised men of color in neighborhoods of concentrated poverty (Franklin 1987; Lee 2000; Staples 1995), athletes desperate to perform as champions (Dworkin and Messner 1999; Klein 1995; White, Young, and McTeer 1995), or ordinary men expressing rage through violence (Harris 2000) and refusing to consult physicians when ill (Courtenay 2000), all manner of men undercut themselves and endanger their lives in the pursuit of their ideals. Harris (2000, 782), for instance, referred to the violence as part of the "doing" of manhood, in line with the sociological theory of gender as accomplishment (Fenstermaker and West 2002). Injury in the pursuit of masculinity extends to social networks, which men more often than women neglect to the point of near isolation and desolation (Courtenay 2000). For those not killed outright, the accumulated damage results in debilitating injury and chronic disease leading to depression (White, Young, and McTeer 1995; Charmaz 1995), fatal heart disease (Helgeson 1995), and high rates of suicide born of lonely despair (Stack 2000). The effect of all of this on old manhood is tremendous, with men experiencing higher death rates than women at every age except after age ninety-five (Federal Interagency Forum on Aging-Related Statistics 2000), at which point few men remain alive.

More important to this discussion, however, than the results of such self-abuse on old age are the effects of age relations on this doing of manhood. To be sure, criminal combat and bone-crunching sports decline with age (much earlier in life, actually) such that old men commit few assaults and play little rugby. The increasing fragility of their bodies leads to relatively sedate lifestyles. Nevertheless, the recent anti-aging boom sells the implicit notion that relaxation equals death or at least defeat and that, once he retires, only high-priced recreation keeps a man a man. Age and gender ideals to which any man can be held accountable shift from careerism to consumption, from sport to milder recreation, but maintain notions of performance all the while.

The theoretical gain here lies in recognizing the historical (and very recent) shift to old manhood as a social problem solved through the consumption of market goods. Men throughout history and across the globe appear always to feel defensive about manhood, in danger of losing or being stripped of it (Solomon-Godeau 1995). This theme takes different forms in different periods, however, and in our own appears as the notion that old men lose their hardness if they relax but can buy it back from leisure companies and medical experts.

"Staying Hard"

Given the importance of heterosexuality to hegemonic masculinity, we should consider the ways in which age and gender interact with sexuality, so often equated for men with "the erect phall[us]" (Marsiglio and Greer 1994, 126). Although graceful acceptance by men of their declining sexual desire had previously served as a hallmark of proper aging (Marshall and Katz 2002), current depictions of old men's masculinity focus on virility as expressed in a (hetero) sexuality enabled by medical products. "Staying hard" goes hand-in-hand with playing hard in the construction of age-appropriate gender ideals in this consumer economy.

Examples of the link among continued sexual functioning, manhood, and resistance to aging, in a context of individual responsibility and control, appear throughout the anti-aging industry, which has been growing as a part of our popular culture through the proliferation of Web sites, direct-mail brochures, journal and magazine advertisements, blurbs in academic newsletters, appearances on talk shows and infomercials, self-help paperbacks, and pricey seminars designed to empower the weakening old. For instance, a few passages from *Newsweek* (Cowley 1996) on the movement toward the use of human growth hormone (HGH) and testosterone draw connections among virility, aging, masculinity, individual control, and consumerism.

> Five years ago, on the eve of his 50th birthday, Ron Fortner realized that time was catching up with him. . . . His belly was soft, his energy and libido were lagging and his coronary arteries were ominously clogged up. After his advancing heart disease forced him into a quintuple bypass operation, Fortner decided he wasn't ready to get old. He . . . embarked on a hormone-based regimen designed to restore his youthful vigor. . . . [H]e started injecting himself with human growth hormone. . . . He claims the results were "almost instantaneous." First came a general sense of well-being. Then within weeks, his skin grew more supple, his hair more lustrous and his upper body leaner and more chiseled. . . . Awash in all these juices, he says he discovered new reserves of patience and energy, and became a sexual iron man. "My wife would like a word with you," he kids his guru during on-air interviews, "and that word is stop." (Cowley 1996, 68, 70)

Significantly, a yearlong supply of HGH in 1996 ran between $10,000 and $15,000, making it most accessible to elite men.

Another "success story" from the article concerns

> Robert, a 56-year-old consultant who wore a scrotal patch [for testosterone] for two and a half years. . . . Since raising his testosterone level from the bottom to the top of the normal range, Robert has seen his beard thicken, his body odor worsen and his libido explode. "Whether it's mental or physical, you start feeling older when you can't do physical things like you could," he says. "Sexually, I'm more comfortable because I know I'm dependable." His only complaint is that he's always covered with little rings of glue that won't come off without a heavy-duty astringent. (Cowley 1996, 71–72)

Finally, the story concludes by noting that

> as the population of aging males grows, the virility preservation movement is sure to grow with it.

> "Basically, it's a marketing issue," says epidemiologist John McKinley, director of the New England Research Institute. . . . "The pharmaceutical industry is going to ride this curve all the way to the bank" (Cowley 1996, 75)

Scientific discourse and practice equate, especially for men, sex with "not aging," and propose technology to retain and restore sexual "functionality" (Katz and Marshall 2003). Indeed, as anti-aging guru Dr. Karlis Ullis, author of *Age Right* and *Super T* (for testosterone), proclaims on his Web site, "Good, ethical sex is the best anti-aging medicine we have" (2003). The appearance of such chemical interventions as sildenafil (Viagra) and the widespread advertising campaigns to promote them have also helped to reconstruct old manhood. A recent ad shows an old, white, finely dressed couple dancing a tango, with the man above and the woman leaning back over his leg. The strenuous dance combines with the caption to convey his virility: "Viagra: Let the dance begin." (*Good Housekeeping* April 1999, 79). Here is a man who likes to be on top and has the (newly enhanced) strength to prove it. Still another ad affirms the role of phallic sex in marital bliss. The bold letters next to a black man with visibly graying hair state, "With Viagra, she and I have a lot of catching up to do." And, at the bottom: "Love life again" (*Black Enterprise* March 2000, 24–5).

Such ideals of virility appear in age-defying ads for active leisure—such as one for Martex towels, which features the caption, "Never, ever throw in the towel." Below this line, three old men stand, towels around their waists, in front of three surfboards that stand erect, stuck in the beach sand. Beneath, one reads that the towels are "for body and soul" *(Oprah* April 2001, 118). An Aetna financial planning ad shows an old white man paddling in the surf, his erect board standing upward between his legs. The caption reads, "A Rocking Chair Is a Piece of Furniture. Not a State of Mind" *(Newsweek* October 27, 1997, 15). In the ideal world of these ads, age is a state of mind, one to be conquered through public displays of a phallic, physical prowess. One accomplishes old manhood, then, by at least appearing to try to live up to some of the ideals pictured in these magazines. The resulting widespread doing of old manhood as consumption of the right products and maintenance of the right activities serves in turn to render natural the ideals toward which men strive.

Masculinity and sexual functioning have long been linked to aging in our popular culture, but the nature of his relationship has shifted as age relations have transformed and come under medical authority. Contemporary drug marketers build on an ancient quest but market it in new ways.

By the 1960s, therapists blamed psychological factors for male impotence and suggested that "to cease having sex would hasten aging itself" (Katz and Marshall 2003, 7). They later redefined male impotence as a physiological event— "erectile dysfunction"—to be addressed through such technologies as penile injections and sildenafil (Viagra)—and declared intercourse vital to successful aging (Marshall and Katz 2002; Potts 2000). More recently, advertisers have catered to a popular notion of "male menopause"—an umbrella label for the consequences of the fears of loss that expectations of high performance, in the context of women's rising status, can engender (Featherstone and Hepworth 1985). Marketers have built their depictions of old manhood on these links among sex, success, and masculinity. Sexual functioning now serves as a vehicle for reconstructions of manhood as "ageless," symbolizing the continued physical vigor and attractiveness derived from the experiences of younger men. To the extent that men can demonstrate their virility, they can still be men and stave off old age and the loss of status that accrues to that label.

To be sure, this shift in advertising imagery toward the phallic can work to the benefit of old men, convincing people to take them seriously as men full of potency as well as consumer power. To stop our analysis there, however, leaves unquestioned the ageism on which these assertions rest, the fact that we root these ideals of activity and virility in the experiences of the younger men. The ads avoid sexuality based on attributes other than hard penises and experiences other than heterosexual intercourse, and these are hegemonic sexual symbols of the young. The little research available suggests that orgasm and intercourse recede in importance for some old men, who turn to oral sex and other expressions of love (Wiley and Bortz 1996). But these phallic ads value men only to the extent that they act like younger, heterosexual (and wealthy) men. Their emphases on both playing and staying hard reveal some of the ways in which gender and other inequalities shape old age. Old men are disadvantaged in relation to younger men, no matter how elite they may be.

The renewed emphasis on sexual intercourse among old men also reinforces the gender inequalities embedded in phallic depictions of bodies and sexuality. Historically, women's bodies and sexualities have been of only peripheral interest in part because they did not fit the "scientific" models based on men's physiologies. For example, rejuvenators were uncomfortable touting sex gland surgery for women (one variation promoted grafting the ovaries of chimpanzees to those of female patients) partly because they knew that they could not restore fertility in women. Thus, when they did speak of women, they tended to focus instead on the "mental" fertility that might result. Part of the problem was that women's "losses" in terms of sexuality (i.e., menopause) occurred much earlier in life. Those women were often "young," which confounded the equation of "loss of sexuality" with "old" (Hirshbein 2000).

People continue to define old women's sexuality in relation to old men's, assessing it in terms of penile-vaginal penetration. An old woman, in such popular imagery, remains passive and dependent on her man's continued erection for any pleasure of her own. Research on old women's accounts of their experiences, however, makes clear that these models represent little of what they want from their sex lives. These popular definitions also ignore that many old women have no partners at all. Even if old women "accept" and try to live up to the burden of being sexual and "not old" in male-defined terms, there are not enough old men for them to be partnered (and our age-based norms do not allow them to date younger men).

Finally, the ageism implicit in the demand to emulate the young is self-defeating and ignores the reality that even with technology and unlimited resources, bodies still change. Ultimately, individuals cannot control this; it is a "battle" one cannot win.

Theorizing Age, Class, and Gender Relations

The rewards for the inclusion of a marginalized group into research extend beyond the satisfaction of listening to oft-ignored voices. The study of old manhood stands to enrich our theories of masculinity as social problem, as disciplinary consumer object, as the accomplishment of heterosexuality, and as the "crisis"-torn struggle to achieve or resist the hegemonic ideals spread through our popular culture.

Studying age relations can render insights into ways that we theorize gender. For instance, Judith Kegan Gardiner (2002) suggested that we clarify gender relations by making an analogy to age relations. This would help reconstruct thinking about gender in our popular culture, she argued, because many people already recognize *continuity* in age categories while they still see gender as dichotomous. People already see themselves as *performing* age-appropriate behavior ("acting their ages") while continuing to take for granted the doing of gender (Fenstermaker and West 2002). And popular culture more fully recognizes enduring group *conflicts* (over divisions of resources) between generations than between sexes. Gardiner (2002) suggested that a fuller theorizing of age relations has much to offer the study of men, that scholars may move beyond their polarization of biological and social construction, and that our popular culture may more fully appreciate the power struggles that govern gender relations.

We recommend just this view—of age and gender, race and class, and other dimensions of inequality—as accomplished by social as well as biological actors; as accountable to ever-changing ideals of age- and sex-appropriate behavior; as constructed in the context of a popular culture shaped by consumer marketing and technological change; and as imposing disciplinary regimens in the names of good health, empowerment, beauty, and success.

Taken together, the mass media reviewed above posit ideals of old manhood to which most if not all men find themselves held accountable. To the men fortunate enough to have been wealthy or well paid for their careerism, corporations (often with the support of those gerontologists who implicitly treat old age as a social problem) sell regimens through which those old men may live full lives, working, playing, and staying hard. If careerism kept the attention of these men from their families and leisure lives, constricting their social networks and degrading their physical health, then this high-priced old age serves as a promised payoff. Once retired, those few wealthy enough to do it can enjoy a reward: high-energy time with a spouse and some friends, enjoyment of tourism, surfing, and sex. Men sacrificed much, even their lives, in their pursuits of hegemonic masculine status. Those who survive face a rougher time with old age as a result: few sources of social support and bodies weakened by self-abuse. Thus, the accomplishment of manhood comes to require some response to the invitation to strain toward middle-age activities. Some men reach with all of their strength for the lifestyle ideals broadcast so loudly, whereas many give up for lack of means to compete, and still others deliberately resist. In a cruel irony, the ideals move all the further out of reach of the men who pursued them with such costly vigor in younger years and damaged their health beyond repair. The final push for hegemonic masculinity involves spending money and enjoying health that many old men do not have to pursue the recreation and phallic sex that the ads tell them they need.

Certainly, the study of old men offers striking views of a popular struggle over heterosexuality (although the study of old gay men will surely be as transforming, the near total lack of research on them prevents us from speculating how). Widely held views of old men's sexuality suggest dominance over women as a form of virility. But, as bodies change, outright predation recedes as an issue and impotency moves to the center of concern. A popular (consumer) culture that figures old manhood in terms of *loss* hardly departs from any trend in images of masculinity. Men have always felt that they were losing their manhood, their pride, and their virility, whether because their penises actually softened or because women gained status and so frightened them. But the study of this transition—from the feelings of invincibility that drive the destructiveness of youth to the growing expectation of vulnerability —throws old masculinity into a valuable relief. For instance, theories that center on violence and predation capture little of the realities of old men's lives, just as scholarly emphasis on coercion and harassment of women excludes most of the experiences of old women. For old women, the more important sexual theme may be that of being *cast aside* (Calasanti and Slevin 2001, 195). For old men, *impotence* in its most general sense, leading to many responses ranging from suicidal depression to more graceful acceptance, may be a more productive theme. It serves as both positive and negative ideal in a classic double-bind: old men should, so as not to intrude on the rights of younger men, retreat from the paid labor market; but they should also, so as to age successfully, never stop consuming opportunities to be active. They should, so as not to be "dirty," stop becoming erect; but they should also, so as to age successfully, never lose that erection. Old men fear impotence to the point that many suffer it who otherwise would not. Anxieties drain them at just the moment when expectations of aggressive consumption, of proving themselves younger than they are, reach their heights.

The notion that men accomplish age just as they do gender has much to offer, with its sensitivity to relations of inequality, its moment-to-moment accountability to unreachable but hegemonic ideals, and the perpetually changing nature of such accomplishments. Never have erections been so easily discussed in public, and never has this "dirty"/"impotent" double bind been tighter, than since the rise of this consumer regimen. Nor have old men, before now, lived under such pressure to remain active further into their lengthening life spans. The ideals of manhood that tempted so many to cripple themselves in younger years now loom large enough to shame those who cannot play tennis or waltz the ballrooms of fancy resorts. The study of manhood should take careful notice of the ways in which men do old manhood under such tight constraints. The popular images that we have reviewed provide ideals of old manhood, but they do not necessarily describe the lives of very many old men. Given how little we know of the ways in which old men respond to such ideals, the research task before us seems clear.

Conclusion

Scholars tend to ignore age relations in part because of our own ageism. Most are not yet old, and even if we are, we often deny it (Minichiello, Browne, and Kendig 2000). Most people know little about the old because we seldom talk to them. Family and occupational segregation by age leave the old outside the purview of the work that most young people do.

Resulting in part from such segregation, the study of men, although no more than any other social science and humanist scholarship, has focused on the work, problems, sexuality, and consumption patterns of the young. This neglect of the old results in theories of masculinity that underplay the lengths to which men go to play and stay hard, the long-term effects of their strenuous accomplishment of manhood, and the variety of ways in which men remain masculine once their appetites for self-destruction begin to wane. Research on the old can reveal much about the desperate struggle for hegemonic masculinity and the varied ways in which men begin to redefine manhood. At the same time, it also uncovers the young and middle-aged biases that inhere in typical notions of masculinity that tend to center on accomplishments and power in the productive sphere, for instance. Few researchers have considered the reality of masculinities not directly tied to the fact of or potential for paid labor.

To leave age relations unexplored reinforces the inequality that subordinates the old, an inequality that we unwittingly reproduce for ourselves. Unlike other forms of oppression, in which the privileged rarely become the oppressed, we will all face ageism if we live long enough. As feminists, scientists, and people growing old, we can better develop our sense of interlocking inequalities and the ways in which they shape us, young and old. Our theories and concepts have too often assumed rather than theorized these age relations. The study of men and masculinity and the scholarship on age relations are just beginning to inform each other.

References

Andrews, M. 1999. The seductiveness of agelessness. *Ageing and Society 19*(3): 301–18.

Calasanti, T. M. 1993. Bringing in diversity: Toward an inclusive theory of retirement *Journal of Aging Studies 7*(2): 133–50.

Calasanti, T. M., and K. F. Slevin. 2001. *Gender, social inequalities, and aging.* Walnut Creek, CA: Alta Mira.

Charmaz, K. 1995. Identity dilemmas of chronically ill men. In *Men's health and illness: Gender, power, and the body,* edited by D. Sabo and D. F. Gordon, 266–91. Thousand Oaks, CA: Sage.

Courtenay, W. H. 2000. Behavioral factors associated with disease, injury, and death among men: Evidence and implications for prevention. *Journal of Men's Studies 9*(1): 81–142.

Cowley, Geoffrey. 1996. Attention: Aging Men. *Newsweek,* September 16, 68–75.

Cruikshank, M. 2003. *Learning to be old: Gender, culture. and aging.* Lanham, MD: Rowman & Littlefield.

Davidson, K. 2001. Later life widowhood, selfishness, and new partnership choices: A gendered perspective. *Ageing and Society* 21:297–317.

Dworkin, S. L., and M. A. Messner. 1999. Just do . . . what? Sport, bodies, gender. In *Revisioning gender,* edited by M. M. Ferree, J. Lorber, and B. B. Hess, 341–61. Thousand Oaks. CA: Sage.

Estes. Carroll L., and Elizabeth A. Binney. 1991. The biomedicalization of aging: Dangers and dilemmas. In *Critical perspectives on aging: The political and moral economy of growing old,* edited by Meredith Minkler and Carroll L Estes, 117–34. Amityville, NY: Baywood.

Featherstone, M., and M. Hepworth. 1985. The male menopause: Lifestyle and sexuality. *Maturitas* 7:235–46.

———. 1995. Images of positive aging: A case study of *Retirement Choice magazine.* In *Images of aging: Cultural representations of later life,* edited by M. Featherstone and A. Wernick, 29–47. London: Routledge.

Federal Interagency Forum on Aging-Related Statistics. 2000. *Older Americans 2000: Key indicators of well-being.* http://www.agingstats.gov.

Fenstermaker, S., and C. West, eds. 2002. *Doing gender, doing difference: Inequality, power, and institutional change.* New York: Routledge.

Franklin, C. 1987. Surviving the institutional decimation of black males: Causes, consequences, and intervention. In *The making of masculinities: The new men's studies,* edited by H. Brod, 155–69. Winchester, MA: Allen and Unwin.

Gardiner, J. K. 2002. Theorizing age and gender: Bly's boys, feminism, and maturity masculinity. In *Masculinity studies & feminist theory: New directions,* edited by J. K. Gardiner, 90–118. New York: University of Columbia Press.

Haraway, D. 1992. The promises of monsters: A regenerative politics for inappropriate/d others. In *Cultural studies,* edited by L. Grossberg, C. Nelson, and P. Treichler, 295–337. New York: Routledge.

Harris, A. P. 2000. Gender, violence, and criminal justice. *Stanford Law Review* 52:777–807.

Helgeson, V. S. 1995. Masculinity, men's roles, and coronary heart disease. In *Men's health and illness: Gender, power and the body,* edited by D. Sabo and D. F. Gordon, 68–104. Thousand Oaks, CA: Sage.

Hirshbein, L. D. 2000. The glandular solution: Sex, masculinity and aging in the 1920s. *Journal of the History of Sexuality* 9(3): 27–304.

Holstein, Martha. 1999. Women and productive aging: Troubling implications. In *Critical gerontology: Perspectives from political and moral economy,* edited by Meredith Minkler and Carroll L. Estes, 359–73. Amityville, NY: Baywood.

Hooyman, N. R. 1999. Research on older women: Where is feminism? *The Gerontologist* 39:115–18.

Katz, S. 2000. Busy bodies: Activity, aging, and the management of everyday life. *Journal of Aging Studies 14*(2): 135–52.

———. 2001–2002. Growing older without aging? Positive aging, anti-ageism, and anti-aging. *Generations 25*(4): 27–32.

Katz, S., and B. Marshall. 2003. New sex for old: Lifestyle, consumerism, and the ethics of aging well. *Journal of Aging Studies 17*(1): 3–16.

Klein, A. M. 1995. Life's too short to die small: Steroid use among male bodybuilders. In *Men's health and illness: Gender power and the body,* edited by D. Sabo and D. F. Gordon, 105–21. Thousand Oaks, CA: Sage.

Lee, M. R. 2000. Concentrated poverty, race, and homicide. *Sociological Quarterly 41*(2): 189–206.

Marshall, B., and S. Katz. 2002. Forever functional: Sexual fitness and the ageing male body. *Body & Society 8*(4): 43–70.

Marsiglio, William, and Richard A. Greer. 1994. A gender analysis of older men's sexuality. In *Older men's lives,* edited by Edward H. Thompson, Jr., 122–40. Thousand Oaks, CA: Sage.

McHugh, K. 2000. The "ageless self"? Emplacement of identities in sun belt retirement communities. *Journal of Aging Studies 14*(1): 103–15.

Minichiello, V., J. Browne, and H. Kendig. 2000. Perceptions and consequences of ageism: Views of older people. *Ageing and Society 20*(3): 253–78.

Potts, A. 2000. The essence of the "hard on": Hegemonic masculinity and the cultural construction of "erectile dysfunction." *Men and Masculinities 3*(1): 85–103.

Slevin, K. F., and C. R. Wingrove. 1998. *From stumbling blocks to stepping stones: The life experiences of fifty professional African American women.* New York: New York University Press.

Solomon-Godeau, A. 1995. Male trouble. In *Constructing masculinities,* edited by M. Berger,

B. Wallis, and S. Watson, 69–76. New York: Routledge.

Stack, S. 2000. Suicide: A 15-year review of the sociological literature. *Suicide & Life-Threatening Behavior 30*(2): 145–76.

Staples, R. 1995. Health among Afro-American males. In *Men's health and illness: Gender, power, and the body,* edited by D. Sabo and D.F. Gordon, 121–38. Thousand Oaks, CA: Sage.

Ullis, Karlis. 2003. *Agingprevent.com.* http://www.agingprevent.com/flash/index.html.

White, P. G., K. Young, and W. G. McTeer. 1995. Sport, masculinity, and the injured body. In *Men's health and illness: Gender, power, and the body*, edited by D. Sabo and D. F. Gordon, 158–82. Thousand Oaks, CA: Sage.

Wiley, D., and W. M. Bortz. 1996. Sexuality and aging—usual and successful. *Journal of Gerontology* 51A (3): M142–M146.

Woodward, K., ed. 1999. *Figuring age: Women, bodies, generations.* Bloomington: Indiana University Press.

Introduction to Reading 24

This article examines the intersection of race, gender, and nativity in consumption culture. The authors focus their analysis on advertising campaigns that repackage old racial stereotypes of Asian American women under the guise of corporate multiculturalism. Asian American women are sexually objectified, culturally misrepresented, and offered up as bodies to be visually consumed. The authors maintain that "Orientalism" has become an object to consume and a vehicle to stimulate consumption.

1. Define Orientalism and describe its history in the United States.
2. How do corporations use multiculturalism to market their products?
3. How are Asian American women positioned in relationship to White males in the cultural schemata of corporate advertisements?

Consuming Orientalism

Images of Asian-American Women in Multicultural Advertising

Minjeong Kim and Angie Y. Chung

Research studies have long challenged the ways in which advertising and marketing campaigns employ gendered imagery that objectify women and reinforce power differences between the sexes in order to sell their products (Berger 1977; Betterton 1987; Bordo 1993; Cortese 1999; Goffman 1979; Kilbourn 1999, 2000; Manca and Manca 1994; Williamson 1978, 1986). Among other things, print advertising has been shown to promote images that distort

From Minjeong, K., & Chung, A. Y. 2005. "Consuming Orientalism: Images of Asian/American women in multicultural advertising" *Qualitative Sociology 28*(1). Reprinted with permission of Springer.

women's bodies for male pleasure, condone violence against women, or belittle the women's movement itself as a playful prank. From a historical perspective, however, women of color rarely figured into the marketing campaigns of these companies—partly because of their small numbers as well as their racialized invisibility to mainstream American society. As a result, aside from research on racial stereotypes in the TV and movie entertainment industries (Gee 1988; Hamamoto 1994; Lee 1999), few scholars have fully examined the commodified images of Asian/American women[1] that have come to play an integral role in today's consumer culture industries.

Recent trends in the global economy have transformed the cultural content and marketing strategies of corporate advertising campaigns today as we demonstrate in this study. In particular, these advertising campaigns have sought to diversify their cultural repertoire through the greater inclusion of Asian and Latino/American characters and the invocation of global imageries. However, we will argue that representations of ethnic minority groups in such advertising campaigns are usually based on gendered and racialized reflections of global culture that draw on resurrected themes of colonialism and American Orientalism. This particularly holds true in their depictions of Asian/American women (and the implicit absence or rarity of Asian/American men). On the one hand, it is important to note that images of Asian/American women in advertising are not ahistorical in origin. Oftentimes, they selectively emulate and modify popular images of Asian/American women in the U.S. culture that have been shaped throughout American history. At the same time, this study aims to show how such representations also emerge from the specific "multicultural" and globalized context of post-Civil Rights America that have destabilized and transformed the identities of White males.

This paper will discuss the dynamics of American Orientalism in advertising and its role in reconstructing Asian/American women in relation to White Americans within the globalizing multicultural context of U.S. society. First, it will provide a theoretical context for understanding gendered and racial representations of women in the print media in post-industrial American society. Second, we will show how stereotypical imageries of Asian/American women and commodified Orientalism have evolved in American media culture over time. Third, we will analyze advertisements taken from various magazines that have included Asian/American female characters with specific focus on three multicultural advertising campaigns. In this section, we will show how the marketing of Orientalist images and meanings take shape under the guise of multiculturalism with more detailed explanations of specific race/gender imagery. Based on this analysis, the paper will conclude by showing how Orientalist ideologies have been rearticulated within the context of today's globalized economy.

Consuming Culture in Post-Industrial America

Much of scholarly attention has focused on the construction of corporate marketing and advertising campaigns through a gendered lens (Cortese 1999; Kilbourn 1990, 2000; Manca and Manca 1994; Williamson 1978, 1986), yet most studies oddly leave out an important racial and nativist element of today's global capitalist culture that feeds on the visual consumption of women's bodies. The early representations of America's consumption culture relied heavily on images of middle-class White women whose idealistic roles were defined within the context of the modern domestic economy.

Although many studies have examined the gendered dimensions of this consumption culture, there has been a noticeable lack of research that analyzes today's capitalist culture through the intersections of race, gender and nativity. Various trends in the current post-industrial global economy have underscored not only the consumptive aspects of traditional gender roles, but also the exploitative international machinery upon which this consumption economy is built. The high standards of living that sustain the growing white-collar sector of the American economy are made possible by the employment and exploitation of cheap immigrant labor, particularly women and children from Asia and Latin America. As more and more white-collar workers are integrated into the expanding highly-skilled and professional labor force, there has been a growing need for immigrant labor to take their place in the home as nannies, housekeepers, lawnmowers and even shopping consumers (Chang and Abramovitz 2000; Hothschild and Ehrenreich 2003; Sassen-Koob 1984). At the same time, the steady growth of low-skilled immigrant workers has also been accompanied by an influx of highly-skilled workers and professionals, particularly from Asia—a pattern that marks the polarized nature of the global economy.

But even beyond the realm of professional service, the mainstream cultural economy as a whole has come to rely increasingly on the cheap labor of immigrants in order to sustain its mass production of cheap goods in large-scale industries like Walmart, Gap, and Nike. Immigrant women from Third World countries have figured greatly into the new economic structure, because of their cheaper labor and greater vulnerability to subcontractors who must drive down labor costs in order to maintain their competitive relations with large-scale corporations. The greater flexibility of production in the new era of technology has allowed corporations to export these jobs as well to Third World countries where such workers are abundant and labor regulation laws are poor. Innovative research by Sassen (Sassen-Koob 1984) and other scholars have shown how the feminization of cheap Third World wage labor and the related rise in female immigration to the U.S. have acted as integral cogs in the corporate machinery of post-industrial capitalism. In this manner, the cultural and structural foundations of today's cultural economy still feed on the colonization of the "Other." The gendered impact of the globalized economy is best exemplified by the coinciding expansion of the Asian sex industry, which has opened its doors to businessmen traveling to Asia (Hochschild and Ehrenreich 2003; Jeffreys 1999).

The rising significance of immigration from Asia and Latin America and America's role in the new global economy will inevitably have an effect on the multicultural representations that advertising and marketing campaigns will promote, particularly among their white-collar, professional audience. Multiculturalism is one of the clever marketing strategies that corporations have recently used to market their products. "Multiculturalism" evokes artificial images of racial unity and harmony among the various cultural groups of America and celebrates the general openness of "color-blind" Americans to the rich cultural traditions of different racial groups. The multicultural approach allows corporations to achieve two things: While allowing them to expand their market share to a racially diversifying population of consumers, corporations have also used the visual consumption of women's bodies—and the bodies of women of color in particular—to re-package and obscure the exploitative labor machinery that produces them.

However, the various manifestations of this new consumption culture represent more than just the hegemonic forces of capitalism. Analysis of this post-industrial global culture must also take into account the historical and cultural context within which this system has taken shape in America. The fantastic imagery of a happy, multicultural society has been a key step for Americans who not too long ago eliminated the last remnants of legalized segregation and discrimination during the 1960s Civil Right era. The series of politically tumultuous struggles that led to its ultimate demise left a deep impression on the White American psyche by calling into question its strong belief in the meritocracy and humanity of American democracy and highlighting the ambiguity of its own identity in an era that rallied cultural pride and self-empowerment for non-White groups. One way that White Americans have established a cultural passageway for themselves has been by laying claim to the birth of this new multicultural world and by establishing their role within it. As best exemplified by the commodification of African American hip-hop, corporations are able to disassociate everyday Americans from the structural context of oppression and the historical context of struggle that define the post-industrial world by laying claim to the bodies and cultures of the "Other"

(Giroux 1994; Rose 1994). The cultural landscapes of post-Civil Rights White America in many ways depend largely on this vision of the American melting pot—a trend that has sustained recent political backlash against "anti-color-blind" policies such as affirmative action (Omi 1991).

Within this context, the article examines the cultural representations of corporate marketing campaigns within the contemporary global era with specific attention to their hegemonic outlooks on race, nativity and gender. The article will argue how the multicultural imagery of specific advertising campaigns, while expanding its campaign to include multi-racial characters, relies on the "foreign" and "seductive" appeal of Asian/American women in order to highlight the supremacy and positionality of White men within the global order. As the next section will show, many of the earlier themes of commodified orientalism are replicated in contemporary depictions of Asian/American women; at the same time, our analysis of corporate campaigns will show how they have now been re-contextualized within the multicultural, global setting of post-industrial American culture.

The History of American Orientalism

Discursive images of American Orientalism have been profoundly shaped by the historical context of race relations in the domestic homefront, as well as the nation's diplomatic relations with Asian countries abroad (Gee 1988; Lee 1999, pp. 8–9). In his influential book, *Orientalism,* Edward W. Said argues that "the essence of Orientalism is the ineradicable distinction between Western superiority and Oriental inferiority" (1979, p. 42). Westerners' knowledge about the East imagines the Orient in a way that polarizes the Orient from the Occident and places the Occident higher than the Orient in the world hierarchy. The West is depicted as developed, powerful, articulate, and superior, while the East is seen as undeveloped, weak, mysterious, and inferior. Although Said focuses mainly on Europe's relations with the Middle East and South Asia, the political ideologies and cultural imageries implicit in such hegemonic dichotomies help to shed light on the internal dynamics of Orientalism in America. Specifically, American Orientalism has been sustained by this notion of Western/White power as a means to justify and exert its cultural domination over Asia and Asian America.

While European Orientalism was purported to justify the colonization and domination of Third World people, early American Orientalism was first invented to exclude Asian immigrants from entering or making a home on American soil. To this end, the mass media began its long history of cultivating insidious stereotypes of Asian/Americans for the visual consumption of the White American public—everything from the aggressive, ominous images of Japanese and Chinese immigrants during the "yellow peril" to more modern depictions of Asian/Americans as the passive "model minority" (Espiritu 1997; Hamamoto 1994; Lee 1999; Moy 1993; Taylor and Stern 1997). In all these stereotypes, the assimilability of Asian/Americans has always been at question (Palumbo-Liu 1999; Yu 2001). Robert G. Lee's book, *Orientals: Asian Americans in Popular Culture* (1999), shows how Orientalist images during the Gold Rush era depicted Asian/Americans as "pollutants" in the free land of California and Chinese immigrant workers as potential threats to the stability of the White immigrant working class. In movies like *The Bitter Tea of General Yen* (1933) and Fu Manchu films, the image of emasculated, asexual Asians co-existed with the image of Orientals as licentious beasts that threatened to undermine the economic and moral stability of the U.S. nation and the American family. Such cultural representations help set the ideological backdrop for anti-Chinese fervor, which led to the outbreak of anti-Chinese rioting and the implementation of the first Chinese Exclusion Act in 1882.

Within this context, it is important to note that the practice of "consuming Orientalism" evolved long before the advent of the post-industrial era. Even in the early twentieth century, Americans supported Orientalism in their day-to-day purchasing and consumption practices. Advertising cards for various products like soaps, dentifrice, waterproof collars and cuffs, clothes wringers, threads, glycerin, hats, and tobacco drew on Sinophobic themes, such as Chinese queues, porcelain doll-like Chinese women, and hyper-feminized Asian men, to market the distinctive appeal of their products (see Chan http://www.chsa.org/features/ching/ching_conf.htm). These cultural representations reinforced White America's moral and masculine superiority over the foreign elements of the East and allowed them to lay both physical and sexual claim to the bodies of Orientals at home and abroad.

During the 1950s and 1960s, the concept of Oriental inassimilability began to give way to the assimilation-oriented Model Minority myth—that is, the belief that Asian/Americans have achieved the American Dream through hard work and passive obedience. After World War II and the Korean War, movies like *Flower Drum Song* (1961) evolved their plots around less threatening, passive versions of Asian/American characters who happily shed their backwards ancestral culture in order to embrace the American lifestyle. However, as Gina Marchetti argues, "Hollywood used Asians, Asian Americans, and Pacific Islanders as signifiers of racial otherness to avoid the far more immediate racial tensions between blacks and whites or the ambivalent mixture of guilt and enduring hatred toward Native Americans and Hispanics" (1993, p. 6). For one, the media's obsession with the model minority arose within the political context of the Civil Rights era (Lee 1999; Suzuki 1989). Images of effeminate Asian men and submissive Asian women were used to counter images of violent and vociferous African Americans and feminists and to demonstrate that familial stability, social mobility, and ethnic assimilation could be achieved without militant social activism. Thus, the Asian American model minority became the symbolic antithesis of militant Civil Rights activists and feminist groups.

Nonetheless, focus on the assimilability of Asian Americans as "honorary whites" did not exempt them from the whims of racial antagonism and continued to co-exist with the image of Asian Americans as "forever foreigners" (Tuan 1998). For instance, America's bitter experiences during the prolonged Vietnam War simultaneously revived cultural images of Asians as villains and "gooks." Countless war movies repeatedly invoked images of the faceless, merciless and destructive Viet Cong instigating unmentionable travesties against brave, White U.S. soldiers. In the 1970s and 1980s, American society, once again threatened by surging economic development in Japan, projected its fears through the cultural resurrection of sinister Fu Manchu-like villains in movies such as *Blade Runner* (1982) or *Rising Sun* (1993) The massive influx of Asian immigrants in the post-1965 era has only helped to sustain the identification of Asian/Americans with mystical beings from the Orient. Compared to African Americans whose activists had been vigilant enough to protest racist movies like *Birth of the Nation* (1915), Asian/Americans were considered to be politically acquiescent and indifferent to misrepresentations in popular culture—a view that seemed to justify Hollywood's all-too-familiar messages of anti-miscegenation and White superiority.

Throughout the evolution of American Orientalism, the notion of the Orient as the culturally-inferior Other has also converged with the concept of women as the gender-inferior Other. Orientalist romanticism in the West synchronized White men's heterosexual desire for (Oriental) women and for Eastern territories through the feminization of the Orient (Kang 1993; Lowe 1991). American Orientalism in many ways depended on the masculine, superior image of White men juxtaposed with the emasculation of Asian/American men. By portraying Asian/American men as sexually excessive or asexually feminine, such cultural themes reaffirmed Orientals' deviance from "normal" heterosexual gender norms implicit in White middle-class families (Espiritu 1997; Lee 1999).

Aside from projecting stereotypes associated with the yellow peril, the model minority, and the gook, cultural representations of Asian/American women in the media have played on specific characteristics that derive from their peculiar status at the crossroads of race and gender (Degabriele 1996; Espiritu 1997; Gee 1988; Hagedorn 1994; Kang 1993; Lee 1996; Lu 1997;

Marchetti 1993; Tajima 1989). Typical representations of Asian/American women have been embodied in what Renee E. Tajima calls, "the Lotus Blossom Baby (e.g. China Doll, Geisha Girl, and the shy Polynesian beauty), and the Dragon Lady (e.g. prostitutes and devious madams)" (1989, p. 309). Although distinctive in many ways, both images have served to stimulate the sexual voyeurism of White American males and the objectification of foreign, exotic Oriental women as their rightful property. As an example of this, the most prevailing image of Asian/American women in movies like *The World of Suzie Wong* (1960) fixated on their shameless sexual desire, their aggressive and manipulative traits, and their inability to resist White men. The storylines of more contemporary movies such as *Year of the Dragon* (1985), and *Heaven and Earth* (1993) continue to focus on Asian/American female characters who are betrayed or exploited by men of their own race but are later saved by White male heroes. Thus, Orientalism in all its guises has been an underlying feature of American culture.

Advertising Multiculturalism

Historically, Asian Americans were never targeted as a significant consumer base for many of these marketing campaigns. However, as ethnic minorities grew in numbers throughout the 1980s and 1990s, marketers and advertisers began to rapidly acknowledge their potential impact as consumers (Cortese 1999; Cui 1997; Reese 1997). As one of the fastest-growing racial groups in the United States, Asian Americans have offered a very attractive market to advertisers because of their high levels of income and education. Furthermore, the steady global expansion of corporate branches into modernizing economies in Asia and Latin America and the growing sector of Asian professionals within the U.S. and abroad have also increased the need to re-conceptualize advertisement campaigns in a multicultural fashion (Ong, Bonacich and Cheng 1994). As a result, ethnic-based marketing strategies have become an indispensable terminology in the area of marketing (Cortese 1999; Cui 1997).

Two decades after the Civil Rights movements, companies and advertisers began to integrate minority consumers into their main marketing strategies (Cui 1997, p. 123) and initiated what Anthony Cortese calls the "copycat ad"—that is, traditional advertisements reproduced with models of different races. However, the copycat ad was perceived to be problematic because it assumed that "African Americans and Latinos are simply dark-skinned white people" (Cortese 1999, p. 96) and ignored the specific consumer needs and ethnic identities of their target population. In response, companies introduced a new style of marketing focused on promoting a corporate brand of multiculturalism.

Some pundits argue that multicultural marketing is more sensitive to the needs of minority consumers and helps to update or abandon traditional stereotypical representations of these populations (Cui 1997, p. 124). However, this study will demonstrate how such campaigns merely re-package long-standing racial stereotypes in their efforts to promote multicultural, globalized settings. Although old advertising cards were explicitly meant to appeal to the native-born White American population, our argument is that contemporary advertising campaigns have tried to re-invent the world in all its multicultural glories without threatening culturally-embedded hierarchies of the past. Images of Asian/Americans in multicultural ad series often employ traditional themes of American Orientalism with a new global twist. In the words of Williamson, "capitalism's constant search for new areas to colonize" (Williamson 1986, p. 116) has permeated the realms of advertisement in terms of the way they portray social movements, feminism, the gay and lesbian movement (Cagan 1978; Clark 1995; Cortese 1999), and now multiculturalism. Under the guise of multiculturalism, Orientalism has evolved into an object to consume and a vehicle to stimulate consumption. Examples of this include recent trends in Asian meditation and spa products and youth-oriented clothing lines that have incorporated "Oriental" paraphernalia like dragons and happy Buddhas into their apparel.

To this end, the research highlights . . . advertising campaigns that were chosen from six different magazines, including *Newsweek, Business Week, Vogue, In Style, Premiere,* and *Entertainment Weekly.*[2] We looked at all issues from these six magazines from September 1999 to December 2000. There were noticeable changes in all magazines in terms of the number of ads showcasing Asian/American models.

Notably, Asian/American males make rare appearances in magazine advertisements we have examined as compared with their female counterparts, which intimates to the racial and gender dynamics of advertisement cultures. And this is not an unprecedented trend. Following the popularity of Connie Chung, Asian American women anchorpersons have been very visible, whereas Asian American men anchors are nearly completely absent (Espiritu 1997). This gender imbalance not only sustains the construction of Asian American women as more desirable candidates to be assimilated when paired with White men but also reinforces the "ownership" of White American males over the bodies and spirits of Asian/American women by negating the potential physical and sexual threat imposed by Asian/American men. This point will be emphasized later.

Our purview of advertisements in the selected magazines also reveals a diversity of targeted Asian/American consumers—from lower to middle-range customers who shop at *Target* to high-class clientele at *Neiman Marcus,* but significantly, a new interest in marketing campaigns has been their attention to white-collar professional clientele, mainly businessmen. For instance, out of the 51 companies that advertised in these widely-read magazines, 8 were prestigious insurance/financial management corporations (e.g. Morgan Stanley Dean Witter), 4 were hotel/travel-related industries (e.g. Hilton, Northwest Airlines), and 5 were high-end designer brands (e.g. Dolce and Gabbanna)—all of which cater to higher-income yuppie or professional consumers. This deliberate attention to the consumption patterns of white-collar professionals stems from Asian Americans' rising purchasing power of this rapidly expanding occupational sector in today's post-industrial economy.

Although these corporations are targeting a multi-racial consumer base, the perspectives of these marketing campaigns implicitly preserve the centrality of White men by re-packaging Orientalist themes of Asian/American women under the guise of multiculturalism. A multicultural ad by Charles Schwab (featured *in Newsweek* and *Business Week*) . . . contextualizes these emerging themes of globalization, enterprise and professionalism through the incorporation of Asian/American models. The ad shows three people sitting on a bench, each holding a book with a title relevant to each person. The White woman to the left is holding a book, "Keep Ahead of the Sharks," the Asian American woman sitting next to her is holding a book "How to Get Rich Overnight," and the White man is holding a book titled "Boy, Am I Happy." The main character is the White man with the blissful look. The advertisement conveys the message that you, the consumer, can be just as happy as this man with help from the consultation company, Charles Schwab.

More interestingly, the White woman is fixated on the book she is reading, but in stark contrast, the Asian American woman is glaring at the White man askance with feelings of resentful anger and jealousy because of his happiness and financial success—a theme that draws on the image of Asia as an economic adversary of the United States (Espiritu 1997; Lee 1999; Suzuki 1989). But also, the unusually large lettering of the word "RICH" on her book highlights her obsession with making money and reaffirms the cultural perception of Asian/Americans as greedy money-mongers and threats to the overall well-being of real (White) "Americans."

The ad's portrayal of the Asian/American woman revolves around the conniving and hostile nature of her role in the global market and more importantly, takes shape *in relation to* her White female and White male counterparts. While these ads speak to a racially-diverse audience of professionals, it is important to note the ways in which relationality is a key component in Orientalism and that multiculturalism arises "within the context of White males within this global market. Importantly, this image that evokes that of the "Dragon Lady," the epithet for belligerent, cunning, and untrustworthy Asian/American women, is placed to represent this societal locality of Asian Americans. Hence, the new wave of advertisement campaigns underscores some of the inherent contradictions of multiculturalism itself: while opening its doors to the new Asian consumer, it does so by representing them within traditional White patriarchal frameworks of Orientalism.

The following . . . corporate campaigns were mainly selected because they are part of a larger multicultural ad series, as opposed to a single-themed advertisement frame. In the last couple of years, the multicultural ad *series* has become a popular way of marketing brand-name products with a racially diverse cast. The multicultural ad series is an advertisement campaign that features each model in various ad copies, poses, or appearances within a series of thematic frames. The frames are either featured all at the same time or in isolation from one another in different magazines. The multicultural ad series tries to include diverse racial groups and has thus generated an increase in Asian/American representations in advertising. They are particularly interesting to our research, because they allow us not only to analyze the cultural undertones of an individual ad but also, compare the thematic representations that come out of each racial/gender frame.

The first series of multicultural advertisements comes from the "Find Your Voice" campaign by Virginia Slims, which depicts different images of women from diverse racial backgrounds expressing ways to "find your voice" in life. Virginia Slims is a brand of cigarettes produced by Philip Morris, the third largest cigarette company (2.4% in 2003) in retail share performance, which specifically targets a racially-diverse audience of female consumers. Established in 1968, Virginia Slims first played on themes of female empowerment through campaign slogans like "You've Come a Long Way Baby," which elicited angry responses from various feminist organizations because of the way it distorted and trivialized feminist issues in order to profit on women's addictions (Cagan 1978; Kilbourn 2000). Following the multicultural marketing trends of the time, Virginia Slims then introduced a new advertisement series in 1999–2000 called the "Find Your Voice" campaign, which again promoted themes of female liberation but this time with attention to a broader multicultural and global consumer base. As one feminist newsletter proclaims, the Virginia Slims campaign now equated smoking their cigarettes with the liberating influence of Western culture through advertisements targeting vulnerable Third World populations (Batchelor 2003). *Ms.* magazine also expressed its indignation to the company's efforts to "globalize addiction and equalize smoking-related illnesses" through multicultural campaigns like these (Comments Please 2000, p. 96).

A glimpse at four different ads from this campaign, each featuring models of different races. The first photo shoot shows a blonde-haired White woman next to the words "I look temptation right in the eye and then make my own decisions"; an African woman proclaiming "No single institution owns the copyright for BEAUTY"; a Latino woman stating "Dance around naked with a rose between your teeth if you want . . . but do it like you mean it"; and finally an Asian woman with the words "My voice reveals the hidden power within." In another ad, the same Asian woman is juxtaposed next to the words "In silence I see. With WISDOM, I speak."

Focusing on the Asian model, we can see blatant references to time-old themes of Oriental feminine exoticism perpetuated by numerous Hollywood films (e.g., *Sayonara* (1957), *The Teahouse of the August Moon* (1956) and *Japanese War Bride* (1952)) as well as western literature (e.g., *Memoir of the Geisha* (1999)) and musicals (e.g., *Madam Butterfly*) in the past century. Stepping away from the feminist undertones of Virginia Slim campaigns, the posture of the Asian woman in two different ads is one of femininity and sexual invitation. She is looking down and sideways, and her head is tilted as well, with a cryptic smile. Her hands are curled in front of her in an "Oriental-like" gesture as if she is dancing. She appears as an entertainer, a Madam Butterfly, a courtesan, a geisha, and "a Lotus Blossom baby." Historically, Lotus

Blossom images represented Asian women as exotic, enticing, subservient, pampering, self-effacing, self-sacrificing and sensual. In a similar manner as this ad, Asian women in Lotus Blossom images throw sexually suggestive smiles and gazes but hesitate to speak. Renee E. Tajima states "Asian women . . . are interchangeable in appearance and name, and are joined together by the common language of non-language—that is, uninterpretable chattering, pidgin English, giggling, or silence" (1989, p. 309). References to "Hidden power" and "In silence I see" again reaffirm the "non-language" embodiment of Asian women. Furthermore, although the ads make no explicit references to men, it is important to note that Lotus Blossom images were traditionally used to obliterate Asian women's subjectivity by validating their role as the objects of White men's sexual fantasy.

What is more interesting about this ad series is the way the "exotic," "feminine" and "mysterious" allure of the Orientalized character becomes accentuated by the projected normalcy of the White characters. The ads that feature women of color consistently promote the strongest cultural references in the series: an African woman in a colorful headdress,[3] a dancing Latino woman in a light cotton weave and wooden beads, an Asian woman in heavy makeup and traditional Chinese dress. In the case of the Asian woman, this aura of foreignness is highlighted by the antediluvian attire and posture. The Asian woman in both ads is wearing dresses and makeup that are modified renditions of traditional Chinese dresses and hairdos that are no longer worn today. This theme derives from Orientalist depictions of Asia—that is, the unchangeable and undeveloped portrayal of a colonized Orient (Said 1979; Nochlin 1989). Furthermore, this particular series invokes feminine and hyper-sexualized stereotypes of Asian women (as well as Latino women) in stark contrast to the themes of liberation and empowerment associated with the White and African American characters.

At the same time, the Westernized version of Orientalism reified by the ad serves to commodify Asian culture. Westerners' indulgence in Asian culture has been often understood as a signifier of their wealth and the broadened purview of their ability to consume (Marchetti 1993, p. 27). The consumption of Asian culture has never required an accurate comprehension of Asian cultures and histories and empathy with Third World experiences of colonialism, imperialism, and economic exploitation. In the ad, the costume of the Asian woman looks Chinese but it is in actuality pseudo-authentic at best. Her hairdo is not done in a traditional Chinese style and her makeup is modern not traditional in fashion. Furthermore, her posture is pan-Asian, drawing on gestures, expressions and stances that stem from various Asian cultures. One of the ads . . . even features messages written in Chinese that make no sense in interpretation. Despite the corporation's attempts to address a multicultural audience, the cultural references in the ads end up perpetuating Orientalist meanings that reaffirm the dominant status of White Americans. The Orientalist depiction of Chinese customs and written characters were there not to be understood but to be objectified by viewers in their visual consumption of the Asian female model. The ad provided neither identification with nor education about Asian cultures, but only the commodification and "objectification" of their people.

Ofoto (Featured in Newsweek and InStyle)

A subsidiary of Eastman Kodak Company, Ofoto is an online photography service (www.ofoto.com) that gives consumers a virtual space where they can store as well as modify and edit their pictures. The company was founded in July 1999 and Internet Service was launched on December 1999, accompanied with an extensive ad campaign. Each of the four ads . . . were found in different magazines or different issues of the same magazine but take on greater meaning when we place them next to one another. The general schemata of the Ofoto advertisement series features an individual sitting on a chair, looking at an unseen picture of himself/herself with someone else. A caption below the scene describes what the person is looking at.

Thus in the illustration, we see four different frames, each featuring models of different races. The White man is supposedly looking at a photo labeled "Tom Gilmartin with the star of the kindergarten play, Hannah Gilmartin, the purple rabbit princess"—that is, a father and daughter scenario. The photo held by an older African American man is captioned "Daryl Lamar Edwards II with Daryl Lamar Edwards IV"—or grandfather with grandson. The White woman is looking at "A rare nose-to-nose meeting between Carol McBride's cat, Manny, and her dog, Marley"—a picture of family pets. And last but not least, the Asian/American version of the series—a woman holding a picture of herself or "Tia Fong with a 'friend' on her hotel room balcony in Prague."

Several points stand out in our analysis of these four different ads—the first being that while both the White and African American men are seen to have a connection to their families and lineage, the female characters are not. Although Carol McBride's wide-legged posture in the ad manifests her power and confidence, her beloved family is her pets, not her family or child. The picture of Tia Fong, the Asian American woman, is even more problematic, because first, Tia's "friend," unlike Carol's pets, does not even have a name and second, she is seeing herself in a hotel room situated in the distant, exotic land of Prague. In essence, Tia Fong is not to be associated with the comforts of home and family but rather, the erotic setting of foreign lands and forbidden pleasures. The fact that Tia's photo includes a mysterious "friend" and that she is located in Prague evokes mystic images of the Orient. This aspect of positionality is also interesting in its racial implications, because it symbolizes not only Asian Americans' detachment from both home and lineage but also their dislocation from American society itself. In essence, these advertisements once again hint to the unassimilability of Asians in America.

The illustration of Tia Fong is imbued with other gendered meanings as well—the most obvious of which is the sexual connotation behind the faceless "friend" and the hotel room where she is staying with this friend. Unlike the other two ads that conjure up feelings of familial belonging (albeit a surrogate family in the case of Carol), the Asian/American ad is replete with references to sexual liaisons within a Prague hotel room—perhaps the modern equivalent of a geisha teahouse. The quotations around the word "friend" and the mysterious smile of the Asian American character is meant to imply the forbidden pleasures associated with this trip. Furthermore, her legs are crossed in a demure, intimate manner that is meant to evoke images of Orientalized sitting postures, while her fingers delicately hold the photo in a gesture that strongly resembles the dancing hand movements of the Virginia Slims Asian model. Again, Ofoto's advertising campaign is drawing on traditional representations of Asian/American women as exotic and erotic objects of White men's sexual adventure.

Conclusion

The article has analyzed . . . advertisements which best exemplify the diverse ways in which Asian/American women are sexually objectified, culturally misrepresented and visually consumed in contemporary American Orientalism. The Virginia Slims campaign was perhaps the most blatant in its resurrection of Asian/American exotica through its mish-mash of simulated Orientalist paraphernalia. However, even when the physical appearances of Asian/American female characters were normalized in relation to other White American characters, the other advertisements series were shown to make more subtle but powerful messages about their inherent cultural and behavioral unassimilability to American society. The Ofoto campaign on the other hand draws on the mystical aspects of Asian/American female sexuality, detaching its characters from the all-American setting of family and home and placing them in the exotic spaces of a hotel room in Prague.

The subtle ingenuity of the multicultural advertisement campaign is the way it is able to profit off a multi-racial consumer base through greater inclusion while maintaining White male supremacy through the visual consumption of

Asian/American women's bodies. By highlighting the ascribed "foreign" nature of Asian/American women, the cultural schemata of corporate advertisements aim to profit off the sense of identity and place they provide for White males in the U.S. through their products, while simultaneously targeting an increasingly diverse global audience. Thus, the illustrations presented in this study would have had less meaning if they included only two White people or even White and African American characters without the added "foreign" Asian/American presence. Furthermore, corporations market on the physical embodiments of sex and pleasure that take the form of Asian/American women in these advertisements. That is, they are not just selling their liquor or their cigarettes or their services to American society, but they are also selling the *bodies* of Asian/American women and the forbidden pleasures that come with them. The ads also displayed racialized and gendered images of other figures (i.e. White females and African American characters), but the presence of Asian/Americans seemed to be most central in highlighting the multicultural nature of today's global society vis-a-vis their exoticization/eroticization, as well as re-affirming White normalcy and supremacy within this global hierarchy.

As our research study has shown, the emerging global culture has been packaged, commodified and marketed by multi-national global corporations in a form that can be sold to dominant White groups attempting to disengage from a historical legacy of racism, segregation and Anglo-conformity. Cultural representations of multiculturalism in corporate advertising campaigns have a more concrete impact on the lived experiences of Asian/American women by reaffirming the complex racial and gender hierarchies underlying the new global order and legitimating the physical domination of their bodies through rape, abuse, exploitation and prostitution. In this way, the perception of multiculturalist advertisements as the symbolic site for cultural diversity and equality overlooks the subtle complexities with which Asian/American bodies are presented and represented to White America.

Notes

1. As Palumbo-Liu explains, the inclusion of the slash in the word "Asian/American" conveys the same meaning as in the construction "and/or." That is, it represents "a choice between two terms, their simultaneous and equal status, and an element of indecidability, that is, as it at once implies both exclusion and inclusion (Palumbo-Liu 1999:1)." This element of "indecidability" is an important factor in this word choice, because Asian Americans are still considered to be "foreigners," or Asians. In this paper, we use "Asian American" only for specific situations related to Asian Americans.

2. Magazines were chosen based on two conditions—to cover different types of magazines (Taylor, Lee, and Stern 1995) and accessibility. They include general interest magazine (*Newsweek*), business press (*Business Week*), women's magazines (*Vogue* and *In Style*), and entertainment magazines (*Premiere* and *Entertainment Weekly*); and, the first author subscribed to four magazines out of six. *Newsweek, Business Week,* and *Vogue* were used in the previous research. According to Ulrich's International Periodical Directories (www.ulrichsweb.com), the numbers of paid circulations are 3.1 million (*Newsweek),* 1.2 million (*Business Week*), 1.3 million (*Vogue*), 1.1 million (*In style*), 1.5 million (*Entertainment Weekly*), and 0.6 million (*Premiere*).

3. Notably, the Virginia Slims campaign also includes ads featuring an African American woman in contemporary attire without such cultural references—a strategy that is used to differentiate the two audiences to whom they are speaking, unlike the case of the Asian and Latino ads which feature exoticized women only.

References

Appadurai, A. (1993). Disjuncture and difference in the global cultural economy. In B. Robbins (Ed.),

The phantom public sphere (pp. 269–295). Minneapolis: University of Minnesota Press.

Batchelor, S. (2003). Traditional women are tobacco's new global market. *Women's Enews.* <http://www.womensenews.org/article.cfm/dyn/aid/1491> (Accessed: June 9, 2003).

Berger, J. (1977). *Ways of seeing.* London: Penguin.

Betterton, R. (1987). *Looking on: Images of femininity in the visual arts and media.* London and Boston: Pandora Press.

Bordo, S. (1993). *Unbearable weight. Feminism, Western culture, and the body.* Berkeley and London: University of California Press.

Cagan, E. (1978). The selling of the women's movement." *Social Policy,* May/June, 4–12.

Chan, J. (2003) 'Rough on rats': Racism and advertising in the latter half of the nineteenth century. <http://www.chsa.org/features/ching/ching_conf.htm> (Accessed: February 25, 2003).

Chang, G., & Abramovitz, M (2000). *Disposable domestics: Immigrant women workers in the global economy.* Cambridge: South End Press.

Clark, D. (1995). Commodity lesbianism. In G. Dines & J. M. Humez (Eds.), *Gender, race and class in media: A text-reader* (pp. 142–151). Thousand Oaks, London and New Delhi: Sage.

Comments please. (2000). *Ms.,* June/July, 96–97.

Cortese, A. J. (1999). *Provocateur. Images of women and minorities in advertising.* Lanham: Rowman & Littlefield Publishers.

Cui, G. (1997). Marketing strategies in a multi-ethnic environment. *Journal of Marketing Theory and Practice, 5,* 122–134.

Degabriele, M. (1996). From Madame Butterfly to Miss Saigon: One hundred years of popular Orientalism. *Critical Arts, 10,* 105–118.

Espiritu, Y. L. (1997). *Asian American women and men: Labor, laws, and love.* Thousand Oaks, CA: Sage.

Gee, D. (Director). (1988). *Slaying the dragon* [Film]. San Francisco: Cross Current Media and National Asian American Telecommunication Association.

Giroux, H. A. (1994). Consuming social change: The 'United Colors of Benetton.' *Cultural Critique, 26,* 5–31.

Goffman, E. (1979). *Gender advertisements.* Cambridge: Harvard University Press.

Hagedorn, J. (1994). Asian women in film: No joy, No luck. *Ms.,* January/February, 74–79.

Hamamoto, D. Y. (1994). *Monitored peril: Asian Americans and the politics of TV representation.* Minneapolis and London: University of Minnesota Press.

Hochschild, A., & Ehrenreich, B. (2003). *Global woman: Nannies, maids and sex workers in the new economy.* New York: Metropolitan.

Jeffreys, S. (1999). Globalizing sexual exploitation: Sex tourism and the traffic in women. *Leisure Studies, 18,* 179–186.

Kang, L. H. (1993). The desiring of Asian female bodies: Interracial romance and cinematic subjection. *Visual Anthropology Review. 9,* 5–21.

Kilbourne, J. (1999). *Deadly persuasion: Why women and girls must fight the addictive power of advertising.* New York: Free Press.

Kilbourne, J. (2000). *Killing us softly 3* [Film]. Northampton, MA: Media Education Foundation.

Lee, J. (1996). Why Suzie Wong is not a lesbian: Asian and Asian American lesbian and bisexual women and femme/butch/gender identities. In B. Beemyn & M. Eliason (Eds.), *Queer studies: A lesbian, gay, bisexual & transgender anthology* (pp. 115–132). New York and London: New York University Press.

Lee, R. G. (1999). *Orientals: Asian Americans in popular culture.* Philadelphia: Temple University Press.

Lowe, L. (1991). *Critical terrains: French and British orientalisms.* Ithaca and London: Cornell University Press.

Lu, L. (1997). Critical visions: The representation and resistance of Asian women. In S. Shah (Ed.), *Dragon ladies: Asian American feminists breathe fire* (pp. 17–28). Boston: South End Press.

Manca, L., & Manca, A. (1994). *Gender & utopia in advertising: A critical reader.* Lisle, IL and Syracuse, NY: Procopian Press and Syracuse University Press.

Marchetti, G. (1993). *Romance and the "yellow peril": Race, sex, and discursive strategies in Hollywood fiction.* Berkeley. Los Angeles, and London: University of California Press.

Moy, J. S. (1993). *Marginal sights: Staging the Chinese in America.* Iowa City: University of Iowa Press.

Nochlin, L. (1989). *The politics of vision.* New York: Harper and Row.

Omi, M. (1991). Shifting the blame: Racial ideology and politics in the post-Civil Rights era. *Critical Sociology, 18.* 77–98.

Ong, P., Bonacich E., & Cheng, L. (1994). The new Asian immigration in Los Angeles and global restructuring. Philadelphia: Temple University Press.

Palumbo-Liu, D. (1999). *Asian/American: Historical crossings of a racial frontier.* Stanford: Stanford University Press.

Reese, S. (1997). A world of differences. *Marketing Tools, 4,* 36–41.

Rose, T. (1994). *Black noise: Rap music and Black culture in contemporary America.* Middletown, CT: Wesleyan University Press.

Said, E. W. (1979). *Orientalism.* New York: Vintage.

Sassen-Koob, S. (1984). Notes on the incorporation of Third World women into wage-labor through immigration and off-shore production. *International Migration Review, 18,* 1144–1167.

Suzuki, B H. (1989). Asian Americans as the 'model minority": Outdoing Whites? Or media hype? *Change,* November/December, 12–19.

Tajima, R. E. (1989). Lotus blossoms don't bleed: Images of Asian women. In Asian Women United of California (Ed.), *Making waves: An anthology of writings by and about Asian American women* (pp. 308–317). Boston: Beacon Press.

Taylor, C. R., & Stern, B. B. (1997). Asian-Americans: Television advertising and the 'model minority' stereotype. *Journal of Advertising, 26,* 47–31.

Tuan, M. (1998). *Forever foreigners or honorary whites?: The Asian ethnic experience today.* New Brunswick, NJ: Rutgers University Press.

Williamson, J. (1978). *Decoding advertisements. Ideology and meaning in advertising.* London: Boyars.

Williamson, J. (1986). Woman is an island: Femininity and colonization. In T. Modleski (Ed.), *Studies in entertainment: Critical approaches to mass culture* (pp. 99–118). Bloomington and Indianapolis: Indiana University Press.

Yu, H. (2001). *Thinking Orientals: Migration, contact, and exoticism in modern America.* Oxford and New York: Oxford University Press.

Introduction to Reading 25

Kirsten Firminger's research on representations of males in five different popular teenage girls' magazines reveals that girls are encouraged to become informed consumers of boys, who are presented as shallow, highly sexual, emotionally inexpressive, and insecure, but also as potential sources of romance, intimacy, and love. Boys appear as products, much like other products and services being sold to girls. Girls are represented as responsible for good shopping, including selecting the right guy.

1. How do teenage girls' magazines function as a guide to selecting boys?

2. What are the links between girls' beauty and fashion products and the presentation of boys as products?

3. Discuss the author's final sentences, "Bottom line: look at dating as a way to sample the menu before picking your entrée. In the end, you'll be much happier with the choice you make! Yum!"

Is He Boyfriend Material?

Representation of Males in Teenage Girls' Magazines

Kirsten B. Firminger

On the pages of popular teenage girls' magazines, boys are presented (in)congruently as the providers of potential love, romance, and excitement and as highly sexual, attracted to the superficial, and emotionally inexpressive. The magazines guide female readers toward avoiding the "bad" male and male behavior (locking up their feelings tighter than Fort Knox) and obtaining the "good" male and male behavior (setting you apart from other girls). Within girls' magazines, success in life and (heterosexual) love is girls' responsibility, tied to their ability to self-regulate, make good choices, and present themselves in the "right" way. The only barriers are girls' own lack of self-esteem or limited effort (Harris 2004). While the "girl power" language of the feminist movement is used, its politics and questioning of patriarchal gender norms are not discussed. Instead, the magazines advocate relentless surveillance of self, boys, and peers. Embarrassing and confessional tales, quizzes, and opportunities to rate and judge boys and girls on the basis of their photos and profiles encourage young women to "fashion" identities through clothes, cosmetics, beauty items, and consumerism.

Popular teenage girls' magazines. In the United States, teenage girls' magazines are read by more than 75 percent of teenage girls (Market profile: Teenagers 2000). The magazines play an important role in shaping the norms and expectations during a crucial stage of identity and relationship development. Currie (1999) found that some readers consider the magazines' content to be more compelling than their own personal experiences and knowledge. Magazines are in the business of both selling themselves to their audience and selling their audience to advertisers (Kilborne 1999). Teenage girls are advertised as more loyal to their favorite magazines than to their favorite television programs, with magazines touted as "a sister and a friend rolled into one" (Market profile: Teenagers 2000). Magazines attract and keep advertisers by providing the right audience for their products and services, suppressing information that might offend the advertiser, and including editorial content saturated in advertiser-friendly advice (Kilborne 1999).

In this textual environment, consumerist and individualist attitudes and values are promoted to the exclusion of alternative perspectives. Across magazines, one relentless message is clear: "the road to happiness is attracting males for successful heterosexual life by way of physical beautification" (Evan et al. 1991; see also Carpenter 1998, Currie 1999, Signorelli 1997). Given the clarity of this message, little work has been done focusing on the portrayal of males that the girls are supposed to attract. I began my research examining this question: how are males and male behavior portrayed in popular teenage girls' magazines?

Method

To explore these questions, I designed a discursive analysis of a cross-section of adolescent girls' magazines, sampling a variety of magazines and

From Firminger, K. B. 2006. "Is he boyfriend material? Representation of males in teenage girls' magazines" *Men and Masculinities 8*(3), p. 298. Reprinted with permission of Sage Publications, Inc.

analyzing across them for common portrayals of males. *Seventeen* and *YM* are long-running adolescent girls' magazines. *Seventeen* has a base circulation of 2.4 million while *YM* has a circulation of 2.2 million *(Advertising Age* 74: 21). As a result of the potential of the market, the magazines that are directed at adolescent girls have expanded to include the new *CosmoGirl* (launched in 1999) and *ELLEgirl* (in 2001). Very successful, *CosmoGirl* has a base circulation of 1 million. *ELLEgirl* reports a smaller circulation of 450,000 *(Advertising Age* 74: 21). Chosen as an alternative to the other adolescent girls' magazines, *Girls' Life* is directed at a younger female audience and is the winner of the 2000, 1999, and 1996 Parents' Choice Awards Medal and of the 2000 and 1998 Parents' Guide to Children's Media Association Award of Excellence. The magazine reports it is the number one magazine for girls ages 10 to 15, with a circulation of 3 million (http://www.girlslife.com/infopage.php, retrieved May 23, 2004).

I coded two issues each of *Seventeen, YM, CosmoGirl, ELLEgirl,* and *Girls' Life,* for a total of ten issues. Magazines build loyalty with their readers by presenting the same kinds of material, in a similar form, month after month (Duke 2002). To take into account seasonal differences in content, I purchased the magazines six months apart, once during December 2002 and once during July 2003. While the magazines range in their dates of publication (for instance, Holiday issue, December issue, January issue, etc.), all the magazines were together at the same newsstand at the singular time of purchase.

Results

Within the pages of the magazines, articles and photo layouts focus primarily on beauty, fashion, & celebrities and entertainment, boys and love, health and sex, and self-development. The magazines specialize, with emphasis more or less on one of these topics over the other: *ElleGirl* presents itself as more fashion focused, while self-development is the emphasis for *Girls' Life*'s younger audience. Within the self-development sections, one can find articles focusing on topics such as activities, school, career aspirations, volunteering, sports, and politics. However, even in these articles, focus is on the social, interpersonal aspects of relationships and on consumption instead of the actual doing and mastery of activities.

Advertising permeates the magazines, accounting for 20.8 percent to 44.8 percent of the pages. Additionally, many of the editorial articles, presumably noncommercial, are written in ways that endorse specific products and services (see Currie 1999, for more information on "advertorials"). For instance, one advice column responded to a reader's inquiry about a first kiss by recommending ". . . [having] the following supplies [handy] for when the magical moment finally arrives: Sugarless mints, yummy flavored lip gloss (I dig Bonne Bell Lip Smackers) . . . ."

Male-focused content. On average, 19.7 percent of the pages focused on males,[1] ranging from a minimum of 13.6 percent in *ELLEGirl* to a maximum 26.6 percent in *Seventeen.* Articles on boys delve into boys' culture, points of view, opinions, interests, and hobbies, while articles on girls' activities focus more pointedly on the pursuit of boys. Girls learn "Where the boys are," since the "next boyfriend could be right under your nose." They are told,

> Where to go: Minor-league ballparks. Why: Cute guys! . . . Who'll be there: The players are just for gazing at; your targets are the cuties in the stands. And don't forget the muscular types lugging soda trays up and down the aisles. What to say: Ask him what he thinks about designated hitters (they're paid just to bat). He'll be totally impressed that you even brought up the subject.

Males are offered up to readers in several different formats. First we read profiles, then we meet "examples," we are allowed question and answer, we are quizzed, and then we are asked to judge the males. Celebrity features contain in-depth interviews with male celebrities, while personal short profiles of celebrities or "regular" guys include a photo, biographical information, hobbies, interests, and inquiries such as his "three big requirements for a girlfriend" and "his perfect date." In question-and-answer articles, regular columnists answer selected questions that the readers have submitted.[2] Some columns consistently focus on boys, such as "GL Guys by Bill and Dave" and *YM's* "Love Q and A," while others focus on a variety of questions, for instance *ELLEgirl*'s "Ask Jennifur" profiles of noncelebrity males are presented and judged in rating articles. The magazines publish their criteria for rating boys, via rhetorical devices such as "the magazine staffs' opinions" or the opinion polls of other teenage girls.

Ratings include categories such as "his style," "dateable?," and "style factor." For example, in *CosmoGirl*'s Boy-a-Meter article, "Dateable?: I usually go for dark hair, olive skin, and thick eyebrows. But his eyes make me feel like I could confide in him," or *ELLEgirl's* The Rating Game, "He's cute, but I don't dig the emo-look and the hair in the face. It's girlie." Readers can then assess their opinions in relation to those of other girls' and the magazine staff.

Romantic stories and quotes enable readers to witness "real" romance and love and compare their "personal experiences" to those presented in the magazines. For instance, "Then one day I found a note tucked in my locker that said, 'You are different than everyone else. But that is why you are beautiful.' At the bottom of the note it said, 'From Matt—I'm in your science class.' We started dating the next day." These can also be rated, as the magazine staff then responded, "Grade: A. He sounds like a very smart boy."

Finally, the readers can then test their knowledge and experiences through the quizzes in the magazines, such as *Seventeen*'s quiz, "Can your summer love last?" with questions and multiple-choice answers:

> As he's leaving for a weeklong road trip with the guys, he: A) tells you at least 10 times how much he's going to miss you. B) promises to call you when he gets a chance. C) can't stop talking about how much fun it will be to "get away' with just his buddies for seven whole days.

Over the pages, boys as a "product" begin to merge with the [other] products and services being sold to girls in "training" as informed consumers, learning to feel "empowered" and make good "choices." While a good boy is a commodity of value, the young women readers learn that relationships with boys should be considered disposable and interchangeable like the other products being sold, "Remember, BFs come and go, but best friends are forever! Is he worth it? Didn't think so."

Embarrassing and confessional stories. Short embarrassing or confessional stories submitted by the readers for publication provide another textual window through which young women can view gender politics; one issue of *YM* included a special pull-out book focused exclusively on confessionals.[3] Kaplan and Cole (2003) found in their four focus groups that the girls enjoy the embarrassing and confessional stories because they reveal "what it is like to be a teenage girl."

On average, two-thirds of the confessional/embarrassing stories were male focused; in 42–100 percent of the stories males were the viewing audience for, or participant in, a girl's embarrassing/confessional moment. Often these stories involve a "cute boy," "my boyfriend," or "my crush." For example:

> My friends and I noticed these cute guys at the ice cream parlor. As we were leaving with our cones, the guys offered to walk with us. I was wearing my chunky-heeled shoes and feeling pretty awesome . . . until I tripped. My double scoop flew in the air and hit one of the guys. Oops.

Teenage girls within these stories are embarrassed about things that have happened, often accidentally, with males typically as the audience. While this may allow the female readers to see that they are not the only ones who have experienced embarrassing moments, it also reinforces the notion of self-surveillance as well as socializes girls to think of boys as the audience and judges of their behavior (Currie 1999).

Representations of males. To assess how males are represented, I coded content across male-focused feature articles and "question and answer" columns.[4] These articles contained the most general statements about boys and their behaviors, motivations, and characteristics (for example, "Guys are a few steps behind girls when it comes to maturity level"[5]).

A dominant tension in the representations of boys involves males' splitting of intimacy from sexuality. The magazine advises girls as they negotiate these different behaviors and situations, trying to choose the "right" guy (who will develop an intimate relationship with a girl), reject the "bad" guy (who is focused only on sex), or if possible, change the "bad" guy into the "good" guy (through a girl's decisions and interactions with the male).

> My boyfriend and I were together for 10 months when he said he wanted to take a break—he wasn't sure he was ready for such a commitment. The thought of him seeing other people tore me apart. So every day while we were broken up, I gave him something as a sign of my feelings for him: love sayings cut out from magazines, or cute comics from the paper. Eventually he confessed that he had just been confused and that he loved me more than ever.

As girls are represented as responsible for good shopping, they are represented also as responsible for selecting/changing/shaping male behavior. If girls learn to make the right choices, they can have the right relationship with the right guy, or convert a "bad"/confused boy into a good catch.

The tension is most evident in stories about males' high sex drive, attraction to superficial appearances, emotional inexpressiveness, and fear of rejection and contrasted with those males who are "keepers": who keep their sex drive in check, value more than just girls' appearances, and are able to open up. The articles and advice columns blend the traditional and the feminist; encompassing both new and old meanings and definitions of what it means to be female and male within today's culture (Harris 2004).

The males' sex drive. The "naturally" high sex drive of males rises as the most predominant theme across the magazines. Viewed as normal and unavoidable in teenage boys, girls write to ask for an explanation and advice, and they are told:

> You invited a guy you kind of like up to your room (just to talk!) and he got the wrong idea. This was not your fault. Guys—especially unchaperoned guys on school trips—will interpret any move by a girl as an invitation to get heavy. And I mean any move. You could have sat down next to him at a lab table and he would have taken that as a sign from God that you wanted his body.

When it comes to the topic of sexuality, traditional notions surround "appropriate behavior" for young women and men. Girls learn that males respect and date girls who are able to keep males' sex drive in check and who take time building a relationship. Girls were rarely shown as being highly sexual or interested only in sexual relationship with a boy. Girls are supposed to avoid potentially dangerous situations (such as being alone with a boy) and draw the line (since the males frequently are unable to do so). If they don't, they can be labeled sluts.

> Don't even make out with someone until you're sure things are exclusive. When you hook up with him too early, you're giving him the message that you are something less than a goddess (because, as you know, a goddess is guarded in a temple, and it's not easy to get to her). Take it from me when I tell you that guys want to be with girls they consider goddesses. So treat your body as a temple—don't let just anyone in.

Valuing superficial appearances. Driven by sex, males were shown as judging and valuing girls based on their appearance.

> That's bad, but it's scarier when combined with another sad male truth: They're a lot more into looks than we are.
>
> Okay, I'm the first to admit that guys can be shallow and insipid and *Baywatch* brainwashed to the point where the sight of two balloons on a string will turn them on.

Since males are thought to be interested in the superficial, girls sought advice on how to be most superficially appealing, asking what do guys prefer, including the size of a girl's breasts, hair color, eye color, height, and weight. Girls are portrayed as wanting to know how to present themselves to attract boys, demonstrating an interaction between girls' ideas and understanding of what males want and girls' own choices and behaviors.

Boys are emotionally inexpressive. Across features, readers learn about boys' inability or unwillingness to open up and share their feelings. However, the articles suggest also that if a girl is able to negotiate the relationship correctly, she could get a guy to trust her.

> Let's say you go to the pet store and see a really cute puppy you'd like to pet but, every time you try, he pulls away because he was treated badly in the past. People aren't much different. Move very slowly, and build up trust bit by bit. Show this guy you're into him for real, and he'll warm up to you. Puppy love is worth the wait.

Girls are responsible for doing the emotional work and maintenance and for being change agents in relationships, not allowing room for or even expecting males to take on any of these tasks (see also Chang 2000).

Boys' insecurity and fear of rejection. Boys are displayed as afraid of rejection. Reflecting the neoliberal ideology of "girl power," girls were urged by the magazines to take the initiative in seeking out and approaching boys. This way they are in control of and responsible for their fate, with only lack of confidence, self-esteem, and effort holding them back from finding romance and love.

> So in the next week (why waste more time?), write him a note, pull him aside at a party, or call him up with your best friend by your side for support. Hey, he could be psyched that you took the initiative.
>
> So I think you may have to do the work. If there's a certain guy you're feelin' and you think he's intimidated by you, make the first move. Say something to relax him, like, "What's up? My name is Chelsea." After that, he'll probably start completing sentences.

Males' potential—the "keepers." "Consider every guy to be on a level playing field—they all have potential." Boys were shown to have "potential" and girls were advised to search out the "right" guys.

> He does indeed sound dreamy. He also sounds like a total gentleman, considering he hasn't attempted to jump your bones yet, so the consensus is: He's a keeper.
>
> Most guys are actually smarter than you think and are attracted to all sorts of things about the female species. Yes, big boobs definitely have their dedicated fan base, but so do musical taste, brains, a cute laugh, style and the ability to throw a spiral football (to name just a few). What's turn-on or dealbreaker for one guy is a nonevent for another.

These boys become the center of the romantic stories and quotes about love and relationships. Resulting from and sustained by girls' self-regulation, personal responsibility, effort, and good choices (as guided by the tools and advice provided by the magazines), these boys are for keeps.

Discussion

Within the magazines, girls are invited to explore boys as shallow, highly sexual, emotionally inexpressive, and insecure and boys who are potential boyfriends, providing romance, intimacy, and love. Males' high sex drive and interest in superficial appearances are naturalized and left unquestioned in the content of the magazines; within a "girl power" version of compulsory heterosexuality, girls should learn the right way to approach a boy in order to get what they want—"the road to happiness is attracting males for successful heterosexual life by way of physical beautification" (Evans et al. 1991). Girls walk the fine line of taking advantage of males' interest in sex and appearance, without crossing over into being labeled a slut. Socialized to be purchasers of beauty and fashion products that promise to make them attractive to boys, girls are "in charge" of themselves and the boys they "choose." It's a competitive market so they better have the right understanding of boys, as well as the right body and outfit to go with it.

The magazines' portrayals, values, and opinions are shaped by their need to create an advertiser-friendly environment while attracting and appealing to the magazines' audience of teenage girls. Skewing the portrayal of males and females to their target audience, magazine editors, writers, and, though I have not highlighted it here—advertisers take advantage of gender-specific fantasies, myths, and fears (Craig 1993). Boys become another product, status symbol, and identity choice. If girls' happiness requires finding romance and love, girls should learn to be informed consumers of boys. By purchasing the magazines, they have a guide to this process, guaranteed to help them understand "What his mixed signals really mean." In addition, if boys are concerned with superficial appearances, it is to the benefit of girls to buy the advertised products and learn "The best swimsuit for [their] bod[ies]."

As girls survey and judge themselves and others, possessions and consumption become the metric for assessing status (Rohlinger 2002; Salamon 2003), the cultural capital for teenagers in place of work, community, and other activities (Harris 2004). The feminist "girl empowerment" becomes personal, appropriated to sell products. The choice and purchase of products and services sold in the magazines promise recreation and transformation, of not only one's outward appearance but also of one's inner self, leading to happiness, satisfaction, and success (Kilborne 1999). Money is the underlying driving force in magazine content. However, while the magazines focus on doing good business, girls are being socialized by the magazines' norms and expectations.

"Bottom line: look at dating as a way to sample the menu before picking your entrée. In the end, you'll be much happier with the choice you make! Yum!"

Notes

1. Percentage of male-focused pages was taken out of total editorial pages, not including advertising pages. Confessional/embarrassing stories did not count toward the total number of pages because of inconsistencies in unit of analysis, with the confessional stories having a variable number of male-focused stories. I analyzed those separately. Feature articles (unique, nonregular) counted if the article focused on or if males significantly contributed to the narrative in the article (for instance, "Out of bounds: A cheerleader tells the story of how the coach she trusted attacked her"). If the feature was equally balanced with focused sections on both boys and girls (for example, if the article is sectioned into different topics or interviews), only pages that focused on males were counted.

Because of the limited nature of the study, I chose to focus purely on the content that was decided upon by the editorial/writing (called "editorial content" within this article) staff of the magazines, since they establish the mission and tone of the content across all of the issues of the magazine. While I acknowledge the influential presence of advertising, I did no analysis of the content of the advertising pages or photographs. The analysis consists only of the written content of the magazines.

2. The magazines report that the question-and-answer columns and embarrassing/confessional tales are "submitted by readers." However, they do not report how they choose the questions and stories that are published, or whether the magazine staff edits this content.

3. *ELLEgirl* did not contain embarrassing or confessional stories.

4. The unit of analysis was the smallest number of sentences that contained a complete thought, experience, or response, ranging from one sentence to a paragraph. For example, "The fact is you can't change other people. He has to change himself—but perhaps your concern will convince him to make some changes." I took this approach so that the meaning and context of a statement was not lost in the coding. Whole paragraphs could not always be used because they sometimes contained contrasting or multiple themes.

5. The other articles that were not included in the coding focused predominantly on a specific boy or a celebrity male and his interests/activities, or on stories including a boy, or activities to do with a boy, rather than making broad statements about how all boys act (for example, "When he was in kindergarten, his mom enrolled Elijah [Wood] in a local modeling and talent school" or "One time, my boyfriend dared me to sneak

out of the house in the middle of the night while my parents were sleeping and meet him at a park.").

References

Carpenter, L. M. 1998. From girls into women: Scripts for sexuality and romance in *Seventeen* Magazine, 1974–1994. *The Journal of Sex Research* 35: 158–168.

Chang, J. 2000. Agony-resolution pathways: How women perceive American men in *Cosmopolitan's* agony (advice) column. *The Journal of Men's Studies* 8: 285–308.

Currie, D. H. 1999. *Girl talk: Adolescent magazines and their readers*. Toronto: University of Toronto Press.

Craig, S. 1993. Selling masculinities, selling femininities: Multiple genders and the economics of television. *The Mid-Atlantic Almanack* 2: 15–27.

Duke, L. 2002. Get real!: Cultural relevance and resistance to the mediated feminine ideal. *Psychology and Marketing* 19: 211–233.

Evans, E., J. Rutberg, and C. Sather. 1991. Content analysis of contemporary teen magazines for adolescent females. *Youth and Society* 23: 99–120.

Girls' Life magazine: About us. Retrieved May 23, 2004 from http://www.girlslife.com/infopage.php.

Kaplan, E. B. and L. Cole. 2003. "I want to read stuff on boys": White, Latina, and Black girls reading *Seventeen* magazine and encountering adolescence. *Adolescence* 38: 141–159.

Market profile: Teenagers. 2000. Magazine Publishers of America.

McRobbie, A. 1991. *Feminism and Youth Culture*. London: Macmillan.

Rohlinger, D. 2002. Eroticizing men: Cultural influences on advertising and male objectification. *Sex Roles: A Journal of Research* 46: 61–74.

Salamon, S. 2003. From hometown to nontown: Rural community effects of suburbanization. *Rural Sociology* 68: 1–24.

Signorelli, N. 1997. A content analysis: Reflections of girls in the media, a study of television shows and commercials, movies, music videos, and teen magazine articles and ads. Children Now and Kaiser Family Foundation Publication.

Tolman, D. L., R. Spencer, T. Harmon, M. Rosen-Reynoso, and M. Stripe. 2004. Getting close, staying cool: Early adolescent boys' experiences with romantic relationships, In *Adolescent boys: Exploring diverse cultures of boyhood*, edited by N. Way and J. Chu. New York: NYU Press.

Tolman, D. L., R. Spencer, M. Rosen-Reynoso, and M. Porche. 2002. Sowing the seeds of violence in heterosexual relationships: Early adolescents narrate compulsory heterosexuality. *Journal of Social Issues* 59: 159–178.

Van Roosmalen, E. 2000. Forces of patriarchy: Adolescent experiences of sexuality and conceptions of relationships. *Youth and Society* 32: 202–227.

Where the girls are. 2003. *Advertising Age* 74: 21.

⚜ Topics for Further Examination ⚜

- Find articles and Web sites that offer critiques of gender stereotypes in the mass media, popular culture, and consumer culture. For example, visit the following Web site: http://womensissues.about.com/cs/genderstereotypes/.
- Locate research on the impact of American media images of masculinity and femininity on the self-perceptions of women and men in non-Western cultures such as India, Thailand, and Kenya.
- Google Hummer ads for women and Hummer ads for men. Compare and contrast the gendered "marketing" strategies. Select other products and compare, contrast, and critique.
- Check out song lyrics by artists who dare to criticize hegemonic masculinity and emphasized femininity. For example, Google Pink lyrics for the tune titled "Stupid Girls."

6

TRACING GENDER'S MARK ON BODIES, SEXUALITIES, AND EMOTIONS

This chapter explores the ways in which gender patterns weave themselves into three of the most intimate aspects of the self: body, sexuality, and emotion. The readings we have selected make the general sociological argument that there is no body, sexuality, or emotional experience independent of culture. That is, all cultures sculpt bodies, shape sexualities, and produce emotions. One of the most powerful ways in which a gendered society creates and maintains gender difference and inequality is through its "direct grip" on these intimate domains of our lives (Schiebinger, 2000, p. 2). Gender ideals and norms require work to be done in, and on, the body to make it appropriately feminine or masculine. The same ideals and norms regulate sexual desire and expression, and require different emotional skills and behaviors of women and men.

At first glance, it may seem odd to think about body, sexuality, and emotion as cultural and gendered products. But consider the following questions: Do you diet, lift weights, dehair your legs or face, or use makeup? If you were interviewing for a job, would you feel more comfortable sitting with legs splayed or legs crossed at the ankles? How do you feel and move your body through public spaces when you are alone at night? If you answered yes to any of these questions, you do gendered body work. Now think about the gendering of sexuality in the United States. The mark of gender on sexual desire and expression is clear and deep, tied to gendered body ideals and norms. Whose breasts are eroticized and why? Are women who have many sexual partners viewed in the same way as men who have many partners? Why are so many heterosexually identified men afraid of being perceived as homosexual? Why do women shoulder the major responsibility for contraceptive control? Like sexuality, emotions are embodied modes of being. And like sexuality and body, emotions are socially regulated and constructed. They are deeply gendered.

Consider these questions: Do you associate emotionality with women or men? Is an angry woman taken as seriously as an angry man? Why do we expect women's body language (e.g., their touch and other gestures) to be more affectionate and gentle than men's? What is your reaction to these word pairs—tough woman/soft man? (See Carter in Chapter 4.) The readings in this chapter explore the complex and contradictory ways in which bodies, sexualities, and emotions are brought into line with society's gender scheme. Two themes unite the readings. First, they demonstrate how the marking of bodily appearance, sexual desire and behavior, and emotional expression as masculine or feminine reinforces Western culture's insistence on an oppositional gender binary pattern. Second, the readings show how patterns of gender inequality become etched into bodies, sexualities, and emotions.

Gendered Bodies

All societies require body work of their members (Black & Sharma, 2004; Lorber & Moore, 2007). But not all societies insist on the molding of men's and women's bodies into visibly oppositional and asymmetrical types—for example, strong male bodies and fragile female bodies. Only societies constructed around the belief and practice of gender dualism and hierarchy require the enactment of gender inequality in body work. To illustrate, consider the fact that the height ideal and norm for heterosexual couples in America consists of a man who is taller and more robust than his mate (Gieske, 2000). This is not a universal cultural imperative. As Sabine Gieske (2000) states, the tall man/short woman pattern was unimportant in eighteenth century Europe. In fact, the ideal was created in the Victorian era under the influence of physicians and educators who defined men as naturally bigger and stronger than women. The expectation that men be taller than their female partners persists today, even though the average height gap between men and women is closing (Schiebinger, 2000). The differential height norm is so strong that many contemporary Americans react to the pairing of a short man and a tall woman with the kind of shock and disapproval typically directed at interracial couples (Schiebinger, 2000).

How are bodies made feminine and masculine as defined by our culture? It takes work, and lots of it. Well into the twentieth century, American gender ideology led men and women alike to perceive women as frail in body and mind. Boys and men were strongly exhorted to develop size, muscle power, physical skills, and the "courage" to beat each other on the playing field and the battlefield, while girls and women were deeply socialized into a world of distorted body image, dangerous dieting, and physical incompetence (Dowling, 2000). Throwing, running, and hitting "like a girl" was a common cultural theme that we now understand to be a consequence of the cultural taboo against girls developing athletic stature and skills. As Collette Dowling (2000) notes, "there is no inherent biological reason for girls not to throw as far, as fast, or as hard as boys do" (p. 64). But there is a cultural reason: the embodiment of the belief in gender difference and inequality. We literally translate the "man is strong, woman is weak" dictum into our bodies. This dictum is so powerful that many people will practice distressed and unhealthy body routines and regimens to try to emulate the images of perfect male and female bodies. The mere threat of seeming masculine or mannish has kept a lot of girls and women from developing their own strength, and the specter of seeming effeminate or sissy has propelled many boys and men into worlds in which their bodies are both weapons and targets of violence (Dowling, 2000).

American men, especially young men, are expected to express stereotyped masculinity through their bodies by the way they move, sit, gesture, eat, and so forth. However, men's bodies are not held to equally severe ideals of bodily attractiveness and expectations of body work as are women's bodies. Gender inequality is etched into women's flesh in more debilitating and painful ways.

American cultural definitions of femininity equate attractiveness with a youthful, slim, fit body that, in its most ideal form, has no visible "flaws"—no hair, pores, discolorations, perspiration marks, body odors, or trace of real body functions. In fact, for many women, the body they inhabit must be constantly monitored and managed—dehaired, deodorized, denied food—so that it doesn't offend. At the extreme, girls and women turn their own bodies into fetish objects to which they devote extraordinary amounts of time and money. What are the models of feminine bodily perfection against which girls and women measure themselves and are evaluated by others? The images and representations are all around us in magazine ads, on TV commercials and programs, in music videos, and in toy stores. The Barbie doll is an excellent example. As Urla and Swedlund argue in Chapter 5, Barbie is the "ever-thin, ever-plastic, ever-wholesome," excessively feminine American icon. Her body is, of course, impossible for real women to achieve.

The embodiment of women's subordination in gender-stratified societies takes some extreme forms. For example, in highly restrictive patriarchies, women's bodies may be systematically deformed, decimated, and restricted. Foot-binding in traditional China is one of the most dramatic examples of an intentionally crippling gender practice. Although foot-binding may seem to be a practice that has no parallels in Western societies, in this chapter sociologist Fatema Mernissi challenges the ethnocentric tendency to dismiss practices such as foot-binding in China, and the veiling of women in some nations today, as alien or primitive. She does so by revealing the symbolic violence hammered directly on the Western female body by fashion codes and cosmetic industries that place Western women in a state of constant anxiety and insecurity. Mernissi's work suggests the usefulness of comparing practices such as foot-binding in China to cosmetic surgery, extreme dieting, and body sculpting among contemporary Western girls and women. Consider the following questions. What do these seemingly different forms of body work have in common? Do they serve similar functions? How do they replicate gender inequality?

The story of the gendering of bodies in the United States is not only about oppositional and asymetrical masculinity and femininity. Prisms of difference and inequality come into play. Let's consider one prism—race—to illustrate this point. What's the answer to the question, "Mirror, mirror on the wall, who's the fairest of them all?" (Gillespie, 1998). You know what it is. The beauty standard is white, blonde, and blue-eyed. It is not Asian American, African American, Latin American, or Native American. In other words, there is a hierarchy of physical attractiveness, and the Marilyn Monroes and Madonnas of the world are at the top. Yes, there are African and Asian American models and celebrities. But, they almost always conform to white appearance norms. When they don't, they tend to be exoticized as the "Other." Just consider the fact that eyelid surgery, nasal implants, and nasal tip refinement procedures are the most common cosmetic surgery procedures undergone by Asian American patients, largely women (Kaw, 1998). The facial features that Asian American women seek to alter, including small, narrow eyes and a flat nose, are those that define them as racially different from white norms (Kaw, 1998). Racial and gender ideologies come together to reinforce an ethnocentric and racist beauty standard that devalues the "given" features of minority women. The medical system has cashed in on it by promoting a beauty standard that requires the surgical alteration of features that don't fit the ideal.

The reading by Peter Hennen examines another prism of difference, sexual orientation. His focus is on the resistance of men in a gay subculture, Bear culture, not only to stereotypical effeminacy and phallic sex but also to adoration of the young, slender body. Hennen's research describes a world of men who have developed standards of physical attractiveness that embrace and elevate the large, hirsute male body as desirable. This study reminds us that although American culture in general emphasizes very narrow standards of gendered physical attractiveness, people can and do construct social worlds in which a greater range of bodies are appreciated and celebrated.

Gendered Sexualities

Like the body, sexuality is shaped by culture. The sexualization process in a gendered society such as our own is tightly bound to cultural ideas of masculinity and femininity. In dominant Western culture, a real man and a real woman are assumed to be opposite human types, as expressed in the notion of the "opposite sex." In addition, both are assumed to be heterosexual as captured in the notion that "opposites attract." Conformity to this gendered sexual dichotomy is strictly enforced.

The term *compulsory heterosexuality* refers to the dominance of heterosexual values and the fact that the meanings and practices of non-hetero(sexuality) as well as hetero(sexuality) are shaped by the dominant heterosexual script. For example, "real sex" is generally conceived of as coitus or penile-vaginal intercourse. This "coital imperative," as Nicola Gavey, Kathryn McPhillips, and Marion Doherty call it in their reading, limits the control individuals have in determining what counts as sexual activity. The coital imperative frames women's sexuality as passive/receptive and men's as active/penetrative. That is, this imperative defines sex as something men "do" to women. Gavey and her coauthors also analyze the negative impact that privileging of men's sexual needs above women's has on the capacity for men and women to have truly reciprocal, safe sexual relations.

Interestingly, the Western obsession with the homosexual/heterosexual distinction is relatively new. Created in the nineteenth century, it became a mechanism by which masculinity and femininity could be further polarized and policed (Connell, 1999). Gay masculinity and lesbian femininity came to be defined as abnormal and threatening to the "natural gender order." Consequently, for many Westerners, the fear of being thought of as homosexual has a powerful impact on presentation of self. Men and boys routinely police each other's behavior and mete out punishment for any suggestion of "effeminacy" (Connell, 1999; Plante 2006), while women and girls who engage in masculinized activities such as military service and elite sports risk being labeled lesbians.

Contemporary Western gender and sexuality beliefs have spawned stereotypes of lesbians as "manly" women and gay men as "effeminate." The reality is otherwise. For example, research shows that many gay men and lesbians are gender conformists in their expression of sexuality (Kimmel, 2000). Gay men's sexual behavior patterns tend to be masculine—oriented toward pleasure, orgasm, experimentation, and many partners—while lesbian sexuality is quite womanly in a Western sense, emphasizing sexual intimacy within romantic relationships. However, this account of nonheterosexuality is incomplete. Recent research suggests that compared to heterosexuals, nonheterosexuals have more opportunities to reflect upon and experiment with ways of being sexual. Although the meanings attached to sexuality in society at large shape their erotic encounters, they are also freer to challenge the dominant sexual scripts (Weeks et al., 2001).

The imposition of gender difference and inequality on sexuality in societies such as the United States is also reflected in the sexual double standard. The double standard, which emphasizes and normalizes male pleasure and female restraint, is a widespread product and practice of patriarchy. Although Western sexual attitudes and behaviors have moved away from a strict double standard, the sexual lives of women remain more constrained than those of men. For example, girls and women are still under the control of the good girl/bad girl dichotomy (see the article by Tanenbaum in Chapter 4), a cultural distinction that serves to pit women against each other and to produce sexual relations between women and men that can be confusing and dissatisfying. Men grow up with expectations to embrace a sexual script by which they gain status from sexual experience (Kimmel, 2000). Women grow up with expectations to believe that to be sexually active is to compromise their value; but at the same time, to be flirtatious and sexy as well. Doesn't this seem confusing? Imagine the relationship misunderstandings and disappointments that can emerge out of the meeting of these "opposite sex" sexual scripts.

The prisms of difference and inequality alter the experience and expression of gendered sexuality in many ways. For example, Mary Nell Trautner's article explores the ways in which social class differences are scripted and represented as sexual differences in exotic dance clubs catering to clientele who come from either the middle class or the working class.

Dancers in working-class clubs are more likely to perform stereotypes of bad girl sexuality while those who dance in middle-class clubs enact good girl sexuality. If you were to review research on the intersections of race and sexuality, you would discover a world of negative sexual stereotypes of members of racial minorities. Rebecca Plante (2006), an expert on the sociology of sexuality, notes, for example, that African Americans "do not enter the discourse and debate about sexualities from the same place where white, middle-class people enter" (p. 231). Black sexuality was historically stereotyped as perverted and predatory, a theme that continues today in pornographic and other representations of African American men in particular.

Gendered Emotions

Masculinity and femininity are emotionally opposed in Western culture (Bendelow & Williams, 1998). This opposition expresses itself in both obvious and subtle ways. The obvious opposition is that the emotions that can be expressed, for example anger and love, and how they are expressed, are tied to gender. Boys and men must appear "hard" by hiding or shutting down feelings of vulnerability, such as fear, while girls and women are encouraged to be "soft"; that is, emotionally in touch, vulnerable, and expressive. Jennifer Lois's analysis of the emotional culture of a volunteer search and rescue team shows how the cultural definition of women and men as emotional opposites plays itself out. Although the men and women in Lois's study were effective in both low- and high-risk crisis situations, women were more likely to be perceived as emotional deviants and inferiors compared to the men. Consider the impact of learning and enacting different, even oppositional, emotional scripts on intimate relationships. If men are not supposed to be vulnerable, then how can they forge satisfying affectionate bonds with women and other men? Additionally, men who embrace the stereotype of hard masculinity may pay a price in well-being by concealing their own pain, either physical or psychological (Real, 2001). Also, it is important to recognize the negative consequences of gendered emotionality for girls and women. Although girls and women receive cultural encouragement to be "in touch with" themselves and others emotionally, this has strong associations with weakness and irrationality. When women express emotion according to cultural rules, they run the risk of being labeled hypersensitive, temperamental, and irrational. The stereotype of the emotionally erratic and unstable woman has been widely used in efforts to undermine the advancement of women in politics, higher education, the professions, business, and other realms of public life. You know how it goes, "We can't risk having a moody, irritable, irrational woman at the helm."

How do the prisms of difference and inequality interact with gendered emotions? Looking through a number of prisms simultaneously, we can see that the dominant definition of hard, stiff-upper-lip masculinity is White, heterosexual, and European (Seidler, 1998). The dominant definition of masculinity assigns rationality and reason to privileged adult men. All other people, including women, minorities, and children, are assumed to be more susceptible to influence by their bodies and emotions, and, as a consequence, less capable of mature, reasoned decision making (Seidler, 1998).

We'd like to conclude this chapter by asking you to think about individual and collective strategies to reject conformity to patterns of gendered body work, sexual expression, and emotionality that demean, disempower, and prove dangerous to women and men. What would body work, emotional life, and sexual desire and experience be like if they were not embedded in and shaped by structures of inequality? How would personal growth, self-expression, and communication with others change if we were not under the sway of compulsory attractiveness, compulsory heterosexuality, the sexual double standard, and gendered emotional requirements? How would your life change? What can we do, as individuals and collectively, to resist and reject the pressure to bring our bodies, sexual experience, and emotional life into line with oppressive and dangerous ideals and norms?

REFERENCES

Bendelow, G., & Williams, S. J. (1998). Introduction: Emotions in social life. In G. Bendelow & S. Bendelow (Eds.), *Emotions in social life* (pp. xv–xxx). London: Routledge.

Black, P., & Sharma, U. (2004). Men are real, women are "made up." In J. Spade & C. Valentine (Eds.), *The kaleidoscope of gender* (pp. 286–296). Belmont, CA: Wadswroth.

Connell, R. W. (1999). Making gendered people. In, M. Feree, J. Lorber, & B. Hess (Eds.), *Revisioning gender* (pp. 449–471). Thousand Oaks, CA: Sage.

Dowling, C. (2000). *The frailty myth.* New York: Random House.

Gieske, S. (2000). The ideal couple: A question of size? In L. Schiebinger (Ed.), *Feminism and the body* (pp. 375–394). New York: Oxford University Press.

Gillespie, M. A. (1998). Mirror mirror. In R. Weitz (Ed.), *The politics of women's bodies* (pp. 184–188). New York: Oxford University Press.

Herdt, G. (1997). *Same sex, different cultures.* Boulder, CO: Westview.

Kaw, E. (1998). Medicalization of racial features: Asian-American women and cosmetic surgery. In R. Weitz (Ed.), *The politics of women's bodies* (pp. 167–183). New York: Oxford University Press.

Kimmel, M. (2000). *The gendered society.* New York: Oxford.

Lorber, J., & Moore, L. J. (2007). *Gendered bodies: Feminist perspectives.* Los Angeles, CA: Roxbury.

Plante, R. (2006). *Sexualities in context.* Cambridge, MA: Westview Press.

Real, T. (2001). Men's hidden depression. In T. Cohen (Ed.), *Men and masculinity* (pp. 361–368). Belmont, CA: Wadsworth.

Schiebinger, L. (2000). Introduction. In L. Schiebinger (Ed.), *Feminism and the body* (pp. 1–21). New York: Oxford.

Seidler,V. J. (1998). Masculinity, violence and emotional life. In G. Bendelow & S. Williams (Eds.), *Emotions in social life* (pp. 193–210). London: Routledge.

Weeks, J., Heaphy, B., & Donovan, C. (2001). *Same sex intimacies: Families of choice and other life experiments.* New York: Routledge.

Weitz, R. (1998). A history of women's bodies. In R. Weitz (Ed.), *The politics of women's bodies* (pp. 3–11). New York: Oxford University Press.

Introduction to Reading 26

Fatema Mernissi is a well-known Moroccan sociologist and Islamic scholar. She has done research in a number of Western nations, including the United States. This article is a chapter from her book titled *Scheherazade Goes West.* Mernissi challenges Westerners to think about the ways in which their feminine beauty images and practices can be as hurtful and humiliating to women as the enforced veiling of women in nations such as Iran and Saudi Arabia. In fact, she argues that the "size 6" ideal is a more violent restriction than the Muslim veil.

1. How do definitions of feminine beauty differ in Morocco compared to the United States?
2. What does Mernissi mean when she states the Western man establishes male domination by manipulating time and light?
3. Read the final paragraph of this article carefully. Why does Mernissi end on this note?

Size 6

The Western Woman's Harem

Fatema Mernissi

It was during my unsuccessful attempt to buy a cotton skirt in an American department store that I was told my hips were too large to fit into a size 6. That distressing experience made me realize how the image of beauty in the West can hurt and humiliate a woman as much as the veil does when enforced by the state police in extremist nations such as Iran, Afghanistan, or Saudi Arabia. Yes, that day I stumbled onto one of the keys to the enigma of passive beauty in Western harem fantasies. The elegant saleslady in the American store looked at me without moving from her desk and said that she had no skirt my size. "In this whole big store, there is no skirt for me?" I said. "You are joking." I felt very suspicious and thought that she just might be too tired to help me. I could understand that. But then the saleswoman added a condescending judgment, which sounded to me like an Imam's fatwa. It left no room for discussion:

> "You are too big!" she said.
>
> "I am too big compared to what?" I asked, looking at her intently, because I realized that I was facing a critical cultural gap here.
>
> "Compared to a size 6," came the saleslady's reply.
>
> Her voice had a clear-cut edge to it that is typical of those who enforce religious laws. "Size 4 and 6 are the norm," she went on, encouraged by my bewildered look. "Deviant sizes such as the one you need can be bought in special stores."

That was the first time that I had ever heard such nonsense about my size. In the Moroccan streets, men's flattering comments regarding my particularly generous hips have for decades led me to believe that the entire planet shared their convictions. It is true that with advancing age, I have been hearing fewer and fewer flattering comments when walking in the medina, and sometimes the silence around me in the bazaars is deafening. But since my face has never met with the local beauty standards, and I have often had to defend myself against remarks such as zirafa (giraffe), because of my long neck, I learned long ago not to rely too much on the outside world for my sense of self-worth. In fact, paradoxically, as I discovered when I went to Rabat as a student, it was the self-reliance that I had developed to protect myself against "beauty blackmail" that made me attractive to others. My male fellow students could not believe that I did not give a damn about what they thought about my body. "You know, my dear," I would say in response to one of them, "all I need to survive is bread, olives, and sardines. That you think my neck is too long is your problem, not mine."

In any case, when it comes to beauty and compliments, nothing is too serious or definite in the medina, where everything can be negotiated. But things seemed to be different in that American department store. In fact, I have to confess that I lost my usual self-confidence in that New York environment. Not that I am always sure of myself, but I don't walk around the Moroccan streets or down the university corridors wondering what people are thinking about me. Of course, when I hear a compliment, my ego expands like a cheese soufflé, but on the whole, I don't expect to hear much from others. Some mornings, I feel ugly because I am sick or tired; others, I feel wonderful because it is sunny out or I have written a good paragraph. But suddenly, in that peaceful American store that I had entered so triumphantly, as a sovereign consumer ready to spend money, I felt savagely attacked. My hips, until then the sign of a relaxed and uninhibited maturity, were suddenly being condemned as a deformity.

"And who decides the norm?" I asked the saleslady, in an attempt to regain some self-confidence by challenging the established rules. I never let others evaluate me, if only because I remember my childhood too well. In ancient Fez, which valued round-faced plump adolescents, I was repeatedly told that I was too tall, too skinny, my cheekbones were too high, my eyes were too slanted. My mother often complained that I would never find a husband and urged me to study and learn all that I could, from storytelling to embroidery, in order to survive. But I often retorted that since "Allah had created me the way I am, how could he be so wrong, Mother?" That would silence the poor woman for a while, because if she contradicted me, she would be attacking God himself. And this tactic of glorifying my strange looks as a divine gift not only helped me to survive in my stuffy city, but also caused me to start believing the story myself. I became almost self-confident. I say almost, because I realized early on that self-confidence is not a tangible and stable thing like a silver bracelet that never changes over the years. Self-confidence is like a tiny fragile light, which goes off and on. You have to replenish it constantly. "And who says that everyone must be a size 6?" I joked to the saleslady that day, deliberately neglecting to mention size 4, which is the size of my skinny twelve-year-old niece.

At that point, the saleslady suddenly gave me an anxious look. "The norm is everywhere, my dear," she said. "It's all over, in the magazines, on television, in the ads. You can't escape it. There is Calvin Klein, Ralph Lauren, Gianni Versace, Giorgio Armani, Mario Valentino, Salvatore Ferragamo, Christian Dior, Yves Saint-Laurent, Christian Lacroix, and Jean-Paul Gaultier. Big department stores go by the norm." She paused and then concluded, "If they sold size 14 or 16, which is probably what you need, they would go bankrupt." She stopped for a minute and then stared at me, intrigued. "Where on earth do you come from? I am sorry I can't help you. Really, I am." And she looked it too. She seemed, all of a sudden, interested, and brushed off another woman who was seeking her attention with a cutting, "Get someone else to help you, I'm busy." Only then did I notice that she was probably my age, in her late fifties. But unlike me, she had the thin body of an adolescent girl. Her knee length, navy blue, Chanel dress had a white silk collar reminiscent of the subdued elegance of aristocratic French Catholic schoolgirls at the turn of the century. A pearl-studded belt emphasized the slimness of her waist. With her meticulously styled short hair and sophisticated makeup, she looked half my age at first glance.

"I come from a country where there is no size for women's clothes," I told her. "I buy my own material and the neighborhood seamstress or craftsman makes me the silk or leather skirt I want. They just take my measurements each time I see them. Neither the seamstress nor I know exactly what size my new skirt is. We discover it together in the making. No one cares about my size in Morocco as long as I pay taxes on time. Actually, I don't know what my size is, to tell you the truth." The saleswoman laughed merrily and said that I should advertise my country as a paradise for stressed working women. "You mean you don't watch your weight?" she inquired, with a tinge of disbelief in her voice. And then, after a brief moment of silence, she added in a lower register, as if talking to herself: "Many women working in highly paid fashion-related jobs could lose their positions if they didn't keep to a strict diet." Her words sounded so simple, but the threat they implied was so cruel that I realized for the first time that maybe "size 6" is a more violent restriction imposed on women than is the Muslim veil. Quickly I said good-bye so as not to make any more demands on the saleslady's time or involve her in any more unwelcome, confidential exchanges about age-discriminating salary cuts. A surveillance camera was probably watching us both. Yes, I thought as I wandered off, I have finally found the answer to my harem enigma. Unlike the Muslim man, who uses space to establish male domination by excluding women from the public arena, the Western man manipulates time and light. He declares that in order to be beautiful, a woman must look fourteen years old. If she dares to look fifty, or worse, sixty, she is beyond the pale. By putting the spotlight on the female child and framing her as the ideal of beauty, he condemns the mature woman to invisibility. In fact, the modern Western man enforces Immanuel Kant's nineteenth-century theories: To be beautiful, women have to appear childish and brainless. When a woman looks mature and self-assertive, or allows her hips to expand, she is condemned as ugly. Thus, the wars of the European harem separate youthful beauty from ugly maturity.

These Western attitudes, I thought, are even more dangerous and cunning than the Muslim ones because the weapon used against women is time. Time is less visible, more fluid than space. The Western man uses images and spotlights to freeze female beauty within an idealized childhood, and forces women to perceive aging—that normal unfolding of the years—as a shameful devaluation. "Here I am, transformed into a dinosaur," I caught myself saying aloud as I went up and down the rows of skirts in the store, hoping to prove the saleslady wrong—to no avail. This Western time-defined veil is even crazier than the space-defined one enforced by the Ayatollahs.

The violence embodied in the Western harem is less visible than in the Eastern harem because aging is not attacked directly, but rather masked as an aesthetic choice. Yes, I suddenly felt not only very ugly but also quite useless in that store, where, if you had big hips, you were simply out of the picture. You drifted into the fringes of nothingness. By putting the spotlight on the prepubescent female, the Western man veils the older, more mature woman, wrapping her in shrouds of ugliness. This idea gives me the chills because it tattoos the invisible harem directly onto a woman's skin. Chinese foot-binding worked the same way: Men declared beautiful only those women who had small, childlike feet. Chinese men did not force women to bandage their feet to keep them from developing normally—all they did was to define the beauty ideal. In feudal China, a beautiful woman was the one who voluntarily sacrificed her right to unhindered physical movement by mutilating her own feet, and thereby proving that her main goal in life was to please men. Similarly, in the Western world, I was expected to shrink my hips into a size 6 if I wanted to find a decent skirt tailored for a beautiful woman. We Muslim women have only one month of fasting, Ramadan, but the poor Western woman who diets has to fast twelve months out of the year. *"Quelle horreur,"* I kept repeating to myself, while looking around at the American women shopping. All those my age looked like youthful teenagers.

According to the writer Naomi Wolf, the ideal size for American models decreased sharply in the 1990s. "A generation ago, the average model weighed 8 percent less than the average American woman, whereas today she weighs 23 percent less. . . . The weight of Miss America plummeted, and the average weight of Playboy Playmates dropped from 11 percent below the national average in 1970 to 17 percent below it in eight years."[1] The shrinking of the ideal size, according to Wolf, is one of the primary reasons for anorexia and other health-related problems: "Eating disorders rose exponentially, and a mass of neurosis was promoted that used food and weight to strip women of . . . a sense of control."[2] Now, at last, the, mystery of my Western harem made sense. Framing youth as beauty and condemning maturity is the weapon used against women in the West just as limiting access to public space is the weapon used in the East. The objective remains identical in both cultures: to make women feel unwelcome, inadequate, and ugly. The power of the Western man resides in dictating what women should wear and how they should look. He controls the whole fashion industry, from cosmetics to underwear. The West, I realized, was the only part of the world where women's fashion is a man's business. In places like Morocco, where you design your own clothes and discuss them with craftsmen and women, fashion is your own business. Not so in the West. As Naomi Wolf explains in *The Beauty Myth,* men have engineered a prodigious amount of fetishlike, fashion-related paraphernalia: "Powerful industries—the $33-billion-a-year diet industry, the $20-billion cosmetic industry, the $300-million cosmetic surgery industry, and the $7-billion pornography industry—have arisen from the capital made out of unconscious anxieties, and are in turn able, through their influence on mass culture, to use, stimulate, and reinforce the hallucination in a rising economic spiral."[3]

But how does the system function? I wondered. Why do women accept it? Of all the possible explanations, I like that of the French sociologist, Pierre Bourdieu, the best. In his latest book, *La Domination Masculine,* he proposes something he calls *"la violence symbolique":* "Symbolic violence is a form of power which is hammered directly on the body, and as if by magic, without any apparent physical constraint. But this magic operates only because it activates the codes pounded in the deepest layers of the body."[4] Reading Bourdieu, I had the impression that I finally understood Western man's psyche better. The cosmetic and fashion industries are only the tip of the iceberg, he states, which is why women are so ready to adhere to their dictates. Something else is going on on a far deeper level. Otherwise, why would women belittle themselves spontaneously? Why, argues Bourdieu, would women make their lives more difficult, for example, by preferring men who are taller or older than they are? "The majority of French women wish to have a husband who is older and also, which seems consistent, bigger as far as size is concerned," writes Bourdieu.[5]

Caught in the enchanted submission characteristic of the symbolic violence inscribed in the mysterious layers of the flesh, women relinquish what he calls "les signes ordinaires de la hiérarchie sexuelle," the ordinary signs of sexual hierarchy, such as old age and a larger body. By so doing, explains Bourdieu, women spontaneously accept the subservient position. It is this spontaneity Bourdieu describes as magic enchantment.[6] Once I understood how this magic submission worked, I became very happy that the conservative Ayatollahs do not know about it yet. If they did, they would readily switch to its sophisticated methods, because they are so much more effective. To deprive me of food is definitely the best way to paralyze my thinking capabilities.

Both Naomi Wolf and Pierre Bourdieu come to the conclusion that insidious "body codes" paralyze Western women's abilities to compete for power, even though access to education and professional opportunities seem wide open, because the rules of the game are so different according to gender. Women enter the power game with so much of their energy deflected to their physical appearance that one hesitates to say the playing field is level. "A cultural fixation on female thinness is not an obsession about female beauty," explains Wolf. It is "an obsession about female obedience. Dieting is the most potent political sedative in women's history; a quietly mad population is a tractable one."[7] Research, she contends, "confirmed what most women know too well—that concern with weight leads to a 'virtual collapse of self-esteem and sense of effectiveness' and that . . . 'prolonged and periodic caloric restriction' resulted in a distinctive personality whose traits are passivity, anxiety, and emotionality."[8]

Similarly, Bourdieu, who focuses more on how this myth hammers its inscriptions onto the flesh itself, recognizes that constantly reminding women of their physical appearance destabilizes them emotionally because it reduces them to exhibited objects. "By confining women to the status of symbolical objects to be seen and perceived by the other, masculine domination . . . puts women in a state of constant physical insecurity. . . . They have to strive ceaselessly to be engaging, attractive, and available."[9] Being frozen into the passive position of an object whose very existence depends on the eye of its beholder turns the educated modern Western woman into a harem slave. "I thank you, Allah, for sparing me the tyranny of the 'size 6 harem,'" I repeatedly said to myself while seated on the Paris-Casablanca flight, on my way back home at last. "I am so happy that the conservative male elite does not know about it. Imagine the fundamentalists switching from the veil to forcing women to fit size 6."

How can you stage a credible political demonstration and shout in the streets that your human rights have been violated when you cannot find the right skirt?

Notes

1. Naomi Wolf, The Beauty Myth: How Images of Beauty Are Used Against Women (New York Anchor Books, Doubleday, 1992), p. 185.
2. Ibid., p. 11.
3. Ibid., p. 17.
4. Pierre Bourdieu: "La force symbolique est une forme de pouvoir qui s' exerce sur les corps, directement, et comme par magie, en dehors de toute contraine physique, mais cette magie n'opère qu'en s'appuyant sur des dispositions déposées, tel des ressorts, au plus profond des corps." In La Domination Masculine (Paris: Editions du Seuil, 1998), Ibid. p. 44. Here I would like to thank my French editor, Claire Delannoy, who kept me informed of the latest debates on women's issues in Paris by sending me Bourdieu's book and many others. Delannoy has been reading this manuscript since its inception in 1996 (a first version was published in Casablanca by Edition Le Fennec in 1998 as "Êtes-Vous Vacciné Contre le Harem").
5. La Domination Masculine, Ibid. p. 41.
6. Bourdieu, Ibid. p. 42.
7. Wolf, Ibid. p. 187.
8. Wolf, quoting research carried out by S. C. Woolly and O. W. Woolly, Ibid. pp. 187–188.
9. Bourdieu, La Domination Masculine. p. 73.

Introduction to Reading 27

Peter Hennen ends his provocative study of the meanings of masculinity, sexuality, and the male body in Bear subculture by asking if Bears cause gender trouble. As you read this article, keep that question in mind. Bears are gay men who define the large, hirsute body as desirable and attractive. They present an image of white, working-class masculinity, even though most Bears are middle class. While Bear culture embraces a number of sexually innovative practices that challenge phallic-oriented sex, the power of these practices to undermine hegemonic masculinity is questionable.

1. How does Bear masculinity differ from the type of masculinity that predominates in other gay cultures?

2. Discuss the social and historical context of Bear culture in relationship to social class and race.

3. What are the three defining functions of the masculinized Bear body?

Bear Bodies, Bear Masculinity

Recuperation, Resistance, or Retreat?

Peter Hennen

One of the most intriguing features to appear on the queer cultural landscape in the past 20 years is the Bear subculture. During that time many gay men seeking to resist the stereotypical association of homosexuality with effeminacy have found the hirsute, masculine image of the Bear enormously attractive. For a significant cohort of men who came out in the late 70s and spent their youth reveling in the freewheeling post-Stonewall sexual culture, the Bear movement's emphasis on the appeal of the husky man provides an enticing antidote to the heartbreak of a slowing metabolism. Consequently, Bear culture has flourished in this country and expanded internationally. A resource Web site for Bears (http://www.resourcesforbears.com/CLUBS/US.html) lists 60 active clubs in cities across the United States, 6 in Canada, 14 in Europe, 5 in Central America, and 6 in New Zealand/Australia. Bear culture has spawned a number of popular books (*The Bear Book, The Bear Book II, Tales from the Bear Cult, Bearotica, The Bear Handbook*) and magazines (*BEAR Magazine, American Bear, American Grizzly*) and dozens of Bear-related Web sites. Several dozen Bear organizations sponsor social events, runs, or camping weekends every year, with the most popular attracting as many as 800 visitors from around the world. Interestingly, although people active in a variety of queer communities are likely to know something about Bears or Bear culture, the phenomenon is not widely recognized outside of these communities.

Just what is a Bear? Responses to this question reveal a variety of answers but almost all reference the Bear body in an attempt either to describe what the typical Bear looks like or to refute the idea that Bears can be defined exclusively by their bodies. As Travis, one of my interview participants, put it, "You know, physical attributes such as stockiness, height, weight, how much facial fur you have, things along those lines. But other people see it as being 90 percent attitude, 10 percent looks." What constitutes Bear attitude? Responses I encountered ranged from "natural, down-to-earth, easy going, likes to have fun" (Larry), "closer to the heterosexual community in their tastes" (Brian), "a sense of independence" (Burt), and finally "an easiness with the body" and "the masculinity thing" (Grant). "The masculinity thing" within Bear culture is complex and inextricably tied to the workings of hegemonic masculinity outside of it. "I think some of what is really appealing to me about the Bear group is that if you saw these guys on the street, they could just as easily be rednecks as gay guys," says Franklin. This suggests that the Bear image not only is conventionally gendered but includes a specifically classed presentation of self.

Bear culture was born of resistance. According to historian and founding figure Les Wright, in the early 1980s men frequenting leather bars in San Francisco and other cities began placing a small teddy bear in their shirt or hip pocket as a way of "refuting the clone colored-hanky code," whereby gay leathermen place different colored hankies in their back pockets to signal their interest in a variety of sex practices. Not willing to be objectified and reduced to an interest in one specific sexual activity, these men sported teddy bears to emphasize

From Hennen, P. 2005. "Bear bodies, Bear masculinity: Recuperation, resistance, or retreat?" *Gender & Society, 19*(1), p. 25. Reprinted with permission of Sage Publications, Inc.

their interest in "cuddling" (1997b, 21). According to Wright, this was a way of saying, "I'm a human being. I give and receive affection" (1990, 54).

Bears reject the self-conscious, exaggerated masculinity of the gay leatherman in favor of a more "authentic" masculinity. This look includes (but is not limited to) jeans, baseball caps, T-shirts, flannel shirts, and beards. To the uninitiated, Bears seem above all to be striving for "regular-guy" status. "The Bear look is all-natural, rural, even woodsy," noted Silverstein and Picano; "full beards are common, as are bushy moustaches. . . . They're just regular guys—only they're gay" (1992, 128–30). But are Bears "just regular guys"? Feminist scholars Kelly and Kane (2001, 342) saw subversive potential in this community: "Is there perhaps something radically subversive of orthodox masculinity at work here, despite all the butch trappings? Might not bears represent the sort of "marginalized men" that Susan Bordo describes as "bearers of the shadow of the phallus, who have been the alchemical agents disturbing the (deceptively) stable elements" of orthodox masculinity in a newly percolating social psyche?'

With the Bears' emphasis on camaraderie instead of competition, the rejection of "body fascism" (as evidenced by the acceptance of heavier and older men), and by popularizing cuddling and "the Bear hug," one finds ample evidence that this is not the type of masculinity that predominates in other gay cultures. As Wright remarked, "Competition with other gay men for sex partners and the depersonalizing effects of a steady stream of sexually consumed bodies is balanced by the humanizing effort to . . . establish contact with the person inside of each of those bodies" (1997c, 10). But at the same time, one finds signs of a recuperative current, a rejection of the insights of feminism, even outright hostility. As Lucie-Smith noted, "There is a challenge to aggressive feminism, which not only seeks female equality, but often tries to subject men to the tastes and standards imposed by women. To be a "Bear" is to assert a homosexual masculinism which rejects this" (1991, 8).

Thus, in staking their claim to gay masculinity, Bears challenge hegemonic assumptions about male sexuality by introducing what feminists have identified as an "ethic of care" (Gilligan 1982) into an objectified sexual culture perceived as alienating. On the other hand, insofar as their rejection of effeminacy signals a broader devaluation of the feminine, Bear masculinity recuperates gendered hierarchies central to the logic of hegemonic masculinity. Furthermore, the pastoral fantasy encoded in Bear semiotics can be linked with earlier movements aimed at revitalizing an "essential" masculinity under assault from the feminizing effects of civilization by retreating to the wilderness, if only symbolically. How then, from a feminist perspective, is one to adjudicate these simultaneously resistant and recuperative features of Bear culture? In this research, I draw on ethnographic and historical evidence as I attempt to make sense of these conflicting currents, with a special emphasis on the way that Bear masculinity is embodied and the effect this has on Bear sexual culture.

Method

Using a case study approach, I have designed this research in response to Stein and Plummer's (1994, 184) call for "a new paradigm for conceptualizing "identity in culture,'" and "developing an understanding of how sexuality, along with gender, race, ethnicity, class, and generation, is articulated and experienced within a terrain of social practices."

My case study community is a Bear club in a major American city, hereinafter referred to as the Friendly Bears.[1] As Bear organizations go, the Friendly Bears are somewhat atypical in that they do not hold regular meetings, do not charge membership fees, and operate with a relatively informal administrative structure. Like other Bear clubs, the Friendly Bears have a board of directors and a slate of officers, but their work is very low profile. The vitality of the club is maintained through monthly social events, all organized on a pay-as-you-go basis by volunteers and various fund-raising events for local charities. Most other clubs have regular meetings and membership dues, and the executive officers of those clubs tend to be much more visible. There is, however, a great deal of communication and interaction between Bear clubs. Some of this happens over the Internet, between individual members of different clubs via Bear chat rooms and message boards, but also through a series of regularly scheduled weekend events sponsored by clubs across the United States. These events, billed primarily as social opportunities, are often closely affiliated with local charities and typically draw men from a wide geographical area. In addition to the half dozen such events that may attract as many as 800 participants from all over the world, there are numerous events sponsored by clubs in midsized cities. These draw anywhere from 100 to 200 out-of-town guests. Thus, despite the peculiarities of my particular case study community, the extraordinary level of interaction between groups suggests a certain degree of national homogeneity across clubs in the United States.

My ethnographic data are drawn from approximately 300 cumulative hours of participant observation at various Friendly Bear sites during 2001 and 2002. As a graduate student and gay man in my early 40s, I had lived in "Friendly town" for more than 10 years before beginning this research. Thus, I had already developed a number of informal relationships with Friendly Bears, greatly facilitated by my own expanding middle-aged frame and natural hirsuteness. My response to these interactions was overwhelmingly positive, following along two distinct dimensions. In the beginning, the hedonic appeal of having my aging body recast in a significantly sexier social frame, through the approving glances of the Friendly Bears, provided the overwhelming appeal. But as time went on, I found myself engaging intellectually. What does the rapid growth of Bear culture mean, and how is it that these men manage to collectively reinterpret and eroticize the very physical attributes stigmatized by the larger gay community (extra weight, body hair)?

As a result of my situation, I found that gaining access to appropriate research sites was relatively easy. In addition to attending a number of semiprivate functions, I attended two Bear summer camping trips (each with more than 100 men attending), a smaller camping trip in the fall of 2001 (approximately 30 attendees), numerous "Bear Bar Nights" (typically hosting more than 150 men), and many casual face-to-face encounters. Observation sites also included "play parties" where sex happened, but this was never the sole purpose of the gathering.[2] The men I studied were overwhelmingly white (approximately 96 percent), and while there were a range of social classes represented, the majority of the men I observed would be most accurately classified as middle class (see below).

I conducted in-depth interviews with a total of seven men for this research, with interviews ranging in length from two to four hours. All interview participants were white and self-identified as middle class. Interview participant selection was guided by a theoretical sampling logic. I made an effort to recruit men who presented a more or less typical Bear image in terms of body size and appearance, but I also sought out participants who decidedly did not fit this profile (i.e., smooth-skinned, thinner men who nevertheless considered themselves members of the Bear community). I also attempted to sample a range of sexual attitudes and styles among my interview participants. At one end of this spectrum, my youngest participant spoke proudly of his sexual conservatism and devotion to monogamy. At the other end was a man of substantial experience who genially spoke of his desire to "have sex with as many men as possible." Finally, I tried to take account of participants' activity profile within the group. While I interviewed several Friendly Bears who had served as officers of the club, I also included men who were less active.

I also accessed historical materials pertaining to Bear culture, as well as contemporary commentary from various writers across the United States. These data include published accounts of Bear history, previously published interviews from key figures in the national Bear culture (particularly men who played formative roles in the late 1980s, as Bear culture was being forged in California's Bay Area), narratives chronicling the establishment of Bear clubs and organizations in other parts of the country, photographs, magazines targeted to Bears, and other related documents.

Bears in Social and Historical Context

In addition to its appeal as a hedge against effeminacy and its eroticization of the heavier body, there are at least two factors contributing to the emergence of the Bear phenomenon during the 1980s. One was, unquestionably, the AIDS pandemic and the effect of AIDS-related wasting syndrome on the erotic imagination of gay men. In an era when thinness could be linked with disease and death, the fleshier body was reinterpreted as an indicator of health, vigor, strength, and virility. The second contributing factor was the Bear movement's ability to co-opt an existing subculture that had been operating on an informal basis for decades prior to the "Bears" arrival on the scene. In 1976, a national network of "chubbies" (big men) and "chasers" (men who were sexually attracted to them) emerged as a new national organization called Girth and Mirth. A dozen years later, as the Bears became a recognizable subculture within the gay community, an uneasy relationship developed between the two groups. Interestingly, in many cities, Girth and Mirth chapters went into decline just as Bear organizations were cropping up (Suresha 2002). One reason for the out-migration from Girth and Mirth may be the more appealing imagery employed by the Bears. The iconic figure of the bear was enormously successful in linking the bigger body with nature, the wilderness, and more conventional notions of masculinity.

With respect to social class, Wright observed that "Bears' 'naturalness' registers in the key of 'blue collar'" (1997c 11). Bears present an image of working-class masculinity, yet many, if not most, are middle class, as Brian observes in this anecdote:

> I will never forget going to a—in fact I was a judge at—International Bear Rendezvous in San Francisco in—when was that?—' 97. And uhm, you'd see these guys, and they were all dressed like, you know, in the Bear drag, bubba drag, you know the, uh, flannel shirt and the ripped jeans, ripped flannel shirts, working boots, and all this sort of stuff, and they were all like—systems analysts at Sun Microsystems [laughs loudly]! I mean, they were all like these, they were all like computer geeks. Not one of them was—you know, like I was saying—a bricklayer, a plumber, a fireman, a policeman.

Because of their purported impatience with abstractions and their daily trials with the harsh realities of material life, working-class men have often been understood as more authentically masculine than their middle-class counterparts. As Connell observed, "Hard labor in factories and mines literally uses up the workers' bodies; and that destruction, a proof of the toughness of the work and the worker, can be a method of demonstrating masculinity" (1995, 36). Furthermore, working-class bodies have long held an erotic fascination for the middle class, as Wray suggested: "Any cursory reading of popular representations of lower-class whites suggests that the middle classes seem obsessed with what lower-class whites do or threaten to do with their sexual bodies" (1994, 1). For all these reasons, it is perhaps not surprising that middle-class Bears, in their revision of gay masculinity, would find working-class images appealing. What is surprising is the silence surrounding these issues, not only the unexamined, underproblematized acceptance of the equation of masculinity with working-class men but also the lack of reflection as to what it means when middle-class men do working-class drag. In this context, Brian's commentary is the exception that proves the rule.

Bear culture advertises itself as racially inclusive but remains overwhelmingly white. For example, my field notes indicate that on a typical Friendly Bear bar night, with more than 100 men attending, I saw 2 African American and 2 Asian American men. Similarly, at Bear Camp 2001, with an enrollment of nearly 120, 1 African American man and 1 Latino man attended. Fall Bear Camp attracted 54 white Bears and 1 African American Bear. My sense is that this is not simply a local problem. According to most of the printed discourse, the Bear body has nothing to do with white skin. To their credit, most Bear organizations actively seek to diversify their ranks, and racially inclusive language can be found on many Bear Web sites. Yet several writers mention the conspicuous absence of Bears of color in their communities. In two separate content analyses of Bear erotic magazines, both Locke (1997) and McCann (2001) commented on the predominance of white bodies. Kelly and Kane (2001, 344) asked why Bears "feel the need to adopt a rhetoric of racial inclusivity when the iconography of the texts before us is so overwhelmingly white."

The whiteness of Bear culture is probably due at least in part to the foundational image of the community (the bear itself) and how this image is perceived across racial lines. For most white men who join the Bear community, the appeal of the bear image is based on its association with masculinity and strength while at the same time signaling a capacity for tenderness and conviviality. But when, in the early 80s, the forerunners of the Bear movement sought to humanize the impersonality of the leather community by wearing teddy bears in their pockets, they were unwittingly drawing on a raced cultural history of white American masculinity. As Bederman (1995, 44) demonstrated, the inspiration for the teddy bear, Teddy Roosevelt, possessed a "talent for embodying two contradictory models of manhood simultaneously—civilized manliness and primitive masculinity." Civilized "manliness," she explained, was a character model that "comprised all the worthy, moral attributes which the Victorian middle class admired in a man" (p. 18). As such, manliness was intimately linked with whiteness. By contrast, "masculinity" was understood in essentialist terms, referring to "any characteristics, good or bad, that all men had" (p. 18). But here again, this primal masculinity was understood to be threatened by the feminizing effects of civilization. On Rotundo's (1993, 228) reading, white masculinist anxieties were further fueled by fears of domination by the more "authentic" masculinity of the tribesmen of "Darkest Africa," the "savage" Indian. Such descriptions of the recuperative back-to-nature narratives of the period reveal their racialized character.

Consequently, as the heirs of a raced cultural dynamic that equates the return to nature with whiteness, Bears may be unintentionally reproducing the raced appeal of the bear image. Exacerbating these effects is the racialized history of identification with animals. While many gay white men revel in their identification with the bear (this extends to purchasing Bear T-shirts, caps, vanity license plates, and other items of "Bearphernalia"), men of color may be much less eager to do so, in light of historically racist comparisons between animals and people of color (Becker 1973; Plous and Williams 1995).

Finally, in addition to class, race, and the dynamics of hegemonic masculinity working from outside of gay cultures, Bear culture is shaped by competing masculinities within gay cultures. The regular-guy masculinity of the typical Bear is a response to the hypermasculine clone phenomenon of the 1970s. The clone look emphasized a muscled, toned body and a presentation of self that was heavily influenced by certain iconic figures of masculinity (The Village People, a popular band during this time, delighted gay audiences by taking this impulse to the extreme). The Bear look was a reaction not against the clone's masculinity per se but rather against his hypermasculinity and the particular way that the clone displayed the body to signal that masculinity—hard, lean, muscled, toned, and smooth. If this is true, it seems to indicate that Bears are interested not so much in revising conventional masculinity but in resignifying it. Wright conceded as much when he acknowledged that "Bears are fully engaged with hegemonic masculinity, seeking an alternative answer, both accepting some of the trappings while rejecting others" (1997c, 6).

Given the context established in this section, what possibilities does Bear culture open up and close off in terms of gender resistance? How are the particular inflections of Bear masculinity manifest in the community's sexual culture? I turn now to an exploration of these questions.

Embodied Bear Masculinity

Bourdieu (1977, 45) proposed habitus, the deeply interiorized and embodied set of mental and physical dispositions that guide social action, as relatively durable but not impervious to change. He allowed that individual experience, or on a societal level, "times of crisis, in which the routine adjustment of subjective and objective structures is brutally disrupted," may indeed affect the habitus in profound ways. I want to argue that men who come to understand themselves as Bears experience just such a time of crisis. However, I first want to examine several of Bourdieu's more specific concepts as they relate to the embodiment of Bear masculinity.

Bourdieu called the first principle that serves to naturalize embodied masculinity "necessitation through systematicity" (1997, 194). Here he acknowledges the . . . primacy of gender as a "master binary": "The limit *par excellence,* that between the sexes, will not brook transgression" (Bourdieu 1990, 211); "[the] binary opposition between male and female appears founded in the nature of things because it is echoed virtually everywhere" (Bourdieu 1997, 194). Thus, the arbitrary "nature" of gender is obscured by virtue of its richly homologous relationship with other already gendered binaries: Hot (masculine)/cold (feminine), hard (masculine)/soft (feminine), outside (masculine)/inside (feminine) (Bourdieu 2001,13–18). By this method, the "arbitrary of the social *nomos*" is transmuted into "a necessity of nature" (Bourdieu 2001, 13). The critical point here is that masculinity is defined relationally, against the feminine. In Bear culture, this pattern is reproduced when Bears define their masculinity not only against the feminine but more specifically against the feminized, hairless, and gym-toned body of the dominant ideal of gay masculinity—"the twink," as he is dismissively known in Bear culture. Wright suggested, "When a Bear makes such a counter-statement, that he is not a 'woman,' not a 'twink,' not a 'heterosexual,' he is using his body to participate in changing social practice and challenging hegemonic power" (1997c, 9). I would argue that with respect to embodied masculinity, this statement obscures the fact that Bear masculinity simultaneously challenges and reproduces hegemonic masculinity.

Bourdieu's concept of "hexis" is instructive here. Closely related to habitus, but more specifically focused on deportment (i.e., ways of presenting and moving the body in social situations) as the physical instantiation of objective political and social relationships.

When Bears refuse to "do submission" or "do effeminacy" with their bodies, they in fact exercise a kind of embodied agency, insofar as the Bear body is perceived by heterosexual men as both "not heterosexual" and "not effeminate." Moreover, this is an agentic deployment of the Bear body that may act to radically destabilize the reified hegemonic narrative linking femininity with male homosexuality. However, this possibility is significantly complicated by the way that Bear masculinity operates *within* gay culture and how this is deployed against other homosexual men, I strongly suspect that of the three defining functions of the masculinized Bear body (not woman, not heterosexual, not twink), it is the twink that provides the real oppositional anchor for most Bears. In their virulent rejection of the effeminate stereotype and female drag, Bears certainly wish to convey that they are "not women," but in practice, this is accomplished indirectly, through an attack on the feminized, narcissistic body of the twink. Furthermore, while Bears may proudly acknowledge that they are "not heterosexual," this should not be read as a rejection of heterosexual masculinity. On the contrary, it seems that the vast preponderance of Bear discourse seeks to minimize the difference between Bear and heterosexual masculinity. On this reading, the Bears' challenge to hegemonic power is negligible, and the power relations reflected in the embodied hexis of Bear masculinity reproduce the hierarchical assumptions of hegemonic masculinity. Both assign lower status to bodies perceived as feminized.

Furthermore, despite their use of the twink as oppositional anchor, the "natural confirmation" (Bourdieu 1997, 194) that is the desired consequence of this "systematicity" remains problematic within Bear culture. This is because, in contrast with heterosexual masculinities, there exists no "rich homology" of binaries to obscure the arbitrary features of gay masculinities. Thus, Bear masculinity must be developed and sustained intersubjectively, within the community itself, an interactive process that is greatly facilitated by the symbol of the bear. The bear operates to link this new form of gay masculinity to the natural; it provides an opportunity for rich elaboration (through the designation of various types of Bear men as sexually submissive "cubs," sexually dominant "grizzlies," gray or white-haired "polar bears," etc.); and most important, through the nostalgic wilderness imagery it evokes, it links Bear masculinity with heteronormative masculinity.

But this construction remains unstable, its arbitrary nature easily revealed, as in this scathing assessment by Harris:

> Its hirsute ideal of rugged masculinity is ultimately as contrived as the aesthetic designer queen. While Bears pretend to oppose the "unnatural" look of urban gay men, nothing could be more unnatural, urban, and middle class than the pastoral fantasy of the smelly mountaineer in long johns, a costume drama that many homosexuals are now acting out as self-consciously as Marie Antoinette and her entourage dress up as shepherds and shepherdesses. (1997, 106)

My time in the field leads me to speculate that this fragility probably works to increase (rather than undermine) group solidarity among Bears.

Bourdieu identifies the other critical process that naturalizes embodied masculinity as "gendered socialization and the somatization of domination" (1997, 195). This describes the various practices that inculcate a gendered habitus during childhood, and Bourdieu further divides the process into four subcomponents.[3] Here I want to apply these processes to the revision of a gendered habitus in adult gay men and apply his ideas to the reconstruction of masculinity in Bear culture. The first practice is identified as "rites of institutions." These rites, such as ritual circumcision in many cultures, serve to underscore the difference between those who participate—men—and those who do not—women (p. 195). Participation, of course, keys directly off of the body. Local Bear organizations such as the Friendly Bears serve the same institutional purpose, and the Bear body becomes the point of reference for those who participate in Bear clubs, organizational planning, and activities. It is worth noting here that membership in these clubs is not strictly limited to men who self-identify as Bears. Most clubs welcome "Bears and their admirers," a phrase familiar to anyone active in this community. The inclusive description serves at least two purposes. First of all, it expands the possible membership beyond those who exhibit the typical Bear physical traits. But even as it does this, it underscores the centrality of the Bear body and its existence as an object of desire. Slim men, hairless men, younger men—all are welcome *provided they identify as Bear admirers.* I observed one such admirer at numerous Friendly Bear events. He was a relatively young, tautly muscled, smooth-skinned ex-gymnast. While he fit the physical description of a twink, his enthusiastic sexual interest in older "Daddy Bear" types meant that he greatly reinforced, rather than undermined, the intersubjectively sustained erotic of Bear sexual culture. As such, he was warmly welcomed in the club, and his interest in larger men was enlisted as supporting evidence of the "natural" appeal of Bears. Thus, the inclusive membership policy contributes significantly to a key agentic function of the Bear clubs—the embodied reassignment of the fleshier, hairier frame from stigmatized to desired object.

The next important process is the "symbolic remaking of anatomical differences." Here Bourdieu (1997, 195) explained that "the socially constructed body serves as an ideological foundation for the arbitrary opposition through which it was itself constructed." Bourdieu used the example of the interpretation of "swelling" and all its various analogies as based on a taken for granted association with the male erection and phallic swelling (2001, 13). In the case of the Bears, the association can (again) not be taken for granted; it must be actively constructed in community and applied to the swelling of the Bear's phallic body. On this reading, the Bear's generous frame, contrasted with the more compact frame of the twink, becomes a kind of homage to phallic power and masculinity.

Bourdieu identifies the third and final process I want to apply as "differential usages of the body and rites effecting the virilization of boys and the feminization of girls" (1997, 198).

Practices for Bears are instructed . . . by the normative structure of the entire group. The self-conscious attempt to dress and groom oneself like a "real" man approaches but never quite registers consciously as drag in the typical Bear consciousness. Nevertheless, this is an ongoing project among Bears, one requiring active construction and constant vigilance. This is perhaps best indicated by those attempts that are perceived as falling short of the prescribed mark. Fritscher (2002) complained, "There's nothing worse than seeing a big brute doing all this standing and posing at a Bear convention or in a Bear bar, only to then watch him pirouette out the door."

Returning to the concept of hexis, it would seem that these discordant displays of improperly masculinized "corporeal dispositions" are upsetting precisely because they reveal the constructed nature of what Bourdieu referred to as "the doxic experience of masculine domination as inscribed in the nature of things, invisible, unquestioned" (1997, 195).

Don's case is particularly interesting with respect to the social construction of the Bear body. Don grew up on a farm and attended high school in a small town, which I quickly surmised was a painful experience for him. He told me, "I came out to myself back when I was 9, 10—I knew I liked what I liked." In high school, Don weighed more than 350 pounds and was ridiculed for being heavy. During his senior year, things got uglier when he was outed by his classmates in a particularly public way. "They were chasing me down the hall with a video camera because they were putting together, like, this news footage and . . . they just outed me, and the next thing you know, I was the gay guy in school."

After graduation, Don wasted little time, waiting only three weeks before moving to the nearest big city. He also managed to drop a considerable amount of weight, and while he was still big, for the first time in his life, he began to feel good about his body: "I had a 45-year-old woman stop me on [Metro] Mall when I was about 20, and [she] said that if she were 20 years younger she'd make me her husband. I was just having my lunch—a chicken salad sandwich—and she came up storming, and, and she wasn't nuts. I mean she was a business-professional-type woman, and it was like she just said that I was an attractive young man. And I went, 'Well thank you,' and then it dawned on me that—well fine—I must be attractive."

But even as this encounter bolstered Don's self-confidence, it also highlighted his same-sex interests. As revealing as this encounter was for Don, it was not something he could pursue. His sexual self-confidence did not really blossom until he found the Friendly Bears, at which time he felt like he had "found family." "I mean, I'm big and hairy, it's obvious. . . . I found my niche, where I was welcome to be who I was and don't have to hide anything."

But Don makes it clear that finding the Friendly Bears was not just about finding interested sex partners. "I never really had a hard time finding sex," he tells me. "When I found the Bears, I found a lot more of what I liked in a man, within that culture. I sort of like knocked out the nellyisms. . . . I knocked out the, you know, the flaming drag queens." On discovering the club, Don was able to quickly parlay his good looks, stocky build, and tall stature into Bear social capital. He is currently very active in the group, both socially and sexually, and often makes a gregarious show of his affection for the community. At the campfire, according to my Bear Camp field notes, "[Don] seems to be running the show, making various bad jokes and, interestingly, using a variety of voices that incorporate growls and grunts into his speech. He is, I realize, talking like a bear. (10/12/01)"[4] His association with the

Friendly Bears has allowed him to come to terms with his traumatic high school experiences, albeit in a way that is obviously informed by hegemonic masculinity: "I still, I see five or six classmates that I went to high school with down at the bar now. So it's sort of like . . . I was sort a like, 'Uh, what was this—I'm the gay one and you're not? On your knees!' [laughter]."

Don's complex and contradictory journey is perhaps best summarized by again returning to Bourdieu's (1977) concept of hexis. Don's acceptance within the Bear community is reflected in a new understanding of his body and the way it can be deployed in social and sexual situations. Prior to finding his way into the Bear community, he wrestled with feelings of shame and inadequacy. From the Bears, he has learned to adjust his gait, posture, and gaze in a way that now signals strength, dominance, and virility. These traits are in turn read by other members of the community as evidence not only of a positive attitude toward his newly discovered Bear self but of an unselfconscious, natural disposition. Perhaps this is why, despite the obvious prominence of a specific body type among Bears, many members (e.g., Larry, quoted earlier) continue to insist that what distinguishes a Bear is "90 percent attitude." This emphasis on attitude may serve to underscore the importance of some of the explicit applications I have made in this section, as well as the utility of Bourdieu's work. As Bourdieu (2001) himself reminded us, it is the presentation of the gendered self as "natural" that obscures its constructed nature in the first place.

Sexual Culture

Given its paradoxical relationship to hegemonic masculinity, how distinctive is Bear sexual culture? What distinguishes the erotic imagination of Bears from that of other gay men? Kelly and Kane looked at Bear erotic fiction and noted the refreshing emphasis on support, nurturance, and playfulness included in the descriptions of sex that they analyzed, albeit with some caution: "I'm wondering whether this discourse of nurturance has to be presented through a discourse of sex in order to make it OK for men to participate? Or is it a way of reclaiming the whole body for eroticism and thereby dephallicizing the cock? And besides the nurturance, what about the playfulness? I think that really mitigates my discomfort with the wild man myth's seeming to reproduce old time sexism" (2001, 341). The concept of "dephallicizing the cock" is not a new one and speaks to the process Bourdieu referred to as the "symbolic coding of the sexual act" (1997, 197).

To what extent does Bear masculinity enable this recoding of the sexual act? In the previous section, I introduced the idea that Bears exhibit agency insofar as the embodiment of Bear masculinity simultaneously resists and complies with hegemonic masculinity. In this section, I discuss the implications of this paradox for Bear sexual culture, as I present evidence that Bear masculinity both challenges and reproduces an emphasis on genital sexuality.

"Sexuality among Bears is sensuality first," Burt tells me. I have seen enough in the field to appreciate what he is talking about. There is a great deal of emphasis on physical touch, both affectionate and sexual, between Bears. On this reading, institutionalized practices such as the Bear hug provide strong evidence of sexual innovation among Bears. My field notes from Bear Camp include an especially vivid example of this. I observed a spontaneous group Bear hug in the middle of the mess hall, wherein six men alternately engaged in kissing, fondling, and massaging each other. After several minutes, another man joined the group:

> I am surprised to hear him introducing himself to one of the other men kissing him, rubbing his body and "nuzzling" his beard. I can't help but notice how "bear like" the men's movements are, especially the rubbing and "nuzzling" of the face. This goes on for some time. Throughout the "hugging" people come in and out of the mess hall and take very little notice of the activity. The "hug" group is momentarily interrupted as new arrivals come in wondering if they can still get some supper. The guy in orange doesn't miss a beat, he takes them

back into the kitchen, the younger guy who had been giving him a backrub stands back from the group looking a bit bereft, but then he starts working on another guy and is drawn into a "group hug" with the remaining four. (7/5/01)

What impressed me here was the absence of any sharp division between the sexual/sensual activity and the practical activity in this scene. This strikes me as a way of claiming space, of sexualizing and sensualizing the everyday—an almost territorial ritual that seems to say, "This is Bear space now." This kind of sexualizing of space, for a variety of reasons, is not possible for these men in the outside world. I am also struck by the fact that other than a brief episode of genital fondling that I saw initially, this activity did not seem to be very genitally centered. All members of the group seemed to know the Bear hug "script," as evidenced by the easy accommodation of the newcomer and the casual introductions during the hug. There is also quite clearly something going on here beyond instrumental sexual "scoring." Franklin told me that this kind of contact has a very special meaning for him: "There was one bar night where about eight guys were all just kind of glumped together . . . like a litter of puppies—some feeling each other up, some hugging, just feeling good to be alive that way."

Burt observes that the emphasis on sensuality helps to foster a more responsible attitude toward safer sex practices. As a long-time member of the community, he has observed this among HIV-positive Bears. "I mean, think of all the varieties of sexual practices that we have to draw on. And we can enter in the 'not so safe' with a few men, but we don't have to. I mean, we can still pleasure total strangers if we want to without ever getting into the unsafe category or even close to it." This is significant because it demonstrates that at least for some Bears, fostering a sexual culture that decenters penetrative intercourse is a conscious and deliberate choice.

Burt clearly articulates this in his critique of what he calls "dick-oriented" sex. "You will find that the language amongst a lot of straight men that indicates subservience surrounds a quick sexual encounter. To fuck you is really meant to say, 'I'm gonna get my rocks off you and leave,' or 'She's a whore.'" When I ask Burt whether Bears would be more or less likely than other gay men to emphasize the importance of fucking in their sex lives, his initial response is equivocal. Eventually he settles the matter by telling me, "Only in the Bear group will you get the, the idea that there are other parts of the body that really bring the intense pleasure. And you can do it in many different ways." He smiles warmly and concludes, "I think we're damn good at sex, to be perfectly honest."[5]

Beyond this, however, my time in the field yielded little that distinguished Bear sexual culture from others' in terms of specific sex practices. When I ask Larry about this, he relates the question directly to the Bear body. "I'd say there's some things I only do with a guy who's hairy . . . like nuzzling chest hair—I can do that for a long time—I love it." Given the obvious emphasis on sensuality and increased attention to touch among Bears, I was surprised when several of my participants explicitly rejected the idea of distinctive sexual practices:

> Sex is sex—one form, shape, or another. (Don)

> I think in some respects, when it comes down to it, the sex is sex. . . . When it comes down to the basic sexual practices, it's all the same. (Travis)

Clearly, not all Bears understand their sexual activity in Burt's more expansive terms. At a Bear play party, I witnessed a scene suggesting that some Bears understand sex in fairly narrow terms, centered on penetrative intercourse. My field notes describe "a brief but enthusiastic fuck session" involving a Bear visiting from out of state. Afterward, the visitor proudly proclaims to the small group of men watching the scene, "*That's* the way we do it in Texas!" After a brief pause, I hear another onlooker wryly reply, "That's the way we do it in [this state] too."

When I ask Brian how he regards intercourse, he smiles and admits, "I mean, honestly . . . everything else is an appetizer [laughs], you know?" On the other hand, Larry seems to concur with Burt, while once again directly referencing the Bear body. "I just base it on more the enthusiasm, the enjoyment of the touching, the feeling, the nipple play, the kissing." Don tells me that intercourse itself is not important to him, but because of his large frame and aggressive personality, and because it is important for other men, he finds that he is often asked to play the top (insertive partner). He responds to these requests with a curious mixture of care and machismo: "If they want it, they'll get it. They'll get it good." Finally, Travis responds to my question about the importance of intercourse in a way that equates it with "real" sex, a definition shared by all "real" men. "You know, guys are guys. They're gonna have sex, you know, whether you're, whether you're in the Bear community or if you're in the gay community in general." Thus, it seems that while Bear culture does yield a number of sexually innovative practices that disperse pleasure across the body and disrupt genitally centered, phallus-and-receptacle interpretations of sex, these innovations coexist with (rather than displace) an attendant set of practices that sustain the phallic emphasis on insertive intercourse.

It seems clear that Bear sexual culture has been heavily influenced by hegemonic masculinity and, to a lesser extent, heteronormativity. But there is also ample evidence of resistance and sexual innovation within this subculture. Perhaps the most accurate way to conclude my observations with respect to Bear sexual culture would be to say that the practices that disperse pleasure across the body coexist with, rather than displace, the phallic emphasis on insertive intercourse.

Conclusion

Do Bears make gender trouble (Butler 1990)? What does it mean when Silverstein and Picano observe of Bears, "They're just regular guys—only they're gay" (1992, 128)? Clearly, there is a move toward normalization here, as well as an identification with heterosexual men, a move that may ironically turn out to be profoundly disruptive of hegemonic masculinity. When Franklin remarks, "Some of what is really appealing to me about the Bear group is that if you saw these guys on the street, they could just as easily be rednecks as gay guys," he speaks for many men who identify as Bears. Herein lies the possibility of subversion, as Bears have been largely successful in divorcing effeminacy from same-sex desire and creating a culture that looks like a bunch of "regular guys." The subversive implications, however, have everything to do with reorganizing sexuality and very little to do with challenging gendered assumptions. Most of these men would like nothing more than to have their masculinity accepted as normative, something that is largely accomplished within the group but remains problematic outside of it.

How is it that Bears come to understand their particular brand of masculinity as natural? It seems clear that this is accomplished quite deliberately, through the appropriation of back-to-nature masculinity narratives that are sustained intersubjectively, as group members reinforce these meanings and associations through their day-to-day interactions. Thus, Bear culture seems currently disposed toward renaturalizing rather than denaturalizing gender relations. It seems far more likely, then, that increasing acceptance of Bear masculinity will encourage greater investment in a heteronormative sexual culture, less experimentation with new pleasures, less dispersal of pleasure across the body, and a renewed appreciation for insertive intercourse as "doing what comes naturally." In this case, the perceived naturalness of the Bear body may be extended to naturalized understandings of sex practices that are increasingly compliant with norms of hegemonic masculinity.

Notes

1. The name I have given my case study community is a pseudonym, as are the names of all of the individual members of the Friendly Bears mentioned in this article. Sources or commentary not specifically attributed to a member of the Friendly Bears should be understood as applying to or coming from the broader national community of Bears or writers commenting on the same.

2. Again, my prior association with the Friendly Bears made the observation of sexual activity relatively

unproblematic. I was an occasional participant in these parties before I began my research, so formal observation entailed simply introducing a higher degree of methodological rigor to a familiar activity. As was the case at other (nonsexual) observation sites, I recorded extensive field notes as soon after leaving the site as possible. The sexual nature of observations made confidentiality an especially important issue, but I saw little reason for additional concern. As a sex-positive community, "play parties" like the ones observed in this study have been a central component of Bear culture since its inception (Wright 1997a), and many Bears are quite frank about participating in them. However, I found a range of attitudes about this type of activity among the Friendly Bears. Two of my interview participants characterized themselves as "vanilla" (conservative) in their sexual tastes. In addition, in his study of Bear erotica, McCann (2001) characterized Bear culture as sexually conservative. The conservative characterization did not go unchallenged, however. One of my participants (Travis) described Bear culture as "almost no holds barred." Another (Franklin) confessed his bewilderment at the popularity of open (nonmonogamous) relationships among Bears.

3. I deal with only three of these in this section: "Rites of institutions," the "symbolic remaking of anatomical differences," and "the rites affecting the masculinization of boys and feminization of girls." The significance of a fourth subprocess, the "symbolic coding of the sexual act" is elaborated in the next section on Bear sexual culture.

4. During my time in the field, I also observed more than one man who quite literally growled during sexual activity, something I have not observed outside of the Bear community.

5. Burt's status among the Friendly Bears is significant here, given his resistant sexual philosophy. In addition to being a long-time member, Burt served for several years as an officer of the club. The overwhelming impression I received from my time in the field is that he is an admired, highly respected, and beloved member of this community.

References

Becker, J. 1973. Racism in children's and young people's literature in the Western world. *Journal of Peace Research 10*(3): 295–303.

Bederman, G. 1995. *Manliness and civilization.* Chicago: University of Chicago Press.

Bourdieu, P. 1977. *Outline of a theory of practice.* Cambridge, UK: Cambridge University Press.

———. 1990. *The logic of practice.* Cambridge, UK: Polity.

———. 1997. The Goffman Prize lecture: Masculine domination revisited. *Berkeley Journal of Sociology* 41:189–203.

———. 2001. *Masculine domination.* Stanford, CA: Stanford University Press.

Butler, J. 1990. *Gender trouble.* London: Routledge.

Connell, R. W. 1995. *Masculinities.* Berkeley: University of California Press.

Fritscher, J. 2002. Bearness's beautiful big blank: Tracing the genome of ursomasculinity. In *Bears on Bears,* edited by R. Suresha. Los Angeles: Alyson Books.

Gilligan, C. 1982. *In a different voice.* Cambridge, MA: Harvard University Press.

Harris, D. 1997. *The rise and fall of gay culture.* New York: Hyperion.

Hocquenghem, G. [1972] 1996. *Homosexual desire.* Durham, NC: Duke University Press.

Kelly, E. A., and K. Kane. 2001. In Goldilocks's footsteps: Exploring the discursive construction of gay masculinity in Bear magazines. In *The Bear book II,* edited by L. K. Wright. New York: Harrington Park Press.

Locke, Philip. 1997. Male images in the gay mass media and Bear-oriented magazines. In *The Bear book,* edited by L. K. Wright. New York: Harrington Park Press.

Lucie-Smith, E. 1991. The cult of the bear. In *The bear cult.* Swaffen, UK: GMP.

McCann, T. 2001. Laid Bear: Masculinity with all the trappings. In *The Bear book II,* edited by L. K. Wright. New York: Harrington Park Press.

Moon, M. [1972] 1996. New introduction. In *Homosexual desire,* by G. Hocquenghem, Durham, NC: Duke University Press.

Plous, S., and T. Williams. 1995. Racial stereotypes from the days of American slavery: A continuing legacy. *Journal of Applied Social Psychology* 25:795–817.

Rotundo, E. A. 1993. *American manhood.* New York: Basic Books.

Silverstein, C., and F. Picano. 1992. *The new joy of gay sex.* New York: Harper Perennial.

Stein. A., and K. Plummer. 1994. "I can't even think straight": "Queer" theory and the missing sexual revolution in sociology. *Sociological Theory* 12 (2): 178–86.

Weeks, J. [1972] 1996. Preface to the 1978 edition. In *Homosexual desire,* by G. Hocquenghem. Durham, NC: Duke University Press.

Wray, M. 1994. Unsettling sexualities and white trash bodies. Association paper presented at the American Studies Working Group, University of California, Berkeley, 10 October.

Wright, L. K. 1990. The sociology of the urban Bear. *Drummer* 140: 53–55.

———. 1997a. The Bear hug group: An interview with Sam Ganczaruk. In *The Bear book,* edited by L. K. Wright. New York: Harrington Park Press.

———. 1997b. A concise history of self-identifying Bears. In *The Bear book,* edited by L. K. Wright. New York: Harrington Park Press.

———. 1997c. Introduction: Theoretical Bears. In *The Bear book,* edited by L. K. Wright. New York: Harrington Park Press.

Introduction to Reading 28

Mary Nell Trautner's study of strip clubs offers intersectional analysis of gender, social class, and sexuality. Her data show that exotic dance clubs create and present performances of gender and sexuality that differ according to the social class of customers. Class differences are represented as sexual differences in very obvious ways: the appearance of dancers and staff, dancing styles, interactions between dancers and customers, and the physical characteristics of the clubs themselves.

1. What were the research challenges faced by the author and how did she "overcome" the challenges?
2. Discuss the role of "good girl" and "bad girl" stereotypes in the images of attractiveness created by the dancers. How are those stereotypes linked to social class?
3. Why are middle-class clubs associated with voyeuristic sexuality while the working-class clubs are more openly sexual?

Doing Gender, Doing Class

The Performance of Sexuality in Exotic Dance Clubs

Mary Nell Trautner

One of the key findings of contemporary feminist scholarship is that organizations and occupations are often gendered—that is, they draw on notions of femininity or masculinity that are hegemonically defined. Building on the idea of gender as a performance (Butler 1990; Moloney and Fenstermaker 2002; West and Zimmerman 1987), scholars find that

Trautner, Mary Nell. 2005. "Doing gender, doing class: The performance of sexuality in exotic dance clubs" *Gender & Society* *19*(6), p. 771. Reprinted with permission of Sage Publications, Inc.

workers in a wide range of occupations and organizations "do gender" in particular ways, based on assumptions about what customers like, motivations, and "normal" interactive behaviors (Acker 1990). Particularly in service-oriented occupations, women work as women, as femininity is constructed and reified in ways that reinforce heterosexuality and male dominance and "naturalize" stereotypical images of women (Dellinger 2002; Dellinger and Williams 2002; Hall 1993; Leidner 1991; Loe 1996; Williams, Giuffre, and Dellinger 1999). Through the continual performance and institutionalization of gender and gendered behaviors and rituals, gender and sexuality become central features of organizational culture—those shared understandings, beliefs, behaviors, and symbols that emerge through interactions between organizational actors (Dellinger 2004; Dellinger and Williams 2002; Gherardi 1995; Hallett 2003; Trice 1993).

While the concept of the gendered organization has been critical to our understanding of how and why sexuality and gender are core features of many jobs, what has received less attention is why some organizations—particularly those that are very similar to one another—exhibit different forms of gender and sexuality (Britton 2003; Dellinger 2004). To examine this question, this article builds on the idea of gendered organizations. I argue that gender in organizations interacts with other major features of stratification—such as class and race—to construct unique organizational cultures that project distinctive images of gender and sexuality that are fitted to their particular organizational settings. I show that the activities and practices of strip clubs construct forms of sexuality that are not only gendered but also distinctively classed—that is, they articulate ideas and presentations of gender that are mediated by class position. I explore this idea of organizations as gendered and classed through a comparative ethnographic analysis of the performance of sexuality in four exotic dance clubs in the southwestern United States.

In exotic dance clubs, women at work must act like women by embodying traditionally female behavior and roles as well as by dressing and behaving femininely. Because the central features of the organizational culture within exotic dance clubs are the commodification and commercialization of women's sexuality, the clubs are premised on the consumption of women's bodies and the presence of those bodies in hegemonic male fantasies. Thus, women work not only as women but as sexualized women. Yet despite having similar underlying institutional logics, clubs offer noticeably different presentations and performances of gender and sexuality. My data demonstrate that exotic dance clubs have different organizational cultures based on distinctions made by the perceived social class of customers. Clubs construct sexuality to be consistent with client class norms and assumptions and with how the clubs and dancers think working-class or middle-class sexuality should be expressed. Those clubs that cater to a middle-class audience present one version of sexuality, while a quite different type of display can be found at working-class clubs. As a result, women in exotic dance clubs work not only as sexualized women but as classed women.

Method

To explore the ways in which social class and organizational culture influence the performance of sexuality within strip clubs, I made a total of five visits each to four exotic dance clubs in Pueblo,[1] resulting in more than 40 hours spent in the field. The advantage of a prolonged direct observation technique in this setting is that I was able to experience the club settings and routines as both a first-time club-goer and a more seasoned customer, familiar with the settings, members, and activities. These four clubs, The Oasis, The Hourglass, The Treasure Chest, and Perfections Showclub, are the busiest, most well-known, and most popular clubs in town. Each of these clubs serves alcohol, which by state law means that they are topless only, as opposed to fully nude. Because my sample is derived from just one city in one state, my findings highlight differences within the boundaries of this particular state's laws. While clubs in other locations would no doubt be responsive to variations in state and city laws, I believe that variations in social class norms would continue to be as salient as I found them to be in Pueblo.

Most clubs in Pueblo allow a woman to enter as a customer only when accompanied by a man. Although it is commonly believed that rules governing the admittance of women were created to prohibit prostitutes and lesbians from entering the club, a private conversation with a club manager in Texas (Trautner 1998) revealed that an additional function of these rules may be to prevent jealous wives and girlfriends from entering the clubs and physically harming dancers and/or customers. Consequently, I presented myself not only as a paying customer but also as either the girlfriend or friend of my male escort(s) to observe naturally occurring interactions and club routines. Like many women researchers who enter strip clubs, my presence in the club did seem to be noticeable to both customers and dancers. To minimize the intrusiveness of my presence, I followed the techniques outlined by Wood (2000): I visited each club frequently (five times each) and for long periods of time (at least two hours each visit), which allowed me to blend into the scene and become less conspicuous to those around me.

At each site, I assumed the role of the naïve stranger to blend in with the crowd as much as possible by looking and acting much like the typical woman customer and also to learn as much as possible about how each club operated. This role involves acting naïve, curious, and responsive but very unknowledgeable about the setting, unspoken rules, and activities taking place, which encouraged members to explain and elaborate on the customs and expectations of the club (Morrill 1995; Snow, Benford, and Anderson 1986). I paid particular attention to interactions between dancers and customers, appearances of dancers, and styles of stage and table dances. A dance is the length of one song, which is usually about three minutes long. This means that on any particular stage, about 40 dances occur in the space of two hours. Field data were collected between January and July of 2001.

The drawback to my covert position is my lack of insight into the club employees' thoughts and feelings toward, and explanations of, the routines in which they participate. In an attempt to sort out this issue, I conducted supplemental in-depth interviews with three exotic dancers in the summer of 2002. These women work at the clubs I observed while in the field and were selected based on prior personal contacts. That is, these particular women were not observed in the field and were thus not selected for their typicality nor their unusualness compared to other dancers at their respective clubs. I was interested in learning the extent to which the patterns and trends I observed in the field were reflective of participants' experiences and actual organizational strategies. Thus, I asked them questions about their style of dancing, the dance styles of other dancers, management involvement, how they interact with customers, the kinds of customers that frequent their club, and their perceptions of and experiences with other clubs in town. Interviews were conducted in respondents' homes and lasted approximately two hours each.

Sexuality in the Strip Club

The four clubs I visited serve two distinct clientele: Perfections Showclub and The Oasis market themselves to a middle- and business-class clientele, while The Hourglass and The Treasure Chest serve primarily working-class and military audiences. While the focus of this article is on how these clubs do class through the performance of sexuality, I should note that they do class in other ways as well. Generally, the middle-class clubs price everything higher than do the working-class clubs, although the differences are often minimal (i.e., a bottle of beer costs fifty cents more at the middle-class clubs, and the cost of admission is $6 and $7, in contrast to $3 and $4 for the working-class clubs).

Clubs also do class in large part through their physical characteristics. These characteristics, such as the state of the parking lot, quality of the lighting and sound systems, club furnishings, amenities offered to customers, and physical layout of the club, signal to potential customers what kind of club it is and what kinds of sexual experiences customers might expect. In doing so, they also encourage customers to become middle-class or working-class "performers" (Bettie 2003). That is, regardless of their own class background, customers can experience a middle- or working-classed event and be seen as a middle- or working-classed individual through their consumption of sexuality as organized by the exotic dance club. By providing their customers with cigars, gourmet meals, soundproof phone booths (presumably to call home or the workplace without revealing the nature of their location), and plush, relaxing arm chairs, the middle-class clubs make the club-going experience about more than just sex, more than just viewing unclothed women. They appear to make every effort to insulate customers from everyday reality by providing them with a safe haven in which they can desire and appreciate women and act and be treated like "gentlemen" (Edgley 1989). These clubs, as I will demonstrate, are characterized by performances of desire and gazing at the female form from a distance, constructed to appear as admiration and respect. I refer to this as "voyeuristic sexuality."

In contrast, the working-class clubs create an atmosphere conducive to pure physical pleasure and lust. Gone are the amenities, high-quality equipment, and soft, comfortable furniture. Customers, who are mostly working-class performers, are able to come to these clubs for vicarious sexual experiences and little else, as the sexuality that is on display is often more interactive than is seen at middle-class clubs. These clubs are havens for the viewing of women as sex objects, for the imagining of these women as sexual partners, and for the enactment of male power (Liepe-Levinson 1998; Wood 2000). This form of sexuality I call "cheap thrills" sexuality.

In addition to the physical characteristics of the clubs, these two forms of sexuality and gender—voyeuristic and cheap thrills—are constructed and institutionalized in various performative aspects of the clubs as well. I argue that these performative aspects—the appearances of the dancers and other staff, the dancing and performance styles, and the interactions that take place between dancers and customers—are as indicative of class and classed expectations as they are of sexuality.

Images of Attractiveness

There is a general difference in the appearance of the women at the middle-class and working-class clubs. Dancers conform much more closely to the hegemonic cultural ideals of attractiveness at Perfections and The Oasis than do dancers at the working-class clubs. In these middle-class clubs, there is a narrowly restricted range of women's body types. For example, there are very few overweight dancers, women with short hair, older women, women with strong musculature, or non-white women. About half of the dancers at each middle-class club appear to have breast implants, and most of the others have naturally large breasts. In fact, one woman I interviewed, Mandy, commented that The Oasis, where she works, "is known for the most . . . for all the girls having them. They call it 'Silicone Valley.'" While there are some small-breasted women, there are comparatively fewer working in these clubs, approximately 10 to 15 percent (as compared to approximately 40 percent in the working-class clubs). Most of the women wear their hair styled in some way (i.e., curled, gelled, sprayed), but all wear their hair loose, flowing down their shoulders and back. Only occasionally will a dancer wear her hair in pigtails to match a schoolgirl costume. All of the women wear makeup, and the majority of the dancers heavily accentuate their eyes with glitter, eyeliner, or eye shadows. Most have long fingernails painted in light or neon shades that reflect the black lights of the club. Mandy says these features describe women who are "classy looking" (at least in this context) and that this is a look that develops over time, as the women become accustomed to the ways in which "Oasis girls" look and "take care of themselves": "Some girls start out and they are so ghetto looking, but then she works and she starts to look better and starts to take more appreciation in what she looks like—tanning, and doing her hair more, and just . . . *changing.* As they are around the other girls and see how the girls keep up themselves, they start to change themselves, because they have the money to also take care of themselves."

According to Mandy, one reason the dancers pay such attention to makeup use and hairstyle is that the amount of money they make from customers is contingent on how sexy, attractive, and feminine they appear to their audience (see Price 1998). While this is true of any club, for middle-class clubs, there is a heavy emphasis on conforming to middle-class cultural ideals. Marina, a dancer at The Treasure Chest, one of the working-class clubs, also recognizes this feature of the middle-class clubs as she comments that "The Oasis and Perfections are probably the best for girls that have ideal bodies according to standards by society today . . . like mainstream standards, pop culture standards."

Fixing their hair, tanning, wearing perfume, and applying particular kinds and shades of makeup not only symbolize doing heterosexuality and femininity—practices that reproduce and naturalize the dominant cultural norms of heterosexuality (Dellinger and Williams 2002; Giuffre and Williams 1994)—they symbolize doing class as well, as the performers distance themselves from women who are "ghetto looking." By using their appearances to simultaneously do gender, heterosexuality, and class, these women increase their financial gains while at the same time conforming to, legitimating, and perpetuating dominant cultural ideals.

At The Treasure Chest and The Hourglass, however, there is a much broader spectrum of female bodies on display. There are several overweight women, as well as some women who are so thin their entire skeletal structure is visible through their skin. There are a few older women working at each club (40-ish), and there is a greater diversity of dancers in terms of race. In contrast to the middle-class clubs, which feature predominantly white women, each of the working-class clubs employs a (relatively) large proportion of Latina dancers, with a few Black and Asian women. On a typical night with 30 dancers, approximately 15 are women of color, compared to about 5 of 30 in the middle-class clubs, thus more accurately reflecting the racial composition of Pueblo. Most women also tend to have long hair, and nearly all the white women have bleached-blond hair. Women are also more creative with their hairstyles. One woman at The Hourglass has a completely shaved head, and other women wear their hair in braids or pulled back into ponytails or barrettes. Marina notes that at "blue-collar" clubs such as The Treasure Chest, "you can see more of the personalities of all these people, which is what I am really interested in. The girls can do whatever they want, and do." Women apply heavy makeup that accentuates their mouths, rather than their eyes as in the middle-class clubs. Most wear dark or bright red lipstick and paint their long fingernails to match, styles typically associated with working-class women (Bettie 2003).

Another aspect of attractiveness is the types of clothing that dancers wear. While by law all dancers must wear a G-string, there are considerable differences in the other types of clothing and accessories worn. Dancers at the middle-class clubs tend to wear outfits—either themed costumes such as a dominatrix or a Catholic schoolgirl outfit or pieces of lingerie like a satin chemise or teddy that covers both the breasts and the buttocks. A few of the dancers wear much more elaborate outfits, such as bodysuits or minidresses. Some dancers choose to wear accessories to appeal to particular members of the audience, like cowboy hats or baseball caps, and some accessorize with thigh-hi stockings or a garter.

All the dancers wear high heels, but they wear the chunky heeled platform shoes that were in fashion at the time, rather than traditional stilettos, as worn in the working-class clubs. Dancers at these clubs wear more jewelry than seen at other places, and many even wear wristwatches. Mandy and Veronica both revealed that there was a time when all the dancers were required to wear evening gowns. Veronica says, "When [the present owner] first bought The Oasis, he really wanted to target like age 30 and up, white-collared-class businessmen. So he said that all the girls had to wear gowns or dresses on stages, and we're only going to play 80s music. That was his way to appeal to that age group."

Few of the dancers at The Hourglass or The Treasure Chest wear costumes or anything that could be dubbed an outfit. Most of the dancers wear a bra-like top and their G-string, with nothing else. Occasionally, a dancer will wear some type of lingerie or dress that covers both the buttocks and the breasts, but it is rare. As one of my escorts commented, "It feels like we're at the beach!" Marina agrees but notes that she dresses a little differently than do most of the other dancers at The Treasure Chest, emphasizing how clothing—"covering up"—reinforces and reproduces classed expectations: "A lot of girls will just walk around in just a bikini," she says, "but I'll wear dresses, I'll cover up more and be more like . . . classy. Not that I *am* more classy, but I have that facade." Note also that Marina is highlighting the difference between class as a material location ("not that I *am* more classy") and class as a symbolic performance ("but I have that façade").

What these body images amount to is a complete catering to the cultural ideals and perceived fantasies in the middle-class clubs and a wider array of images of women's sexuality and appearance in the working-class clubs. That the dancers at The Oasis and Perfections draw attention to their eyes suggests an invitation to look and an aura of mystery—they are meant to see and be seen. The red lipstick that accentuates the lips of the dancers at the working-class clubs oozes sensuality, fire, and excitement. In a sense, the dancers draw from two different yet equally stereotyped images of femininity: the good girl (who looks but does not touch, "innocent" in her sexualized schoolgirl outfit) and the bad girl (who falls outside the hegemonic beauty ideals and flaunts her exaggerated sexuality). Yet as Ortner (1991), Bettie (2000, 2003), and others have pointed out, these stereotypes are as much about class as they are about femininity.

Stage Performances: Constructing the Gaze

The stage is perhaps the most visible and obvious place in which sexuality is performed. On a main, center stage of a club, nearly everyone in the audience has a view of the show, and as such, performing on the main stage is a dancer's main method of being seen, making eye contact with customers, and finding people for whom she may perform table dances later in the evening. (Ronai and Trautner 2001). For many customers, the stage area is their only experience with dancers, as many patrons never purchase table dances at all, preferring to watch the constant and varied entertainment provided on the stage. This is where dancers can show off their bodies, show of their athletic and dance abilities, and show customers, through their dance style and choice of music, the personalities they have constructed for their performances.

There is a distinct dissimilarity between the styles of music featured at each of the clubs. While all the women are allowed to choose their music for dancing, there is much less variation than there would be if organizational cultures had no effect on music style. The songs heard at Perfections and The Oasis, the middle-class clubs, are generally contemporary pop music, such as one would find on the Billboard Top-40 or on the television station VH-1 (recall that the club owner at The Oasis required all dancers to perform to hits from the 1980s a few years ago). The songs are for the most part slower, with lyrics that are decipherable. Most of the songs feature male vocalists (such as songs by Third Eye Blind or Matchbox 20), although some instrumental techno songs, as well as some music by Madonna and Janet Jackson, are played. Veronica reveals that a few years ago, "no rap was allowed. None. No rap. But now they play some. You really gotta appeal to everyone." A typical evening, however, features only one or two rap songs.

In contrast, the music at The Hourglass and The Treasure Chest is remarkably different from the middle-class clubs. Most of the songs played are rap songs, heavy metal, and classic rock. There are very few pop songs played, and even fewer songs with women vocalists. Marina describes the music in similar terms, stating that "there's girls who dance to rap music and they'll have more of a 'tough' side. There's girls who do country, and there's girls that do really angry, evil music. And it doesn't necessarily mean that the girl *is* that way, it's just the style that she dances. She'll, like, appeal to the S&M crowd." At the working-class Treasure Chest, where they "can do whatever they want," dancers exercise even more control over the music in the club, while simultaneously playing up their classed, sexualized, "bad girl" image. Marina describes scenes in which "girls will go up on stage and if they don't like the music the DJ is playing, they might tell him 'fuck you,' or they might just lay there and not dance. It is *crazy.*" In other words, dancers exercise social control over each other, DJs, and managers by not dancing to songs that fall outside the club's regular style of music.

The dancers at the middle-class clubs take a very passive approach to their stage performances, meaning that the dances are noninteractive. Most of the dancers get on stage fully clothed about 10 to 15 seconds after their song has begun to play and slowly walk around the stage, stopping every few seconds to strike a pose, in which they arch their back, lean their head back, lift their arms up to their head, and lift their hair up off their shoulders and then let it cascade down their back, shaking their head from side to side ever so slowly. About halfway through the song, the dancer will remove her clothing while on stage and let the clothes fall gently to the floor. She then will continue to saunter around the stage, occasionally lightly touching the sides of her breasts with her fingertips as she moves her arms upward to play with her hair. The aspect of the performances that struck me the most on my first visit to the middle-class clubs was that only three dancers on stage ever touched their breasts while dancing. Of these, none touched their nipples. Instead, they ran their hands across the top of their breasts, almost like a shelf they were checking for dust. One of these women lightly ran her hands up and down the sides of her torso, just barely touching the sides of her breasts with her fingertips. This was a pattern repeated on every visit to the middle-class clubs. Few of the dancers incorporate actual dance moves, and even fewer attempt to move to the rhythm of the music. The focus appears to be on showing off, presenting a sensual and delicate image of sexuality, and making sure that all eyes follow her as she strolls around the stage (creating an audience of voyeurs). Before the song is even over, she picks up her clothes and gets off stage, putting her clothes back on in the process. Consequently, there are about 20 to 30 seconds with no dancer on stage, as the women arrive on stage late and leave early.

To move into one of the working-class clubs is to move into an entirely different world. The music is fast, and the dancers are very active and somewhat rowdy during their performances. While some women do perform the slow, noninteractive dance routines characteristic of the middle-class clubs, these performances are rare, and only 3 or 4 (of approximately 120) will be witnessed over the course of a two-hour time period on all three stages. Most of the dancers use the poles on stage as props to hang onto as they spin around, to lean against as they bend over to give the audience a view of their buttocks, as phallic symbols on which they gyrate, or as climbing devices that they shimmy up and then grab onto with their legs as they suspend their bodies upside-down from the poles. Many of the dancers also display their flexibility by performing the splits on stage, or in the case of a few dancers at each club, lying flat on their backs with their legs up in the air and moving them around from side to side. All of a sudden, a dancer would spread her legs widely and slam them down in a center-splits position onto the stage, causing her shoes to hit the stage with a loud thumping sound. She would repeat this move two or three times before finding other ways to demonstrate her flexibility.

Many of the women in these clubs dance as if they are having sex without a partner. An excerpt from my field notes shows how one woman, Crystal, had a routine in which this style of dance was captured perfectly:

> She kept her eyes closed through most of the performance and rolled her head with a slight look of pleasure and intensity on her face. She would grab the outsides of her breasts and cup them upwards and together, and then arch her back, lean her head backwards, and smile. Soon after she began doing this, she pushed her bikini top up over her breasts so that it was all smooshed up at the top of her chest, and fell to her knees. She leaned backwards, back still arched, with one hand on the floor behind her to prop herself up, the other hand now rubbing the insides of her thighs as she thrusted her body up and

> down, back and forth. She began to move faster and harder, breasts bouncing up and down, hand moving towards her head where she would clench a piece of her hair up by her forehead, then let go and drag her open-faced hand down her neck, over her breasts, down the side of her stomach to her outer thigh, ending up back at her inner thigh.

Crystal is giving an explicitly sexual performance— she is showing the audience what it would be like to have sex with her, or what she wants them to think it would be like. The focus is still on watching her, but she invites looks of lust and desire rather than cool contemplation and distanced admiration. Other women would position themselves on stage on all fours with their buttocks facing the crowd. These women would sometimes arch their backs or turn their heads to also have a view of the audience while they shook their bottom and thrust their body back and forth.

Tipping and Table Dances: Interacting with Customers

Dancers expect to receive tips while performing on stage. The stage tip is a customer's way of communicating that he likes what he sees and perhaps that he may be willing to purchase a table dance from the dancer later in the evening. For the dancer, accepting a stage tip simultaneously functions as a way for her to advertise herself to the entire audience and for them to imagine what getting a table dance from her might be like (Ronai and Ellis 1989). Much like the dancing routines themselves, the styles by which dancers accepted tips were characteristic of very different forms of sexuality.

Dancers at The Oasis and Perfections, the middle-class clubs, were much more likely to accept tips without permitting any sort of touching between themselves and the customers, consistent with the passive dancing style. Even when customers attempted to touch dancers, the women enforced voyeuristic sexuality. One customer at The Oasis, for instance, held his dollar between his teeth for the dancer (presumably so that she would use her mouth or breasts to retrieve the bill, a method often seen at working-class clubs), but the woman, with perhaps a smidge of disdain, simply used her hands to pluck the bill out of his mouth.

Many dancers at the middle-class clubs, instead of touching or allowing contact between themselves and patrons, will perform a "mini-show" for the tipping customer to view, as Kayla at Perfections does (as illustrated by my field notes): "A man wearing dark slacks and a polo-style shirt approached the stage during Kayla's dance, with a crisp one-dollar bill held firmly in his hand. He stood patiently at the side of the stage, waiting for her to notice him. She saw him as she turned around, and strolled over to where he was standing, smiling and making eye contact. She lay down on the stage, and for 10–15 seconds, rolled around on the stage, moving her legs in the air in a scissors-like fashion before lifting the side of her G-string for dollar placement." Other dancers, instead of giving the customer a special performance, will lightly place their hands on the man's shoulders and lean over him, throwing their hair over his head and then whisper something in his ear, before removing their hair, stepping back, and lifting the side of their G-strings. Several times I asked my male escorts to tip these women and let me know what kinds of things the dancers were whispering to the patrons. By those reports, they are whispering light conversational fare: "What's your name?" "Are you having a good time?" "Is that your girlfriend with you?" None of these comments are inherently sexy, but in this context, the way in which these questions are whispered leaves men feeling somewhat special, like the dancer thinks he is interesting enough to have been noticed, or that she wants to know his name.[2] If the trick works, perhaps he will purchase a table dance from her after she leaves the stage (Ronai and Ellis 1989).

Like the stage tips, table dances at The Oasis and Perfections, the middle-class clubs, are typically characterized by slow sensuality and distance between dancers and customers. Most of the women spend a great deal of their time standing in front of the customer, in between his spread legs, swaying back and forth, lightly touching her breasts and flipping her hair around. The Oasis in particular has been characterized in club reviews and Web discussions as the "home of the air dance," meaning that the dancers rarely make contact with the customers, even during table dances. Even when dancers do make physical contact with customers, they do so in ways that encourage voyeuristic sexuality. Dancers will lean backwards, sitting on the customer's lap, resting her head on one of his shoulders, and squeeze her breasts together. In this position, the man can look down the front of a dancer's body but not see her face. Both Veronica and Mandy describe their styles of dancing as sensual and voyeuristic, yet devoid of explicit sexual contact. Veronica says, "Some girls, like me, dance very slowly, and it's not even really dancing. It's just massaging, looking, a lot of eye contact. I'm trying to enchant the guy or put him in a . . . trance almost, and it works! I mean, women do have a very . . . we have a power that men are . . . it just works. We have a power that works over men. And when you learn how to use it, then you can take their money [laughs]. In that environment. Now, I haven't figured out how to use it in real life, but in that environment, yeah." Mandy's description is nearly identical: "My style is more exotic, like rubbing their hands and playing with their hair and touching their face. And most of the time, when I'm in the right mood, and I got my crap together like last night, I can actually have guys in trances where they're actually closing their eyes while I'm dancing for them."

In contrast, the dancers at the working-class clubs accept tips and perform table dances in very different ways. In these clubs, stage tips and table dances appear to be driven by the desire to give a "cheap thrill," a term that merges sexuality and social class. There is much contact between dancer and customer, and the seemingly most popular dancers are those who touch the customers in some way with their breasts or genital region. One widespread method of accepting tips is demonstrated by Sara at The Hourglass (from my field notes): "She bent down so that her breasts were directly in front of his face, moved close to him, and then grabbed the back of his head, pulling his face into the space between her breasts, and shook her body from side to side. There was not enough space between the man's face and her breasts to see how the man responded, so he might have kissed or licked her breasts and no one would have seen." Another popular method of receiving tips from customers is to grab the back of the man's head and shove it into the dancer's genital region while shaking his head from side to side. Trina, a dancer at The Treasure Chest, thanked a man for his $1 tip by holding the back of his neck with her left hand, whispering in his ear, and grabbing his bottom with her right hand. She then slapped his rear end with an open hand and poked him with her index finger in the space right between his cheeks.

Likewise, table dances simulate sex and sexual acts in dramatic ways. I observed several dancers at both working-class clubs who spent at least one-third of each song sitting on her knees facing the customer, with her face directly inside the man's legs, simulating oral sex. Marina is one dancer who uses this trick, and she describes in detail just what it is she is doing down there: "If the guy seems clean and seems all right, I will go down like I'm going to give him a blow job, and I'll do this thing where I'll blow warm air through his pants and I'll roll my Rs so that it feels like a vibrator." It also appears that many of the dancers will very lightly run one of their hands over the man's genital region. Most dancers sit on a customer's lap, facing away from him, grinding their buttocks in a circular motion over his genitals. Dancers also will stand with one foot firmly on the floor, the other resting on the man's shoulder, leaning in toward the man's face so as to give him a close view of her crotch. About a year ago, Marina got in trouble with regulatory agents for "flashing," quickly pushing her G-string over to the side to reveal her genitalia to the customer, a practice she asserts is common (and even condoned) at The Treasure Chest: "When I got in trouble for flashing, the manager told me, 'Well, 95 percent of the girls here flash, so don't feel bad. They just happened to catch you. You should probably just be more careful.'"

These examples make it clear that very different styles of dancing and interacting with customers are taking place at these strip clubs. The middle-class clubs are characterized by performances of passive desire and distance, while the working-class clubs are marked with explicit allusions to sex and sexuality, physical activity and exertion, and contact between patron and dancer.

Staff Attire

Although dancers are the primary signals of sexuality within these organizations, the attire and mannerisms of waitresses, managers, bouncers, and bartenders also support the images of sexuality the clubs are constructing. Staff can be (and are) dressed in such a way as to contribute to an environment of raw sex and casualness or to one of distance, unapproachability, and composure.

The managers and bouncers at the middle-class clubs dress in what might be perceived as a more inaccessible and intimidating manner than do similar types of employees at the working-class clubs. The managers (all men) wear suits, complete with tie and vest, while the bouncers are dressed in all black—slacks, button-down collared shirt, and vest. These outfits send a signal of restraint, distance, and formality to the patrons of the club. They seem to be telling customers that businessmen belong there. As businessmen themselves, in their suits and ties, they signal that they understand and identify with the needs and expectations of their middle- and business-class clientele. The managers at The Hourglass and The Treasure Chest, however, dress in polo-style shirts with khaki pants, while the bouncers dress in jeans and T-shirts. They seem both casual and approachable, signaling an easygoing, "anything goes" attitude. "We're no better than you," their outfits communicate to their patrons. "We're workers, too."

Perhaps one of the most striking differences between the two types of clubs is the ways in which the waitresses are dressed. At Perfections and The Oasis, the middle-class clubs, the waitresses have the choice of wearing either black shorts or black pants, with a white shirt of any style, which varies greatly from waitress to waitress. The clothing, typical of waitstaff at many restaurants nationwide, lends the club an aura of legitimacy, and this constructed boundary makes it clear that it is the dancers who are on display, not the waitresses.

It is almost hard to tell the waitresses and the dancers apart at the working-class clubs. Like the performers, the waitresses wear string-bikini tops and G-strings, but they also wear a mesh sarong that only slightly covers their buttocks. The women bartenders wear the same outfits, and all the women employees are available to give table dances, although the waitresses do not go on stage, and the dancers do not serve drinks. The attire of the waitresses makes it clear that these clubs are places in which all the women present are legitimate and permissible sex objects, that there are no boundaries placed on men's desires and curiosities.

Conclusion

In this article, I have emphasized the ways that exotic dancers perform gendered and classed sexuality. Of course, women's performances are both a reflection and an interpretation of other core features of organizational culture such as management styles and organizational rules. Had I interviewed managers, more dancers, or customers, or had I become an employee of these clubs, my perspective on the performance of class and the performance of sexuality might perhaps be more focused on those features that constrain or enable women's agency. Yet the strength of my observational approach and small sample is that as an outsider, I am able to make sense of the variations in one city's club experiences much as any customer or potential customer might. Consumers, when faced with an array of seemingly similar services, such as those found in exotic dance clubs, make distinctions based on the frontstage of organizational culture (i.e., the performance of sexuality and class) not based on the backstage constraints that produce them.

Expressions and performances of sexuality, I have argued, are not homogeneous. Clubs construct distinctive working-class and middle-class performances and performers of sexuality that are consistent with popular ideas of how class and sexuality intersect. In this way, my analysis shows that social class is a central feature of organizational culture.

Class differences are . . . represented as sexual differences in very concrete ways: in the appearance of dancers and other staff, dancing and performance styles, and the interactions that take place between dancers and customers. Middle-class clubs are associated with a sexuality that is voyeuristic, characterized by distance, gazing, and a formal sexual atmosphere, while working-class clubs are associated with cheap thrills, contact, and a casual sexual atmosphere. These types of sexuality are consistent with and further specify popular images of femininity and masculinity that are also mediated by social class (and race): sexually restrained middle-class white men and women and sexually promiscuous working-class men and women of color. Sex as performance, class, race, power, and gender are thus all intertwined.

Thus, sexuality is much more than an individual attribute, and social class is much more than simply a material location. As women and men construct, perform, and consume gender and sexuality in exotic dance clubs, they are simultaneously constructing, performing, and consuming social class.

Notes

1. Pueblo is a mid-sized city in southwestern United States (population approximately 500,000). Pseudonyms have been used in place of the names of the city, clubs, and dancers throughout.

2. Most of my male escorts reported feeling this way, even though they also said they "knew better." That is, they were aware that the dancers were interested in a potential financial opportunity, not in a date, yet each still left the stage with a rush of excitement and feeling "special."

References

Acker, Joan. 1990. Hierarchies, jobs, bodies: A theory of gendered organizations. *Gender & Society* 4:139–58.

Bettie, Julie. 2000. Women without class: Chicas, cholas, trash, and the presence/absence of class identity. *Signs: Journal of Women in Culture and Society, 26*: 1–35.

———. Julie. 2003. *Women without class: Girls, race, and identity.* Berkeley: University of California Press.

Britton, Dana. 2003. *At work in the iron cage: The prison as a gendered organization.* New York: New York University Press.

Butler. Judith. 1990. *Gender trouble.* New York: Routledge.

Dellinger, Kirsten. 2002. Wearing gender and sexuality "on your sleeve": Dress norms and the importance of occupational and organizational culture at work. *Gender Issues* 20:3–25.

———. 2004. Masculinities in "safe" and "embattled" organizations: Accounting for pornographic and feminist magazines. *Gender & Society* 18:545–66.

Dellinger, Kirsten, and Christine L. Williams. 2002. The locker room and the dorm room: Workplace norms and the boundaries of sexual harassment in magazine editing. *Social Problems* 49:242–57.

Edgley, Charles. 1989. Commercial sex: Pornography, prostitution, and advertising. In *Human sexuality: The societal and interpersonal context,* edited by Kathleen McKinney and Susan Sprecher. Norwood, NJ: Ablex.

Frank, Katherine. 2002. *G-strings and sympathy: Strip club regulars and male desire.* Durham, NC: Duke University Press.

Gherardi, Sylvia. 1995. *Gender, symbolism and organizational culture.* London: Sage.

Giuffre, Patti A., and Christine L. Williams. 1994. Boundary lines: Labeling sexual harassment in the workplace. *Gender & Society* 8:378–401.

Hall, Elaine J. 1993. Smiling, deferring, and flirting: Doing gender by giving "good service." *Work and Occupations* 20:452–71.

Hallett, Tim. 2003. Symbolic power and organizational culture. *Sociological Theory* 21:128–49.

Leidner, Robin. 1991. Serving hamburgers, and selling insurance: Gender, work, and identity in interactive service work. *Gender & Society* 5:154–77.

Liepe–Levinson, Katherine. 1998. Striptease: Desire, mimetic jeopardy, and performing spectators. *The Drama Review: TDR* 42:9–29.

Loe, Meika. 1996. Working for men: At the intersection of power, gender, and sexuality. *Sociological Inquiry* 66:399–421.

Moloney, Molly, and Sarah Fenstermaker. 2002. Performance and accomplishment: Reconciling feminist conceptions of gender. In *Doing gender, doing difference: Inequality, power, and institutional change,* edited by Sarah Fenstermaker and Candace West. New York: Routledge.

Morrill, Calvin. 1995. *The Executive Way: Conflict Management in Corporations.* Chicago: University of Chicago Press.

Morrill, Calvin, Christine Yalda, Madalaine Adelman, Michael Musheno, and Cindy Bejarano. 2000. Telling tales in school: Youth culture and conflict narrative. *Law & Society Review* 34:521–65.

Ortner, Sherry B. 1991. Reading America: Preliminary notes on class and culture. In *Recapturing Anthropology: Working in the Present,* edited by Richard G. Fox. Santa Fe, NM: School of American Research Press.

Price, Kim. 1998. Stripping work: Women's labor power in strip clubs. Paper presented at the annual meetings of the American Sociological Association, San Francisco, August.

Ronai, Carol Rambo, and Carolyn Ellis. 1989. Turn-ons for money: Interactional strategies of the table dancer. *Journal of Contemporary Ethnography* 16:271–98.

Ronai, Carol Rambo, and Mary Nell Trautner. 2001. Table and lap dancing. *Encyclopedia of criminology and deviant behavior,* vol. 3, edited by Nanette Davis and Gilbert Geis. Philadelphia: Taylor and Francis.

Snow, David A., Robert Benford, and Leon Anderson. 1986. Fieldwork roles and informational yield: A comparison of alternative settings and roles. *Urban Life* 15:377–408.

Trautner, Mary Nell. 1998. When you are a scholar and a sex object: The conflicting identities of female college students who strip for a living. Sociology honors thesis, Southwestern University, Georgetown, TX.

Trice, Harrison M. 1993. *Occupational subcultures in the workplace.* Ithaca. NY: ILR Press.

West, Candace, and Sarah Fenstermaker. 1995. Doing difference. *Gender & Society* 9:8–37.

West, Candace, and Don H. Zimmerman. 1987. Doing gender *Gender & Society* 1:125–51.

Williams, Christine, Patti Giuffre, and Kirsten Dellinger. 1999. Sexuality in the workplace: Organizational control, sexual harassment, and the pursuit of pleasure. *Annual Review of Sociology* 25:73–93.

Wood, Elizabeth Anne. 2000. Working in the fantasy factory: The attention hypothesis and the enacting of masculine power in strip clubs. *Journal of Contemporary Ethnography* 29:5–31.

Introduction to Reading 29

Nicola Gavey, Kathryn McPhillips, and Marion Doherty conducted in-depth interviews with fourteen New Zealand women, ages twenty-seven to thirty-seven, who came from diverse work experiences and educational backgrounds. The authors questioned the women about their past and current experiences with condoms, their heterosexual relationships and practices, their personal views of condoms, and how they thought others viewed condoms. The authors focus on the conflict between public health campaigns, which encourage women to demand condom use by male partners, and dominant heterosexual scripts that limit the ways in which women may control the course and outcomes of heterosexual encounters.

1. What are the two main "discursive arenas" or organizing principles of heterosexual sex that constrain women's sexual pleasure and activity?
2. Why do the authors believe that advice to women that encourages them to be assertive in sexual relationships with men is problematic?
3. Why did many of the women in this study have experiences of engaging in unwanted sex with men? How is this study's analysis of engaging in unwanted sex helpful to understanding the phenomenon of date rape?

"If It's Not On, It's Not On"—Or Is It?

Nicola Gavey, Kathryn McPhillips, and Marion Doherty

If it's not on, it's not on!" Slogans such as these exhort women not to have sexual intercourse with a man unless a condom is used. Health campaigns targeting heterosexuals with this approach imply that it is women who should act assertively to control the course of their sexual encounters to prevent the spread of HIV/AIDS and other sexually transmitted infections (STIs). Researchers and commentators, too, have sometimes explicitly concluded that it is women in particular who should be targeted for condom promotions on the basis of assumptions such as "the disadvantages of condom use are fewer for girls" (Barling and Moore 1990). . . .

Aside from the obvious question of whether women should be expected to take greater responsibility for sexual safety, this approach relies on various assumptions that deserve critical attention. For example, what constraints on women's abilities to unilaterally control condom use are overlooked in these messages? What assumptions about women's sexuality are embedded in the claim that the disadvantages of condoms are fewer for women? That is, is safer sex simply a matter of women deciding to use condoms at all times and assertively making this happen, or do the discursive parameters of heterosex work to subtly constrain and contravene this message? Moreover, are condoms as unproblematic for women's experiences of sex as the logic offered for targeting women implies? These questions demand further investigation given that research has repeatedly shown the reluctance of heterosexuals to consistently use condoms despite clear health messages about their importance. There is now a strong body of feminist research suggesting that condom promotion in Western societies must compete against cultural significations of condom*less* sexual intercourse as associated with commitment, trust, and "true love" in relationships (e.g., Holland et al. 1991; Kippax et al. 1990; Willig 1995; Worth 1989; see also Hollway 1989). Here, we contribute to this body of work, which collectively highlights how women's condom use needs to be understood in relation to some of the complex gender dynamics that saturate heterosexual encounters. In particular, we critically examine (1) the concept of women's control over condom use that is tacitly assumed in campaigns designed to promote safer sex and condoms to women and (2) the foundational assumption of such campaigns that condoms are relatively unproblematic for women's sexual experiences. . . .

We suggest there are two main discursive arenas that need to be considered in critically evaluating strategies that assume women's ability to control condom use and the appropriateness of targeting women in particular for messages of (hetero)sexual responsibility. These are (1) a male sex drive discourse (Hollway 1984, 1989) and the corresponding scarcity of discourse around female desire (Fine 1988), and (2) the constraints of a "coital imperative" on how much control women (or men) have in determining what sexual activity counts as "real sex." We suggest that there are two important points where these kinds of discursive influences converge for women in ways that mediate the possibilities for condom use, that is, the sites of identity and pleasure.

Assertion, Identity, and Control

* * *

Public sexual health education fervently promotes the rights of women to demand condom use by male partners. Against a backdrop of implied male aversion to condoms, both the Planned Parenthood

From Gavey, N., McPhillips, K., & Doherty, M. 2001. "'If it's not on, it's not on'—or is it?" *Gender & Society 15*(6), p. 917. Reprinted with permission of Sage Publications, Inc.

Federation of America (1998–2001) and the Family Planning Association (2000) in New Zealand offer mock scripts that despite an air of gender neutrality suggest ways women can assert their right to insist on a condom being used for sexual intercourse. In this section, we consider what these kinds of hard-line "calls to assertiveness" might look like in practice. The excerpt below is taken from the interview with Rose, a young woman in her early 20s. At the very beginning of the interview, in response to a question about her "current situation, in terms of your relationships," she said, "My last experience was fairly unpleasant and so I thought I'll really try and um devote myself to singledom for a while before I rush into anything." In this one-night stand, approximately three weeks prior to the interview, Rose did manage to successfully insist that her partner use a condom. Her account of this experience is particularly interesting for the ways in which it graphically demonstrates the kinds of interactional barriers that a woman might have to overcome to be assertive about using a condom; moreover, it demonstrates how even embodiments of the male sex drive discourse that are not perceived to be coercive can act out levels of sexual urgency that provide a momentum that is difficult to stop. Although the following quote is unusually long, we prefer to present it intact to convey more about the flavor of the interaction she describes. We will refer back to it throughout this article and want to keep the story intact.

Nicola: So were condoms involved at all in that—

Rose: Yeah um, actually that's quite an interesting one because we were both very drunk, but I still had enough sense to make it a priority, you know, and I started to realize that things were getting to the point where he seemed to be going ahead with it, without a condom, and I was—had to really push him off at one point and—'cause I kept saying sort of under my breath, are you going to get a condom now, and—and he didn't seem to be taking much notice, and um—

Nicola: So you were actually saying that?

Rose: Yeah. I was—

Nicola: In a way that was audible for him to—

Rose: Yeah. (NICOLA: Yeah) And um—at least I think so. (NICOLA: Yeah) And—and it got to the point where [laughing] I had to push him off, and I think I actually called him an arsehole and—I just said, look fuck, you know, and—and so he did get one then but it seemed—

Nicola: He had some of his own that he got?

Rose: Yeah. (NICOLA:Yeah) Yeah it was at his flat and he had them. In fact that was something I'd never asked beforehand. I presumed he would have some. And um he did get one and then you know—and that was okay, but it was sort of like you know if I hadn't demanded it, he might've gone ahead without it. And I am fairly sure that he's pretty, um—you know he has had a pretty dubious past, and so that worried me a bit. I know he was very drunk and out of control and, um—otherwise the whole situation would never have occurred I'm sure, but still I had it together enough, thank God, to demand it.

Nicola: And you actually had to push him off?

Rose: Yeah, I think—I'm pretty sure that I did, you know it's—it's all quite in a bit of a haze, [laughing] (NICOLA: laughing) but um, it wasn't that pleasant, the whole experience. It was—he was pretty selfish about the whole thing. And um—yeah.

Nicola: How much old—how—You said that he was a bit older than you.

Rose: Ohh, he's only twenty-eight or nine, but that's quite a big difference for me. I usually see people that are very much in my own age-group. Mmm.

Nicola: And um, you said that it was kind of disappointing sexually and otherwise and he was quite selfish, (ROSE: laughter) like at the point where you know you were saying, do you want to get—are you going to get a condom, at that point were you actually wanting to have sexual intercourse?

Rose: Um, mmm, that's a good question. I can't remember the whole thing that clearly. And I seem to remember that I was getting um—I was just getting sick of it, or—[laughing] or—I might have been, but my general impression was that it was quite a sort of fumble bumbled thing, and—and it was quite—he just didn't—he didn't have it together. He wasn't—he possibly would've been better if he was less drunk, but he was just sort of all over the place and um—and I was just thinking, you know, I want to get this over and done with. Which is not the [laughing] best way to go into—into that sort of thing and um—yeah I—I remember at points—just at points getting into it and then at other points it just being a real mess. Like he couldn't—he wasn't being very stimulating, he was trying to be and bungling it 'cause he was drunk. And being too rough and just too brutish, just yeah. And um—

Nicola: When you say he was trying to [laughing] be, what do you—

Rose: Ohh, he was just like you know, trying to use his hands and stuff and just it—it was just like a big fumble in the dark (NICOLA: Right) type thing. I mean I'm sure h—I hope he's not usually that bad it was just like—I was in—sometimes—most of the time in fact I was just saying, look don't even bother. [laughing] You obviously haven't got it together to make it pleasurable. So in some ways—

Nicola: You—you said you were thinking that or you said that?

Rose: I just sort of—I did push his hand away and just say, look don't bother. Because—and I thought at that point that probably penetration would be, um, more pleasurable, yeah. But—and then—yeah. And I think I was actually wanting to just go to sleep and I kept—I think I um mentioned that to him as well and said, you know why don't we—we're we're not that capable at the moment, why don't we leave it. But he wasn't keen on [laughing] that idea.

Nicola: What did he—is that from what he said, or just the fact that he didn't—

Rose: I think he just said, ohh no, no, no we can't do that. It was kind of like we had to put on this big passionate spurt but um it just seemed quite farcical, considering the state we were both in. And um, so I—yeah I thought if—yeah I thought it would probably be the best idea to [laughing] just get into it and like as I—as I presumed—ohh, actually I don't know what I expected, but he didn't last very long at all, which was quite a relief and he was quite—he was sort of a bit apologetic and like ohh you know, I shouldn't have come so soon. And I was just thinking, oh, now I can go to sleep. [laughter] Yes, so I just um—since then I've just been thinking, um one-night stands ahh don't seem to be the way to go. They don't seem to be much fun and [laughing] I'm not that keen on the idea of a relationship either at the moment. So I don't know. Who—I—it's hard to predict what's going to happen but—mmmm.

This unflattering picture of male heterosexual practice painted by Rose's account classically illustrates two of the dominant organizing principles of heterosexual sex—a male sex drive discourse and a coital imperative. The male sex drive discourse (see Hollway 1984, 1989) holds that men are perpetually interested in sex and that once they are sexually stimulated, they need to be satisfied by orgasm. Within the terms of this discourse, it would thus not be right or fair for a woman to stop sex before male orgasm (normatively through intercourse). This discursive construction of male sexuality thus privileges men's sexual needs above women's; the absence of a corresponding discourse of female desire (see Fine 1988) or drive serves to indirectly reinforce these dominant perceptions of male sexuality.

The extent to which male behavior, as patterned by the male sex drive discourse, can constrain a woman's attempts to insist on condom use are graphically demonstrated in Rose's account. The man she was with behaved with such a sense of sexual urgency and unstoppability that although she was able to successfully ensure a condom was used, it was only as a result of particularly determined and persistent efforts. He was unresponsive to her verbal requests and did not stop proceeding with intercourse until Rose became more directly confrontational—calling him "an arsehole" and physically pushing him off. At this point, he eventually did agree to wear a condom and did not use his physical strength to resist and overcome her actions to retain or take control of the situation. We will come back to analysis of Rose's account later in the article. . . .

Pleasing a Man

The male sex drive discourse constructs masculinity in ways that directly affect women's heterosexual experiences as evidenced in the above description . . . of Rose['s] . . . experience. However, as we alluded to above, this discursive framework can also constitute women's sexual subjectivity in complex and more indirect ways. For example, for some women situated within this discursive framework, their ability to "please a man" may be a positive aspect of their identity. In this sense, a woman can be recruited into anticipating and meeting a man's "sexual needs" (as they are constituted in this discursive framework) as part of her ongoing construction of a particular kind of identity as a woman.

Sarah reported not liking condoms. She said she did not like the taste of them; she did not like the "hassle" of putting them on, taking them off, and disposing of them; and that they interfered with her sexual pleasure. However, it seemed that her reluctance to use them was also related to her sexual identity and her taken-for-granted assumptions about men's needs, desires, and expectations of her during sex. Sarah traced some of her attitudes toward sex to her upbringing and her mother's attitudes in which the male sex drive discourse was strongly ingrained.[1] She directly connected the fact that "when I've started, I never stop" to her difficulty in imagining asking a man to use a condom. As she said in an ironic tone, "If a man gets a hard-on you've gotta take care of it because he gets sick. You know these are the mores I was brought up with. So you can't upset his little precious little ego by asking to use a condom, or telling if he's not a good lover."

Sarah explained her own ambivalence toward this male sex drive discourse by drawing on a psychoanalytic distinction between the conscious and unconscious mind:

> I mean I've hopefully done enough therapy to have moved away from that, but it's still in your bones. You know there's my conscious mind can say, that's a load of bullshit, but my unconscious mind is still powerful enough to drive me in some of these moments, I would imagine.

Thus, despite having a rational position from which she rejected the male sex drive discourse, Sarah found that when faced with a man who wanted to have sex with her, her embodied response would be to acquiesce irrespective of her own desire for sex. Her reference to this tendency being "in [her] bones" graphically illustrates how she regarded this as a fundamental influence. It is evocative of Judith Butler's suggestion "that discourses do actually live in bodies. They lodge in bodies, bodies in fact carry discourses as part of their own lifeblood" (Meijer and Prins 1998, 282). Sarah recalled an incident where her own desire for sex ceased immediately on seeing the man she was with undress, but she explained that she would not be prepared to stop things there:

> *Sarah*: He's the hairiest guy I ever came across. I mean—but no way I would stop. I mean, as soon as he took his shirt off I just kind of

> about puked [laughing], but I ain't gonna say—I mean having gone to all these convolutions to get this thing to happen there's no way I'm gonna back down at that stage.

Nicola: So when he takes his shirt off he's just about the most hairiest guy you've ever met, which you find really unappealing.

Sarah: Terribly.

Nicola: Um, but you'd rather go through with it?

Sarah: Well I wouldn't say rather, but I do it.

Sarah also described another occasion where she met a man at a party who said, "do you want to get together? and I said, well, you know, I just want to cuddle, and he said, well, that's fine, okay." She explained that it was very important to her to make her position clear before doing anything. However, they ended up having sexual intercourse, because as she said,

Sarah: I was the one. I mean we cuddled, and then I was the one that carried it further.

Nicola: And what was the reason for that?

Sarah: As I said, partly 'cause I want to and partly 'cause if ohh he's got a hard-on you have to.

For women like Sarah, it seemed that an important part of their identity involved being a "good lover." This required having sexual intercourse to please a man whenever he wanted it and, in Sarah's case, to the point of anticipating this desire on the basis of an erect penis. . . . Given these expectations, and her belief that most men do not like condoms, it is not surprising that she had developed an almost fatalistic attitude toward her own risk of contracting HIV, such that she could say, "There's a part of me that also says, as long as I'm clear and not passing it on, I'm not gonna worry. You know, and if I get it, hey, it was meant to be."

For understanding the actions of Sarah and women like her, the assertiveness model is not at all helpful. In these kinds of sexual encounters, Sarah's lack or possession of assertiveness skills is beyond the point. What stops her from acting assertively to avoid undesired sex are deeply inscribed features of her own identity—characteristics that are not related to fear of assertion so much as the production of a particular kind of self. Well before she gets to the point of acting or not acting assertively, she is motivated by other (not sexual) desires, about what kind of woman she wants to *be*.

In a similar way to Sarah, an important part of Sally's identity was having "integrity about sexuality." She explained her position in relation to not "leading somebody on" in terms of a desire for honesty:

> And so—and I—that really was clinched somehow that you didn't lead somebody on and so that's part of—that's one of the sort of ways in which I understand that contract notion, really. And so maybe that had something to do with how I see myself about being a person with a reasonable—with integrity about sexuality. I won't go into something with false promises kind of thing.

Prior to these comments, Sally talked about the origins of her beliefs about this kind of contractual notion where it was not possible to be physically intimate with someone unless you were prepared to have intercourse. She remembered feeling awful about touching a man's penis when she was younger:

> I had been unfair because of the—I suppose the feeling of sort of cultural value of—of the belief that men somehow you know it's tormenting to them to leave a cock unappeased (N:[laughter]) [laughing] basically or something like that.

Like Sarah, Sally too reflected on her experiences in a way that highlights the limitations of attempting to influence sexual behavior based on understandings of people as unitary rational actors. Sally discussed a six-month relationship with a past lover in which she had not used condoms. She said that she had made the decision not to use condoms because she already had an intrauterine device (IUD) and because they were seen to connote a more temporary rather than long-lasting relationship (a view ironically reinforced by the advice of a nurse at the Family Planning Clinic):

Sally: It's like condoms are about more casual kinds of encounters or I mean—I mean, I'm kind of—um they are kind of anti-intimacy at some level.

Nicola: And so if you'd used condoms with him, that would've meant—

Sally: Maybe it would have underscored its temporariness or its—yeah, its lack of permanence. I don't understand that. What I've just said really particularly. It doesn't [seem] very rational to me. [laughter]

Nicola: [laughter] No it wasn't very rational and I—

Sally: [indistinguishable] it seems to be coming out of you know, somewhere quite deeper about um—I think it goes back to that business about ideals stuff. And I think that's one of the things about not saying no, you know. And that the ideal woman and lover—the ideal woman is a good lover and doesn't say no. Something like that. And it is incredibly counterproductive [softly] at my present time in life. [sigh/laugh]

Sally's reference to the ideal woman who is a good lover (because) she does not say no implicitly recognizes the strength of male sex drive discourse and its effect on her sexual experiences. In a construction that is similar to Sarah's, Sally refers to this kind of influence as "deeper" than her "rational" views. She presents an appreciation of this kind of cultural ideal as internalized in some way that is capable of having some control over her behavior despite her assessment of this as "counterproductive."

Engaging in Unwanted Sex

Many of the women in this study recounted experiences of having sex with men when they did not really want to, for a variety of reasons. This now common finding (e.g., Gavey 1992) underscores the extent to which women's control of sex with men is limited by various discursive constraints in addition to direct male pressure, force, or violence. As Bronwyn said, it is part of "the job:"

Nicola: You said that you enjoy intercourse up to a point. Um, beyond that point, um, what are your reasons for continuing, given that you're not enjoying it?

Bronwyn: Ohh I just think it's part of my function if you like [laughter] that sounds terribly cold-blooded, but it is, [laughing] you know, it's part of the job.

Nicola: The job of—

Bronwyn: Being a wife. A partner or whatever.

Rose's account of her confrontational one-night stand, discussed previously, can be seen to be influenced in complex and subtle ways by the discursive construction of normative heterosexuality in which male sexuality and desire are supreme. She described the encounter as being quite unpleasant and disappointing, both sexually and in the way that he treated her—with respect to condoms and more generally throughout the experience. She described his actions during the encounter as clumsy and not the least bit sexually arousing ("He wasn't being very stimulating, he was trying to be and bungling it because he was drunk. And being too rough and just too brutish") and at one stage she suggested that they give up and go to sleep but he rejected this idea very strongly. Despite the extremely unsatisfactory nature of the sexual interaction and despite Rose's demonstrated skills of acting assertively, she did continue with the encounter until this man had had an orgasm through vaginal intercourse. It is difficult to understand why she would have done this without appreciating the power of the male sex drive discourse and the coital imperative in determining the nature of heterosexual encounters.

Although she did not define herself as a victim of the experience, Rose's account also makes clear that somehow she did not feel it was an option to end sex unless he gave the okay:

> It was sort of like—and I guess in that case he didn't have my utmost respect by that point. But it was sort of like, um, you know, he'd—he seemed to be just going for it, and I really really—I was drunk and sort of dishevelled and—and pretty resigned to having a bit of loose un—and unsatisfactory time which I didn't have a lot of control over.

She partially attributed this lack of control over the situation to the fact that the encounter took place at his flat, but this seems to be reinforced by underlying assumptions about sexuality and the primacy of his desires:

> Um, but it was kind of like he had his idea of what was going to happen and um I sort of realized after a while that he was so intent on it and the best thing to do was to just comply I guess, and make it as pleasurable as possible. Try and get into the same frame of mind that he was in. And—and yeah. Mmm, get it over with. It sounds really horrible in retrospect, it wasn't that bad it was just lousy, you know. It was just sort of a poor display of [laughing] everything. I suppose. Of intelligence and—and good manners. [laughter]

Part of the reason for the ambivalent nature of her account (swinging from describing what happened in very negative terms to playing down the experience as a poor display of manners) can be argued to originate from her positioning within a kind of liberal feminist discourse about sex. She had made a point of saying that she did not think things went in stages and defining herself in opposition to a model of female sexuality as fragile and in need of protection. The result of the connection between this liberal discourse of sexuality and that of the male sex drive is that she is left with no middle ground from which to negotiate within a situation of this kind. The absence of an alternative discourse of active female sexuality leaves her in a position of no return once she has consented to heterosexual relations and when certain minimal conditions are fulfilled (for her, this was the use of condoms). If heterosexuality were instead discursively constructed in such a way that women's sexual pleasure was central rather than optional, it makes sense that Rose may have felt able to call an end to this sex—which was, after all, so unpleasant that it led to her resolve to "devote [herself] to singledom for a while."

The Question of Pleasure

"I'm probably atypical from what I read of women in the fact that I personally don't like condoms." Sarah made this comment in the first minutes of the interview, and then much later she shared her assumption about how men regard using condoms: "Most men hate it. I've never asked them, but . . . that's the feeling I have from what I've read or heard." Sarah's generalized views about how women and men regard condoms echo dominant commonsense stories in Western culture. That is, most men do not like using condoms—a view shared by 80.5 percent of women in one large U.S. sample (Valdiserri et al. 1989), while women do not mind them. To explain her own dislike of condoms, Sarah was forced to regard herself as atypical. In the following section, we will discuss evidence that challenges the tacit assumption that condoms are relatively unproblematic for women.

Enforcers of the Coital Imperative: Condoms as Prescriptions for Penetration

Research on how men and women define what constitutes "real sex" has repeatedly found that a coital imperative exists that places penis-vagina intercourse at the center of (hetero)sex (Gavey, McPhillips, and Braun 1999; Holland et al. 1998; McPhillips, Braun, and Gavey 2001). The strength of this imperative was also reflected in the current research—as one women explained, "I don't think I worked out a model of being with someone like naked intimate touching which doesn't have sex at the end of it" (Sally defined sex as penis-vagina intercourse during the interview). Although this coital imperative could be viewed as forming part of the male sex drive discourse examined above, it is addressed separately here as it has particular consequences for safer sex possibilities.

Condoms seem to reinforce the coital imperative in two interconnected ways, both in terms of their symbolic reinforcement of the discursive construction of sex as *coitus* and through their material characteristics that contribute at a more practical level to rendering sex as finished after coitus. In the analyses that follow, we will be attending to the material characteristics of condoms as women describe them. As discussed earlier, we adopt a realist reading of women's accounts here. What the women told us about the ways in which condoms help to structure the material practice of heterosex casts a shadow over the assumption that condoms are unproblematic for women's sexual experience. These accounts illuminate how the male sexual drive discourse can shape not only the ways people speak about and experience heterosex, but also the ways in which a research lens is focused on heterosexual practice to produce particular ways of seeing that perpetuate commonsense priorities and silences (which, in this case, privilege men's pleasure above women's). That is, the ways in which condoms can interfere with a woman's sexual pleasure are relatively invisible in the literature, which tends, at least implicitly, to equate "loss of sexual pleasure" with reduced sensation in the penis, or disruption of desire and pleasure caused by the act of putting a condom on. As the following excerpts show, the material qualities of condoms have other particular effects on the course of sex. These effects are especially relevant both to the question of a woman's pleasure and to the way in which the coital imperative remains unchallenged as the definitive aspect of heterosex.

While many of the women said they like (sometimes or always) and/or expect sexual intercourse (i.e., penis-vagina penetration) when having sex with a man, many of the women noted how condoms operated to enforce intercourse as the finale of sex. Several women found this to be a disadvantage of condoms, in that they tend to limit what is possible sexually, making sex more predictable, less spontaneous, playful, and varied. That is, once the condom is on, it is there for a reason and one reason only—penile penetration. It signals the beginning of "the end." Women who identified this disadvantage tended to be using condoms for contraceptive purposes and so were comparing sex that involved condoms unfavorably to intercourse with some less obtrusive form of contraception such as the pill or an IUD, or with no contraception during a "safe" time of the month.

For example, Julie found that condoms prescribed penetration at a point where she could be more flexible if no condom was involved:

> That's what I mean about the condom thing. It's like this is *the act* you know, and you have to go through the whole thing. Whereas if you don't use condoms, you know, like he could put it in me and then we could stop and then put it in again, you know, you can just be a bit more flexible about the whole thing.

The interconnection between the material characteristics of condoms (its semen-containing properties require "proper use") and the discursive construction of the encounter (coitus is spoken about as "*the act*") has the effect of constraining a woman's sexual choices and leaving this generally unspoken coital imperative unchallenged. The discursive centrality of coitus within heterosex is materially reinforced by the practical difficulties associated with condoms, as Deborah said,

> Once you've put on the condom . . . that limits you. Once you've got to the stage in sex that you put a condom on, you then— it's not that you can't change plans, but it's a hassle if you then decided that you might like to move to do um—you might like to introduce oral sex at this stage, as opposed to that stage, then you have to take it off, or you don't, or—

* * *

Thus, condoms not only signaled when penetration would take place, but their use served to reinforce the taken-for-granted axiom of heterosexual practice, that coitus is the main sexual act. Furthermore, one woman (Sally) described how the need for a man to withdraw his penis soon after ejaculation when using a condom disrupted the postcoital "close feeling" she enjoyed. These excerpts can be seen to represent a form of resistance to the teleological assumptions of the coital imperative; women's accounts of desiring different forms of sexual pleasure (including, but by no means limited to, emotional pleasures) may provide rich ground for exploring safer sex options. This potentially productive area has yet to be fully exploited by traditional health campaigns, which perhaps reflects the lack of acknowledgment given to discourses of female sexual desire and pleasure in Western culture in general (see also Fine 1988).

Some women also talked about the effect using condoms had on their sexual pleasure by using the language of interruption and "passion killing" more commonly associated with men. Sarah, who rarely used condoms, said that "it breaks the flow":

> I have a lot of trouble reaching an orgasm anyway, and it's probably one of the reasons I don't like something that's interrupting, because I do go off the boil very quickly. Um, once it's on and it's sort of decided that penetration tends to be what happens. I don't suppose it's a gold rule, but it seems to be the way it is. So you know there's no more warm-up.

Unlike health campaigns directed at the gay community, which have emphasized the range of possible sex acts carrying far less risk of HIV infection than penile penetration, campaigns aimed at heterosexuals have done little to challenge the dominant coital imperative.[2] Health campaigns that promote condoms as the only route to safer sex implicitly reinforce this constitution of heterosexuality and the dominance of the male sex drive discourse. As the responses of the women in this study demonstrate, this reluctance to explore other safer sex possibilities may be a missed opportunity for increasing erotic possibilities for women at the same time as increasing opportunities for safer sex. The fact that all of these women spent some time talking about the ways in which condoms can operate to enforce the coital imperative or reduce their desire indicates the importance of taking women's pleasure into account when designing effective safer sex programs. That is, it may simply not be valid to assume that "the disadvantages of condom use are fewer for girls" (Barling and Moore 1990) if we expect women's sexual desires and pleasures to be taken as seriously as men's. Special effort may be required to ask different questions to understand women's experiences in a way that doesn't uncritically accept a vision of heterosex as inherently constrained through the lens of the coital imperative and male sex drive discourses.

Discussion

* * *

Holland et al. (1998) have argued that heterosexual relations as they stand are premised on a construction of femininity that endangers women. Evidence for this position can be drawn from the current study as the interaction of discourses determining normative heterosexuality produces situations in which women are unable to always ensure their safety during sexual encounters. Holland et al. (1998) have argued that a refiguring of femininity is needed to ensure that women have a greater chance of safer heterosexual encounters. One of the prerequisites for change of this kind would be acknowledgment of the discourses of active female desire, which have traditionally been repressed (Fine 1988). Indeed, our research here suggests that the claim that condoms are "relatively unproblematic" for women is based on a continued relegation of the importance of women's sexual pleasure, relative to men's. Without challenging the gendered nature of dominant representations of desire, and more critically examining the coital imperative, condom promotion to women is likely to remain a double-edged practice. As both a manifestation and a reinforcement of normative forms of heterosex, it may be of limited efficacy in promoting safer heterosex.

Notes

1. It should be emphasized that Sarah's mother's attitudes would have been in line with contemporary thought at the time. Take, for example, the advice of "A Famous Doctor's Frank, New, Step-by-Step Guide to Sexual Joy and Fulfillment for Married Couples" (on front cover of Eichenlaub 1961, 36), published

when Sarah was nearly an adolescent: Availability: If you want good sex adjustment as a couple, you must have sexual relations approximately as often as the man requires. This does not mean that you have to jump into bed if he gets the urge in the middle of supper or when you are dressing for a big party. But it does mean that a woman should never turn down her husband on appropriate occasions simply because she has no yearning of her own for sex or because she is tired or sleepy, or indeed for any reason short of a genuine disability. (Eichenlaub 1961, 36)

2. While some sexuality education directed at teenagers might be more likely to encourage alternatives to coital sex, and hence broader definitions of safer sex (e.g., Family Planning Association 1998; see Burns 2000), this is still less evident in material designed for the "mature sexuality" of heterosexual adults.

References

Barling, N. R., and S. A. Moore. 1990. Adolescents' attitudes towards AIDS precautions and intention to use condoms. *Psychological Reports* 67: 883–90.

Burns, M. 2000. "What's the word?" A feminist, post-structuralist reading of the NZ Family Planning Association's sexuality education booklet. *Women's Studies* 16:115–41.

Crawford, M. 1995. *Talking difference: On gender and language.* London: Sage.

Eichenlaub, J. E. 1961. *The marriage art.* London: Mayflower.

Family Planning Association. 1998. *The word—On sex, life & relationships* [Booklet]. Auckland, New Zealand: Family Planning Association. [ISBN 0–9583304–8-4]

———. 2000. *Condoms* [Pamphlet]. Aukland, New Zealand: Family Planning Association. (Written and produced in 1999. Updated 2000).

Fine, M. 1988. Sexuality, schooling, and adolescent families: The missing discourse of desire. *Harvard Educational Review* 58:29–53.

Gavey, N. 1992. Technologies and effects of heterosexual coercion. *Feminism & Psychology* 2:325–51.

Gavey, N., K. McPhillips, and V. Braun. 1999. Interruptus coitus: Heterosexuals accounting for intercourse. *Sexualities* 2:37–71.

Holland, J., C. Ramazonaglu, S. Scott, S. Sharpe, and R. Thomson. 1991. Between embarrassment and trust: Young women and the diversity of condom use. In *AIDS: Responses, interventions, and care,* edited by P. Aggleton, G. Hart, and P. Davies. London: Falmer.

Holland, J., C. Ramazonaglu, S. Sharpe, and R. Thomson. 1998. *The male in the head: Young people, heterosexuality and power.* London: The Tufnell.

Hollway, W. 1984. Gender difference and the production of subjectivity. In *Changing the subject: Psychology, social regulation and subjectivity,* edited by J. Henriques, W. Hollway, C. Urwin, and V. Walkerdine. London: Methuen.

———. 1989. *Subjectivity and method in psychology: Gender, meaning, and science.* London: Sage.

Kippax, S., J. Crawford, C. Waldby, and P. Benton. 1990. Women negotiating heterosex: Implications for AIDS prevention. *Women's Studies International Forum* 13:533–42.

McPhillips, K.,V. Braun, and N. Gavey. 2001. Defining heterosex: How imperative is the "coital imperative"? *Women's Studies International Forum* 24:229–40.

Meijer, I. C., and B. Prins. 1998. How bodies come to matter: An interview with Judith Butler. *Signs: Journal of Women in Culture and Society* 23:275–86.

Planned Parenthood Federation of America, Inc. 1998—2001. *Condoms.* Retrieved 18 June 2001 from the World Wide Web: http://www.planned-parenthood. org

Statistics New Zealand. 2001. *Statistics and information about New Zealand.* Retrieved 4 July 2001 from the World Wide Web:http://www.stats.govt .nz/default.htm

Valdiserri, R. O.,V. C. Arena, D. Proctor, and F. A. Bonati. 1989. The relationship between women's attitudes about condoms and their use: Implications for condom promotion programs. *American Journal of Public Health* 79:499–501.

Willig, C. 1995. "I wouldn't have married the guy if I'd have to do that. "Heterosexual adults' accounts of condom use and their implications for sexual practice. *Journal of Community and Applied Social Psychology* 5:75–87.

Worth, D. 1989. Sexual decision-making and AIDS: Why condom promotion among vulnerable women is likely to fail. *Studies in Family Planning* 20:297–307.

Introduction to Reading 30

Jennifer Lois explores the "gendered emotional culture of high-risk takers," specifically a volunteer search and rescue group made up of women and men, with whom she conducted six years of ethnographic research. Edgework engages people in many physically and emotionally threatening situations, including conditions in which individuals must negotiate the boundary between life and death. Lois discovered four stages of edgework during which rescue team members had to deal with intense emotions. Most significantly, the feelings and emotion management techniques of the men and women differed.

1. What were the major differences in the emotional experiences and management techniques of the women and men rescuers? How about similarities?
2. How does "doing gender" shape the experience of adrenaline rushes?
3. Why were men who were not willing to take risks subject to more criticism than women who were concerned with safety?

Gender and Emotion Management in the Stages of Edgework

Jennifer Lois

If you let those concerns bother you—"Oh God, there's three kids out there freezing to death"—you're losing sight of your task. And you're jeopardizing your own safety and your team's safety by not being focused on what your task is. If you're out there searching for a plane crash, the odds are you have a [radio locator] signal. You're not looking for those three little kids, you're looking for that signal. And if you start thinking about those three little kids, you're going to get off your task. Emotions just get in the way on [search and rescue] missions.

—Gary, eight-year member of Peak Volunteer Search and Rescue

Emergency situations call for effective action in the face of potentially overwhelming emotions. As Gary states above, strong feelings can "get in the way" of performing an important task. This article is about how male and female members of "Peak," a volunteer search and rescue group, "managed" their emotions (Hochschild, 1983) before, during, and after their most dangerous, stressful, or gruesome rescues. In the mountains of the western United States, Peak's

From Lois, J. 2005. "Gender and emotion management in the stages of edgework" in Stephen Lyng (Ed.), *Edgework: The Sociology of Risk-Taking*. Reprinted with permission of Routledge.

rescuers engaged in a wide variety of risky activity during the searches and rescues (called "missions"). Missions presented many physically and emotionally threatening situations for rescuers, such as when they searched for missing skiers in avalanche-prone terrain, negotiated the rapids of rushing rivers to reach stranded rafters, were lowered down cliff faces to evacuate injured rock climbers, and extracted mutilated bodies from planes that crashed in the wilderness. These extreme conditions—the most crucial life-and-death circumstances—called for members to engage in what Lyng (1990) has termed "edgework": negotiating the boundary between life and death during voluntary risk taking.

I introduce four stages of edgework by tracing members' specific feelings and the corresponding management techniques they employed before, during, and after urgent rescues. The data reveal two gendered ways rescuers prepared for, engaged in, and reflected on edgework.

Setting and Method

These data are drawn from a six-year ethnographic study of Peak, a volunteer search and rescue group in a Rocky Mountain resort town. Peak County consisted of 1700 square miles, 1300 of which were undeveloped national forest or wilderness area lands. Local residents and tourists alike used this "backcountry" land year-round for various recreational purposes such as hiking, camping, rock climbing, whitewater rafting/kayaking, snowmobiling, and backcountry skiing. Occasionally, recreational enthusiasts became lost or injured in these vehicle-inaccessible regions. Because the county sheriff's deputies did not have the skills or resources to venture into these remote areas, the sheriff commissioned Peak, a volunteer group of local citizens, to act as the public safety agent in the backcountry. Because these emergencies could happen at any time of the day or night, members were given pagers so that they could be notified immediately when they were needed, and respond if they were available. Frequently, this meant getting out of bed in the middle of the night to search for an overdue snowshoer or rescue an injured camper.

I became interested in Peak after reading several local newspaper accounts of their rescues. With no specialized backcountry skill or experience, I joined the group in 1994 to study it sociologically. I began attending the biweekly business meetings, weekly training sessions, post-training social hours at the local bar, and a few missions. Through these initial interactions, I became intrigued by how members defined their participation in rescue work, as well as how these definitions affected their lives (Blumer, 1969).

Over the next year, I developed strong friendships with several group members, and was able to discuss some of my observations with them. I occasionally asked them for their interpretations of certain events, which enhanced my analysis through the perspective of everyday life (Jorgensen, 1989). During this time period, I became an "active" member (Adler and Adler, 1987) of the setting; I was given deeper access to members' thoughts and feelings as they began to trust me, both as a researcher and a rescuer.

For six years, I kept detailed field notes of the group activities in which I participated, including the business meetings, training sessions, social hours, and missions. It was my participation in the missions, however, that most helped me to identify with other rescuers' experiences. I was struck by the intense emotions I felt during searches and rescues, and once I realized that other rescuers experienced some of the same emotional patterns, I began to ask about them specifically during casual conversations. In addition to taking extensive field notes, I conducted 23 in-depth, semi-structured interviews with rescuers, focusing the questions loosely around their motivation for participating in Peak and their experiences on missions.

Gender, Emotions, and the Stages of Edgework

The levels of difficulty, danger, and stress varied greatly among Peak's missions. At times, members were asked to perform only slightly demanding, low-urgency tasks such as hiking a short distance up a trail to carry a hiker with a twisted ankle out to the parking lot. Other times they were asked to perform very difficult, dangerous, or gruesome tasks such as rappelling down a 200-foot rock face to recover the body of a fallen hiker, traversing known avalanche terrain to reach a hypothermic snowshoer, and bushwhacking for fifteen hours over miles of treacherous terrain in search of a lost hunter. It was these physically and emotionally demanding situations that most threatened rescuers' sense of control, requiring them to engage in edgework—to negotiate the boundary between order and chaos—not only physically, but emotionally as well.

There were four stages of edgework in Peak's missions. These stages were distinctly marked not only by the flow of rescue events, but also by the feelings members experienced in each stage. Women and men experienced edgework differently, interpreting and managing feelings in gender-specific ways while they prepared for missions, performed high-risk activities, reflected on their participation, and made sense of their actions.

Preparing for Edgework: Establishing Confidence Levels

Missions were unpredictable, sometimes chaotic events, and members were often required to use whatever resources they had to accomplish their task. Many of Peak's members, like other edgeworkers, found it exciting not to know what to expect from a rescue, and they felt challenged by the prospect of relying on their cognitive and technical skills to quickly solve any puzzle that suddenly presented itself. Yet other rescuers viewed the missions' unpredictability as stressful, and they worried in anticipation of performing under trying conditions.

One common worry was that they might be physically unable to perform a task because they did not have the skills to handle a difficult situation, even if others trusted their abilities. Maddie, a thirty-four-year-old, ten-year member and mission coordinator told me about first achieving "leader" status after being in the group for two years. The old-time members in the group had been evaluating her skills for weeks and, after a mock rescue scenario, told her she was ready to advance to this higher group position. She, however, was unsure of her abilities, and thus, of their decision:

> I kind of questioned it at the time because I didn't think my avalanche skills were there yet. I knew my knots, and I could make a [rope and pulley] system, and I could get other people to make one, but I asked Roy and Jim both, you know, "Are you sure I'm ready to do this?" They're like, "Yeah, I mean, you can definitely take a group of people and lead them through a rescue or a mission, and we know you'll get the job done, even though [those people] might not know a thing." And I went, "Oh, okay, I guess that's what it means to be [leader status]. *But please don't ever put me in that spot [laughs]!"* Only because you look at the rock jocks we had, or the really knowledgeable systems people, and it was just hard to think that I was considered their equal. For me, anyway; I didn't think that.

Members also worried about their ability to maintain emotional control, realizing that they could encounter a particularly upsetting scene on a mission. For example, Maddie told me that one situation she dreaded was encountering a dead victim whom she knew. She expected that this situation would be one that most threatened her emotional control, the one in which she would be most likely to go over the edge.

Worrying about what could arise on a future mission led rescuers to plan their actions ahead of time, anticipating their potential reactions to stressful events. Preparing for edgework by imagining numerous different scenarios gave them some sense of control over the unpredictable future, and through such planning, they were able to manage their uncomfortable anticipatory feelings about the unknown, a dynamic found in other research on high-risk takers (see Holyfield, 1997; Lyng, 1990).

Maddie's statement also typified another technique some rescuers used along with planning and rehearsing future scenarios: They set low expectations for themselves. Part of their planning process was to prepare for the most demanding possible situations, the ones in which they were most likely to fail. This emotion-management strategy served two functions: First it made members acutely aware of their progress toward the edge on missions. For example, Cyndi, a four-year member in her late twenties, told me that as a rescuer, "You need to know your limits. . . . If you start to do something and [find] 'I can't do this,' you shouldn't push yourself to do it just because everyone's like, 'Yes you can.' You should be like, 'No. I can't do this,' and work on other things that you can do." The second function of setting low expectations was that rescuers would probably perform beyond them. In this way, they set themselves up for success, remaining within their limits on a mission while feeling good about surpassing their expectations.

Anticipating a poor performance and uncontrollable conditions were common ways the women on the team prepared for the variety of situations they could encounter on a mission. Underestimating their ability was not a very common practice among the men in the group, however. When I talked to Martin, a fifty-three-year-old construction supervisor and five-year member, he told me that he never worried about his performance. He used an example from one of the team's most critical missions, a car accident with four casualties. A van had driven off the side of a dirt road and tumbled to the bottom of a 400-foot ravine. Search and rescue was called because the accident was inaccessible to the paramedics, who needed ropes to get down to the victims and a hauling system to get them out. Martin was the third rescuer to arrive on the scene and was immediately directed to join Gary (a rescuer and EMT) down at the accident site. Despite the stress and danger he experienced in this situation, Martin said he was never afraid of getting on a rescue scene and losing control:

Martin: I've never had that worry at all. I've had thoughts that maybe I'd get on scene and somebody'd be hurt so bad, and I wouldn't have the training to be able to help them. But I don't think that's losing control. What if I'd been the first one down to that van accident? That was way beyond anything I could've known what to do.

Jen: What do you think would've happened if you had been the first one down there?

Martin: Actually, thrown into the situation, I probably would've handled it. I think, thrown into a situation, I think I could react well. I would've handled it somehow.

In contrast to many women's strategy of trepidation, most of the men in Peak used the opposite technique—building confidence—to prepare for emergency action, as Martin did.

Brooke, a mid-twenties student, commented . . . on rescuers' confidence levels, referring specifically to Gary and Nick. She claimed they were able to perform at very high levels *because* of their high expectations for themselves.

According to Brooke, extreme confidence was effective for Gary and Nick, yet she did not think that it was a viable emotion management technique for her to employ. She felt safer being wary of an uncertain situation, but acknowledged that Gary and Nick were safer charging into the same situation. She saw their confidence as enabling them to perform at high levels.

When I asked Gary himself about his experience in extreme situations, he responded with incredible certainty in his ability, supporting Brooke's perception of him:

Gary: I like being thrown knee-deep [into challenging situations]. I like it when the shit hits the fan, and having to get my way out of it.

Jen: Don't you get nervous that you might not be able to do that?

Gary: Nope.

Jen: Do you think you'll always be able to do that?

Gary: Yup. I am a cocky, young, think-I-can-do-it-all kid.

Gary highlighted the mutually reinforcing relationship between confidence and ability: Not only did confidence enhance performance, but past performance enhanced confidence. Lyng (1990) noted a similar tautological relationship between skill and success in his study of skydivers. Successfully performing at high levels (surviving) served as "proof" that a skydiver had "the right stuff," while failing (dying or becoming critically injured) served as "proof" that one never really had it. Lyng designated this belief the "illusion of control," and showed how the skydivers he studied relied on it to enhance their confidence. This same form of ex post facto validation helped Peak's male rescuers interpret the relationship between their confidence and skill.

Many of Peak's members, furthermore, saw such confidence as a necessary quality in rescuers, largely because it allowed them to take risks and to be aggressive. Some linked these traits to gender. Nick, a twenty-eight-year-old, five-year member, and construction worker, told me that one reason the men in the group tended to be better at riding the snowmobiles was because they had "more guts" and "less fear" than the women.

Other male rescuers also explained how confidence and a "big ego" were desirable features in rescuers, and often equated these traits with testosterone, thus linking them closely to men and masculinity via biology.

Several members expressed similar beliefs about men's willingness to take risks and about women's reticence to do so. However, women's lack of confidence and their desire to minimize risk was much more accepted than men's (perhaps because men's trepidation was seen as "unnatural"). Thus, the men in the group who did not possess confidence and were not willing to take risks were ridiculed and disparaged for allegedly not having what it took to accomplish a mission. Denigration was most severe when these men insisted on staying involved in a mission, despite their unwillingness to take (what others saw as) the necessary risks.

Martin's opinion was typical of several others' when he criticized Russ, a fifty-year-old doctor and member of five years, who was fairly experienced, yet always unwilling to take the risks other men were:

> I think he's been trained as a doctor to be ultraconservative in everything he does. And I think there may be a fear factor in there. I think the members we need and we're not getting—this is also what Shorty thinks—are people who are willing to lay it on the line. There's gotta be a certain bit of cockiness. There's gotta be a certain bit of "I'll-take-a-chance" kind of thing. It's just because that's what we do. I mean, whether it's a man or woman, the kind of things we do are masculine things, they're testosterone things.

Russ' unwillingness to engage in edgework was unacceptable to many group members, who, as Martin noted, did not trust him. It was ironic that Russ was considered "too safe" for rescue work, yet that was how Peak's rescuers judged him. He did not engage in edgework, which painted him ineffective as a rescuer. Yet I rarely heard this degree of criticism about women who were unwilling to take risks, like Brooke (who dubbed herself "Miss Safety") or Cyndi (who thought rescuers should never try to do something they didn't think they could do). Thus, two very different emotional strategies were condoned for women and men during the edgework preparation stage. Women were expected to show trepidation, while men (by virtue of their "hormones," in many cases), were expected to display confidence and a willingness to take risks.

These gendered differences in preparation strategies can be explained in several different ways. First, in general, men were more experienced than women. Through their own recreation and group-related activities, men's exposure to risk was both more frequent and more hazardous than women's. Yet this gendered confidence pattern was not totally explained by differential risk exposure. For example, when I talked to equally experienced men and women, apprehension and lack of confidence still dominated women's anticipatory feelings (like Maddie's and Alex's), and most men still tended to feel confident and want to get involved. Furthermore, even when women performed well on missions, it did not seem to boost their confidence for future situations, while conversely, men's poor performances did not erode theirs.

A second factor in explaining this pattern was that Peak's group culture constructed rescue work as masculine, or "testosterone-filled." This enabled men to feel more at ease in the setting, and thus most tended to display unwavering certainty that they could handle any situation in which they might find themselves. Like the female ambulance workers in Metz's (1981) study, Peak's women felt disadvantaged by the "masculine" nature of rescue work and, taking this into account, set low expectations for themselves. In one way, their feelings were based in reality: They were aware that, on the whole, the men in the group were physically stronger, and thus able to perform harder tasks than they.

In another way, though, women's insecurities were due to cultural and group stereotypes about men's superior rescue ability. For example, the prevailing belief that men are emotionally stronger than women made women question whether they would be able to perform edgework in potentially upsetting situations, while the same stereotype enabled men to have confidence that they would maintain control in those situations (recall Maddie's fear that she would "lose it" if she encountered someone she knew and Gary's love for the "hot seat"). Yet, my observations (discussed later in this article) yielded no gendered pattern of emotional control during missions. Another stereotype that made women worry about their rescue ability was the belief that men were more mechanically and technically inclined than they were. This stereotype came into play during trainings and missions when the group used any kind of mechanization, such as rope and pulley systems, helicopters, snowmobiles, or whitewater rafting equipment. Cyndi told me she felt "hugely" intimidated in her first year by the technical training, yet she later became quite adept in setting up and operating rope and pulley systems. Elena told me that during her first training she looked around at "all the guys" and thought, "What am I doing here? I'm not even qualified for any of this." Cyndi and Elena's feelings of inferiority acted as "place markers" where "the emotion conveys information about the state of the social ranking system" (Clark, 1990, p. 308).

Because women were often marginalized in these ways, both their own and others' expectations of them were lower than they were for most men in the group. By remaining trepidacious and maintaining low expectations that they would often exceed, the women reaffirmed their place in the group as useful. Although they feared admitting when they would be unable to complete a task, because as Cyndi said, it meant "you're admitting to everyone else that you're not as good as them," the women in the group felt that bowing out early was preferable to failing. Cyndi said others would think, "at least they didn't fuck the mission up. They stayed, and they helped, and they did something." Thus, trepidation and confidence emerged as gendered emotional strategies used in preparing for edgework.

Performing Edgework: Suppressing Feelings

During Peak's urgent missions, effective action, a core feature of edgework (Lyng, 1990; see also Mannon, 1992; Mitchell, 1983), was seen as especially crucial. However, in such demanding situations, members' capacity for emotional and physical control was seen as more tenuous; emotions threatened to push them over the edge, preventing them from physically performing at all.

Rescuers who were easily scared, excited, or upset by a mission's events were considered undependable. There were several strategies members employed to control these feelings during missions, which allowed them to perform under pressure.

Members were particularly wary of the onset of adrenaline rushes because such potent physiological reactions threatened their composure; they felt that the emotions they experienced as a result of adrenaline rushes could "get in the way" of their performance. Although Schachter and Singer (1962) demonstrated that the physiological arousal associated with adrenaline does not signify a particular emotion in the absence of other situational information, Peak's members used the term "adrenaline rush" to refer to two distinct (and potentially problematic) emotional states: fear and urgency. Yet adrenaline was not totally undesirable; in fact, at lower levels rescuers welcomed it because it helped them focus and heightened their awareness. Mostly, though, rescuers were encouraged to see adrenaline rushes as an important physiological cue, one they should heed as a warning that they were approaching the edge and at risk of losing control.

One way rescuers talked about adrenaline impeding performance was through paralyzing fear (see Holyfield and Fine, 1997), which rendered them ineffective, increasing risk for their teammates and for the victim.

Interestingly, Cyndi equated adrenaline with fear. She used these terms interchangeably, nothing the edge between a useful and detrimental physiological reaction that she experience as fear.

Another problem that members attributed to adrenaline was that it could cause them to misinterpret routine situations as urgent. Maddie, the ten-year member, explained how feelings of urgency and excitement were problematic and had to be suppressed:

> [Members need to] realize the urgency, and manage that urgency. And that really is the big part [of participating in a mission]. I know I still get it every now and again, that adrenaline rush is really going as you're walking in to the [victim], and that can really get in the way, big time, out there. Because most of the time we're not in a rush to get that person out. We can't be, or we're gonna injure them.

While excessive fear and urgency were seen as potentially dangerous overreactions to adrenaline, loss of control due to fear was more often associated with women's reactions, while becoming too excitable was considered more of a male phenomenon. One reason men and women might have experienced adrenaline rushes differently was because men were confident at the prospect of undertaking risk, which may have caused them to interpret their adrenaline during the missions as pleasurable and exciting. Since women reported more cautious mindsets in preparing for missions, worrying about their ability to exercise emotional and physical control in risky situations, perhaps they were more likely to define their adrenaline rushes as fearful and unpleasant.

Men and women, however, managed their feelings of urgency and fear similarly: They suppressed them.

When I asked women how they suppressed their emotions, they gave me long and detailed descriptions of what they did and how they did it. Yet men's answers were brief. Some said that since they did not think about the risks they were taking, dangerous emotions did not arise. Nick said, "The majority of the time I just don't think about the worst part, like getting hurt or dying. I just think about getting the job done." Another experienced member commented on the phenomenon as well, saying, "Are you gonna stop and look at the risk factor every time you get paged out for a mission? Or are you just gonna go and do it?"

When men did give detailed responses to questions about suppressing emotions, their explanations still differed from women's in that they tended to describe suppressing their emotions as a process that required little conscious control, one that came almost "naturally" to them.

Another way emotions interfered with performance was when members were disturbed by the graphic sight of the accidents they encountered. Recovering the body of a dead victim, for example, held great potential for negative feelings, especially if the death was violent or gruesome, leaving the body in pieces, excessively bloody, or positioned unnaturally (such as having the legs bent backwards or an arm missing). Such situations could cause extreme reactions in rescuers, possibly preventing them from doing the job they were assigned. On the whole, men were assumed to be better suited for these gruesome or graphic jobs because they were perceived to be emotionally stronger than women.

Yet men, experienced or not, were not immune to the potentially disturbing effects of gruesome rescues. In fact, Martin told me that when he came upon a particularly gruesome plane crash, "I barfed my guts out." Meg, a resort worker and ten-year member in her mid-forties, told me that despite stereotypes of masculine emotional strength, she had seen experienced men who had trouble dealing with dead bodies, even though they were willing to assist in the recovery task.

Thus, even for members who were expected to be emotionally tough, emotions also "got in the way" in dead body situations, because they could cause some members to go over the edge, losing their ability to perform under stressful circumstances.

One emotion management strategy members used to combat these upset feelings was to depersonalize the victims, which Meg alluded to by saying a body is not a person because the "who" part is gone.

Jim, the founding member, also felt very strongly that emotions not be a part of recovering a dead body. He described one occasion when his team was extracting a drowned victim from a river:

> There is nothin' glamorous about taking somebody's human remains, stuffing 'em in a black bag, hauling 'em up the hill, and throwing 'em in the back of the sheriff's van. There is *nothin'* glamorous about that. And when I'm in those types of situations, there's a space that I have to go to in my head. And it's real no-nonsense; it's time to say, "Let's get the job done. Let's roll up our sleeves," if you will. You know, that is not the time to reflect on "What's it all about?" or "Why we're all here." It's a time to roll this *carcass* into a bag and drag it up the hill. And people on our team started with that [being philosophical during one extrication]. [So I said], "I don't wanna hear it! I want the guy *off* the log, *in* the bag, to the *top* of the hill. Are we ready to go?" And we should be, *[snaps],* ready to go.

Jim's "space" in his head helped him perform because it allowed him to depersonalize the victim (the "carcass"). In his view, the other members' choice to reflect on larger philosophical questions was poorly timed, interfering not only with the mission's efficiency, but with his own ability to suppress emotion as well. He clearly demonstrated edgeworkers' ability to filter out part of the social self when he explained that he had a special "space" in his head, and he showed how edgework was impeded when others reflected on their place in the world.

Fear, urgency, and emotional upset were some of the powerful feelings that threatened rescuers' control during missions. As a result, members worked to suppress them (women more actively than men), maintaining a demeanor of "affective neutrality" (Parsons, 1951) by focusing on their task and depersonalizing the victims. This group norm of displaying affective neutrality signified a cool-headedness that was thought to be safe and effective, and the group considered those who could achieve it in the most critical of circumstances to be those who could work closest to the edge. Thus, the best edgeworkers, and by extension, Peak's most valuable members, were those who could reliably suppress their emotions and get the job done.

Completing Edgework: Releasing Feelings

Immediately after the missions, members' suppressed feelings began to surface. They viewed the sensations they got from successful mission outcomes, like reuniting victims with their families, as the ultimate reward, and I often witnessed them expressing these positive feelings upon hearing the news of a saved victim. They instantly discarded their objective demeanors and became jovial, slap-happy, and chatty.

Occasionally, rescuers would realize that they had been in a risky situation that could have gone awry during the mission. Reflecting back on the hazards they had undertaken and overcome during a mission made them feel ambivalent. On the one hand, they felt energized, which they generally regarded as a positive feeling of control and competence.

Alex, the British botanist, told me more about the time she and Roy, the fifteen-year-member and respected mountaineer, were chosen to ice-climb up the 800-foot, avalanche-prone gully to reach the stranded hot-air balloonists whose balloon had crashed at the top of a cliff. About an hour after they were dispatched into the field, the mission coordinator was able to contact a helicopter, which was going to be able to reach the stranded subjects sooner than Roy and Alex, who were still only halfway up the gully. The mission coordinator radioed Alex and Roy and told them to get to a stable, safe place because the wind from the incoming helicopter above them could stir up the snow, causing an avalanche into their gully. They assessed their location at the base of a seventy-five-foot frozen waterfall and thought they might be safe if they positioned themselves close to the ice (the momentum of an avalanche would propel the snow in an arc, providing a safety zone between the stream of snow and the vertical icefall). At the last minute, Alex saw an ice cave twenty feet up, behind the frozen waterfall. They decided they would be safer there so she climbed up and then pulled Roy (whose climbing footgear was falling off) up to her. She told me what happened next:

> The two of us landed in this snow cave, and just then we hear this *roar,* and we're like "Oh my God, an avalanche!" And we like, peer out from behind the ice fall, and the *balloon,* with the basket and all of the fuel tanks and everything, just came thundering down the gully, [and the basket] just *exploded* [broke into hundreds of pieces] at the bottom where we'd been standing, and just carried on down the gully. And we were just like [*shocked expression: jaw dropped, eyes wide, speechless*]. I mean, we just sort of lay there like "Oh my God." I mean, we were expecting an avalanche, not the *balloon.* . . . If we had not moved out of that position and got out of the gully, I mean, we'd've just gotten smack-a-roonied by the balloon. We were both pretty shaken by that one. I don't know, maybe we felt more vulnerable or something, but we were definitely shaken up. . . . Roy kept grabbing me and saying [*yelling*], "I can't believe that happened!" and "Oh my God!" I'd say it was definitely the closest I've come in a rescue to getting snookered.

Even when rescuers emerged from missions safely, these types of close calls could disturb them for days or weeks afterward, bringing their mortality into sharp relief. It was difficult when they realized that they were vulnerable, even if they believed it was only to freak accidents. Such events made it hard for them to maintain the "illusion of control" that many edgeworkers rely on (Lyng, 1990).

Missions with negative outcomes frequently left rescuers with even stronger and more unpleasant feelings than rescues that ended successfully, and members often reported being at home alone when they first encountered these feelings. One source of upset feelings was recurring memories of emotionally disturbing scenes, a common reaction for rescuers in other settings as well (see Moran, Britton, and Correy, 1992; Oliner and Oliner, 1988).

One member told me that he once assisted in the body recovery of a fallen rock climber, after which he didn't sleep for three weeks. He had many upsetting flashbacks of both the sights and sounds from the scene: seeing the victim's legs broken backwards from the fall, and hearing them crack as rescuers straightened them to fit him in the body bag. Alex, the British rescuer, remembered feeling the weight of one dead body she helped carry out from the four-passenger van rollover: "I just remember how *heavy* it was. You know, they say 'dead weight'? That was one of the most memorable [missions]. That one lived with me for a while."

These upsetting flashbacks could be compounded by knowing personal information about the victim. When the rescue hit too close to home, members' confrontation with the stressful emotions was more intense.

While not all rescuers were equally affected by the negative impact of dead body recoveries or of "failed" missions, it did not appear that men were immune to these feelings in the post-mission period, as many rescuers had imagined they would be. Both women and men reported having trouble dealing with feelings of vulnerability in these instances, which diluted the emotional charge they got from edgework. As a result, rescuers tried to manage the uncontrolled flow of conflicting emotions in the immediate post-mission period. In the most intense cases, they reported feeling overwhelmed with emotion, unable to control it, and needing to release it in some way, a prevalent pattern for other types of emergency workers as well (see Mannon, 1992; Metz, 1981).

There were several ways members released these feelings. One way was by crying. I talked to Elena, a hotel manager and five-year member in her early thirties, shortly after her first (and only) dead body recovery. She told me that she thought she was "okay" until she got home and was in the shower, where she started to cry.

This pattern of releasing emotions by crying was gendered: Men never reported crying as a means of dealing with the emotional turmoil of missions, while women occasionally did. Although it is possible that men and women did cry with equal frequency and masculinity norms prevented men from reporting it, it was more likely that women saw this as a more acceptable emotion management technique and coped in this way more often than men.

After the most traumatic missions, such as one occasion when members extricated the charred remains of several forest fire fighters caught in a "fire storm" (an extremely hot, quick-moving, and dangerous type of forest fire), the group provided a professionally run "critical incident debriefing" session where they could talk about their feelings after the mission. While these sessions encouraged men (who were the ones most often involved in such intense missions) to express their feelings, there were only two of these sessions offered in my six years with Peak. As a general rule, Peak's culture did not encourage men to express their feelings after emotionally taxing rescues, a phenomenon that is common to American culture in general: Women tend to cope with emotionally threatening feelings by crying, while men tend to cope with stress by withdrawing, becoming angry, and using drugs and alcohol (Gove, Geerken, and Hughes, 1979; King, Delaronde, Dinoi, and Forsberg, 1996; Mirowsky and Ross, 1995; Moran, Britton, and Correy, 1992; Roehling, Koelbel, and Rutgers, 1996; Thoits, 1995).

It was common for Peak's members to drink alcohol after both positive- and negative-outcome missions, but they generally consumed more after negative outcomes, such as dead body recoveries, as a way of coping with anxiety and unpleasant feelings. After Tyler, Nick, and Shorty recovered the trapped kayaker's body in another county, Tyler told me that they bought a twelve-pack of beer for Nick and Shorty to drink while Tyler drove them home. In the three-hour drive, Shorty drank two of the beers, and Nick drank the remaining ten. When I asked Nick about this, he told me that he drank beer after missions to try to "calm down, to relax. . . . [I was tense] because I didn't think the body was gonna be that beat up. It's kinda like if you had a rough day at work, you drink a couple beers. . . . I think [it's] just part of releasing any tension, even if it's just adrenaline that you have stored up." In this way members used "bodily deep acting," manipulating their physiological state to change their emotional state (Hochschild, 1990), by relaxing themselves with alcohol in an effort to dampen the chaos of their surfacing feelings.

Maddie also talked about members' alcohol use after emotionally stressful missions. In her ten years with Peak, she noticed that more men than women used alcohol to release their emotions, attributing this difference to gender socialization.

Maddie explained men's higher alcohol consumption rate over women's in terms of cultural expectations for men to hide their feelings and appear to remain emotionally unaffected, an observation supported by social research (see Mirowsky and Ross, 1995; Thoits, 1995). Maddie also believed alcohol use to be a dysfunctional, ineffective strategy for some of the men she knew, an observation that has received inconclusive support in coping research (see Robinson and Johnson, 1997; Roehling, Koelbel, and Rutgers, 1996; Patterson and McCubbin, 1984; Sigmon, Stanton, and Snyder, 1995).

Rescuers often reported feeling more easily overwhelmed in the period after a mission, attributing it to their failure to release all of their pent-up emotions, much like the disaster volunteers Gibbs and her colleagues (1996) studied. Thus, another way they dealt with their feelings was to leave the group temporarily or quit altogether. After recovering the trapped kayaker, Nick turned off his pager so that he did not hear any calls for missions, saying that during the several weeks that followed, "I didn't go on any rescues. . . . I just wanted a break."

Maintaining the "Illusion of Control": Redefining Feelings

The fourth stage of edgework was marked by members' ability to regain control of their feelings and cognitively process them, retrospectively redefining and shaping their experiences, a process Kitsuse (1962) termed "retrospective interpretation." Often this involved neutralizing their post-mission negative feelings, which was important to edgework because, if left unresolved, negative emotions could destroy the positive elements of edgework, shake members' confidence, and impede their performance on future missions. In the long term, positive-outcome missions allowed rescuers to maintain the "illusion of control"; members' success served as evidence that they could push their limits next time, too. Negative outcomes, however, threatened the illusion of control, leaving members wondering if they were capable, and unsure of the risk they were willing to assume in the future. In this stage, rescuers employed another type of "deep acting" where they "visualiz[ed] a substantial portion of reality in a different way" (Hochschild, 1990, p. 121), and their emotion management techniques were aimed at cognitively changing the meaning of what happened, which transformed their feelings about it. This helped them to maximize their ability to conduct future edgework.

Guilt was a stressful emotion for rescuers in the wake of unsuccessful missions. Members could feel personally responsible for the outcome, for example, if they failed to save a victim. On one occasion, rescuers felt bothered by a mission where a kayaker died in a river race. Many of Peak's members were at the race, volunteering to act as safety agents on the riverbanks, throwing lines to any kayakers in trouble. One racer's kayak flipped upside down, and he was unable to right himself. Although many tried to reach him—fellow racers chased him down, people standing on the banks threw safety lines—no one could get to him until he floated through the finish line four minutes later. Many bystanders speculated that he must have been knocked unconscious while he was inverted and subsequently drowned. Jim told me that he went over and over the incident in his mind that night, trying to think of something he and the team could have done to reach the boater more quickly. He could find no flaws in the team's response, and yet found it difficult to accept that the boater was killed.

Two days later the local newspaper reported that the kayaker had died when, due to a genetic defect, his heart "exploded." Many members were relieved by this news because it confirmed the conclusions they had come to through their careful reanalysis: They could not have saved him.

One way members neutralized their guilt was by redefining their part in missions. One technique was "denying responsibility" (Sykes and Matza, 1957) for the victim's fate. Members were often reminded that their first concern on a mission was to protect themselves, second, their teammates, and third, the victim.

Another way members avoided feeling responsible for not saving people was by attributing control to a higher power. Meg told me that she relied on her "personal philosophical beliefs" to help her deal with the potentially negative feelings that could result from recovering a dead body. She used this technique to accept the reality that people die, as well as to deny any responsibility for it.

Members also denied responsibility by blaming the victims themselves. Gary told me that he often said to himself, "'God, I'm glad that's not me.' It sounds selfish but it's really not. . . . If they're dead, they might have done something stupid to get there; that's not our fault. Through these methods, rescuers were able to temper their feelings of guilt and vulnerability, which, in turn, helped them to maintain a positive self-image as well as to maintain the illusion of control, to reassure themselves about their own ability to survive edgework.

Members also redefined their part in "failed" missions by emphasizing the positive side of negative events. For instance, although dead body recoveries were very unpleasant experiences for everyone in the group, most members were prepared to voluntarily assist in them, pointing out how their part in retrieving bodies helped the grieving families.

Another technique members used to counter the stress of emotionally taxing missions was to weight the successes more that the failures. Although they took great pains to separate themselves personally from failed missions, denying responsibility and downplaying the significance of those situations, members actively sought a personal connection with the successful missions, acknowledging their role in them, and allowing their participation to be meaningful and important reflections of their abilities. Personally accepting credit for successes protected rescuers by bolstering their confidence and making them feel that they had control over risky conditions, a strategy used by other edgeworkers who strive to maintain the illusion of control (see Lyng, 1990; Mannon, 1992).[1]

Conclusion

There were four stages of edgework that were marked not only by the flow of rescue events, but also by the corresponding emotions they evoked. Rescuers risked both their physical and emotional well-being before, during, and after the missions, and maintaining a sense of an ordered reality was a key concern in each stage. Because each of these four stages was characterized by different threatening feelings, members utilized several types of emotion management strategies as they prepared for, performed, completed, and redefined edgework. Moreover, these feelings and management techniques were gender specific. The men in the group tended to feel confident and excited on missions and to display emotional stoicism about negative outcomes. Conversely, the women tended to feel trepidacious and fearful on the critical missions and to express their feelings in the mission's aftermath. Both genders, though, tended to maintain a positive self-image by using cognitive strategies to redefine their feelings in retrospect.

Note

1. This strategy, known more broadly to social psychologists as the "self-serving bias," is one of the most common ways that people protect their self-esteem.

References

Adler, P. A., and Adler, P. *Membership Roles in Field Research.* Newbury Park, CA: Sage, 1987.

Blumer, H. *Symbolic Interactionism.* Englewood Cliffs, NJ: Prentice-Hall, 1969.

Clark, C. "Emotions and Micropolitics in Everyday Life: Some Patterns and Paradoxes of 'Place'" in *Research Agendas in the Sociology of Emotions,* edited by T. D. Kemper. Albany, NY: SUNY Press, 1990, 305–333.

Gibbs, M., Lachenmeyer, J. R., Broska, A., and Deucher, R. "Effects of the AVIANCA Aircrash on Disaster Workers." *International Journal of Mass Emergencies and Disasters* 14 (1996): 23–32.

Gove, W. R., Geerken, M., and Hughes, M. "Drug Use and Mental Health Among a Representative National Sample of Young Adults." *Social Forces* 58 (1979): 572–590.

Hochschild, A. R. *The Managed Heart.* Berkeley: University of California Press, 1983.

———. "Ideology and Emotion Management: A Perspective and Path for Future Research" in *Research Agendas in the Sociology of Emotions,* edited by T. D. Kemper. Albany, NY: SUNY Press, 1990, 117–142.

Holyfield, L. "Generating Excitement: Experienced Emotion in Commercial Leisure" in *Social Perspectives on Emotion,* Vol. 4, edited by R. J. Erickson and B. Cuthbertson-Johnson. Greenwich, CT: JAI Press, 1997, 257–281.

Holyfield, L., and Fine, G. A. "Adventure as Character Work: The Collective Taming of Fear." *Symbolic Interaction* 20 (1997): 343–363.

Jorgensen, D. L. *Participant Observation.* Newbury Park, CA: Sage, 1989.

King, G., Delaronde, S. R., Dinoi, R., and Forsberg, A. "Substance Use, Coping, and Safer Sex Practices Among Adolescents with Hemophilia and Human Immunodeficiency Virus." *Journal of Adolescent Health* 18 (1996): 435–441.

Kitsuse, J. "Societal Reactions to Deviant Behavior: Problems of Theory and Method." *Social Problems* 9 (1962): 247–256.

Lyng, S. "Edgework: A Social Psychological Analysis of Voluntary Risk Taking." *American Journal of Sociology* 95 (1990): 851–886.

Mannon, J. M. *Emergency Encounters: A Study of an Urban Ambulance Service.* Sudbury, MA: Jones and Bartlett, 1992.

Metz, D. L. *Running Hot.* Cambridge, MA: Abt Books, 1981.

Mirowsky, J., and Ross, C. E. "Sex Differences in Distress: Real or Artifact?" *American Sociological Review* 60 (1995): 449–468.

Mitchell, R. G., Jr. *Mountain Experience: The Psychology and Sociology of Adventure.* Chicago: University of Chicago Press, 1983.

Moran, C., Britton, N. R., and Correy, B. "Characterising Voluntary Emergency Responders: Report of a Pilot Study." *International Journal of Mass Emergencies and Disasters.* 10 (1992): 207–216.

Oliner, S. P., and Oliner, P. M. 1988. *The Altruistic Personality: Rescuers of Jews in Nazi Europe.* New York: Free Press.

Parsons, T. *The Social System.* New York: Free Press, 1951.

Patterson, J. M., and McCubbin, H. I. "Gender Roles and Coping." *Journal of Marriage and the Family* 46 (1984): 95–104.

Robinson, M. D., and Johnson, J. T. "Is It Emotion or Is It Stress? Gender Stereotypes and the Perception of Subjective Experience." *Sex Roles* 36 (1997): 235–258.

Roehling, P. V., Koelbel, N., and Rutgers, C. "Codependence and Conduct Disorder: Feminine Versus Masculine Coping Responses to Abusive Parenting Practices." *Sex Roles* 35 (1996): 603–618.

Schachter, S., and Singer, J. E. "The Interactions of Cognitive and Physiological Determinants of Emotional State." *Psychological Review* 69 (1962): 379–399.

Sigmon, S. T., Stanton, A. L., and Snyder, C. R. "Gender Differences in Coping: A Further Test of Socialization and Role Constraint Theories." *Sex Roles* 33 (1995): 565–587.

Syres, G. "Techniques of Neutralization: A Theory of Delinquency." *American Sociological Review, 22:* 664–670.

Thoits, P. A. "Identity-Relevant Events and Psychological Symptoms: A Cautionary Tale." *Journal of Health and Social Behavior* 36 (1995): 72–82.

✤ Topics for Further Examination ✤

- Look up research and Web sites on patterns of eating disorders and cosmetic surgery procedures among women and men in the United States today. Check out this Web site: http://www.pbs.org/wgbh/nova/thin.
- Explore the world's largest online library of books and articles on homosexuality: http://www.questia.com/Index.jsp?CRID=homosexuality&OFFID=se1.
- Compare and contrast the messages about gendered emotions in ads, articles, columns and other features in popular women's magazines and popular men's magazines.

7

Gender at Work

Throughout this book we emphasize the social construction of gender, a dominant prism in people's lives. This chapter explores some of the ways that the social and economic structures within capitalist societies create gendered opportunities and experiences at work, and how work and gender affect life choices, particularly as they relate to family and parenting. The gendered patterns of work that emerge in capitalist systems are complex, like those of a kaleidoscope. These patterns reflect the interaction of gender with other social prisms such as race, sexuality, and social class. Readings in this chapter support points made throughout the book. First, women's presence, interests, orientations, and needs tend to be diminished or marginalized within occupational spheres. Second, one can use several of the concepts we have been studying to understand the relationships of men and women at work, including hegemonic masculinity, "doing gender," the commodification of gender, and the idea of separate spaces for men and women.

In this chapter, we explore the construction and maintenance of gender within both paid and unpaid work in the United States, including an article that looks at gender in the global workplace. We begin with a discussion of work and gender inequality. The history of gender discrimination in the paid labor market is a long one (Reskin & Padavic, 1999), with considerable social science research that documents gendered practices in workplace organizations. The first reading, by Joan Acker, discusses what she calls "inequality regimes," or the ways in which work organizations create and maintain inequality across the intersections of gender, race, and social class. In this piece, she looks beneath the surface to almost invisible institutional practices that maintain unequal opportunities within organizations (see also Acker, 1999). Care work is another aspect of work discussed in this chapter, in the reading by Paula England. Care work is gendered. Care work is also undervalued and either unpaid or underpaid relative to other types of work. In this reading, England discusses five theories that explain both why care work is devalued and why women continue to do it.

Consider the various ways that the workforce in the United States is gendered. Think about different jobs (e.g., nurse, engineer, teacher, mechanic, domestic worker) and ask yourself if you consider them to be "male" or "female" jobs. Now, take a look at Table 7.1, which lists job categories used by the Bureau of Labor Statistics (2006a). You will note that jobs tend to be gender typed; that is, men and women are segregated into particular jobs. The consequences for men and women workers of this continuing occupational gender segregation are significant in the maintenance of gendered identities. Included in Table 7.1 are jobs predominantly held by men (management, architecture and engineering, and construction) and those predominantly held by women (education, health care support, and office and administrative support).

Gender segregation of jobs is linked with pay inequity in the labor force. As you look through Table 7.1, locate those jobs that are the highest paid and determine whether they employ more men or more women. Also, compare women's to men's salaries across occupational categories. Clearly a "gender wage gap" is evident in Table 7.1. Even in those job categories that are predominantly filled by women, men earn more than women. For example, going beyond the data in Table 7.1 and looking specifically at elementary and middle school teachers, a traditionally female job, the 2006 median weekly earnings for men regardless of race or ethnicity is $920 compared to $824 for women (Bureau of Labor Statistics, 2006b). However, in a traditionally male job such as financial managers, men also outearn women $1,421 to $894 (Bureau of Labor Statistics, 2006b), a situation discussed in the reading by Louise Marie Roth in this chapter.

The pattern you see does not deny that *some* women are CEOs of corporations, such as Xerox and Hewlett Packard, and that today we see women workers everywhere including on construction crews. However, although a few women crack what is often called "the glass ceiling," getting into the top executive or hypermasculine jobs is not easy for women and minority group members. The glass ceiling refers to the point at which women and others, including racial minorities, are "blocked from any further upward movement" in organizations (Baxter & Wright, 2000, p. 275). Informal networks generally maintain the impermeability of glass ceilings, with executive women often isolated and left out of "old boys' networks," finding themselves "outsiders on the inside" (Davies-Netzley, 1998, p. 347). Similar internal mechanisms within union and trade-related organizations also keep women out because "knowing" someone often helps to get a job in the higher paid, blue-collar occupations.

Gender, Race, and Social Class at Work

When we incorporate the prism of race, segregation in the work force and pay inequality become more complex. Another look at Table 7.1 indicates that Hispanic or Latino and African American men and women earn less than White, non-Hispanic men and women, although minority men often make more than White women. In addition, Hispanic and African American women and men are much less likely to be found in the job categories with higher salaries than their percentages in the labor force would suggest. The continuing discrimination against African Americans, Hispanics, and other ethnic minority groups (as indicated in Table 7.1) shows patterns similar to the discrimination against women, both in the segregation of certain job categories as well as in the wage gap that exists within the same job category. These processes operate to keep African American, Hispanic, and other marginalized groups "contained" within a limited number of occupational categories in the labor force.

The inequities of the workplace carry over into retirement (Calasanti & Selvin, 2001). Women and other marginalized groups are at a disadvantage as they leave the workforce and retire because their salaries are lower during their paid work years (see England in this chapter). Calasanti and Slevin find considerable inequalities in retirement income, which indicate that the inequalities in the labor force have a long-term effect for women and racial/ethnic minorities. They argue that only a small group of the workforce—privileged White men—are able to enjoy their "golden years" and the reasons for this situation are monetary.

Efforts to change inequality in the workplace by combating wage and job discrimination

Table 7.1 2006 Median Weekly Salary and Percentages of Men and Women in Selected Occupational Categories by Gender and Race/Ethnic Group[1]

Occupational Category		*White Women*	*White Men*	*Black Women*	*Black Men*	*Hispanic Women*	*Hispanic Men*
		Median Weekly Wages					
	Total Number Employed in Category (16 years and older)	*% in Category*	*% in Category*	*% in Category*	*% in Category*	*% in Category*	*% in Category*
All occupations	$671 106,106	$609 34.3%	$761 46.8%	$519 6.3%	$591 5.7%	$440 5.4%	$505 9.4%
Management, Professional, and Related Occupations							
Management occupations	$1,127 10,661	$931 33.4%	$1,295 53.8%	$886 3.8%	$955 3.5%	$836 3.0%	$944 4.3%
Business and financial operations occupations	$930 4,786	$827 44.1%	$1,143 35.9%	$778 7.4%	$1,016 3.6%	$696 4.2%	$907 2.8%
Computer and mathematical occupations	$1,166 2,935	$1,022 17.9%	$1,219 55.5%	$918 2.8%	$1,095 4.5%	$739[2] 1.1%	$1,058 3.8%
Architecture and engineering occupations	$1,155 2568	$935 10.3%	$1,173 72.6%	$864[2] 1.2%	$996 4.8%	$815[2] 0.9%	$993 5.5%
Life, physical and social science occupations	$984 1,220	$880 32.1%	$1,125 46.1%	$689[2] 3.0%	$1,033[2] 2.7%	$850[2] 1.5%	$1,066[2] 2.8%
Community and social services occupations	$740 1,816	$716 42.4%	$841 30.6%	$674 13.3%	$699 7.9%	$689 5.5%	$768 3.6%
Legal occupations	$1,144 1,156	$901 49.4%	$1,750 39.3%	$766 4.6%	$1,344[2] 2.0%	$715 4.9%	$1,484[2] 1.5%
Education, training and library occupations	$819 6,158	$773 61.2%	$986 23.4%	$686 8.4%	$760 2.2%	$698 5.4%	$800 1.7%

(Continued)

Table 7.1 (Continued)

	Median Weekly Wages						
	Total Number Employed in Category (16 years and older)	*% in Category*	*% in Category*	*% in Category*	*% in Category*	*% in Category*	*% in Category*
Arts, design, entertainment, sports, and media occupations	$841 1,476	$739 38.6%	$943 47.2%	$748[2] 2.8%	$772 4.3%	$604[2] 3.1%	$838 5.4%
Healthcare practioner and technical occupations	$905 5,048	$865 57.2%	$1,127 19.5%	$746 10.0%	$812 2.7%	$732 4.1%	$985 2.0%
Service Occupations							
Healthcare support occupations	$423 2,231	$425 58.2%	$508 6.2%	$397 25.3%	$483 3.4%	$397 12.7%	$502[2] 1.8%
Protective service occupations	$693 2,633	$584 13.2%	$775 62.4%	$512 6.8%	$571 13.3%	$498 2.1%	$718 8.6%
Food preparation and serving related occupations	$371 4212	$354 39.2%	$390 37.3%	$350 7.3%	$373 6.6%	$317 10.1%	$371 18.0%
Building and grounds cleaning and maintenance occupations	$406 3,594	$359 26.1%	$447 51.4%	$364 7.5%	$416 9.6%	$338 13.8%	$392 22.1%
Personal care and service occupations	$407 2,079	$390 52.7%	$534 18.3%	$369 13.7%	$425 4.5%	$333 11.2%	$510 3.7%
Sales and Office Occupations							
Sales and related occupations	$628 10,336	$501 35.8%	$786 48.7%	$406 5.6%	$579 4.1%	$396 5.7%	$565 6.0%
Office and administrative support occupations	$572 15,351	$558 59.3%	$633 19.8%	$541 10.6%	$556 4.2%	$505 8.4%	$525 4.4%

	Median Weekly Wages						
	Total Number Employed in Category (16 years and older)	% in Category	% in Category	% in Category	% in Category	% in Category	% in Category
Natural Resources, Construction, and Maintenance Occupations							
Farming, fishing, and forestry occupations	$387 716	$323 16.3%	$406 71.8%	$396[2] 2.4%	$338[2] 3.1%	$305 8.8%	$352 36.7%
Construction and extraction	$619 7,166	$520 1.9%	$624 87.0%	$675[2] 0.3%	$552 6.8%	$493 0.8%	$495 32.3%
Installation, maintenance, and repair occupations	$742 4,630	$701 3.5%	$752 84.0%	$689[2] 0.8%	$671 6.6%	$582[2] 0.6%	$606 13.2%
Production, Transportation, and Material Moving Occupations							
Production occupations	$559 8,391	$429 21.5%	$637 58.8%	$448 4.4%	$519 7.9%	$344 7.5%	$481 13.6%
Transportation and material moving occupations	$556 6,942	$417 9.0%	$595 68.8%	$404 2.9%	$524 14.5%	$361 2.8%	$481 18.0%

[1] Data for this table were taken from the Current Population Survey, Table A2. "Usual weekly earnings of employed full-time wage and salary workers by intermediate occupation, sex, race, and Hispanic or Latino ethnicity and Non-Hispanic ethnicity, Annual averages 2006 (Bureau of Labor Statistics, unpublished data, 2007)

[2] These estimates do not meet BLS standard for statistical reliability (50,000 cases), therefore, they must be used cautiously.

through legislation have included both gender and race. In 1963, Congress passed the Equal Pay Act, prohibiting employment discrimination by sex, but not by race. Men and women in the same job, with similar credentials and seniority, could no longer receive different salaries. Although this legislation was an important step, Blankenship (1993) cites two weaknesses in it. First, by focusing solely on pay equity, this legislation did not address gender segregation or gender discrimination in the workplace. Thus, it was illegal to discriminate by paying a woman less than a man who held the same job, but gender segregation of the workforce and differential pay across comparable jobs was legal. As Blankenship (1993) notes, this legislation saved "men's jobs from women" (p. 220) because employers could continue to segregate their labor force into jobs that were held by men and those held by women, and then pay the jobs held by men at a higher rate. Second, this legislation did little to help minority women, as a considerable majority of employed women of color were in occupations such as domestic workers in private households or employees of hotels/motels or restaurants that were not covered by the act (Blankenship, 1993).

In 1964, Congress passed Title VII of the Civil Rights Act. They drafted this Act to address racial discrimination in the labor force. This act prohibited discrimination in "hiring, firing, compensation, classification, promotion, and other conditions of employment on the basis of race, sex, color, religion, or national origin" (Blankenship, 1993, p. 204). Sex-based discrimination was not originally part of this legislation, but was added at the last minute, an addition that some argue was to assure the bill would not pass Congress. However, the Civil Rights Act did pass Congress and women were protected along with the other groups. Unfortunately, the enforcement of gender discrimination was much less enthusiastic than that of race discrimination (Blankenship, 1993).

Blankenship (1993) argues that the end result of these two pieces of legislation to overcome gender and race discrimination was to "protect white men's interests and power in the family" (p. 221) with little concern about practices that kept women and men of color out of higher-paying jobs. Sadly, these attempts seem to have had little impact on race and gender discrimination (Strum & Guinier, 1996). In the reading in this chapter, Roth describes the effects of this legislation in her discussion of women on Wall Street and, unfortunately, shows how much still needs to change in this high-paced, high-paying occupational field. Take another look at Table 7.1 and think about the ways the different allocations of jobs and wages affect women's and men's lives across race and social class—their ability to be partners in relationships and their ability to provide for themselves and their families.

As you think about the differences that remain in wage inequality, consider what still needs to be accomplished. Pay equity may seem like a simple task to accomplish. After all, now we have laws that should be enforced. However, the process by which most companies determine salaries is quite complex. They rank individual job categories based upon the degree of skill needed to complete job-related tasks. Ronnie Steinberg (1990), a sociologist who has studied comparable worth of jobs for over thirty years, portrays a three-part process for determining wages for individual jobs. First, jobs are evaluated based upon certain job characteristics such as "skill, effort, responsibility, and working conditions" (p. 457). Second, job complexity is determined by applying a "value to different levels of job complexity" (p. 457). Finally, the values determined in the second step help to set wage rates for the job. As England notes in her reading in this chapter, care work as well as other types of work typically performed by women, are undervalued in this wage-setting process.

On the surface, this system of determining salaries seems consistent and "compatible with meritocratic values," where each person receives pay based upon the value of what he or she actually does on the job (Steinberg, 1987, p. 467). What is recognized as "skill," however, is a matter of debate and is typically decided by organizational leaders who are predominately White, upper-class men. The gender and racial bias in the system of determining skills is shocking. Steinberg gives an example from the State of Washington in 1972 in which two job categories, legal secretary and heavy equipment operator, were evaluated as "equivalent in job complexity," but the heavy equipment operator was paid $400 more per month than the legal secretary (Steinberg, 1990, p. 456). Although it appears that all wages are determined in the same way based upon the types of tasks they do at work, Steinberg (1987, 1990, 1992) and others (including England in this chapter) argue that the processes used to set salaries are highly politicized and biased.

Gender Discrimination at Work

One way of interpreting why these gendered differences continue in the workforce is to examine workplaces as gendered institutions, as discussed in the Introduction to this book. Acker, in the first reading in this chapter, as well as other researchers examine work as a gendered institution (Acker, 1999; Reskin, McBrier, & Kmec, 1999). For example, Martin (1996) studied managerial styles and evaluations of men and women in two different organizations: universities and a multinational corporation. She found that when promotions are at stake, male managers mobilized hegemonic masculinity to benefit themselves, thus excluding women. Understanding the processes and patterns by which hegemonic masculinity is considered "normal" within organizations is one avenue to understanding how organizations work to maintain sex segregation and pay inequity. These "inequality regimes" disadvantage all but a few and, as Acker notes in this chapter, things are likely to get worse, rather than better. Unfortunately, as corporations are spreading across the globe, rather than across nations, these patterns of inequality are also spreading. In a reading in this chapter, Steven McKay describes how, as high-tech production goes global, multinational corporations spread cultural and business practices that disadvantage women in their own countries to other countries where they relocate. What is interesting, in McKay's reading, is that the three multinational corporations each brings a different approach to labor force relations, with different consequences for women workers in the Phillipines.

Looking at some of the top wage jobs in the United States, Roth describes some of the subtle and not so subtle mechanisms of discrimination in Wall Street firms. Unfortunately, even the women who succeeded on Wall Street ended up making less than their male counterparts. The "inequality regimes," surrounding career trajectories and compensation patterns make meritocracy a myth and discourage women from trying because it is clear that they are on an uneven playing field. As you read the articles by Acker, McKay, England, and Roth, consider those mechanisms and others where gender segregates workplaces and keeps women from advancing into particular jobs.

Gender discrimination at work is much more than an outcome of cultural or socialization differences in women's and men's behaviors in the workplace. Corporations have vested interests in exploiting gender labor. The exploitation of labor is a key element in the global as well as the U.S. economy, particularly as companies seek to reduce labor costs. Women, in particular, are likely targets for large, multinational corporations. In developing nations, companies exploit poor women's desires for freedom for themselves and responsibilities to their families. The reading by Steven McKay illustrates these points as he describes local factories in the Phillipines owned by three multinational corporations based in three different countries and three different cultures. The gendered assumptions, policies, and practices these corporations bring to their factories in the Phillipines reflect the culture of their home country and illustrate the various ways "inequality regimes" are created in the workplace. As you read McKay's description of how corporations structure their employees work experiences, ask yourself how other structures and patterns within the workplace reinforce gender difference and inequality in men's and women's identities, relationships, and opportunities.

Sexual Harassment

In the reading in Chapter 9, Beth Quinn describes situations that are considered to be sexual harassment. The definition of sexual harassment includes two different types of behaviors (Welsh, 1999). Quid pro quo harassment is that which involves the use of sexual threats or bribery to make employment decisions. Supervisors who threaten to withhold a promotion to someone they supervise unless that person has sex with them engage in quid pro quo harassment. The second form of sexual harassment is termed "hostile environment" harassment, which the EEOC defines as consisting of behavior that creates an "intimidating, hostile, or offensive working environment" (Welsh, 1999, p. 170). A hostile environment occurs when a company allows situations that interfere with an individual's job performance.

Unfortunately, we do not know exactly how prevalent sexual harassment is. Reports of the incidence of sexual harassment are not consistent, primarily due to the various ways sexual harassment is measured across studies (Welsh, 1999). Although various studies report that anywhere from 16 to 90 percent of women experience sexual harassment at their workplace, an attempt to summarize eighteen different surveys estimated the prevalence rate at 44 percent (Welsh, 1999). Any incidence of sexual harassment, however, creates an uncomfortable workplace for the target and perpetuates gender inequality.

The effects of sexual harassment and other forms of gender discrimination create a unwelcoming and hostile workplace for women workers, one that tends to overlook talents and emphasizes normative gender characteristics.

The Effect of Work on Our Lives

The work we do shapes our identities and affects people's expectations for themselves and others (Kohn, Slomczynski, & Schoenbach, 1986) and their emotions (as indicated in the reading on emergency rescue workers by Lois in Chapter 6). It is not just paid work that affects our orientations toward self and others (Spade, 1991), but also work done in the home. In Western societies, work also defines leisure, with leisure related to modernization and the definition of work being "done at specific times, at workplaces, and under work-specific authority" (Roberts, 1999, p. 2). Although the separation of leisure from work is much more likely to be found in developed societies, work is not always detached from leisure, as evidenced by the professionals who carry home a briefcase at the end of the day, or the beepers that summon individuals to call their workplaces.

Women's Work

Care work is one gendered pattern that restricts women's leisure more so than men's leisure, as England notes in the reading in this chapter. Women's leisure is often less an escape from work and more a transition to another form of work, domestic work. In an international study using time budgets collected from almost 47,000 people in ten industrialized countries, Bittman and Wajcman (2000) found that men and women have a similar amount of free time; however, women's free time tends to be more fragmented by demands of housework and caregiving. Another study using time budgets found that women spent 30.9 hours on average in various different family care tasks such as cooking, cleaning, repairs, yard work, and shopping, while men spent 15.9 hours per week performing those tasks (Robinson & Godbey, 1997, p. 101). Women also reported more stress in the Bittman and Wajcman (2000) study, which the authors attributed to the fact that "[f]ragmented leisure, snatched between work and self-care activities, is less relaxing than unbroken leisure" (p. 185).

Domestic work, while almost invisible and generally devalued, cannot be left out of a discussion of work and leisure (Gerstel, 2000). England describes the devaluation of care work in her reading, particularly unpaid care work, which rests largely on women's shoulders. Gerstel (2000) refers to the contribution of women to care work as "the third shift." As a result, domestic labor and caregiving, being unpaid, are done by people least valued in the paid market. As England points out, the undervaluation of care work carries over to the paid market as well. Look again at Table 7.1 and identify those job categories that encompass care work such as health support workers, and personal care and service workers. Now, compare salaries and percentages of men and women in these care giving jobs. As you start to consider these issues, ask why we undervalue care work—the unpaid care work in the home as well as care work in the workplace? Why are men encouraged not to participate in care work and why are women the default caregivers? How is it that the work of the home is undervalued and how is it related to the workplace and the amount of leisure time available to men and women?

Work, Family, and Parenting

With care work perceived as a "feminine" activity, it is not surprising that women's lives are more likely to be focused around, or expected to be focused around, care work activities. As Kathleen Gerson discusses in her reading, couples have started to modify expectations for marriage and marital responsibilities in response to changes in educational attainment and workforce participation of women. Gerson describes the dilemmas faced and strategies tried by young people as they attempt to balance commitment to a relationship and autonomy in their lives. Care giving and parenting have been transformed and both men and women must respond to the constraints on their time as they balance what they want in their lives with idealized gender roles and the relationships of their own parents.

Decisions about care giving and parenting are much more difficult for young couples, however, when society and work organizations assume they are a private (read "woman's") responsibility. Contrast the work/family policies of other industrialized nations to that of the United States as described in the article by Gwen Moore in this chapter. She looks at the lives of elite men and women in a comparative analysis of how these leaders of industrialized nations balance their work lives and family lives. Unfortunately, considerable gender differences remain, but other countries provide a much broader support network for dual-worker families. What work/family policies would you like to see the United States adopt?

We can only illustrate a few patterns of work in this chapter. The rest you can explore on your own as you take the examples from the readings and apply them to your own life. When you read through the articles in this chapter, consider the consequences of maintaining gendered patterns at work for yourself and your future. While you are at it, consider why these patterns still exist and what these patterns of inequality look like in your life.

References

Acker, J. (1999). Gender and organizations. In J. S. Chafetz (Ed.), *The handbook of the sociology of gender* (pp. 171–194). New York: Kluwer Academic/Plenum.

Baxter, J., & Wright, E. O. (2000). The glass ceiling hypothesis: A comparative study of the United States, Sweden, and Australia. *Gender & Society, 14* (2), 275–294.

Bittman, M., & Wajcman, J. (2000). The rush hour: The character of leisure time and gender equity. *Social Forces, 79* (1), 165–189.

Blankenship, K. M. (1993). Bringing gender and race in: U.S. employment discrimination policy. *Gender & Society, 7* (2), 204–226.

Bureau of Labor Statistics (2006a). *Current Population Survey, Bureau of Labor Statistics*, unpublished data retrieved February 23, 2007.

Bureau of Labor Statistics (2006b). *Current Population Survey, Bureau of Labor Statistics*. Retrieved April 19, 2007 from http://www.bls.gov/cps/cpsaat39.pdf

Calasanti, M., & Selvin, K. F. (2001). *Gender, social inequalities, and aging*. Walnut Creek, CA: AltaMira Press.

Davies-Netzley, S. A. (1998). Women above the glass ceiling: Perceptions on corporate mobility and strategies for success. *Gender & Society, 12* (3), 339–355.

Gerstel, N. (2000). The third shift: Gender and care work outside the home. *Qualitative Sociology, 23* (4), 467–483.

Kohn, M. L., Slomczynski, K. M., & Schoenbach, C. (1986). Social stratification and the transmission of values in the family: A cross-national assessment. *Sociological Forum, 1*, 73–102.

Martin, P.Y. (1996). Gendering and evaluating dynamics: Men, masculinities, and managements. In D. Collinson & J.Hearn (Eds.), *Men as managers, managers as men* (pp. 186–209). Thousand Oaks, CA: Sage.

Reskin, B. F., McBrier, D. R., & Kmec, J. A. (1999). The determinants and consequences of workplace sex and race composition. *Annual Review of Sociology, 23*, 335–361.

Reskin, B. F., & Padavic, I. (1999). Sex, race, and ethnic inequality in United States workplaces. In J. S. Chafetz (Ed.), *Handbook of the sociology of gender* (pp. 343–374). New York: Kluwer Academic/Plenum Publishers.

Roberts, K. (1999). *Leisure in contemporary society.* Oxon, England: CABI Publishing.

Robinson, J. P., & Godbey, G. (1997). *Time for life: The surprising ways Americans use their time.* University Park: Pennsylvania State University Press.

Spade, J. Z. (1991). Occupational structure and men's and women's parental values. *Journal of Family Issues, 12* (3), 343–360.

Steinberg, R. J. (1987). Radical changes in a liberal world: The mixed success of comparable worth. *Gender & Society, 1* (4): 446–475.

Steinberg, R. J. (1990). Social construction of skill: Gender, power, and comparable worth. *Work and Occupations, 17* (4), 449–482.

Steinberg, R. J. (1992). Gendered instructions: Cultural lag and gender bias in the Hay System of job evaluation. *Work and Occupations, 19* (4), 387–423.

Sturm, S., & Guinier, L. (1996). Race-based remedies: Rethinking the process of classification and evaluation: The future of Affirmative Action. *California Law Review.*

Welsh, S. (1999). Gender and sexual harassment. *Annual Review of Sociology, 25*, 169–190.

Introduction to Reading 31

Joan Acker draws from her vast research on gender, class, work, and organizations to describe the structure of organizations that maintain gender, class, and race disparities in wages and power in organizations. She also explores why inequalities in organizational structures and practices are not likely to change. She describes "inequality regimes," or practices and policies embedded in the organization itself, and shows how they work to create and maintain inequality across gender, race, and class. In this article, Acker provides detailed examples of how organizations maintain the gender inequalities in wages described in Table 7.1 and also why individuals seem powerless to overcome these gender inequalities.

1. Using your own life, think about whether you can identify any "inequality regimes" in the organizations you have worked in.
2. How does her description of inequality regimes explain the data in Table 7.1?
3. Look for examples in this reading of how inequality regimes influence the decisions individuals make relative to autonomy and commitment as discussed in the last reading of this chapter by Kathleen Gerson.

INEQUALITY REGIMES

GENDER, CLASS, AND RACE IN ORGANIZATIONS

Joan Acker

All organizations have inequality regimes, defined as loosely interrelated practices, processes, actions, and meanings that result in and maintain class, gender, and racial inequalities within particular organizations. The ubiquity of inequality is obvious: Managers, executives, leaders, and department heads have much more power and higher pay than secretaries, production workers, students, or even professors. Even organizations that have explicit egalitarian goals develop inequality regimes over time, as considerable research on egalitarian feminist organizations has shown (Ferree and Martin 1995; Scott 2000).

I define inequality in organizations as systematic disparities between participants in power and control over goals, resources, and outcomes; workplace decisions such as how to organize work; opportunities for promotion and interesting work; security in employment and benefits; pay and other monetary rewards; respect; and pleasures in work and work relations. Organizations vary in the degree to which these disparities are present and in how severe they are. Equality rarely exists in control over goals and resources, while pay and other monetary rewards are usually unequal. Other disparities may be less evident, or a high degree of equality might exist in particular areas, such as employment security and benefits.

Inequality regimes are highly various in other ways; they also tend to be fluid and changing. These regimes are linked to inequality in the surrounding society, its politics, history, and culture. Particular practices and interpretations develop in different organizations and subunits. One example is from my study of Swedish banks in the late 1980s (Acker 1994). My Swedish colleague and I looked at gender and work processes in six local bank branches. We were investigating the degree to which the branches had adopted a reorganization plan and a more equitable distribution of work tasks and decision-making responsibilities that had been agreed to by both management and the union. We found differences on some dimensions of inequality. One office had almost all women employees and few status and power differences. Most tasks were rotated or shared, and the supervision by the male manager was seen by all working in the branch as supportive and benign. The other offices had clear gender segregation, with men handling the lucrative business accounts and women handling the everyday, private customers. In these offices, very little power and decision making were shared, although there were differences in the degrees to which the employees saw their workplaces as undemocratic. The one branch office that was most successful in redistributing tasks and decision making was the one with women employees and a preexisting participatory ethos.

* * *

From Acker, Joan "Inequality Regimes: Gender, Class, and Race in Organizations," *Gender & Society* 2006 Vol. *20*(4):441–464. Reprinted with permission of Sage Publications, Inc.

What Varies? The Components of Inequality Regimes

Shape and Degree of Inequality

The steepness of hierarchy is one dimension of variation in the shape and degree of inequality. The steepest hierarchies are found in traditional bureaucracies in contrast to the idealized flat organizations with team structures, in which most, or at least some, responsibilities and decision-making authority are distributed among participants. Between these polar types are organizations with varying degrees of hierarchy and shared decision making. Hierarchies are usually gendered and racialized, especially at the top. Top hierarchical class positions are almost always occupied by white men in the United States and European countries. This is particularly true in large and influential organizations. The image of the successful organization and the image of the successful leader share many of the same characteristics, such as strength, aggressiveness, and competitiveness. Some research shows that flat team structures provide professional women more equality and opportunity than hierarchical bureaucracies, but only if the women function like men. One study of engineers in Norway (Kvande and Rasmussen 1994) found that women in a small, collegial engineering firm gained recognition and advancement more easily than in an engineering department in a big bureaucracy. However, the women in the small firm were expected to put in the same long hours as their male colleagues and to put their work first, before family responsibilities. Masculine-stereotyped patterns of on-the-job behavior in team-organized work may mean that women must make adaptations to expectations that interfere with family responsibilities and with which they are uncomfortable. In a study of high-level professional women in a computer development firm, Joanne Martin and Debra Meyerson (1998) found that the women saw the culture of their work group as highly masculine, aggressive, competitive, and self-promoting. The women had invented ways to cope with this work culture, but they felt that they were partly outsiders who did not belong.

Other research (Barker 1993) suggests that team-organized work may not reduce gender inequality. Racial inequality may also be maintained as teams are introduced in the workplace (Vallas 2003). While the organization of teams is often accompanied by drastic reductions of supervisors' roles, the power of higher managerial levels is usually not changed: Class inequalities are only slightly reduced (Morgen, Acker, and Weigt n.d.).

The degree and pattern of segregation by race and gender is another aspect of inequality that varies considerably between organizations. Gender and race segregation of jobs is complex because segregation is hierarchical across jobs at different class levels of an organization, across jobs at the same level, and within jobs (Charles and Grusky 2004). Occupations should be distinguished from jobs: An occupation is a type of work; a job is a particular cluster of tasks in a particular work organization. For example, emergency room nurse is an occupation; an emergency room nurse at San Francisco General Hospital is a job. More statistical data are available about occupations than about jobs, although "job" is the relevant unit for examining segregation in organizations. We know that within the broad level of professional and managerial occupations, there is less gender segregation than 30 years ago, as I have already noted. Desegregation has not progressed so far in other occupations. However, research indicates that "sex segregation at the job level is more extensive than sex segregation at the level of occupations" (Wharton 2005, 97). In addition, even when women and men "are members of the same occupation, they are likely to work in different jobs and firms" (Wharton 2005, 97). Racial segregation also persists, is also complex, and varies by gender.

Jobs and occupations may be internally segregated by both gender and race: What appears to be a reduction in segregation may only be its reconfiguration. Reconfiguration and differentiation have occurred as women have entered previously male-dominated occupations. For example, women doctors are likely to specialize in pediatrics, not surgery, which is still largely a male domain. I found a particularly striking example of the internal gender segregation of a job category in my research on Swedish banks (Acker 1991). Swedish banks all had a single job classification for beginning bank workers: They were called "aspiranter," or those aspiring to a career in banking. This job classification had one description; it was used in banking industry statistics to indicate that this was one job that was not gender segregated. However, in bank branches, young women aspiranters had different tasks than young men. Men's tasks were varied and brought them into contact with different aspects of the business. Men were groomed for managerial jobs. The women worked as tellers or answered telephone inquiries. They had contact only with their immediate supervisors and coworkers in the branch. They were not being groomed for promotion. This was one job with two realities based on gender.

The size of wage differences in organizations also varies. Wage differences often vary with the height of the hierarchy: It is the CEOs of the largest corporations whose salaries far outstrip those of everyone else. In the United States in 2003, the average CEO earned 185 times the earnings of the average worker; the average earnings of CEOs of big corporations were more than 300 times the earnings of the average worker (Mishel, Bernstein, and Boushey 2003). White men tend to earn more than any other gender/race category, although even for white men, the wages of the bottom 60 percent are stagnant. Within most service-sector organizations, both white women and women of color are at the bottom of the wage hierarchy.

The severity of power differences varies. Power differences are fundamental to class, of course, and are linked to hierarchy. Labor unions and professional associations can act to reduce power differences across class hierarchies. However, these organizations have historically been dominated by white men with the consequence that white women and people of color have not had increases in organizational power equal to those of white men. Gender and race are important in determining power differences within organizational class levels. For example, managers are not always equal. In some organizations, women managers work quietly to do the organizational housekeeping, to keep things running, while men managers rise to heroic heights to solve spectacular problems (Ely and Meyerson 2000). In other organizations, women and men manage in the same ways (Wacjman 1998). Women managers and professionals often face gendered contradictions when they attempt to use organizational power in actions similar to those of men. Women enacting power violate conventions of relative subordination to men, risking the label of "witches" or "bitches."

Organizing Processes That Produce Inequality

Organizations vary in the practices and processes that are used to achieve their goals; these practices and processes also produce class, gender, and racial inequalities. Considerable research exists exploring how class or gender inequalities are produced, both formally and informally, as work processes are carried out (Acker 1989, 1990; Burawoy 1979; Cockburn 1985; Willis 1977). Some research also examines the processes that result in continuing racial inequalities. These practices are often guided by textual materials supplied by consultants or developed by managers influenced by information and/or demands from outside the organization. To understand exactly how inequalities are reproduced, it is necessary to examine the details of these textually informed practices.

Organizing the general requirements of work. The general requirements of work in organizations vary among organizations and among organizational levels. In general, work is organized on the image of a white man who is totally dedicated to the work and who has no responsibilities for children or family demands other than earning a living. Eight hours of continuous work away from the living space, arrival on time, total attention to the work, and long hours if requested are all expectations that incorporate the image of the unencumbered worker. Flexibility to bend these expectations is more available to high-level managers, predominantly men, than to lower-level managers (Jacobs and Gerson 2004). Some professionals, such as college professors, seem to have considerable flexibility, although they also work long hours. Lower-level jobs have, on the whole, little flexibility. Some work is organized as part-time, which may help women to combine work and family obligations, but in the United States, such work usually has no benefits such as health care and often has lower pay than full-time work (Mishel, Bernstein, and Boushey 2003). Because women have more obligations outside of work than do men, this gendered organization of work is important in maintaining gender inequality in organizations and, thus, the unequal distribution of women and men in organizational class hierarchies. Thus, gender, race, and class inequalities are simultaneously created in the fundamental construction of the working day and of work obligations.

Organizing class hierarchies. Techniques also vary for organizing class hierarchies inside work organizations. Bureaucratic, textual techniques for ordering positions and people are constructed to reproduce existing class, gender, and racial inequalities (Acker 1989). I have been unable to find much research on these techniques, but I do have my own observations of such techniques in one large job classification system from my study of comparable worth (Acker 1989). Job classification systems describe job tasks and responsibilities and rank jobs hierarchically. Jobs are then assigned to wage categories with jobs of similar rank in the same wage category. Our study found that the bulk of sex-typed women's jobs, which were in the clerical/secretarial area and included thousands of women workers, were described less clearly and with less specificity than the bulk of sex-typed men's jobs, which were spread over a wide range of areas and levels in the organization. The women's jobs were grouped into four large categories at the bottom of the ranking, assigned to the lowest wage ranges; the men's jobs were in many more categories extending over a much wider range of wage levels. Our new evaluation of the clerical/secretarial categories showed that many different jobs with different tasks and responsibilities, some highly skilled and responsible, had been lumped together. The result was, we argued, an unjustified gender wage gap: Although women's wages were in general lower than those of men, women's skilled jobs were paid much less than men's skilled jobs, reducing even further the average pay for women when compared with the average pay for men. Another component in the reproduction of hierarchy was revealed in discussions with representatives of Hay Associates, the large consulting firm that provided the job evaluation system we used in the comparable worth study. These representatives would not let the job evaluation committees alter the system to compare the responsibilities of managers' jobs with the responsibilities of the jobs of their secretarial assistants. Often, we observed, managers were credited with responsibility for tasks done by their assistants. The assistants did not get credit for these tasks in the job evaluation system, and this contributed to their relatively low wages. But if managers' and assistants' jobs could never be compared, no adjustments for inequities could ever be made. The hierarchy was inviolate in this system.

In the past 30 years, many organizations have removed some layers of middle management and relocated some decision making to lower organizational levels. These changes have been described as getting rid of the inefficiencies of old bureaucracies, reducing hierarchy and inequality, and empowering lower-level employees. This happened in two of the organizations I have studied—Swedish banks in the late 1980s (Acker 1991), discussed above, and the Oregon Department of Adult and Family Services, responsible for administration of Temporary Assistance to Needy Families and welfare reform (Morgen, Acker, and Weigt n.d.). In both cases, the decision-making responsibilities of frontline workers were greatly increased, and their jobs became more demanding and more interesting. In the welfare agency, ordinary workers had increased participation in decisions about their local operations. But the larger hierarchy did not change in either case. The frontline employees were still on the bottom; they had more responsibility, but not higher salaries. And they had no increased control over their job security. In both cases, the workers liked the changes in the content of their jobs, but the hierarchy was still inviolate.

In sum, class hierarchies in organizations, with their embedded gender and racial patterns, are constantly created and renewed through organizing practices. Gender and sometimes race, in the form of restricted opportunities and particular expectations for behavior, are reproduced as different degrees of organizational class hierarchy and are also reproduced in everyday interactions and bureaucratic decision making.

Recruitment and hiring. Recruitment and hiring is a process of finding the worker most suited for a particular position. From the perspectives of employers, the gender and race of existing jobholders at least partially define who is suitable, although prospective coworkers may also do such defining (Enarson 1984). Images of appropriate gendered and racialized bodies influence perceptions and hiring. White bodies are often preferred, as a great deal of research shows (Royster 2003). Female bodies are appropriate for some jobs; male bodies for other jobs.

A distinction should be made between the gendered organization of work and the gender and racial characteristics of the ideal worker. Although work is organized on the model of the unencumbered (white) man, and both women and men are expected to perform according to this model, men are not necessarily the ideal workers for all jobs. The ideal worker for many jobs is a woman, particularly a woman who, employers believe, is compliant, who will accept orders and low wages (Salzinger 2003). This is often a woman of color; immigrant women are sometimes even more desirable (Hossfeld 1994).

Hiring through social networks is one of the ways in which gender and racial inequalities are maintained in organizations. Affirmative action programs altered hiring practices in many organizations, requiring open advertising for positions and selection based on gender- and race-neutral criteria of competence, rather than selection based on an old boy (white) network. These changes in hiring practices contributed to the increasing proportions of white women and people of color in a variety of occupations. However, criteria of competence do not automatically translate into gender- and race-neutral selection decisions. "Competence" involves judgment: The race and gender of both the applicant and the decision makers can affect that judgment, resulting in decisions that white males are the more competent, more suited to the job than are others. Thus, gender and race as a basis for hiring or a basis for exclusion have not been eliminated in many organizations, as continuing patterns of segregation attest.

Wage setting and supervisory practices. Wage setting and supervision are class practices. They determine the division of surplus between workers and management and control the work process and workers. Gender and race affect assumptions about skill, responsibility, and a fair wage for jobs and workers, helping to produce wage differences (Figart, Mutari, and Power 2002).

Wage setting is often a bureaucratic organizational process, integrated into the processes of creating hierarchy, as I described above. Many different wage-setting systems exist, many of them producing gender and race differences in pay. Differential gender-based evaluations may be embedded in even the most egalitarian-appearing systems. For example, in my study of Swedish banks in the 1980s, a pay gap between women and men was increasing within job categories in spite of gender equality in wage agreements between the union and employers (Acker 1991). Our research revealed that the gap was increasing because the wage agreement allowed a small proportion of negotiated increases to be allocated by local managers to reward particularly high-performing workers. These small increments went primarily to men; over time, the increases produced a growing gender gap. In interviews we learned that male employees were more visible to male managers than were female employees. I suspected that the male managers also felt that a fair wage for men was actually higher than a fair wage for women. I drew two implications from these findings: first, that individualized wage-setting produces inequality, and second, that to understand wage inequality it is necessary to delve into the details of wage-setting systems.

Supervisory practices also vary across organizations. Supervisory relations may be affected by the gender and race of both supervisor and subordinate, in some cases preserving or reproducing gender or race inequalities. For example, above I described how women and men in the same aspiranter job classification in Swedish banks were assigned to different duties by their supervisors. Supervisors probably shape their behaviors with subordinates in terms of race and gender in many other work situations, influencing in subtle ways the existing patterns of inequality. Much of this can be observed in the informal interactions of workplaces.

Informal interactions while "doing the work." A large literature exists on the reproduction of gender in interactions in organizations (Reskin 1998; Ridgeway 1997). The production of racial inequalities in workplace interactions has not been studied so frequently (Vallas 2003), while the reproduction of class relations in the daily life of organizations has been studied in the labor process tradition, as I noted above. The informal interactions and practices in which class, race, and gender inequalities are created in mutually reinforcing processes have not so often been documented, although class processes are usually implicit in studies of gendered or racialized inequalities.

As women and men go about their everyday work, they routinely use gender-, race-, and class-based assumptions about those with whom they interact, as I briefly noted above in regard to wage setting. Body differences provide clues to the appropriate assumptions, followed by appropriate behaviors. What is appropriate varies, of course, in relation to the situation, the organizational culture and history, and the standpoints of the people judging appropriateness. For example, managers may expect a certain class deference or respect for authority that varies with the race and gender of the subordinate; subordinates may assume that their positions require deference and respect but also find these demands demeaning or oppressive. Jennifer Pierce (1995), in a study of two law firms, showed how both gendered and racialized interactions shaped the organizations' class relations: Women paralegals were put in the role of supportive, mothering aides, while men paralegals were cast as junior partners in the firms' business. African American employees, primarily women in secretarial positions, were acutely aware of the ways in which they were routinely categorized and subordinated in interactions with both paralegals and attorneys. The interaction practices that re-create gender and racial inequalities are often subtle and unspoken, thus difficult to document. White men may devalue and exclude white women and people of color by not listening to them in meetings, by not inviting them to join a group going out for a drink after work, by not seeking their opinions on workplace problems. Other practices, such as sexual harassment, are open and obvious to the victim, but not so obvious to others. In some organizations, such as those in the travel and hospitality industry, assumptions about good job performance may be sexualized: Women employees may be expected to behave and dress as sexually attractive women, particularly with male customers (Adkins 1995).

The Visibility of Inequalities

Visibility of inequality, defined as the degree of awareness of inequalities, varies in different organizations. Lack of awareness may be intentional or unintentional. Managers may intentionally hide some forms of inequality, as in the Swedish banks I studied (Acker 1991). Bank workers said that they had been told not to discuss their wages with their coworkers. Most seem to have complied, partly because they had strong feelings that their pay was part of their identity, reflecting their essential worth. Some said they would rather talk about the details of their sex lives than talk about their pay.

Visibility varies with the position of the beholder: "One privilege of the privileged is not to see their privilege." Men tend not to see their gender privilege; whites tend not to see their race privilege; ruling class members tend not to see their class privilege (McIntosh 1995). People in dominant groups generally see inequality as existing somewhere else, not where they are. However, patterns of invisibility/visibility in organizations vary with the basis for the inequality. Gender and gender inequality tend to disappear in organizations or are seen as something that is beside the point of the organization. Researchers examining gender inequality have sometimes experienced this disappearance as they have discussed with managers and workers the ways that organizing practices are gendered (Ely and Meyerson 2000; Korvajärvi 2003). Other research suggests that practices that generate gender inequality are sometimes so fleeting or so minor that they are difficult to see.

Class also tends to be invisible. It is hidden by talk of management, leadership, or supervision among managers and those who write and teach about organizations from a management perspective. Workers in lower-level, nonmanagement positions may be very conscious of inequalities, although they might not identify these inequities as related to class. Race is usually evident, visible, but segregated, denied, and avoided. In two of my organization studies, we have asked questions about race issues in the workplace (Morgen, Acker, and Weigt n.d.). In both of these studies, white workers on the whole could see no problems with race or racism, while workers of color had very different views. The one exception was in an office with a very diverse workforce, located in an area with many minority residents and high poverty rates. Here, jobs were segregated by race, tensions were high, and both white and Black workers were well aware of racial incidents. Another basis of inequality, sexuality, is almost always invisible to the majority who are heterosexual. Heterosexuality is simply assumed, not questioned.

The Legitimacy of Inequalities

The legitimacy of inequalities also varies between organizations. Some organizations, such as cooperatives, professional organizations, or voluntary organizations with democratic goals, may find inequality illegitimate and try to minimize it. In other organizations, such as rigid bureaucracies, inequalities are highly legitimate. Legitimacy of inequality also varies with political and economic conditions. For example, in the United States in the 1960s and 1970s, the civil rights and the women's movements challenged the legitimacy of racial and gender inequalities, sometimes also challenging class inequality. These challenges spurred legislation and social programs to reduce inequality, stimulating a decline in the legitimacy of inequality in many aspects of U.S. life, including work organizations. Organizations became vulnerable to lawsuits for discrimination and took defensive measures that included changes in hiring procedures and education about the illegitimacy of inequality. Inequality remained legitimate in many ways, but that entrenched legitimacy was shaken, I believe, during this period.

Both differences and similarities exist among class, race, and gender processes and among the ways in which they are legitimized. Class is fundamentally about economic inequality. Both gender and race are also defined by inequalities of various kinds, but I believe that gender and racial differences could still conceivably exist without inequality. This is, of course, a debatable question. Class is highly legitimate in U.S. organizations, as class practices, such as paying wages and maintaining supervisory oversight, are basic to organizing work in capitalist economies. Class may be seen as legitimate because it is seen as inevitable at the present time. This has not always been the case for all people in the United States; there have been periods, such as during the depression of the 1930s and during the social movements of the 1960s, when large numbers of people questioned the legitimacy of class subordination.

Gender and race inequality are less legitimate than class. Antidiscrimination and civil rights laws limiting certain gender and race discriminatory practices have existed since the 1950s. Organizations claim to be following those laws in hiring, promotion, and pay. Many organizations have diversity initiatives to attract workforces that reflect their customer publics. No such laws or voluntary measures exist to question the basic legitimacy of class practices, although measures such as the Fair Labor Standards Act could be interpreted as mitigating the most severe damages from those practices. In spite of antidiscrimination and affirmative action laws, gender and race inequalities continue in work organizations. These inequalities are often legitimated through arguments that naturalize the inequality (Glenn 2002). For example, some employers still see women as more suited to child care and less suited to demanding careers than men. Beliefs in biological differences between genders and between racial/ethnic groups, in racial inferiority, and in the superiority of certain masculine traits all legitimate inequality. Belief in market competition and the natural superiority of those who succeed in the contest also naturalizes inequality.

Gender and race processes are more legitimate when embedded in legitimate class processes. For example, the low pay and low status of clerical work is historically and currently produced as both a class and a gender inequality. Most people take this for granted as just part of the way in which work is organized. Legitimacy, along with visibility, may vary with the situation of the observer: Some clerical workers do not see the status and pay of their jobs as fair, while their bosses would find such an assessment bizarre. The advantaged often think their advantage is richly deserved. They see visible inequalities as perfectly legitimate.

High visibility and low legitimacy of inequalities may enhance the possibilities for change. Social movements may contribute to both high visibility and low legitimacy while agitating for change toward greater equality, as I argued above. Labor unions may also be more successful when visibility is high and legitimacy of inequalities is low.

Control and Compliance

Organizational controls are, in the first instance, class controls, directed at maintaining the power of managers, ensuring that employees act to further the organization's goals, and getting workers to accept the system of inequality. Gendered and racialized assumptions and expectations are embedded in the form and content of controls and in the ways in which they are implemented. Controls are made possible by hierarchical organizational power, but they also draw on power derived from hierarchical gender and race relations. They are diverse and complex, and they impede changes in inequality regimes.

Mechanisms for exerting control and achieving compliance with inequality vary. Organization theorists have identified many types of control, including direct controls, unobtrusive or indirect controls, and internalized controls. Direct controls include bureaucratic rules and various punishments for breaking the rules. Rewards are also direct controls. Wages, because they are essential for survival in completely monetized economies, are a powerful form of control (Perrow 2002). Coercion and physical and verbal violence are also direct controls often used in organizations (Hearn and Parkin 2001). Unobtrusive and indirect controls include control through technologies, such as monitoring telephone calls or time spent online or restricting information flows. Selective recruitment of relatively powerless workers can be a form of control (Acker and Van Houten 1974). Recruitment of illegal immigrants who are vulnerable to discovery and deportation and recruitment of women of color who have few employment opportunities and thus will accept low wages are examples of this kind of control, which preserves inequality.

Internalized controls include belief in the legitimacy of bureaucratic structures and rules as well as belief in the legitimacy of male and white privilege. Organizing relations, such as those between a manager and subordinates, may be legitimate, taken for granted as the way things naturally and normally are. Similarly, a belief that there is no point in challenging the fundamental gender, race, and class nature of things is a form of control. These are internalized, often invisible controls. Pleasure in the work is another internalized control, as are fear and self-interest. Interests can be categorized as economic, status, and identity interests, all of which may be produced as organizing takes place. Identities, constituted through gendered and racialized images and experiences, are mutually reproduced along with differences in status and economic advantage. Those with the most powerful and affluent combination of interests are apt to be able to control others with the aim of preserving these interests. But their self-interest becomes a control on their own behavior.

* * *

Globalization, Restructuring, and Change in Inequality Regimes

Organizational restructuring of the past 30 years has contributed to increasing variation in inequality regimes. Restructuring, new technology, and the globalization of production contribute to rising competitive pressures in private-sector organizations and budget woes in public-sector organizations, making challenges to inequality regimes less likely to be undertaken than during the 1960s to the 1980s. The following are some of the ways in which variations in U.S. inequality regimes seem to have increased. These are speculations because, in my view, there is not yet sufficient evidence as to how general or how lasting these changes might be.

The shape and degree of inequality seem to have become more varied. Old, traditional bureaucracies with career ladders still exist. Relatively new organizations, such as Wal-Mart, also have such hierarchical structures. At the same time, in many organizations, certain inequalities are externalized in new segmented organizing forms as both production and services are carried out in other, low-wage countries, often in organizations that are in a formal, legal sense separate organizations. If these production units are seen as part of the core organizations, earnings inequalities are increasing rapidly in many different organizations. But wage inequalities are also increasing within core U.S.-based sectors of organizations.

White working- and middle-class men, as well as white women and all people of color, have been affected by restructuring, downsizing, and the export of jobs to low-wage countries. White men's advantage seems threatened by these changes, but at least one study shows that white men find new employment after layoffs and downsizing more rapidly than people in other gender/race categories and that they find better jobs (Spalter-Roth and Deitch 1999). And a substantial wage gap still exists between women and men. Moreover, white men still dominate local and global organizations. In other words, inequality regimes still seem to place white men in advantaged positions in spite of the erosion of advantages for middle- and lower-level men workers.

Inequalities of power within organizations, particularly in the United States, also seem to be increasing with the present dominance of global corporations and their free market ideology, the decline in the size and influence of labor unions, and the increase in job insecurity as downsizing and reorganization continue. The increase in contingent and temporary workers who have less participation in decisions and less security than regular workers also increases power inequality. Unions still exercise some power, but they exist in only a very small minority of private-sector organizations and a somewhat larger minority of public-sector unions.

Organizing processes that create and recreate inequalities may have become more subtle, but in some cases, they have become more difficult to challenge. For example, the unencumbered male worker as the model for the organization of daily work and the model of the excellent employee seems to have been strengthened. Professionals and managers, in particular, work long hours and often are evaluated on their "face time" at work and their willingness to put work and the organization before family and friends (Hochschild 1997; Jacobs and Gerson 2004). New technology makes it possible to do some jobs anywhere and to be in touch with colleagues and managers at all hours of day and night. Other workers lower in organizational hierarchies are expected to work as the employer demands, overtime or at odd hours. Such often excessive or unpredictable demands are easier to meet for those without daily family responsibilities. Other gendered aspects of organizing processes may be less obvious than before sex and racial discrimination emerged as legal issues. For example, employers can no longer legally exclude young women on the grounds that they may have babies and leave the job, nor can they openly exclude consideration of people of color. But informal exclusion and unspoken denigration are still widespread and still difficult to document and to confront.

The visibility of inequality to those in positions of power does not seem to have changed. However, the legitimacy of inequality in the eyes of those with money and power does seem to have changed: Inequality is more legitimate. In a culture that glorifies individual material success and applauds extreme competitive behavior in pursuit of success, inequality becomes a sign of success for those who win.

Controls that ensure compliance with inequality regimes have also become more effective and perhaps more various. With threats of downsizing and off-shoring, decreasing availability of well-paying jobs for clerical, service, and manual workers, and undermining of union strength and welfare state supports, protections against the loss of a living wage are eroded and employees become more vulnerable to the control of the wage system itself. That is, fear of loss of livelihood controls those who might challenge inequality.

Conclusion

* * *

Greater equality inside organizations is difficult to achieve during a period, such as the early years of the twenty-first century, in which employers are pushing for more inequality in pay, medical care, and retirement benefits and are using various tactics, such as downsizing and outsourcing, to reduce labor costs. Another major impediment to change within inequality regimes is the absence of broad social movements outside organizations agitating for such changes. In spite of all these difficulties, efforts at reducing inequality continue. Government regulatory agencies, the Equal Employment Opportunity Commission in particular, are still enforcing antidiscrimination laws that prohibit discrimination against specific individuals (see www.eeoc.gov/stats/). Resolutions of complaints through the courts may mandate some organizational policy changes, but these seem to be minimal. Campaigns to alter some inequality regimes are under way. For example, a class action lawsuit on behalf of Wal-Mart's 1.3 million women workers is making its way through the courts (Featherstone 2004). The visibility of inequality seems to be increasing, and its legitimacy decreasing. Perhaps this is the opening move in a much larger, energetic attack on inequality regimes.

References

Acker, Joan. 1989. *Doing comparable worth: Gender, class and pay equity.* Philadelphia: Temple University Press.

———. 1990. Hierarchies, jobs, and bodies: A theory of gendered organizations. *Gender & Society* 4:139–58.

———. 1991. Thinking about wages: The gendered wage gap in Swedish banks. *Gender & Society* 5:390–407.

———. 1994. The gender regime of Swedish banks. *Scandinavian Journal of Management* 10:117–30.

Acker, Joan, and Donald Van Houten. 1974. Differential recruitment and control: The sex structuring of organizations. *Administrative Science* Quarterly 19:152–63.

Adkins, Lisa 1995. *Gendered work.* Buckingham, UK: Open University Press.

Barker, James R. 1993. Tightening the iron cage: Concertive control in self-managing teams. *Administrative Science Quarterly* 38:408–37.

Burawoy, Michael 1979. *Manufacturing consent.* Chicago: University of Chicago Press.

Charles, Maria, and David B. Grusky. 2004. *Occupational ghettos: The worldwide segregation of women and men.* Stanford, CA: Stanford University Press.

Cockburn, Cynthia. 1985. *Machinery of dominance.* London: Pluto.

Ely, Robin J., and Debra E. Meyerson. 2000. Advancing gender equity in organizations: The challenge and importance of maintaining a gender narrative. *Organization* 7:589–608.

Enarson, Elaine. 1984. *Woods-working women: Sexual integration in the U.S. Forest Service.* Tuscaloosa, AL: University of Alabama Press.

Featherstone, Lisa. 2004. *Selling women short: The landmark battle for workers' rights at Wal-Mart.* New York: Basic Books.

Ferree, Myra Max, and Patricia Yancey Martin, eds. 1995. *Feminist organizations.* Philadelphia: Temple University Press.

Figart, D. M., E. Mutari, and M. Power. 2002. *Living wages, equal wages.* London: Routledge.

Forsebäck, Lennart. 1980. *Industrial relations and employment in Sweden.* Uppsala, Sweden: Almqvist & Wiksell.

Glenn, Evelyn Nakano. 2002. *Unequal freedom: How race and gender shaped American citizenship and labor.* Cambridge, MA: Harvard University Press.

Hearn, Jeff, and Wendy Parkin. 2001. *Gender, sexuality and violence in organizations.* London: Sage.

Hochschild, Arlie Russell. 1997. *The time bind: When work becomes home & home becomes work.* New York: Metropolitan Books.

Hossfeld, Karen J. 1994. Hiring immigrant women: Silicon Valley's "simple formula." In *Women of color in U.S. society,* edited by M. B. Zinn and B. T. Dill. Philadelphia: Temple University Press.

Jacobs, Jerry A., and Kathleen Gerson, 2004. *The time divide: Work, family, and gender inequality.* Cambridge, MA: Harvard University Press.

Korvajärvi, Päivi. 2003. "Doing gender"—Theoretical and methodological considerations. In *Where have all the structures gone? Doing gender in organizations, examples from Finland, Norway and Sweden,* edited by E. Gunnarsson, S. Andersson, A. V. Rosell, A. Lehto, and M. Salminen-Karlsson. Stockholm, Sweden: Center for Women's Studies, Stockholm University.

Kvande, Elin, and Bente Rasmussen. 1994. Men in male-dominated organizations and their encounter with women intruders. *Scandinavian Journal of Management* 10:163–74.

Martin, Joanne, and Debra, Meyerson. 1998. Women and power: Conformity, resistance, and disorganized coaction. In *Power and influence in organizations,* edited by R. Kramer and M. Neale. Thousand Oaks. CA: Sage.

McIntosh, Peggy. 1995. White privilege and male privilege: A personal account of coming to see correspondences through work in women's studies. In *Race, class, and gender: An anthology,* 2nd ed., edited by M. L. Andersen and P. H. Collins. Belmont, CA: Wadsworth.

Mishel, L., J. Bernstein, and H. Boushey. 2003. *The state of working America 2002/2003.* Ithaca, NY: Cornell University Press.

Morgen, S., J. Acker, and J. Weigt. n.d. *Neo-Liberalism on the ground: Practising welfare reform.*

Perrow, Charles. 2002. *Organizing America.* Princeton, NJ: Princeton University Press.

Pierce, Jennifer L. 1995. *Gender trials: Emotional lives in contemporary law firms.* Berkeley: University of California Press.

Reskin, Barbara. 1998. *The realities of affirmative action in employment.* Washington, DC: American Sociological Association.

Ridgeway, Cecilia. 1997. Interaction and the conservation of gender inequality. *American Sociological Review* 62:218–35.

Royster, Dierdre A. 2003. *Race and the invisible hand: How white networks exclude Black men from blue-collar jobs.* Berkeley: University of California Press.

Salzinger, Leslie. 2003. *Genders in production: Making workers in Mexico's global factories.* Berkeley: University of California Press.

Scott, Ellen. 2000. Everyone against racism: Agency and the production of meaning in the anti racism practices of two feminist organizations. *Theory and Society* 29:785–819.

Spalter-Roth, Roberta, and Cynthia Deitch. 1999. I don't feel right-sized; I feel out-of-work sized. *Work and Occupations* 26:446–82.

Vallas, Steven P. 2003. Why teamwork fails: Obstacles to workplace change in four manufacturing plants. *American Sociological Review* 68: 223–50.

Wacjman, Judy. 1998. *Managing life a man.* Cambridge, UK: Polity.

Wharton, Amy S. 2005. *The Sociology of gender.* Oxford, UK: Blackwell.

Willis, Paul. 1977. *Learning to labor.* Farmborough, UK: Saxon House.

Introduction to Reading 32

In this excerpt from her book *Selling Women Short: Gender Inequality on Wall Street,* Louise Marie Roth provides a detailed description of why women's and men's salaries are so unequal on Wall Street. She explores the idea of "meritocracy," that is, those who are most worthy will get the most rewards, based upon "merit" or achievement. She interviewed and followed the careers of 1991, 1992, and 1993 graduates of elite MBA programs, interviewing these men and women some ten years after they began working on Wall Street. These men and women came from similar backgrounds, worked in Wall Street firms that were similar, and worked in similar conditions, including a booming market, which means more money for traders. This reading provides a detailed examination of what Acker called "inequality regimes."

1. What "inequality regimes" do you see described in this article?
2. Why doesn't meritocracy dictate similar salaries for men and women in these Wall Street firms?
3. How much difference does inequality on Wall Street mean for these women interviewed?

Selling Women Short

Gender Inequality on Wall Street

Louise Marie Roth

Women first entered Wall Street as professionals in the mid-1970s. At that time they faced blatant discrimination: they were excluded or ejected from meeting in all-male clubs and were often told outright that they would receive less money than men. The glass ceiling blocking their promotions was more obvious, as was their clear underpayment relative to their male peers—this despite sometimes superior performance [Fisher 1989, Herera 1997]. These women had only two options: to endure compensation discrimination or to leave the profession.

This type of barefaced discrimination became less prevalent in the 1980s, largely because it clearly violated Title VII, the federal law that prohibits sex discrimination in hiring and promotion, and because the most flagrant incidences of discrimination had been attacked in court. In the early 1980s, recruiters for Goldman Sachs, for instance, asked female MBAs if they were willing to have abortions to stay on the fast track, which violated the Pregnancy Discrimination Act (PDA) of 1978 (discrimination on the basis of pregnancy is illegal sex discrimination) [Fisher 1989].

A group of Stanford MBAs complained in 1984, producing a public outcry and leading to changes in the firm's interview questions. Views of women as reproductive time bombs did not completely disappear, but the most obvious discrimination on that basis has.

Legal challenges motivated Wall Street firms to develop internal policies and procedures to comply with equal employment opportunity guidelines and to address discrimination in hiring and promotion, sexual harassment, and maternity leave. At least on the surface, these firms made efforts to hire and retain qualified women. The most obvious barriers to women's success eroded in the late 1980s, when most investment banks promoted their first female managing directors and partners. But at the same time, the sex discrimination cases of the last decade imply that discrimination and a culture of sexual harassment remained defining features of the industry.

Having won court cases to institute formal policies against discrimination, by the 1990s women on Wall Street started to fight back against the male-dominated culture of the industry, filing class action suits against major securities firms. In some cases, like the infamous "boom-boom room" case against Smith Barney, these suits focused partly on verbal and physical sexual harassment [Antilla 2002; *New York Times,* February 27, 1999; *Wall Street Journal,* June 22, 1999]. In 1995, after enduring years of put-downs, lewd remarks, sexual gestures, and practical jokes, as well as being denied promotions and access to the best accounts, three women at Smith Barney's Garden City brokerage office filed a class action suit charging the firm with sex discrimination and sexual harassment. They alleged that "the office maintained a fraternity-house culture" that was hostile and degrading to women [*New York Times,* February 27, 1999]. (The "boom-boom room" was a party room in the basement of the Garden City office where male brokers held parties that excluded women or subjected women to sexual innuendo and pranks.) The suit claimed that the firm had tolerated a hostile and discriminatory work environment, and Smith Barney settled before it went to court. While the firm accepted no culpability, the settlement led to an overhaul of its sexual harassment and diversity policies.

A similar class action suit against Merrill Lynch focused primarily on patterns of economic disparity, although sexual harassment was also an issue [*New York Times,* February 27, 1999; Antilla 2002]. Female brokers at Merrill entered the class action suit en masse because of discrimination in wages, promotions, account distributions, and maternity leaves. They were especially concerned about the allocation of accounts from departing brokers, walk-ins, leads, and referrals, which they claimed went disproportionately to male brokers. When a panel of arbitrators awarded Hydie Sumner $2.2 million on April 19, 2004, it was the first legal ruling to find that a Wall Street firm engaged in systematic discrimination.

In addition to these class actions, individual women also filed suits. In 1999, a former employee of J. P. Morgan sued the firm for barring her from working as a trader because of her gender [*New York Post,* August 31, 1999]. Around the same time, Allison Schieffelin filed her highly publicized gender discrimination lawsuit with the EEOC after she was passed over for promotion to managing director in Morgan Stanley's international equity sales division [*Wall Street Journal,* July 29, 1999, July 30, 1999, September 24, 1999; *New York Times,* August 3, 1999, June 6, 2000]. Morgan Stanley only reluctantly, and under duress, cooperated with the investigation. The EEOC decided to pursue the case, in their first ruling against Wall Street in over a decade. The day before the case was scheduled to go to court, on July 13, 2004, Morgan Stanley settled for a total of $54 million [Ackman 2004; Gibson 2004; Martinez 2004; McGeehan 2004]. The fact that these lawsuits have found some Wall Street firms liable for gender discrimination reveals that obstacles for women remain entrenched, even after past legislative actions have removed the most blatant displays of gender bias. The question remains, however, why such barriers exist and how they persist.

Equal employment legislation may have paved the way for women's entrance into the securities industry, but more than half of all employees in the industry are still white men (even when female-dominated positions like administrative assistants and sales assistants are included). The figure rises to over 70 percent among investment bankers, traders, and brokers, and to over 80 percent among executive managers [McGeehan 2004]. While the number of women in professional positions has risen, and securities firms have a heightened awareness of equal employment opportunity laws, Wall Street remains a male-dominated environment. And while the most obvious forms of differential treatment have been squelched, more subtle forms of discrimination still flourish on Wall Street.

* * *

The Myth of Meritocracy: Gender and Performance-Based Pay

[S]ome women succeeded on Wall Street despite the subtle discrimination and prevalent workaholism. Some might think that the existence of successful women supports the notion of a Wall Street meritocracy and therefore that less successful women deserved lower pay because they made personal choices that derailed their careers. But merit alone cannot explain the differences in men's and women's career trajectories.

The fact remains that women who were equal to their male counterparts as a group made less money. This is even more surprising given that Wall Street firms compete for the most talented individuals; this is, after all the path to making the most money for both the firm and the client. The long bull market of the 1990s should have created more opportunity regardless of biases. With business booming, Wall Street firms should have tried to retain anyone who could skillfully manage the volume of work and contribute to their huge bottom lines, leading them to be gender blind. The performance-based pay system should have, theoretically, created equity by compensating workers on their worth rather than according to characteristics like gender. But men and women encountered different environments in this industry, and bonus system encouraged unequal results independent of skill, effort, or other aspects of merit. The result is a larger-than-average gender gap in pay.

Gender and Career Trajectories

[G]ender shapes Wall Street careers right from the start. At the entry level, Wall Street firms made efforts to hire women, but women applied in smaller numbers than men. Women represented only 20 percent of incoming associates. Women also started in different areas of the securities industry, either because of their own preferences or because managers hired workers who resembled them or their clients. As a result, hiring practices had a disparate impact on women, who tended to be tokens and who were especially isolated in the highest-paying jobs, which had the highest proportion of men. As tokens, women were highly visible and vulnerable to performance pressures, stereotyping, and discrimination. They had to fight cultural assumptions that they were less competent than their male peers. Because of these assumptions, many of women's contributions to the productivity of their work groups were overlooked or devalued by their managers and colleagues, especially if indicators of performance were not clear-cut. Managers' and coworkers' preferences for coworkers similar to themselves gave further advantages to those in the majority. In most areas, these preferences favored men.

The processes that obstructed women's early careers meant that very few women made it into the senior ranks, and those who did had to overcome contradictory standards—those that applied to them as workers and those that applied to them as women. As one informant commented, most women who made it in finance had a "Stepford wife" quality to them; they were not necessarily strong role models or advocates for women in the junior ranks. As a result, there were few real role models for new female hires, especially in the higher-paying areas. And the few pockets where there was greater diversity were usually job ghettos with lower pay.

Another significant barrier to many women's success was the intense time commitment that Wall

Street required. This generated a culture of workaholism that precluded involvement in family life and defined the ideal worker as one with no extra-work responsibilities and preferably with a stay-at-home spouse. This definition of the ideal worker was based on the average life shared by the male majority and contradicted women's typical family situation, posing additional obstacles for women who did not want to, or could not, emulate male patterns.

Approximately one quarter of the women managed to succeed despite these obstacles. But women had a higher rate of attrition than men in the early part of their careers, were promoted at a lower rate, and often moved into lower-paying positions where there were more women. Given that fewer women continued to pursue their careers and were at higher levels, and those who remained continued to encounter obstacles to success, gender inequality on Wall Street probably worsened as their careers progressed further.

Markets and Meritocracy

Those who believe in the justice of competitive markets might argue that this type of inequality must be justified, otherwise firms that permit subtle discrimination would be at a disadvantage compared to those that do not [Becker 1971]. That is, they would argue that discrimination must make firms more competitive or they would fail. But securities firms do not necessarily pay a high economic price for offering privileges to white men. They can distribute workers across work groups, or work within groups, in a variety of ways without changing the firm's profits. In some cases managers appeared to believe that distributing fewer good accounts to women was economically rational because they expected clients to prefer men or believed that investments in women's careers were a waste of resources because they would eventually leave the labor force. Firms also had substantial leeway when they allocated bonuses. Bonuses provided incentives for future performance, but pay that did not exactly match performance might have little impact on workers' motivation or subsequent performance, especially if performance was hard to measure and the fair bonus amount was unknown. Most of the influences on gender differences in access to accounts and performance evaluations were also subtle enough to be unconscious. Meanwhile, the ideology of market efficiency was used to justify inequalities that are incorporated into and perpetuated by the bonus system, defining them as meritocratic [Nelson and Bridges 1999; Frank 2004].

A meritocracy is a system in which advancement is based on individual ability or achievement. While many touted Wall Street's compensation system as meritocratic, *Selling Women Short* [2006] shows how it is instead a self-reproducing system of structural inequality. The men and women I interviewed were similar, and yet a large gender gap remained after accounting for other influences. Wall Street's bonus system permits nonmerit influences to affect pay decisions, many of which have a disparate impact on women. While firm revenues and group revenues affected men and women alike, women were disproportionately funneled into groups with lower revenue potential. Perceptions of performance within teams of workers adversely affected women because of universal tendencies to prefer similar others and to view men as more competent than women. The lack of fit between Wall Street's workaholic culture and typical family arrangements had negative effects not only on women's performance but on how they were perceived and thus on evaluations of their performance, even in the face of contradictory information. Thus ends the myth of meritocracy.

If this kind of subtle discrimination can persist on Wall Street, which is supposedly driven only by money, then it's likely that similar processes operate in other settings with performance-based rewards. This isn't a "blame me first" polemic, rather the bonus system itself reproduces inequality by concealing information about criteria for determining bonus amounts and about others' pay and by holding women to higher standards than men. Recognizing this, 29 percent of the workers I interviewed viewed meritocracy on Wall Street as a myth.

Some of these workers disavowed meritocracy on Wall Street because bonuses seemed arbitrary. Large pay differences within work groups and disconnections between performance and pay made bonus amounts seem random. Vicki, who left Wall Street to work at a commercial bank, compared the pay at her current job with that of her first job in equity research.

> Working at [the Wall Street firm where I was an analyst], there were some people there who were making huge amounts of money, and there was a very large discrepancy between the highest-paid analyst and the lowest-paid analyst. One of the things that is just very obvious is that the pay scale where I am now is much more compressed. There is not nearly the same differentiation in compensation between people of a similar rank. That is just a lot better to be around. You're not making fifty thousand dollars, while somebody down the hall from you doesn't seem to be working as hard as you but is making five million.

From her perspective, pay on Wall Street did not correspond to effort or productivity. The lack of evidence that differences were merit based led her to view the bonus system as arbitrary and unfair.

For some workers, subjective evaluations even contradicted more objective measures of performance, leading to especially strong rejections of the myth of meritocracy. Tracy, a trader who left the industry after her flexible work arrangement led to a poor bonus, remarked,

> It's very subjective whereas you think in trading, "Okay, you made your P&L and you get X." But it's not. . . . The team review process, while it sounds good on paper, is difficult because it is very subjective and anyone can write a review on you. So if somebody doesn't like you because your trade affected them in some way . . . they can write a scathing review, unsolicited, on you and that can show up on your [evaluation form] or your comprehensive review that your manager puts together and suddenly you're faced with, "I don't like her because she wears really loud pink shoes." . . . Is it fair? I don't know. Because you can make your P&L and do your job and go to work every day but if you don't kiss everyone's butt or be everyone's little favorite you're not going to get your team review to be stellar.

Because she had profit and loss indicators to gauge whether or not her bonus accurately corresponded to her performance, she was particularly disturbed by subjective influences on her evaluations. When evaluations contradicted tangible evidence of performance, workers viewed meritocracy on Wall Street as a myth.

Women were much more likely than men to disbelieve the myth of meritocracy because, . . . women were more subject to nonmerit influences having a negative impact on their bonuses. Forty-three percent of women, compared to 9 percent of men, rejected the myth of meritocracy [Roth 2006]. Could Wall Street be a true meritocracy despite this? Theories and research on women's experiences in the workplace suggest that it is unlikely.

Theories About Stratification and Meritocracy

Theories in social psychology imply that a true meritocracy in a setting like Wall Street is improbable because of nearly universal preferences for others who resemble oneself and cultural tendencies to view men as more competent than women, even when men's performance is equal to or worse than that of women. Because men are the incumbents of Wall Street and have more status in general, women are at a disadvantage—it's as simple as that. This is common to many, if not most, occupations, but my findings reveal how they perpetuate gender inequalities in a performance-based system. When evaluations of performance are partly subjective, they will inevitably be swayed by attraction to and beliefs about others. Men's preference for other men and a cultural gender hierarchy that awards men higher status than women led to higher evaluations for men. In fact, most people, both male and female, unconsciously hold biases that favor men in this setting. That fact, coupled with the trend toward subjective performance reviews more generally, bodes ill for meritocracy anywhere.

Gender theories also predict that men will receive advantages on Wall Street even when their performance is identical to that of women because the industry is male dominated. Being in the minority can have detrimental effects beyond those created by the majority's attraction to similar others, as Kanter [1977] has pointed out. Women's increased visibility as tokens forced them into stereotyped roles, heightening performance pressure on them and demanding that they represent all women while also meeting standards set by the male majority—a tall order for anyone. Even the most successful women, who sometimes believed that standing out from their peers had helped them, noted that they were held to different standards as women. For example, Julie noted,

> I actually think that being a woman has been an advantage. And I would be surprised if most of the women at my level didn't say that. I think that Wall Street desperately wants to hire and retain qualified women. And I think that if you are good, the senior management at these firms, especially the senior management at my last firm, they know who every single woman is at the VP level. They don't know every guy. To be able to get in the elevator and have them talk to me is an advantage. But if you are not performing well then it is probably going to hit you harder. I think that a weak woman stands out more than a weak man.

As this suggests, women who were exceptional could receive an advantage from their higher visibility, but mediocre men could pass through the ranks more easily than mediocre women. This type of double standard permeated the securities industry and prevented it from producing a true meritocracy.

Thirty-two percent of the women observed that women's performance was subject to different standards than men's, which affected their evaluations. Jacqueline, a research analyst, said that men held women to higher standards in their reviews.

> [The head of the group] who still runs research at [my first firm] probably had this deep-rooted belief that women belonged at home. And I think you could prove that women were not compensated at the same level men were for doing equal work. . . . We all had our year-end reviews and you get reviewed from all different constituencies. And there was one woman who was a great analyst. You couldn't possibly find fault with anything she did and she goes in there and he has something to say about her performance whereas the guys go in there and there's no big deal. Nothing. So I think he probably demonstrated some bias toward men in the performance reviews.

From her perspective, meritocracy must be a myth because subtle gender biases influenced the reviews on which compensation was based, leading to greater scrutiny of women with impeccable performance than of average-performing men.

Theories about gender and skill also suggest similar biases against women in performance evaluations because skill is an ideological category that is defined by the typical characteristics of workers who perform a particular task [Collins 2002; England 1992; Nelson and Bridges 1999; Phillips and Taylor 1980; Steinberg 1990]. In other words, definitions of skill in a male-dominated setting like Wall Street are biased toward expertise and characteristics that are typically held by men, while those that most women possess are devalued—even if they might be readily applicable to the situation at hand. When pay is based on performance evaluations, cultural and job-specific definitions of skill and performance lead to systematic inequality rather than meritocracy. Illustrating this, Maureen said that performance evaluations rewarded men for stereotypical male behavior and devalued qualities more typical among women even if they improved performance.

> This woman is 5'10," played field hockey and lacrosse. She's not thin, she's not fat but she's not—she went to Harvard. She's not shy and timid. She can be really aggressive. She holds it back. She was told, like apparently every other analyst woman in her class, that she was very timid. She was like, "Oh my God! I've never in my life been called timid." I don't think she's timid. I think she has a manner about her which isn't—she doesn't shoot from the hip and make stuff up when she goes to a meeting, and she definitely kind of has different style but women's style is categorized as "timid." . . . And guys who mouth off when they

> don't know what they're talking about, I find in the end, have been typically rewarded as opposed to being penalized and they . . . do huge damage.

In her view, men were rewarded for behaviors that negatively affected their performance, while women were penalized despite better performance. She added, "You're not out there beating your chest and doing this stuff, and you're not out there making non-credible threats, and so you get paid less because you're 'timid.'" From her point of view, compensation discrimination occurred partly because the value of women's contributions was not perceived, while stereotypically male behavior was idealized.

Fletcher [1999] found that women engineering executives had relational skills that added value in the workplace but that remained invisible to their managers and colleagues—largely because they are acquired through gender socialization. Similarly, Gutek's [1985] theory of sex-role spillover argued that as occupations become gender typed, the gender role becomes part of the work role. Work in female-dominated jobs is then structured to take advantage of women's stereotyped traits while not rewarding them as skills, and male-dominated jobs incorporate men's stereotyped traits and define them as essential for performance—even when they actually detract from it, as Maureen observed. On Wall Street, the skills that were rewarded were those that fit with stereotypical masculinity, while women's contributions were discounted.

Accordingly, one-third of the women believed that the bonus system perpetuated gender inequality by holding women to different and higher standards and by not rewarding them for attributes that contributed to their productivity. These women rejected the idea that Wall Street was a free market paradise, and their observations exposed the effects of entrenched cultural biases on the performance review system. But the "myth of meritocracy" still provides a potent interpretation of bonus pay for workers across pay levels. If you didn't get a good bonus, so the thinking goes, then you must just not be that good: it's not the system, it's you. Over two-thirds (71 percent) of Wall Street workers believed that the compensation system was a meritocracy, sometimes despite experiences that clearly contradicted the myth. They believed in it because they were high earners, they did not know how much others were paid, they compared their earnings only to others in the exact same job, and/or they had clear measures of performance that corresponded to their pay. In fact, a belief in meritocracy persisted, even among many workers at the bottom of the ladder.

* * *

Selling Women Short [2006] has demonstrated that gender inequality coexisted with a compensation system that aimed to pay for performance. Men and women on Wall Street believed that the securities industry had mellowed since the era when blatant sex discrimination and sexual harassment were tolerated, as described in *Liar's Poker, The Bonfire of the Vanities,* and *Tales from the Boom-Boom Room.* But gender inequality remained, and many of the processes that maintained a boys' club in the higher echelons of Wall Street were subtle and operated through the performance review system. Economically irrational gender inequality persisted in Wall Street's highly rationalized compensation system because performance evaluations were vulnerable to gender double standards and subtle discrimination.

Wall Street workers reveal that organizations can maintain a myth of rationality, efficiency, and meritocracy while producing inequality that is not economically rational or based on merit. Gender stereotypes powerfully, if subtly, influenced evaluations of performance and this prevented Wall Street from being a genuine meritocracy. The fact that Wall Street created a gender hierarchy rather than a meritocracy should encourage other types of organizations, which are increasingly moving toward flat hierarchies and variable incentive structures like those of securities firms, to examine their own practices.

Many people find performance-based incentives appealing because they imply that people receive what they deserve based on their efforts. There are many longstanding performance-based reward systems, like the system of awarding grades based on academic merit. But any teacher can tell you that academic merit may not always be the only criterion for grades and that grading is not an exact science. There are, of course, better and worse methods for evaluating students' performance, and Wall Street might take lessons from the field of education—but first there must be recognition that no performance-based system produces perfectly just results. This challenges the entrenched assumption that performance-based pay structures produce a meritocracy. But because the system's winners support this ethos and the losers leave, the subjective influences that produce systematic inequality in opportunities and evaluations are often hidden.

References

Ackman, Dan. 2004. "Morgan Stanley: Big Bucks for Bias." *Forbes,* July 13.

Antilla, Susan. 2002. *Tales from the Boom-Boom Room: Women vs. Wall Street.* Princeton, NJ: Bloomberg Press.

Becker, Gary M. 1971. *The Economics of Discrimination.* Chicago: University of Chicago Press.

Collins, Jane L. 2002. "Mapping a Global Labor Market: Gender and Skill in the Globalizing Garment Industry." *Gender & Society* 16, 6 (December): 921–40.

Edelman, Lauren B. 1990. "Legal Environments and Organizational Governance: The Expansion of Due Process in the American Workplace." *American Journal of Sociology* 95, 6 (May): 1401–40.

———. 1992. "Legal Ambiguity and Symbolic Structures: Organizational Mediation of Civil Rights Law." *American Journal of Sociology* 97: 1531–76.

England, Paula. 1992. *Comparable Worth: Theories and Evidence.* New York: Aldine de Gruyter.

Fisher, Anne B. 1989. *Wall Street Women: Women in Power on Wall Street Today.* New York: Alfred P. Knopf.

Fletcher, Joyce K. 1999. *Disappearing Acts: Gender, Power, and Relational Practices at Work.* Cambridge, MA: MIT Press.

Frank, Thomas. 2004. *What's the Matter with Kansas? How Conservatives Won the Heart of America.* New York: Metropolitan Books.

Gibson, Gail. 2004. "2 Wins for Working Women." *Baltimore Sun,* July 19.

Gutek, Barbara A. 1985. *Sex and the Workplace.* San Francisco: Jossey-Bass Publishers.

Herera, Sue. 1997. *Women of the Street: Making It on Wall Street—The World's Toughest Business.* New York: John Wiley.

Kanter, Rosabeth Moss. 1977. *Men and Women of the Corporation.* New York: Basic Books.

Martinez, Jose. 2004. "The Woman Who Won $12M Settlement from Morgan Stanley." *New York Daily News,* July 14.

McGeehan, Patrick. 2004. "Merrill Lynch Ordered to Pay for Sexual Bias." *New York Times,* April 21.

Nelson, Robert L., and William P. Bridges. 1999. *Legalizing Gender Inequality: Courts, Markets, and Unequal Pay for Women in America.* New York: Cambridge University Press.

New York Post. "Suit Charges Sex Bias at J. P. Morgan." August 31, 1999, p. 18.

New York Times. "Bias at the Bull: Merrill Lynch's Class-Action Settlement Draws a Crowd." February 27, 1999, pp. C1, C14.

———. "If Wall Street Is a Dead End, Do Women Stay to Fight or Go Quietly?" August 3, 1999, pp. D1, D6.

———. "Morgan Stanley Is Cited for Discrimination Against Women." June 6, 2000, pp. C1, C2.

Phillips, Anne, and Barbara Taylor. 1980. "Sex and Skill: Notes Towards a Feminist Economics." *Feminist Review* 6 (Winter): 79–88.

Roth, Louise Marie. 2006. *Selling Women Short: Gender Inequality on Wall street.* Princeton, NJ: Princeton University Press.

Steinberg, Ronnie J. 1990. "Social Construction of Skill: Gender, Power, and Comparable Worth." *Work and Occupations* 17, 4 (November): 449–82.

Wall Street Journal. "Morgan Stanley Executive Files Charges with the EEOC Alleging Gender Bias." July 29, 1999.

———. "Morgan Stanley Agrees to Discuss Policies on Hiring with EEOC." July 30, 1999, p. B8.

———. "Morgan Stanley Is Told by Judge to Produce All Sex Complaints." September 24, 1999, p. C12.

Introduction to Reading 33

In this review of literature, Paula England, a sociologist who has studied workplace policies and practices for some time, explores both theories and research findings related to "care work." One topic England explores is whether the gendering of care work can explain salary differentials for women and men. In this reading, England proposes five different ways of thinking about care work and supports her arguments with research in this area. Most people agree that care work is important. However, you will soon discover that there is considerable debate about how care work is incorporated into the work world and society in general.

1. What is care work and why is it paid in some instances and not in others?
2. Is care work undervalued or devalued? In what ways?
3. Do you think care work should be paid?

Emerging Theories of Care Work

Paula England

Some jobs involve providing care for pay; child care providers, teachers, nurses, doctors, and therapists all provide care. Some care is provided without pay; for example, parents rear their children and adults care for their disabled kin. This review surveys emerging scholarship on paid and unpaid care work. Most of it comes from gender scholars. They take an interest because women do such a high proportion of paid and unpaid care work, so that how well a society rewards care work impacts gender inequality. But gender arrangements also affect how care is provided; increasing women's employment means that more of the care of children and disabled elders is provided by paid workers rather than unpaid female family members.

I review both empirical and theoretical work, but organize my discussion around five conceptual frameworks deployed in the literature. I evaluate their logic as well as how well they fit available empirical evidence. In some cases, these frameworks offer different (competing or complementary) answers to the same questions. In other cases, they address distinct questions. . . .

Gender Bias and the Devaluation of Care Work

. . . [S]ome argue that female-dominated jobs involving care are especially devalued because care is the quintessentially female-identified activity (Cancian & Oliker 2000, England & Folbre 1999, England et al. 2002). To test this, researchers examined whether those in care work earn less than other workers after controlling for jobs' requirements for education, skill, and working conditions, and even their sex composition. For example, England (1992,

From England, P., "Emerging Theories of Care Work," *Annual Review of Sociology* 2005 Vol. 31: 381–99. Reprinted with permission.

chapter 3) examined the relative pay of a broader category called nurturant work. In addition to including the things called care work, such as child care work, teaching, nursing, and therapy, this category included all jobs involving giving a face-to-face service to clients or customers of the organization for which one works. . . . England (1992) found that occupations involving interactive service work had a pay penalty; a 1990 replication found the same results (England et al. 2001). These penalties are net of the sex composition of the occupation. Other research has examined the returns to the kinds of social skills used in care work. In an analysis of the New York State civil service jobs, Steinberg et al. (1986, p. 152) found that jobs involving communication with the public and group facilitation paid less than other jobs, net of skill demands. Kilbourne et al. (1994) developed a scale to measure nurturant skill, largely from measures in the *Dictionary of Occupational Titles,* assessing whether jobs involve dealing with people and communication. They found that, other things being equal, workers in such occupations suffered a wage penalty.

In more recent work, England and colleagues (2002) operationalized care work as those occupations providing a service to people that helps develop their capabilities. The main categories of jobs termed care work were child care, all levels of teaching (from preschool through university professors), and health care workers of all types (nurses aides, nurses, doctors, physical and psychological therapists). Controlling for skill demands, educational requirements, industry, and sex composition, we found a net penalty of 5%–10% for working in an occupation involving care (one exception was nursing, which did not seem to experience the pay penalty of other care work). Thus, overall the evidence suggests that care work pays less than we would expect, given its educational and other requirements. This finding is consistent with the devaluation framework, although there is no direct evidence that the mechanism is the cultural devaluation of jobs because they are filled largely with women and subsequent institutionalization of this devaluation in wage structures.

The devaluation perspective can be applied to race as well as to gender. Although paid care work requiring a college degree is done largely by white women, much care work without such requirements is done by women of color, some of whom are immigrants (Hondagneu-Sotelo, 2001, Misra 2003, Romero 1992). The work done by these women is the lowest paid. Is the relative pay of this work influenced by racist assumptions that devalue work associated with people of color?. . .

But what about unpaid care work, the most time-intensive example of which is parenting? Women do the lion's share of parental work. Recent data show that women spend about twice as much time as men in childrearing in married couple families (Sayer et al. 2004). How are women economically supported while they are raising children and either not employed or employed less fully for pay than they otherwise would be? The traditional answer is that they are supported by their husbands. But what about mothers without husbands or cohabitational partners? This is a growing group in all industrial societies owing to increased divorce and nonmarital births. If these women are not supported by their own earnings or child support voluntarily supplied by the fathers of their children, then they are supported by the state or state-mandated child support.

Scholarship on gender and the welfare state focuses on how much the state provides public support for such women and their children through direct payments, child allowances (received by married couples as well in affluent nations other than the United States), and state-supported child and health care (O'Connor et al. 1999, Orloff 1996, Sainsbury 2001). The conceptual framework of devaluation of activities associated with women is present in this literature as well, although the term devaluation is typically not used. But as gender scholars have pointed out, gendered assumptions are built into welfare states—that men can be relied upon to support their families and thus that the forms of economic insecurity the state needs to address are those that occur to men, such as unemployment because of disability or economic downturns and the need to retire in old age. Benefits are often conditioned on prior employment. They often offer little for the mother without prior employment because she has been caring for her children. Or if she has a claim to retirement benefits, it is based on marriage rather than care. Payments to lone mothers are not only smaller, but they are much more controversial, especially in the United States. Indeed, the Personal Responsibility and Work Opportunity Reconciliation Act of 1996, usually called welfare reform, eliminated a federal entitlement of those meeting the means test to welfare, and allowed states to institute lifetime time limits on welfare receipt. The fact that welfare recipients are disproportionately women of color may further erode public support for welfare. Overall, it seems that what men do is seen as a basis of citizenship rights more than what women do (Glenn 2000). This difference in treatment is consistent with the devaluation perspective, suggesting that the same cultural processes of devaluing activities associated with women and persons of color are reflected in political decisions of the state.

Care as Public Good Production

Care work, whether paid or unpaid, often includes investment in the capabilities of recipients. At issue is not only how care imbues cognitive skills that increase earnings, but more broadly that receiving care also helps recipients develop skills, values, and habits that benefit themselves and others (England & Folbre 2000). Care helps recipients develop capabilities for labor market success as well as for healthy relationships as a parent, friend, or spouse. Care contributes to the intellectual, physical, and emotional capabilities of recipients. These capabilities contribute to recipients' own and others' development and happiness. The benefits that accrue to the direct recipient also benefit indirect recipients. The direct beneficiaries of care are the student who is taught, the patient of the nurse or doctor, the client of the therapist, and the child cared for by a parent or child care worker. But when a direct recipient of care learns cognitive skills, stays or gets healthy, learns how to get along with others, or learns habits of self-control, others also benefit.

The many benefits of care to indirect beneficiaries make it arguably a public good. But how do the benefits of care diffuse to indirect beneficiaries? Education is an obvious example. Schooling makes people more productive, increasing their later productivity in a job, which benefits the owner and customers of the employing organization. As another example, if a client in psychotherapy learns to listen deeply and articulate his wants in a nonblaming way, this is likely to benefit his spouse, children, friends, and coworkers.

In the 1970s, Marxist feminists made a similar but narrower point (Dalla Costa & James 1972). They argued that homemakers were among those exploited by capitalists because their caretaking of their husbands and children made the current and next generation of workers more productive. Thus, in making profits, capitalists extract surplus value from homemakers as well as from paid workers. Those proposing the broader public good framework for care work do not necessarily subscribe to the Marxist labor theory of value. They see the indirect beneficiaries of care to be all of us, not merely capitalist employers. If children given love and taught patience and trustworthiness turn out to be better spouses when they grow up, their spouses benefit. If they are better parents, their children benefit. If they are better neighbors, the social capital of the community increases. If they become good Samaritans rather than predators, safety goes up, and the costs of building and maintaining prisons go down, benefiting their fellow citizens. Benefits to all these indirect recipients accrue because care workers help develop the capabilities of direct beneficiaries, and these beneficiaries spread them through social interaction. The extent to which benefits of caring labor will go beyond the direct beneficiary to others depends, in part, on how altruistic the beneficiary is—which is often a function of the kind of care she or he received.

The claim that care work, more than other kinds of work, produces public goods hinges mainly on the fact that care work involves a higher ratio of investment in capabilities than production of items immediately consumed. For example, the manager of a toy manufacturing plant, as well as the secretaries, janitors, and assembly workers in the firm, and the sales people selling toys contribute to providing something (toys) that consumers enjoy. But it is unclear that providing toys to a child leads him or her to later provide benefits to others. You could substitute those who produce clothes, makeup, furniture, and so forth for those providing toys. By contrast, the care-giving functions of teaching a child discipline and reading and of providing her with healthcare are much surer to lead to benefits for others.

On the central claim of the public good framework for understanding care—that paid or unpaid care work creates diffuse social benefits beyond its immediate beneficiaries—there is no direct confirmatory or disconfirmatory evidence, nor has a relevant research strategy been proposed. The evidence is largely indirect. First, there is some evidence for a public good aspect to fertility (Lee & Miller 1990). That is, the costs of rearing children, especially in the United States, are borne mostly privately (except for public education). But, with the pay-as-you-go Social Security system in the United States, as soon as children reach the age to have earnings, they contribute to the retirement of their parents' generation through the Social Security payroll tax (Lee & Miller 1990). Thus, because few of the costs (except school) of childrearing have been collectivized, but one of the major economic benefits of having children (their support for their elders in retirement) has been collectivized, having and rearing children has benefits for the viability of the Social Security system (Folbre 1994). Second, there is evidence of social benefits of education—benefits to other individuals that go beyond the earning power of those who are educated (Bowen 1977, Wolfe & [Haveman] 2003).

The low wages of care work can also be seen as indirect evidence that care produces public goods. In the previous section, research documenting a care penalty was presented as evidence for the devaluation perspective, arguing that care work pays less than we would otherwise expect because of its association with women. Another possible explanation for the wage penalty in care work is the public good aspect of the work. The standard economic argument is that public goods will be underprovided by markets because there is no way to capture and turn into profits (or wages, we might note) the benefits that come through social interaction. . . . Those using the public good framework have pointed to the low net wages of care work as evidence that care work creates public benefits not reflected in the wage received (England & Folbre 1999, England et al. 2002).

The public good framework has also been used to interpret policy implications of the wage penalty for motherhood. Several recent studies find a wage penalty for motherhood in the United States (Budig & England 2001; Lundberg & Rose 2000; Neumark & Korenman 1994; Waldfogel 1997, 1998a,b). A motherhood penalty has also been found in the United Kingdom (Harkness & Waldfogel 1999, Joshi & Newell 1989) and Germany (Harkness & Waldfogel 1999). Why do mothers earn less? First, although mothers have very high rates of employment today [for example, over 40% of women with children under one year of age are in the labor force (Klerman & Leibowitz 1999)], many women still lose at least some employment time to childrearing (Cohen & Bianchi 1999, Klerman & Leibowitz 1999). Women have no earnings while they are not employed, reducing their lifetime earnings and affecting their pensions. Intermittent employment also affects women's wage level when they return to work because employers reward experience and seniority. Budig & England (2001) found that about 40% of the motherhood wage penalty results from moms losing experience and seniority. Another portion of the motherhood penalty comes from the minority of moms who work part-time and that part-time work generally pays less per hour (Budig & England 2001, Waldfogel 1997). After experience, seniority, part-time status, and many job characteristics are controlled, there is still a residual penalty for being a mother (Budig & England 2001, Waldfogel 1997). This residual penalty could be an effect of motherhood on productivity. But a recent experimental study provides evidence that some of it is discrimination by employers against mothers. Correll & Benard (2004) asked students to help screen applications for a job. Subjects were told that a company was hiring for a mid-level marketing position in a telecommunications company, that the company wanted feedback from younger adults because they are heavy users of communications technology, and that the company would incorporate their rating when making hiring decisions. They evaluated resumes of fictitious applicants (presented to them as real applicants), indicating whom they would hire and at what salary. Compared with female applicants whose resume mentioned no children, those mentioning small children were less often recommended for hire and, if recommended, were offered lower starting salaries (Correll & Benard 2004).

What does the relative pay of mothers compared with other women have to do with public goods? The research reviewed above clarifies how various labor market processes in our economy—the return to continuous labor market experience, the lower hourly pay of part-time work, and employers' discrimination between mothers and nonmothers—disadvantage mothers. But Budig & England (2001) argue that if the unpaid care work that goes with motherhood is creating a public good, then the inequity is more unjust and the state should intervene to lessen the penalty. For example, the state could prohibit discrimination based on motherhood (Williams & Segal 2003). Another possibility is for the state to mandate that employers hold the jobs of workers who take parental leave after the birth or adoption of a child. Although the Family and Medical Leave Act of 1993 requires this, the mandated leave is unpaid, it is only for six weeks, and small firms are exempted. Gornick & Meyers (2003) argue for public policies that provide state payments for (gender-neutral) parenthood leave for a few months. In their scheme, employers are required to hold jobs (so that previously accrued seniority rights are preserved), and the state provides replacement of a certain proportion of the wage. Their proposals are based, in part, on the public good argument, as well as on evidence of how such policies have worked in European nations.

Caring Motives and Prisoners of Love

What's love got to do with it? Does genuine care or altruism motivate care work and provide some intrinsic reward for those who do it? Feminist writings on care contain both an insistence that care work really is hard work, as well as a concern for the negative consequences for society if we lose truly caring motivations for care work. Sometimes the word care itself is used to describe a motive or a moral imperative. . . . According to Abel & Nelson (1990, p. 4), "caregiving is an activity encompassing both instrumental tasks and affective relations. Despite the classic Parsonian distinction between these two modes of behavior, caregivers are expected to provide love as well as labor" Cancian & Oliker (2000, p. 2) define caring as a combination of feelings and actions that "provide responsively for an individual's personal needs or well-being, in a face-to-face relationship." Folbre has defined caring labor as work that provides services based on sustained personal (usually face-to-face) interaction, and is motivated (at least in part) by concern about the recipient's welfare (Folbre 1995, Folbre & Weisskopf 1998). Stone (2000) talks about how professional care workers (e.g., nurses) often want to talk to patients and show them real love but are frustrated by bureaucratic requirements that make this difficult. Implicit in much of this discussion is the idea that the recipients of care will be better off if the person giving care really cares about them than if they are motivated strictly by money.

What is the effect on the care workers' wage of having some altruism as one of the motivations for doing care work? When neoclassical economists confront evidence of the pay penalty in care work, they generally suggest that the correct explanation lies in the theory of compensating differentials (e.g., Filer 1989). (See

England 1992, pp. 69–73, and Jacobs & Steinberg 1990 for criticisms of the claim that this theory explains the low pay of most female jobs.) The theory calls attention to differences between jobs in their intrinsic rewards or penalties. . . . [S]imply put, the low pay may be made up for by the intrinsic fulfillment of the jobs. Indeed, this is the common economists' alternative to the claim that care is paid less (relative to skill) because of devaluation. Because neither the tastes of the marginal worker nor employers' processes of devaluation are observed, empirical evidence cannot adjudicate between the two views. In this orthodox neoclassical view, there is no policy problem with the low wages of care work; if women do not find the intrinsic rewards to make up for the low pay, they will enter other jobs. If they cannot find other types of jobs because of hiring discrimination, then economists see that as the problem policy should address, rather than the relative pay of care work.

Folbre (2001) has coined the term prisoner of love for this effect of care workers' caring motives on their pay. But her model differs from the standard economic view of compensating differentials in seeing altruistic preferences as at least in part endogenous to doing the work. Rational choice theorists, including economists, generally assume preferences to be exogenous and unchanging. But paid care workers may become attached to care recipients after they start the job, and this may make it difficult for them to withhold their services in order to demand more remuneration for them (England & Folbre 2003, Himmelweit 1999). . . . Although I know of no evidence of this, it makes sense that child care workers become attached to the toddlers they see every day, nurses empathize with their patients, and teachers worry about their students. These emotional bonds put care workers in a vulnerable position, discouraging them from demanding higher wages or changes in working conditions that might have adverse effects on care recipients. A kind of emotional hostage effect occurs.

Owners, employers, and managers are less likely to have direct contact with clients or patients than are care workers. Therefore, they can generally engage in cost-cutting strategies without feeling their consequences. Sometimes they can even be confident that adverse effects of their decisions on clients will be reduced by workers' willingness to make personal sacrifices to maintain high-quality care. For instance, workers may respond to cutbacks in staffing levels by intensifying their effort or agreeing to work overtime. This perspective suggests an equity problem of taking advantage of altruistic motives. It also suggests that if the motives are endogenous to doing the work, and people realize this, women may increasingly forego such work because they know they will become prisoners of love. It is like the decision not to have a child because one knows it is too taxing to be a good parent. As more women have the option to choose non-care work to avoid becoming prisoners of love, the supply of care work may be jeopardized.

The prisoner of love phenomenon applies not only to paid work, but also to struggles between mothers, fathers, and the state. Although mother love (and father love) may spring partly from nature, they are undoubtedly cultivated by the experience of providing care to one's child. If this is true, then gendered practices that assign child care to mothers mean that mothers will develop greater caring for their children than either their male partners or others. In the case of divorce, one way to understand men's failure to pay child support is that they know they can count on the mother's willingness to care for the child anyhow and share her money with the child—rather than to retaliate by abandoning the child. In matters of welfare reform, we can also see mothers as prisoners of love. If they did not care about their children, they could bargain more successfully with the state for higher welfare payments by threatening to put their children in foster care. The state pays foster parents much more than it pays welfare mothers—precisely because mothers' love and sense of obligation to their children can be counted on even in the absence of pay. Thus, state actors would not see such a threat by mothers as credible; they are prisoners of their love for their children.

The Commodification of Emotion

What happens when care is a commodity? The idea that the provision of care through markets may harm those who do the work is identified largely with Hochschild (1983), who coined the term emotional labor in *The Managed Heart.* She emphasized how being recruited by capitalists to sell one's emotion is harmful to workers. *The Managed Heart* was largely a study of the occupation of flight attendants, based on interviews with flight attendants and those who devise their training, and on observation of the training. Hochschild was struck by how flight attendants were taught to display feelings that they did not actually feel—to be cheerful even when sad, deferential to passengers even when furious at their disrespectful behavior. She argued that many jobs in the new service economy require workers to act emotions they do not really feel. Sometimes they even require deep acting, where the actor comes to feel the feelings prescribed. (See Smith-Lovin 1998 and Steinberg & Figart 1999 for reviews of research following the lead of Hochschild.) Whereas the prisoner of love view focuses on how care work is emotionally satisfying—so much so that workers will take a lower wage—Hochschild worried about psychological distress from deep acting. That is, one theory sees nonpecuniary amenities and the other disamenities of care work. Wharton has conducted a number of empirical tests to see if those in jobs requiring emotional labor have lower job satisfaction or worse mental health. In general, she did not find this (Wharton 1993, 1999), and found some evidence that many workers like the social interaction their jobs afford them. Under certain circumstances, however, emotional labor was taxing, as when workers had to combine emotional labor with low control and autonomy (Wharton 1999).

In more recent work, Hochschild (2000; 2003, especially chapter 14) has focused on the global penetration of capitalist market forces, and how they have special consequences for women and their families in poor countries. Her focus is on women who migrate from poor to rich countries to take jobs as nannies (see also Hondagneu-Sotelo 2001, Romero 1992). Some of these women have left their own children at home with kin to come to work in a richer country for a better economic future for their families. It is very poignant that they care for affluent, usually white, American children while their own children are left back home to experience their mother's love only through memory. The situation is similar to that of African American women in earlier periods who cared for white children in white homes, while leaving their own children at home, except that here children are not across town but across the world, only seldom visited. This situation is encouraged by the increased employment of well-educated American women (who then need child care for their children) and the large disparity between the wages available in poor and rich nations (one motivation for migration). One could use Hochschild's earlier work on emotional labor to analyze how taxing it is to have to feign love for someone else's children. But Hochschild finds even more poignant the cases in which nannies come to really love their American charges and feel closer to them than to their own children, given the distance. She describes the First World as extracting love from the Third World, and sees it as analogous to extraction of raw materials by colonial powers. She worries that Third World children pay the price, although she does not provide evidence that the children whose mothers come are worse off than they would be if their mothers stayed—that is, that the trade-off between losing their mothers' time and gaining the money their mothers earned was not worth it. . . .

Rejecting the Dichotomy Between Love and Money

The love *and* money perspective rejects the idea of an oppositional dichotomy between the realms of love and self-interested economic action. Leading voices promoting this view are Nelson (1999; 2004; J. Nelson, unpublished manuscript) and Zelizer (2002). . . . Nelson sees dichotomizing habits of thought as rooted in tacit assumptions about gender. Because male and female are seen as opposite, and because gender schema organize so much of our thinking, we develop a dualistic view that "women, love, altruism, and the family are, as a group, radically separate and opposite from men, self-interested rationality, work, and market exchange" (Nelson & England 2002). Zelizer (2002) calls this the "hostile worlds" view.

In this dichotomizing scheme that Nelson and Zelizer are contesting, we cannot pay care workers well and still get people doing the work who bring genuine, felt care to the work. Moreover, profit-making firms or waged labor can only contaminate or erode love. . . . Consider the question of whether it is possible for people to have access to adequate care by people whose motives are caring, or the question of whether we could change wage structures to get rid of the penalty for doing care work. Analysis should try to ascertain which particular structural or cultural features of behavior in markets, families, or states have which consequences, rather than assuming that solving these problems is impossible as long as care is done as waged work in private-sector firms.

Experimental psychologists and economists have studied the effects of payment on intrinsic motivation—willingness to expend effort on a task without extrinsic reward. Because doing care work for love or out of altruism is one example of intrinsic motivation, this line of research may apply to whether paying (more) for care work increases or decreases the supply of and quality of care. It should be noted, however, that none of the tasks in the experimental literature involved care work. . . . These results suggest that the more that pay is combined with trust and appreciation, the less it drives out genuine intrinsic motivation—especially important in care work. Furthermore, the experimental research shows that unexpected rewards increase intrinsic motivation more than expected rewards. This line of research exemplifies the love *and* money framework—it looks at specific mechanisms of achieving desirable results in care work, rather than assuming that the world is divided into two opposite systems.

Conclusion

Serious research on the care sector is just beginning. Several empirical generalizations have come out of the research: that an increasing amount of care is done by paid workers (rather than at home by women without pay); that women's unpaid care for their families is a more controversial basis for state support than men's employment or military service; that those who do care work for pay often report intrinsic motivations; and that paid care work pays less than would be predicted by its skill level, and even less than other predominantly female jobs at its skill level. There is a fair degree of consensus on these empirical generalizations. More challenging has been conceptualizing care with a theoretical apparatus that explains the source of these empirical regularities. Why do care workers earn less than those in similarly skilled jobs? Why is welfare controversial? Which method of organizing care best combines caring motives, an adequate supply of care, and an erasure of the economic penalty for care? Do love and money drive each other out?

References

Abel E, Nelson M. 1990. *Circles of Care: Work and Identity in Women's Lives*. Albany: SUNY Press

Bowen HR. 1977. *Investment in Learning: The Individual and Social Value of American Higher Education.* San Francisco: Jossey-Bass

Budig MJ, England P. 2001. The wage penalty for motherhood. *Am. Sociol. Rev.* 66:204–25

Cancian FM, Oliker SJ. 2000. *Caring and Gender.* Thousand Oaks, CA: Pine Forge

Cohen PN, Bianchi SM. 1999. Marriage, children, and women's employment: What do we know? *Mon. Labor Rev.* 122:22–31

Correll S, Benard S. 2004. *Getting a job: Is there a motherhood penalty?* Work. Pap., Dep. Sociol., Cornell Univ.

Dalla Costa M, James S. 1972. *The Power of Women and the Subversion of the Community.* Bristol, UK: Falling Wall

England P. 1992. *Comparable Worth: Theories and Evidence.* New York: Aldine de Gruyter

England P, Budig MJ, Folbre N. 2002. Wages of virtue: the relative pay of care work. *Soc. Probl.* 49:455–73

England P, Folbre N. 1999. The cost of caring. *Ann. Am Acad. Polit. Soc. Sci.* 561:39–51

England P, Folbre N. 2000. Reconceptualizing human capital. In *The Management of Durable Relations,* ed. W Raub, J Weesie, pp. 126–28. Amsterdam, The Neth.: Thela Thesis

England P, Folbre N. 2003. Contracting for care. In *Feminist Economics Today,* ed. J Nelson, M Ferber, pp. 61–80. Chicago: Univ. Chicago Press

England P, Thompson J, Aman C. 2001. The sex gap in pay and comparable worth: an update. In *Sourcebook of Labor Markets: Evolving Structures and Processes,* ed. I Berg, AL Kalleberg, pp. 551–65. New York: Kluwer Academic/Plenum

Filer R. 1989. Occupational segregation, compensating differentials, and comparable worth. In *Pay Equity: Empirical Inquiries,* ed. RT Michael, HI Harmann, B O'Farrell, pp. 153–70. Washington, DC: Natl. Acad. Press

Folbre N. 1994. *Who Pays for the Kids? Gender and the Structures of Constraint.* New York: Routledge

Folbre N. 1995. "Holding hands at midnight": the paradox of caring labor. *Fem. Econ.* 1:73–92

Folbre N. 2001. *The Invisible Heart: Economics and Family Values.* New York: New Press

Folbre N, Weisskopf T. 1998. Did father know best? Families, markets and the supply of caring labor. In *Economics, values and organization,* ed. A Ben-Ner, L Putterman, pp. 171–205. New York: Cambridge Univ. Press

Glenn EN. 2000. Creating a caring society. *Contemp. Sociol.* 29:84–94

Gornick JC, Meyers MK. 2003. *Families That Work: Policies for Reconciling Parenthood and Employment.* New York Russell Sage Found.

Harkness S, Waldfogel J. 1999. *The family gap in pay: evidence from seven industrialized counties.* CASE Paper 30, Cent. Anal. Soc. Exclusion, London Sch. Econ., London. http://sticerd.Ise.ac.uk/dps/case/cp/CASEpaper30.pdf

Himmelweit S. 1999. Caring labor. *Ann. Am. Acad. Polit. Soc. Sci.* 561:27–38

Hochschild AR. 1983. *The Managed Heart: Commercialization of Human Feeling.* Berkeley: Univ. Calif. Press

Hochschild AR. 2000. The nanny chain. *Am. Prospect* 11:32–36

Hochschild AR. 2003. *The Commercialization of Intimate Life: Notes from Home and Work.* Berkeley: Univ. Calif. Press

Hondagneu-Sotelo P. 2001. *Domestica: Immigrant Workers Cleaning and Caring in the Shadows of Affluence.* Berkeley: Univ. Calif. Press

Jacobs JA., Steinberg RJ. 1990. Compensating differentials and the male-female wage gap: evidence from the New York comparable worth study. *Soc. Forces* 69:439–68

Joshi H, Newell M-L. 1989. *Pay Differentials and Parenthood: Analysis of Men and Women Born in 1946.* Coventry, UK: Univ. Warwick Inst. Employ. Res.

Kilbourne BS, England P, Farkas G, Beron K, Weir D. 1994. Return to skill, compensating differentials, and gender bias: effects of occupational characteristics on the wages of white women and men. *Am. J. Sociol.* 100:689–719

Klerman JA, Leibowitz A. 1999. Job continuity among new mothers. *Demography* 36:145–55

Lee RD, Miller T. 1990. Population policy and externalities to childbearing. *Ann. Am. Acad. Polit. Soc. Sci.* Spec. Iss. *World Population: Approaching the Year* 2000, ed. SH Preston, 510:17–32

Lundberg S, Rose E. 2000. Parenthood and the earnings of married men and women. *Labour Econ.* 7:689–710

Misra J. 2003. Caring about care. *Fem. Stud.* 29:387–401

Nelson J. 1999. Of markets and martyrs : It is OK to pay well for care? *Fem. Econ.* 5:43–59

Nelson J. 2004. *Feminist economists and social theorists: Can we talk?* Work. Pap., Glob. Dev. Environ. Inst., Tuffs Univ.

Nelson JA, England P. 2002. Feminist philosophies of love and work. *Hypatia* 17: 1–18

Neumark D, Korenman S. 1994. Sources of bias in women's wage equations: results using sibling data. *J. Hum. Resourc.* 29:379–405

O'Connor JS, Orloff AS, Shaver S. 1999. *States, Markets, Families: Gender, Liberalism and Social Policy in Australia, Canada, Great Britain, and the United States.* New York: Cambridge Univ. Press

Orloff AS. 1996. Gender in the welfare state. *Annu. Rev. Sociol.* 22:51–78

Romero M. 1992. *Maid in the U.S.A.* London: Routledge

Sainsbury D. 2001. Social welfare policies and gender. In *International Encyclopedia of the Social & Behavioral Sciences,* ed. N. Smelser, PB Baltes, pp. 14476–81. Amsterdam: Elsevier

Sayer LC, Bianchi SM, Robinson JP. 2004. Are parents investing less in children? Trends in mothers' and fathers' time with children. *Am. J. Sociol.* 110:1–43

Smith-Lovin L. 1998, On Arlie Hochschild, *The Managed Heart:* emotion management as emotional labor. In *Required Reading: Sociology's Most Influential Books,* ed. D Clawson, pp. 113–19. Amherst: Univ. Mass. Press

Steinberg RJ, Figart DM, 1999. Emotional labor since *The Managed Heart. Ann. Am. Acad. Polit. Soc. Sci.* 561:8–26

Steinberg RJ, Haignere L, Possin C, Chertos CH. Trieman D. 1986. *The New York State Pay Equity Study: A Research Report.* Albany, NJ: Cent. Women Gov., SUNY Press

Stone DA. 2000. Caring by the book. In *Care Work: Gender, Labor and the Welfare State,* ed. MH Meyer, pp. 89–111. London: Routledge

Waldfogel J. 1997. The effects of children on women's wages. *Am. Sociol. Rev.* 62:209–17

Waldfogel J. 1998a. The family gap for young women in the United States and Britain: Can maternity leave make a difference? *J. Labor Econ.* 16:505–45

Waldfogel J. 1998b. Understanding the family gap in pay for women with children. *J. Econ. Perspect.* 12:137–56

Wharton AS. 1993. The affective consequences of service work: managing emotions on the job. *Work Occup.* 20:205–32

Wharton AS. 1999. The psychological consequences on emotional labor. *Ann. Am. Acad. Polit. Soc. Sci.* 561:158–76

Williams J, Segal N. 2003. Beyond the maternal wall: relief for family caregivers who are discriminated against on the job. *Harvard Women's Law J.* 26:77–162

Wolfe BL, Haveman RH. 2003. Social and nonmarket benefits from education in an advanced economy. In *Education in the 21st Century: Meeting the Challenges of a Changing World,* ed. Y Kodrzychi, pp. 207–34. Boston: Fed. Res. Bank Boston

Zelizer VA. 2002. How care counts. *Contemp. Sociol.* 31:115–19

Introduction to Reading 34

With the globalization of the economy, it is interesting to read Steven McKay's examination of the workplace policies and practices of three different high-tech companies in the Philippines. These factories are owned by three different multinational companies from America, Europe, and Japan. Each multinational corporation brings its own culture, but it also brings gendered assumptions to the factories they own in the Philippines. While the basic needs of workers for decent wages, reasonable hours, and safe working conditions are the same around the world, these companies used gendered assumptions to attract and maintain their workforce, not always to the benefit of the women and men working for them.

1. Which of the three factories has the most female-friendly practices?
2. What workplace practices and gender-based assumptions about work disadvantage women in these factories?
3. Why was it so difficult for women to be promoted from "operators" of machines to technical positions?

Hard Drives and Glass Ceilings

Gender Stratification in High-Tech Production

Steven C. McKay

Rapid industrialization and foreign direct investment in developing countries since the 1960s touched off what has become a three-decades-old debate about the impact of technology and "the global assembly line" on women and the division of labor (see Mills 2003). Recent trends in globalization, technological change, and flexible work reorganization have reawakened the debate (Acker 2004; Kelkar and Nathan 2002). In the electronics industry, where women have dominated production work, technological upgrading has been held up as a potential avenue for increased women's employment, empowerment, and socioeconomic mobility (Kuruvilla 1996; Yun 1995). Yet others argue that globalization and more flexible work organization may only intensify the exploitation of women workers and exacerbate occupational gender segmentation (Fox 2002; Joekes and Weston 1994).

While many recent studies document global restructuring, they—like earlier studies of women workers—tend to frame their analysis in terms of whether employment in advanced electronics represents either empowerment or exploitation. Thus, Chhachhi (1999) poses but cannot possibly answer the question of whether women electronic workers in India represent "dormant volcanoes," with a new, structural potential to explode, or "fresh green vegetables," that remain naïve, docile, and dominated by global capital.

However, focusing on either empowerment or exploitation becomes an analytical cul-de-sac, particularly given the wide variations in how firms organize high-tech production, how workers struggle over and make sense of their own employment, and the gendered meanings that emerge from these intersecting processes within a specific sociopolitical context. In this article, I concentrate on the myriad ways gender and inequality are constituted—even within a single industry and in a single national context—analyzing at multiple levels the links between gender ascription and the division of labor (Mills 2003). At the workplace level, I focus on management's structuring of the shop floor and key sources of variation that influence how advanced technologies and flexible work are organized, stratified, and gendered (Salzinger 2003). Equally important, I also examine the extra-organizational contexts—particularly the role of the state and persistent gender ideologies—that shape firm-level decision making as well as workers' assessments and action. Finally, to better assess the meaning of flexible production for women workers, I analyze high-tech work in terms of its impact on a range of issues concerning the gender division of labor. Specifically, I focus on the degree to which the three firms' gender and flexibility strategies address women's more "practical" everyday responsibilities and individual survival needs within the existing gendered division of labor versus their impact on the more "strategic" gender interest of women in "extending the conditions for choices to be made *about* the gender division of labor" (Chhachhi and Pittin 1998, 71, emphasis added; Molyneux 1985). In this sense, I hope to gauge to what extent changes in the organization of high-tech work directly challenge deeper gender inequality or "the structural roots of unequal access to resources and control" (Chow and Lyter 2002, 41). . . .

From McKay, S. C., "Hard Drives and Glass Ceilings: Gender Stratification in High-Tech Production," *Gender & Society*: 207–235. Reprinted with permission of Sage Publications, Inc.

Data and analysis are based on 11 months of interview and observation-based field research conducted in the Philippines in 1999 and follow-up research conducted during several weeks in 2003. Following interviews with human resource managers or staff at 20 firms, I selected three cases, representing the three dominant types of electronics products produced in the Philippines—disk drives, discrete semiconductors, and integrated chips—and the diverse nationalities of the home corporation.

* * *

THE STATE, THE ELECTRONICS INDUSTRY, AND GENDER

The Philippines has been a site of low-cost electronics assembly since the early 1970s, in large part because it offers—and the Philippine state has helped guarantee—an investment context favored by foreign high-tech manufacturers: political and economic stability; inexpensive, educated, and disciplined labor; good infrastructure; few operational restrictions; and investment/tax incentives tailored to the electronics industry (Austria 1999; Ernst 2002). State-led labor control has gone hand in hand with foreign investment policies at least since the early 1970s, when then-president Ferdinand Marcos launched an export-oriented industrialization program while also declaring martial law, rewriting the labor code, banning all strikes in export manufacturing, and cracking down on militant labor (Bello, Kinley, and Elinson 1982). The debt crisis of the 1980s only increased the Philippines's dependence on foreign investment, as structural adjustment packages attached to International Monetary Fund loans forced the Philippines to further open its economy and compete directly with other developing countries—particularly Malaysia, Thailand, and Indonesia—for mobile investment capital (Bello et al. 2004). By the 1990s, the Philippines had fully embraced a neoliberal economic growth model, centered on courting foreign investment through reregulating employment and production conditions, especially in its booming export-processing zones (World Bank 1997)....

As it has since the 1970s, the electronics industry continues to hire primarily women for factory operator positions largely because women can be paid less and are viewed as more patient, docile, and detail oriented (Chant and McIlwaine 1995). Management's desire to develop such a "cheapened" labor force is directly connected to the intensity of the labor process: production workers still make up 82 percent of the electronics workforce. Overall, 74 percent of the electronics industry workers are women and 78 percent younger than 30. But the workforce is increasingly well educated: More than 60 percent of the workers are high school graduates, and 37 percent have college degrees (Bureau of Labor Relations 1999).

Although women have made strides through their electronics employment, they still face a host of challenges, particularly given prevailing gender ideologies and complex gender relations in the Philippines. Women in the Philippines are sometimes viewed as having relatively more power than women in other Asian countries since women in the Philippines traditionally control domestic budgets and make most household decisions (Eviota 1992). And the preference for women in two key industries, electronics and garments, has contributed to overall employment and wage gains made by women in recent decades. While women's labor force participation rates have remained stable since 1975 at about 47 percent, women make up nearly 70 percent of the export industry workforce, women's unemployment rate (9.9 percent) is slightly below the men's rate (10.3 percent), and the wage differential between women and men, at 87 percent, is one of the lowest for a developing country (National Statistical Coordination Board 2001; Seguino 2000).

Despite these gains, employers' preference for female labor both taps into and reproduces gender ideologies and gendered labor market segmentation that systematically undervalues the labor of young educated women (Chant and McIlwaine 1995). As Parrenas (2001) points out, although women dominate the workforces of the two largest foreign exchange earning sectors—overseas contract labor and electronics—women are nevertheless constructed through Philippine families, churches, and the state as "secondary earners," whose proper roles of "mother," "daughter," and/or caregiver tie them primarily to the home, particularly after marriage (Eviota 1992). Thus, although many women have secured formal waged work through electronics assembly jobs, they nevertheless still confront a rigid gender division of labor, gendered labor control strategies, and gendered associations with technology. These issues, which only intensify as firms upgrade production technologies, will be discussed in each of the cases below.

Storage Ltd.

Storage Ltd., a subsidiary of a leading Japanese hard disk drive producer, demonstrates that while new flexibilities do lead to different labor practices than those documented in earlier studies of transnational assembly, such restructuring may also sharpen gendered labor control regimes. The firm's $124 million plant, operational since 1996, assembles and tests high-end hard disk drives for computer servers, requiring expensive semi-automated machinery, sophisticated quality control systems, and enormous startup costs. To meet these demands, Storage Ltd. has chosen a labor process and control regime that diverge somewhat from past practices in electronics and other export industries. For example, the firm does not subcontract out its labor-intensive assembly nor rely on temporary workers to boost its external flexibility (Standing 1999). Instead, to meet quality standards and enhance internal flexibility, the firm develops multi-skilled workers, trained to operate from one to three machines or work stations. Such training can take up to three months and considerable firm investment, so the firm tries hard to keep turnover down, making all employees permanent after six months and providing some positive incentives. For example, although base pay is only minimum wage, the company does pay more for overtime, pushing take-home pay up to 300 pesos per day (about US$6) or 50 percent more than the legal minimum. The company also provides transportation, free uniforms, subsidized meals, emergency medical insurance, and 12 days of sick and vacation leave a year—benefits relatively high by local standards.

Yet in other ways, the demanding character of disk drive production coupled with cut-throat cost competition has led the firm to develop more sophisticated forms of gendered labor control focused on stability, cost saving, and "disciplinary management." First, the actual production process remains quite labor intensive: The plant employs more than 9,000 workers, 87 percent of whom are shop floor operators, who churn out more than 1.5 million disk drives per month. To maintain such high productivity, the firm uses both high-tech strategies and more traditional direct control. On the technical side, the firm relies on a computerized assembly line that automatically gathers data for statistical process control, providing managers and customers with real-time productivity and quality data down to the individual workstation. On the social side, there is strict enforcement of company rules. As one human resources staff member admitted, "For most operators, they have no control. They are treated like robots; told what to do, when, where and how. Every movement is controlled. . . . They are told where to go, when to eat. And everything is timed, even going to the bathroom." While officially, there are three 8-hour shifts, in practice, there are only two 12-hour shifts with 4 hours of daily forced overtime. The workweek is six and often seven days a week. The long hours clearly take a toll on workers. Irma, a production operator noted, "The overtime is just too much. I only sleep five hours a day. I get home (from the night shift) at nine a.m., I'm asleep by ten. Then, I get up again at three or four p.m. just so I can do my laundry and clean up before I go in again."

In addition to such in-plant intensity, the firm utilizes and helps construct gender stereotypes to dampen its labor costs and extend its labor control far beyond the factory. The process begins in hiring. The company selectively recruits rural women, assuming—to some degree correctly—that they have fewer labor market options and will thus be cheaper, more dependent on the firm, and willing to put up with the demanding labor process. The profile of the company's 9,000 workers is remarkably uniform: 88 percent are women; 83 percent between 16 and 22, with an average age of 21; and 97 percent single. Fifty-one percent have some college or vocational training, and the rest are high school graduates. The reasons given for this rigid hiring profile follow and reinforce the gender stereotypes and "public narratives" about the nature of women's work that have become the norm in export factories around the world (Mills 2003). Managers say they prefer women because the work requires "patience, dexterity, and attention to detail," while men are viewed as lazy, sloppy, impatient, and generally *magulo* or disruptive. The company also wants intelligent, English-speaking workers, but ones without high job, pay, or promotion expectations. For this reason, management again prefers women—stereotyped as loyal but not ambitious—and does not hire anyone at the operator level with more than two years of college. This gendered hiring and job allocation essentially cuts women operators off from promotions to the more lucrative technical jobs, since these positions require a three-year technical or four-year college degree. Women operators can thus aspire only to a horizontal move to office clerk or to a climb up a short job ladder to line leader or quality control inspector. Technicians and engineers in the plant are almost all men.

Finally, Storage Ltd. also tries to localize its labor control, tapping into gender subordination at the family and community level. For example, when screening potential workers, recruiters ask extensively about an applicant's family and her role in the household. While seemingly innocent, the line of questioning aims to judge an applicant's level of dependence and reflects another of the firm's gendered strategies: to hire oldest daughters. Since the eldest are usually responsible for aiding parents and younger siblings, they are considered less likely to quit. The firm also maintains a Background Investigation Unit to scrutinize both workers and their families and make links with village-level officials to head off potential unionizing efforts. Before workers are made permanent following six months' probationary employment, the Background Investigation Unit conducts a home visit to interview workers' parents, inspect their residence, and contact the lowest-level officials. Primary targets for investigations are those reported to have gotten married since being hired and who may be hiding families back home.

The firm's sophisticated labor control strategies have important implications for both women's practical needs and strategic gender interests. In a practical sense, employment at Storage Ltd. does benefit women workers by providing permanent work paying above-minimum wages in a high-status industry. Ironically, because the firm can legally and explicitly target women with little experience and few labor market options, it has created a workforce that is grateful for its jobs despite the demanding working conditions, yet one with little collective or individual bargaining power to directly confront the hierarchical divisions of labor. Although workers complained of the overtime and lack of rest days, they also referred to the work as "clean" and "light," particularly in contrast to the physically demanding and low-status agricultural- or informal-sector work in their rural hometowns. For many, the job also gave them their first experience of living away from parents, increased status, control over income, and a means to fulfill independently their basic survival needs. As one operator noted, "This is my first job and I'm satisfied with the pay and benefits. . . . Working for 12 hours is not so bad. The work's pretty light . . . my parents treat me better now that I already have my own job." . . .

Integrated Production

Integrated Production, a subsidiary of an American semiconductor firm producing advanced integrated circuits, shows a different gendered strategy to promote labor stability, as well as the possible—but unrealized—potential of high-tech flexible production for women. Production at the firm's $200 million plant opened in late 1996 is less standardized, less labor intensive, and more automated than at Storage Ltd. At the core of the labor process are engineers and technicians, who make up more than 20 percent of the workforce and are crucial for ramping up production, programming in flexibility, and keeping the lines running. Due to the high level of automation, only 55 percent of the 1,373 permanent employees are production workers. These operators play only a minor role in flexible production: They work under strict supervision and are told to simply check the computer readouts from the quality control programs and report any problems to engineers.

Yet the automated machinery is extremely expensive and critical to production, so Integrated Production, like the other firms in the study, wants to minimize turnover or production disruption and thus provides training and employs only permanent workers. However, unlike Storage Ltd., Integrated Production combines its "hard" engineering-intensive labor process with a "soft" strategic labor relations approach—modeled after a Silicon Valley variant of the American human resources model—that stresses positive incentives, loyalty, and "empowerment" through ritual participation (Katz and Darbishire 2000). First, although take-home pay is actually less than at the other two firms, production operators are monthly-paid salaried employees, considered a notch up in status from daily-paid laborers. In addition, workers receive other benefits such as stock options, education reimbursement, and a free computer after two years. The firm also provides an assortment of morale-boosting activities, such as outings, sports fests, and pizza parties. The employee relations manager boasted, "It is really like an American company [in America]—there's really toilet paper in the bathroom, really free coffee all the time." Finally, the company "empowers" individual workers through an employee suggestion program and open intercompany communication. Yet the firm also studiously avoids—and actively disrupts—any collective worker action through anti-union trainings and surveillance.

The organization of work and labor relations is also gendered, in ways both similar to and different from that at Storage Ltd. Overall, the firm has one of the most balanced gender ratios of the companies visited: 38 percent men and only 63 percent women. But when broken down by position, the gender segregation typical of the industry becomes clearer: All nine directors and 90 percent of the engineers and technicians are men, while 85 percent of shop floor workers are women. As we will see below, this strict gender division of labor makes upward mobility for current women operators into new technical positions extremely difficult.

Nevertheless, gender strategies differ. First, unlike Storage Ltd., the firm is less concerned with trying to maximize women's multiple labor market vulnerabilities by recruiting only from rural areas and screening out experienced or married workers. The human resources manager complained that, local rural residents are poorly educated. . . . Rather, the firm puts a premium on educated workers, requiring a minimum of two years of college or technical training and preferring college graduates. Its emphasis on education has led to a frontline workforce dominated by women yet demographically distinct from other plants: Only 60 percent of workers are younger than 24 years of age, and only 70 percent are single. Operators are more likely from urban areas, 95 percent have finished at least two-year technical degrees, and a large portion are college graduates.

However, the firm's gendered strategy to hire educated women but to maintain a strict gender division of labor has also created problems. In part because these women are well educated, they also have higher career aspirations, more labor market options, and more individual bargaining power. Thus, the firm must rely more on positive rather than negative incentives to reduce turnover. In fact, some women took jobs in production in the hopes that they could eventually move up in the company. . . . In terms of internal promotion, the technical job ladder has far better prospects for earning and security but requires increasingly specific qualifications and is thus generally blocked to nontechnically trained (if formally educated) operators. . . . The mismatch between women's high education and blocked mobility has led to a turnover rate of more than 9 percent, the highest among the three companies. . . .

As the plant moves toward full automation, the new technical operators, who are slated to replace other shop floor workers, will still handle basic production but also take on responsibilities of current technicians, performing all minor maintenance and repairs. These enhanced operator positions would then have a direct promotional ladder leading to full technicians and possibly engineers—an avenue previously closed to operators. If women operators were to gain access to these jobs, their increased skills and mobility could provide more decision-making power over the division of labor and help break down gendered notions associated with technology and technical positions that have helped keep women out of technical work.

Unfortunately, such potential remains unrealized. First, the firm requires these new technical operators to have at least two year's training from an accredited electronics training school. In fact, managers, operating under gendered assumptions about women's technical (in)capacities, do not fill these new positions by training current operators. Rather, the firm recruits directly from engineering or technical schools, where almost all the trainees, and thus all the new technical operators, are men.

In other ways, technological upgrading may already be shifting management's gendered notions about electronics assembly, effectively masculinizing the work and the workforce. The human resources manager explained that, in the early days, women dominated semiconductor assembly in the Philippines because "everyone followed the dictum that women work better with small things." But with the increase in the number of technicians and technical operators, he noted, "Guys are beginning to gain ground." Specifically, men have been gaining ground by masculinizing the new machines and processes. For example, the manager noted that the new automated machines are not as "gender specific" as old machines and that production has become "less dependent on the fine skills of women. . . . The machines have large screens instead of microscopes, so [operators] don't need 20/20 vision or the patience to sit at a scope." Thus, there is a regendering process as managers and skilled male workers try to capture a new technology or area of production and revalue previously feminized assembly labor. The human resources manager went on to explain why women do not make "ideal workers": "Women get sick, they need maternity leave, have monthly periods." Men, on the other hand, were seen as "stronger" and able to work longer shifts. Here, the manager referred to the 12-hour day as an important reason for the shifting gender balance. According to the manager, "with the long shifts, we need more stamina and not agility. . . . Guys are strong and can work for 24 hours. With the right pay, they will do anything." Thus, the demands of an essentially still-standardized labor process become regendered: from the stereotyped call for feminized patience into a new demand for masculine stamina. In such cases, where the gendering of technology is widening the divide between male technical workers and female shop floor operators, women operators facing mass layoffs and blocked paths to promotion ladders may find that upgrading serves neither their practical survival needs nor their strategic gender interests.

Discrete Manufacturing

Discrete Manufacturing, a branch of a leading European electronics multinational that produces discrete semiconductors, refutes arguments that electronics work is necessarily exploitative of women workers, particularly if workers act collectively. Yet the introduction of new technologies may threaten the hard-fought gains won through unionization. The $110 million plant, opened in 1994, employs nearly 3,500 permanent employees, 85 percent of whom are factory workers and almost half of whom transferred from an older assembly plant. Relocation allowed management to reorganize production with more automated equipment, new line layouts, increased multiskilling, and the introduction of teamwork. But the firm was constrained from taking full advantage of an otherwise greenfield investment since workers at Discrete Manufacturing are unionized and have been able to check management's ability to introduce flexibility in the same manner as at the other two firms.

Discrete Manufacturing also differs in how its gendered employment practices have played out over time. Ironically, Discrete Manufacturing—like other foreign electronics firms—originally recruited women with only high school educations because it too sought "cheap assembly hands" for its labor-intensive plant. However, workers at the older plant successfully unionized during the height of the Philippine labor movement's organizing surge in the turbulent 1980s. When production and the workforce shifted to the new plant, management initially hired a human resources director to bust the union. But when the union successfully resisted these attempts, the firm shifted to a more cooperative stance. This crucial development helped women workers, who traditionally have little bargaining power, to collectively shape an agenda for fulfilling both their practical needs and, to some extent, their strategic gender interests.

The most immediate areas in which unionization has made a difference are in pay and benefits. Although new workers' base pay is similar to that at other firms, experienced production workers earn nearly three and a half times the starting rate. The high pay is a big reason so many women have stayed with the company. Thus, while the firm's overall gender balance is similar to other electronics firms—75 percent of the factory workers are women—the workforce is quite a bit older. Among factory workers, 77 percent are 26 years old or older, with an average age of 29, and less than 1 percent are younger than 20. In part a reflection of their older age, 35 percent of operators are married. Interestingly, 72 percent of married women had spouses who were unemployed, meaning that a large percentage of women workers, 26 percent, were sole breadwinners for their families. It is also interesting to note that the older women remain largely unmarried. While managers joked that there are "a lot of *sultera* [old maids] in this company," this pattern may also reflect women's ability to delay marriage, since earning a decent wage has allowed them some measure of independence (Chant and McIlwaine 1995; Wolf 1992).

The breadth of benefits—which are negotiated rather than simply defined by management—also reflects workers' own priorities as both workers and women. First, all union members enjoy general provisions such as seniority-based pay, family health insurance, and protection against layoffs and subcontracting. Second, workers have negotiated gender-sensitive benefits not found in most other companies. For example, maternity leave and childbirth subsidies take into account the kind of delivery (caesarian, normal, or miscarriage). Since women in the Philippines are usually responsible for attending to sick family members or emergencies, workers have negotiated benefits such as bereavement assistance and a total of 26 days of leave a year. Women are also often responsible for family finances, so they have bargained for an array of low-interest loans and subsidies for housing, education, and emergencies to finance expenses.

With the unionized status and negotiated benefits, many workers recognize their jobs as some of the best in the industry and are committed to staying with the firm. Carmen, a production operator, stated, "The work is hard. . . . Two machines I have to watch over and I'm standing the whole shift," but, "I'm not planning to look for anything else. I figure I'll be here at Discrete Manufacturing for the next 10 years. There's a CBA [collective bargaining agreement]. My projection . . . is that it's really lifetime employment." Indeed, the average tenure of the 24 operators interviewed was 4.75 years, and the overall yearly turnover rate was between just 3 and 5 percent. That these women remain in the workforce and consider themselves "lifetime" employees is again in contrast to other studies that have focused on the temporary nature of women's factory work.

But despite wide gains by the union, workers now face a dual challenge: technological upgrading and a restive management trying to reassert its command over the labor process. Like Integrated Production, Discrete Manufacturing is also experimenting with a new, fully automated assembly line run by new technical operators. And also like Integrated Production, the technical operator positions have the potential for increasing upward mobility. Previously, when new equipment was introduced, the union had been able to negotiate technical training for its frontline workers, ensuring gradual upgrading of skills. Initially, when the new technical operator positions were developed, union officers tried to get these positions filled by senior workers, who are primarily women. However, management has taken this opportunity of more radical redesign to enhance its control over workers. Invoking their claim to "management prerogative" over production issues, managers have refused to promote existing operators into the technical operator positions. Thus, the firm, like Integrated Production, is not filling these positions with its women employees but with new "fresh graduates" of four-year engineering or two-year electronics training programs, who all happen to be men. Thus, it is clear that while women workers have made many gains through their employment and collective bargaining, the industry is also changing in ways that may not bode well for the future.

* * *

Conclusions: Realizing Practical and Strategic Gender Interests

The three case studies demonstrate that the impact of high-tech employment and technological upgrading on women workers varies: It is impossible to speak of exploitation or empowerment. But how then to assess the variation and what it might mean for the gendering of high-tech production in the future? To conclude, it will be useful to return to the issue of practical needs versus strategic gender interests.

Focusing first on the practical material needs of women, the cases confirm what other studies have also pointed out: that multinational electronics firms tend to pay higher wages and provide higher job security and better working conditions than local factories or the informal sector (Elson 1999; Mills 2003). In this sense, work in the electronics sector does seem to help meet women's daily needs and improve their lives. Workers themselves often have positive assessments of their jobs, as waged work helps increase their mobility, independence, and control over income. Equally important for interviewees has been the higher status that their "clean" and "modern" employment in the electronics industry provides. As others have argued, these gains from formal-sector jobs are significant and may help women workers renegotiate power and gender relations within their families, particularly regarding child care and household spending decisions (Chant and McIlwaine 1995). Nevertheless, these "practical" gains are made within existing gender divisions of labor—both in the firms and in wider Philippine society—which still slot women into lower-level, secondary positions (Eviota 1992; Parrenas 2001). Workers' positive views of seemingly punishing factory work are also relative to the limited options these women face in the informal economy or service sector due to the persistence of gender-stratified labor markets. Thus, while women workers do benefit from their factory jobs, such labor is embedded in and subjected to persistent gender ideologies that employers have long used to feminize and cheapen assembly work. Therefore, work in the electronics sector, in and of itself, is unlikely to lead to a transformation of traditional gender divisions of labor.

The dialectics of gendered work organization are most extreme in the case of Storage Ltd. While the firm may serve its women employees' survival needs by hiring them, it also has by far the most exploitative gendered labor control strategy. By consciously leveraging wider gender inequalities in the labor market, the firm creates a kind of asymmetrical agency, which allows a sense of individual worker autonomy and empowerment yet only within the confines of existing organizational and gender hierarchies that help (re)produce rather than challenge management's authority in production and broader gendered social relations.

Nevertheless, the variation across the three firms in terms of individual versus collective bargaining power suggests that the most potent vehicle for maximizing the benefits of industrial work is through collective organization (Elson 1999; Hutchinson and Brown 2001). At Discrete Manufacturing, a group of initially marginal women workers—selected along stereotypical lines for their cheapness and docility—have in fact increased their income security, class mobility, and autonomy by negotiating a gender-income sensitive contract that addresses worker-defined needs. This is similar to Fernandez's (2001) findings that although technological upgrading can lead to greater wage inequality, an active union can at least help mitigate the most polarizing gender and racial effects. Unfortunately, the conditions at Discrete Manufacturing are not widely shared: Despite similar structural conditions of permanent employment across most of the industry, fewer than 10 percent of electronics firms in the Philippines are unionized.

But more important, even impressive collective bargaining agreements such as the one at Discrete Manufacturing do not directly address strategic gender interests, as they focus primarily on improving wages and benefits and not on challenging the gender division of labor. Thus, the development in the industry that has the most potential for transforming workplace gender hierarchies and expanding women's choices about the gender division of labor is the emergence of the technical operator. As shown most vividly at Integrated Production—but also at Discrete Manufacturing—the new technical operator positions fuse the jobs, responsibilities, and promotional ladders of production workers (traditionally women) and technicians (traditionally men). These positions, which are likely to dominate high-end production facilities not only in the Philippines but across the global industry could allow current women operators to breach both the glass ceiling for women that has predominated in high-tech production work and the gendered associations with technology that have helped reproduce and sustain it. Yet despite the real potential that technological upgrading holds for realizing strategic gender interests, management at both firms chose to extend—rather than challenge—traditional associations between masculinity and technology, demonstrating the power and durability of gendered ideologies and frameworks.

Thus, while the gendering of work is always an ongoing, contested, and negotiated process, the ongoing trope of productive femininity, low unionization rates, and the complicity of the Philippine government means that there are few checks on management's power to implement and gender the new technology in ways it sees fit. And given the polarizing character of high-tech production and the current trends in the global industry toward automation, downsizing, and the masculinization of production, women electronics workers may witness—but have little access to—more skilled high-tech manufacturing jobs that the industry is finally producing.

References

Acker, Joan. 2004. Gender, capitalism and globalization. *Critical Sociology 30*(1): 17–41.

Austria, Myrna S. 1999. Assessing the competitiveness of the Philippine IT industry. PIDS discussion paper. Makati: Philippine Institute for Development Studies.

Bello, Walden, David Kinley, and Elaine Elinson. 1982. *Development debacle: The World Bank in the Philippines.* San Francisco: Institute for Food and Development Policy.

Bureau of Labor Relations. 1999. Labor management schemes and workers' benefits in the electronics industry. Unpublished draft. Manila, Philippines: Bureau of Labor Relations

Chant, Sylvia, and Cathy McIlwaine. 1995. *Women of a lesser cost: Female labor, foreign exchange and Philippine development.* London: Pluto.

Chhachhi, Amrita. 1999. Gender, flexibility, skill and industrial restructuring: The electronics industry in India. Working paper series no. 296. The Hague, the Netherlands: Institute of Social Studies.

Chhachhi, A., and R. Pittin. 1998. Multiple identities, multiple strategies: Confronting state, capital and patriarchy. In *Labor worldwide in the era of globalization,* edited by P. Waterman and R. Munck. London: Macmillan.

Chow, E. N., and D. M. Lyter. 2002. Studying development with gender perspectives: From mainstream theories to alternative frameworks. In *Transforming gender and development in East Asia,* edited by E. N. Chow. New York: Routledge.

Elson, Diane. 1999. Labor markets as gendered institutions: Equality, efficiency and empowerment issues. *World Development 27*(3): 611–27.

Ernst, Dieter. 2002. Digital information systems and global flagship networks: How mobile is knowledge in the global network economy? East West Center working papers, economics series no. 48. Honolulu, HI: East West Center.

Eviota, Elizabeth U. 1992. *The political economy of gender: Women and the sexual division of labor in the Philippines.* London: Zed Books.

Fernandez, Roberto. 2001. Skill-biased technological change and wage inequality: Evidence from a plant retooling. *American Journal of Sociology, 107*(2): 273–320.

Fox, J. 2002. Women's work and resistance in the global economy. In *Labor and capital in the age of globalization,* edited by B. Berberoglu. Lanham, MD: Rowan and Littlefield.

Hutchinson, Jane, and Andrew Brown, eds. 2001. *Organizing labour in globalizing Asia.* London: Routledge.

Joekes, Susan, and Ann Weston. 1994. *Women and the new trade agenda.* New York: UNIFEM.

Katz, Harry C., and Owen Darbishire. 2000. *Converging divergences.* Ithaca, NY: Cornell University Press.

Kelkar, G., and D. Nathan. 2002. Gender relations and technological change in Asia. *Current Sociology 50*(3): 427–41.

Kenney, M., R. Goe, O. Contreras, J. Romero, and M. Bustos. 1998. Learning factories or reproduction factories? Labor-management relations in the Japanese consumer electronics maquiladoras in Mexico. *Work and Occupations 25*(3): 269–304.

Kuruvilla, Sarosh. 1996. Linkages between industrialization strategies and industrial relations/human resource policies: Singapore, Malaysia, the Philippines and India. *Industrial & Labor Relations Review 49*(4): 635–58.

Lin, V. 1987. Women electronics workers in Southeast Asia: The emergence of a working class. In *Global restructuring and territorial development,* edited by J. Henderson and M. Castells. London: Sage.

Mills, Mary Beth. 2003. Gender and inequality in the global labor force. *Annual Review of Anthropology* 32:41–62.

Molyneux, Maxine. 1985. Mobilization without emancipation? Women's interests, the state and revolution in Nicaragua. *Feminist Studies* 11:225–54.

National Statistical Coordination Board. Various years. Statistics on men and women in the Philippines. Government of the Philippines. Available from http://www.nscb.gov.ph/stats/ wmfact.htrm.

Ong, Aiwa. 1991. The gender and labor politics of post modernity. *Annual Review of Anthropology* 20:279–309.

Parrenas, R. S. 2001. Breaking the code: Women, migration, and the 1987 family code of the Republic of the Philippines. Paper presented at the Workshop on Globalization and the Asian "Migrant" Family, National University of Singapore, 16 April.

Safa, Helen. 1981. Runaway shops and female employment: The search for cheap labor. *Signs: Journal of Women in Culture and Society* 7:418–33.

Salzinger, Leslie. 2003. *Gender in production: Making workers in Mexico's global factories.* Berkeley: University of California Press.

Seguino, Stephanie. 2000. Accounting for gender in Asian economic growth. *Feminist Economics 6*(3): 27–58.

Standing, Guy. 1999. Global feminization through flexible labor: A theme revisited. *World Development 27*(3): 583–602.

Tzannatos, Zafiris. 1999. Women and labor market changes in the global economy: Growth helps, inequalities hurt and public policies matter. *World Development 27*(3): 551–69.

Wolf, Diane L. 1992. *Factory daughters: Gender, household dynamics, and rural industrialization in Java.* Berkeley: University of California Press.

World Bank. 1997. *Philippines: Managing global integration.* Vol. 2, report no. 17024-PH. Poverty Reduction & Economic Management Sector, East Asia and the Pacific Office. Washington, DC: World Bank.

Yun, Hing Ai. 1995. Automation and new work patterns: Cases from Singapore's electronics industry. *Work, Employment & Society 9*(2): 309–27.

Introduction to Reading 35

Sociologist Gwen Moore is known for her research on men and women elites in the most prestigious and visible careers in industrialized countries. This article compares the experiences of men and women elites in twenty-seven industrialized nations on their orientations toward careers and families, and reports upon the various social support structures available. She explores interpersonal dynamics in the family—who does the unpaid work of the household including child care—and the social systems that support work and family in these countries.

1. How is the work experience of women elites different from men's?
2. What social support systems made the work lives of these elite workers easier?
3. Is it easier for men to "have it all" in terms of elite jobs and families?

Mommies and Daddies on the Fast Track in Other Wealthy Nations

Gwen Moore

Social and cultural contexts, as well as public policies, shape the experiences of women and men in demanding occupations. Women's employment in influential positions in the public and private sectors varies cross-nationally. Taking politics as an example, in the 2003 Swedish election, more than 45 percent of the parliamentarians who were elected are women. This contrasts sharply with the United States, where women hold just 14 percent of congressional seats, a smaller proportion than in fifty-nine other nations (InterParliamentary Union Web site, www.ipu.org, accessed January 16, 2004). Still, men hold nearly all top economic posts in all countries throughout the world (Adler and Izraeli 1994: Wirth 2001). These variations across countries and sectors demonstrate the importance of placing the topic of work and family in an international perspective.

In this comment, I will compare some work-family themes to research outside of the United States, especially to the findings of a mid-1990s survey of approximately twelve hundred women and men who held the highest positions in elected politics and private business in twenty-seven capitalist industrial nations, including twenty-one European nations, Australia, New Zealand, and others in North America and Asia (Vianello and Moore 2000). This collaborative project—the Comparative Leadership Study—gathered information on the leaders' backgrounds and careers, experiences in office, and gender and career attitudes as well as current family characteristics (Vianello and Moore 2000).

From Moore, G., "Mommies and Daddies on the Fast Track in Other Wealthy Nations" from *Annals, AAPSS, 596,* November, 2004, pp. 208–213. Reprinted with permission of Sage Publications, Inc.

Consistent with findings on lawyers, professors, scientists, and finance managers, the majority of the women and men in the Comparative Leadership Study were or had been married or cohabiting. Marriage was more common for the men leaders than for their women counterparts, as has been found in previous research. Eighteen percent of the women and 8 percent of the men had never married. Virtually all of the men (94 percent) and most (75 percent) of the women leaders were currently married or cohabiting (Neale 2000, 158–59).

Most leaders of both sexes in the international study were also parents. Just over one-fourth of the women and less than 10 percent of the men had no children (Kuusipalo, Kauppinen, and Nuutinen 2000, Table 15.1). Among parents, women more often had just one child (Neale 2000, 158–59). The vast majority of men (88 percent) and three-fifths of women were living with both a partner and at least one child at the time of the survey (Kuusipalo, Kauppinen, and Nuutinen 2000, Table 15.1).

Most top business and political leaders in the comparative study, as well as those in professional and managerial positions in the United States, marry and have at least one child. These leaders face the dilemma of combining a demanding career and family life.

How do they manage this work-family conflict? Rarely by cutting back on working time. Even in countries with relatively shorter workweeks and longer vacations than the United States, the national leaders in our study reported far longer working hours than the general workforce in their country (Woodward and Lyon 2000; also see Jacobs and Winslow 2004). Politicians average more than sixty-five hours per week and business leaders about ten hours fewer, with little gender difference (Woodward and Lyon 2000, Table 8.1). In addition, few female leaders and almost none of their male colleagues had worked part-time or interrupted their careers for care work (Neale 2000; also see Epstein et al. 1998). Especially among managerial elites, the typical career proceeds without interruption from its beginning (Blair-Loy 2003).

In dual-career families, both partners work hard and long in paid employment. Taking time from paid work for child care or household labor is difficult. Some leaders—mostly men—have spouses who were not in paid employment, worked part-time, or had less demanding occupations. Dual-career families—with both partners in senior positions in the labor force—were far more common for the women leaders in our study than for the men (see Boulis 2004 for similar patterns among physicians). More than 70 percent of the men—and just more than one-third of the women—had partners who were not in paid employment or who worked in nonprofessional jobs with no supervisory responsibilities (Kuusipalo, Kauppinen, and Nuutinen 2000, Table 15.2; see Stone and Lovejoy 2004).

Most leaders earn high enough salaries to pay others to perform some of the household labor and child care. Yet most women in our research did some of these tasks themselves in spite of their long workweeks. The amounts done differ considerably between men and women. Well more than half of the men and none of the women reported that their spouse had cared for their children when they were preschoolers (Kuusipalo, Kauppinen, and Nuutinen 2000, Table 15.4). Likewise, one in five women (nearly all politicians) and no men had cared for their preschool children themselves (Neale 2000, Tables 13.1 and 13.2; Kuusipalo. Kauppinen, and Nuutinen 2000, Table 15.4). Child care responsibilities clearly fell disproportionately on women, even for those holding top economic and political positions.

In response to a question about the division of household labor, few women (9 percent) and about a quarter of the men reported doing none (Kuusipalo, Kauppinen, and Nuutinen 2000, Table 15.2). Nearly a third of the women business and political leaders said they did more than half of the household labor themselves, including 13 percent of the women business leaders who reported doing all the housework (Neale 2000, 164–65; Kuusipalo, Kauppinen, and Nuutinen 2000, Table 15.2; Esseveld and Andersson 2000, Table 16.2).

Family responsibilities, including care of young children and completion of household labor, fall disproportionately on women, even among those in top leadership positions in the professions, management, and politics. Marriage and, even more, parenthood impinge on women's careers to a far larger extent than they do on similarly situated men's. Compared to women, fast-track men are more frequently married and parents. And men also more often have wives who are not in a demanding career and are thus more available for child care, housework, and involvement in building his career (see Stone and Lovejoy 2004).Women are frequently married to highly placed men who have as many time constraints as they do and thus are less available for sharing family labor and focusing on advancing the woman's career.

Research beyond the United States generally paints a similar picture to that portrayed. . . . But this broad picture obscures variations in patterns within regions or countries. When one looks more closely at work-family conflicts in European and other industrialized countries variations appear. National cultural norms and social policies provide contexts for the employment and family lives of workers in demanding occupations (see Wax 2004).

The Nordic countries stand out as models of (relatively) woman- and family-friendly societies. Gender equality norms and government policies support women's equal participation in public life, and women's rates of paid employment are high. According to Kuusipalo, Kauppinen, and Nuutinen (2000), the Nordic countries have replaced the male breadwinner model with the dual-earner model. They wrote. "Parental leave, flexible working hours and state childcare support women's right to work and men's right to fatherhood" (p. 178). Data from the Nordic countries in the Comparative Leadership Study (Denmark, Finland, Norway, and Sweden) do show fewer gender differences in family status and duties than are seen in other areas of Europe, North America, Asia, Australia and New Zealand (Vianello and Moore 2000, pt. III).

Public child care is available in the Nordic countries, and about 30 percent of leaders placed their preschool children in it. Nordic male leaders were more involved in household labor than men in the other regions: 20 percent reported doing at least half of the housework themselves. Nordic men's higher rate of household labor may be partly due to Nordic norms against employing private household workers.

Despite men's greater participation in family care in the Nordic countries, women do even more. Twice as many women leaders reported caring for their preschool children (26 percent of women vs. 14 percent of men), and nearly two-thirds of women reported doing at least half of the housework (Esseveld and Andersson 2000). Primarily mothers, not fathers, apparently use the generous parental leave policies available in the Nordic countries.

Career pathways and occupational settings also affect work-family conflicts. . . . Careers in management and science, for instance, require an early and steady commitment from aspirants, often beginning in high school. Dropping in and out or beginning a career later in life is hardly an option for these fields. In the Comparative Leadership Study, business-women had the lowest rates of marriage and childbearing (Vianello and Moore 2000). By contrast, some top careers are more flexible and more easily entered at an older age. Elected politicians, for example, often enter politics in their late twenties or even thirties. Politics, then, is not closed to women who have reared children or begun in other careers. Some strategies to improve the work-family balance for women and men in demanding careers seem unrealistic for those in senior leadership positions. Jerry Jacobs (forthcoming) has called for institutions to clearly state that tenure-track faculty are expected to work no more than fifty hours per week to create a "family-compatible workstyle." Reduced working hours for national leaders in politics, business, or voluntary associations is not a workable solution. Top leaders in key institutions are expected to show total devotion to their work (e.g., Kanter 1977; Blair-Loy 2003). Possibly more feasible is Phyllis Moen's (2004) advocacy of social expectations that careers include "second acts" and "time-outs" allowing men and women to develop integrated career and family lives. For those aspiring to national elite positions, a time-out early in their career to have and rear children—as taken by many women politicians—could facilitate a more compatible career and family life.

An international perspective on the work-family conflicts helps to show in what ways the United States is similar to and different from other countries. In many ways, the United States differs little, as I have shown above. Yet the Nordic countries appear more successful in lessening work-family conflicts, even for women and men in top positions. These countries have made extraordinary progress in opening political decision-making positions to women. Women are prominent among prime ministers, cabinet members, and members of parliament in Denmark, Finland, Norway, and Sweden. Women constitute a critical mass in their parliaments, erasing their token status and normalizing the image of politics as a woman's game as well as a man's game (Kanter 1977; Epstein 1988). Scholars and policy makers would benefit from a closer examination of these models promoting gender equality.

References

Adler, Nancy J., and Dafna N. Izraeli, eds. 1994. *Competitive frontiers: Women managers in a global economy.* Cambridge, MA: Blackwell.

Blair-Loy, Mary. 2003. *Competing devotions: Career and family among women executives.* Cambridge, MA: Harvard University Press.

Boulis, Ann. 2004.The evolution of gender and motherhood in contemporary medicine. *Annals of the American Academy of Political and Social Science* 596: 172–206.

Epstein, Cynthia Fuchs. 1988. *Deceptive distinctions: Sex, gender, and the social order.* New Haven, CT: Yale University Press.

Epstein, Cynthia Fuchs, Carroll Seron, Bonnie Oglensky, and Robert Saute. 1998. *The part-time paradox: Time norms, professional lives, family, and gender.* New York: Routledge.

Esseveld, Johanna, and Gunnar Anderson. 2000. Career life-forms. In *Gendering elites: Economic and political leadership in 27 industrialised societies,* edited by M. Vianello and G. Moore, 189–204. London: Macmillan.

Jacobs, Jerry A., and Sarah E. Winslow. 2004. Overworked faculty: Job stresses and family demands. *Annals of the American Academy of Political and Social Sciences, 596*: 104–129.

Jacobs, Jerry A. Forthcoming. The faculty time divide. *Sociological Forum.*

Kanter, Rosabeth. 1977. *Men and women of the corporation.* New York: Basic Books.

Kuusipalo, Jaana, Kaisa Kauppinen, and Iira Nuutinen. 2000. Life and career in north and south Europe. In *Gendering elites: Economic and political leadership in 27 industrialised societies,* edited by M. Vianello and G. Moore, 177–88. London: Macmillan.

Moen, Phyllis. 2004. Integrative careers: Time in, time out, and second acts. Presidential Address, Eastern Sociological Society meetings, February 20.

Neale, Jenny. 2000. Family characteristics. In *Gendering elites: Economic and political leadership in 27 industrialised societies,* ed. M. Vianello and G. Moore, 157–68. London: Macmillan.

Stone, Pamela, and Meg Lovejoy. 2004. Fast-track women and the "choice" to stay home. *Annals of the American Academy of Political and Social Science* 596: 62–83.

Vianello, Mino, and Gwen Moore, eds. 2000. *Gendering elites: Economic and political leadership in 27 industrialised societies.* London: Macmillan.

Wax, Amy L. 2004. Family-friendly workplace reform: Prospects for change. *Annals of the American Academy of Political and Social Science* 596: 36–61.

Wirth, Linda. 2001. *Breaking through the glass ceiling: Women in management.* Geneva, Switzerland: International Labour Office.

Woodward, Alison, and Dawn Lyon. 2000. Gendered time and women's access to power. In *Gendering elites: Economic and political leadership in 27 industrialised societies,* edited by M. Vianello and G. Moore, 91–103. London: Macmillan.

Introduction to Reading 36

Sociologist Kathleen Gerson has studied work and family patterns for some time. Her first book examining women's and men's roles in families, published in 1985, examined the decisions women make about work, career, and motherhood. This book was aptly titled *Hard Choices*. In this reading, she considers what she has learned from studying men's and women's choices about work and family, exploring the moral dilemmas that she has observed and considering those that young people face as they negotiate familial relationships and work responsibilities. This article uses more recent data she collected from life history interviews with 120 men and women from ages eighteen to thirty-two. These individuals lived in New York City and surrounding areas and represent a diverse group of young people, both in race and social class.

1. What does Gerson mean by "the tension between autonomy and commitment"?
2. What changes does she observe in heterosexual relationships?
3. What is neotraditionalism and how does it affect men and women in heterosexual relationships?

Moral Dilemmas, Moral Strategies, and the Transformation of Gender

Lessons from Two Generations of Work and Family Change

Kathleen Gerson

Choosing between self-interest and caring for others is one of the most fundamental dilemmas facing all of us. To reconcile this dilemma, modern societies in general—and American society in particular—have tried to divide women and men into different moral categories. Since the rise of industrialism, the social organization of moral responsibility has expected women to seek personal development by caring for others and men to care for others by sharing the rewards of independent achievement.

Although labeled "traditional," this gendered division of moral labor represents a social form and cultural mandate that rose to prominence in the mid-twentieth century but reached an impasse as the postindustrial era opened new avenues for work and family life. . . . At the outset of the twenty-first century, women and men face rising conflicts over how to resolve the basic tensions between family and work, public and private, autonomy and commitment. They are searching for new strategies for reconciling

From Gerson, K., "Moral Dilemmas, Moral Strategies, and the Transformation of Gender: Lessons from Two Generations of Work and Family Change, "*Gender & Society 16*(1): 8–28.

an "independent self" with commitment to others.

While the long-term trajectory of change remains unclear, new social conditions have severely undermined the link between gender and moral obligation. The young women and men who have come of age amid this changing social landscape face risks and dangers, but they also inherit an unprecedented opportunity to forge new, more egalitarian ways to balance self-development with commitment to others. To enable them to do so, however, we must reshape work and family institutions in ways that overcome beliefs and practices that presume gender differences in moral responsibility. . . .

While it is important to assert that it is just as valuable to pursue emotional connection and provide care as it is to create an independent self or provide economically for a family, it is also critical to question the premise that women and men can be separated into distinct, opposed, or unchanging moral categories. As Epstein argues, any vision of dichotomous gender distinctions is not only inaccurate; it is also an ideological construct that justifies and reinforces inequality. Connell points out that "masculinities" and "femininities" vary across historical time and space. Lorber and Risman, among others, question the concept of gender itself, pointing to the social paradoxes and cultural contradictions to which all human actors must respond in constructing their public and private selves. These theorists recognize that gender is a social institution, not an inherent trait, and that it shapes organizations and opportunity structures as well as personal experiences (Connell 1987, 1995; Epstein 1988; Lorber 1994; Risman 1998).

There are good analytic and empirical reasons to reject the use of gender to resolve the knotty moral conflicts between public and private, work and family, self and other. It is difficult to avoid the conclusion that using gender in this way is more prescriptive than descriptive. Such approaches may depict how women and men *should* behave, but they do not provide an accurate description or explanation of how women and men actually *do* behave or how they *would* behave if alternative options were available. Certainly, the proportion who have conformed to gendered injunctions about appropriate moral choices has varied substantially across societies, subcultures, and historical periods. Countless women and men have been labeled "deviant" for their reluctance or inability to uphold idealized conceptions of gender. A framework of gendered moralities helps justify inequalities and stigmatize those who do not conform.

* * *

New Dilemmas, Ambiguous Strategies

How does this generation view its moral choices? As adult partnerships have become more fluid and voluntary, they are grappling with how to form relationships that balance commitment with autonomy and self-sufficiency. As their mothers have become essential and often sole breadwinners for their households, they are searching for new ways to define care that do not force them to choose between spending time with their children and earning an income. And in the face of rising work-family conflicts, they are looking for definitions of personal identity that do not pit their own development against creating committed ties to others. As young women and men wrestle with these dilemmas, they are questioning a division of moral responsibility that poses a conflict between personal development and caring for others.

Seeking Autonomy, Establishing Commitment

The decline of permanent marriage has raised new and perplexing questions about how to weigh the need and desire for self-sufficiency against the hope of creating an enduring partnership. In wrestling with this quandary, young women and men draw on lessons learned in their families and personal relationships. Yet, they also recognize that past experiences and encounters can provide, at best, a partial and uncertain blueprint for the future.

Few of the women and men who were interviewed reacted in a rigidly moralistic way to their parents' choices. Among those whose parents chose to divorce (or never marry), about 45 percent viewed the breakup as a prelude to growing difficulty, but the other 55 percent supported the separation and felt relief in its aftermath. Danisha, a 21-year-old African American, concluded that conflict would have emerged had her parents stayed together. . . .

And at 26, Erica, who grew up in a white middle-class suburb, supported her parents' decision to separate and received more support from each of them in its aftermath:

> I knew my parents were going to get divorced, because I could tell they weren't getting along. They were acting out roles rather than being involved. They were really drifting apart, so it was something perfectly natural to me. In the new situation, I spent more valuable time with my parents as individuals. So time with my father and mother was more meaningful to me and more productive.

Among those whose parents stayed together, almost 60 percent were pleased and, indeed, inspired by, their parents' lifelong commitment, but about 40 percent concluded that a breakup would have been better than the persistently unhappy, conflict-ridden relationship they watched unfold. Amy, a 24-year-old Asian American, explains:

> I always felt my parents would have divorced if they didn't have kids and didn't feel it was so morally wrong. They didn't really stick together because they were in love. I know all couples go through fights and stuff, but growing up, it seemed like they fought a lot, and each of them has made passing comments—like. "Oh, I would have divorced your mom by now" or "I would have left your dad a thousand times." [So] I wouldn't have broken down or been emotionally stressed if my parents divorced. I didn't want to hear the shouting, and I didn't want to see my mom cry anymore. And I was also afraid of my dad, because he would never lay a hand on my mom, but he's scary. He could be violent.

Whether their parents stayed together or parted, most concluded that neither steadfast commitment nor choosing to leave has moral meaning in the abstract. The value of enduring commitment depends on the quality of the relationship it embodies.

When considering their own aspirations, almost everyone hopes to establish a committed, lasting relationship with one partner. Yet, they also hold high standards for what a relationship should provide and anticipate risks in sustaining such a commitment. Across the divides of gender, race, and class, most agree that a satisfying and worthwhile relationship should offer a balance between autonomy and sharing, sacrifice and support. . . .

Amy imagines a partnership that is equal and fluid, capable of adapting to circumstances without relinquishing equity:

> I want a fifty-fifty relationship, where we both have the potential of doing everything. Both of us working, and in dealing with kids, it would be a matter of who has more flexibility with regard to their career. And if neither does, then one of us will have to sacrifice for one period, and the other for another.

Most acknowledge, however, that finding a lasting and satisfying relationship represents an ideal that is hard to reach. If it proves unattainable, they agree that being alone is better than remaining in an unhappy or destructive union. Building a full life thus means developing the self in multiple ways. . . . Across the range of personal family experiences, most also agree that children suffer more from an unhappy home than from separated parents. . . .

Women and men both wonder if it is possible to establish relationships that strike a good balance between self-affirmation and commitment, providing and receiving support. Having observed their parents and others struggle with varying degrees of success against the strictures of traditional gender categories, they are hopeful but guarded about the possibilities for resolving the tension between autonomy and commitment in their own lives. . . .

Care as Time, Care as Money

If the rise of fluid adult partnerships has heightened the strains between commitment and autonomy, then the rise of employed mothers and the decline of sole male breadwinners have made the meaning of care ambiguous. Now that most children—whether living in single-parent or two-parent households—depend on the earnings of their mothers, parents face conflicts in balancing the need to provide economic support with the need to devote time and attention.

Rigid notions of gendered caring do not fit well with most family experiences, and the majority express support for parents who transgressed traditional gender categories. Among those who grew up in two-earner households, four out of five support such an arrangement, most with enthusiasm. Across race, class, and gender groups, they believe that two incomes provided the family with increased economic resources, more flexibility against the buffeting of economic winds, and greater financial security. . . .

Of course, this means they see a mother's employment as largely beneficial. Whether in a two-parent or single-parent home, women and men agree that an independent base enhanced a mother's sense of self, contributed to greater parental equality, and provided an uplifting model. Rachel, 24 and from a white, working-class background, explains, "I don't think that I missed out on anything. I think it served as a more realistic model. . . ."

Kevin, 25 and from a middle-class, white family, agrees:

> For quite a while, my mom was the main breadwinner. She was the one who was the driving force in earning money. My mother's persona was really hard working, and that's something I've strived to be with and to emulate. I didn't think it was wrong in any way. I actually feel it's a very positive thing. Whatever my relationships, I always want and appreciate people who work, and I'm talking about female involvement. It's part of who I am, and it makes me very optimistic knowing that hard work can get you somewhere.

They also deemed highly involved fathers, whether in two-earner or single-parent households, as worthy examples. Daniel, now 23, describes his Irish father's atypical working hours and parental involvement:

> My father was always around. He's a fire fighter, so he had a lot of free time. When he was home, he was usually coaching me and my brother or cooking dinner or taking us wherever we wanted to go. He was the only cook up until me and my brother started doing it. So I want to make sure that, if I get married and have kids, I'm there for my kids.

In contrast, those who grew up in a largely traditional household expressed more ambivalence. Although half felt fortunate to have had a mother devoted primarily to their care, the other half would have preferred for their mothers to pursue a more independent life. At 21, Justin, who grew up in a white, largely middle-class suburb, looks back on his mother's domestic focus with a strong conviction that it took its toll on the whole household. . . .

Breadwinning fathers may also elicit mixed reactions. Their economic contributions are appreciated but not necessarily deemed sufficient. A good father, most concluded, takes time and offers emotional support as well. At 29, Nick, who grew up in a white working-class neighborhood and remembers feeling frustrated by his own father's distance, is seeking joint custody of his own young daughter:

> I have seen a lot of guys who have kids and have never changed a diaper, have never done anything for this child. Don't call yourself daddy. Even when she was saying, "Oh, she might not be yours," it didn't matter to me. This child is counting on me.

In this context, care becomes a slippery concept. Across family circumstances, these young adults judge an ideal parent—whether mother or father—to be one who supports her or his children both economically and emotionally. . . .

If fathers should resemble traditional conceptions of mothers, then mothers should resemble fathers when it comes to work outside the home.

Gabriel, a white 25-year-old who was raised by his father after his parents divorced when he was in grade school, explains,

> In terms of splitting parental stuff, it should be even. Kids need a mother and a father. And I'm really not high on the woman giving up her job. I have never wanted to have a wife who didn't make a salary. But not for the sake of leeching off of her, but so that she was independent. . . .

If such an ideal proves beyond reach, as many expect it will be, women and men agree that families should apportion moral labor however best fits their circumstances—whether or not this means conforming to classic notions of gender difference. Mothers can and often do demonstrate care through paid work and fathers through involvement. Now 26 and raising a child on her own, Crystal, an African American, rejects a natural basis for mothering:

> I don't really believe in the mother instinct. I don't believe that's natural. Some people really connect with their children, and some people just don't. I think it should be whoever is really going to be able to be there for that child.

In the end, the material and emotional support a child receives matters more than the type of household arrangement in which it is provided. . . .

Identity Through Love, Identity Through Work

In a world where partnerships are fragile and domesticity is devalued, young women and men are confronting basic questions about identity and self-interest. Do they base their personal wellbeing and sense of self on public pursuits or be struck between them?

In pondering their parents' lives, most could find no simple way to define or measure self-interest. While a minority uphold traditional gendered identities, most do not find such resolutions viable. Women are especially likely to conclude that it is perilous to look to the home as the sole source of satisfaction or survival. Reflecting on the many examples of mothers and other women who languished at home, who were bereft when marriages broke up, or who found esteem in the world of paid work, 9 out of 10 express the hope that their lives will include strong ties to the workplace and public pursuits. . . .

On the other side of the gender divide, many men have also become skeptical of work-centered definitions of masculine identity. As traditional jobs have given way to unpredictable shifts in work prospects, they are generally guarded about the prospect of achieving stable work careers. Having observed fathers and friends who found work either dissatisfying or too demanding, two-thirds of the men concluded that, while important, work alone could not provide their lives with meaning. These young men hope to balance paid work and personal attachments without having to sacrifice the self for a job or paycheck. Traditional views persist, but they increasingly compete with perspectives that define identity in more fluid ways. Widely shared by those who grew up in different types of families, these outlooks also transcend class and race differences. They cast doubt on some post-feminist assertions that a "new traditionalism" predominates among young women and men (Crittenden 1999). . . .

Yet, beyond the apparent similarities, a gender divide emerges. With one-third of men—but far fewer women—preferring traditional arrangements over all others, women are more likely to uphold flexible views of gender for themselves and their partners. More important, women and men both distinguish between their ideals and their chances of achieving them. If most hope to integrate family and work—and to find partners with whom to share the rewards and burdens of both—far fewer believe they can achieve this lofty aspiration. It is difficult to imagine integrating private with public obligations when most workplaces continue to make it difficult to balance family and job. And it is risky to build a life dependent on another adult when relationships are unpredictable. . . .

An Emerging Gender Divide: Autonomy and Neotraditionalism as Fallback Positions

The ideal of a balanced self continues to collide with an intransigent social world. New generations must thus develop contingent strategies for less than ideal circumstances. If egalitarian aspirations cannot be reached, what options remain? Here, women and men tend to diverge. Indeed, even as they are developing similar ideals, they are preparing for different outcomes. If an egalitarian commitment proves unworkable, most men would prefer a form of "modified traditionalism" in which they remain the primary if not sole family breadwinner and look to a partner to provide the lion's share of domestic care. Women, in contrast, tend to look toward autonomy as preferable to any form of traditionalism that would leave them and their children economically dependent on someone else.

As young women and men consider the difficulties of building balanced, integrated lives, they move from ideals to consider the fallback positions that would help them avert worst-case scenarios. Here, as we see below, the gender gap widens. Women, in hoping to avoid economic and social dependence, look toward autonomy, while men, in hoping to retain some traditional privileges, look toward modified forms of traditional arrangements. Yet, both groups hope to resolve these conflicts as they construct their lives over time.

Women and Autonomy

Among the women, 9 out of 10 hope to share family and work in a committed, mutually supportive, and egalitarian way. Yet, most are skeptical that they can find a partner or a work situation that will allow them to achieve this ideal. Integrating caretaking with committed work remains an uphill struggle, and it seems risky to count on a partner to sustain a shared vision in the long run. Even a modified version of traditionalism appears fraught with danger, for it creates economic vulnerability and constricted options in the event that a relationship sours or a partner decides to leave. Four out of five women thus prefer autonomy to a traditional marriage, concluding that going it alone is better than being trapped in an unhappy relationship or being abandoned by an unreliable partner. . . .

Autonomy for women means, at its core, economic self-sufficiency. A life that is firmly rooted in the world of paid work provides the best safeguard against being stuck in a destructive relationship or being left without the means to support a family. Healthy relationships, they reason, are based on a form of economic individualism in which they do not place their economic fate in the hands of another. Rachel declares,

> I'm not afraid of being alone, but I am afraid of being with somebody who's a jerk. I can spend the rest of my life alone, and as long as I have my sisters and my friends, I'm okay. I want to get married and have children, but I'm not willing to just do it. It has to be under the right circumstances with the right person.

Men and Neotraditionalism

Young men express more ambivalence about the choice between autonomy and traditionalism. If a committed, egalitarian ideal proves out of reach, about 40 percent would opt for independence, preferring to stress the autonomous self so long associated with manhood and now increasingly affirmed by women as well. But six out of 10 men would prefer a modified traditionalism in which two earners need not mean complete equality. This split among men reflects the mix of options they confront. Work remains central to constructing a masculine identity, but it is difficult to find work that offers either economic security or good opportunities for family involvement. Without these supports, men are torn between avoiding family commitments and trying to retain some core advantages provided by traditional arrangements.

From men's perspective, opting for the autonomy conferred by remaining unmarried, unattached, or childless relieves them of the economic burden of earning a family wage in an uncertain economy, but it also risks cutting them off from close, committed, and lasting intimate connections. A neotraditional arrangement, in contrast, offers the chance to create a family built around shared breadwinning but less than equal caretaking. In this scenario, men may envision a dual-earner arrangement but still expect their partner to place family first and weave work around it. Josh, a white 27-year-old who was raised by his father after his mother was diagnosed with severe mental illness, asserts,

> All things being equal, it should be shared. It may sound sexist, but if somebody's gonna be the breadwinner, it's going to be me. First of all, I make a better salary. If she made a much better salary, then I would stay home, but I always feel the need to work, even if it's in the evenings or something. And I just think the child really needs the mother more than the father at a young age.

Modified traditionalism provides a way for men to cope with economic uncertainties and women's shifting status without surrendering some valued privileges. It collides, however, with women's growing desire for equality and rising need for economic self-sufficiency.

Resolving Moral Dilemmas Over Time

In the absence of institutional supports, postponing ultimate decisions becomes a key strategy for resolving the conflicts between commitment and self-development. For women as much as men, the general refrain is, "You can't take care of others if you don't take care of yourself." Michael wants to be certain his girlfriend has created a base for herself at the workplace before they marry, hoping to increase the chances the marriage will succeed and to create a safety net if it fails:

> There are a lot of problems when two people are not compatible socially, economically. When Kim gets these goals under her belt, and I have my goals established, it'll be a great marriage. You have to nurture the kind of marriage you want. You have to draw it out before you can go into it.

For Jennifer, 19 and white, autonomy also comes first. Commitment may follow, but only when she knows there is an escape route if the relationship deteriorates:

> I will have to have a job and some kind of stability before considering marriage. Too many of my mother's friends went for that—let him provide everything—and they're stuck in a relationship they're not happy with because they can't provide for themselves or the children they now have. The man is not providing for them the way they need, or he's just not a good person. Most of them have husbands who make a lot more money, or they don't even work at all, and they're very unhappy, but they can't leave. So it's either welfare or putting up with somebody else's crap.

Establishing an independent base becomes an essential step on the road to other goals, and autonomy becomes a prerequisite for commitment. This developmental view rejects the idea that individualism and commitment are in conflict by defining the search for independence as a necessary part of the process of becoming able to care for others. To do that, women as well as men tend to look to work, and its promise of autonomy, to complete the self. For those with children as well as those who are childless, lifelong commitments can be established when "you feel good enough about yourself to create a good relationship." . . .

These strategies are deeply felt and intensely private responses to social and personal conflicts that seem intractable. More fundamental solutions await the creation of systematic supports for balancing work and family and for providing women and men with equal opportunities at the workplace and in the home. Without these supports, new generations must cope as best they can, remaining both flexible and guarded. . . .

Conclusion: Toward a New Moral Order?

Deeply rooted social and cultural changes have created new moral dilemmas while undermining a traditional gendered division of moral labor. The widespread and interconnected nature of these changes suggests that a fundamental, irreversible realignment is under way. Less clear is whether it will produce a more gender-equal moral order or will, instead, create new forms of inequality. The long-term implications are necessarily cloudy, but this ambiguity has created some new opportunities along with new risks.

While large-scale social forces are propelling change in a general direction, the specific forms it takes will depend on how women and men respond, individually and collectively, to the dilemmas they face. Those who have come of age during this period are adopting a growing diversity of moral orientations that defies dichotomous gender categories. Their experiences point to a growing desire for a social order in which women and men alike are afforded the opportunity to integrate the essential life tasks of achieving autonomy and caring for others.

Yet, persistent inequalities continue to pose dilemmas, especially for those who aspire to integrate home and work in a balanced, egalitarian way. To understand these processes, we need to focus on the social conditions that create such dilemmas and can transform, and potentially dissolve, the link between gender and moral responsibility. Of course, eradicating this link might only mean that women are allowed to adopt the moral strategies once reserved for men. We also need to discover how to enable everyone, regardless of gender, class, or family situation, to balance care of others with care of self.

The possibilities have never been greater for creating humanistic, rather than gendered, conceptions of moral obligation. New moral dilemmas have prompted women and men to develop innovative strategies, but the long-term resolution of these dilemmas depends on reorganizing our social institutions to foster gender equality and a better balance between family and work. Freud once commented that a healthy person is able "to love and to work." Achieving this vision depends on creating a healthy society, where all citizens are able to combine love and work in the ways they deem best.

References

Connell, R. W. 1987. *Gender and power.* Stanford, CA: Stanford University Press.

———. 1995. *Masculinities.* Berkeley and Los Angeles: University of California Press,

Crittenden, Danielle. 1999. *What our mothers didn't tell us: Why happiness eludes the modern woman.* New York: Simon & Schuster.

Epstein, Cynthia F. 1988. *Deceptive distinctions: Sex, gender and the social order.* New Haven, CT: Yale University Press.

Gerson, Kathleen. 1985. *Hard choices: How women decide about work, career, and motherhood.* Berkeley and Los Angeles: University of California Press.

———. 1993. *No man's land: Men's changing commitments to family and work.* New York: Basic Books.

———. 2001. Children of the gender revolution: Some theoretical questions and findings from the field. In *Restructuring work and the life course,* edited by Victor W. Marshall, Walter R. Heinz, Helga Krueger, and Anil Verma. Toronto, Canada: University of Toronto Press.

Hochschild, Arlie R. 1997. *The time bind: When work becomes home and home becomes work.* New York: Metropolitan Books.

Jacobs, Jerry A., and Kathleen Gerson. Forthcoming. *The time divide: Work, family, and social policy in a hurried era.* Cambridge. MA: Harvard University Press.

Lorber, Judith. 1994. *Paradoxes of gender.* New Haven, CT: Yale University Press.

Putnam, Robert. 2000. *Bowling alone: The collapse and revival of American community.* New York: Simon & Schuster.

Risman, Barbara J. 1998. *Gender vertigo: American families in transition.* Hew Haven, CT: Yale University Press.

Schor, Juliet. 1992. *The overworked American: The unexpected decline of leisure.* New York: Basic Books.

Webster's dictionary of the English language. 1992. Chicago: J. G. Ferguson.

West, Candace, and Don H. Zimmerman. 1987. Doing gender. *Gender & Society* 1:125–51.

⚜ Topics for Further Examination ⚜

- Check out the most recent research on women and work done by the Institute for Women's Policy Research at http://www.iwpr.org and the current activism underway at 9 to 5 National Association of Working Women at http://www.9to5.org/.
- Using the Web, find a list of the top executives in a sample of the largest firms in this country and calculate a gender ratio of women to men (Hint: Fortune 500 is one such list).
- Find information on work/family policies in different countries. You should be able find articles in an academic database, or just searching the Web.

8

Gender in Intimate Relationships

Although social institutions and organized activities such as work, religion, education, and leisure provide frameworks for our lives, it is the relationships within these activities that hold our lives together. What surprises many people is that these everyday relationships are patterned. We don't mean "the daily routine" kind of patterns, for we are referring to patterns across individuals and relationships related to gender. For example, sociologists consider the family to be more than just a personal relationship; they view it as a social institution, with relatively fixed roles and responsibilities that meet some basic needs in society such as caring for dependent members and providing emotional support for its members. As you read through this chapter, you will come to realize how social norms influence all gendered relationships, including intimate relationships. This introduction and the readings in this chapter illustrate two key points. First, gendered intimate relationships always evolve, often in response to social changes unrelated to the relationships themselves. Second, gender embeds itself in an idealized version of intimacy—the traditional family—that is not the reality in the United States and most parts of the world today, as illustrated in readings in Chapters 1 and 3.

Before going on about these details of intimacy, let's stop for a moment and look at relationships in general. The word *relationship* takes on many different meanings in our lives. We can have a relationship with the teller at our bank because he or she is usually there when we deposit our paycheck. We have relationships with our friends; some we may have known most of our lives, others we have met more recently. And, we have relationships with our family and with people who are like family. Then, there are those relationships that surround us with love, economic support, intimacy, and almost constant engagement. All of these relationships are shaped by gender. You have already read about relationships at work in Chapter 7; in this chapter you will read about how gender shapes more intimate relationships from friendships to partnering to parenting.

Gendered Relationships

Consider the impact of gender on our relationships. We can have many friends, with whom we love or share affection. In the past, researchers often argued that friendships varied by gender in predictable and somewhat stereotypical ways. That is, they described women's friendships as more intimate, or focused on sharing feelings and private matters, while describing men's friendships as more instrumental or focused around doing things. To illustrate, Cancian (1990) argued that men were more instrumental or task-oriented in their love relationships, whereas women expected emotional ties. For example, when asked how he expressed his love to his wife in Cancian's study, one man told the interviewer that he washed her car.

While Cancian's earlier study is revealing, it is important to remember that other social prisms influence the gendering of our relationships, putting social constraints and expectations on even our most intimate times. For example, Walker (1994) found that while both men and women hold stereotypical views of behavior in men's and women's friendships, actual friendship patterns were more complex and related to the social class of the individual. Even though they recognized what was gender appropriate behavior for friendships, working-class men tended to describe their friendships in ways that would be defined as more stereotypically female (disclosure and emotional intimacy), while professional women tended to describe more masculine friendship patterns with less reported intimacy. These exceptions to stereotypical views of gendered friendships, however, also are patterned; that is, they appear to vary by social class and reflect the constraints of work lives. Both readings by Shirley Hill and Karen Pyke in this chapter contribute to our understanding of how prisms of class and race influence gendered patterns in relationships.

Gender and Changing Living Arrangements

One of the strongest gendered influences on relationships is the expectation that men and women are supposed to fall in love and marry. As we have emphasized throughout this book, American culture assumes idealized intimate relationships to be heterosexual, accompanied by appropriate gendered behaviors, and, of course, based in nuclear families. You will notice that we did not include the word "family" in the title of this chapter. We did this because the stereotypical vision of family—mom, dad, two kids and a dog all living in a house behind a white picket fence—is only a small percentage of households today and never was the predominant form of relationships.

As we explore relationships in this chapter, it is important to begin by examining how households actually are patterned. In Table 8.1, we list the various household configurations and the percentage in each category in the United States today. These data may surprise you. Relationships in the United States are changing, as indicated by the diversity noted in Table 8.1. Only 53.1 percent of all Americans age fifteen or older were married in 2005 (U.S. Census Bureau, 2006), with almost half of the population living separate from a spouse. These percentages are much different for Blacks, with only 31.4 percent of Blacks married with a partner present and 70 percent of Black women not in a marital relationship. Clearly, these demographics influence the relationships of Black men and women which Shirley Hill describes in the reading in this chapter.

The increase in single and non-family households reflects both a trend toward postponing marriage and a longer life expectancy. A White boy born in 2003 can expect to live 75.3 years on average, and a White girl, 80.5 years (Arias, 2006). These life expectancies are shaped by other social factors such as race and social class as well as gender. For example, the average life expectancy for a Black boy born in 2003 is 69.0 years and a Black girl 76.1 years (Arias, 2006).

Increased life expectancy means we have more "time," thus postponing first marriage makes sense. The median age at first marriage has risen for both men and women. In 1980, the median age at first marriage for women was twenty-two and twenty-five for men, whereas from 2000 to 2003 the estimated median age at first marriage increased to twenty-five for women and twenty-seven for men (U.S. Census Bureau, 2005). Also,

Table 8.1 Household Composition in the United States, 2005

Both Sexes	*Percentage*			
	White Non-Hispanic	Black	Hispanic	Asian
Married				
Spouse Present	55.8	31.4	46.3	57.7
Spouse Absent	1.1	2.2	3.2	3.1
Widowed	6.6	6.2	3.5	4.0
Divorced	10.1	10.9	7.4	5.0
Separated	1.5	4.4	3.6	1.5
Never Married	25.0	45.0	36.0	28.5
Women Only				
	White Non-Hispanic	Black	Hispanic	Asian
Married				
Spouse Present	54.0	27.9	47.9	59.0
Spouse Absent	1.0	2.2	2.0	2.9
Widowed	10.3	9.2	5.7	6.4
Divorced	11.2	12.5	9.0	6.2
Separated	1.7	4.8	4.7	1.8
Never Married	21.7	43.4	30.6	23.8
Men Only				
	White Non-Hispanic	Black	Hispanic	Asian
Married				
Spouse Present	57.7	35.6	44.7	56.4
Spouse Absent	1.1	2.2	4.4	3.4
Widowed	2.7	2.5	1.4	1.5
Divorced	8.8	9.0	5.9	3.8
Separated	1.3	3.8	2.5	1.3
Never Married	28.4	46.9	41.1	33.7

U.S. Census Bureau, 2005

because we live longer, we may be less inclined to stay in a bad relationship, since it could last for a very long time. However, there are clear differences in the patterning of marriage across the race and ethnic groups described in Table 8.1, clearly reflecting the influence of other "prisms" in constructing relationships—points made clear in the Hill reading in this chapter.

The "Idealized" Family

The growing diversity of household and family configurations in the United States has reshaped and challenged the rigid gender roles that pattern the ways we enter, confirm, maintain, and envision long-term intimate relationships. Not surprisingly, the rigid gender roles associated with that mythical little home behind the white picket fence with mom staying at home to care for children and dad heading off to work is not the reality for households in the United States. Instead, we have many different household patterns, with some single head of households, and others, unrelated by blood or family bonds, living together. Some new household patterns arise from choice, including an increased tolerance for lifestyle diversity, such as single-person households. Others, such as single parents (typically mothers) who raise children alone, often arise out of divorce and poverty. Whatever the reasons, the idealized, traditional family with its traditional gendered relationships never was (Coontz, 2000) and certainly is not now, the norm in American households.

To illustrate how rare that idealized family is, only 44.6 percent of all married couples (from 2000 to 2003) were living with their own children under eighteen years of age (U. S. Census Bureau, 2006). Furthermore, the idealized image of the father at work and the mother and the children at home (Coontz, 2000) comprises a very small percent of all households in the United States. In 2005, of the 13,970,000 couples with children under age eighteen, just 1.5 percent indicated that only the father was employed (U.S. Census Bureau, 2006). Over half of the women with children under the age of six are in the labor force (62.2 percent of all women), with considerable variation by race (61.1 percent of White women, 89.6 percent of Black women, 57.9 percent of Asian women, and 51.0 percent of Hispanic women [U. S. Department of Labor, 2005]). When you consider that most women are employed, only a small percentage of households fit the "Ozzie and Harriet" model for families, with mom at home taking care of the children. Like gender, "the family" is a culturally constructed concept that often bears little resemblance to reality. The idealized, traditional family with separate and distinct gender roles does not exist in most people's lives.

Historically, families changed considerably in terms of how they are formed and how they function. While enduring relationships typically involve affection, economic support, and concern for others, marriage vows of commitment are constructed around love in the United States today. In previous generations, most marriage vows promised commitment and love "until death do us part." Marriages in the 1800s, even when rooted in love, often were based on economic realities. These nineteenth century marriages were likely to evolve into fixed roles for men and women linked to the economy of the time, roles that reinforced gender difference but not necessarily gender inequality. For example, farm families developed patterns that included different, but not always unequal, roles for men and women (Smith, 1987). In the latter half of the nineteenth century and into the early 1900s, families—particularly immigrant families—worked together to earn enough for survival. Women often worked in the home, doing laundry and/or taking in boarders, and many children worked in the factories (Bose, 2001; Smith, 1987).

What changed and how did the current idealized roles of men and women within the stereotypical traditional family come to be? Martha May (1982) argues that the father as primary wage earner was a product of early industrialization in the United States. She notes that unions introduced the idea of a family wage in the 1830s to try to give men enough wages so their wives and children were not forced to work. In the early 1900s, Henry Ford expanded this idea and developed a plan to pay his male workers $5.00 per day if their wives did not work for money (May, 1982). The Ford Motor Company then hired sociologists to go into workers' homes to make sure that the wives were not working for pay either outside or inside the home (i.e., taking boarders or doing laundry) before paying this family wage to male workers (May, 1982). In fact, very few men actually were paid the higher wage (May, 1982). You might ask, why was Ford Motor Company so interested in supporting the family with an adequate wage? At that time, factories faced high turnover because work on these first assembly lines was unpleasant, paid very little, and the job was much more rigid and unpleasant than the farm work that most workers were accustomed to. Ford enacted this policy to reduce this turnover and lessen the threat of unionization. Thus, one reason behind the social construction of the "ideal family" was capitalist motivation to tie men to their jobs for increased profits, not individual men wanting to control their families (May, 1982).

These historical and structural changes in the family affect our interpersonal relationships both in and outside of marriage. A key factor in elationships is power, as Veronica Tichenor discusses in this chapter. She examines different explanations for the gendered distribution of power in relationships and proposes a new way to think about power.

Same-Gender Couples

A change that is hidden in Table 8.1 is the growing number of same-gender couples. In the 2000 Census, a category "unmarried partner" was added to the questionnaire, allowing a count of same-gender couples. With this change, one percent of the total households in Table 8.1 were identified as same-gender partner households, while 8.1 percent of the total households were opposite-gender partner households (U.S. Census Bureau, 2003). The legal parameters surrounding same-gender households are still changing in the United States, although the possibility of marriage for same-gender couples has met considerable resistance. Other countries, such as the Netherlands, England, Germany, and Canada allow same-gender couples to legally marry. Even if they cannot do so legally in the United States many lesbian and gay couples choose to profess their love and commitment formally in ceremonies that are both similar and different from traditional heterosexual weddings. Same-gender couples also create families and redefine gender, parenthood, and family as Gillian Dunne describes in her article in this chapter. Because gender cannot pattern parenting relationships in same-gender couples, Dunne's article helps us to understand the constraints and concessions needed to parent children.

Feminization and Juvenilization of Poverty

Another reality of today's families that differs from the idealized, traditional family is that 38.9 percent of children under the age of nineteen were at or below 200 percent of the poverty level in 2003 to 2005 (U. S. Census Bureau, 2005). Many of these children are living with single parents, primarily their mothers. This change in household composition challenges the traditional gendered relationships expected in families, but also relates to social class inequalities tied to gender. The increase in the number of households headed by poor women raising children has been called the feminization or juvenilization of poverty (Bianchi, 1999; McLanahan & Kelly, 1999). Diana Pearce coined the term, "the feminization of poverty" in 1978 (Bianchi, 1999, p. 308), at a time when the number of poor women-headed families rapidly increased. The rate of women's poverty relative to men's fluctuates over time and is about 50 to 60 percent higher (Bianchi, 1999, p. 311), a reflection of the inequality in wages for men and women discussed in the last chapter.

The juvenilization of poverty refers to an increase in the poverty rates for children that began in the early 1980s, whether in single or two-parent families (Bianchi, 1999). The Urban Institute (2001) published 1997 data indicating that only 47 percent of children from low-income households live with two parents, whereas 75 percent of children from families with higher incomes do so. Children's poverty rates in the 1980s and 1990s have stayed above 20 percent; during the 1980s and 1990s, children were 20 to 23 percent of those officially categorized as poor in the United States (Bianchi, 1999, pp. 314–315). The feminization and juvenilization of poverty are serious problems and both are gender as well as race and social class issues.

Government Policies and Family Relationships

Governmental policies play a role in shaping families in the form of tax laws, health and safety rules, and other legislation. A multitude of policies frame parents' decisions to have children and shape how children are expected to act in societies. Early in the twentieth century, health and safety laws increased the age of employment for factory workers (Bose, 2001) and are still in effect in terms of what age a child can begin to work for pay. These laws related to the creation of adolescence, with the expectation that children would remain in school throughout high school. Another act of legislation, the dependent deduction on tax returns was first intended to encourage families to have more children at a time when politicians worried about the declining birth rate in this country. At the same time and for the same reason, Canada instituted a policy that is still in place in which women are given a payment each month for every child under the age of eighteen. The government makes this payment directly to the mother, based on the assumption that it is the mother in a two-parent family who assumes responsibility for the children. The forms, however, allow for the payment to go to the father, if written documentation is submitted to indicate that he is the primary caregiver. While only a nominal sum, it was distributed across social classes as an incentive to bear and raise children. Canada also recently instituted the Universal Child Care Benefit, which provides $100 per month for child care for each child under age six in 2007.

Policies and laws affect all relationships because they often idealize the women's role in two-parent families and reinforce hegemonic masculinity and emphasized femininity, while ignoring other choices and life situations, such as same-gender couples. The impact of policies and practices on relationships in poor families is also considerable, making it difficult to "do" idealized gender because living, in and of itself, is challenging, as illustrated in the reading by Karen Pyke.

Changing Relationships

Heterosexual, marital, and parenting relationships have changed considerably over time. In addition to government regulations, the feminization of the workforce has affected men's and women's roles in marriages (Blackwelder, 1997). The need for middle-class families to have two workers is intricately connected to the economic structure of the country and has changed considerably relationships in the home. As Kathleen Gerson discusses in her reading in Chapter 7, couples have started to modify expectations for marriage and marital responsibilities in response to changes in educational attainment and workforce participation of women. Gerson describes the dilemmas faced, and strategies tried, by young people as they attempt to balance commitment and autonomy in their lives. Parenting has thus been transformed and both men and women must respond to the constraints on their time as they consider the idealized gender roles of mothers and fathers. Scott Coltrane, Ross Parke, and Michele Adams discuss the relationship between gender stereotypes and the reality of meeting children's and household's needs for low-income Mexican American men in their article in this chapter.

Gender continues to permeate and frame ever-evolving relationships even though fathers are more involved in household labor, particularly caring for children (Bianchi, Robinson, & Milkie, 2006). However, while there may be more equal distribution of household labor today, mothers continue to feel the pressure of caring for children and the household (Bianchi et al., 2006). Thus, the idealized, traditional gendered responsibility for women in the United States to care for children and the household remains, even though most women work outside of the home. Jacobs and Gerson (2001) argue that the changes in family composition and gender relations have created situations in which members of families, particularly women and most particularly single women, are overworked with little free time left for themselves or their families. Jacobs and Gerson (2001) describe the situation as particularly acute for those couples whose work weeks are 100 hours or more and who tend to be highly educated men and women with prestigious jobs. However, as described by Pyke, relationships across social class face different constraints. Young men and women face many choices in terms of how they will form relationships, including the moral dilemma between commitment and autonomy that Gerson discussed in the last chapter.

The readings in this chapter provide a fuller understanding of how our most intimate relationships are socially constructed around gender. As friends, lovers, parents, and siblings, we are defined in many ways by our gender. Compare, for instance, the dilemmas young people face in Gerson's reading in the last chapter to the realities of the couples in the selections by Veronica Tichenor, Karen Pyke, Shirley Hill, Scott Coltrane et al., and Gillian Dunne. Ask yourself what choices you have made or wish to make as you consider how gender influences what you expect in your intimate relationships.

References

Arias, E. (2006). United States life tables, 2003. *National Vital Statistics Reports.* 54 (14). Retrieved January 25, 2007, from http://www.edc.gov/nchs/data/nvsr/nvsr54/nvsr54_14.pdf

Bianchi, S. M. (1999). Feminization and juvenilization of poverty: Trends, relative risks, causes, and consequences. *Annual Review of Sociology, 25,* 307–333.

Bianchi, S. M., Robinson, J. P., & Milkie, M. A. (2006). *Changing rhythms of American family life.* New York: Russell Sage Foundation.

Blackwelder, J. K. (1997). *Now hiring: The feminization of work in the United States, 1900–1995.* College Station: Texas A&M University Press.

Bose, C. E. (2001). *Women in 1900: Gateway to the political economy of the 20th century.* Philadelphia: Temple University Press.

Cancian, F. M. (1990). The feminization of love. In C. Carlson (Ed.), *Perspectives on the family: History, class and feminism* (pp. 171–185). Belmont, CA: Wadsworth.

Clark, G. (1999). Mothering, work, and gender in urban Asante ideology and practice. *American Anthropologist, 101*(4), 717–729.

Coontz, S. (2000). *The way we never were: American families and the nostalgia trap.* New York: Basic Books.

Freiberg, P. (1999). *Gay couples identify as married.* Retrieved June 6, 2002, from http://www.gfn.com/archives/story/phtml?sid=1259

Gray, J. (1992). *Men are from Mars, women are from Venus: A practical guide for improving communication and getting what you want in your relationships.* New York: Harper Collins.

Jacobs, J. A., & Gerson, K. (2001). Overworked individuals or overworked families? Explaining trends in work, leisure, and family time. *Work and Occupations, 28*(1), 40–63.

May, M. (1982). The historical problem of the family wage: The Ford Motor Company and the five dollar day. *Feminist Studies, 8* (2), 399–424.

McLanahan, S. S., & Kelly, E. L. (1999). The feminization of poverty: Past and future. In J. S. Chafetz (Ed.), *Handbook of the sociology of gender* (pp. 127–145). New York: Kluwer Academic/Plenum Publishers.

Smith, D. (1987). Women's inequality and the family. In N. Gerstel & H. E. Gross (Eds.), *Families and work* (pp. 23–54). Philadelphia: Temple University Press.

U.S. Census Bureau. (2003). *Married-couple and unmarried-partner hHouseholds: 2000.* Retrieved January 28, 2007 from http://www.census.gov

U.S. Census Bureau. (2005). *Indicators of marriage and fertility in the United States from the American Community Survey: 2000 to 2003.* Retrieved January 25, 2007, from http://www.census.gov/population/www/socdemo/fertility/mar-fert-slides.html

U.S. Census Bureau. (2006). *America's families and living arrangements.* Retrieved January 23, 2007, from http://www.census.gov/population/www/socdemo/hh-fam/cps2005.html

U.S. Census Bureau. (2007). *Health insurance: Low-income uninsured children by state: 2003, 2004, and 2005.* Retrieved January 28, 2007, from http://www.census.gov/hhes/www/hlthins/liuc05.html

U.S. Department of Labor Bureau of Labor Statistics. (2005). *Women in the labor force: A databook.* Retrieved January 25, 2007, from http://www.bls.gov/cps/wlf-databook2005.htm

Walker, K. (1994). Men, women, and friendship: What they say and what they do. *Gender & Society 8*(2), 246–265.

Introduction to Reading 37

In this selection, Veronica Tichenor, a sociologist, develops a theoretical critique of previous explanations for power distributions in marriage. She uses this critique to argue for the development of a new conceptualization of power and a different way of explaining why power is gendered in heterosexual marriages. The gendered distribution of power in relationships can shape virtually every aspect of our daily lives. This reading makes us think about where power in relationships comes from and how power is used in relationships that are supposed to be based on love. So, as you read this article, think about the theories explaining power discussed herein and ask why men and women might accept gendered power in our relationships and why such relationships are so difficult to change.

1. Compare some of the explanations Tichenor gives for gender and power in marriage to the theories discussed in the Introduction to this book.
2. If you accept Tichenor's explanation of gendered power as hidden power, what would we have to change to create equality in heterosexual relationships?
3. Think about people you know. What does a gendered power relationship look like? What does an equal power relationship look like?

Thinking About Gender and Power in Marriage

Veronica Jaris Tichenor

The balance of power in most marriages reflects the ideology of separate spheres in the conventional marital contract. Of course, this contract is not a written document; it consists of cultural understandings of the reciprocal rights and obligations that each spouse has within the institution of marriage. According to this unwritten contract, these rights and obligations are divided along gender lines, which construct men as breadwinners and women as mothers and homemakers. The man's main responsibilities are to provide for the family economically and to represent the family to the community or the world at large. The woman's main responsibility is to care for the home, husband, and children. If spouses hold up their end of the bargain, this exchange is considered both reasonable and fair.

While this model may seem overstated and outdated, Joan Williams (2000) argues that the basic assumptions of this contract persist. This complementary organization of market work and family life exists as a system that Williams calls "domesticity." Under this system, men are entitled and encouraged to perform as "ideal workers" in the marketplace, unencumbered by the demands of family life. Women, whether engaged in paid labor or not, are marginalized in the workplace by their domestic responsibilities. They continue to be seen and treated by employers as mothers or potential mothers, which limits their options and opportunities at work. That women maintain responsibility for domestic labor and child care hampers their ability to engage in paid labor as ideal workers (i.e., men). So while the assumption that women will be engaged only in domestic labor has changed in recent years, the underlying contract that delegates breadwinning responsibility to men and domestic responsibility to women remains largely unchallenged.

The conventional marital contract does not simply divide responsibilities between spouses; it also reinforces men's power within marriage. This is because the responsibilities and tasks of husbands and wives are valued differently. Within most U.S. families, the income that the husband earns is the most highly valued asset. It confers a higher status on the husband, both within and outside the relationship, and has been used to justify men's greater power in marriage, especially in terms of decision-making practices and control over the family's financial resources. Historically, men have wielded power based on their greater incomes, and wives were expected to defer to their husbands' authority. By contrast, women's caring work at home has not been accorded the same status as breadwinning. That it is unpaid work signifies its lesser value, and the ability to refuse to do such work is one of the privileges men typically enjoy in marriage (Hochschild 1989 [also England in Chapter 7]). The conventional marital contract, then, underscores the greater value of the man's contributions (income), while devaluing those of the woman (domestic services). In short, the bargain implied by the conventional marital contract is the key to continued gender inequality in marriage (Williams 2000).

Admittedly, life has changed dramatically for married couples, especially in the last several decades. This model of husband as sole breadwinner and wife as homemaker describes the reality of only 25 percent of married couples in the United States today (Raley et al. 2003). Transformations in the economy have made it impossible for all but a comparative handful of families to enjoy a moderate standard of living on only one income. This makes it tempting to think of the man-as-breadwinner/woman-as-homemaker model of married life as outdated and irrelevant.

However, marriages are still constructed against the backdrop of the conventional marital contract. Culturally, we still hold men accountable for breadwinning and women for mothering, regardless of whatever additional responsibilities they may take on. In most circles, men are still revered and respected based on the kind of work they perform and the standard of living their families are able to enjoy because of it. Remember that most of the men with higher-earning wives profiled in recent news articles felt like outcasts or failures for not being the major earners. Women are still expected to keep neat homes and present clean, well-adjusted, and well-mannered children to the outside world. Poorly behaved children might still be asked, "Didn't your mother teach you any better than that?"

This means that men and women get more "credit," both inside and outside the marital relationship, for engaging in activities that are consistent with conventional gender identities. While a wife's income may be important to the family, her employment lacks the social legitimacy accorded her husband's work. Women's paid work is typically seen as an option, rather than a duty. Since social convention does not obligate a woman to provide for her family, she is not protected from housework or other domestic intrusions on her breadwinning activities as a man would be. Women typically retain responsibility for the household and simply add the role of worker onto those of mother and homemaker (Hochschild 1989; Rubin 1994). Women also receive less social approval than do men for engaging in paid work and may even face condemnation for "neglecting" domestic duties (Popenoe 1989). Similarly, while men may help out with the workload at home, and in middle- to upper-middle-class circles might receive a great deal of social approval for doing so, their domestic labor is not a substitute for breadwinning; even the most involved father rarely opts out of providing altogether (Coltrane 1996; Deutsch 1999). In short, the meanings attached to paid work and domestic labor are fundamentally different for men and women, and tend to reinforce the identities of breadwinner and mother/homemaker embedded in the conventional marital contract.

The continued distinction between mothering and breadwinning as gendered activities means that we can think of these identities enduring "gender boundaries"—ways to mark the difference between women and men (Connell 1987; Potuchek 1997). While women may work outside the home, men still have the responsibility to provide that makes them breadwinners. Though men may help with housework or child care, it is still a woman's duty to provide the level of attention and care associated with mothering. The lines dividing these gendered responsibilities are still clearly drawn (Williams 2000).

If it is true that breadwinning is the central identity for men, and mothering is the central identity for women, this could be problematic for dual-earner couples. If wives are also providing an income, what distinguishes them from their husbands? In most dual-earner couples, men still outearn their wives, often by a large margin (Raley et al. 2003). Couples typically respond to this shift in behavior by thinking of husbands as the primary breadwinners, with wives as secondary earners simply "helping out" (Potuchek 1997; Willinger 1993). Similarly, couples see women's mothering and domestic responsibilities as primary, and their work commitments are often organized around the needs of the family (Hochschild 1989; Williams 2000). In this way, men can see themselves as still meeting the masculine imperative of providing for their families, and wives can see themselves as good mothers, despite being employed outside the home (Coltrane 1996).

Of course, having both spouses in the workforce could represent an opportunity to change the gendered expectations and meanings surrounding breadwinning by rejecting the idea of separate spheres embedded in the conventional marital contract. Husbands and wives could think of themselves as co-providers with a joint responsibility to meet the financial obligations of the family. They could then share the responsibilities for maintaining a clean, orderly home and raising healthy children. Sharing all family work (both paid and unpaid) more equally could break down these rigid gender boundaries.

We know that some couples have worked successfully to erode these boundaries (Coltrane 1996; Deutsch 1999; Risman and Johnson-Sumerford 1998; Schwartz 1994). These couples consciously share the work of providing and caring for a family in ways that begin to undermine the breadwinner/mother identities, as well as the power imbalance associated with them. However, even some of these partnerships contain rumblings of gender unease. For example, men and women in these relationships often collaborate to maintain some gender specialization; women want to guard part of the domestic domain as their own (Hertz 1989) or want to feel like "I'm still the mom" (Coltrane 1996; Deutsch 1999), and men still think of providing as their own responsibility (Wilkie 1993). Such expectations are so strong that even couples with higher-earning wives continue to cling to them (Brennan, Barnett, and Gareis 2001). These results suggest that spouses are often more comfortable with a certain level of conventional gender asymmetry in their relationships.

Williams (2000) describes this pull toward the conventional as being caught in a "gender force field." While conventional gender expectations do not determine behavior, they can exert a strong pull that can be difficult to resist and can wear people down over time. For example, after couples have their first child, the call of traditional roles and expectations can be particularly loud. With a new life depending on them, men often feel a greater need to be a good provider, and even women who had planned to continue working after their child's birth can feel unexpectedly drawn toward staying at home (Cowan and Cowan 1992; Rexroat and Shehan 1987).

In other words, even though spouses' behavior is changing, as women continue to be a strong presence in the workforce and some men become more engaged in domestic labor, it may be too threatening to give up all their conventional gender expectations—it does not feel right. This means that as men and women are engaged in similar activities, such as providing for the family, the breadwinning and mothering boundaries can take on great importance; that is, these boundaries become a crucial way for husbands and wives to create and maintain a sense of gender difference (Potuchek 1997).

So even though the conventional marital contract no longer describes the reality of most U.S. couples, by maintaining the gender boundaries of mothering and breadwinning, dual-earner couples reinforce the bargain implied by the old contract (Brennan et al. 2001; Coltrane 1996; Potuchek 1997; Wilkie 1993). This finding is significant because of the power dynamics embedded in the contract. As we have seen, the activities associated with mothering and breadwinning are differentially valued, with breadwinning generally conferring more privileges than does mothering. If employed women are not defined as breadwinners, they may lose access to these privileges. In other words, by maintaining the gender boundaries written into the conventional marital contract, spouses may undercut women's power within marriage.

However, all of this assumes that men's power within marriage is truly rooted in their greater economic resources. While this assumption has driven much of the research on marital power, the accumulating evidence suggests that it is flawed. If money is the key to the power dynamics within marriage, we would expect the balance of power to shift as women have begun to share breadwinning with their husbands. In fact, earning an income has done little to increase women's power, which undermines the fair exchange of income earned by male breadwinners for the domestic services of their wives implied by the conventional marital contract. This means that we need to rethink the money/power link within marriage.

Marital Power as the Exchange of Resources

Early attempts to talk about the balance of power within marriages rested on the assumptions embedded in the conventional marital contract. This research (beginning with Blood and Wolfe 1960) was driven by resource and exchange theories that link the balance of power in marriages to the relative contributions, or resources, of spouses. Resources are anything of value, tangible or intangible, that partners bring to a relationship. They include money, occupational or social status, education, love and affection, physical attractiveness, special knowledge or expertise, services (such as performing domestic labor or giving back rubs), and so on. Under the conventional marital contract, men contribute their incomes, as well as the status attached to their occupations, in exchange for domestic labor and child-care services from their wives.

The resource/exchange model views power as the ability to prevail in a variety of household decisions, ranging from how much to spend each week on groceries to when and if the family should move. Since men and women both reported that husbands had more control over most decisions, Blood and Wolfe concluded that husbands had more power in their marriages. They also concluded that this power came, not from the influence of patriarchal ideology, but from husbands' contributing the more socially valued resources (income and status) to the marriage. Thus, resources such as income and status represent the potential for exercising power. And while both spouses have access to some resources, men have more power in marriages because they contribute the more important resources to the relationship.

This logic is compelling and has held sway both inside and outside the academy: The more you give, the more you should receive in return. However, if this conceptualization of power within marriage were accurate, we would expect to see a shift in the balance of power between spouses over the last several decades as women have moved into the paid labor force in great numbers. According to resource and exchange theory assumptions, women who contribute economically to the relationship should be able to exercise greater control over finances and decision making, and buy a certain amount of relief from domestic labor and childrearing responsibilities.

However, the marital power literature over the past few decades demonstrates that this is not happening (see, for example, Bianchi et a1. 2000; Blumberg 1984; Blumstein and Schwartz 1983, 1991; Hochschild 1989; Perry-Jenkins and Folk 1994; Pleck 1985; Wright et a1. 1992). Women may gain a greater measure of control over finances or household decisions, but few couples report patterns that could be characterized as egalitarian. Husbands continue to exercise greater control in financial matters and decision making.

Similarly, women's employment has done little to alter the division of domestic labor. Husbands may help a little more with household chores and child care, but much of the research argues that these changes reflect shifts in the proportion of work being done by each spouse. In other words, it looks like men are doing more because women, who are now further crunched for time, are doing less. . . . It is clear that merely earning a wage does not significantly enhance a woman's power in most marriages.

As we have said, one reason for this continued imbalance of power may be that men typically outearn their wives, often by a large margin. This income advantage continues to lend legitimacy to the husband's authority within the marriage. He may not be earning all the money, but he is still earning most of it. This circumstance may allow spouses to continue to think of their economic assets as largely his and to justify his continued control over them. However, this is only a partial explanation for the enduring imbalance of power within marriage. If women's income buys them so little, then power is not about money—or at least, not entirely about money. Gender is also a factor.

Two examples from studies of marital power suggest that this is the case. The first example comes from Blood and Wolfe's (1960) work. One of their most interesting findings, given their reliance on resource and exchange theory, is that the wives in their sample who worked outside the home full time got the least amount of help with domestic labor from their husbands. It is not just that these wives could not exchange their income for more help with domestic labor from their husbands, but that they got the worst deal overall of any group of wives when they were contributing the most (in terms of paid and domestic labor) to their relationships. The second example comes from Arlie Hochschild's *The Second Shift* (1989). Hochschild reported that, while substantial sharing of domestic duties was not common in her sample of dual-earner couples, among couples where women earned more than their husbands, none of the men shared the housework and child-care duties. These results directly challenge the exchange of resources implied by the conventional marital contract, since these wives got no credit for the substantial incomes they contributed.

More recent quantitative research has produced results similar to Hochschild's and demonstrates that there is a curvilinear relationship between income contributed and the amount of housework each spouse performs (Brines 1994; Greenstein 2000; Bittman et al. 2003). Husbands who are sole (or major) breadwinners successfully trade their income for domestic labor, but wives who are the major breadwinners in their families are unable to negotiate a similar deal. That is, wives who earn all a family's income perform about the same amount of housework as wives who earn no income at all. Their husbands seem to receive domestic services, rather than to compensate for their wives' unusually high earnings by taking on more household labor. Their wives' earnings disrupt a balance of power that feels culturally right, and either these men attempt to restore that balance by asserting their right as men to their wives' domestic labor, or wives take on more household work voluntarily to avoid further assaulting their husbands' masculinity. Couples engage in "gender display" (Brines) or "deviance neutralization" (Greenstein) to restore a sense that spouses are meeting their conventional obligations. However, the exact dynamic by which gender overrides the money-equals-power equation among these couples is unclear.

That higher-earning wives cannot trade their income for a reduction in their domestic labor burden undermines the theoretical assumptions of the bulk of research on power in marriages. These results demonstrate that men's power in marriage does not come from their income or their role as (primary) breadwinner—or at least, it does not come from these resources alone. Husbands in dual-earner families retain and enjoy some rights or privileges as men. Thus, it makes sense to talk about gender as exerting an influence on marital power dynamics that is independent of income earned by spouses. We can then think of gender as a separate structure that shapes the balance of power within marriage.

Gender as Structure

Because gender so thoroughly pervades social life, it is often conceptualized as a cultural dynamic that is "woven into" other institutions, meaning that beliefs about gender and gender differences are used to maintain and justify other social practices. However, gender also exists as a separate entity (Lorber 1994). That is, while gender is indeed embedded in and shapes the practices of other social institutions, it also exerts an influence that is separable from all other institutions. In the case of marital power, men's ability to retain their control and privileges, even in the absence of the economic dominance that has legitimated this power advantage, suggests that gender exerts an influence on marital power dynamics that is distinct from men's successful enactment of breadwinner responsibilities. In short, that men retain their power advantages in these circumstances points to gender as both a separate and a stable entity.

This means that we can talk about gender as structure (Risman 1998). This structure exists and operates on multiple levels within social life: institutional, interactional, and individual. At the institutional level, gender exists as beliefs about what men and women are or should be and as organizational practices that serve to reinforce these beliefs. The typing of women and men into particular occupations, the gap in wages between men and women, and the glass ceiling in organizations are all examples of gendered organizational practices. But gender at the institutional level also exists as ideology. Beliefs that men should be stronger and rational and women weaker and emotional are part of conventional gender ideology. The conventional marital contract, with the expectation that men are breadwinners and women are mothers and homemakers, is also part of gendered ideology. These institutional-level beliefs and practices organize the behavior of men and women at the interactional and individual levels.

At the interactional level, gender shapes face-to-face communication. That is, we interact with others as men and women, drawing on the cultural expectations for behavior that exist at the institutional level. Perhaps our best conceptualization of how the gender structure operates at the interactional level comes from West and Zimmerman (1987). They argue that men and women "do gender" as they interact with others in ordinary settings. Within a given context, individuals must clearly demonstrate to others that they are appropriately masculine or feminine. Men and women are aware of the gendered expectations for dress, speech, and behavior that exist at the institutional level, and they manage their conduct in light of the possibility that they may be held accountable to these standards by others.

For example, aside from cassocks or ceremonial kilts, men in Western societies do not wear skirts, but not because of any inherent property of the garment. Skirts are actually quite practical in terms of allowing freedom of movement and can be more comfortable than pants, particularly in warm weather when they generate their own breeze. However, since skirts have been successfully typed as women's clothing, no self-respecting (conventional) male would be caught dead in one. Those rare males who choose to adopt this style of dress risk social sanctions ranging from disapproval to ridicule to physical violence. They also risk being labeled inappropriately feminine and, therefore, not men. This example demonstrates that although conventional gender expectations may not completely determine behavior, even the smallest rules can be quite compelling. So while men and women do not always conform to the expectations for their sex categories, all know that deviations may come at a cost.

While "doing gender" encompasses a wide range of behaviors, the primary activities for men and women are the doing of dominance and submission (Berk 1985; West and Zimmerman 1987). This means that gender differences are not neutral but tied to larger power structures. At the institutional level, men enjoy greater economic and political power. At the interactional level, men assert their authority and women defer to this authority. This gender imperative is particularly salient in the context of heterosexual love relationships and marriage. Cultural notions of a man as "head of the household" or "king of his castle" continue to resonate, even if more subtly than in the past. Correspondingly, being a wife has typically entailed a certain level of service to one's husband. All spouses have to negotiate their relationships against the backdrop of these expectations, whether in congruence with or in opposition to them. That is, while spouses may choose to challenge conventional expectations or practices, others may still hold spouses accountable to conventional standards. These conventional assumptions regarding gender continue to shape the interactions of spouses and remain a central part of how men and women think about themselves as husbands and wives.

Gender is also a fundamental component of identity construction; it is impossible to think of ourselves separately from our identity as a man or woman. At the individual level, the gender structure constrains men and women as they attempt to construct meaningful identities. That is, doing gender is an internal process as well (Acker 1992). Individuals often hold themselves accountable to conventional conceptions of gender-appropriate behavior, regardless of the standards imposed by those around them. That is, it is important to feel that one is behaving in a way consistent with one's identity as a man or woman.

The research reviewed thus far suggests that, within marriage, the gendered identities of breadwinner and mother may still resonate for spouses and represent an important touchstone for constructing individual identities. Still, for husbands and wives, this identity construction occurs in the context of the couple; doing gender is a team performance. One spouse's failure to engage his or her part appropriately may reflect negatively on the partner. For example, having a wife who makes substantially more money may represent a significant threat to a man's gender identity, given that breadwinning is such a fundamental component of masculinity in U.S. culture, and the couple must find a way to manage this tension. Spouses must construct gender identities in tandem to find a balance that feels right to them, both as a couple and as individuals.

Although we can conceptualize these three levels of the gender structure as distinct from one another, they operate simultaneously and reinforce each other. While institutional practices and ideologies shape microlevel behavior, behavior on the interactional and individual levels has an impact on gender at the institutional level, for it is through microlevel dynamics that the larger gender structure is either challenged or reproduced. For example, at the microlevel, couples with higher-earning wives could represent a potential site of gender change. They could challenge the conventional link between breadwinning and power by sharing domestic labor, decision-making power, and the responsibility to provide equally. They could rewrite the gender scripts of the conventional marital contract and expand the possibilities for both men and women in marriage.

In spite of this opportunity, it seems that women's incomes have bought them little in their marriages because the gender structure assures men certain privileges within the marital relationship. As we have seen, the gender structure seems to have accommodated women's paid labor by constructing men as primary breadwinners, and therefore still due the privileges attached to this activity. However, couples with higher-earning wives present a more serious challenge to the gender structure. These couples disrupt the cultural link between gender, money, and power more profoundly and create new tensions for spouses to manage. These couples' efforts to preserve men's authority and interpersonal dominance despite women's economic advantage highlight the difficulty of rewriting conventional gender scripts and demonstrate the resilience of the gender structure in the face of potential challenges to it.

Conceptual Issues in Analyzing Power

Love: What's Power Got to Do With It?

Conducting research on marital power raises some sticky questions, because we are not used to thinking that power operates in our most intimate relationships. In U.S. culture, marriage is supposed to be based on romantic love. This notion requires that our relationships be ruled by our hearts and emotions, as in "I'm so crazy about her" or "He just swept me off my feet." This overpowering emotion often puts us beyond the reach of reason. We do not rationally calculate whether we should be in love with someone; we act largely on the basis of feelings. In fact, thinking rationally about a relationship, or weighing its pros and cons, opens us up to the charge that we are not really in love.

Being involved in a romantic relationship also means being focused on the other. That is, love requires a certain degree of altruism. One's personal desires cannot always be primary. Romantic love often requires putting the needs of one's partner first to make her or him happy or even to preserve the relationship. This selfless giving is often the standard used to assess the nature of our feelings. Only when we can place the interests of the other before our own is our caring and commitment seen as genuine.

This emphasis on affection and altruism leaves little room for power considerations in a love relationship. In fact, Western culture sets up love and power as opposites. If love means a denial of self for the sake of the other, power implies a calculation of one's rational self-interest above the interests of the other. Power means forcing another to do something she or he would prefer not to do, or even taking advantage of another for personal gain. Power considerations, then, seem anathema to the kind of blissful relationship idealized by our cultural emphasis on romantic love.

It is the equation of power with this kind of domination that makes us reluctant to admit that power plays a role in our love relationships. Few people want to think of having power over (or being subordinate to) their beloved. However, power is not simply domination; power also refers to autonomy—the ability to act according to one's wishes and desires. Using this conceptualization of power, the need for love and the need for power are no longer mutually exclusive dynamics in a love relationship but are intimately connected. Great acts of love depend on a free and autonomous self capable of both feeling and action. In this way, love and power are both fundamental dynamics in any healthy relationship (Nyberg 1981).

Even if we think of power as autonomy rather than domination, we can see how power and love exist in tension with one another in any relationship. Fulfilling one's own desires can sometimes mean thwarting those of one's partner, raising important questions, such as, How much should I give up to make my partner happy? and What do I have a right to expect from my partner in return? Conflicting desires or needs between spouses mean that at some point one person's autonomy will be sacrificed to meet the desires of the other. Often these issues exist on an unconscious level and surface in a relationship only in the context of "fairness," as in, Why do *I* have to do all the cleaning? or Why do you always get your way? Despite the clear power implications of these issues, they are rarely framed as evidence of differential power between spouses.

Conceptualizing and Measuring Power Within Marriage

The widespread cultural denial of the presence of power in love relationships makes it difficult to conceptualize and measure power within marriage. Most obviously, one spouse's exerting his or her will over the other (overt power) gives us a concrete way to examine the balance of power in a relationship, and much of the work on marital power has taken this factor as its starting point. Spouses have been asked to report on how much "say" each spouse has in a wide range of decisions commonly made by married couples. One's ability to influence or control decision making is a fundamental indicator of one's power within the relationship.

However, power dynamics, particularly within marriage, can be much more nuanced and subtle, and decision-making outcomes tell only part of the story. The process of making decisions, including the various ways in which partners can influence negotiations, also reveals much about who exerts more control within the relationship. For example, while one spouse may be making what appear to be important decisions, it is possible that these are merely tasks delegated by the other partner. Paying the bills could put one spouse in a position of power by allowing her or him to closely monitor the family's finances and make at least some monetary decisions unilaterally. However, in other circumstances, paying the bills may be a task dumped on a spouse by a partner who considers this job menial and stressful. Therefore, knowing how couples come to decisions can be a more important indicator of the balance of power between spouses than the actual outcome of the decision-making process.

The ability to suppress issues, or "non-decision making," is also an important indicator of power. This kind of power would show up in the successful resolution of conflict or resistance in the past in ways that keep similar conflicts from reemerging (latent power). In this case, the partner who "lost" the first round on a particular issue might fear open confrontation if she or he attempts to renegotiate the outcome, and therefore lives with the past decision rather than actively pursuing her or his desires. Power then lies with the partner who is able to avoid discussion or conflict over an issue once it has been settled to her or his satisfaction.

Considering overt and latent power, as well as power processes, gives us a number of ways to assess the balance of power in a relationship. But even these approaches do not examine all the possible avenues for exercising power. Steven Lukes (1974, 1986) has conceptualized power in a way that allows us to explore power dynamics that are embedded in larger cultural assumptions or ideologies. He advocates what he calls a three-dimensional view of power, which examines overt and latent power but also attempts to uncover power that is "hidden." Lukes argues that the ability to keep particular issues from entering the arena of conflict is a more thorough-going exercise of power than any overt struggle for dominance. The most effective exercise of power draws on prevailing ideological constructions, so that an individual's or group's domination seems beneficial, reasonable, or natural. In this way, the most adept uses of power are largely hidden.

Hidden power can be exercised in a variety of ways: through individual decisions, institutional procedures, or by dominant values that shape interaction (e.g., conventional gender expectations). Consensus may seem to exist, as ideology masks the contradictions in lived experience; but uncovering these contradictions reveals the exercise of power. We can easily see the distinction between the various conceptualizations of power by looking at the issue of domestic labor. If a husband and wife struggle over domestic labor, and the husband successfully resists his wife's request for him to do more around the house, he has exercised overt power. If his wife then accepts the situation and avoids raising the issue again out of fear of renewed conflict, he has exercised latent power. But even if this issue is never raised between the two spouses because the wife accepts it as her duty to bear the domestic labor burden—even when she is employed outside the home—her husband has benefited from the hidden power in prevailing gendered practices and ideology.

Aafke Komter (1989) adapted Lukes's framework to her examination of the hidden power in Dutch marriages. She found that husbands benefit from the implicit hierarchy of cultural worth that values men over women, and that couples rely on conventional gender expectations to explain inequities in their relationships. For example, men explained that their wives perform more housework because they were "better at it" or "enjoyed it more." These explanations reinforce conventional gender expectations and obscure men's power advantage in these relationships.

This conceptualization of hidden power. . . . allows us to assess how cultural expectations regarding gender at the institutional level affect both the interactions between spouses and their attempts to construct meaningful identities. Attention to hidden power can sensitize us to the subtle ways in which gender expectations shape the power dynamics within marriage.

While this conceptualization of power moves us forward theoretically, it represents only half the battle. Measuring power can be equally challenging because power cannot generally be measured in any direct way. Rather, theoretically driven indications of power are measured. Within the context of a marriage, the ability to prevail in the face of conflict is an obvious reflection of one's power. The relative level of control spouses exercise over financial and other family decisions has also been seen as indicative of the relative power of each spouse. More recently, especially as women have moved into the workforce, the division of domestic labor has been used to reflect the balance of power between spouses. The assumption here is that performing such labor is onerous and undesirable, and that a spouse will avoid this labor if she or he has the power to do so. And men's strong resistance to performing household chores, despite women's labor-force participation (as well as their continued efforts to get men to help), has been viewed as a successful expression of men's power or privilege (Hochschild 1989).

References

Acker, Joan. 1992. "Gendered Institutions: From Sex Roles to Gendered Institutions." *Contemporary Sociology* 21:565–569.

Berk, Sarah Fenstermaker. 1985. *The Gender Factory: The Apportionment of Work in American Households.* New York: Plenum Press.

Bianchi, Susan, Melissa Milkie, Liana Sayer, and John Robinson. 2000. "Is Anyone Doing the Housework? Trends in the Gender Division of Household Labor." *Social Forces* 79:191–228.

Bittman, Michael, Paula England, Liana Sayer, Nancy Folbre, and George Matheson. 2003. "When Does Gender Trump Money? Bargaining and Time in Household Work." *American Journal of Sociology* 109:186–214.

Blood, Robert O., and Donald M. Wolfe. 1960. *Husbands and Wives.* Glencoe, Ill.: Free Press.

Blumberg, Rae Lesser, 1984. "A General Theory of Gender Stratification." In *Sociological Theory 1984,* edited by R. Collins. San Francisco: Jossey-Bass.

Blumstein, Philip, and Pepper Schwartz. 1983. *American Couples: Money, Work, Sex.* New York: Morrow.

———.1991. "Money and Ideology: Their Impact on Power and the Division of Household Labor." In *Gender, Family, and Economy,* edited by R. L. Blumberg. Newbury Park, Calif.: Sage.

Brennan, Robert, Rosalind C. Barnett, and Karen Gareis. 2001. "When She Earns More Than He Does: A Longitudinal Study of Dual Earner Couples." *Journal of Marriage and the Family* 63:168–182.

Brines, Julie. 1994. "Economic Dependency, Gender, and the Division of Labor at Home." *American Journal of Sociology* 100:652–688.

Coltrane, Scott. 1996. *Family Man: Fatherhood, Housework, and Gender Equity.* New York: Oxford University Press.

Connell. R. W. 1987. *Gender and Power: Society, the Person, and Sexual Politics.* Stanford, Calif.: Stanford University Press.

Cowan, Carolyn Pape, and Philip A. Cowan. 1992 *When Partners Become Parents: The Big Life Change for Couples.* New York: Basic Books.

Deutsch, Francine. 1999. *Halving It All: How Equally Shared Parenting Works.* Cambridge: Harvard University Press.

Greenstein, Theodore, 2000. "Economic Dependence, Gender, and the Division of Labor at Home: A Replication and Extension." *Journal of Marriage and the Family* 62:322–335.

Hertz, Rosanna. 1989. "Dual Career Corporate Couples: Shaping Marriages through Work." In *Gender in Intimate Relationships: A Microstructural Approach,* edited by B. Risman and P. Schwartz. Belmont, Calif.: Wadsworth.

Hochschild, Arlie. 1989. *The Second Shift.* New York: Viking.

Komter, Aafke. 1989. "Hidden Power in Marriage." *Gender & Society* 3:187–216.

Lorber, Judith. 1994. *Paradoxes of Gender.* New Haven: Yale University Press.

Lukes, Steven. 1974. *Power: A Radical View.* London: Macmillan.

———. 1986. *Power.* Oxford: Basil Blackwell.

Nyberg, David. 1981. *Power over Power.* Ithaca, N.Y.: Cornell University Press.

Perry-Jenkins, Maureen, and Karen Folk. 1994. "Class, Couples, and Conflict: Effects of the Division of Labor on Assessments of Marriage in Dual-Earner Families." *Journal of Marriage and the Family* 56:165–180.

Pleck, Robert. 1985. *Working Wives, Working Husbands.* Beverly Hills: Sage.

Popenoe, David. 1989. *Disturbing the Nest: Family Change and Decline in Modern Society.* New York: Aldine de Gruyter.

Potuchek, Jean. 1997. *Who Supports the Family? Gender and Breadwinning in Dual-Earner Families.* Stanford, Calif.: Stanford University Press.

Raley, Sara, Marybeth Mattingly, Suzanne Bianchi, and Erum Ikramullah. 2003. "How Dual Are Dual-Income Couples? Documenting Change from 1970–2001." Presented at the annual meeting of the American Sociological Association, Atlanta, Georgia, August.

Rexroat, Cynthia, and Constance Shehan. 1987. "The Family Life Cycle and Spouses' Time in Housework." *Journal of Marriage and the Family* 49:737–750.

Risman, Barbara. 1998. *Gender Vertigo: American Families in Transition.* New Haven: Yale University Press.

Risman, Barbara, and Danielle Johnson-Sumerford. 1998. "Doing It Fairly: A Study of Post-Gender Marriages." *Journal of Marriage and Family* 60:23–40.

Rubin, Lillian Breslow. 1994. *Families on the Faultline.* New York: Harper-Collins.

Schwartz, Pepper. 1994. *Love between Equals: How Peer Marriage Really Works.* New York: Free Press.

West, Candace, and Donald Zimmerman. 1987. "Doing Gender." *Gender & Society* 1:125–151.

Wilkie, Jane Riblett. 1993. "Changes in U.S. Men's Attitudes toward the Family Provider Role." *Gender & Society* 7:261–279.

Williams, Joan. 2000. *Unbending Gender: Why Family & Work Conflict and What to Do about It.* New York: Oxford University Press.

Willinger, Beth. 1993. "Resistance and Change: College Men's Attitudes toward Family and Work in the 1980's." In *Men, Work, and Family,* edited by Jane Hood. New York: Sage.

Wright, Eric Olin, Karen Shire, Shu-Ling Hwang, Maureen Dolan, and Janeen Baxter. 1992. "The Non-Effects of Class on the Gender Division of Labor in the Home: A Comparative Study of Sweden and the U.S." *Gender & Society* 6:252–281.

Introduction to Reading 38

Sociologist Shirley Hill describes findings from her interviews with seventy African American men and women and other research on the topic of intimacy in relationships. This excerpt looks at relationships through the prism of gender, and what it means to form, maintain, and be in intimate relationships for African American men and women. These relationships reflect the marital patterns described in the Introduction to this chapter. They involve power; intimacy both psychologically and physically; support, psychological as well as economic; and much more. This reading helps us to understand how men and women think about relationships and the social factors that influence what are our most intimate moments.

1. What social factors shape the relationships of African American men and women?
2. What changes does Hill observe in heterosexual relationships?
3. Why are African American men and women less likely to marry, as indicated in Table 8.1?

BLACK INTIMACIES

LOVE, SEX, AND RELATIONSHIPS—THE PURSUIT OF INTIMACY

Shirley A. Hill

Historically, scholars gave only sporadic attention to what P. H. Collins calls the "love and trouble" tradition among black men and women, as necessity demanded a sustained focus on the broader forces of inequality, such as institutionalized racism, segregation, and economic exclusion. In light of such issues, explorations of relationship troubles between black women and men were seen as nothing more than inappropriately airing dirty linen or, even worse, contributing to literature that impugned and stereotyped African Americans. . . .

Yet for the most part, this surge of interest in analyzing the intimate relationships of African American women and men, sparked more by popular artists and authors than scholars, had been stunted by a lack of contextual and structural analysis of the issue. In this chapter, I explore the search for intimacy among African Americans in the broader context of race and class. I contend that the very heart of the troubles that proliferate in black intimate relationships is the inability to resolve the tension between their own cultural traditions and resources and Eurocentric notions of love, gender, courtship, and marriage.

African Americans often embrace dominant societal definitions of masculinity and femininity, ideals about female attractiveness, and gendered prescriptions of male-female relationships, despite the fact that their race and class position preclude conformity to such norms. Dominant societal rules about courtship, for example, sanction men as the aggressors, protectors, proposers, and providers and cast women as passive recipients of male attention that is gained largely through their physical attractiveness, sexuality, and coyness. The courtship process is seen as logically culminating in romantic love and marriage, despite the fact that marriage is essentially a legal contract based on property rights, legitimate heirs, and gender-defined roles for men and women. The historically marginalized class and racial status of African Americans has compromised their ability to meet the economic and gender expectations of marriage, thus reducing the relevance and stability of marriage. Heroic efforts aside, black men have often found themselves unable to shield women from the hostilities of the public arena, foot the entire bill for dating, or provide economic support for their families. Black women have had to fend for themselves, work outside the home, and rely on female kinship networks for support, and they have thus developed a strong sense of independence. This disjuncture between the historic experiences of African Americans and the gender norms of courtship and legal requisites of marriage have pitted black women and men against each other in patterns of resentment, mutual blame, and denigration, each blaming the other for failing to create successful relationships.

African American women criticize black men for being unfaithful and exploitative in relationships and often deride them for being "less than a man" when it comes to earning money and supporting their families. Black women are also criticized and penalized by their men for violating White female codes of behavior, especially for

---

From Shirley A. Hill, "Love, Sex, and Relationships: The Pursuit of Intimacy," pp. 93–117 in *Black Intimacies: A Gender Perspective on Families and Relationships,* copyright © 2005. Reprinted with permission of AltaMira Press.

spoiling intimacy by failing to be submissive and depriving men of the dominance they deserve. In their reflective moments, African Americans understand that race and racism shaped their dilemma, but such insights do not produce healing or peace, nor do they exempt African Americans from the stigma of having disordered gender relations. Although explicitly invoking the black matriarch concept has become almost verboten in scholarly literature on African Americans, its enduring legacy is evident in the frequent charge that black women are "simply too strong, too independent, and too self-sufficient for their own good or for the good of their relationships" [Franklin 2000]. Contention over the strong women/weak men premise remains a central theme among blacks in their everyday talk, with many embracing and idealizing the very models of family and gender that they have never been able to implement and that are now being rapidly abandoned by other races. Black couples who have experienced enough class mobility and assimilation to carve out a comfortable niche for themselves between dominant societal and black cultural norms are seen as exceptions to the rule, but even they find themselves playing out their relationships against the backdrop of palpable tension between race/gender myths and realities. . . .

Pursuing Gendered Intimacy

Delia Williams is a fifty-six-year-old African American divorced mother of three children who has recently earned a doctorate degree and teaches at a predominantly black community college. Concerned about the prevalence of early sex, childbearing, and welfare dependency among her young, mostly female students, as well as the way the men in their lives seem to exploit and abandon them, she offers her students a few rules on how to remedy the problem. According to the advice she shares with her students, male-female relationships should go this way: The man asks for a date, picks you up at home, meets your family, pays for the date, does the pursuing, proposes marriage, and provides economically for the family. Her advice is only partially tongue-in-cheek and strictly at odds with her own marital and class background: She has always been in the labor force, is divorced from a husband who did not support her values and career ambition, and has worked hard to raise her children as a single mother while advancing her education. Yet her advice reflects attitudinal support among blacks for dominant norms in love, courtship, and marriage—despite her own inability and even unwillingness to live such norms. Even blacks who proffer strong support for gender equality at work and in marriage often embrace traditional norms of dating and courtship. . . .

[I]n buying into a gendered scenario of love and marriage, they place unrealistic expectations not only on themselves but also on men, as traditional values tend to measure their masculinity by their economic achievements. Black women's weariness of male partners who do not measure up in the economic arena has made money a central issue in relationships. A recent study of college students found that, compared to their white counterparts, black female students were ten times more likely to say that a man should pay for the date (regardless of who initiated it), twice as likely to say that it was insulting for a man to even expect a women to share the cost of a date, and twice as likely to say that it was improper for women to initiate intimacy. Black men and women both indicated that upward mobility was an important trait in their partners, but women placed much more value on the social status of their partners than did men [Ross 1996].

Support for traditional gender expectations also runs high among African American men who, deprived of other legitimate sources of power, often cling tenaciously to the ideology of male dominance and see controlling black women as crucial to their claim to masculinity. Stung by the "black matriarchy" thesis, many black men appear to have become even more blatantly sexist in their demand for female subordination. Even Black Power militants missed the irony of advocating violence to overthrow racial injustice, while insisting on the subordination of women. "What makes a woman appealing is femininity," says black nationalist Maulana Ron Karenga, "and she can't be feminine without being submissive" [Quoted in Marable 2000

[1983]:97]. All too often, the physical and personality characteristics associated with black women were denigrated as deviant and destructive, and they found themselves pushed aside by black male leaders who were frequently accused of "talking black but sleeping white." As one black female law student pointed out, "We black women are always being reminded of how marginal and unworthy we are. We're never smart enough or beautiful enough or supportive, sexy, understanding, and resourceful enough to deserve a good black man" [Quoted in Bell 2001].

Endorsing dominant societal norms fosters the dismaying idea that relational harmony hinges on the economic success of men and the ability of women to assume submissive, secondary, and dependent roles. While black women are applauded for the active, strong, vital roles they have played in their families and communities, they are contradictorily told that such roles are improper in their intimate relationships with men. Moreover, dominant societal norms of femininity uphold exclusionary, Eurocentric images of female attractiveness that are defined in terms black women rarely can meet. A recent study found that African American males (33 percent) were nearly twice as likely as black females (17 percent) to say they preferred to date a person with light skin and even more likely to want to marry such a person [Ross 1997]. These attitudes help account for the sixfold increase in interracial marriage between blacks and whites between 1960 and 2000, with black men twice as likely as black women to marry outside their race [Franklin 2000]. While less than 1 percent of marriages are between blacks and whites, this low rate of marriage may mask the growing rate of interracial dating and relationships: A 2001 survey found that two-thirds of black men and half of black women have dated someone of another race [Romano 2003].

Revisioning Love and Gender

The dilemma facing African Americans is scarcely addressed in theories of heterosexual love and partner selection. Most emphasize the values of reciprocity, equity, and homogamy in forming successful relationships, although social exchange theorists (and some feminists) have seen marriage as essentially an exchange of a woman's sexuality and physical attractiveness for the economic resources of a man. Theorists such as Afrocentrist Molefi Asante have addressed the strain between black women and men by providing an African-based vision of the ideal relationship. Asante argues that four major value components—sacrifice, inspiration, vision, and victory—shape successful black unions, yet his model is clearly more prescriptive than analytical. Most black cultural theorists have fallen short of reconciling dominant societal traditions of love with the material realities and racial status of African Americans, often even insisting on the propriety of patriarchal relationships.

The civil rights era ushered in a burgeoning of black pride and support for black cultural traditions, such as an emphasis on the strength and self-reliance of women, alternative standards of beauty, attire, and attractiveness, and oppositional expressions of masculinity and womanhood. At least publicly, strength was an appreciated and embraced characteristic of black women, and many devised styles of beauty and sexuality that challenged the emphasis on long, straight hair, small bodies, and sexual passivity. Black magazines are filled with images of women donning strikingly black cultural hairstyles and clothing, and popular singers often depict the sexual freedom of black women by depicting them as "independent, strong, and self-reliant agents of their own desires, masters of their own destiny" [Emerson 2002:116]. The black cultural tradition of valuing women with big bodies has also been noted in publications like *Ebony* magazine, whose cover recently featured three plus-size African American female celebrities above an article title that read, "The full-figured revolution: If you got it, flaunt it." Yet, most efforts by black women to embrace the traditions that arise from their cultural, material, and gender experiences are rebuffed by white society as reflecting class deficits, sexual promiscuity, a lack of self-discipline, or behaviors that are detrimental to health and well-being. Thus, images of a distinctively black womanhood have often failed to gain currency in the dominant society and have been discredited and devalued by many black people who embrace white standards of style and the dominant society's almost fanatical dieting culture.

Forging alternative expressions of black masculinity has been equally difficult, especially for the vast majority of men who are not celebrity-status athletes, successful rap artists, or members of the inner-city underclass. Unique styles of dress and behavior are often cultivated as reflecting authentic black masculinity, and they often revolve around displays of power and sexuality.

Hip-hop artists are often the trendsetters for performances of black masculinity, but most of the penalties that accrue from such behaviors fall to the poor, young black men who imitate them. Such displays of manhood are less available for black middle-class men, who do not fit or embrace these versions of masculinity. Their class achievement brings prestige but also deprives them of being seen as either authentically black or particularly masculine. Performances of black masculinity by low-income men are associated with pathology and failure in the dominant society and are seen as socially threatening. Moreover, they enact the very stereotypes social activists have condemned as erroneous portrayals of black men as violent, criminal, and hypersexual. bell hooks [2004:155] argues that although "popular culture has made the black male body and presence stand for the apex of "cool," it is a death-dealing coolness, not one that is life-enhancing." These dangerous behaviors further isolate young men from mainstream society, and the emphasis on hypersexuality, cool pose, game playing, and violence renders them ill equipped for stable relationships with women.

Courtship and Marriage Class Context

Analyses of the intimacy dilemma faced by African Americans has often focused on the plight of young, economically marginal women and men, whose fragile ties to marriage are most evident. Young black females living in poor, urban areas, for example, are especially vulnerable when it comes to hanging all of their hopes for a better life on finding the right man. The paucity of fatherly love and male attention in their lives, coupled with their own dismal prospects for escaping poverty through educational or career success, heightens their search for a male partner to validate their self-worth and provide them with a home, a family, and adult status. Nothing in their background prepares them to demand respect from their potential suitors, and most of the males they encounter lack the emotional maturity, material resources, or inclination to marry or form stable relationships. These young women use their sexuality to negotiate the lives they yearn for—freedom from parental control, womanhood, and a secure relationship with their male partners—yet they rarely fully realize their dreams. An exclusive relationship with a woman who they are accountable to but unable to support holds little appeal for disadvantaged young men, who are more likely to gain respect among their peers by being seen as "players." Elijah Anderson's [1990:113–4] insightful work explains the pattern of relationships between young men and women in inner-city areas this way:

> The girls have a dream, the boys a desire. The girl's dream of being carried off by a Prince Charming who will love them, provide for them, and give them a family. The boys often desire sex without commitment or babies without responsibility for them. . . . Yet the boy knows what the girl wants and plays that role to get sex. In accepting his advances, she may think she is maneuvering him toward a commitment or that her getting pregnant is the nudge he needs to marry her and give her the life she wants. What she does not see is that the boy, despite his claims, is often incapable of giving her that life. For in reality he has little money, few prospects for earning much, and no wish to be tied to a woman who will have a say in what he does.

Black teenage girls in poor families *do* face relentless pressure from their male partners for sex, but Anderson's analysis falls short by portraying them as helpless victims who are easily duped by their male counterparts, as this view ignores their sexual desires and efforts to garner power through their sexuality. Nearly all research reports a racial gap in the sexual patterns of blacks and whites, although class may hold as much explanatory power as race in understanding it. For example, although the percentage of sexually active high school students has declined in recent years, it still remains much higher among blacks (73 percent) than among whites (44 percent) [Risman and Schwartz 2002]. Studies show that the onset of sexuality is earlier for blacks than for whites and that being African American is associated with having more sex partners and higher rates of marital infidelity [Scott and Sprecher 2000; Franklin 2000]. Nonetheless, exploring the racial gap in sexual behavior is generally thwarted by concern over reinforcing negative views of black sexuality, and most researchers steer clear by focusing on the outcomes of sexuality, such as pregnancy, teenage motherhood, and sexually transmitted diseases.

The pursuit of intimacy among older, more affluent African Americans is also shaped by economic factors, class-based relationship expectations, and gender factors. The traditional norm of women "marrying up"—or marrying men who have more status, education, and income than they do—is complicated among blacks and increases the fragility of their relationships. Black men as a group have higher earnings than black women, mostly because they are concentrated in male-typed jobs but also because they are more likely to hold high-paying professional positions. Nonetheless, during the 1950s and 1960s, the actual educational, economic, and employment gains of black women began to exceed those of black men, as did their entry into white-collar jobs (albeit mostly female-typed jobs, such as teaching and social work) [Franklin 2002]. The status of African American women is also based on the fact that they have held jobs that have more fully integrated them into mainstream society. This situation was exacerbated during the economic decline of the 1980s, intensifying what D. King describes as a "low-level gender war" between black men and women ["Marriage dilemma . . . 1997]. Black women today hold more bachelor's degrees and especially more graduate degrees than black men and more jobs described as managerial/professional ["Marriage dilemma . . . 1997]. Black men often argue that black women are less threatening to the dominant culture and have an easier time getting good jobs—fueling dissension with the suggestion that their "double minority status" (rather than their merit) allows them to get ahead. Some African American women feel ambivalent about their own success, but others contend that black men are simply insufficiently motivated. As one women I interviewed asserted, black women "are just more go-getters than black men are. . . . [A black woman] has more of a driving force to succeed in what she wants to do, because black women overall are like that compared to black men."

The economic advantage of black men often does not compensate for the status and educational advantage of black women, who often find themselves "marrying down." In 1996, only 46 percent of college-educated black women were married to college-educated men, compared to 70 percent of white women, and college-educated black women were twice as likely as other black women to be interracially married ["Marriage dilemma . . . 1997]. The higher education of women, especially in relation to their spouses, increases marital instability, as it plays itself out as key differences in values about family life. The class background of women is likely to predict certain attitudes about child socialization, how money is spent, and how gender is organized; moreover, educated women are more likely to demand companionship, equity, and emotional intimacy in their marriages. As Robert Staples admits, meeting such needs "is not the forte of American males," and this may be even more likely to be true for less educated and economically disadvantaged men. Men and women thus often find themselves in separate and hostile camps; indeed, as a black woman with a doctorate degree and married to a white man said: "I never ran across a lot of black men who liked women, who really liked women and appreciated women. Perhaps because of upbringing, because of society, black men haven't been allowed to, or just don't know how to like women and how to value them" [Cose 1995:51–2].

Although they are still considerably less likely to marry interracially, the actual increase in black-white interracial marriages involving black women has grown faster than that of black men. Still, "A good man is hard to find" is virtually a mantra among black women, as high rates of underemployment, low-wage work, incarceration, and mortality have diminished the supply of African American men. Moreover, the number of available women has decreased black men's incentive for fidelity and marriage and increased the vulnerability of women in relationships. As Robert Staples admits in his 1994 book, the gender ratio favors men and reduces the bargaining power of women: "In a sense," he writes, "black women often find themselves in the position of sexually auditioning for a meaningful relationship. After a number of tryouts, they may find a black male who is willing to make a commitment to them" [Quoted in Zinn and Eitzen 2002]. To rectify the shortage of black men, the notion of polygamy has been offered as a way to organize and institutionalize extramarital and other relationships.

* * *

Marriage: The Culture, Contract, and Reality

Modern marriage has been seen as the natural outcome of heterosexual love, as proof of commitment and value conformity, and as the only legitimate arena for sexuality, reproduction, and rearing children. Despite undergoing significant revolutions in the past few decades, Karla Hackstaff [1999] argues that the "marriage culture" remains particularly strong in the United States. At least 90 percent of Americans plan to marry, and most still believe that married people are happier than single people. The dominant culture romanticizes marriage as an expression of committed love, free choice, intimacy, sexuality, and security. As the highly idealized nucleus of the family, it is entered into with elaborate ceremonies, celebrations, and solemn vows. The vision of a lifelong relationship with a soul mate, with someone to love and share life with, is probably universally appealing—even among those who are divorced, more than 80 percent of whom say marriages ought to be permanent. The marriage culture is inspired by Western Judeo-Christian religion, notes Hackstaff [1999], which regulates marriage and divorce, gives meaning to family practices, reinforces the notion of male dominance, and accepts divorce only as a last resort. The cultural romanticizing of the value of marriage is also heralded by scholars who document its numerous benefits, including better health, greater sexual satisfaction, more financial affluence, and better social, academic, and psychological outcomes for children [Waite 2000].

The idealized view of marriage as the locus of love and sexuality sharply contrasts with the reality of marriage as a legal contract that upholds male domination and female dependency, assures property rights, and sanctions the legitimacy of children. Traditional marriage has been described as the linchpin of gender inequality in families, rigidly defining the roles of husbands as legal heads of their households and economic providers and obliging women to provide unpaid domestic work and child care. Women were seen as owing their husbands complete obedience, as common law held that "the husband and father had nearly absolute authority over his wife and children, including the right to administer physical correction" [Walsh 1985:4]. The concept of marital rape did not exist, as women were expected to accept the sexual overtures of their husbands under all circumstances, and husbands could sue a third party who interfered with his exclusive right to the sexuality and domestic work of his wife. Marriage was also a property arrangement based on legitimizing childbirth and inheritances, offering significant economic benefits to women and children. As such, it has been less relevant to African Americans.

Although the majority of blacks historically have gotten married, economic strains and female independence made it difficult for them to reap the usual benefits of marriage—that is, household authority and exemption from domestic work for men and economic solvency and the ability to engage in full-time domesticity for women. Black Americans have not, of course, been completely exempt from such marital arrangements. The working-class family in which I was born and reared, for example, consisted of a blue-collar, wage-earning father and a full-time homemaker mother who had full responsibility for the housework and caring for six children. My father embodied the strong male head of household: He earned the money, laid down the law, demanded obedience, and meted out punishment, all of which was uncritically accepted by us as normal. My mother deferred to his authority, and the marriage lasted until her death (nearly forty years), although I never thought much about whether she was happily married or satisfied with the relationship. Two-parent working-class families like my own remain fairly invisible in most scholarship, which tends to dichotomize African American families as either poor and headed by single mothers or as dual-income, middle-class families. In neither case is gender viewed as much of an issue, as men were absent in the former and the latter were held out as models of gender egalitarianism. The notion that married black couples had achieved gender equality in the domestic arena has now been challenged by systematic research, but few studies have offered much insight about other factors that might affect marital satisfaction among African Americans.

Feminist research implies gender equality is the key to marital stability and happiness. . . . For the most part scholars have settled on studying the division of housework and child care as the most important determinant of gender equality, documenting repeatedly that women of all races do much more of it than men. Age and level of education appear best at predicting male partners' contributions to family work, with young, educated men being more involved than other men. In 1997, John and Shelton found that black men spent only about half as much time doing housework as their partners; moreover, although black and white women spent a comparable number of hours each week on housework, black women worked longer hours on their jobs and had more children. The myth of gender equality between black spouses may stem from the fact that African American men have been shown to be more accepting of their wives' employment than are white men; yet, they are also more conservative in other gender beliefs [Blee and Tickamyer 1995] and experience more conflict over work-family responsibilities, especially when the wife has a career [Bridges and Orza 1996]. At best, the handful of studies conducted during the 1990s suggest that blacks are moving toward a more egalitarian pattern, since in some cases black women do less of the housework than white women [Veroff, Young and Coon 2000].

African Americans often hold seemingly incongruent gender ideologies; my research, for example, finds strong support for the idea that men and women should share housework and child care equally *and* that mothers should be in the home and fathers in the labor market [Hill 1999]. For the most part, gender ideologies are probably less potent predictors of behavior than economic factors. There is no conflict between believing in gender equality and being a stay-at-home wife and mother, but women whose husbands earned enough money to exempt them from the labor force did most of the domestic work and were often quite satisfied with the relationship. . . .

Middle-class married African American men are often quite *happily* involved in the care of children—in many cases eager to become the fathers they never had—although there tends to be a gender division in the work they perform. The owner of a day care center, who was married to a physician, described her husband this way: "My husband is incredible . . . he's just a great help. He loves to help the kids with their homework, and as they get older he takes more responsibility. And he loves to plan our vacations and go on field trips with the kids at school."

Beyond the sharing of family work, other factors undoubtedly shape marital satisfaction among African Americans, although little research has explored the issue. One interesting (and contested) article by Ball and Robbins [1986] reported that being single produced more happiness among blacks than being married. Black men who were single, divorced, or separated were happier in life than those who were married, and among black women, widows had the highest level of life satisfaction. A more recent study found that, compared to white women, black women felt they were receiving fewer benefits from being married; they also expressed less trust in their spouses and had lower levels of marital well-being [Goodwin 2003]. In their study of divorced black men, Erma Jean Lawson and Aaron Thompson found financial strain and male unemployment to be the most frequently mentioned causes of divorce, and that these factors were also associated with wife abuse. But while an overall higher standard of living predicts marital satisfaction among blacks [Orbuch and Custer 1995], higher earnings by wives than husbands place some black marriages at risk [Staples and Johnson 1993].

Frequent participation in religion is apt to strengthen black marriages, despite the fact that churches support patriarchy in principle and often disperse antifeminist rhetoric. Yet such polemical teachings by Christian fundamentalists are frequently accompanied by marriage enrichment classes that teach men to sacrifice practically every male privilege in supporting their wives, and they define male headship in seemingly benign ways [Veroff et al. 2000]. Thus, understanding how religion affects marriage and gender will require looking beyond the polemical literature supporting patriarchal norms.

Race translates into cultural patterns that also affect the quality of black women and men's relationships, often in class-specific ways. African American women have historically been known to place more value on blood than marriage relationships, and the bonds of mutual aid and support formed by low-income black women often militate against marriage and long-term relationships. Research has found that wives in kin-centered units are less likely to discuss their problems with their husbands and more likely to experience marital dissatisfaction [Milardo and Graham 2000]. The strong emphasis on extended family relationships may explain why black wives' closeness with their in-laws correlates with marital happiness for both spouses but has no impact on white marriages [Veroff et al. 2000]. Family bonds also seem to matter in that husbands' close contact with their fathers is associated with their having greater love for their wives and with greater marital satisfaction among wives [Berger and Milardo, cited in Milardo and Graham 2000].

The challenges facing African Americans historically did not keep most of them from marrying and trying to create viable relationships, yet the fragility of those marriages is seen in the fact that even in the 1950s and 1960s, their rate of divorce was four times higher than that of whites [Franklin 2000]. Since then, it has been about two times higher, but the reduction in the racial disparity is largely the result of high rates of nonmarriage among both blacks and whites. In 1998, 80 percent of white adults had been married at least once, compared to only 61 percent of blacks [Oropesa and Gorman 2000]. Combining those who have never married and those who are separated, divorced, or widowed reveals about 65 percent of black women are single. Black men have a higher rate of marriage than black women but a lower rate than white men [Hemmons 1996]. These trends in black marriage rates parallel and amplify patterns of singleness, delayed marriage, cohabitation, and divorce evident in the broader society. Yet while whites are often seen as delaying marriage until they acquire adequate resources to sustain families, blacks are seen as irresponsibly making lifestyle choices that undermine family stability and increase rates of crime, poverty, and welfare dependency. In reality, the same forces that have weakened black marriages are increasingly affecting white marriages.

* * *

Living Single

. . . The ideological connection among intimacy, love, and marriage has been weakened by growing rates of singleness, delayed marriage, and nonmarital cohabitation. Black adults are more likely to be single than whites, but just as marriage does not assure intimacy, being single does not exclude it. . . .

African American women have a long tradition of relying on each other for assistance in raising families, and the deepening of those relationships today both includes and transcends the swapping of child care and economic resources. Black women pull together in religious and myriad other ways to bond and support each other—whether through prayer groups, community uplift, and recreation activities. *Ebony* magazine wrote on the growing importance of the "sister circles" that black women are engaged in, and they are proliferating informally and through organized single groups, offering women much of the companionship and intimacy of marriage. Most of the single women I know are quick to say they still may marry—if they find the right man—but they find themselves willing to make fewer and fewer concessions to do so. . . .

Terry McMillan's popular 1992 novel *Waiting to Exhale* vividly portrays the search for sexual satisfaction and intimacy among a group of accomplished black women, who engage in one-night stands, have affairs with married men, and even tolerate drug-abusing partners—only to find themselves relying on each other in the end. Many criticized the movie for its negative depiction of black men, but few questioned the portrayal of successful, intelligent black women who were unable to figure out why they could not find true love despite giving generously of their sexuality to men who were adulterers, abusive, and sometimes virtual strangers. Today, more black women (and some black men) are considering the benefits of celibacy as a way to achieve greater respect for themselves and control over their bodies. Efforts to convince young African American girls of the virtues of sexual abstinence have been widespread in churches and communities for decades, and a spate of new books—mostly by Christian authors—now argue that many older single women should be more circumspect in their sexual choices, since relationships have often left them feeling used and abandoned.

Many African American women, of course, form sexual relationships with other women, although doing so openly often leads to criticism and stigma. Omolade [1995] argues that the historic homosocial ties between black women have often had a sexual dimension, as they sought to free themselves from sexual exploitation by white men and the patriarchal leanings of black men. In other cases, black women take heterosexual orientation for granted rather than exploring their sexuality. . . .

Black Americans today are depicted as especially homophobic [Battle and Bennett 2000], and, while such attitudes seem consistent with their educational and religious background, it appears that they may have grown more openly intolerant of gays and lesbians since the civil rights era. The black nationalist movement, for example, was especially hostile to gay men, even arguing that homosexuality was unknown in Africa and was a European strategy for destroying black people. Beverly Greene argues that internalized racism also heightens homophobia among blacks, who see it as another basis for their denigration. Black religious leaders today align themselves more closely with the new Christian Right and their fundamentalist views of the Bible. Together, black nationalism and conservative religious thought may have fostered more anti-gay and anti-lesbian attitudes. Battle and Bennett [2000] find that blacks are less likely to disclose their homosexuality and that those who do have more negative mental health outcomes than whites. Still, black leaders like Al Sharpton, Coretta Scott King, and Carol Moseley Braun are among those who support the rights of gays and lesbians as a civil rights issue, and polls show that issues of social justice and jobs are more likely to sway black voters than politicians' stances on gay issues [Boykin 2004].

Cohabitation, another alternative to marriage, rose dramatically among singles during the 1990s, increasing nearly 50 percent to more than 4 million couples. For poor people, cohabitation has been described in some studies as the "budget way" to start a family [Furstenberg 1996] and it can produce relatively stable relationships. As one black mother said, "I'm not married. I got three kids. But their father is there with the kids.

He's been there since I was 16. . . . I been with the same guy since I was sixteen years old and I'm still with him now. I only had really one man in my life" [Jarrett 1994:39].

Since the 1960s, cohabitation can no longer be the explicit basis for denying welfare eligibility, and states have become fairly lenient about relationships between welfare recipients and their partners [Moffitt, Reville, and Winkler 1998]. Those in long-term cohabiting relationships often feel they have less to gain from marriage, especially if they are unable to meet the financial obligations of marriage. Moreover, cohabitation among single mothers improves their economic standing and decreases the rate of childhood poverty by about 30 percent [Seltzer 2000]. Edin [2000] finds that low-income women place much emphasis on the financial stability and respectability of potential mates in considering marriage, and they may even have a "pay and stay" rule for co-residing male partners to ensure that they contribute to household expenses. Their mistrust of men, fear of domestic violence, and concern over losing control of their own households also affect their marital decisions.

Despite the risk poor women face in their intimate relationships, the past few years have seen the emergence of a new campaign to solve poverty and single motherhood by getting them to marry. Welfare reform policies designated $300 million to promote marriage among the poor, in many cases through premarital counseling and making marriage education a part of the high school curriculum [Campbell 2002]. Whether welfare reform or marriage will reduce poverty and births to young, single mothers remains to be seen. A recent study by Lichter and his colleagues [2003] found that poverty rates would still be more than twice as high among black women if they had the same family background and rates of marriage and unwed childbirth as whites. Moreover, they noted that among economically disadvantaged black women, marriage is associated with downward educational mobility, and those who marry and then divorce have even higher rates of poverty than never-marrieds. Hao and Cherlin [2004] found that welfare reform has failed to decrease teenage births for girls in welfare families; in fact, rates of unwed pregnancy and high school dropout had increased slightly in those families. More important, marriage promotion policies seem to miss the significance of economic solvency and gender equity in creating viable unions.

Summary

This chapter has explored the clash between the black cultural traditions of love and marriage and those of the dominant society. For African Americans, racism and poverty have stripped marriage of much of its facade, exposing its economic and patriarchal underpinnings. Still, many have accepted the gendered scenario of love and marriage, although doing so has been especially perilous, as few black men have achieved the economic prowess sufficient to warrant female subordination, and black women have a long tradition of employment and independence. The gender dilemma, coupled with racial and economic oppression, has made marriages fragile and left African Americans leading the way in creating alternative family systems that are now becoming common in the broader society. While social conservatives champion marriage as compatible with family values and as a strategy for strengthening the economic base of poor families, research so far has not supported the efficacy of this strategy. Families need a solid economic base if they are to succeed and outmoded traditions of notions of male dominance must be put to rest.

References

Anderson, E. 1990. *Streetwise: Race, Class, and Change in an Urban Community.* Chicago: University of Chicago Press.

Ball, R. E., and L. Robbins. 1986. "Marital status and life satisfaction among black Americans." *Journal of Marriage and the Family* 48:389–94.

Battle, J., and M. Bennett. 2000. "Research on lesbian and gay populations within the African American community: What we have learned." *African American Research Perspectives* 6:35–47.

Bell, D. 2001. "The sexual diversion: The black man/black woman debate in context." pp. 168–76 in *Traps: African American Men on Gender and Sexuality,* ed. B. Guy-Sheftall and R. P. Byrd. Bloomington: Indiana University Press.

Blee, K. M., and A. R. Tickamyer. 1995. "Racial differences in men's attitudes about women's gender roles." *Journal of Marriage and the Family* 57:21–30.

Boykin, K. 2004. "Your blues ain't like mine: Blacks and gay marriage." *Crisis* 111:23–25.

Bridges, J. S., and A. M. Orza. 1996. "Black and White employed mothers' role experience." *Sex Roles* 35:337–85.

Campbell, K. 2002. "Can marriage be taught? The state of matrimony." *Christian Science Monitor,* pp. 1, 9.

Carolan, M. T., and K. R. Allen. 1999. "Commitments and constraints to intimacy for African American couples at midlife." *Journal of Family Issues* 20:3–24.

Cose, E. 1995. *A Man's World: How Real Is Male Privilege and How High Is Its Price.* New York: HarperCollins.

Edin, K. 2000. "What do low-income single mothers say about marriage?" *Social Problems* 47:112–13.

Emerson, R. A. 2002. "'Where my girls at?' Negotiating black womanhood in music videos." *Gender & Society* 16:115–35.

Fitch, C. A., and S. Ruggles. 2000. "Historical trends in marriage formation: The United States 1850–1990." In *The Ties That Bind: Perspectives on Marriage and Cohabitation,* ed. L. J. Waite, with C. Bachrach, M. Hindin, E. Thomson, A Thornton. New York: Aldine de Gruyter.

Foston, N. A. 2004. "Is celibacy the new virginity?" *Ebony* (January):120–24.

Franklin, D. L. 2000. *What's Love Got to Do with It? Understanding and Healing the Rift between Black Men and Women.* New York: Touchstone.

Furstenberg, F. F. 1996. "The future of marriage." *American Demographics* (June), 34–40.

Goodwin, P. Y. 2003. "African American and European American women's marital well-being." *Journal of Marriage and the Family* 65:550–60.

Hackstaff, K. B. 1999. *Marriage in a Culture of Divorce.* Philadelphia: Temple University Press.

Hao, L., and A. Cherlin. 2004. "Welfare reform and teenage pregnancy, childbirth, and school dropout." *Journal of Marriage and the Family* 66:179–94.

Hemmons, W. M. 1996. *Black Women in the New World Order: Social Justice and the African American Female.* Westport, CT: Praeger

Hill, S. A. 1999. *African American Children: Socialization and Development in Families.* Thousand Oaks, Calif.: Sage.

hooks, b. 2004. *We Real Cool: Black Men and Masculinity.* New York: Routledge.

Hunter, A. G. 2002. "(Re)Envisioning cohabitation: A commentary on race, history, and culture." pp. 41–50 in *Just Living Together: Implications of Cohabitation on Families, Children, and Social Policy,* ed. A. Booth and A. C. Crouter. Mahwah, N.J.: Erlbaum.

Jarrett, R. L. 1994. "Living poor: Family life among single parent, African-American women." *Social Problems* 41:30–50.

King, A. O. 1999. "African American females' attitudes toward marriage: An exploratory study." *Journal of Black Studies* 29:416–37.

Leggon, C. B. 1983. "Career, marriage, and motherhood: 'Copping out' or coping?" pp. 113–34 in *Black Marriage and Family Therapy,* ed. C. E. Obudho. Westport, Conn.: Greenwood.

Lichter, D. T., D. Roempke Graefe, and J. B. Brown. 2003. "Is marriage a panacea? Union formation among economically disadvantaged unwed mothers." *Social Problems* 50:60–86.

Marable, M. 2000. "Groundings with my sisters: Patriarchy and the exploitation of black women." pp. 119–52 in *Traps: African American Men on Gender and Sexuality,* ed. R. P. Byrd and B. Guy-Sheftall. Bloomington: Indiana University Press.

"Marriage dilemma of college-educated women." 1997. *Journal of Blacks in Higher Education* (Autumn):52–57.

Milardo, R. M., and A. Graham. 2000. "Social networks and marital relationships." Chap. 3 in *Families as Relationships,* ed. R. M. Milardo and S. Duck. New York: Wiley.

Moffitt, R. A., R. Reville, and A. E. Winkler. 1998. "Beyond single mothers: Cohabitation and marriage in the AFDC program." *Demography* 35:259–78.

Omolade, B. 1995. "Hearts of darkness." pp. 362–77 in *Words of Fire: An Anthology of African-American*

Feminist Thought, ed. B. Guy-Sheftall. New York: New Press.

Orbuch, T. L., and L. Custer. 1995. "The social context of married women's work and its impact on black husbands and white husbands." *Journal of Marriage and the Family* 57:333–45.

Oropesa, R. S., and B. K. Gorman. 2000. "Ethnicity, immigrations, and beliefs about marriages as a 'tie that binds.'" pp. 188–211 in *The Ties That Bind: Perspectives on Marriage and Cohabitation,* ed. L. J. Waite, with C. Bachrach, M. Hindin, E. Thomson, and A. Thornton. New York: Walter de Gruyter.

Risman, B., and P. Schwartz. 2002. "After the sexual revolution: Gender politics in teen dating." *Contexts* 1:16–24.

Romano, R. C. 2003. *Race Mixing: Black–White Marriage in Postwar America.* Cambridge, Mass.: Harvard University Press.

Ross, L. E. 1996. "Black–white college student attitudes and expectations in paying for dates." *Sex Roles* 35:43–56.

———. 1997. "Male selection preferences among African American college students." *Journal of Black Studies* 27:554–69.

Scott, C. F., and S. Sprecher. 2000. "Sexuality in marriage, dating, and other relationships: A decade review." *Journal of Marriage and the Family* 62:999–1017.

Seltzer, J. A. 2000. "Families formed outside of marriage." *Journal of Marriage and the Family* 62:1247–68.

Staples, R., and L. B. Johnson. 1993. *Black Families at the Crossroads: Challenges and Prospects.* San Francisco: Jossey-Bass.

Veroff, J., A. Young, and H. Coon. 2000. "The early years of marriage." Chap. 2 in *Families as Relationships,* ed. R. M. Milardo and S. Duck. New York: Wiley.

Waite, L. J., with C. Bachrach, M. Hindin, E. Thomson, and A. Thornton (Eds.). 2000. *The Ties That Bind: Perspectives on Marriage and Cohabitation.* New York: Aldine de Gruyter.

Walsh, L. 1985. "The experiences and status of women in the Chesapeake." pp. 1–18 in *The Web of Southern Social Relations: Women, Family and Education,* ed. W. Fraser, R. F. Saunders, and J. Wakelyn. Athens: University of Georgia Press.

Zinn, M. B., and D. S. Eitzen. 2002. *Diversity in Families.* Boston: Allyn & Bacon.

Introduction to Reading 39

This research study by Scott Coltrane, Ross Parke, and Michele Adams examines the cultural factors that affect how men father their children. They analyze responses of 167 Mexican American fathers, a sample taken from a larger longitudinal study which they conducted of working-class families in Los Angeles. In this selection, the authors examine how cultural expectations shape involvement with children, illustrating the powerful role of culture in framing expectations for fathering and mothering. Their findings also suggest that attitudes toward parenting and equality in the household are shaped by circumstances of individuals lives.

1. When are Mexican American fathers more likely to be involved in children and household tasks?
2. How does the economic situation of these families affect fathers' involvement in their families? Note the subtle differences between the argument in this paper and that of Tichenor in this chapter.
3. What role does culture play in the involvement of these men in their families?

Complexity of Father Involvement in Low-Income Mexican American Families

Scott Coltrane, Ross D. Parke, and Michele Adams

When the economy falters as it has in the past few years, American parents must struggle to make ends meet and maintain families. Even when the economy is doing well, as it was in the late 1990s, such struggles are common among low-income families. Among the groups most at risk for economic stress, adolescent problems, and school dropout are Latinos (Hispanics), who are projected to comprise nearly one-quarter of the U.S. population by the year 2050. Two-thirds of U.S. Latinos are Mexican Americans, a population disproportionately composed of two-parent, working-poor families with unique needs and cultural resources. Though previous research has tended to ignore this group, in this article, we draw on this population to investigate multiple components of father involvement and offer an analysis of predictor variables associated with higher levels of men's participation in family life.

When job markets do not support full employment, the family lives of marginalized ethnic groups are adversely affected. Although traditional cultural ideals call for men to be sole breadwinners and women to be stay-at-home mothers, contemporary labor markets increasingly require households to have two earners. The individualism and gender ideals commonly associated with women sharing breadwinning and men sharing child rearing are not prevalent among Latino families. Instead, such families are commonly described as having high levels of family cohesion and cooperation ("familism") but also being governed by traditional gender ideals (see Buriel, 1986; Cauce & Rodriguez, 2001; Gonzales, Knight, Morgan-Lopez, Saenz, & Sirolli, 2002; Segura, 1992; Vega, Kolody, Valle, & Weir, 1991). How, then, have recent economic and social changes affected parenting practices in working-class Latino families? We explore some of the tensions and changes facing Mexican American families and identify the cultural and labor market conditions associated with father involvement in low-income and working-poor communities.

Contradictory Cultural Expectations

National surveys report that the vast majority of American men rank marriage and children among their most precious goals, and most American fathers say they value their families over their jobs (Coltrane, 1996). Similarly, most Americans agree that women should have equal rights to men and that job discrimination on the basis of gender should be prohibited. . . . Ironically, surveys also show that American husbands and fathers perform relatively little housework or child care (compared to their wives) and that American husbands and wives continue to judge unbalanced divisions of family labor as "fair" (Coltrane, 2000). Thus, Americans embrace equal parental involvement at the same time that they accept minimal family work contributions from men. Elsewhere, we have argued that such contradictions reflect longstanding tensions in American culture dating back to the founding of the nation (Coltrane & Adams, 2001, 2003; Coltrane & Parke, 1998; Parke & Tinsley, 1984). Tensions between individual rights and family obligations have surfaced most often in debates about women's employment, but as we move into the 21st century, these tensions are

---

increasingly played out in debates about men, marriage, and fatherhood (Coltrane, 2001).

Typically missing from these debates is explicit consideration of men who are not White and middle class. In this study, we move beyond the simple observation that poor men from minority communities are less likely to be married and to contribute money to their children, and document how Mexican American men in two-parent households contribute to their families in a multitude of ways. In particular, we investigate the frequency with which they interact with their school-aged children, document whether they participate in "feminine" gender-typed activities as well as "masculine" activities, assess whether they monitor their children's whereabouts and activities, and ascertain the extent to which they participate in direct child supervision and routine household labor. In addition, we specify the personal, family, social, economic, and cultural factors that predict father involvement in each of these activities and invoke theories that might help us understand differential levels of involvement.

* * *

Conceptualizing and Measuring Father Involvement

The leading strategy for measuring fathers' involvement with children posits three components: engagement or interaction with the child, accessibility or availability to the child, and responsibility for the care of the child (as distinct from the direct performance of care; Lamb, Pleck, Charnov, & Levine, 1985). Although some activities include aspects of more than one of these components, specifying father involvement in these terms allows for refinement in theories and research on family functioning and child development (Parke, 1995). Father involvement is most frequently measured in terms of the first component of engagement or interaction, exemplified by holding, talking to, and especially playing with younger children (McBride, 1989; Parke, 2000). Studies show that fathers have increased the amount of time they spend interacting with their children, though they still spend less time interacting with children than do mothers (Coltrane, in press; Levine & Pittinsky, 1997; Pleck & Masciadrelli, 2003).

Some recent attempts to quantify fathers' interaction or engagement with children include more careful consideration of the specific types of contact involved, especially distinguishing between childcare activities and involvement in play, leisure, or affiliative activities with the child (Parke, 1995). When compared to mothers, fathers are likely to spend a much greater proportion of their interaction time with children in play or leisure, and such activities make unique contributions to children's emotional self-regulation and social competence (Parke, 1996). Especially when considering older children, the types of activities associated with father-child interaction have implications for the potential effects of such interaction on the child's understanding of gender (see Adams & Coltrane, in press). For example, when a father plays sports with his sons (but not with his daughters), or when he directs his daughters to cook for him (but not his sons), his children learn different messages about the family obligations and entitlements of men and women. To capture differences in types of father-child interaction, we separated masculine-typed interaction (e.g., outdoor games) from feminine-typed interaction (e.g., cooking together).

Fathers' availability is measured less often than interaction, but it is a necessary precondition for more active forms of parenting—like monitoring and rule enforcement—that predict positive child outcomes. Availability is sometimes measured by hours of employment, although more accurate assessments include the actual time that a parent is with a child (but not necessarily interacting), in the vicinity of the child (e.g., in the house or yard with the child), or reachable via phone or other means (especially for older children). One recent study found that a father's availability now has a greater impact on his assumption of childrearing duties than it did in past decades (Brewster, 2000).

In the late 1970s and early 1980s, fathers tended to use their discretionary nonworking hours for other activities, whereas in the late 1980s and 1990s, they were far more likely to use those hours for child care. In this study, we measure parent availability by asking how many hours per week each parent supervises the child, allowing for the computation of the father's percentage of child supervision. In addition, we measure mothers' and fathers' employment hours and use them as predictors of fathers' availability.

In addition, researchers have begun to stress the importance of the responsibility dimension of father involvement, which includes the more hidden managerial aspects of child care, as well as household maintenance activities such as housework. Managerial parenting functions include organizing and arranging the child's environment, thereby regulating the child's access to social contacts (Parke, 1995, 2000). In infancy, management includes setting boundaries for play, taking the child to the doctor, arranging day care, or setting up opportunities for interacting with others. In middle childhood, managerial responsibility includes regulating meals, baths, clean-up, and monitoring or arranging for play with other children. Parke (2000) suggested that the monitoring and managerial role that parents play may be just as important for child development as time spent in face-to-face interaction, because the amount of time children spend in social environments far exceeds their time interacting directly with parents. As Pleck (1997) noted, mothers remain childcare managers in the vast majority of households, but evidence from some studies indicates that at least some fathers are taking a more active role in this domain (Coltrane, 1996). In this study, we measured one aspect of responsibility by ascertaining the extent to which each parent monitors the child's whereabouts and activities. Monitoring reflects knowledge about children's behaviors, as well as setting and enforcing behavioral rules.

Although often neglected in studies of parenting, another way that fathers (and mothers) take responsibility for children is by participating in routine household labor. National surveys and time-diary studies show that the average American woman performs about three times as much routine housework (i.e., cooking, meal clean-up, house cleaning, shopping, laundry) as the average man and that divisions of labor become more gender segregated after couples become parents (Coltrane, 2000; Robinson & Godbey, 1997; Thompson & Walker, 1989). Although most studies looked at either housework or parenting, the few studies that include both suggest that when men share more of the daily parenting, they also do more housework. One study using the National Survey of Families and Households (Coltrane & Adams, 2001) showed that when men participated in nurturing and supportive activities serving children, they also were more likely to share in the housework. In contrast, when fathers enacted fatherhood based on masculine recreation or family headship, they were less likely to share domestic work with their wives. Such findings suggest that it is important to disaggregate fathering behaviors into various categories (e.g., housework, transportation, play or sports, rule enforcement, supervision, monitoring, emotional support, shared activities of various types) and to evaluate whether the activities have similar or different predictors.

Predictors of Father Involvement

As Pleck's comprehensive reviews reveal (Pleck, 1997; Pleck & Masciadrelli, 2003), father involvement is multiply determined. No single factor is responsible for all types of father involvement, and studies often report contradictory effects for factors like family size, birth timing, socioeconomic status, or gender attitudes. For example, some studies found that fathers are more involved if they believe in gender equality (Baruch & Barnett, 1981; Blair, Wenk, & Hardesty, 1994), whereas others find no significant association between involvement and gender equality ideals (Pleck). A few factors, like education level, are consistently associated with higher reports of father involvement, though questions remain about how to interpret such results, or how much they are influenced by self-reporting biases and social desirability. Mothers with more education tend to report higher levels of father participation in interaction and responsibility, in part because they are more likely to endorse the view that fathers should be involved in children's lives (Coltrane, 1996; Pleck, 1997). One of the most consistent findings in studies of fatherhood is that men are more involved with sons than with daughters (e.g., Harris & Morgan, 1991; McBride, Schoppe, & Rane, 2002), especially among older children (Pleck, 1997). These findings, like most in family studies and social science, are based primarily on studies of White middle-class families.

Most research also finds that fathers' work hours are a strong predictor of interaction with children or participation in child supervision, especially when considered in conjunction with mothers' work hours. To illustrate, one study found that fathers spend about 2 hours interacting with children on weekdays, but over 6 hours on Sundays (McBride & Mills, 1993). When mothers of preschool children are employed, a father's employment hours predict his participation in routine child care (Casper & O'Connell, 1998), and nonoverlapping work shifts are one of the best predictors of fathers sharing routine child care (Presser, 1988). Fathers who work fewer hours and have more flexible work schedules tend to be more actively involved in child care than those who work long hours or who have inflexible schedules (Pleck, 1993). Fathers also tend to participate more to the extent that they view their wives' career prospects more positively, suggesting that mothers' human capital (e.g., education) and relative contributions to the household economy are predictive of greater father involvement (Pleck, 1997).

Predictors of men's participation in housework usually include relative resources, gender attitudes, and time availability (Coltrane, 2000). Relative resource predictions suggest that people with higher relative incomes or higher status jobs will do less housework than their spouses and that those with fewer resources will not be able to avoid doing the housework. Gender attitude predictions suggest that people socialized to believe in gender-segregated work will conform to those beliefs. Time availability predictions, as noted for child care above, suggest that when people spend more time in paid work, they spend less time in housework. Although methodological and conceptual issues abound, support for the first two explanations (resource and attitudes) often is mixed, whereas the third factor (time availability) usually is found to explain significant amounts of variance in the sharing of household labor (Coltrane, 2000). In general, wives who make more money do less housework, and those who earn more of the household income do proportionately less housework (though still more than their partners; see Blair & Lichter, 1991; Greenstein, 1996; Silver & Goldscheider, 1994). Of the time availability variables, women's employment hours have demonstrated the strongest and most consistent effects on women's absolute levels of family work and men's share of that work (e.g., Almeida, Maggs, & Galambos, 1993; Demo & Acock, 1993; Robinson & Godbey, 1997). Using national samples, researchers typically find that men who are employed fewer hours also do a greater share of the housework and child care (Brines, 1993; Greenstein, 1996; Waite & Goldscheider, 1992). In contrast, some studies find no relationship between men's employment hours and their housework (Almeida et al., 1993; John & Shelton, 1997; Sullivan, 1997). Once again, data for these conclusions are drawn from samples composed primarily of White middle-class couples.

Studies evaluating the relative contributions of men and women to household labor usually control for such variables as education, family income, financial stress, and family size. Although interpretation of education effects is complicated by conceptual confusion about whether years of education should be considered a measure of human capital accumulation, a relative resource, a component of social class, an indicator of attitudes, or something else, studies suggest that women with more education do less housework (Orbuch & Eyster, 1997; Pittman & Blanchard, 1996; Presser, 1994). In contrast, men with more education generally do more housework (Haddad, 1994; Orbuch & Eyster; Pittman & Blanchard, 1996; Presser, 1994;

South & Spitze, 1994). Attitudes about women being suited to tend homes and raise children remain linked to the allocation of labor in most American homes. Women's egalitarian gender ideology is a consistent predictor of housework sharing, and some studies also show that more egalitarian men share more housework (or child care; e.g., Almeida et al., 1993; Perry-Jenkins, & Crouter, 1990). Findings from parenting and household labor studies rarely consider the issue of social class, except to use income as a general control variable (with few consistent findings). In contrast, the economic stress literature tends to consider the effects of job loss or poverty, as noted above, but generally focuses on parental engagement, monitoring, and responsibility rather than other forms of household labor (Elder & Conger, 2000).

Although studies find that women still perform most of the housework and parenting, when men are more involved, women report that their division of labor is more fair, they are less depressed, and they enjoy higher levels of marital satisfaction (Coltrane, 2000). These findings, like those identifying predictors of men's housework and parenting, are based on samples that rarely include significant numbers of ethnic minority men [see Hill in this chapter]. . . . Some find that common predictor variables work somewhat differently for Blacks, in part because of more egalitarian attitudes and greater employment/earnings equality between spouses (Orbuch & Eyster, 1997). Findings are contradictory concerning the sharing of family work in Latino families, with some suggesting that there is slightly more sharing than among White families (Mirandé, 1997; Shelton & John, 1993) and some suggesting that there is less (Coltrane & Valdez, 1993; Golding, 1990). Most studies show similar patterns of association between predictor and outcome variables regardless of whether the couples are Latino or White (Coltrane & Valdez; Golding; Herrerra & Del Campo, 1995), but findings remain tentative, primarily because of the small number of Latino men in most studies and the lack of differentiation among different subgroups (i.e., Mexican, Latin American, South American, Cuban, Puerto Rican). In the analysis that follows, we assume that predictor variables found to be associated with greater father involvement in White families are likely candidates to be associated with greater father involvement in Mexican American families as well (e.g., more education, more liberal gender attitudes, more equal husband-wife employment and earnings). In addition, we examined the extent to which acculturation and familism might influence Mexican American men's involvement in routine parenting and housework. Following earlier research, we expected that higher levels of acculturation would lead to more egalitarian housework arrangements.

* * *

Although this study was exploratory in nature and was not strictly intended to test specific theoretical perspectives, our analytic strategy was premised on several abstract hypotheses. We posited that the greater the resources available to mothers, the more they will be able to entice fathers into doing "practical" family work, such as household labor and direct child supervision. Likewise, mother's and father's time availability were both anticipated to affect father's participation in "practical" family work, her greater work hours leading to more father involvement and his greater work hours leading to less father involvement. In addition to these relative resource and time availability predictions, we expected gender ideology to play a significant role in predicting the amount and type of father involvement. In particular, we expected that more egalitarian gender attitudes would be associated with more participation by fathers in feminine-typed activities and in child monitoring and that more traditional gender attitudes would be associated with more father involvement in masculine-typed activities. We also predicted that greater levels of mothers' education would be associated with more sharing of routine family work. Finally, contrary to straight-line assimilation models, we expected that, after controlling for multiple independent variables, low levels of acculturation and high levels of familism might be associated with greater father involvement among these low- to moderate-income Mexican American families.

FINDINGS*

Components of father involvement generally were correlated with each other. For example, men who performed housework also were likely to supervise children ($r = .549$, $p < .001$) and men who interacted with children in feminine-typed activities (e.g., shopping, cooking, reading, indoor games) also interacted with them in masculine-typed activities (e.g., hobbies, outdoor games, spectator entertainment; $r = .675$, $p < .001$). Child supervision and participation in feminine-typed activities were correlated with each other and with all other components of father involvement. Though positively related, monitoring children and interaction in masculine-typed activities were not correlated with participating in housework for fathers. This suggests that doing "guy stuff" like sports is not related to doing more service-oriented aspects of parenting like cooking for, cleaning up after, or doing laundry for children. Similarly, the monitoring scale, which includes items on rule setting (e.g., "I limit the amount of time my child can spend watching TV," "I expect my child to be in bed by a certain time"), was associated with being a parental authority figure and congruent with expectations for male household heads. Both of these components may reflect a more traditional approach to fathering (play, discipline), whereas the other three components of father involvement entail more quiet verbal interaction and domestic support (e.g., read or enjoy a book together, bake or cook a meal together, clean up after child, supervise child). Although most forms of father involvement appear to be mutually reinforcing, this analysis provides some support for claims in the literature that there are distinct styles of fathering combining various components in different measure. In this case, one style focused on the components of outdoor play and child discipline, and the other style focused on indoor care and child service (Coltrane, 1996; Parke, 1996).

Although previous depictions of Latino fathers portrayed them as aloof and uninvolved in daily family life (Madsen, 1973), our results suggest a more complex portrait of father involvement. Although not included in this analysis, compared to White (non-Latino) fathers in the larger study, the Mexican American fathers in this sample were more involved in both masculine-typed and feminine-typed interactions with their children, although they were less involved in housework. In the multivariate models presented here, father's Mexican identification was associated with higher levels of interaction in feminine-typed activities with children ($\beta = .183$, $p < .05$), and with more supervision of children ($\beta = .315$, $p < .001$), net of other independent variables (see Table 8.2). The Mexican identification variable reflects preference for speaking, reading, and watching TV and movies in Spanish, as well as having more contact with friends and relatives who identify as

---

*Authors' (KV and JS) Note: To help you in interpreting the data in this paper, you can follow these general guidelines. First, r and regression coefficients indicate the strength of the relationship between variables in the study, with r indicating the strength of the relationship between two variables and the regression coefficient indicating the strength of that variable in predicting the dependent variable (listed along the top of Table 8.2). R^2 and R^2 adjusted in Table 8.2 indicate the strength of all of the variables in predicting the dependent variable. All of these indicators range from 0 to 1 with 1 being a perfect relationship, that is the variable or variables predict the other or the dependent variable in every instance studied. For example, a .05 relationship is weak, but a .50 relationship is much stronger. These indicators can also be either negative or positive. If a number is positive, the higher the value a person has on one variable in the study, the higher that person is on the other variable. If the number is negative, the higher one variable is, the lower the other will be. For example, if education and father's involvement is .50, the higher father's level of education, the higher their involvement for this study. If, in the same example, it is -.50 then the higher the level of father's education, the lower the level of involvement.

The p value of <.05 to <.001 indicates something entirely different. It represents the likelihood that what you found could have occurred by chance through the process of randomly selecting the sample. In this case, lower is better. For example, you wouldn't want to rely on data that could have happened randomly one out of every two times you drew a sample (.5). In this case, the less likely you would find these results, the better (.001 is better than .05); however, there is general agreement that anything less than .05 (meaning 5 times out of 100 your results could be attributed to sampling error alone) is statistically significant and not likely to have occurred by sampling error alone.

Table 8.2 Standardized Regression Coefficients for Variables Predicting Different Components of Father Involvement ($n = 139$)

	Components of Father Involvement				
Independent Variable	*Proportion of household labor hours*	*Proportion of child supervision hours*	*Feminine-typed interaction with children*	*Masculine-typed interaction with children*	*Monitoring children*
Child sex (dummy-coded, female – 1)	.037	.008	.093	–.076	–.049
Mother's percent couple income	.415***	.080	–.141	–.162	–.131
Father's employment hours/day	.083	–.116	.081	.103	–.016
Mother's employment hours/day	.032	.189*	.100	.076	–.111
Number of children in home	–.090	–.197*	–.214**	–.164	–.152
Family rituals (father perception)	.074	.025	.286***	.195*	.212**
Mother's education	.272***	.223**	.268**	.248**	.174*
Father gender traditionalism	–.147*	–.165*	–.202**	–.156	–.192*
Father's level of Mexican identification	.016	.315***	.183*	.009	.093
Financial stress	.199**	.034	–.019	.007	–.038
Couple income	–.066	–.093	.081	.014	.154
F value	7.30***	3.87***	5.60***	3.40***	3.15***
R^2	.387	.251	.327	.227	.214
Adjusted R^2	.334	.186	.268	.161	.146

$*p < .05$. $**p < .01$. $***p < .001$. (two-tailed tests)

Mexican. Because all men in this analysis are of Mexican descent, the comparison group is one of more acculturated Mexican Americans (i.e., those who speak, read, and watch TV and movies in English and have more contact with English speakers and non-Latino Whites). This finding provides support for Mirandé's (1997) suggestion that Mexican men are labeled by the majority culture as macho and uninvolved in family life, when in fact they often exhibit high levels of commitment to family and spend considerable time interacting with their children in nurturing and emotional ways.

One of the factors pulling Mexican American men into more involved fathering is familism. In this study, the family rituals variable signifies family cohesion and commitment to family-level interaction. In our multivariate models, higher levels of family rituals predicted more monitoring of child by the father ($\beta = .212$, $p < .01$), and more interaction in both masculine-typed ($\beta = .195$, $p < .05$) and feminine-typed ($\beta = .286$, $p < .001$) activities (see Table 8.2). These families spent significant amounts of time and energy sustaining family-level group activities, and family cohesion appeared to spill over into more frequent father-child interactions.

With most families reporting incomes in the working-poor range, it is notable that the mean responses to the questions on economic stress reflect only *little* to *some* difficulty paying bills and having just enough money to make ends meet. These families are managing on limited resources, but their financial situations are not strongly influencing patterns of father involvement, except in the area of housework. Income was not related to any component of father-child interaction or parental monitoring. However, housework was more likely to be performed by fathers when the family was undergoing financial stress ($\beta = .199$, $p < .01$; see Table 8.2). Mother's percent of couple income also was significantly associated with father's share of housework ($\beta = .415$, $p < .001$), suggesting that performing domestic labor is something from which people symbolically "buy out." That is, even though women tend to do more housework than men, and even though men tend to earn more than women, when women's earnings increase, the burden of doing housework is more equally shared between them in these Mexican American families, just as it is in families of other ethnic groups. In contrast, other forms of father involvement are not associated with mother's relative earnings, with interaction and monitoring components negatively (but nonsignificantly) associated with relative earnings.

Employment variables also were infrequently associated with the different aspects of father involvement. Father's employment hours were not significant predictors in any of the models, and mother's employment hours reach significance only for father's proportion of child supervision hours ($\beta = .189$, $p < .05$). Number of children in the home was negatively correlated with all components of fatherhood, and coefficients reach significance for child supervision ($\beta = -.197$, $p < .05$) and feminine-typed interaction ($\beta = -.214$, $p < .01$). This finding can be interpreted as fathers doing more with each child when there are fewer of them (the interaction variables refer to contact with target child only), but it could reflect a tendency for mothers to become the childcare specialists in families with more children. The Mexican American families in this study had relatively large families, with an average of over three children ($M = 3.4$, $SD = 1.3$).

Mother's education was the most consistent predictor of father involvement, correlating with all measured components. This can be interpreted in multiple ways. Because the mean level of education among this group was *completing the ninth grade,* this does not necessarily signify advanced academic attainment. Rather, it captures literacy, sometimes high school graduation, and more rarely, some college attendance. Higher levels of education do signify more exposure to "expert" advice about child development, as opposed to sole reliance on one's own experiences and one's kin relations. Such exposure probably entails more acceptance of the idea that child development is enhanced by the participation of fathers in multiple aspects of parenting. Higher levels of education also can signify women's higher expectations for men's involvement in domestic activities and direct child supervision. In short, Mexican American women with more education may expect their husbands to do more parenting and housework.

Discussion

The macho stereotypes applied to Mexican American men can be misleading (Mirandé, 1997). Not only did the Mexican American men in this study share in many aspects of family work, but controlling for a host of demographic, economic, and family variables, Mexican-identified men were more likely to supervise their children and engage in feminine-typed interactions with them than were the more acculturated Mexican American men. We interpret these findings as confirmation of the influence of familism in Mexican American families—that is, high levels of family cohesion, cooperation, and reciprocity may encourage these men to focus on the health and well-being of their children and to interact with them in intimate ways. Parenthetically, we suspect that the familistic orientation of Mexican American fathers (and mothers) buffers their families from various risk factors associated with living in a low-income neighborhood, but this association remains to be tested. The emphasis in these families on eating evening meals together and participating in collective activities on the weekends provides frequent opportunities for fathers to interact with their children and monitor their activities. Based on the research of others (e.g., Patterson & Stouthamer-Loeber, 1984), we know that monitoring is a key to healthy adolescent development in high-risk environments.

The ways that fathers participate in family life are many and varied. Higher levels of participation in one form often generalized to other forms. However, it is important to specify the type and quality of involvement, as well as frequency. There are also multiple pathways to father involvement in these families, with some common and some distinct predictors. In general, men are more involved when their wives are more educated and when they believe that parenting and housework should be shared. These findings support gender theories focusing on the constraints of cultural ideals that dictate a strict separation of gender spheres. If men and women believe that tasks should be shared, they do share more of them, even if women remain the primary caregivers and housekeepers, and men remain the primary breadwinners.

Among the different components of father involvement, routine household labor is most influenced by a combination of gender beliefs and economics. Our models consistently explained more variance in this dependent variable than other aspects of fathering. If women earn more of the money and have more education, men are more likely to share the work of shopping, cooking, meal clean-up, housecleaning, and laundry—all tasks that increase greatly when children are in the home. When families experience financial hardship, men also are more likely to assume responsibility for household chores. The allocation of routine household labor in these working-poor Mexican-origin families, as in middle-class White families, responds to financial stress, relative resources, and ideology. These findings support a power-dependence model of household labor allocation (Coltrane, 2000) in which spouses with greater earnings can avoid onerous tasks. In particular, they suggest that when Mexican American families are dependent on wives' earnings for economic survival, wives may be more successful in recruiting husbands to share in the more mundane and routine household chores that typically fall to women.

Household labor was the only area strongly influenced by financial considerations in this analysis; the other forms of father involvement did not vary significantly as a function of family income, economic stress, or mother's share of earnings. Child supervision by the father was more likely when mothers were employed longer hours, but other forms of father involvement did not vary according to either fathers' or mothers' employment. These findings run counter to theories suggesting that time availability and work schedules are the primary determinants of men's involvement in family life. In these Mexican American families, time availability was much less important to father involvement than factors associated with familism, gender ideology, and education.

One conclusion is that fathering needs to be contextualized. By isolating similarities and differences between different forms of father involvement in this group of mostly low-income Latino families, we highlight the multidimensional nature of fatherhood and attempt to place fathering behaviors in their social contexts. We view simplistic attempts to reduce fathering to breadwinning, role modeling, or family headship as detrimental to scholarship on fatherhood and misleading for public debates about potential family policies. Fathers and mothers do many different things with and for families and children, and we need to pay attention to all of them—for both men and women. Further, fathering cannot be understood without paying serious attention to the social, economic, cultural, and family contexts in which it occurs. Our finding that family rituals were associated with more father-child monitoring and interaction suggests that low-income Latino men's parenting contributions should be studied in the context of the family system. Because these men shared more housework when their wives made more money and supervised children when their wives worked longer hours, we need to conceive of fathers' family work as sometimes directly substituting for mothers.' Because men did more housework when families were under economic stress, we also need to pay attention to the impact of larger economic forces. When families struggle to earn a livable wage, they also face numerous stressors associated with neighborhood safety, lack of access to health care, poor nutrition, and inadequate community services. The stressors associated with low income create pressures on men and women to share more family work, but such sharing does not alleviate the multiple adverse effects of living on the edge of poverty.

Mothers in these Mexican American families, like mothers in other families, did more parenting and housework than their husbands, but spouses with more egalitarian gender attitudes shared more of all but the most masculine-typed household and parenting duties. In other words, belief in the equality of men and women in marriage may be an important component of shared parenting in low-income Latino families, at least insofar as sharing parenting and household labor requires the performance of similar activities. Adding support to this interpretation is the finding that father involvement was associated with higher levels of education for women. Although the men in this sample tended to hold more traditional gender ideals than men in the general population, those who espoused a belief in the equality of men and women were more likely to be involved fathers. Because gender ideology was a consistent predictor of sharing parental duties in these families, and because second- and third-generation Mexican Americans tend to hold more liberal gender attitudes than their parents (Zinn & Wells, 2000), we might expect even more sharing of family work in the future.

Our findings lend support to a pluralistic model of father involvement in Mexican American families that simultaneously acknowledges unique cultural influences, economic opportunities, human capital, and changing gender relations. As researchers and agenda setters for public policy debates, scholars and practitioners ought to examine the multitude of ways that men can share in the parenting and domestic labor needed to raise healthy children. To that end, we advocate further research into the many ways that fathers of all ethnicities and income levels contribute to their families and children.

References

Adams, M., & Coltrane, S. (in press). Boys and men in families. In M. Kimmel, R. W. Connell, & J. Hearn (Eds.), *Handbook on men and masculinities.* Thousand Oaks, CA: Sage.

Almeida, D. M., Maggs, J. L., & Galambos, N. L. (1993). Wives' employment hours and spousal participation in family work, *Journal of Family Psychology, 7,* 233–244.

Baruch, G., & Barnett, R. (1981). Fathers' participation in the care of their preschool children. *Sex Roles, 7,* 1043–1055.

Blair, S. L., & Lichter, D. T. (1991). Measuring the division of household labor: Gender segregation of housework among American couples. *Journal of Family Issues, 12,* 91–113.

Blair, S. L., Wenk, D., & Hardesty, C. (1994). Marital quality and paternal involvement: Interconnections of men's spousal and parental roles. *Journal of Men's Studies, 2,* 221–237.

Brewster, K. L. (2000, March). *Contextualizing change in fathers' participation in child care.* Paper presented at Work and Family: Expanding the Horizons conference, San Francisco, CA.

Brines, J. (1993). The exchange value of housework. *Rationality and Society, 5,* 302–340.

Buriel, R. (1986). *Latino value systems and their educational implications.* Unpublished manuscript, Pomona College, Claremont, CA.

Casper, L. M., & O'Connell, M. (1998). Work, income, the economy, and married fathers as childcare providers. *Demography, 35,* 243–250.

Cauce, A. M., & Rodriguez, M. D. (2001). Latino families: Myths and realities. In J. M. Contreras, K. A. Kerns, & A. M. Neal-Barnett (Eds.), *Latino children and families in the United States.* Westport, CT: Greenwood.

Coltrane, S. (1996). *Family man: Fatherhood, housework, and gender equity.* New York: Oxford University Press.

Coltrane, S. (2000). Research on household labor: Modeling and measuring the social embeddedness of routine family work. *Journal of Marriage and the Family, 62,* 1209–1233.

Coltrane, S. (2001). Marketing the marriage "solution." *Sociological Perspectives, 44,* 387–422.

Coltrane, S., & Adams, M. (2001). Men's family work: Child centered fathering and the sharing of domestic labor. In R. Hertz & N. Marshall (Eds.), *Work and family: Today's realities and tomorrow's vision* (pp. 72–99). Berkeley: University of California Press.

Coltrane, S., & Adams, M. (2003). The social construction of the divorce "problem": Morality, child victims, and the politics of gender. *Family Relations, 52,* 363–372.

Coltrane, S., & Parke, R. D. (1998). *Reinventing fatherhood: Toward an historical understanding of continuity and change in men's family lives.* Philadelphia: National Center on Fathers and Families (WP 98–12A).

Coltrane, S., & Valdez, E. (1993). Reluctant compliance: Work-family role allocation in dual earner Chicano families. In J. Hood (Ed.), *Men, work, and family* (pp. 151–174). Newbury Park, CA: Sage.

Demo, D. H., & Acock, A. C. (1993). Family diversity and the division of domestic labor: How much have things really changed? *Family Relations, 42,* 323–331.

Elder, G. H., Jr., & Conger, R. D. (2000). *Children of the land: Adversity and success in rural America.* Chicago: University of Chicago Press.

Golding, J. M. (1990). Division of household labor, strain, and depressive symptoms among Mexican Americans and Non–Hispanic Whites. *Psychology of Women Quarterly, 14,* 103–117.

Gonzales, N. A., Knight, G. P., Morgan-Lopez, A., Saenz, D. S., & Sirolli, A. (2002). Acculturation, enculturation and the mental health of Latino youths: An integration and critique of the literature. In J. M. Contreras, K. A. Kerns, & A. M. Neal-Barnett (Eds.), *Latino children and families in the United States* (pp. 45–74). Westport, CT: Greenwood.

Greenstein, T. N. (1996). Husbands' participation in domestic labor: Interactive effects of wives' and husbands' gender ideologies. *Journal of Marriage and the Family, 58,* 585–595.

Haddad, T. (1994). Men's contribution to family work: A re-examination of "time-availability." *International Journal of Sociology of the Family, 24,* 87–111.

Harris, K. M., & Morgan, S. P. (1991). Fathers, sons and daughters: Differential paternal involvement in parenting. *Journal of Marriage and the Family, 53,* 531–544.

Herrera, R. S., & Del Campo, M. (1995). Beyond the superwoman syndrome: Work satisfaction and family functioning among working-class, Mexican American women. *Hispanic Journal of Behavioral Sciences, 17,* 49–60.

John, D., & Shelton, B. A. (1997). The production of gender among Black and White women and men:

The case of household labor. *Sex Roles, 36,* 171–193.

Lamb, M. E., Pleck, J., Charnov, E., & Levine, J. (1985). Paternal behavior in humans. *American Zoologist, 25,* 883–894.

Levine, J., & Pittinsky, T. L. (1997). *Working fathers: New strategies for balancing work and family.* New York: Harcourt Brace.

Madsen, W. (1973). *Mexicans-Americans of South Texas.* New York: Holt, Rinehart and Winston.

McBride, B. A. (1989). Stress and fathers' parental competence: Implications for family life and parent educators. *Family Relations, 38,* 385–389.

McBride, B., & Mills, G. (1993). A comparison of mother and father involvement with their preschool age children. *Early Childhood Research Quarterly, 8,* 457–477.

McBride, B., Schoppe, S., & Rane, T. (2002). Child characteristics, parenting stress, and parental involvement: Fathers versus mothers. *Journal of Marriage and Family, 64,* 998–1011.

Mirandé, A. (1997). *Hombres y machos: Masculinity and Latino culture.* Boulder, CO: Westview.

Orbuch, T. L., & Eyster, S. L. (1997). Division of household labor among Black couples and White couples. *Social Forces, 76,* 301–332.

Parke, R. D. (1995). Fathers and families. In M. H. Bornstein (Ed.), *Handbook of Parenting* (Vol. 3, pp. 27–63). Hillsdale, NJ: Erlbaum.

Parke, R. D. (1996). *Fatherhood.* Cambridge, MA: Harvard University Press.

Parke, R. D. (2000). Father involvement: A developmental psychological perspective. *Marriage and Family Review, 29,* 43–58.

Parke, R. D., & Tinsley, B. (1984). Fatherhood Historical and contemporary perspectives. In K. A. McCloskey & H. W. Reese (Eds.), *Life-Span developmental psychology: Historical and generational effects* (pp. 429–457). New York: Academic Press.

Patterson, G. R., & Stouthamer-Loeber, M. (1984). The correlation of family management and delinquency. *Child Development, 55,* 1299–1307.

Perry-Jenkins, M., & Crouter, A. C. (1990). Men's provider-role attitudes: Implications for household work and marital satisfaction. *Journal of Family Issues, 11,* 136–156.

Pittman, J. F., & Blanchard, D. (1996). The effects of work history and timing of marriage on the division of household labor: A life-course perspective. *Journal of Marriage and the Family, 58,* 78–90.

Pleck, J. H. (1993). Are "family-supportive" employer policies relevant to men? In J. C. Hood (Ed.), *Men, work, and family* (pp. 217–237). Newburry Park, CA: Sage.

Pleck, J. H. (1997). Paternal involvement: Levels, sources, and consequences. In M. E. Lamb (Ed.), *The role of the father in child development* (3rd ed., pp. 66–104). New York: Wiley.

Pleck, J. H., & Masciadrelli, B. P. (2003). Paternal involvement: Levels, sources, and consequences. In M. E. Lamb (Ed.). *The role of the father in child development* (4th ed.). New York: Wiley.

Presser, H. B. (1988). Shift work and child care among young dual-earner American parents. *Journal of Marriage and the Family, 50,* 133–148.

Presser, H. B. (1994). Employment schedules among dual-earner spouses and the division of household labor by gender. *American Sociological Review, 59,* 348–364.

Robinson, J. P., & Godbey, G. (1997). *Time for life: The surprising ways that Americans use time.* University Park: Pennsylvania State University Press.

Segura, D. (1992). Chicanas in white-collar jobs. *Sociological Perspectives, 35,* 163–182.

Shelton, B. A., & John, D. (1993). Ethnicity, race, and difference: A comparison of White, Black, and Hispanic men's household labor time. In J. C. Hood (Ed.). *Men, work, and family* (pp. 131–150). Newbury Park, CA: Sage.

Silver, H., & Goldscheider, F. (1994). Flexible work and housework: Work and family constraints on women's domestic labor. *Social Forces, 72,* 1103–1119.

South, S. J., & Spitze, G. (1994). Housework in marital and nonmarital households. *American Sociological Review, 59,* 327–347.

Sullivan, O. (1997). The division of housework among "remarried" couples. *Journal of Family Issues, 18,* 205–223.

Thompson, L., & Walker, A. J. (1989). Gender in families: Women and men in marriage, work, and parenthood. *Journal of Marriage and the Family, 51,* 845–871.

Vega, W. A., Kolody, B., Valle, R., & Weir, J. (1991). Social networks, social support and their relationship to depression among immigrant Mexican women. *Human Organization, 50,* 154–162.

Waite, L., & Goldscheider, F. K. (1992). Work in the home: The productive context of family relationships. In S. J. South & S. E. Tolnay (Eds.), *The changing American family* (pp. 267–299). Boulder, CO: Westview.

Washington Post. (1998). With More Equity, More Sweat, by Richard Morin and Megan Rosenfeld (March 22, 1998), and unpublished Gender Poll Results. Washington DC: Kaiser Family Foundation/WashingtonPost/Harvard University.

Zinn, M. B., & Wells, B. (2000). Diversity within Latino families: New lessons for family social science. In D. H. Demo, K. R. Allen, & M. A. Fine (Eds.). *Handbook of family diversity* (pp. 252–273). New York: Oxford University Press.

Introduction to Reading 40

In this reading, Gillian Dunne analyzes data from a study called the Lesbian Household Project. The study employed qualitative and quantitative methods to illuminate the work and parenting experiences of thirty-seven cohabiting lesbian couples with dependent children across England. Dunne notes that both the sensitive nature of the research topic and the invisibility of lesbians in the population make it difficult to claim that the sample is representative. Notwithstanding, Dunne's research into the lives of women parenting together reveals a world in which they challenge the connections between biological and social motherhood, and often put egalitarian ideals into practice.

1. Why does Dunne view heterosexuality and gender as playing a central role in reproducing gender inequality in parenting? How does her argument compare with that of Tichenor and Coltrane's, Parke's, and Adams' readings earlier in this chapter?
2. Why do women parenting with women have a "head start" over heterosexual couples with respect to egalitarianism?
3. How do women parenting together challenge the connections between biological and social motherhood?

Opting Into Motherhood

Lesbians Blurring the Boundaries and Transforming the Meaning of Parenthood and Kinship

Gillian A. Dunne

The extension of educational and employment opportunities for women, together with widening experience of the "plastic" nature of sexualities (Giddens 1992, 57), has enabled increasing numbers of Western women to construct independent identities and lifestyles beyond traditional marriage, motherhood, and indeed, heterosexuality (Dunne 1997). As contemporary women's identities expand to incorporate the expectations and activities that have been traditionally associated with masculinity, there has not been an equivalent shift of male identity, let alone practice, into the traditional domains of women. Exceptions not withstanding (Blaisure and Allen 1995; Doucet 1995; Ehrensaft 1987; VanEvery 1995), a distinctly asymmetrical division of labor remains the majority pattern (Berk 1985; Brannen and Moss 1991; Ferri and Smith 1996; Gregson and Lowe 1995; Hochschild 1989). The intransigent nature of the gender division of labor means that women continue to perform the bulk of domestic work and that mothers bear the brunt of the social and economic penalties associated with caring for children. Men's relative freedom from the time constraints and labor associated with the home and parenting enables them to be more single-minded in the pursuit of employment opportunities and retain their labor market advantages. . . .

While contemporary women begin to see the demands of motherhood as conflicting with their newly won bid for autonomy, there has been a recent shift in attitudes toward parenting among the lesbian population. A rising awareness of alternatives to heterosexual reproduction has led to the growing recognition that their sexuality does not preclude the possibility of lesbian and gay people having children. In Britain and in the United States, we are witnessing the early stages of a "gayby" boom, a situation wherein lesbian women and gay men are opting into parenthood in increasing numbers. According to Lewin, "The 'lesbian baby boom' and the growing visibility of lesbians who became mothers through donor insemination constitute the most dramatic and provocative challenge to traditional notions of both family and of the non-procreative nature of homosexuality" (1993, 19). In this article, I want to address this apparent contradiction between childlessness as resistance and lesbian motherhood as provocative challenge by showing that the mothering experiences that lesbians are opting into are qualitatively different from those that some women seek to avoid.

I take the view that sexuality is socially and materially constructed and that heterosexuality plays a central role in reproducing gender inequality (Dunne 1997, 1998d, 2000). The dominance of heterosexuality is the outcome of institutional processes that render alternatives undesirable and/or unimaginable (Dunne 1997; Rich 1984) and that construct gender difference and gender hierarchies (Butler 1990, 17; Rubin 1975). Consequently, there is a crucial relationship between gender and sexuality. . . .

I wish to support and extend Lewin's observations on single lesbian mothers by drawing on my work on lesbian couples who have become

---

From Dunne, G. "Opting into Motherhood: Lesbians Blurring the Boundaries and Transforming the Meaning of Parenthood and Kinship" *Gender & Society 14*(1): 11–35.

parents via donor insemination. I argue that an attentiveness to the gender dynamic of sexuality illuminates additional challenges that arise when women combine with women to rear children—the possibility of showing what can be achieved when gender difference as a fundamental structuring principle in interpersonal relationships is minimized (see Dunne 1997, 1998a). I suggest a complex and contradictory situation for lesbians who have opted into motherhood via donor insemination. By embracing motherhood, lesbians are making their lives "intelligible" to others—their quest to become parents is often enthusiastically supported by family and heterosexual friends. However, their sexuality both necessitates and facilitates the redefinition of the boundaries, meaning, and content of parenthood. When women parent together, the absence of the logic of polarization to inform gender scripts, and their parity in the gender hierarchy, means that, to borrow Juliet's words, "We have to make it up as we go along." Their similarities as women insist on high levels of reflexivity and enable the construction of more egalitarian approaches to financing and caring for children. In this way, some of the more negative social consequences of motherhood can be transformed. Although not unique in their achievements, nor assured of their success, women parenting with women have a head start over heterosexual couples because of their structural similarities and the way that egalitarianism is in the interests of both partners.

* * *

Parenting Circumstances

The sample includes 8 households where children were from a previous marriage, 1 household where the children were adopted, and 28 (75 percent) where they had been conceived by donor insemination. In the majority of households (60 percent), there was at least one child younger than five; and in 40 percent of households, coparents were also biological mothers of older, dependent, or nondependent children. The research revealed a fairly unique and important opportunity for women parenting together—the possibility of detaching motherhood from its biological roots through the experience of social motherhood. Interestingly, 15 women in the study expressed a long-standing desire to mother as a social experience but a strong reluctance to experience motherhood biologically. These women had often taken responsibility for siblings in their families of origin and for the children of others usually featured in their lives and occupational choices. This social-biological separation also meant that motherhood is not necessarily ruled out for women who have fertility problems. Parenting was depicted as jointly shared in 30 households (80 percent). As we will see in the three case studies, in contrast to men who share mothering (Ehrensaft 1987) yet remain happy with the identity of father, the singularity and exclusivity of the identity of mother represented a major problem for women parenting together.

[A]lmost all of the women who had experienced donor insemination organized this informally—they rarely used National Health or even private fertility services. Respondents tended to want to know the donor, and in 86 percent of households, this was the case. A wide range of reasons was given for this preference. A common feeling related to wanting to know that a good man, in terms of personal qualities, had a role in creating their child. Sometimes more specific ideas about biogenetic inheritance came up in discussions, and for Jewish women there was a preference for Jewish donors. Some employed the metaphor of adoption—the idea that children should have the option of knowing their biological father at some stage in the future. Commonly, donors were located through friendship networks or by advertising. Occasionally, they made use of the informal women's donor networks that exist in many British cities. When organized informally, children were always conceived by self or partner insemination, and the majority became mothers in their current lesbian relationship.

Lesbian motherhood undermines a core signifier of heterosexuality and challenges heterosexual monopoly of and norms for parenting. The social hostility toward those parents and children who transgress the sanctity of heterosexual reproduction is such that the decision to become a mother by donor insemination can never be easily made. Typically, respondents described a lengthy period of soul-searching and planning preceding the arrival of children. For some, this process lasted as long as seven years. Unlike most women, they had to question their motives for wanting children, to critique dominant ideas about what constitutes a "good" mother and family, and to think about the implications of bringing up children in a wider society intolerant of difference. Informing this process was much research—reading the numerous self-help books that are available on lesbian parenting, watching videos on the topic, and attending discussion groups. I would suggest that lesbian parenting via donor insemination is the "reflexive project" par excellence described by Giddens (1992, 30). For respondents in partnerships, a central part of this process was the exploration of expectations in relation to parenting, for example, attitudes to discipline, schooling, and if and how far responsibilities would be shared. Key considerations related to employment situations. Respondents did not expect or desire a traditional division of labor, and thus timing was often influenced by their preference to integrate child care and income generation. In the meanwhile, potential donors were contacted. Respondents described a fairly lengthy process of negotiation with donors that focused on establishing a mutuality in parenting expectations and, if he was previously unknown to the couple, getting to know each other and developing confidence. While recognizing the generosity of potential donors, some were rejected because of personality clashes or concerns about motives, but more usually, rejection was because a donor wanted too much or too little involvement.

Men featured in the lives of most of the children, and it was not unusual for donors to have regular contact with their offspring (40 percent of households); in three households, fathers were actively co-parenting. This involvement was usually justified in terms of providing children with the opportunity to "normalize" their family arrangements by being able to talk to peers about doing things with father. Donors were usually gay men—and all male co-parents were gay. This preference appeared to be based on three main assumptions. First was the respondents' perceptions of gay men as representing more aware, acceptable, and positive forms of masculinity. Their desire to involve men (donors or other male friends) in the lives of children, particularly boys, was often described as being about counteracting dominant stereotypes of masculinity. Second, because of the particularities of gay men's lifestyles, respondents believed that they would be less likely to renege on agreements. Third, they thought that should a dispute arise, a heterosexual donor (particularly if he were married) had greater access to formal power to change arrangements in relation to access and custody. That none expressed any serious difficulties in relation to father and/or donor involvement attests to the value of the careful negotiation of expectations before the arrival of children. It also says much about the integrity and generosity of the men concerned, although it must be noted that most had preschool-age children, and conflicts of interests may come as the children mature.

In situations where children had been conceived in a previous marriage or heterosexual relationship, there was more diversity and conflict regarding fathers' involvement. In several cases, the father had unsuccessfully contested custody on the grounds of the mother's lesbianism. Indeed, two had appeared on daytime television arguing that their ex-wives' sexuality conflicted with their capacity to be good mothers. There were also examples of good relations between mothers and ex-husbands. While there were several examples of fathers having lost contact with their children, in most cases, respondents suggested that the child or children had more quality time with their fathers after divorce than before. Despite tensions and possible conflict between mothers and ex-husbands, these respondents suggested that they worked hard to maintain their children's relationships with the fathers. Thus, ironically, in this group as well as in the donor insemination group of parents, there are examples of highly productive models of cooperation between women and men in parenting.

The role of fathers and/or donors and other male friends in children's lives reminds us that lesbian parenting does not occur in a social vacuum. While generally hostile to the idea of the privatized nuclear family, respondents were keen to establish more extended family networks of friends and kin. Often, respondents described the arrival of children as bringing them closer to or helping repair difficult relations with their families of origin. Typically, they described a wide circle of friends (lesbian "aunties," gay "uncles," and heterosexual friends) and kin supporting their parenting.

I now want to illustrate some of these themes by drawing on the voices of respondents in three partnerships where parenting was shared and where men were involved.

Vivien[1] and Cay's Story

. . . It was not uncommon to find a woman who had been married and had grown-up children who was starting over again with a partner who wanted to have children herself ($n = 4$). Women parenting together was understood as offering the opportunity to experience parenting in new and exciting ways that were tempered by the wisdom that comes from already having raised children. Cay and Vivien are fairly typical of these households. Vivien, age 44, has a grown-up son, Jo, who lives independently. Cay, age 32, is the biological mother of two boys, Frank, age four, and Steve, age two. When we first met, they had been living together for six years in a small terraced house in inner-city Birmingham. Cay, born in North America, is a self-employed illustrator of children's books who supplements her income by working as a cleaner. Vivien, of Irish-Greek descent, recently completed a degree and acquired her "first real job" as a probation officer. Cay told me that she had always wanted to have children and that her sexuality had not changed this desire. Vivien was enthusiastically supportive of the idea although she did not want to go through a pregnancy herself.

Like the vast majority of respondents, Vivien and Cay organized donor insemination informally. They had little difficulty in locating a willing donor—John, an old friend of Vivien's.

> Vivien: It worked out well. He's my oldest friend, and we've known each other since we were teenagers, and he has the same kind of coloring and stuff, he could be my brother in terms of coloring and looks. Originally we asked one of my brothers to donate, and he felt he would maybe want more of an involvement, more of a say in the children's lives, and we wanted somebody who would let us have the responsibility and would take on a sort of a kindly uncle role. And John agreed to do that.

The description "kindly uncle" was frequently used by respondents to describe what was a fairly limited yet enthusiastic relationship between a donor and his child or children. Respondents almost always wanted to retain responsibility for bringing up their children. Like most of the couples in the study, Vivien and Cay regard these responsibilities as shared.

> Cay: It can't be anything but joint I think. The way we've approached it is that if it's not totally agreeable between both of us, it couldn't have really gone forward, given the kind of relationship we have. We've seen other people, you know, where one parent has said, "Well, I want a child and that's it." But the other one says, "Yes, you can have one, but I don't want to have lots of responsibility." That's not our way.

When respondents described their parenting as jointly shared, they meant that each partner took an active role in the routine pleasures, stresses, and labor of child care. . . .

Like Vivien, respondents took great pleasure in child care, and this was reflected in their ordering of priorities. Cay suggested that because she and Vivien had joint responsibility for housework, they were less subject to the tyranny of maintaining high domestic standards—a sentiment reflected across the sample. This, together with their shared approach to doing tasks, she believed, gave them more fun time with their children—this was supported in the time-use data across the sample. For example, comparison of respondents' time use with trends for married parents with young children revealed that regardless of the employment status of married mothers, because they did the bulk of routine domestic work, it occupied far more of their time than child care, while the reverse was the case for respondents (Dunne 1998a).

Vivien and Cay described their roles before and after the arrival of children as interchangeable; earlier, Cay had been the main earner when Vivien was a student. Routinely, birth mothers and co-parents alike spoke of seeking integrated lives—valuing time with children, an identity from the formal workplace, and the ability to contribute financially. Within reason, they were prepared to experience a reduced standard of living to achieve the kind of quality of life desired. Thus, there was an unusually wide range of partner employment strategies in the sample. Like Vivien and Cay, some took turns in who was the main earner, while others (a quarter of the sample) opted for half-time employment for both parents. Rather than the polarization of employment responsibilities that characterizes married couples' parenting experiences, particularly when children are young,[2] few households had extreme partner differences in employment hours, and being the birth mother was a poor predictor of employment hours. . . .

While Vivien and Cay describe themselves as the boys' mothers, in common with most respondents, they struggle over terminology to describe and symbolize that relationship. Because of the singularity and exclusivity of the label *mother* or *mum* and/or their feminist critique of the way the term can eclipse other important aspects of identity, respondents often preferred to encourage the use of first names, special nicknames, or the word *mother* borrowed from another language.

Vivien: Yes, [we are the boys' mothers] absolutely, yes. Very much so.

Interviewer: What do they call you?

Vivien: By our names. . . . They very rarely use the word *mother.*

Cay: In fact [Frank] never used the word *mother* until he started going to school, and then, hearing the other kids saying it, it was just a kind of copying thing.

Here we catch a glimpse of some of the everyday pressures toward social conformity and the dilemmas experienced by parents and children as they negotiate a world hostile to difference. Just as this motivated some to involve donors in their children's lives so that they had the option to pass as relatively "normal" in school, many respondents relented and used the term *mother* to describe the biological mother. . . .

However, the very positioning outside conventionality that enables the construction of more creative approaches to organizing parenting brings also the problem of lack of recognition and validation from the outside world. Vivien speaks for many in the study:

> I think we have to acknowledge that within this house we can sit down and we can talk about the equality that we feel and the experiences that we have and the confidence that we have in our relationship and in our parenting. But very little outside of this house tells us that those things that we're talking about tonight are actually true. . . . I think heterosexual friends that we have tend to probably see our relationship in their own terms. . . . I don't think they've got an insight into how much we really do work together. . . . You know, we have to work at it all the time, we have to forge links with the school, we have to forge links with this and forge links with that, we have to work hard at being good neighbors and making contact with the neighbors so that as the children come along they're not surprised and they can adjust. We're doing the work, we're doing the outreach, we're doing the education, and what we get back is the right to be ourselves, sort of, as long as we're careful.

Again, their experience underscores the difficulties associated with challenging the normative status of heterosexuality in relation to reproduction and the organization of parenting roles. Constantly, these pioneering women feel obliged to justify their alternative families and approaches to parenting to a wider society that cannot see beyond the constraints of heterosexuality and that is informed by media representations that vilify lesbian parents. Their struggle for validation was not confined to the heterosexual world.

> Vivien: Other lesbians I think may see its as trying to repeat some sort of heterosexual relationship, and that's not what we're trying to do. So we have to kind of justify it to our heterosexual friends and justify it to our lesbian friends.

The contradiction illuminated here between being a lesbian and being a mother serves to remind us that while it can be argued that assisted conception is an important expression of the ideologies supporting compulsory motherhood, it is less easy to apply this thinking to lesbian mothers. Within lesbian culture, the absence of children within a relationship does not constitute failure. In fact, research (Sullivan 1996) supports much of what respondents said about their decision to have children going against established societal norms, specifically those of the lesbian and gay community. Until recently, this community, particularly the radical or revolutionary wing, has been suspicious of motherhood because of fears of constraints on women's autonomy and the importation of oppressive family arrangements (see Green 1997).

Thelma and Louise's Story

* * *

It was not unusual for both partners to have experienced biological motherhood as the result of donor insemination while in their relationship. At the time of first contact, four couples were in this situation (this number had risen to seven at the follow-up stage two years later). In these households, children were brought up as siblings, and parenting was equally shared. The experiences of Thelma and Louise are not atypical of mothers in this situation. They have been living together for seven years in an apartment that they own in inner-city Manchester. They have two daughters, Polly, age four, and Stef, age two. Thelma works in desktop publishing, and Louise is a teacher. Like many in the sample, Thelma and Louise operationalize shared parenting by reducing their paid employment to half-time. They both wanted to have children; their decisions about timing and who would go first were shaped by emotional and practical considerations. Thelma needed to build up sufficient clientele to enable self-employment from home, and Louise wanted to gain more secure employment. . . .

By the time Louise was pregnant, two years after Thelma, she was in a much stronger position at work, having undergone retraining. She had secured a permanent position in teaching and, after maternity leave, arranged a job share with a friend. Like women more generally, respondents' careers had rarely progressed in a planned linear manner. Instead, their job histories have a more organic quality (see Dunne 1997)—moving across occupations and in and out of education or training. However, in contrast to married women more generally, where the gender division of labor supports the anticipation of financial dependence on husbands when children are young (Mansfield and Collard 1988), an important consideration in the timing of the arrival of children for most biological mothers in this study was the achievement of certain employment aims that would enable greater financial security and allow time to enjoy the children. Their gender parity and this approach to paid employment meant that there were not major earning differentials between partners. . . .

After several miscarriages with an earlier donor, Thelma finally got pregnant. Again, they used their friendship networks to locate a donor who then took on a "kindly uncle" role. . . .

Interviewer: And what will Polly call her donor?

Thelma: His name—and she calls him Daddy Paul. So I mean she doesn't ever really call him Daddy. Either she calls him Paul or Daddy Paul.

Louise: He is a bit like an uncle [to them both] she'd see now and again, you know, he'd be like this kindly uncle figure,

> who'd take her to the pics and take her to the zoo and that kind of thing. Give her treats.

They originally planned that Paul would be the donor for Louise; however, there were difficulties in conception, so a new donor was found. Hugh, a gay friend of Thelma's brother, who was temporarily living in England, agreed. While Thelma and Louise both wanted to experience motherhood biologically, they viewed parenting as shared, and this situation was legally recognized in their gaining of a joint parental responsibility order. . . .

Again, their interpretation of shared parenting brought them up against the limitations of language to describe a social mother's relationship to a child.

> *Thelma:* They both call us Mum.
>
> *Louise:* It started off that you were going to be Mum and I was going to be Louise, and then coming up to me giving birth to Stef, it just got a bit kind of funny, so we thought it's not really going to work any more because if they're sisters how come?—it just all didn't work, so now we're both Mums. And they just call us Mum.
>
> *Thelma:* Stef says Mummy Louise or Mummy Thelma.
>
> *Louise:* And Polly mostly calls us Louise and Thelma doesn't she?
>
> *Thelma:* Yeah she does. She calls us both Mum when she wants to, but mostly she calls us by our names.
>
> *Louise:* The last couple of years she's started calling me Mum.

Some of the immensity of the creative project in which lesbians engage is revealed in the tensions in the last two extracts and in the next. While they describe the children as having two mothers, Louise reminds us of the contingent nature of this. The rule of biological connection is unquestioned in the assumption that in the event of a breakup each will depart into the horizon with her own child. This next extract illustrates other practical difficulties faced by the couple as they engage with the wider society.

> Louise: It's a lot easier now because we've both had a child. I don't think I had any role models in terms of being a nonbiological mum. There's a thing that if you want to be acknowledged as a parent, you just had to "come out." It's the only way to explain that you're a parent. And even that is a very hard way to explain you're a parent. My inner circle at work would know and it's funny—I nearly wrote it down one day—because it was just like some days I'd be a parent and some days I wasn't. So it would depend on what day of the week it was and who I was talking to. I think I made it harder for us by me not being called Mum [in the early stages]. Because as soon as people found out you weren't the mum, then they'd just—it was like "who the hell are you then?"

Such is the power of ideas about the singularity and the exclusivity of the identity of "Mum" in a social world structured by heterosexual norms that polarize parenting along lines of gender. Respondents had a store of both amusing and uncomfortable stories about other people's confusions about who was the mother of the child or children or the status of social mothers. . . .

Without exception, respondents believed that they approached and experienced parenting in ways that were very different from the heterosexual norm. They were redefining the meaning and content of motherhood, extending its boundaries to incorporate the activities that are usually dichotomized as mother and father. Going against prevailing norms was never without difficulties and disappointments. In joint and individual interviews, respondents usually singled out the ability and commitment to communicate as crucial. They spoke of arrangements being constantly subject to negotiation and the need to check in regularly with each other so that routines that may lead to taking the other for granted could be rethought and sources of conflict discussed.

Bonnie and Claudia's Story

* * *

In three partnerships, donors were actively coparenting from separate households—becoming a "junior partner in the parenting team" as one father described himself. In two cases, the father's parenting was legally recognized in a joint residency order. Bonnie, Claudia, and Philip share the care of Peter, age two. Bonnie and Claudia have lived together for nine years in a terraced house in inner-city Bristol. Bonnie, Peter's biological mother, works full-time in adult education, and Claudia has a half-time teaching post. They describe and contrast their feelings about wanting to have children. . . .

Their experience illustrates another fairly unique advantage for women who want to become mothers in a lesbian relationship—if one partner has fertility problems, the other may agree to go through the pregnancy instead. There were three other examples of partners swapping for this reason, and several others expressed their willingness to do so. . . .

In their negotiations with Philip over the four years that preceded the birth of Peter, they came to the decision that he would be an actively involved father.

> Claudia: Philip wanted a child, and he, I think, was also looking for a kind of extended family relationship, wasn't he?—with us and the children. But he also wants his freedom, I suppose, his lifestyle, a lot of which he needs not to have children around for. Yes, so it fits in the sense that what we get is time without Peter, to have a relationship that needs its own sort of nurturing and stuff, and he gets special time with Peter and a real bonding. I mean he's seen Peter every day since he's been born. So he has become part of the family, hasn't he?—in a sense, or we've become part of his. But we live in two separate homes. People sometimes don't realize that.

Claudia's words alert us to another underlying reason for respondents' confidence in fathers and/or donors retaining a more minor role in children's lives—routine child care does not usually fit in with the lifestyles of most men, gay or heterosexual. The masculine model of employment that governs ideas of job commitment and what constitutes a valuable worker is based on the assumption that employees are free from the constraints of child care.

After extended maternity leave, Bonnie returned to her successful career in adult education. At this point, Claudia, despite being the higher earner, reduced her employment hours to half-time so that she could become Peter's main caregiver. Men's superior earnings are often described by egalitarian-minded heterosexual couples as ruling out opportunities for shared parenting (Doucet 1995; Ehrensaft 1987). However, women parenting together, without access to ideologies that polarize parenting responsibilities, bring fresh insights to this impasse, which supports gender inequalities.

Bonnie: We started in a completely different place [from heterosexual couples]. I think we feel it's just much easier to be cooperative and to be more creative in the way that we share out paid work and domestic work, because that's how we look at it. We're constantly chatting about it, aren't we, over the weeks, and saying, "How does it feel now? Are you still thinking about staying on part-time?" and we've talked about what it would be like if I went part-time as well, and could we manage on less money?

Claudia: Yes, and I think the thing that's part of the advantage is that in a conventional setup, although it may be easier to start with, everyone knowing what they are supposed to be doing, but the men don't know their children so they miss out. . . . I'm having a balanced life really.

Bonnie: I think that's why we've got the space to enjoy our child in a way that a lot of heterosexuals perhaps don't. It's so easy to fall in—the man earns slightly more so it makes sense for him to do the paid work, and women have babies anyway. Because we could potentially each have had the child it's all in the melting pot. Nothing is fixed.

Claudia: And I don't think a lot of women [enjoy mothering]. They think they're going to, but they get isolated and devalued, and lose their self-confidence and self-esteem.

It was not unusual to find the higher earner in a partnership reducing her hours of employment to share care or become the main caregiver. In contradiction to the dictates of rational economic models, this was often justified on the grounds that a person in a higher paid or higher status occupation has more power and may be less penalized for time out than someone in a more marginal position (Dunne 1998a). I would argue that their rationale (like the part-time/part-time solution) can actually make good long-term financial sense. It also illuminates masculine assumptions in relation to value—the idea that market work is superior to caring.

As in the vast majority of households (Dunne 1998a), routine domestic work was fairly evenly divided between Claudia and Bonnie. Their guiding principle was that "neither should be running around after the other." . . .

Peter goes to a private nursery three days a week (the costs are shared with Philip), and the rest of his care is divided between Claudia, Bonnie, and Philip.

Bonnie: Philip lives in the next street, and so he can just come round every day after work or pick Peter up from nursery and bring him back and do his tea, bath and things, and then we'll roll in about 6:30 or whenever, or sometimes one of us is here anyway.

Claudia: Yes, we try to work that one of us is always at home, either with him or working at home. . . . Quite often there's days when we both have to commute, so Philip usually covers. . . . He's the only one of us who works locally and he's got a bleep [beeper] as part of his job and it's ideal because the nursery can call at any time if there's an emergency.

Interviewer: It strikes me you've got the most ideal situation!

Bonnie: Yes, we think so! [laughing] We're the envy of the mother and toddler group.

Their experience with Philip provides a radical alternative model of cooperative parenting between women and men, based on a consensual nonsexual relationship with a father who is interested in being actively involved in his child's life. In effect, Philip is prepared to engage in mothering, and in doing so, he shares some of the social penalties associated with this activity—all three parents collaborate in balancing the demands of employment and child care, and the result is the lessening of its overall impact. . . .

Again, finding the right words to describe their parenting relations was difficult. Bonnie expresses a common feminist critique of the label *mummy,* which is hostile to ways that it can be employed to subsume other aspects of a woman's identity.

> Bonnie: I've always been quite keen that Peter should know what our names are anyway. I think there's something completely depersonalizing about the way women sit around and talk about a child's mummy as if she's got no identity. It's fine if there's a baby in the room and it's your child, but everyone will say, "Ask Mummy, tell Mummy." But you become this amorphous mummy to everybody. All women are sort of mummy, they don't have their own identity. So I've been quite keen that he should grow up knowing that people have roles and names, and that you should be able to distinguish between the two.

Yet, her radicalism is tempered by her recognition and desire to celebrate her special connection with the child, and she becomes swayed by arguments for the best interests of the child.

Bonnie: But I also feel completely contradictory, that there is something very special emotionally about having your own mummy.

Claudia: And then Philip had very strong feelings about it all, didn't he? He'd always been

> clear that he wanted to be Daddy, and while we went on holiday together last summer, he made it very clear that he thought that in some sense you needed to be recognized as Peter's mother, that that was important, an important thing in terms of what the relationship meant, and that it would be wrong to deny Bonnie that. . . . Yes, he [also thought] that Peter would, if we started him calling both of us Mummy, sooner or later he'd be ridiculed by some of the other children, and then he would have a terrible conflict of loyalties, does he go with the crowd or does he protect us? And that we shouldn't put him in that position. So we went for Mummy, Daddy and Claudia. And then he started calling me Mummy anyway. But now he calls me Addie. [laughter]

This Mummy, Daddy, and Claudia configuration that then evolved into Claudia being called Mummy or the nickname Addie is potentially very undermining of the co-mother. Other couples specifically avoided involving biological fathers to this extent because of such complications of status and role. Claudia's confidence in her relationship with Peter was affirmed through her experience of mothering as main caregiver and, hopefully, by their capacity to be aware of the issues, as the discussion above appears to indicate. Philip's desire for recognition as Daddy is at one level less problematic. He earns this validation through his active involvement in parenting, and because he is not attempting to share fatherhood with a partner, there are no additional complications in relation to exclusion. However, the gender dynamics of this are interesting. While much of the social aspect of Philip's parenting involves the activities of mothering, he is content with the identity of dad. Conversely, in common with the rest of the sample, rather than draw upon dominant polarized heterosexual frameworks—mother/father—respondents extend the meaning of motherhood to include so-called fathering activities such as breadwinning. This raises the wider question, What exactly is a father?

It is no simple act, however, for extended family to claim kinship ties in these nontraditional situations that require coming to terms with a relative's sexuality. While part of being lesbian and gay is about learning how to come out to self and others, I think we have given scant attention to the work involved when heterosexual family members, particularly elderly parents, claim kinship ties that require coming out on behalf of others. For Philip's parents, it was easier for them to explain his entry into fatherhood to other family members by inventing a complicated story about Philip and Bonnie being or having been lovers. . . .

Conclusion

These three stories illustrate many common themes that emerged across the sample, particularly the creativity and cooperation that appear to characterize much of the parenting experience of lesbian couples. I have focused on the involvement of fathers and/or donors and on the complexity of kinship to show how like and unlike these families are to other sorts of family formations. I could equally have looked at the important friendship networks that supported their parenting, the presence of lesbian aunties and heterosexual friends. Lesbian families are usually extended families, supported by elaborate networks of friends and kin.

In common with single lesbian mothers in the United States (Lewin 1993, 9), kin occupy an important place in respondents' accounts of their social interaction. My focus on couples in shared parenting situations reveals other interesting dimensions of kinship: the complexity of these relations and the importance respondents placed on having nonbiogenetic ties recognized and validated by family of origin. . . .

Regardless of whether parenting was shared, mothering was usually carried out in a context where mothers experienced a great deal of practical and emotional support from their partners, where routine domestic responsibilities were fairly evenly shared, and where there was a mutual recognition of a woman's right to an identity beyond the home. Beyond the confines of heterosexuality, they had greater scope to challenge the connections between biological and social motherhood and fatherhood. By deprivileging the biological as signifier of motherhood (although this appears to be contingent on the relationship remaining intact) and the capacity to mother, many were actively engaged in extending the meaning, content, and consequence of mothering to include both partners (or even fathers) on equal terms. . . .

They consequently have greater scope to operationalize their egalitarian ideals in relation to parenting. The high value they attached to nurturing, together with their desire to be fair to each other, meant that within reason they were prepared to experience a reduced standard of living (see Dunne 1998a). Their views about what constitutes shared parenting were less distorted by ideologies that dichotomized parenting along lines of gender in such a way that men can be seen and see themselves as involved fathers when they are largely absent from the home (Baxter and Western 1998, Ferri and Smith 1996). Consequently, their solution to the contradiction was to integrate mothering and breadwinning.

In their everyday lives of nurturing, housework, and breadwinning, respondents provide viable alternative models for parenting beyond heterosexuality. While our focus is on lesbian partners, anecdotal evidence suggests that lesbians are also founding parenting partnerships on the basis of friendship—with gay men or other lesbians. By finding a way around the reproductive limitations of their sexuality, they experience their position as gatekeepers between children and biological fathers in an unusual way. Ironically, we find examples of highly productive models of cooperation between women and men in bringing up children. Unhampered by the constraints of heterosexuality, they can choose to include men on the basis of the qualities they can bring into children's lives. It is no accident, I believe, that respondents usually chose to involve gay men. These men were seen as representing more acceptable forms of masculinity, and their sexuality barred them from some of the legal rights that have been extended to heterosexual fathers.

Their positioning outside conventionality and the similarities they share as women enable and indeed insist upon the redefinition of the meaning and content of motherhood. Thus, when choosing to opt into motherhood, they are anticipating something very different from the heterosexual norm. . . .

They challenge conventional wisdom by showing the viability of parenting beyond the confines of heterosexuality. Rather than being incorporated into the mainstream as honorary heterosexuals, by building bridges between the known and the unknown, their lives represent, I believe, a fundamental challenge to the foundation of the gender order.

Notes

1. To maintain confidentiality, the names of participants and their children and their geographical location and occupations have been changed. To give some sense of their employment circumstances, I have assigned similar kinds of occupations.

2. While British mothers are more likely now than in the past to be employed full-time, it is mothers rather than fathers who balance the demands of paid work and child care. It is very unusual for mothers and fathers to have similar lengths of paid-work weeks, even when mothers are employed full-time (Dunne 1998a; Ferri and Smith 1996).

References

Baxter, J., and M. Western. 1998. Satisfaction with housework: Examining the paradox. *Sociology* 1:101–20.

Berk, S. F. 1985. *The gender factory :The appointment of work in American households.* New York: Plenum.

Blaisure, K., and K. Allen. 1995. Feminists and the ideology and practice of marital equality. *Journal of Marriage and the Family* 57:5–19.

Brannen, J., and P. Moss. 1991. *Managing mothers: Dual earner households after maternity leave.* London: Unwin Hyman.

Butler, J. 1990. *Gender trouble: Feminism and the subversion of identity.* New York: Routledge.

Doucet, A. 1995. Gender equality, gender difference and care. Ph.D. diss., Cambridge University, Cambridge, UK.

Dunne, G. A. 1997. *Lesbian lifestyles :Women's work and the politics of sexuality.* London: MacMillan.

———. 1998a. "Pioneers behind our own front doors": Towards new models in the organization of work in partnerships. *Work Employment and Society 12*(2):273–95.

———. 1998b. A passion for "sameness"? Sexuality and gender accountability. In *The new family?* edited by E. Silva and C. Smart. London: Sage.

———. 1998c. Opting into motherhood: Lesbian experience of work and family-life. London School of Economics, Gender Institute Discussion Paper Series 6.

———. 1998d. Add sexuality and stir: Towards a broader understanding of the gender dynamics of work and family life. In *Living "difference": Lesbian perspectives on work and family life,* edited by G. A. Dunne. New York: Haworth.

———. 1999. Balancing acts: On the salience of sexuality for understanding the gendering of work and family-life opportunities. In *Women and work: The age of post-feminism?* edited by L. Sperling and M. Owen. Aldershot, UK: Ashgate.

———. 2000. Lesbians as authentic workers? Institutional heterosexuality and the reproduction of gender inequalities. *Sexualities. 3*(2):133–48.

Ehrensaft, D. 1987. *Parenting together: Men and women sharing the care of the children.* New York: Free Press.

Fenstermaker, S., C. West, and D. H. Zimmerman. 1991. Gender inequality: New conceptual terrain. In *Gender family and economy, the triple overlap,* edited by R. L. Blumberg. London: Sage.

Ferri, E., and K. Smith. 1996. *Parenting in the 1990s.* London: Family Policy Studies Center.

Gartrell, N., A. Banks, J., Hamilton, N., Reed, H. Bishop, and C. Rodas. 1999. The national lesbian family study 2: Interviews with mothers of toddlers. *American Journal of Orthopsychiatry. 69*(3): 362–369.

Giddens, Anthony. 1992. *The transformation of intimacy.* Cambridge, MA: Polity.

Green, S. 1997. *Urban amazons: The politics of sexuality, gender and identity.* Basingstoke, UK: MacMillan.

Gregson, N., and M. Lowe. 1995. *Servicing the middle-classes: Class, gender and waged domestic labor.* London: Routledge.

Hochschild, A. R. 1989. *The second shift.* New York: Avon.

Lewin, E. 1993. *Lesbian mothers.* Ithaca, NY: Cornell University Press.

Mansfield, P., and J. Collard. 1988. *The beginning of the rest of your life: A portrait of newly wed marriage.* London: MacMillan.

McAllister, F., and L. Clarke. 1998. *Childless by choice: A study of childlessness in Britain.* London: Family Policy Studies Centre.

Rich, A. 1984. On compulsory heterosexuality and lesbian existence. In *Desire: The politics of sexuality,* edited by A. Snitow, C. Stansell, and S. Thompson. London:Virago.

Rubin, G. 1975. The traffic in women: Notes on the "political economy" of sex. In *Towards an anthropology of women,* edited by R. R. Reiter. London: Monthly Review Press.

Sullivan, M. 1996. Rozzie and Harriet? Gender and family patterns of lesbian coparents. *Gender & Society 10*(6):747–67.

VanEvery, J. 1995. *Heterosexual women changing the family: Refusing to be a "wife."* London: Taylor Francis.

Introduction to Reading 41

This reading helps us to understand how the prism of social class modifies the privilege and power of males in interpersonal relationships. Karen Pyke uses data from a survey of 215 individuals, along with excerpts from interviews with seventy divorced and remarried men and women who participated in the larger survey. The experiences of the men and women she interviewed illustrate how class privilege combines with hegemonic masculinity to shape the interpersonal power dynamics of men and women, and disadvantages lower-class men as well as women.

1. Is hegemonic masculinity a characteristic of individual men or the system within which gender is shaped?
2. What are the effects of men's class position on wives of upper- and lower-class men?
3. In what ways does class undermine the privilege of hegemonic masculinity for lower-class men?

Class-Based Masculinities

The Interdependence of Gender, Class, and Interpersonal Power

Karen D. Pyke

Men's Jobs and Conjugal Privileges

Leisure and Autonomy

An important indicator of marital power is the ease with which husbands can free themselves from the boundaries of family life to pursue other interests. Roughly half of the 69 husbands in first marriages did not spend an inordinate amount of time away from their families beyond what was required of their work day. Among those who did, however, some interesting class differences emerge. The ideological supremacy of the male career provided a means by which higher-class husbands could absent themselves in the evenings and on weekends. These absences often were due to legitimate business trips, though not always as necessary as portrayed to their wives and sometimes lengthened for the pursuit of leisure. Some higher-class husbands used work as a smoke-screen for leisure time with friends or extramarital affairs. For example, one husband, the owner of a textile firm and father of two, extended his foreign business trips to add some pleasure, which included sexual affairs.

In fact, middle-class wives were more often shocked than were working-class wives to learn of their husbands' sexual affairs, which had been easily obscured by the broad cloak of the male career. When her first husband, a salesperson who spent a few nights a month out of town, confessed that he had been with 10 different women, one respondent said, "'When? When?' I couldn't believe that he even had the time to do that. . . . It was a real shock." Another respondent recalled how she "would go out of [her] way to make sure that [her husband] was ready to go" when he had his weekend business trips. She did so for years before she learned they weren't business trips at all and that he had been having an affair with one woman in particular for the previous two years. "And you'd think that if I was bright enough or something I would've noticed it."

For lower-class men there is less blurring of the line between work and leisure, often delineated by the punch of a time card. Men's time with male friends, often drinking or "tinkering" with cars (sometimes resold for a profit), was viewed by their wives as self-centered leisure

From Pyke, K. D., "Class-Based Masculinities: The Interdependence of Gender, Class, and Interpersonal Power" *Gender & Society 10*(5): 527–549.

(see also Halle 1984, 58). In reality, however, many higher-class husbands also were drinking with other men, but it was associated with "working" and "getting ahead." These varying meanings shaped wives' resentment and acceptance. Working-class wives viewed husbands who spent a lot of time with their male friends as "lazy," "not ambitious," "self-centered," "carefree," "immature," and "irresponsible." But higher-class men who spent a lot of time away from home pursuing leisure that was at least nominally associated with work were more often viewed as "ambitious" and doing so out of necessity, even if their wives wished they could cut back on their hours.

When working-class men, such as self-employed contractors, had jobs that could have provided a smokescreen for leisure, they relied on them as a cover less often. Instead they tended to be more blatant in their pursuit of leisure away from their families. They also were more careless in hiding their extramarital affairs and, consequently, were more likely to get caught. For example, with his wife in the hospital after having given birth, one husband, a truck driver, brought another woman home without concealing her from his wife's visiting brother, who later reported the infidelity to his sister.

Half of all working-class men (20 of 38) did not engage in rebellious behavior or stray outside of the boundaries of "good husbands." Those who did typically moved in a social milieu of like men, often coworkers, who encouraged such behavior (see Connell 1991; Halle, 1984; LeMasters 1975; Rubin 1976; Willis 1977). For example, Nick, who had several jobs in construction and other trades as well as periods of unemployment in his first marriage, drank heavily, often with coworkers. He said, "The people I worked with, that's just what you did, especially on the weekends, on Friday nights, you'd get hammered."

This interplay between social milieu and the construction of a defiant masculinity is evident in Ted and Debbie's 9-year-long marriage. At the age of 15, Debbie married Ted, a self-employed plumber 9 years her senior. She finished high school and, later, stayed home and raised their son. Ted was often away from home drinking with male friends and "running around" with women. Although he sometimes used work as an excuse, he wasn't covering it up very well. For example, as Debbie explained, he would say," 'I'm going out to buy a pack of cigarettes' and wouldn't come back until the next day."

Debbie regards Ted's affairs as having been "quick thrills." She said, "He wasn't emotionally involved. It was more part of the recreation of being drunk, being high, being part of that group of people." She referred to that group as "low lifes," . . .

Lower-class husbands who ostentatiously pursued drugs, alcohol, and sexual carousing are constructing a compensatory form of masculinity. Such behavior was worn like a badge of masculinity in the work and social environments they inhabited. By drinking with other working-class men at the bar and openly engaging in extramarital relationships, they appear to be defying existing power structures, displaying their independence from the control of their wives and "the establishment" (i.e., higher-status men). This exaggerated masculinity compensates for their subordinated status in the hierarchy of their everyday work worlds. It gave them a sense of autonomy and self-gratification, entitlements that higher-status men acquire more easily and with greater impunity, thereby creating the illusion of ascendant masculinity. Although this behavior is characteristic of some and not all working-class men, it reinforces a stereotype of subordinated heterosexual masculinity that higher-class men call on as evidence of their own civility and gender equity, thereby further obscuring their power and privilege and reaffirming their ascendant masculinity.

In sum, lower-class men do not enjoy the same ideological legitimations for personal autonomy and leisure in their marriages that higher-class men acquire as part of their career package. Instead, some working-class husbands engage in defiant behavior and construct a compensatory masculinity (see Collinson 1992; Connell 1991, 1995; Willis 1977). In the next section, I describe this overt form of power in more detail.

Overt Domination of Wives

Because working-class men's jobs do not provide a shortcut to marital power, they must either concede power to wives or maintain dominance by some other means. They were, overall, both more egalitarian (especially in sharing housework and child care) and more *explicitly* domineering in their marriages than were higher-class husbands. Domineering lower-class husbands draw more directly and overtly on personal masculine privilege as their essential right as a means of bolstering their conjugal power (see Collinson 1992; Rubin 1976). The following case provides an example of such overt power and illustrates its link to the denigrated status of working-class men.

Nick, age 38, remarried his first wife, Nina, following 4 years of divorce and a tumultuous 10-year marriage that was marked by his drinking, violence, and chronic depression. In his first marriage, his dissatisfaction with several jobs, mostly in construction, led to his current position as a splicer for a utility company. Nick's description of his transition to a splicer reflects the centrality of his work in affecting his low self-esteem. . . .

Nick's self-esteem as a man also plummeted when Nina returned to work. He recalled, "That probably hit me really hard. . . . I wanted to be the provider. When she went to work, it took that away, it took away my status as the man of the house, I thought."

Nick's heavy drinking in his first marriage often was accompanied by violent attacks on his wife and their house, usually prompted by violations of the traditional and submissive role Nick wanted Nina to fulfill. "Small things would trigger it," Nick explained, such as his wife's "lousy" housekeeping. "Plus I was a real jealous person, and whenever [her] old flames would appear I just couldn't handle that, even though I'm sure she was pretty dedicated." Nick's violent rage also was triggered when Nina challenged his domination. . . . Nick's low self-esteem, alcoholism, and violence eventually gave way to chronic depression and thoughts of suicide that landed him in the hospital. "All I was trying to do was provide for the family and be with the kids, but I was sinking the whole way. . . . Even though I was trying harder, I was still getting violent, and things were getting worse and worse," he explained.

Nick was not exceptional in his abuse. Among the 36 women interviewed, 52 percent married the first time to lower-class men said their first husbands had hit them, compared to 20 percent of those married to higher-class men. The greater incidence of wife abuse (based mostly on self-reports) committed by lower-status, underemployed, and unemployed husbands who cannot fulfill the provider role has been documented elsewhere (Dibble and Straus 1980; Gelles and Cornell 1985; Levinger 1966; O'Brien 1971; Straus, Gelles, and Steinmetz 1980). Other research links economic disadvantage with the husband's increased hostility toward his wife (Conger, Ge, and Lorenz 1994; Liker and Elder 1983).

Working-class husbands' subordinated class status in relation to other men—and women—in the labor force seems to exacerbate their need to use their marriage as a place where they can be superior (Ferraro 1988; O'Brien 1971; Pyke 1994; Rubin 1976). With their power base on shakier ground, they are more likely to resort to explicit and relentless tactics, such as violence, as well as criticism and constant surveillance of their wives.

Some lower-class husbands, particularly those who were violent and adulterous, greatly feared their wives' infidelity. They expressed this fear in baseless accusations, demands for their wives' constant attention, restrictions on their wives' movement and employment, or spying (see also Ferraro 1988). Seemingly irrational, this fear appears to be a reflection of their sense that they offered too little compared to other men to hold onto their wives. These feelings of powerlessness led some to use terrorizing tactics to bolster their control over their wives; however, it had devastating consequences on the marriage, typically pushing wives to leave. . . .

Husbands' Entitlement to Housework

The hidden power of higher-class men and the explicit power of lower-class husbands are also evident by the mechanisms that excuse them, to varying degrees, from performing domestic work. Male careers provide a rationale for higher-class husbands' freedom from family work, whereas working-class men are more likely to rely on blatant and increasingly contested patriarchal ideologies for similar entitlements. In both higher and lower-class marriages, husbands' absences precluded their doing household chores. However, because the absences of higher-class men were more likely due, ostensibly at least, to evening or out-of-town business obligations or night classes, wives viewed this division as fair. In addition, any work higher-class husbands brought home from their jobs also excused them from family tasks. For example, a female accounting manager remarried to a systems analyst said, "I do 75 percent of the housework because he also works at home. . . . He's always on the computer so I don't know if it's work or play."

Higher-class husbands also derived from their careers greater entitlement to a stay-at-home wife. They were more likely than working-class husbands to veto, discourage, or limit their wives' labor force participation, especially in first marriages (Pyke 1994). For example, Jane was married the first time to a prominent attorney who earned $300,000 annually. His position entitled him to limit her teaching to part-time. She said,

> But he made sure that we both knew that his job came first and if I was working too many hours, he made it clear that I should cut back on my hours. . . . He had a very prominent job, a lot of public recognition that came with his job. . . . So I supported him in that and I was sort of content to be in his shadow. . . . His career came before my career and was much more important to me at the time.

Another attorney, who was childless prior to remarriage, gave a description of the kind of wife he sought after his divorce that emphasized her supporting role to his career. He said, "I made good money as a lawyer, so my wife didn't need to work to support the household. I needed someone to take care of my children and my house when I am not there."

This need for a wife to serve as a maid and nanny propelled many higher-status men to look in the same places for a wife as they do for paid domestic help: among the lower class. An interclass mate selection occurred in remarriage between men employed in high-status occupations and women who were unhappy with dead-end, low-skilled jobs and who had worked out of necessity in their first marriages. Among 102 women whom I surveyed whose first and second marriages were identified as being either working or middle to upper class, 25 percent moved from a working-class first husband to a middle-class second husband (70 percent remarried husbands of the same social class as their first husbands, and 5 percent moved from middle to working class). Similarly, Gerson (1985, 1993) found that men are more likely to seek domestically oriented partners as their breadwinning ability increases, and women are more likely to veer toward domesticity when faced with blocked job opportunities and married to men who earn enough to be the sole provider.

Even though both higher and lower-class husbands tended to avoid household labor, they enjoyed varying levels of legitimacy for doing so. Higher-class husbands were more likely to be excused by the priorities granted to their career and provider role (see also Gerson 1993; Hochschild with Machung 1989), which also served as the places they most prominently produced ascendant masculinity. Lower-class husbands, on the other hand, whose jobs and lower earnings provided them with little justification for not sharing chores—especially when wives worked for pay also—more *directly* relied on rigid gender divisions of labor in the home as a means of producing masculinity (see Game and Pringle 1983). However, explicit traditional ideologies about the proper roles of husbands and wives were likely to be challenged and resented by wives. Hence, they were a less reliable basis of men's freedom from housework than was the ideology of the husband's career. This again suggests that the power of lower-class husbands that rests on notions of masculine privilege is likely to be undermined, especially in long-term, stable marriages.

What about the minority of husbands actively embracing egalitarian divisions of labor? Higher-class husbands more often were constrained by their job demands from doing so, even when they professed a sense of obligation or desire. In contrast, there is a greater structural incentive for working-class families to adopt egalitarian practices. Lower-class husbands often cared for children while wives worked a different shift from their own. And, for some, such as Nick in his remarriage, greater involvement with children and family life provided a sense of self-worth and meaning that compensated for the degradation endured on the job (Connell 1991; Gerson 1993; Pyke 1994). These men do not appear to put stock in ascendant or exaggerated masculinities and instead produce an egalitarian masculinity involving expressiveness and high levels of family involvement (referred to as the "New Man"; Messner 1993).

The pressures exerted by the structural conditions of working-class life may lead some men to juggle a Dr. Jekyll and Mr. Hyde existence in which they produce hypermasculinity in male cliques and on the job and an egalitarian masculinity in their family relations. For example, working-class men might use talk of masculine superiority, privilege, and authority as a means of producing hypermasculinity while nonetheless sharing power and family work with their wives on a day-to-day basis. That is, when the situational context changes, the form of masculinity produced, even within categories of social class, can change as well. Furthermore, the consistent construction and maintenance of hypermasculinity across all arenas of social life, including family relations, are so costly as to become less desirable and untenable to individual men. In fact, as they approach midlife, working-class men tend to drop out of the male cliques in which hypermasculinity is, collectively produced (Rubin 1976).

Egalitarian masculinity may not be appreciated by some wives who view it as a threat to their feminine identity. In the next section, I discuss some surprising insights about how the supremacy of the male career leads some middle-class wives to negatively evaluate egalitarian husbands out of preference for male dominance.

Middle-Class Men and Egalitarian Masculinity

Ideological hegemonies present elite interests as everyone's interests. Thus, they lead subordinates, as well as elites, to sanction those who fail to reproduce the dominant group's power advantage. For example, higher-class egalitarian men who violate notions of ascendant masculinity often attract hostility, even from wives who would appear to derive benefits from their husbands' defection but are convinced otherwise. This was evident in a few middle-class first and second marriages in which husbands were unable to live up to the ethic of masculine ambition and high earnings to which their wives felt entitled. It was not that these husbands were poor providers or husbands; on the contrary, they tended to be very family-involved men with moderate earnings. However, their wives, who expected them to have *greater* ambition and earnings, were less likely to view them as "real" men and reacted to their "inadequacy" with disdain. Disappointment in the "failure" of their husbands to live up to the ideal of ascendant masculinity bolstered wives' marital power. Because explicit patriarchal ideologies are not prominent in the discourse of the higher-class cultural milieu, middle-class husbands less often call them up as a means of boosting their power.

The resultant power sharing is not always appreciated by wives. The ideological supremacy of the male career and the "doing" of essential femininity fosters the desire among some wives to be dependent and to "look up" to their husbands. For example, Jean resented her shared power as a further indication of her first husband's failure as a man. Her disappointment in him for not being less "submissive" and more successful like the people she worked with, even though he earned what she considered to be a good income, resulted in the break up of their otherwise happy marriage. They married when she was 19 and he was 24. She finished her education and became a successful accountant. It bothered her that her husband, Phil, an insurance title representative, did not want to finish his bachelor's degree. . . .

Such unhappiness endured despite Phil's instrumental and emotional support of Jean and her career. For example, when she had to work until late at night or on Saturdays, "it was no problem. . . . He would get dinner started." After she had a car accident, he drove her to and from work. "He was very good, generous," she said. The marriage faltered despite all the positive qualities of her husband, which enabled her to advance in her career. Indeed, it was those very qualities that led her to lose respect for him as a man; he was not as "egotistical" as other men. Instead, she said, "He was generally pretty accommodating. I was really the one who forced my decision. I think that's one of the things that bothered me. I had a very strong personality and he would back down sometimes."

The very traits she desired in Phil, such as egotism, ambition, and dominance, would have undermined her marital power, obligating her to more instrumental and expressive support of his career at the potential expense of her own. This illustrates the pressure on men to accomplish ascendant masculinity via a successful career, its connection to male dominance and feminine identity, and the negative effects endured by some higher-class men who are not hyperambitious.

One middle-class man described the strain in both his first and second marriages as relating to his failure to achieve a higher level of success. He said,

> I think from [ex-wife's] perspective, and my current wife is still kind of the same way, it's almost like . . . they both felt that when you get married it's . . . you hit a button and all of a sudden your husband is out there making $75,000 a year. And they [wives] work as a lark.

His second wife pressures him to leave advertising, where his earnings have suffered, and go into commercial real estate like one of her relatives who earns $100,000 a year. Their individual highest annual earnings during their marriage are equal at $40,000. Her resentment, which he gingerly tiptoes around, is a dominant theme in their marriage. He described her as "contentious" and a "battle ax." . . .

Although her husband performs 60 percent of their weekly total household labor hours, her resentment for his failure to live up to the middle-class standard of success overshadows any gratitude she might have for his greater household work.

The previous two examples illustrate how some women's acceptance of the dominant ideology about what it means to be a man reflects and contributes to the hegemony of the male career. These examples suggest that when middle-class men are not very successful in their careers, power can shift to a resentful wife. Because methods of accomplishing femininity often rest on women's subordination to men, some women may resent their power. They may use it, as exemplified here, to steer men toward the production of masculinity in ways that emphasize male power and female subordination. These examples underscore the ways cultural notions of what constitute a "real" man and a "real" woman elicit women's participation in the project of male dominance and female subordination.

Summary and Conclusion

In contrast to oversimplified, gender-neutral or gender-static approaches, the theoretical framework presented here integrates interpersonal power with broader structures of class and gender inequality. I used empirical examples of conjugal power to illustrate how interpersonal powering processes and gender and class relations can be considered components of an interacting system. Specifically, structures of inequality are expressed in ideological hegemonies, which construct gender in ways that reemphasize and normalize the domination of men over women and that of privileged men over lower-class men. Furthermore, the relational constructions of ascendant and subordinated masculinities have different implications for interpersonal power dynamics.

For example, the different conjugal power processes available across social class further feed into the cultural legitimations of higher-class men's superior position. In the absence of legitimated hierarchical advantages, lower-class husbands are more likely to produce hypermasculinity by relying on blatant, brutal, and relentless power strategies in their marriages, including spousal abuse. In so doing, they compensate for their demeaned status, pump up their sense of self-worth and control, and simulate the uncontested privileges of higher-class men. The production of an exaggerated masculinity in some working-class subcultures also serves the interests of higher-class men by deflecting attention from their covert mechanisms of power and enabling them to appear egalitarian by contrast.

The coercive power strategies of lower-class men appear unmatched in degree. This is misleading, however. It is precisely their demeaned status and weak power base that have propelled many working-class men to rely on extreme methods of control as a kind of last resort in asserting power and producing masculinity. Thus, those who study power relations must be careful not to equate brutality of power with quantity of power and examine the ways that power inequalities may be obscured in other seemingly egalitarian relationships by hegemonic cultural ideologies. In addition, brutal styles of masculinity, such as displayed in the workplace or tavern, should not be assumed to be automatically linked to brutal power strategies in marriage. Some men may balance more egalitarian practices in their personal life with more public displays of hypermasculinity and claims to male dominance. Similarly, higher-class men who are mild mannered and civil in the workplace may nonetheless exercise brutal forms of power in their family life.

The omnirelevance of gender to social life and the ways it is taken for granted as essential and inevitable makes it an especially effective ideology in normalizing and mystifying gendered power relations. In doing gender, men and women engage in practices that promote male dominance and female subordination in most social contexts. Specifically, I have shown how some women pursue marital arrangements that contribute to male dominance as a method of accomplishing their gender; that is, they do so to affirm their "essential nature" (West and Fenstermaker 1993). It is not necessarily that they consciously desire male dominance, but the methods they employ in "doing gender" produce conditions that foster such power differentials. It is thus important that researchers studying interpersonal power consider how it is a symbolic artifact of the routine production of gender as well as the structural conditions of men's and women's lives.

Although I have used the case of marital power and the hegemony of the male career, other hegemonic ideologies similarly affect power dynamics in marriage as well as other social relationships and reinforce "essential" gender differences. For example, a white heterosexual masculine ethic pervades capitalist, managerial ideologies that stress rationality, success orientation, impersonality, emotional flatness, and a disregard for family concerns. Because these traits are associated with "essential" masculinity and are antithetical to notions of "essential" femininity, this ethic would appear to exclude women from management positions and undermine the power of those who have successfully acceded to such ranks while (re)constructing "essential" femininity and ascendant masculinity. Similarly, the masculine ethic of management associated with higher-class men embodies traits that reflect and perpetuate the negative evaluation of lower-class men, men of color, young men, and homosexual men. This, in turn, reinforces the construction of compensatory masculinities, such as the "cool pose" associated with African American men (Majors and Billson 1992) and the hypermasculinity described here among lower-class white men and also common among male youths (Messerschmidt 1993). Thus, ideological hegemonies have a different impact on men and women across race, sexuality, age, and social class in ways that reflect and (re)construct relational conceptions of masculinities and femininities with different implications for interpersonal power. The ensuing practices of interpersonal power, in turn, reinforce structures of inequality and their ideological legitimations.

Femininity is also cross-cut into diverse forms by the structural and cultural conditions of race, social class, sexuality, and age. For the sake of clarity, however, I have centered this discussion almost exclusively on white, heterosexual, class-based masculinities. The construction of femininities can perhaps best be understood in relation to men. As Connell observed (1987, 186–87), all forms of femininity are constructed in the context of the overall subordination of women to men. The interplay of diverse femininities does not reemphasize a hierarchy among women as much as intermale hierarchies of dominance as well as gender hierarchies. The degree to which women are accommodating to men provides a useful basis for conceptualizing femininities. "Emphasized femininity" (Connell 1987, 187) is produced among women who view their role as naturally subservient to men. Noncompliant femininity, on the other hand, emphasizes women's independence and desired equality with men. It is displayed by the woman who can do it all: maintain a good job, a clean house, well-behaved children, and a loving marriage. Noncompliant femininity obscures women's subordination to men by associating their paid labor with equality and downplays how their employment benefits elite males who purchase women's discounted labor.

How the construction of femininities reflects and (re)constructs (or resists) the gender order and intermale hierarchies needs to be further explored. An examination of hidden power dynamics might reveal that the key difference in the ways these two forms of femininities are played out has less to do with quantity of male dominance than with quality. For example, women who display greater egalitarianism in some arenas of their marriage or job may feel pressed to accomplish their gender with greater submission in other arenas (see Hochschild with Machung 1989; Pyke 1994). On the other hand, women who emphasize their femininity may be able to wield considerable power from behind a smokescreen of female subservience. By examining the underlying cultural ideologies at play and the actual practices, we can learn how the construction of these and other forms of femininity shapes interpersonal power, plays into the construction of masculinities, and obscures while (re)producing inequality. . . .

References

Collinson, David L. 1992. "Engineering humour": Masculinity, joking, and conflict in shop-floor relations. In *Men's lives,* edited by Michael S. Kimmel and Michael A. Messner, 232–46. New York: MacMillan.

Conger, Rand D., Xiao-Jia Ge, and Frederick O. Lorenz. 1994. Economic stress and marital relations. In *Families in troubled times: Adapting to change in rural America,* edited by Rand D. Conger and Glen H. Elder, Jr., 187–203. New York: Aldine de Gruyter.

Connell, R. W. 1987. *Gender and power.* Stanford, CA: Stanford University Press.

———. 1991. Live fast and die young: The construction of masculinity among young working-class men on the margin of the labour market. *The Australian and New Zealand Journal of Sociology* 27:141–71.

———. 1995. *Masculinities.* Los Angeles: University of California Press.

Dibble, Ursula, and Murray S. Straus. 1980. Some social structure determinants of inconsistency between attitudes and behavior: The case of family violence. *Journal of Marriage and the Family* 42:71–80.

Ferraro, Kathleen J. 1988. An existential approach to battering. In *Family abuse and its consequences,* edited by Gerald T. Hotaling, David Finkelhor, John T. Kirkpatrick, and Murray A. Straus, 126–38. Newbury Park, CA: Sage.

Game, Ann, and Rosemary Pringle. 1983. *Gender at work.* Boston: Allen & Unwin.

Gelles, Richard J., and Claire Pedrich Cornell. 1985. *Intimate violence in families.* Beverly Hills, CA: Sage.

Gerson, Kathleen. 1985. *Hard choices: How women decide about work, career, and motherhood.* Los Angeles: University of California Press.

———. 1993. *No man's land: Men's changing commitments to family and work.* New York: Basic Books.

Halle, David. 1984. *America's working man: Work, home, and politics among blue-collar property owners.* Chicago: University of Chicago Press.

Hochschild, Arlie, with Anne Machung. 1989. *The second shift: Working parents and the revolution at home.* New York: Viking.

LeMasters, E. E. 1975. *Blue-collar aristocrats: Life-styles at a working-class tavern.* Madison: University of Wisconsin Press.

Levinger, George. 1966. Sources of marital dissatisfaction among applicants for divorce. *American Journal of Orthopsychiatry* 36:803–7.

Liker, Jeffrey K., and Glen H. Elder, Jr. 1983. Economic hardship and marital relations in the 1930s. *American Sociological Review* 48: 343–59.

Majors, Richard, and Janet Mancini Billison. 1992. *Cool pose: The dilemmas of black manhood in America.* Lexington, MA: Lexington Books.

Messerschmidt, James W. 1993. *Masculinities and crime: Critique and reconceptualization of theory.* Lanham, MD: Rowman & Littlefield.

Messner, Michael. 1993. "Changing men" and feminist politics in the United States. *Theory and Society* 22:723–37.

O'Brien, John E. 1971. Violence in divorce-prone families. *Journal of Marriage and the Family* 33:692–98.

Pyke, Karen. 1994. Women's employment as a gift or burden? Marital power across marriage, divorce, and remarriage. *Gender & Society* 8:73–91.

Rubin, Lillian. 1976. *Worlds of pain: Life in the working class family.* New York: Basic Books.

Straus, Murray A., Richard J. Gelles, and Suzanne K. Steinmetz. 1980. *Behind closed doors: Violence in the American family.* New York: Doubleday.

West, Candace, and Sarah Fenstermaker. 1993. Power, inequality, and the accomplishment of gender: An ethnomethodological view. In *Theory on gender/feminism on theory.* edited by Paula England, 151–74. New York: Aldine de Gruyter.

Willis, Paul. 1977. *Learning to labor.* New York: Columbia University Press.

⚜ Topics for Further Examination ⚜

- Go to http://www.umbc.edu/wmst and check out their resource sites to find information on families (if you can't find something immediately, try the miscellaneous link).
- Go to the Web and search for "healthy relationships" and examine the first twenty or so to see the focus for the Web site and who is sponsoring the site. What does this data collection exercise tell you about how we are to view relationships in our society?
- Check out the most recent research on gender and relationships using an academic database. How does this research differ from that which you found on the Web?
- Examine marriage and engagement announcements in your local paper. What race, gender, and sexuality patterns do you find in these short announcements? Do the same thing with a listing of personal ads from a local paper. What does this tell us about expectations for relationships?

9

ENFORCING GENDER

Throughout Part II, we have discussed patterns of learning, selling, and doing gender at work, play, and in intimate relationships. In this final chapter of our section on patterns, we look at those surrounding the enforcement of gender. Enforcing gender is about more than just *doing* gender; it is about assault, coercion, and constraints on people's behaviors, as well as more subtle and tacit constraints on identities to enforce gender conformity. Enforcing gender involves people using a range of social control strategies, such as physical abuse and rape, harassment, gossip and name calling, as well as laws and rules created by governments, work organizations, and religions to coerce people to conform to gender norms they might not otherwise wish to obey. Many readings throughout this book are about the enforcement of gender, particularly those in Chapter 4. This chapter extends that prior discussion, explicitly focusing on the different forms of social control used to enforce gender.

The enforcement of gender can have profound effects on women's and men's choices, self-esteem, relationships, and abilities to care for themselves. We argue two main points in this chapter. First, doing gender is not something that we freely choose; rather, there are many times that we are forced to do gender, whether we would wish to do it or not. Second, there are many occasions whereby the very act of maintaining a gendered identity hurts ourselves and others, either physically, emotionally, or both.

SOCIAL CONTROL

Enforcing gender is about the physical and emotional control of everyone. Berger (1963) describes the processes by which we learn to conform to the norms of society as "a set of concentric rings, each representing a system of social control" (p. 73). At the middle of the concentric rings, Berger (1963) places the individual. Social control mechanisms, including family and friends, are in the next ring, and the legal and political system of a society are in the outer ring. He argues that most social control of behavior occurs in the inner rings, which he describes as "broad coercive systems that every individual shares with a vast number of fellow controlees" (Berger, 1963, p. 75). As such, most gendered behavior is enforced in those center rings, with the forms of social control differing slightly by gender. For example, homophobia is a control mechanism that is often used to enforce gender patterns for men.

Friends or peers who call a boy a "fag" or "sissy" can make him uncomfortable and force him to display hegemonic masculinity. The commodification of gender most often enforces emphasized femininity, with young girls comparing themselves to the idealized images of the media, as described in Chapter 5.

Gender Violence

Although most of the enforcement of gender happens as part of normal interaction, some forms of social control are more coercive, including physical assault, rape, sexual harassment, and even murder. We begin by discussing the harm done when violence is used to enforce gender. We hope you have never experienced physical violence, but many individuals have. The actual incidence of gender violence is not clear. The reading by Julia Hall illustrates that much of the violence against women is not reported to the authorities (Centers for Disease Control and Prevention, 2007a; Rennison, 2002), making the incidence of violence for women higher than the statistics suggest. Furthermore, gender violence is often normalized, both in relationships and how we define violence. The reading by Kristen Anderson and Debra Umberson in this chapter describes how some men dismiss their acts of violence against female partners as trivial. Men are also victims of acts of violence by women, but this violence is likely to be less physically harmful and men may be less likely to talk about it, since to claim to be physically harmed by a woman clearly violates gendered expectations.

Victims, particularly women, often do not report acts of gender violence because of fear for themselves and their children, based upon threats of additional physical violence or the withdrawal of economic support and/or the outcome of emotional abuse, which leads them to believe they "asked for it." Abusers exert "coercive control" over their victims, making victims fearful for their lives and the lives of loved ones. As such, victims often become psychologically battered and emotionally dependent on their perpetrators (Mahoney, Williams, & West, 2001). Domestic violence not only lowers the victim's self-esteem, but also a victim's ability to leave an abusive situation is difficult because abusers often control victim's freedom of movement and finances (even for individuals who work for pay outside the home). Abusers also prevent victims from getting the psychological support they need to leave the abusive situation (Mahoney, Williams, & West, 2001). In her reading in this chapter, Julia Hall describes the situation of poor women returning to the men that beat and abuse them—a pattern found across social class relationships. Hall's article illustrates how the effects of violent acts extend beyond the abuser and abused, with the poor, White middle-school girls in her study constructing their lives and planning for their futures based upon the gender violence they experienced. These young girls distrusted men and expressed the wish to never marry and have children because of the violent acts they witnessed.

Even given concerns about the accuracy and underreporting of the data, statistics detailing incidences of gender violence should give us pause. In the year 2000, the U.S. Department of Justice (Rennison, 2003) reported 1,247 women who were murdered by an intimate partner. This equates to almost three and a half women per day. In the same year, 440 men were murdered by an intimate partner, or 1.2 men per day. The rates of intimate partner homicide are three times higher for Black women (3.55 per 100,000 versus 1.11 for White women) and Black men (4.11 per 100,000 versus .49 for White men) and Native American men (1.20 per 100,000) and women (2.26) according to Palozzi, Saltzman, Thompson and Holmgreen (2001).

The Centers for Disease Control and Prevention defines four types of intimate partner violence: physical, sexual, threats of physical or sexual violence, and psychological/emotional violence (2007b), with over 5.3 million incidents each year for women, age eighteen and older, and 3.2 million for men in the United States (Centers for Disease Control and Prevention, 2007a). Using a narrower definition of intimate partner violence, the Bureau of Justice Statistics reports that 588,490 women reported rape/sexual assault, robbery, aggravated assault or simple assault by intimate partners (Rennison, 2003), representing 85 percent of all reports of violence by intimate partners. Clearly, women are more likely than men to be victims of intimate violence, which reflects and reinforces the system of hegemonic masculinity and differential power and control in relationships.

Gender violence often begins in the early years and is a form of intimate abuse that almost entirely affecting women. Studies find that approximately one in four college women have been raped or the victims of attempted rape (Bachar & Koss, 2001; Fisher, Cullen & Turner, 2000). Unfortunately, this situation of date rape evolves in a rape culture which encourages and justifies rape (Boswell & Spade, 1996). The literature on the prevalence of rape has been consistent over the last fifteen years; however, there is little work on prevention of intimate violence. Attempts by educators and activists to work with mixed-gender audiences in preventing rape are not as successful as those directed at helping women to understand that they do not cause rape or trigger intimate violence (Bachar & Koss, 2001). The reading by Elizabeth Armstrong, Laura Hamilton, and Brian Sweeney in this chapter argues that the inordinately high incidence of sexual violence for college students (Fisher, Cullen, & Turner, 2000) is related to organizational structure and interactional dynamics, as well as individual choices—most of which are gendered.

We argue that the underlying reasons for intimate violence and sexual abuse are complicated and relate to the maintenance of hegemonic masculinity. Hegemonic masculinity depends upon the sexual, physical, and emotional degradation of women. Culturally, it is often inherent in "becoming" a man to learn to disdain women. As Beth Quinn suggests in her reading in this chapter, many men are expected to disrespect women and other men congratulate them when they do so. The same patterns and practices we have been describing throughout this book also explain how men learn to instigate and justify the physical and emotional abuse of women as described in Anderson and Umberson's reading in this chapter. These patterns also illustrate what encourages men to be violent. Although there have been some changes in the way police and courts handle intimate violence, there is a long-standing attitude that the "victims ask for it" (Mahoney, Williams, & West, 2001).

Institutionalized Enforcement of Gender

Although the enforcement of gender mainly occurs in daily activities, as Berger (1963) notes, institutionalized settings also sometimes enforce gender. Organizations either participate in or simply ignore more subtle forms of gender enforcement such as "girl watching" in office settings described in the reading by Quinn in this chapter. These patterns can be found in schools (e.g., fraternities and sports teams), the workplace, and men's sports teams (Schact, 1996). In patriarchal societies worldwide, it is men who make the rules (and laws) and write the informal scripts that prescribe behaviors for themselves, other men, and women. These rules become institutionalized in daily patterns of life.

Religion is one institution that often enforces gender difference and inequality. Many religions have rules that exclude or segregate women within the practice of that religion. For example, women are excluded from the priesthood in the Catholic church and marginalized in many other Christian churches (Nesbitt, Baust, & Bailey, 2001). Some Jewish congregations segregate women physically and do not allow women to study the Torah with the same seriousness as men (Rose, 2001). Some Islamic communities also apply rules of gender, such as enforced veiling of women which women negotiate in a variety of ways (Gerami & Lehnerer, 2001).

In addition to explicit practices found in various religions, many rules subtly maintain the domination of women—including the sexual violation of women by men. Politicians, police, and traffickers construct and maintain the rules supporting human trafficking of women for the sex trade (Kempadoo & Doezema, 1998). These rules allow girls and women to enter countries under dubious documentation and fail to provide the help that trafficked girls and women need to escape lives of prostitution and abuse (Kempadoo & Doezema, 1998). The women sold into virtual sexual slavery, and the men and women who take advantage of vulnerable girls and women, are part of a brutal pattern of gender enforcement. For example, one study of trafficking of women for sex/prostitution estimates that Indonesian women are most likely to go to Japan, Malaysia, and probably Singapore. They are lured to these places by the possibility of jobs in karaoke bars as singers or, in Japan, as cultural ambassadors (American Center for International Labor Solidarity, 2005). Unfortunately, they are forced into and kept in the sex trade by withholding of salary or immigration documents, or the need to pay off the "debt" they incurred to get to the country. Other women face similar situations as they are forced into domestic labor. The trafficking of women in both contexts helps to maintain the sexual domination of women, just as fraternities help to maintain the sexual domination of fraternity brothers over women as discussed by Armstrong, Hamilton, and Sweeny in this chapter.

Another, more subtle and often "unnoticed" form of gender enforcement is sexual harassment as discussed in Chapter 7. Sexual harassment begins early in school settings, even among young schoolchildren (American Association of University Women, 2001; Hand & Sanchez, 2000). Although the gender difference in the experience of harassment is only 9 percent greater for girls (76 percent of boys and 85 percent of girls experience sexual harassment), girls endure the more demeaning forms of harassment and at a higher rate than do boys (Hand & Sanchez, 2000). Unfortunately, sexual harassment continues long beyond high school, as noted by Quinn in this chapter and discussed in Chapter 7.

Although it is subtler than sexual assault, sexual harassment has serious consequences for girls and women. In her reading, Quinn helps us to understand how sexual harassment enforces gender for men and women by making women objects of men's masculinity. The "girl watching" that Quinn describes creates a wall between men and women that makes it impossible for them to relate on an equal basis.

The enforcement of gender integrates many prisms, not all of which we can cover in this chapter. For example, concern about intimate partner violence in same-gender relationships is often underreported and hidden because statistics are not collected on same-gender relationships. However, using data collected on same-gender intimate partner homicides collected by the FBI and Bureau of the Census, the rate of violence for these relationships is similar to that for heterosexual relationships, between 22 and 38 percent (Puzone et al., 2000).

Sometimes we only see gender enforcement when we look at someone else's life circumstances, such as the young women in Hall's article or the reports of men's use of intimate partner violence in Anderson and Umberson's article, or Quinn's descriptions of "girl watching." Think about your life as you read this chapter and consider the ways conformity to gender is enforced among the people you know and observe.

References

American Association of University Women. (2001). *Hostile hallways: Bullying, teasing, and sexual harassment in school.* Washington, DC: American Association of University Women Educational Foundation.

American Center for International Labor Solidarity. (2005). *When they were sold.* Retrieved June 6, 2007, from http://solidarity.timberlakepublishing.com/files/when_they_were_sold_chapter2.pdf

Bachar, K., & Koss, M. P. (2001). From prevalence to prevention: Closing the gap between what we know about rape and what we do. In C. M. Renzetti, J. L. Edleson, & R. K. Bergen (Eds.), *Sourcebook on violence against women* (pp. 117–142). Thousand Oaks, CA: Sage.

Berger, P. L. (1963). *An invitation to sociology: A humanistic perspective.* Garden City, NY: Anchor Books.

Boswell, A., & Spade, J. Z. (1996). Fraternities and rape culture: Why are some fraternities more dangerous places for women? *Gender & Society, 10* (2), 133–147.

Centers for Disease Control and Prevention. (2007a). *Intimate partner violence: Fact sheet*. Retrieved February 22, 2007, from http://www.cdc.gov/ncipc/factsheets/ipvfacts/htm

Centers for Disease Control and Prevention. (2007b). *Intimate partner violence: Overview*. Retrieved February 22, 2007, from http://www.cdc.gov/ncipc/factsheets/ipvoverview/htm

Fisher, B. S., Cullen, F. T., & Turner, M. G. (2000). *The sexual victimization of college women*. Washington, DC: Bureau of Justice Statistics. Retrieved February 22, 2007, from http://www.ojp.usdoj.gov

Gagné, P., & Tweksbury, R. (1998). Conformity pressures and gender resistance among transgendered individuals. *Social Problems, 45*(1), 81–101.

Gerami, S., & Lehnerer, M. (2001). Women's agency and household diplomacy: Negotiating fundamentalism. *Gender & Society, 15*(4), 556–573.

Guiffre, P. A., & Williams, C. L. (1994). Boundary lines: Labeling sexual harassment in restaurants. *Gender & Society, 8*(3), 378–401.

Hand, J. Z., & Sanchez, L. (2000). Badgering or bantering? Gender differences in experience of, and reactions to, sexual harassment among U.S. high school students. *Gender & Society, 14*(6), 718–746.

Hochschild, A. (1983). *The managed heart: Commercialization of human feeling*. Berkeley, CA: University of California Press.

Kempadoo, K., & Doezema, J. (1998). *Global sex workers: Rights, resistance, and redefinition*. New York: Routledge.

Mahoney, P., Williams, L. M., West, C. M. (2001). Violence against women by intimate relationship partners. In C. M. Renzetti, J. L. Edleson, & R. K. Bergen (Eds.), *Sourcebook on violence against women* (pp. 143–178). Thousand Oaks, CA: Sage.

Nesbitt, P., Baust, J., & Bailey, E. (2001). Women's status in the Christian church. In D. Vannoy (Ed.), *Gender mosaics: Societal perspectives* (pp. 386–396). Los Angeles: Roxbury Publishing Company.

Paulozzi, L. J., Saltzman, L. E., Thompson, M. P., & Holmgreen, P. (2001). *Surveillance for homicide among intimate partners—United States, 1981–1998*. Washington, DC: U.S. Government Printing Office.

Puzone, C. A., Saltzman, L. E., Kresnow, M., Thompson, M. P., & Mercy, J. A. (2000). National trends in intimate partner homicide. *Violence Against Women, 6*(4), 409–426.

Rennison, C. M. (2002).*Rape and Sexual Assault: Reporting to Police and Medical Attention, 1992–2000*. Retrieved February 22, 2007, from the U.S. Department of Justice Web site: http://www.ojp.usdoj.gov

Rennison, C. M. (2003).*Intimate Partner Violence, 1993–2001*. Retrieved February 22, 2007, from the U.S. Department of Justice http://www.ojp.usdoj.gov

Rose, D. R. (2001). Gender and Judaism. In D. Vannoy (Ed.), *Gender mosaics: Societal perspective* (pp. 415–424). Los Angeles: Roxbury Publishing Company.

Schact, S. P. (1996). Misogyny on and off the "pitch": The gendered world of male rugby players. *Gender & Society, 10*(5), 550–565.

Introduction to Reading 42

The authors of this selection combine a thorough review of the research on sexual assault on college campuses with findings from their own research at a large research university located in a small Midwestern city. The data they use in this study included ethnographic observations of a residence hall "party dorm" conducted in 2004–2005; interviews or in-depth surveys from almost all of the women residing in that dorm; and sixteen group interviews with 24 men and 63 women. The focus of their study was how students understood and participated in the party scene on this campus, particularly as it related to sexual assault. This reading is particularly valuable in illustrating the multiple factors that contribute to and maintain a climate of sexual assault on campuses.

1. What elements of campus structure facilitate sexual assault?
2. How do individuals, in their interactions and expectations, maintain a "rape culture" on campuses?
3. If you were a college president, what recommendations would you make to your campus to reduce sexual assaults on your campus?

Sexual Assault on Campus

A Multilevel, Integrative Approach to Party Rape

Elizabeth A. Armstrong, Laura Hamilton, and Brian Sweeney

A 1997 National Institute of Justice study estimated that between one-fifth and one-quarter of women are the victims of completed or attempted rape while in college (Fisher, Cullen, and Turner 2000).[1] College women "are at greater risk for rape and other forms of sexual assault than women in the general population or in a comparable age group" (Fisher et al. 2000:iii). At least half and perhaps as many as three-quarters of the sexual assaults that occur on college campuses involve alcohol consumption on the part of the victim, the perpetrator, or both (Abbey et al. 1996; Sampson 2002). The tight link between alcohol and sexual assault suggests that many sexual assaults that occur on college campuses are "party rapes." A recent report by the U.S. Department of Justice defines party rape as a distinct form of rape, one that "occurs at an off-campus house or on- or off-campus fraternity and involves . . . plying a woman with alcohol or targeting an intoxicated woman" (Sampson 2002:6).[2] While party rape is classified as a form of acquaintance rape, it is not uncommon for the woman to have had no prior interaction with the assailant, that is, for the assailant to be an in-network stranger (Abbey et a1. 1996).

Colleges and universities have been aware of the problem of sexual assault for at least 20 years, directing resources toward prevention and providing services to students who have been sexually assaulted. Programming has included education of various kinds, support for *Take Back the Night* events, distribution of rape whistles, development and staffing of hotlines, training of police and administrators, and other efforts. Rates of sexual assault, however, have not declined over the last five decades (Adams-Curtis and Forbes 2004:95; Bachar and Koss 2001; Marine 2004; Sampson 2002:1).

Why do colleges and universities remain dangerous places for women in spite of active efforts to prevent sexual assault? While some argue that

From Armstrong, E. A., Hamilton, L., & Sweeney. B., "Sexual Assault on Campus: A Multilevel Integrative Approach to Party Rape" in *Social Problems, Vol. 53*(4) pp. 483–499. Reprinted with permission of University of California Press.

"we know what the problems are and we know how to change them" (Adams-Curtis and Forbes 2004: 115), it is our contention that we do not have a complete explanation of the problem. To address this issue we use data from a study of college life at a large midwestern university and draw on theoretical developments in the sociology of gender (Connell 1987, 1995; Lorber 1994; Martin 2004; Risman 1998, 2004). Continued high rates of sexual assault can be viewed as a case of the reproduction of gender inequality—a phenomenon of central concern in gender theory.

We demonstrate that sexual assault is a predictable outcome of a synergistic intersection of both gendered and seemingly gender neutral processes operating at individual, organizational, and interactional levels. The concentration of homogenous students with expectations of partying fosters the development of sexualized peer cultures organized around status. Residential arrangements intensify students' desires to party in male-controlled fraternities. Cultural expectations that partygoers drink heavily and trust party-mates become problematic when combined with expectations that women be nice and defer to men. Fulfilling the role of the partier produces vulnerability on the part of women, which some men exploit to extract nonconsensual sex. The party scene also produces fun, generating student investment in it. Rather than criticizing the party scene or men's behavior, students blame victims. By revealing mechanisms that lead to the persistence of sexual assault and outlining implications for policy, we hope to encourage colleges and universities to develop fresh approaches to sexual assault prevention.

Approaches to College Sexual Assault

Explanations of high rates of sexual assault on college campuses fall into three broad categories. The first tradition, a psychological approach that we label the "individual determinants" approach, views college sexual assault as primarily a consequence of perpetrator or victim characteristics such as gender role attitudes, personality, family background, or sexual history (Flezzani and Benshoff 2003; Forbes and Adams-Curtis 2001; Rapaport and Burkhart 1984). While "situational variables" are considered, the focus is on individual characteristics (Adams-Curtis and Forbes 2004; Malamuth, Heavey, and Linz 1993). For example, Antonia Abbey and associates (2001) find that hostility toward women, acceptance of verbal pressure as a way to obtain sex, and having many consensual sexual partners distinguish men who sexually assault from men who do not. Research suggests that victims appear quite similar to other college women (Kalof 2000), except that white women, prior victims, first-year college students, and more sexually active women are more vulnerable to sexual assault (Adams-Curtis and Forbes 2004; Humphrey and White 2000).

The second perspective, the "rape culture" approach, grew out of second wave feminism (Brownmiller 1975; Buchward, Fletcher, and Roth 1993; Lottes 1997; Russell 1975; Schwartz and DeKeseredy 1997). In this perspective, sexual assault is seen as a consequence of widespread belief in "rape myths," or ideas about the nature of men, women, sexuality, and consent that create an environment conducive to rape. For example, men's disrespectful treatment of women is normalized by the idea that men are naturally sexually aggressive. Similarly, the belief that women "ask for it" shifts responsibility from predators to victims (Herman 1989; O'Sullivan 1993). This perspective initiated an important shift away from individual beliefs toward the broader context. However, rape supportive beliefs alone cannot explain the prevalence of sexual assault, which requires not only an inclination on the part of assailants but also physical proximity to victims (Adams-Curtis and Forbes 2004:103).

A third approach moves beyond rape culture by identifying particular contexts—fraternities and bars—as sexually dangerous (Humphrey and Kahn 2000; Martin and Hummer 1989; Sanday 1990, 1996; Stombler 1994). Ayres Boswell and

Joan Spade (1996) suggest that sexual assault is supported not only by "a generic culture surrounding and promoting rape," but also by characteristics of the "specific settings" in which men and women interact (p. 133). Mindy Stombler and Patricia Yancey Martin (1994) illustrate that gender inequality is institutionalized on campus by "formal structure" that supports and intensifies an already "high-pressure heterosexual peer group" (p. 180). This perspective grounds sexual assault in organizations that provide opportunities and resources.

We extend this third approach by linking it to recent theoretical scholarship in the sociology of gender. Martin (2004), Barbara Risman (1998; 2004), Judith Lorber (1994) and others argue that gender is not only embedded in individual selves, but also in cultural rules, social interaction, and organizational arrangements. This integrative perspective identifies mechanisms at each level that contribute to the reproduction of gender inequality (Risman 2004). Socialization processes influence gendered selves, while cultural expectations reproduce gender inequality in interaction. At the institutional level, organizational practices, rules, resource distributions, and ideologies reproduce gender inequality. Applying this integrative perspective enabled us to identify gendered processes at individual, interactional, and organizational levels that contribute to college sexual assault.

Risman (1998) also argues that gender inequality is reproduced when the various levels are "all consistent and interdependent" (p. 35). Processes at each level depend upon processes at other levels. Below we demonstrate how interactional processes generating sexual danger depend upon organizational resources and particular kinds of selves. We show that sexual assault results from the intersection of processes at all levels.

We also find that not all of the processes contributing to sexual assault are explicitly gendered. For example, characteristics of individuals such as age, class, and concern with status play a role. Organizational practices such as residence hall assignments and alcohol regulation, both intended to be gender neutral, also contribute to sexual danger. Our findings suggest that apparently gender neutral social processes may contribute to gender inequality in other situations.

* * *

Explaining Party Rape

We show how gendered selves, organizational arrangements, and interactional expectations contribute to sexual assault. We also detail the contributions of processes at each level that are not explicitly gendered. We focus on each level in turn, while attending to the ways in which processes at all levels depend upon and reinforce others. We show that fun is produced along with sexual assault, leading students to resist criticism of the party scene.

Selves and Peer Culture in the Transition from High School to College

Student characteristics shape not only individual participation in dangerous party scenes and sexual risk within them but the development of these party scenes. We identify individual characteristics (other than gender) that generate interest in college partying and discuss the ways in which gendered sexual agendas generate a peer culture characterized by high-stakes competition over erotic status.

Non-Gendered Characteristics Motivate Participation in Party Scenes. Without individuals available for partying, the party scene would not exist. All the women on our floor were single and childless, as are the vast majority of undergraduates at this university; many, being upper-middle class, had few responsibilities other than their schoolwork. Abundant leisure time, however, is not enough to fuel the party scene. Media, siblings, peers, and parents all serve as sources of anticipatory socialization (Merton 1957). Both partiers and non-partiers agreed that one was "supposed" to party in college. This orientation was reflected in the popularity of a poster titled "What I Really Learned in School" that pictured mixed drinks with names associated with academic disciplines. As one focus group participant explained:

> You see these images of college that you're supposed to go out and have fun and drink, drink lots, party and meet guys. [You are] supposed to hook up with guys, and both men and women try to live up to that. I think a lot of it is girls want to be accepted into their groups and guys want to be accepted into their groups.

Partying is seen as a way to feel a part of college life. Many of the women we observed participated in middle and high school peer cultures organized around status, belonging, and popularity (Eder 1985; Eder, Evans, and Parker 1995; Milner 2004). Assuming that college would be similar, they told us, that they wanted to fit in, be popular, and have friends. Even on move-in day, they were supposed to already have friends. When we asked one of the outsiders, Ruth, about her first impression of her roommate, she replied that she found her:

> Extremely intimidating, Bethany already knew hundreds of people here. Her cell phone was going off from day one, like all the time. And I was too shy to ask anyone to go to dinner with me or lunch with me or anything. I ate while I did homework.

Bethany complained to the RA on move-in day that she did not want to be roommates with Ruth because she was weird. A group of women on the floor—including Bethany, but not Ruth—began partying together and formed a tight friendship group, Ruth noted: "There is a group on the side of the hall that goes to dinner together, parties together, my roommate included, I have never hung out with them once. . . . And, yeah, it kind of sucks." Bethany moved out of the room at the end of the semester, leaving Ruth isolated.

Peer Culture as Gendered and Sexualized. Partying was also the primary way to meet men on campus. The floor was locked to non-residents, and even men living in the same residence hall had to be escorted on the floor. The women found it difficult to get to know men in their classes, which were mostly mass lectures, They explained to us that people "don't talk" in class. Some complained they lacked casual friendly contact with men, particularly compared to the mixed-gender friendship groups they reported experiencing in high school.

Meeting men at parties was important to most of the women on our floor. The women found men's sexual interest at parties to be a source of self-esteem and status. They enjoyed dancing and kissing at parties, explaining to us that it proved men "liked" them. This attention was not automatic, but required the skillful deployment of physical and cultural assets (Stombler and Padavic 1997; Swidler 2001). Most of the party-oriented women on the floor arrived with appropriate gender presentations and the money and know-how to preserve and refine them. While some more closely resembled the "ideal" college party girl (white, even features, thin but busty, tan, long straight hair, skillfully made-up, and well-dressed in the latest youth styles), most worked hard to attain this presentation. They regularly straightened their hair, tanned, exercised, dieted, and purchased new clothes.

Women found that achieving high erotic status in the party scene required looking "hot" but not "slutty," a difficult and ongoing challenge (West and Zimmerman 1987). Mastering these distinctions allowed them to establish themselves as "classy" in contrast to other women (Handler 1995; Stombler 1994). Although women judged other women's appearance, men were the most important audience. A "hot" outfit could earn attention from desirable men in the party scene. A failed outfit, as some of our women learned, could earn scorn from men. One woman reported showing up to a party dressed in a knee length skirt and blouse only to find that she needed to show more skin. A male guest sarcastically told her "nice outfit" accompanied by a thumbs-up gesture.

The psychological benefits of admiration from men in the party scene were such that women in relationships sometimes felt deprived.

One woman with a serious boyfriend noted that she dressed more conservatively at parties because of him, but this meant she was not "going to get any of the attention." She lamented that no one was "going to waste their time with me" and that, "this is taking away from my confidence." Like most women who came to college with boyfriends, she soon broke up with him.

Men also sought proof of their erotic appeal. As a woman complained, "Every man I have met here has wanted to have sex with me!" Another interviewee reported that: this guy that I was talking to for like ten/fifteen minutes says, "Could you, um, come to the bathroom with me and jerk me off?" And I'm like, "What!" I'm like, "Okay, like, I've known you for like, fifteen minutes, but no." The women found that men were more interested than they were in having sex. These clashes in sexual expectations are not surprising: men derived status from securing sex (from high-status women), while women derived status from getting attention (from high-status men). These agendas are both complementary and adversarial: men give attention to women en route to getting sex, and women are unlikely to become interested in sex without getting attention first.

University and Greek Rules, Resources, and Procedures

Simply by congregating similar individuals, universities make possible heterosexual peer cultures. The university, the Greek system, and other related organizations structure student life through rules, distribution of resources, and procedures (Risman 2004).

Sexual danger is an unintended consequence of many university practices intended to be gender neutral. The clustering of homogeneous students intensifies the dynamics of student peer cultures and heightens motivations to party. Characteristics of residence halls and how they are regulated push student partying into bars, off-campus residences, and fraternities. While factors that increase the risk of party rape are present in varying degrees in all party venues (Boswell and Spade 1996), we focus on fraternity parties because they were the typical party venue for the women we observed and have been identified as particularly unsafe (see also Martin and Hummer 1989; Sanday 1990). Fraternities offer the most reliable and private source of alcohol for first-year students excluded from bars and house parties because of age and social networks.

University Practices as Push Factors. The university has latitude in how it enforces state drinking laws. Enforcement is particularly rigorous in residence halls. We observed RAs and police officers (including gun-carrying peer police) patrolling the halls for alcohol violations. Women on our floor were "documented" within the first week of school for infractions they felt were minor. Sanctions are severe—a $300 fine, an 8-hour alcohol class, and probation for a year. As a consequence, students engaged in only minimal, clandestine alcohol consumption in their rooms. In comparison, alcohol flows freely at fraternities.

The lack of comfortable public space for informal socializing in the residence hall also serves as a push factor. A large central bathroom divided our floor. A sterile lounge was rarely used for socializing. There was no cafeteria, only a convenience store and a snack bar in a cavernous room furnished with big-screen televisions. Residence life sponsored alternatives to the party scene such as "movie night" and special dinners, but these typically occurred early in the evening. Students defined the few activities sponsored during party hours (e.g., a midnight trip to Wal-Mart) as uncool.

Intensifying Peer Dynamics. The residence halls near athletic facilities and Greek houses are known by students to house affluent, party-oriented students. White, upper-middle class, first-year students who plan to rush request these residence halls, while others avoid them. One of our residents explained that "everyone knows [what the residence hall] is like and people are dying to get in here. People just think it's a total party or something." Students of color tend to live elsewhere on campus. As a consequence, our floor was homogenous in terms of age, race, sexual orientation, class, and appearance. Two women identified as lesbian; one moved within the first few weeks. The few women from less privileged backgrounds were socially invisible.

The homogeneity of the floor intensified social anxiety, heightening the importance of partying for making friends. Early in the year, the anxiety was palpable on weekend nights as women assessed their social options by asking where people were going, when, and with whom. One exhausted floor resident told us she felt that she "needed to" go out to protect her position in a friendship group. At the beginning of the semester, "going out" on weekends was virtually compulsory. By 11 p.m. the floor was nearly deserted.

Male Control of Fraternity Parties. The campus Greek system cannot operate without university consent. The university lists Greek organizations as student clubs, devotes professional staff to Greek-oriented programming, and disbands fraternities that violate university policy. Nonetheless, the university lacks full authority over fraternities; Greek houses are privately owned and chapters answer to national organizations and the Interfraternity Council (IFC) (i.e., a body governing the more than 20 predominantly white fraternities).

Fraternities control every aspect of parties at their houses: themes, music, transportation, admission, access to alcohol, and movement of guests. Party themes usually require women to wear scant, sexy clothing and place women in subordinate positions to men. During our observation period, women attended parties such as "Pimps and Hos," "Victoria's Secret," and "Playboy Mansion"—the last of which required fraternity members to escort two scantily-clad dates. Other recent themes included: "CEO/Secretary Ho," "School Teacher/ Sexy Student," and "Golf Pro/Tennis Ho."

Some fraternities require pledges to transport first-year students, primarily women, from the residence halls to the fraternity houses. From about 9 to 11 p.m. on weekend nights early in the year, the drive in front of the residence hall resembled a rowdy taxi-stand, as dressed-to-impress women waited to be carpooled to parties in expensive late-model vehicles. By allowing party-oriented first-year women to cluster in particular residence halls, the university made them easy to find. One fraternity member told us this practice was referred to as "dorm-storming."

Transportation home was an uncertainty. Women sometimes called cabs, caught the "drunk bus," or trudged home in stilettos. Two women indignantly described a situation where fraternity men "wouldn't give us a ride home." The women said, "Well, let us call a cab." The men discouraged them from calling the cab and eventually found a designated driver. The women described the men as "just dicks" and as "rude."

Fraternities police the door of their parties, allowing in desirable guests (first-year women) and turning away others (unaffiliated men). Women told us of abandoning parties when male friends were not admitted. They explained that fraternity men also controlled the quality and quantity of alcohol. Brothers served themselves first, then personal guests, and then other women. Non-affiliated and unfamiliar men were served last, and generally had access to only the least desirable beverages. The promise of more or better alcohol was often used to lure women into private spaces of the fraternities.

Fraternities are constrained, though, by the necessity of attracting women to their parties. Fraternities with reputations for sexual disrespect have more success recruiting women to parties early in the year. One visit was enough for some of the women. A roommate duo told of a house they "liked at first" until they discovered that the men there were "really not nice."

The Production of Fun and Sexual Assault in Interaction

Peer culture and organizational arrangements set up risky partying conditions, but do not explain *how* student interactions at parties generate sexual assault. At the interactional level we see the mechanisms through which sexual assault is produced. As interactions necessarily involve individuals with particular characteristics and occur in specific organizational settings, all three levels meet when interactions take place. Here, gendered and gender neutral expectations and routines are intricately woven together to create party rape. Party rape is the result of fun situations that shift—either gradually or quite suddenly—into coercive situations. Demonstrating how the production of fun is connected with sexual assault requires describing the interactional routines and expectations that enable men to employ coercive sexual strategies with little risk of consequence.

College partying involves predictable activities in a predictable order (e.g., getting ready, pre-gaming, getting to the party, getting drunk, flirtation or sexual interaction, getting home, and sharing stories). It is characterized by "shared assumptions about what constitutes good or adequate participation"—what Nina Eliasoph and Paul Lichterman (2003) call "group style" (p. 737). A fun partier throws him or herself into the event, drinks, displays an upbeat mood, and evokes revelry in others. Partiers are expected to like and trust partymates. Norms of civil interaction curtail displays of unhappiness or tension among partygoers. Michael Schwalbe and associates (2000) observed that groups engage in scripted events of this sort "to bring about an intended emotional result" (p. 438). Drinking assists people in transitioning from everyday life to a state of euphoria.

Cultural expectations of partying are gendered. Women are supposed to wear revealing outfits, while men typically are not. As guests, women cede control of turf, transportation, and liquor. Women are also expected to be grateful for men's hospitality, and as others have noted, to generally be "nice" in ways that men are not (Gilligan 1982; Martin 2003; Phillips 2000; Stombler and Martin 1994; Tolman 2002). The pressure to be deferential and gracious may be intensified by men's older age and fraternity membership. The quandary for women, however, is that fulfilling the gendered role of partier makes them vulnerable to sexual assault.

Women's vulnerability produces sexual assault only if men exploit it. Too many men are willing to do so. Many college men attend parties looking for casual sex. A student in one of our classes explained that "guys are willing to do damn near anything to get a piece of ass." A male student wrote the following description of parties at his (non-fraternity) house:

> Girls are continually fed drinks of alcohol. It's mainly to party but my roomies are also aware of the inhibition-lowering effects. I've seen an old roomie block doors when girls want to leave his room; and other times I've driven women home who can't remember much of an evening yet sex did occur. Rarely if ever has a night of drinking for my roommate ended without sex. I know it isn't necessarily and assuredly sexual assault, but with the amount of liquor in the house I question the amount of consent a lot.

Another student—after deactivating—wrote about a fraternity brother "telling us all at the chapter meeting about how he took this girl home and she was obviously too drunk to function and he took her inside and had sex with her." Getting women drunk, blocking doors, and controlling transportation are common ways men try to prevent women from leaving sexual situations. Rape culture beliefs, such as the belief that men are "naturally" sexually aggressive, normalize these coercive strategies. Assigning women the role of sexual "gatekeeper" relieves men from responsibility for obtaining authentic consent, and enables them to view sex obtained by undermining women's ability to resist it as "consensual" (e.g., by getting women so drunk that they pass out).

In a focus group with her sorority sisters, a junior sorority woman provided an example of a partying situation that devolved into a likely sexual assault.

Anna: It kind of happened to me freshman year. I'm not positive about what happened, that's the worst part about it. I drank too much at a frat one night, I blacked out and I woke up the next morning with nothing on in their cold dorms, so I don't really know what happened and the guy wasn't in the bed anymore, I don't even think I could tell you who the hell he was, no I couldn't.

Sarah: Did you go to the hospital?

Anna: No, I didn't know what happened. I was scared and wanted to get the hell out of there. I didn't know who it was, so how am I supposed to go to the hospital and say someone might've raped me? It could have been any one of the hundred guys that lived in the house.

Sarah: It happens to so many people, it would shock you. Three of my best friends in the whole world, people that you like would think it would never happen to, it happened to. It's just so hard because you don't know how to deal with it because you don't want to turn in a frat because all hundred of those brothers . . .

Anna: I was also thinking like, you know, I just got to school, I don't want to start off on a bad note with anyone, and now it happened so long ago, it's just one of those things that I kind of have to live with.

This woman's confusion demonstrates the usefulness of alcohol as a weapon: her intoxication undermined her ability to resist sex, her clarity about what happened, and her feelings of entitlement to report it (Adams-Curtis and Forbes 2004; Martin and Hummer 1989). We collected other narratives in which sexual assault or probable sexual assault occurred when the woman was asleep, comatose, drugged, or otherwise incapacitated.

Amanda, a woman on our hall, provides insight into how men take advantage of women's niceness, gender deference, and unequal control of party resources. Amanda reported meeting a "cute" older guy, Mike, also a student, at a local student bar. She explained that, "At the bar we were kind of making out a little bit and I told him just cause I'm sitting here making out doesn't mean that I want to go home with you, you know?" After Amanda found herself stranded by friends with no cell phone or cab fare, Mike promised that a sober friend of his would drive her home. Once they got in the car Mike's friend refused to take her home and instead dropped her at Mike's place. Amanda's concerns were heightened by the driver's disrespect. "He was like, so are you into ménage à trois?" Amanda reported staying awake all night. She woke Mike early in the morning to take her home. Despite her ordeal, she argued that Mike was "a really nice guy" and exchanged telephone numbers with him.

These men took advantage of Amanda's unwillingness to make a scene. Amanda was one of the most assertive women on our floor. Indeed, her refusal to participate fully in the culture of feminine niceness led her to suffer in the social hierarchy of the floor and on campus. It is unlikely that other women we observed could have been more assertive in this situation. That she was nice to her captor in the morning suggests how much she wanted him to like her and what she was willing to tolerate in order to keep his interest.

This case also shows that it is not only fraternity parties that are dangerous; men can control party resources and work together to constrain women's behavior while partying in bars and at house parties. What distinguishes fraternity parties is that male dominance of partying there is organized, resourced, and implicitly endorsed by the university. Other party venues are also organized in ways that advantage men.

We heard many stories of negative experiences in the party scene, including at least one account of a sexual assault in every focus group that included heterosexual women. Most women who partied complained about men's efforts to control their movements or pressure them to drink. Two of the women on our floor were sexually assaulted at a fraternity party in the first week of school—one was raped. Later in the semester, another woman on the floor was raped by a friend. A fourth woman on the floor suspects she was drugged; she became disoriented at a fraternity party and was very ill for the next week.

Party rape is accomplished without the use of guns, knives, or fists. It is carried out through the combination of low level forms of coercion—a lot of liquor and persuasion, manipulation of situations so that women cannot leave, and sometimes force (e.g., by blocking a door, or using body weight to make it difficult for a woman to get up). These forms of coercion are made more effective by organizational arrangements that provide men with control over how partying happens and by expectations that women let loose and trust their party-mates. This systematic and effective method of extracting non-consensual sex is largely invisible, which makes it difficult for victims to convince anyone—even themselves—that a crime occurred. Men engage in this behavior with little risk of consequences.

Student Responses and the Resiliency of the Party Scene

The frequency of women's negative experiences in the party scene poses a problem for those students most invested in it. Finding fault with the party scene potentially threatens meaningful identities and lifestyles. The vast majority of heterosexual encounters at parties are fun and consensual. Partying provides a chance to meet new people, experience and display belonging, and to enhance social position. Women on our floor told us that they loved to flirt and be admired, and they displayed pictures on walls, doors, and websites commemorating their fun nights out.

The most common way that students—both women and men—account for the harm that befalls women in the party scene is by blaming victims. By attributing bad experiences to women's "mistakes," students avoid criticizing the party scene or men's behavior within it. Such victim-blaming also allows women to feel that they can control what happens to them. The logic of victim-blaming suggests that sophisticated, smart, careful women are safe from sexual assault. Only "immature," "naïve," or "stupid" women get in trouble. . . .

[A] floor resident relayed a sympathetic account of a woman raped at knife point by a stranger in the bushes, but later dismissed party rape as nothing to worry about "'cause I'm not stupid when I'm drunk," Even a feminist focus group participant explained that her friend who was raped "made every single mistake and almost all of them had to do with alcohol. . . . She got ridiculed when she came out and said she was raped." These women contrast "true victims" who are deserving of support with "stupid" women who forfeit sympathy (Phillips 2000). Not only is this response devoid of empathy for other women, but it also leads women to blame themselves when they are victimized (Phillips 2000).

Sexual assault prevention strategies can perpetuate victim-blaming. Instructing women to watch their drinks, stay with friends, and limit alcohol consumption implies that it is women's responsibility to avoid "mistakes" and their fault if they fail. Emphasis on the precautions women should take—particularly if not accompanied by education about how men should change their behavior—may also suggest that it is natural for men to drug women and take advantage of them. Additionally, suggesting that women should watch what they drink, trust party-mates, or spend time alone with men asks them to forgo full engagement in the pleasures of the college party scene.

Victim-blaming also serves as a way for women to construct a sense of status within campus erotic hierarchies. As discussed earlier, women and men acquire erotic status based on how "hot" they are perceived to be. Another aspect of erotic status concerns the amount of sexual respect one receives from men (see Holland and Eisenhart 1990: 101). Women can tell themselves that they are safe from sexual assault not only because they are savvy, but because men will recognize that they, unlike other women, are worthy of sexual respect. For example, a focus group of senior women explained that at a small fraternity gathering their friend Amy came out of the bathroom. She was crying and said that a guy "had her by her neck, holding her up, feeling her up from her crotch up to her neck and saying that I should rape you, you are a fucking whore." The woman's friends were appalled, saying, "no one deserves that." On the other hand, they explained that: "Amy flaunts herself. She is a whore so, I mean . . ." They implied that if one is a whore, one gets treated like one.

Men accord women varying levels of sexual respect, with lower status women seen as "fair game" (Holland and Eisenhart 1990; Phillips 2000). On campus the youngest and most anonymous women are most vulnerable. High-status women (i.e., girlfriends of fraternity members) may be less likely victims of party rape. Sorority women explained that fraternities discourage members from approaching the girlfriends (and ex-girlfriends) of other men in the house. Partiers on our floor learned that it was safer to party with men they knew as boyfriends, friends, or brothers of friends. One roommate pair partied exclusively at a fraternity where one of the women knew many men from high school. She explained that "we usually don't party with people we don't know that well." Over the course of the year, women on the floor winnowed their party venues to those fraternity houses where they "knew the guys" and could expect to be treated respectfully.

Opting Out. While many students find the party scene fun, others are more ambivalent. Some attend a few fraternity parties to feel like they have participated in this college tradition. Others opt out of it altogether. On our floor, 44 out of the 51 first-year students (almost 90%) participated in the party scene. Those on the floor who opted out worried about sexual safety and the consequences of engaging in illegal behavior. . . .

Socially, the others simply did not exist. A few of our "misfits" successfully created social lives outside the floor. The most assertive of the "misfits" figured out the dynamics of the floor in the first weeks and transferred to other residence halls. However, most students on our floor lacked the identities or network connections necessary for entry into alternative worlds. Life on a large university campus can be overwhelming for first-year students. Those who most needed an alternative to the social world of the party dorm were often ill-equipped to actively seek it out. They either integrated themselves into partying or found themselves alone in their rooms, microwaving frozen dinners and watching television. A Christian focus group participant described life in this residence hall: "When everyone is going out on a Thursday and you are in the room by yourself and there are only two or three other people on the floor, that's not fun, it's not the college life that you want."

Discussion and Implications

We have demonstrated that processes at individual, organizational, and interactional levels contribute to high rates of sexual assault. Some individual level characteristics that shape the likelihood of a sexually dangerous party scene developing are not explicitly gendered. Party rape occurs at high rates in places that cluster young, single, party-oriented people concerned about social status. Traditional beliefs about sexuality also make it more likely that one will participate in the party scene and increase danger within the scene. This university contributes to sexual danger by allowing these individuals to cluster.

Notes

1. Other studies have found similar rates of college sexual assault (Abbey et al. 1996; Adams-Curtis and Forbes 2004; Copenhaver and Grauerholz 1991; DeKeseredy and Kelly 1993; Fisher et al. 2000; Humphrey and White 2000; Koss 1988; Koss, Gidycz, and Wisniewski 1987; Mills and Granoff 1992; Muehlenhard and Linton 1987; Tjaden and Thoennes 2000; Ward et al. 1991).

2. On party rape as a distinct type of sexual assault, see also Ward and associates (1991). Ehrhart and Sandler (1987) use the term to refer to group rape. We use the term to refer to one-on-one assaults. We encountered no reports of group sexual assault.

References

Abbey, Antonia, Pam McAuslan, Tina Zawacki, A. Monique Clinton, and Philip Buck. 2001. "Attitudinal, Experiential, and Situational Predictors of Sexual Assault Perpetration." *Journal of Interpersonal Violence* 16:784–807.

Abbey, Antonia, Lisa Thomson Ross, Donna McDuffie, and Pam McAuslan. 1996. "Alcohol and Dating Risk Factors for Sexual Assault among College Women." *Psychology of Women Quarterly* 20: 147–69.

Adams-Curtis, Leah and Gordon Forbes. 2004. "College Women's Experiences of Sexual Coercion: A Review of Cultural, Perpetrator, Victim, and Situational Variables." *Trauma, Violence, and Abuse: A Review Journal* 5:91–122.

Bachar, Karen and Mary Koss. 2001. "From Prevalence to Prevention: Closing the Gap between What We Know about Rape and What We Do." Pp. 117–42 in *Sourcebook on Violence against Women,* edited by C. Renzetti, J. Edleson, and R. K. Bergen. Thousand Oaks, CA; Sage.

Boswell, A. Ayres and Joan Z. Spade. 1996. "Fraternities and Collegiate Rape Culture: Why Are Some Fraternities More Dangerous Places for Women?" *Gender & Society* 10:133–47.

Brownmiller, Susan. 1975. *Against Our Will: Men, Women, and Rape.* New York: Bantam Books.

Buchward, Emilie, Pamela Fletcher, and Martha Roth, eds. 1993. *Transforming a Rape Culture.* Minneapolis, MN: Milkweed Editions.

Connell, R. W. 1987. *Gender and Power.* Palo Alto. CA: Stanford University Press.

———. 1995. *Masculinities.* Berkeley, CA: University of California Press.

Copenhaver, Stacey and Elizabeth Grauerholz. 1991. "Sexual Victimization among Sorority Women: Exploring the Link between Sexual Violence and Institutional Practices." *Sex Roles* 24:31–41.

DeKeseredy, Walter and Katharine Kelly. 1993. "The Incidence and Prevalence of Women Abuse in Canadian University and College Dating Relationships." *Canadian Journal of Sociology* 18:137–59.

Eder, Donna. 1985. "The Cycle of Popularity: Interpersonal Relations among Female Adolescents." *Sociology of Education* 58:154–65.

Eder, Donna, Catherine Evans, and Stephen Parker. 1995. *School Talk: Gender and Adolescent Culture.* New Brunswick, NJ: Rutgers University Press.

Ehrhart, Julie and Bernice Sandler. 1987. "Party Rape," *Response* 9:205.

Eliasoph, Nina and Paul Lichterman. 2003. "Culture in Interaction." *American Journal of Sociology* 108:735–94.

Fisher, Bonnie, Francis Cullen, and Michael Turner. 2000. "The Sexual Victimization of College Women." Washington, DC: National Institute of Justice and the Bureau of Justice Statistics.

Flezzani, James and James Benshoff. 2003. "Understanding Sexual Aggression in Male College Students: The Role of Self–Monitoring and Pluralistic Ignorance." *Journal of College Counseling* 6:69–79.

Forbes, Gordon and Leah Adams-Curtis. 2001. "Experiences with Sexual Coercion in College Males and Females: Role of Family Conflict, Sexist Attitudes, Acceptance of Rape Myths, Self-Esteem, and the Big-Five Personality Factors." *Journal of Interpersonal Violence* 16:865–89.

Gilligan, Carol. 1982. *In a Different Voice: Psychological Theory and Women's Development.* Cambridge, MA: Harvard University Press.

Handler, Lisa. 1995. "In the Fraternal Sisterhood: Sororities as Gender Strategy." *Gender & Society* 9:236–55.

Herman Diane. 1989. "The Rape Culture." Pp. 20–44 in *Women: A Feminist Perspective,* edited by J. Freeman. Mountain View, CA: Mayfield.

Holland, Dorothy and Margaret Eisenhart. 1990. *Educated in Romance: Women, Achievement, and College Culture.* Chicago: University of Chicago Press.

Humphrey, John and Jacquelyn White. 2000. "Women's Vulnerability to Sexual Assault from Adolescence to Young Adulthood." *Journal of Adolescent Health* 27:419–24.

Humphrey, Stephen and Arnold Kahn. 2000. "Fraternities, Athletic Teams, and Rape: Importance of Identification with a Risky Group." *Journal of Interpersonal Violence* 15:1313–22.

Kalof, Linda. 2000. "Vulnerability to Sexual Coercion among College Women: A Longitudinal Study." *Gender Issues* 18:47–58.

Koss, Mary. 1988. "Hidden Rape: Incidence and Prevalence of Sexual Aggression and Victimization in a National Sample of Students in Higher Education." Pp. 4–25 in *Rape and Sexual Assault,* edited by Ann W. Burgess. New York: Garland.

Koss, Mary, Christine Gidycz, and Nadine Wisniewski. 1987. "The Scope of Rape: Incidence and Prevalence of Sexual Aggression and Victimization in a National Sample of Higher Education Students." *Journal of Counseling and Clinical Psychology* 55:162–70.

Lorber, Judith. 1994. *Paradoxes of Gender.* New Haven, CT: Yale University Press.

Lottes, Ilsa L. 1997. "Sexual Coercion among University Students: A Comparison of the United States and Sweden." *Journal of Sex Research* 34:67–76.

Malamuth, Neil, Christopher Heavey, and Daniel Linz. 1993: "Predicting Men's Antisocial Behavior against Women: The Interaction Model of Sexual Aggression." Pp. 63–98 in *Sexual Aggression: Issues in Etiology, Assessment, and Treatment,* edited by G. N. Hall, R. Hirschman, J. Graham, and M. Zaragoza. Washington, DC: Taylor and Francis.

Marine, Susan. 2004. "Waking Up from the Nightmare of Rape." *The Chronicle of Higher Education.* November 26, p. B5.

Martin, Karin. 2003. "Giving Birth Like a Girl." *Gender & Society.* 17:54–72.

Martin, Patricia Yancey. 2004. "Gender as a Social Institution." *Social Forces* 82: 1249–73.

Martin, Patricia Yancey and Robert A. Hummer. 1989. "Fraternities and Rape on Campus." *Gender & Society* 3:457–73.

Merton, Robert. 1957. *Social Theory and Social Structure.* New York: Free Press.

Mills, Crystal and Barbara Granoff. 1992. "Date and Acquaintance Rape among a Sample of College Students." *Social Work* 37:504–09.

Milner, Murray. 2004. *Freaks, Geeks, and Cool Kids: American Teenagers, Schools, and the Culture of Consumption.* New York: Routledge.

Mohler-Kuo, Meichun, George W. Dowdall. Mary P. Koss, and Henry Weschler. 2004. "Correlates of Rape While Intoxicated in a National Sample of College Women." *Journal of Studies on Alcohol* 65:37–45.

Muehlenhard, Charlene and Melaney Linton. 1987. "Date Rape and Sexual Aggression: Incidence and Risk Factors." *Journal of Counseling Psychology* 34:186–96.

O'Sullivan. Chris. 1993. "Fraternities and the Rape Culture." Pp. 23–30 in *Transforming a Rape Culture,* edited by E. Buchward, P. Fletcher, and M. Roth. Minneapolis, MN: Milkweed Editions.

Phillips, Lynn. 2000. *Flirting with Danger: Young Women's Reflections on Sexuality and Domination.* New York: New York University.

Rapaport, Karen and Barry Burkhart. 1984. "Personality and Attitudinal Characteristics of Sexually Coercive College Males." *Journal of Abnormal Psychology* 93:216–21.

Risman, Barbara. 1998. *Gender Vertigo: American Families in Transition.* New Haven, CT: Yale University Press.

———. 2004. "Gender as a Social Structure: Theory Wrestling with Activism." *Gender & Society* 18:429–50.

Rosow, Jason and Rashawn Ray. 2006. "Getting Off and Showing Off: The Romantic and Sexual Lives of High-Status Black and White Status Men." Department of Sociology, Indiana University, Bloomington, IN. Unpublished manuscript.

Russell, Diana. 1975. *The Politics of Rape.* New York: Stein and Day.

Sampson, Rana. 2002. "Acquaintance Rape of College Students." Problem-Oriented Guides for Police Series, No. 17. Washington, DC: U.S. Department of Justice, Office of Community Oriented Policing Services.

Sanday, Peggy. 1990. *Fraternity Gang Rape: Sex, Brotherhood, and Privilege on Campus.* New York: New York University Press.

———. 1996. "Rape-Prone versus Rape-Free Campus Cultures." *Violence against Women* 2:191–208.

Schwalbe, Michael, Sandra Godwin, Daphne Holden, Douglas Schrock, Shealy Thompson, and Michele Wolkomir. 2000. "Generic Processes in the Reproduction of Inequality: An Interactionist Analysis." *Social Forces* 79:419–52.

Schwartz, Martin and Walter DeKeseredy. 1997. *Sexual Assault on the College Campus. The Role of Male Peer Support.* Thousand Oaks, CA: Sage Publications.

Stombler, Mindy. 1994. "'Buddies' or 'Slutties': The Collective Reputation of Fraternity Little Sisters." *Gender & Society* 8:297–323.

Stombler, Mindy and Patricia Yancey Martin. 1994. "Bringing Women In, Keeping Women Down: Fraternity 'Little Sister' Organizations." *Journal of Contemporary Ethnography* 23:150–84.

Stombler, Mindy and Irene Padavic. 1997. "Sister Acts: Resisting Men's Domination in Black and

White Fraternity Little Sister Programs." *Social Problems* 44:257–75.

Sweeney, Brian. 2004. "Good Guy on Campus: Gender, Peer Groups, and Sexuality among College Men." Presented at the American Sociological Association Annual Meetings, August 17, Philadelphia, PA.

Swidler, Ann. 2001. *Talk of Love: How Culture Matters.* Chicago: University of Chicago Press.

Tjaden, Patricia and Nancy Thoennes. 2000. "Full Report of the Prevalence, Incidence, and Consequences of Violence against Women: Findings from the National Violence against Women Survey." Washington, DC: National Institute of Justice.

Tolman, Deborah. 2002. *Dilemmas of Desire: Teenage Girls Talk about Sexuality.* Cambridge, MA: Harvard University Press.

Ward, Sally, Kathy Chapman, Ellen Cohn, Susan White, and Kirk Williams. 1991." Acquaintance Rape and the College Social Scene." *Family Relations* 40:65–71.

West, Candace and Don Zimmerman. 1987. "Doing Gender." *Gender & Society* 1:125–51.

Introduction to Reading 43

Julia Hall uses her sociological training here to examine the effects of shifting socioeconomic tides on domestic violence. She studies the effects of domestic violence on the female children of those whose lives have been turned upside down by the closing of factories and loss of jobs. This extensive case study of nine poor, White girls—ages eleven to thirteen, living in a Northeast urban area—describes the effects of domestic violence on future generations. These girls interpreted their lives based upon the anger, abuse, and violence they witnessed.

1. In what ways are incidents of violence normalized in these girls' lives and their communities?
2. What strategies have the girls developed to escape the violence they live with on a daily basis?
3. How have schools failed to deal with the violence these children face?

It Hurts to Be a Girl

Growing Up Poor, White, and Female

Julia Hall

In this investigation, I contend that a group of poor white middle school young women in the postindustrial urban Northeast are living among high concentrations of domestic violence. I refer to this group as "Canal Town" girls.[1] These young women are envisioning lives in which, by charting a course of secondary education, they hope to procure jobs and self-sufficiency. As their narrations indicate, such plans are fueled by the hope that they will live independent lives as single career women and, therefore, will bypass the domestic violence that currently rips through their own and their mothers' lives. . . .

While there are many analyses that focus on the ways in which institutions and the formation of female youth cultures contribute to inequitable futures (Finders 1996; Holland and Eisenhart 1990; McRobbie 1991; Raissiguier 1994; Smith 1988; Valli 1988; Weis 1990), none of this work picks up on the issue of violence. Fine and Weis (1998) examine this theme as it boldly emerges in data on the lives of poor and working-class adult white women and the production of identity. They found that white working-class women experience more abuse, as compared with working-class women from other cultural backgrounds, and are more apt to treat their abuse as a carefully guarded secret (Weis et al. 1997; Weis, Marusza-Hall, and Fine 1998).[2]

Informed by such work, I turn this critical lens on middle school girls. What I found is that their lives are also saturated with domestic abuse. They are not talking about it, not reporting it, and covering their bruises with clothes. They are also hiding it from others and themselves to such an extent that it is not openly dealt with at a critical level at all. Nowhere in their narrations is there any sense that males are accountable for their violent behavior.

This research is contextualized in a postindustrial economy characterized by the systematic dismantling of the basic productive capacity of a nation, a trend sharply experienced in the United States during the 1970s and 1980s. During these decades, the U.S. steel industry had already begun a process of shifting to foreign, less expensive, less regulated markets, as did other areas of manufacturing and production. As a result, smaller businesses that were dependent on industry also closed (Bluestone and Harrison 1982). No longer able economically to support its own populace, the city in this analysis currently relies on shrinking state resources. Left in the wake of global restructuring are empty factories, gutted warehouses, and people who can no longer make a decent living. Canal Town is an urban neighborhood reflecting these changes.

No longer able to find the wage-earning jobs they once enjoyed, today residents are often unemployed (Perry 1996). Many rely on food stamps and Aid to Families with Dependent Children (AFDC). The demography of Canal Town has also shifted from white to racially diverse. This change is reflected in the neighborhood school, which was transformed into a Spanish-English bilingual magnet in the late 1970s. The community center, which has traditionally been staffed by white adults, however, is almost exclusively visited by local white youth. . . .

Although still socially and economically privileged by their whiteness, among most white former workers, a family wage has disappeared. The cushion of wealth that white laborers were often able to amass for their families across generations is quickly eroding. Still, there may be pockets of accumulated resources that are shared among white families in Canal Town, for example, in the form of home ownership or a pension (Fine and Weis 1998; Oliver and Shapiro 1995). . . .

The Canal Town girls' families have historically been working-class, most having fathers and grandfathers who worked in industry while their mothers and grandmothers stayed home. The subordination of women to men within white working-class families has been heavily investigated (Smith 1987). Others explore this subordination through the notion of the *family wage.* The family wage appeared advantageous to all family members, but in reality it supported the notion that women should receive lower wages than men or stay home (May 1987; Woodcock Tentler 1979). Although the young women in this investigation contend that the adult men around them are no longer employed in full-time labor jobs, present-day gender arrangements in the Canal Town community are linked to the ideology prevalent during the days of heavy industry. . . .

Canal Town Girls

To obtain some sense of the nine poor white sixth-, seventh-, and eighth-grade girls who participated in this research, I share information from the individual interviews pertaining to their home life. Out of the nine girls—Anne, 11; Rosie, 11; Sally, 11; Jamie, 12; Elizabeth, 12; Lisa, 12; Katie, 13; Christina, 13; and Lisette, 13—only Jamie says she lives with both parents. Elizabeth, Anne, and Katie maintain they live with their mothers, siblings, and their mothers' steady boyfriends. The rest of the girls—Christina, Lisette, Lisa, Rosie, and Sally—report they live with their mothers, siblings, and on occasion, their mothers' different boyfriends. Only Jamie and Christina said they were in contact with their biological fathers, while the remaining girls contend they have no knowledge of their fathers' whereabouts. In terms of employment, seven of the girls state their mothers are not presently working, nor have they been in the past. Only Elizabeth says that her mother used to work as a secretary before she was born. Of the adult men who contribute to household expenses, Jamie's father holds a part-time job in the trucking industry, while Katie is the only girl to claim her mother's boyfriend earns money for their family. As Katie explains, he collects items on trash day that he sells to pawn shops. Jamie says her family receives food stamps, while all of the other girls say their families rely on food stamps and AFDC. Jamie lives in a house that her parents inherited from her grandparents, while the rest of the girls say they live in apartments.

Dream Jobs

As the Canal Town girls begin to talk about what they want their lives to look like after high school, they stress going to college and/or obtaining a good job, and only mention marriage or family after being asked. Since the girls are only in middle school, their plans for the future may not yet be specific or thought out, but the positioning of a job or career as central to the production of identity is worth noting.

Christina: I want to be a doctor . . . I'll have to go to college for a long time . . . I don't know where I'll go [to college], hopefully around here . . . I'm not sure what type of doctor, but I'm thinking of the kind that delivers babies.

Lisette: I want to be a leader and not a follower . . . I want to be a teacher in [the neighborhood] because I never want to leave here . . . I want to go to [the local] community college, like my sister, learn about teaching little kids . . . I definitely want to be a teacher. . . .

All of these young girls envision further education in their future, but most do not yet have a clear sense of what school they hope to attend or how long they plan to go. . . . Any of these girls may switch ideas about careers a number of times, yet when asked about the future, all of them focus their energies on the single pursuit of furthering their schooling and landing a job. Christina says she worries about being homeless, which is likely a chronic fear among poor youth.

These white young adolescents are the daughters of presently poor adults. None of their parents continued their education beyond high school and a few did not graduate from grade 12. College, they say, is not an option that is really discussed much at home. Perhaps Lisette has the clearest idea of where she would like to go to school because she is the only white girl who I worked with who has an older sibling enrolled in an institution of higher education. Lisette's sister attends a nearby community college and studies early childhood education, a circumstance that likely influenced her little sister's plans. Interestingly, three of these white girls indicate that although they want to break out of cycles of dependence and have careers, they do not want to leave their neighborhood—whether for school or work.

The importance of a job or career is emerging within the identities of these girls, but it is too soon to say whether they will follow through on their plans for further education or training. The outlook is not promising, as all but a few of their older siblings are negotiating lives riddled with substance abuse and early pregnancies. My conversations with the principal of the area high school reveal that very few local teenagers are enrolling in any form of advanced studies.

Even though the Canal Town girls view education as important in obtaining their goals, they both accept and reject academic culture and knowledge. I observed that on a daily basis while in class, these girls copy homework, pass notes, read magazines and/or books, or, in other words, participate in the form rather than the content of schooling. Time spent in school involves passively skipping across the surface of learning.

In only a few instances did the girls actually narrate resentment toward school authority. As part of a tradition of working-class women whose personal choices have been mediated by structural constraints, the Canal Town girls, by virtue of gender, are not part of this legacy of expressed resentment. Animosity toward institutional authority is typically male and is linked to the historical contestation between capital and workers (Everhart 1983; Weis 1990). Since white women generally labored in the private sphere or as marginalized wage workers, they did not directly engage in such struggles.

Family Planning

* * *

[M]ost of the Canal Town girls, with the exception of Lisa, contend they do not wish to have husbands, homes, and families at all. Rather, the Canal Town girls claim they are looking to the life of a single career woman as a way to circumvent the abuse that they see inscribed in future families or relationships with men. It quickly becomes clear that seeking refuge from domestic violence plays a big role in constructing identities. For many of these girls, the future includes avoiding marriage and family altogether and getting a job so they can rely on themselves.

Christina: I don't want to be married because if I was married, my husband would want a kid. I don't want to have a kid because its father may not treat us right . . . hitting and stuff. . . . There's not enough for everybody and the kid shouldn't have to suffer . . . I want to always stay in [this neighborhood] . . . live alone. . . . At least I know trouble here when I see it.

Lisette: I don't want to get married and be told to stay at home . . . and be someone's punching bag . . . I'll get a one-bedroom apartment and live alone and just try to be the best teacher I can be. . . .

The girls are not devising career-oriented plans simply to escape a patriarchal-dominant home. Rather, they specifically say they view a job as the ticket to a life free of abuse. By concentrating energies on the world of work instead of family, some of the Canal Town girls feel they can spare bringing children into the world, whom they feel often bear the brunt of adult problems. . . .

While the girls say they want to live as independent women in the public sphere, they are developing such identities in response to violent men. Nowhere, in more than one year of observations and interviews, did I hear these girls hold men and boys accountable for their abusive behavior. While it may be the case that they hold such a critique, the absence of any such discourse in the data is glaring, especially given the frequency and detail in which abuse was mentioned in the private space of an interview or in hushed conversations with friends.

It Hurts to Be a Girl

When these young girls are asked to describe their neighborhood, they soon begin to tell stories of women being abused at the hands of men. The women in their narrations seemingly work to conceal their abuse from authorities and ultimately end up "going back."

Jamie: It's a pretty good place to live. . . . There's lots of auto crashes, drunk people. Lots of people go to the bars on Friday and Saturday and get blasted. They're always messing with people. Some guy is always getting kicked out of the bar for fighting. Guys are mostly fighting with their girlfriends and are getting kicked out for punching so they continue to fight in the street; I see it from my bedroom window; only the girl mostly gets beat up really bad . . . but later she was saying it was her fault.

Christina: There's lots of violence in this neighborhood. Like there's this couple that's always fighting. When the guy gets mad, he hits her. It happens upstairs in their house. She's thrown the coffee pot at him and the toaster, they [the coffeepot and toaster] landed in the street . . . I saw it while walking by. . . . The guy would show off all the time in front of his friends. One day when he was hitting her, she just punched him back and told him she wasn't going to live with him anymore. He used to hit her hard. She used to cry but she would still go out with him. She said she loved him too much to dump him. A lot of people go back. . . .

Even though the community is seen as "a pretty good place to live" and "overall a nice neighborhood," the girls' descriptions of residency quickly devolve into stories of violence—mostly violence directed toward women by the hands of men in both public and private spaces. . . .

Although these girls may look at abuse differently, they all are quick to recognize violence as a defining feature of their community. Many mention that women "go back" to their abuser as if it were acceptable or normal for men to abuse women, and that it is the women's duty to negotiate their way around this violence. Again, missing in these arguments is the recognition that men are responsible for their abusive behavior. The only critique articulated is raised by Lisette, but it is directed toward neighborhood mothers whom she feels are not adequately putting their children's needs first.

Despite their young ages, the Canal Town girls have heard of abuse in a variety of different contexts and forms and, according to their narrations, women in this neighborhood do not always endure their violence in isolation. Rosie, for example, reveals that a neighbor sought refuge with her mother after a severe beating. Abuse, however, is seemingly concealed within the community—that is, complaints rarely reach a more public forum.

For girls, abuse does not just exist in public places and in the private dwellings of others. Violence also occurs in their own homes. In talking about personal experiences with abuse, they typically contextualize violence as part of the past, as "things are better now." The younger girls, though, are not consistent in packaging such events in history. For instance, Elizabeth and Sally shift from present to past in describing the abuse in their homes. Many women recall chilling vignettes of unbridled rage that pattern their upbringing. . . .

Sally: I used to think of myself as a zero, like I was nothing. I was stupid; I couldn't do anything . . . I don't anymore because we're all done with the violence in my house . . . I've tried to keep it out since I was a kid. . . . My mom and John [her mother's boyfriend] will argue over the littlest things. My mom is someone who is a violent person too. Sometimes she hits us, or he does. Then she would take a shower and we would get all dressed up, and we would all go out somewhere. After something bad would happen, she would try to make it better. She's a real fun person. . . . We're really close. We make cookies together and breakfast together.

Anne: My mom and her boyfriend constantly fight because they drink. When I was little, I remember being in my bed. I was sleeping, only my other sisters came and woke me up because my mom and her boyfriend were fighting. We [Anne and her sisters] started crying. I was screaming. My sisters were trying to calm me down. Our door was above the staircase and you could see the front door. I just had visions of me running out the door to get help because I was so scared. My oldest sister was like nine or ten and I had to go the bathroom and we only have one and it was downstairs. She sneaked me downstairs and into the kitchen and there was glasses smashed all over, there were plants underwater, the phone cord was underwater in the kitchen sink. It was just a wreck everywhere. But most of all, there were streams of blood mixing in with the water, on the floor, on the walls. . . .

As these narrations indicate, domestic violence patterns the lives of these girls. . . . Mom can offer little salvation as she is often drunk, violent herself, or powerless as the man in her life is on an abusive rampage. . . . [T]hese females have little recourse from the extreme and terrifying conditions that govern their lives. In Sally's case, her mother is also violent yet is thought of as making up for that abuse by involving her daughter in family-style activities. Sally learns, therefore, not to see or feel pain. Given these accounts, it is easy for me to conclude that the effects of domestic violence are not something that can be contained at home, and the Canal Town girls indicate that exposure to abuse profoundly shapes their behavior in other places, such as school.

Elizabeth: About twice a month they [her mother and mother's boyfriend] fight. But not that far apart. Last time he [the boyfriend] smacked me, I had a red hand on my face. I walked around with a red hand on my face, only I wouldn't let anybody see it . . . I skipped school and the [community] center for, like, three days so no one would ask me about it . . . I hid in my closet until you could barely see it. Then when I went back to school, I stayed real quiet because I didn't want people to look at me, notice the hand on my face.

Christina: When I had a boyfriend, he [her father] got so mad at me. He told me I wasn't allowed to have a boyfriend. I didn't know that because he never told me. He said that if he ever saw him again, I would get my ass kicked. So one day he heard that Robbie [her boyfriend] walked me to school. Well, he [her father] came over that night and pulled down my pants and whipped me with his belt. I was bloody and the next day full of bruises. But I hurt more from being embarrassed to have my pants pulled down at my age. It hurt to sit all day long at school; that's all I could concentrate on. I couldn't go to the nurse because then she would find out. Nobody knew how I hurt under my clothes. I couldn't go to gym because people would find out, so I skipped. I hid in the bathroom but got picked up by the hall monitor who accused me of skipping gym to smoke. I just got so mad when I heard this, I pushed her [the hall monitor] away from me and yelled. I was out of control with anger when they were dragging me down to the principal's. I got suspended for a week and had to talk to a school psychologist for two weeks about how bad smoking is for your health. . . .

The glimpses into these lives suggest children from violent homes are learning at very young ages how to negotiate lives that are enmeshed inside a web of overwhelming circumstances. Elizabeth talks about how her mother's boyfriend blames her and her mother and sister for all that is wrong. . . . As they devise ways to conceal their bruises, they each face their pain alone. Elizabeth skips school and seeks shelter from the world in the same closet in which she is punished by her mother's boyfriend. Christina is choked on her anger and pain and separates herself from school activity only to become embroiled in another set of problems. . . .

The narrations of the poor white girls in this study reflect findings in much of the existing research (Weis, Marusza-Hall, and Fine 1998). As these girls indicate, abuse at home makes it difficult to concentrate in school, and the hurt, anger, and fear that they harbor inside often render them silent, which also corroborates these studies (Elkind 1984; Jaffe, Wolfe, and Wilson 1990). According to Afulayan (1993), some children blame themselves for the abuse and skip school to protect a parent from the abuser, while other children become ill from worry. Depression, sleep disturbances, suicidal tendencies, and low self-esteem are other symptoms exhibited among children living in violent homes (Hughes 1988; Reid, Kavanaugh, and Baldwin 1987).

All of these girls reveal they spend incredible energy on keeping their abuse a secret while in school. This is likely in response to a number of fears, including fear of public embarrassment, fear of further angering an abuser, or fear that families will be torn apart by authorities. While observing the girls at school, I noticed that some of them sustained bruises that could not be so easily hidden under long sleeves or turtlenecks. One day, for example, Christina came to school wearing an excessive amount of eye makeup, which was noticeable, considering she usually did not wear any. While talking to her outside after school, I realized this was probably an attempt to conceal a black eye, which could clearly be seen in the harsh light of day.

Interestingly, I did not hear any talk of domestic violence at school—critical or otherwise. This finding parallels the poor and working-class white women in the study of Weis et al. (1997) who also were silent about the abuse in their lives, which was similarly not interrupted by schools, the legal system, and so forth. It did not seem to me that any of the girls sought help at school from their white female peers, teachers, or anyone else in coping with abuse. Instead, in the space of the school, a code of silence surrounding domestic violence prevailed, even though the girls articulate an awareness of others' abuse throughout the community. Not once did I hear students or teachers query others about violence, nor was abuse even mentioned as a social problem in classes in which human behavior was discussed. Even on the day that Christina came to school attempting to camouflage a bruised eye, I did not observe a teacher pull her aside to talk, nor did I hear her friends ask her if she was all right. Dragged by their families from one violent situation to the next, it is remarkable that these girls are, for the most part, able to get through the school day, go home, and come back again tomorrow.

Conclusion

The Canal Town girls are from families that had been working class for generations. Born into the snares of a postindustrial economy, today these girls are growing up in poverty. As their narrations on work and family indicate, gender arrangements in their lives echo that of the working class in which women are subordinate to men. Embedded in this subordination is a silencing of domestic violence.

Domestic violence runs painfully deep in the lives of the Canal Town girls. These girls are socialized at an early age to conceal abuse from those outside the community who might take action. As a method of coping, they have learned to work around abuse to such an extent that by envisioning their future lives as financially independent, they hope to sidestep violence. The sting of abuse provides much of the scaffolding for how these girls wish to construct their lives, and men are seemingly not taken to task. In this tight-knit community, it often hurts to be a girl.

During an entire year of fieldwork, I never saw or heard a teacher approach a student concerning domestic violence. I also never witnessed a teacher initiate a discussion on the topic of abuse in class. Throughout the year, I had the opportunity to ask all of the teachers if they had knowledge of the extent of violence in the Canal Town community or the possibility of abuse in the lives of their students. The teachers had little to say on this topic, many indicating they had not thought much about domestic violence, although they "wouldn't be surprised."

By not responding to violence in the home, institutions that structure the lives of these girls, such as schools, arguably contribute to its concealment. The guidelines already in place in some schools, the counselors, and child abuse training for teachers do not typically address the needs of battered youth. Due to shame or punishment that awaits at home, youth do not always visit a counselor. It is often the case that teachers also are afraid to report abuse—afraid of upsetting the students, parents, and school administrators. Perhaps educators feel "unsure" about their suspicions and "wait" to see more evidence.

Indeed, alerting Child Protective Services many times leads to further abuse by an angered parent. Likewise, police investigations and court appearances often prove unproductive and humiliating for women and children (Weis, Marusza-Hall, and Fine 1998).

Educators must come to the conclusion that at least some students in their classes go home to abusive situations. Teachers and policy makers, therefore, are confronted with the task of formulating more tangible responses. In English and History classes, boys and girls can often be led in critical discussions about domestic violence as it relates to classroom material and to daily life. Through these lessons, abuse must be positioned as abnormal behavior, with social and historical roots that can be unraveled. Older kids can also be encouraged to enter internships at domestic violence shelters and hot lines, so youth can learn that abuse is wrong, it is not a personal problem, and there is some recourse (Weis, Marusza-Hall, and Fine 1998).

Educators and social scientists must additionally seek out other safe spaces in students' lives where critical conversations can take place (Fine and Weis 1998; Weis, Marusza-Hall, and Fine 1998). The Canal Town girls, for example, are regular visitors to a neighborhood community center. Places such as community centers, arts programs, and youth groups offer a location that is unbounded by state guidelines where such talk can happen. By conducting workshops in these sites by those who run domestic violence shelters, youth can be led to think critically about abuse and can come to realize they are not alone and that there is a possibility for a different way of life.

Notes

1. This name is based on the fact that during the early 1800s, this area of the city was selected as the last stop on a major canal that was constructed across the state. This opened up the city, transforming it into a formidable site for the production and transport of steel. During the past few decades, however, most of this industry has left the area.

2. Boys living in violent homes may experience more abuse than girls. This has been found to be the case because when angry, boys typically act out more than girls. Because this acting out is more apt to enrage a violent adult, boys often end up as a more primary target (Jouriles and Norwood 1995).

References

Afulayan, J. 1993. Consequences of domestic violence on elementary school education. *Child and Family Therapy* 15:55–58.

Bluestone, B., and B. Harrison. 1982. *The deindustrialization of America: Plant closings, community abandonment, and the dismantling of basic industry.* New York: Basic Books.

Elkind, P. 1984. *All grown up and no place to go.* Reading, MA: Addison-Wesley.

Everhart, R. 1983. *Reading, writing and resistance: Adolescence and labor in a junior high school.* Boston: Routledge and Kegan Paul.

Finders, M. 1996. *Just girls: Hidden literacies and life in junior high.* New York: Teachers College Press.

Fine, M., and L. Weis. 1998. *The unknown city: The lives of poor and working-class young adults.* New York: Beacon.

Holland, D., and M. Eisenhart. 1990. *Educated in romance: Women, achievement, and college culture.* Chicago: University of Chicago Press.

Hughes, H. 1988. Psychological and behavioral correlates of family violence in child witnesses and victims. *American Journal of Orthopsychiatry,* 58:77–90.

Jaffe, P., S. Wolfe, and S. Wilson.1990. *Children of battered women.* Newbury Park, CA: Sage.

Jouriles, E., and W. Norwood. 1995. Physical aggression toward boys and girls in families characterized by the battering of women. *Journal of Family Psychology* 9:69–78.

May, M. 1987. The historical problem of the family wage: The Ford Motor Company and the five dollar day. In *Families and work,* edited by N. Gerstel and H. E. Gross. Philadelphia, PA: Temple University Press.

McRobbie, A. 1991. *Feminism and youth culture: From Jackie to just seventeen.* Boston: Unwin Hyman.

Oliver, M., and T, Shapiro. 1995. *Black wealth, white wealth: A new perspective on racial inequality.* New York: Routledge.

Perry, D. 1996. *Governance in Erie County: A foundation for understanding and action.* Buffalo: State University of New York Press.

Raissiguier, C. 1994. *Becoming women, becoming workers: Identity formation in a French vocational school.* Albany: State University of New York Press.

Reid, J., T. Kavanaugh, and J. Baldwin. 1987. Abusive parents' perception of child problem behavior: An example of paternal violence. *Journal of Abnormal Child Psychology* 15:451–66.

Smith, D. 1987. *The everyday world as problematic: A feminist sociology.* Boston: Northeastern University Press.

———. 1988. Femininity as discourse. In *Becoming feminine: The politics of popular culture,* edited by L. Roman, L. Christian-Smith, and E. Ellsworth. London: Falmer.

Valli, L. 1988. Gender identity and the technology of office education. In *Class, race, and gender in American education,* edited by L. Weis. Albany: State University of New York Press.

Weis, L. 1990. *Working class without work.* New York: Routledge.

Weis, L., M. Fine, A. Proweller, C. Bertram, and J. Marusza-Hall. 1997. I've slept in clothes long enough: Excavating the sounds of domestic violence among women in the white working class. *Urban Review* 30:43–62.

Weis, L., J. Marusza-Hall, and M. Fine. 1998. Out of the cupboard: Kids, domestic violence, and schools. *British Journal of Sociology of Education* 19:53–73.

Woodcock Tentler, L. W. 1979. *Wage earning women: Industrial work and family life in the US, 1900–1930.* New York: Oxford University Press.

Introduction to Reading 44

In this selection, the authors examine domestic violence from the perspective of men. They argue that domestic violence is gendered and the enactment of domestic violence maintains gender. To illustrate these points, Anderson and Umberson conducted interviews with thirty-three men who were in an educational domestic violence program (Family Violence Diversion Network), using men's own voices to explain their role in violence against women. The men are diverse in terms of social class (mean number of years of education is 13 and mean income is \$30,463), age (mean 32), and race and ethnicity (58 percent European American, 18 percent African American, and 21 percent Hispanic). Of the men interviewed, 82 percent were mandated by the courts to attend this program.

1. How does this reading differ in the impact of domestic violence on individual lives from the previous reading by Hall?
2. What do the authors mean when they say that, through violence, the men "naturalized a binary and hierarchical gender system"?
3. Why don't these men feel remorse over their use of domestic violence?

GENDERING VIOLENCE

MASCULINITY AND POWER IN MEN'S ACCOUNTS OF DOMESTIC VIOLENCE

Kristin L. Anderson and Debra Umberson

In the 1970s, feminist activists and scholars brought wife abuse to the forefront of public consciousness. Published in the academic and popular press, the words and images of survivors made one aspect of patriarchy visible: Male dominance was displayed on women's bruised and battered bodies (Dobash and Dobash 1979; Martin 1976). Early research contributed to feminist analyses of battery as part of a larger pattern of male domination and control of women (Pence and Paymar 1993; Yllö 1993). Research in the 1980s and 1990s has expanded theoretical understandings of men's violence against women through emphases on women's agency and resistance to male control (Bowker 1983; Kirkwood 1993); the intersection of physical, structural, and emotional forces that sustain men's control over female partners (Kirkwood 1993; Pence and Paymar 1993); and the different constraints faced by women and men of diverse nations, racial ethnic identities, and sexualities who experience violence at the hands of intimate partners (Eaton 1994; Island and Letellier 1991; Jang, Lee, and Morello-Frosch 1998; Renzetti 1992). This work demonstrates ways in which the gender order facilitates victimization of disenfranchised groups.

Comparatively less work has examined the ways in which gender influences male perpetrators' experiences of domestic violence (Yllö 1993). However, a growing body of qualitative research critically examines batterers' descriptions of violence within their relationships. Dobash and Dobash (1998), Hearn (1998), and Ptacek (1990) focus on the excuses, justifications, and rationalizations that batterers use to account for their violence. These authors suggest that batterers' accounts of violence are texts through which they attempt to deny responsibility for violence and to present nonviolent self-identities.

Dobash and Dobash (1998) identify ways in which gender, as a system that structures the authority and responsibilities assigned to women and men within intimate relationships, supports battery. They find that men use violence to punish female partners who fail to meet their unspoken physical, sexual, or emotional needs. Lundgren (1998) examines batterers' use of gendered religious ideologies to justify their violence against female partners. Hearn (1998, 37) proposes that violence is a "resource for demonstrating and showing a person is a man." These studies find that masculine identities are constructed through acts of violence and through batterers' ability to control partners as a result of their violence.

This article examines the construction of gender within men's accounts of domestic violence. Guided by theoretical work that characterizes gender as performance (Butler 1990, 1993; West and Fenstermaker 1995), we contend that batterers attempt to construct masculine identities through the practice of violence and the discourse about violence that they provide. We examine these performances of gender as "routine, methodical, and ongoing accomplishment[s]" that create and sustain notions of natural differences between women and men (West and Fenstermaker 1995, 9). Butler's concept of performativity extends this idea by suggesting that it is through performance that gendered

From Anderson, K. L., & Umberson, D., "Gendering Violence: Masculinity and Power in Men's Accounts of Domestic Violence," in *Gender & Society, 15*(3)2001: 358–380. Reprinted with permission of Sage Publications, Inc.

subjectivities are constructed: "Gender proves to be performative—that is, constituting the identity it is purported to be. In this sense, gender is always a doing, though not a doing by a subject who may be said to preexist the deed" (1990, 25). For Butler, gender performances demonstrate the instability of masculine subjectivity; a "masculine identity" exists only as the actions of individuals who stylize their bodies and their actions in accordance with a normative binary framework of gender.

In addition, the performance of gender makes male power and privilege appear natural and normal rather than socially produced and structured. Butler (1990) argues that gender is part of a system of relations that sustains heterosexual male privilege through the denigration or erasure of alternative (feminine/gay/lesbian/bisexual) identities. West and Fenstermaker (1995) contend that cultural beliefs about underlying and essential differences between women and men, and social structures that constitute and are constituted by these beliefs, are reproduced by the accomplishment of gender. In examining the accounts offered by domestically violent men, we focus on identifying ways in which the practice of domestic violence helps men to accomplish gender. We also focus on the contradictions within these accounts to explore the instability of masculine subjectivities and challenges to the performance of gender.

* * *

Findings

How do batterers talk about the violence in their relationships? They excuse, rationalize, justify, and minimize their violence against female partners. Like the batterers studied by previous researchers, the men in this study constructed their violence as a rational response to extreme provocation, a loss of control, or a minor incident that was blown out of proportion. Through such accounts, batterers deny responsibility for their violence and save face when recounting behavior that has elicited social sanctions (Dobash and Dobash 1998; Ptacek 1990).

However, these accounts are also about the performance of gender. That is, through their speech acts, respondents presented themselves as rational, competent, masculine actors. We examine several ways in which domestic violence is gendered in these accounts. First, according to respondents' reports, violence is gendered in its practice. Although it was in their interests to minimize and deny their violence, participants reported engaging in more serious, frequent, and injurious violence than that committed by their female partners. Second, respondents gendered violence through their depictions and interpretations of violence. They talked about women's violence in a qualitatively different fashion than they talked about their own violence, and their language reflected hegemonic notions of femininity and masculinity. Third, the research participants constructed gender by interpreting the violent conflicts in ways that suggested that their female partners were responsible for the participants' behavior. Finally, respondents gendered violence by claiming that they are victimized by a criminal justice system that constructs all men as villains and all women as victims.

Gendered Practice

Men perpetrate the majority of violence against women and against other men in the United States (Bachman and Saltzman 1995). Although some scholars argue that women perpetrate domestic violence at rates similar to men (Straus 1993), feminist scholars have pointed out that research findings of "sexual symmetry" in domestic violence are based on survey questions that fail to account for sex differences in physical strength and size and in motivations for violence (Dobash et al. 1992; Straton 1994). Moreover, recent evidence from a large national survey suggests that women experience higher rates of victimization at the hands of partners than men and that African American and Latina women experience higher rates of victimization than European American women (Bachman and Saltzman 1995).

Although the majority of respondents described scenarios in which both they and their partners perpetrated violent acts, they reported that their violence was more frequent and severe than the violence perpetrated by their female partners. Eleven respondents (33 percent) described attacking a partner who did not physically resist, and only two respondents (6 percent) reported that they were victimized by their partners but did not themselves perpetrate violence. The twenty cases (61 percent) in which the participants reported "mutual" violence support feminist critiques of "sexual symmetry":

> We started pushing each other. And the thing is that I threw her on the floor. I told her that I'm going to leave. She took my car keys, and I wanted my car keys so I went and grabbed her arm, pulled it, and took the car keys away from her. She—she comes back and tries to kick me in the back. So I just pushed her back and threw her on the floor again. (Juan)

Moreover, the respondents did not describe scenarios in which they perceived themselves to be at risk from their partners' violence. The worst injury reportedly sustained was a split lip, and only five men (15 percent) reported sustaining any injury. Female partners reportedly sustained injuries in 14 cases (42 percent). Although the majority of the injuries reportedly inflicted on female partners consisted of bruises and scratches, a few women were hospitalized, and two women sustained broken ribs. These findings corroborate previous studies showing that women suffer more injuries from domestic violence than men (Langhinrichsen-Rohling, Neidig, and Thorn 1995). Moreover, because past studies suggest that male batterers underreport their perpetration of violence (Dobash and Dobash 1998), it is likely that respondents engaged in more violence than they described in these in-depth interviews.

Domestic violence is gendered through social and cultural practices that advantage men in violent conflicts with women. Young men often learn to view themselves as capable perpetrators of violence through rough play and contact sports, to exhibit fearlessness in the face of physical confrontations, and to accept the harm and injury associated with violence as "natural" (Dobash and Dobash 1998; Messner 1992). Men are further advantaged by cultural norms suggesting that women should pair with men who are larger and stronger than themselves (Goffman 1977). Women's less pervasive and less effective use of violence reflects fewer social opportunities to learn violent techniques, a lack of encouragement for female violence within society, and women's size disadvantage in relation to male partners (Fagot et al. 1985; McCaughey 1998). In a culture that defines aggression as unfeminine, few women learn to use violence effectively.

Gendered Depictions and Interpretations

Participants reported that they engaged in more frequent and serious violence than their partners, but they also reported that their violence was different from that of their partners. They depicted their violence as rational, effective, and explosive, whereas women's violence was represented as hysterical, trivial, and ineffectual. Of the 22 participants who described violence perpetrated by their partners, twelve (55 percent) suggested that their partner's violence was ridiculous or ineffectual. These respondents minimized their partners' violence by explaining that it was of little concern to them:

> I came out of the kitchen, and then I got in her face, and I shoved her. She shoved, she tried to push me a little bit, but it didn't matter much. (Adam)

> I was seeing this girl, and then a friend of mine saw me with this girl and he went back and told my wife, and when I got home that night, that's when she tried to hit me, to fight me. I just pushed her out of the way and left. (Shad)

This minimizing discourse also characterizes descriptions of cases in which female partners successfully made contact and injured the respondent, as in the following account:

> I was on my way to go to the restroom. And she was just cussing and swearing and she wouldn't let me pass. So, I nudged her. I didn't push her or shove her, I just kind of, you know, just made my way to the restroom. And, when I done that she hit me, and

> she drew blood. She hit me in the lip, and she drew blood. . . . I go in the bathroom and I started laughing, you know. And I was still half lit that morning, you know. And I was laughing because I think it maybe shocked me more than anything that she had done this, you know. (Ed) . . .

Even in the case of extreme danger, such as when threatened with a weapon, respondents denied the possibility that their partners' violence was a threat. During a fight described by Steve, his partner locked herself in the bathroom with his gun:

> We were battering each other at that point, and that's when she was in the bathroom. This is—it's like 45 minutes into this whole argument now. She's in the bathroom, messing with my [gun]. And I had no idea. So I kicked the door in—in the bathroom, and she's sitting there trying to load this thing, trying to get this clip in, and luckily she couldn't figure it out. Why, I don't—you know, well, because she was drunk. So, luckily she didn't. The situation could have been a whole lot worse, you know, it could have been a whole lot worse than it was. I thank God that she didn't figure it out. When I think about it, you know, she was lucky to come out of it with just a cut in her head. You know, she could have blown her brains out or done something really stupid.

This account contains interesting contradictions. Steve stated that he had "no idea" that his partner had a gun, but he responded by kicking down the door to reach her. He then suggested that he was concerned about his partner's safety and that he kicked in the door to save her from doing "something really stupid" to herself. . . .

[The men] described their partners' acts as irrational and hysterical. Such depictions helped respondents to justify their own violence and to present themselves as calm, cool, rational men. Phil described his own behavior of throwing his partner down as a nonviolent, controlled response to his partner's outrageous behavior. Moreover, he suggests that he used this incident to demonstrate his sense of superior rationality to his partner. Phil later reported that a doctor became "very upset" about the marks on his wife's neck two days after this incident, suggesting that he was not the rational actor represented in his account.

In eight other cases (36 percent), respondents did not depict their partner's violence as trivial or ineffectual. Rather, they described their partners' behavior in matter-of-fact terms:

> Then she starts jumping at me or hitting me, or tell me "leave the house, I don't want you, I don't love you" and stuff like that. And I say, "don't touch me, don't touch me." And I just push her back. She keeps coming and hit me, hit me. I keep pushing back, she starts to scratch me, so I push hard to stop her from hurting me. (Mario)

Other respondents depicted their partner's violence in factual terms but emphasized that they perceived their own violence as the greater danger. Ray took his partner seriously when he stated that "she was willing to fight, to defend herself," yet he also mentioned his fears that his own violence would be lethal: "The worst time is when she threw an iron at me. And I'm gonna tell you, I think that was the worst time because, in defense, in retaliation, I pulled her hair, and I thought maybe I broke her neck." Only two respondents—Alan and Jim—consistently identified as victims:

> One of the worst times was realizing that she was drunk and belligerent. I realized that I needed to take her home in her car and she was not capable of driving. And she was physically abusive the whole way home. And before I could get out of the door or get out of the way, she came at me with a knife. And stupidly, I defended myself—kicked her hand to get the knife out. And I bruised her hand enough to where she felt justified enough to call the police with stories that I was horribly abusing. (Jim)
>
> Jim reported that his partner has hit him, stabbed him, and thrown things at him. However, he also noted that he was arrested following several of these incidents, suggesting that his accounts tell us only part of the story. . . .

In contrast to their reported fearlessness when confronted by women wielding weapons, respondents constructed their own capacity for violence as something that should engender fear. These interpretations are consistent with cultural constructions of male violence as volcanic—natural, lethal, and impossible to stop until it has run its course.

Respondents' interpretations of ineffectual female violence and lethal male violence reflect actual violent practices in a culture that grants men more access to violence, but they also gender violence. By denying a threat from women's violence, participants performed masculinity and reinforced notions of gender difference. Women were constructed as incompetent in the practice of violence, and their successes were trivialized. For example, it is unlikely that Ed would have responded with laughter had his lip been split by the punch of another man (Dobash and Dobash 1998). Moreover, respondents ignored their partners' motivations for violence and their active efforts to exert change within their relationships.

The binary representation of ineffectual, hysterical female behavior and rational, lethal male violence within these accounts erases the feminine; violence perpetrated by women and female subjectivity are effaced in order that the respondents can construct masculinities. These representations mask the power relations that determine what acts will qualify as "violence" and thus naturalize the notion that violence is the exclusive province of men.

Gendering Blame

The research participants also gendered violence by suggesting that their female partners were responsible for the violence within their relationships. Some respondents did this by claiming that they did not hit women with whom they were involved in the past:

> I've never hit another woman in my life besides the one that I'm with. She just has a knack for bringing out the worst in me. (Tom) . . .

Respondents also shifted blame onto female partners by detailing faults in their partners' behaviors and personalities. They criticized their partners' parenting styles, interaction styles, and choices. However, the most typically reported criticism was that female partners were controlling. Ten of the 33 respondents (30 percent) characterized their partners as controlling, demanding, or dominating:

> She's real organized and critiquing about things. She wanna—she has to get it like—she like to have her way all the time, you know. In control of things, even when she's at work in the evenings, she has to have control of everything that's going on in the house. And—but—you know, try to get, to control everything there. You know, what's going on, and me and myself. (Adam) . . .

In a few cases, respondents claimed that they felt emasculated by what they interpreted as their partners' efforts to control them:

> You ask the guy sitting next door to me, the guy that's down the hall. For years they all say, "Bill, man, reach down and grab your eggs. She wears the pants." Or maybe like, "Hey man, we're going to go—Oh, Bill can't go. He's got to ask his boss first." And they were right. (Bill)

These representations of female partners as dominating enabled men to position themselves as victims of masculinized female partners. The relational construction of masculinity is visible in these accounts; women who "wear the pants" disrupt the binary opposition of masculinity/femininity. Bill's account reveals that "one is one's gender to the extent that one is not the other gender" (Butler 1990, 22); he is unable to perform masculinity to the satisfaction of his friends when mirrored by a partner who is perceived as dominating.

Moreover, respondents appeared to feel emasculated by unspecified forces. Unlike female survivors who describe concrete practices that male partners utilize to exert control (Kirkwood 1993; Walker 1984), participants were vague about, what they meant by control and the ways in which their partners exerted control. . . . Respondents who claimed that their partners are controlling offered nebulous explanations for these feelings, suggesting that these claims may be indicative of these men's fears about being controlled by a woman rather than the actual practices of their partners.

Finally, respondents gendered violence through their efforts to convince female partners to shoulder at least part of the blame for their violence. . . .

Contemporary constructions of gender hold women responsible for men's aggression (Gray, Palileo, and Johnson 1993). Sexual violence is often blamed on women, who are perceived as tempting men who are powerless in the face of their primal sexual desires (Scully 1990). Although interviewees expressed remorse for their violent behavior, they also implied that it was justified in light of their partners' controlling behavior. Moreover, their violence was rewarded by their partners' feelings of guilt, suggesting that violence is simultaneously a performance of masculinity and a means by which respondents encouraged the performance of femininity by female partners.

"The Law Is for Women": Claiming Gender Bias

Participants sometimes rationalized their violence by claiming that the legal system overreacted to a minor incident. Eight of the 33 interviewees (24 percent) depicted themselves as victims of gender politics or the media attention surrounding the trial of O. J. Simpson:

> I think my punishment was wrong. And it was like my attorney told me—I'm suffering because of O. J. Simpson. Mine was the crime of the year. That is, you know, it's the hot issue of the year because of O. J. Two years ago they would have gone "Don't do that again." (Bill) . . .

These claims of gender bias were sometimes directly contradicted by respondents' descriptions of events following the arrival of the police. Four participants (12 percent) reported that the police wanted to arrest their female partner along with or instead of themselves—stories that challenged their claims of bias in the system. A few of these respondents reported that they lied to the police about the source of their injuries to prevent the arrest of their partners. Ed, the respondent who sustained a split lip from his partner's punch, claimed that he "took the fall" for his partner. . . .

When the police arrived, these respondents were in a double bind. They wanted to deny their own violence to avoid arrest, but they also wanted to deny victimization at the hands of a woman. "Protecting" their female partners from arrest allowed them a way out of this bind. By volunteering to be arrested despite their alleged innocence, they became chivalrous defenders of their partners. They were also, paradoxically, able to claim that "gender bias" led to their arrest and participation in the [Family Violence Diversion Network] FVDN program.

* * *

When batterers "protect" their partners from arrest, their oppressor becomes a powerful criminal justice system rather than a woman. Although even the loser gains status through participation in a fight with another man, a man does not gain prestige from being beaten by a woman (Dobash and Dobash 1998). In addition, respondents who stepped in to prevent their partners from being arrested ensured that their partners remained under their control, as Jim suggested when he described "the thin line between being protected by somebody and possessing somebody." By volunteering to be arrested along with his partner, Jim ensured that she was not "taken into possession" (e.g., taken into custody) by the police.

By focusing the interviews on "gender bias" in the system, respondents deflected attention from their own perpetration and victimization. Constructions of a bias gave them an explanation for their arrest that was consistent with their self-presentation as rational, strong, and nonviolent actors. Claims of "reverse mentality" also enabled participants to position themselves as victims of gender politics. Several interviewees made use of men's rights rhetoric or alluded to changes wrought by feminism to suggest that they are increasingly oppressed by a society in which women have achieved greater rights. . . .

A number of recent studies have examined the increasingly angry and antifeminist discourse offered by some men who are struggling to construct masculine identities within patriarchies disrupted by feminism and movements for gay/lesbian and civil rights (Fine, et al. 1997; Messner 1998; Savran 1998). Some branches of the contemporary "men's movement" have articulated a defensive and antifeminist rhetoric of "men's rights" that suggests that men have become the victims of feminism (Messner 1998; Savran 1998). Although none of our interviewees reported participation in any of the organized men's movements, their allusions to the discourse of victimized manhood suggest that the rhetoric of these movements has become an influential resource for the performance of gender among some men. Like the angry men's rights activists studied by Messner (1998), some respondents positioned themselves as the victims of feminism, which they believe has co-opted the criminal justice system and the media by creating "myths" of male domination. The interviews suggest that respondents feel disempowered and that they identify women—both the women whom they batter and women who lead movements to criminalize domestic violence—as the "Other" who has "stolen their presumed privilege" (Fine et al. 1997, 54): "Now girls are starting to act like men, or try and be like men. Like if you hit me, I'll call the cops, or if you don't do it, I'll do this, or stuff like that" (Juan). Juan contends that by challenging men's "privilege" to hit their female partners without fear of repercussions, women have become "like men." This suggests that the construction of masculine subjectivities is tied to a position of dominance and that women have threatened the binary and hierarchical gender framework through their resistance to male violence.

Discussion: Social Locations and Discourses of Violence

Respondents' descriptions of conflicts with female partners were similar across racial, ethnic and class locations. Participants of diverse socioeconomic standings and racial ethnic backgrounds minimized the violence perpetrated by their partners, claimed that the criminal justice system is biased against men, and attempted to place responsibility for their violence on female partners. However, we identified some ways in which social class influenced respondents' self-presentations.[1]

Respondents of higher socioeconomic status emphasized their careers and the material items that they provided for their families throughout the interviews:

> We built two houses together and they are nice. You know, we like to see a nice environment for our family to live in. We want to see our children receive a good education. (Ted)
>
> That woman now sits in a 2,700 square foot house. She drives a Volvo. She has everything. A brand-new refrigerator, a brand-new washer and dryer. (Bill)

Conversely, economically disenfranchised men volunteered stories about their prowess in fights with other men. These interviewees reported that they engaged in violent conflicts with other men as a means of gaining respect:

> Everybody in my neighborhood respected me a lot, you know. I used to be kind of violent. I used to like to fight and stuff like that, but I'm not like that anymore. She—I don't think she liked me because I liked to fight a lot but she liked me because people respected me because they knew that they would have to fight if they disrespected me. You know I think that's one thing that turned her on about me; I don't let people mess around. (Tony) . . .

The use of violence to achieve respect is a central theme in research on the construction of masculinities among disenfranchised men (Messerschmidt 1993; Messner 1992). Although men of diverse socioeconomic standings valorize fistfights between men (Campbell 1993; Dobash and Dobash 1998), the extent to which they participate in these confrontations varies by social context. Privileged young men are more often able to avoid participation in social situations that require physical violence against other men than are men who reside in poor neighborhoods (Messner 1992).

We find some evidence that cultural differences influence accounts of domestic violence. Two respondents who identified themselves as immigrants from Latin America (Alejandro and Juan) reported that they experienced conflicts with female partners about the shifting meanings of gender in the United States:

> She has a different attitude than mine. She has an attitude that comes from Mexico—be a man like, you have to do it. And it's like me here, it's fifty-fifty, it's another thing, you know, it's like "I don't have to do it." . . . I told her the wrong things she was doing and I told her, "It's not going to be that way because we're not in Mexico, we're in the United States." (Juan)

Juan's story suggests that unstable meanings about what it means to be a woman or a man are a source of conflict within his relationship and that he and his partner draw on divergent gender ideologies to buttress their positions. Although many of the respondents expressed uncertainty about appropriate gender performances in the 1990s, those who migrated to the United States may find these "crisis tendencies of the gender order" (Connell 1992, 736) to be particularly unsettling. Interestingly, Juan depicts his partner as clinging to traditional gender norms, while he embraces the notion of gender egalitarianism. However, we are hesitant to draw conclusions about this finding due to the small number of interviews that we conducted with immigrants.

Conclusions

Many scholars have suggested that domestic violence is a means by which men construct masculinities (Dobash and Dobash 1998; Gondolf and Hannekin 1987; Hearn 1998). However, few studies have explored the specific practices that domestically violent men use to present themselves as masculine actors. The respondents in this study used diverse and contradictory strategies to gender violence and they shifted their positions as they talked about violence. Respondents sometimes positioned themselves as masculine actors by highlighting their strength, power, and rationality compared with the "irrationality" and vulnerability of female partners. At other times, when describing the criminal justice system or "controlling" female partners, they positioned themselves as vulnerable and powerless. These shifting representations evidence the relational construction of gender and the instability of masculine subjectivities (Butler 1990).

Recently, performativity theories have been criticized for privileging agency, undertheorizing structural and cultural constraints, and facilitating essentialist readings of gender behavior: "Lacking an analysis of structural and cultural context, performances of gender can all too easily be interpreted as free agents' acting out the inevitable surface manifestations of a natural inner sex difference" (Messner 2000, 770). Findings from our study show that each of these criticisms is not necessarily valid.

First, although the batterers described here demonstrate agency by shifting positions, they do so by calling on cultural discourses (of unstoppable masculine aggression, of feminine weakness, and of men's rights). Their performance is shaped by cultural options.

Second, batterers' performances are also shaped by structural changes in the gender order. Some of the batterers interviewed for this study expressed anger and confusion about a world with "TV for women" and female partners who are "too educated." Their arrest signaled a world askew—a place where "the law is for "women" and where men have become the victims of discrimination. Although these accounts are ironic in light of the research documenting the continuing reluctance of the legal system to treat domestic violence as a criminal act (Dobash and Dobash 1979), they demonstrate the ways in which legal and structural reforms in the area of domestic violence influence gender performances. By focusing attention on the "bias" in the system, respondents deflected attention from their own perpetration and victimization and sustained their constructions of rational masculinity. Therefore, theories of gender performativity push us toward analyses of the cultural and structural contexts that form the settings for the acts.

Finally, when viewed through the lens of performativity, our findings challenge the notion that violence is an essential or natural expression of masculinity. Rather, they suggest that violence represents an effort to reconstruct a contested and unstable masculinity. Respondents' references to men's rights movement discourse, their claims of "reverse discrimination," and their complaints that female partners are controlling indicate a disruption in masculine subjectivities. Viewing domestic violence as a gender performance counters the essentialist readings of men's violence against women that dominate U.S. popular culture. What one performs is not necessarily what one "is."

Disturbingly, however, this study suggests that violence is (at least temporarily) an effective means by which batterers reconstruct men as masculine and women as feminine. Participants reported that they were able to control their partners through exertions of physical dominance and through their interpretive efforts to hold partners responsible for the violence in their relationships. By gendering violence, these batterers not only performed masculinity but reproduced gender as dominance. Thus, they naturalized a binary and hierarchical gender system.

Note

1. We define high socioeconomic status respondents as those who earn at least $25,000 per year in personal income and who have completed an associate's degree. Seven respondents fit these criteria. We define disenfranchised respondents as those who report personal earnings of less than $15,000 per year and who have not completed a two-year college program. Nine respondents fit these criteria.

References

Bachman, R., and L. E. Saltzman. 1995. *Violence against women: Estimates from the redesigned survey August 1995.* NCJ-154348 special report. Washington, DC: Bureau of Justice Statistics.

Bowker, L. H. 1983. *Beating wife-beating.* Lexington, MA: Lexington Books.

Butler, J. 1990. *Gender trouble: Feminism and the subversion of identity.* New York: Routledge.

———. 1993. *Bodies that matter: On the discursive limits of sex.* New York: Routledge.

Campbell, A. 1993. *Men, women and aggression.* New York: Basic Books.

Connell, R. W. 1992. A very straight gay: Masculinity, homosexual experience, and the dynamics of gender. *American Sociological Review* 57:735–51.

Dobash, R. E., and R. P. Dobash. 1979. *Violence against wives: A case against the patriarchy.* New York: Free Press.

———. 1998. Violent men and violent contexts. In *Rethinking violence against women,* edited by R. E. Dobash and R. P. Dobash. Thousand Oaks, CA: Sage.

Dobash, R. P., R. E. Dobash, M. Wilson, and M. Daly. 1992. The myth of sexual symmetry in marital violence. *Social Problems* 39:71–91.

Eaton, M. 1994. Abuse by any other name: Feminism, difference, and intralesbian violence. In *The public nature of private violence: The discovery of domestic abuse,* edited by M. A. Fineman and R. Mykitiuk. New York: Routledge.

Fagot, B., R. Hagan, M. B. Leinbach, and S. Kronsberg. 1985. Differential reactions to assertive and communicative acts of toddler boys and girls. *Child Development* 56:1499–1505.

Fine, M., L. Weis, J. Addelston, and J. Marusza. 1997. (In)secure times: Constructing white working-class masculinities in the late 20th century. *Gender & Society* 11:52–68.

Goffman, E. 1977. The arrangement between the sexes. *Theory and Society* 4 (3): 301–31.

Gondolf, Edward W., and James Hannekin. 1987. The gender warrior: Reformed batterers on abuse, treatment, and change. *Journal of Family Violence* 2:177–91.

Gray, N. B., G. J. Palileo, and G. D. Johnson. 1993. Explaining rape victim blame: A test of attribution theory. *Sociological Spectrum* 13:377–92.

Hearn, J. 1998. *The violences of men: How men talk about and how agencies respond to men's violence against women.* Thousand Oaks, CA: Sage.

Island, D., and P. Letellier. 1991. *Men who beat the men who love them: Battered gay men and domestic violence.* New York: Harrington Park.

Jang, D., D. Lee, and R. Morello-Frosch. 1998. Domestic violence in the immigrant and refugee community: Responding to the needs of immigrant women. In *Shifting the center: Understanding contemporary families,* edited by S. J. Ferguson. Mountain View, CA: Mayfield.

Kirkwood, C. 1993. *Leaving abusive partners: From the scars of survival to the wisdom for change.* Newbury Park, CA: Sage.

Langhinrichsen-Rohling, J., P. Neidig, and G. Thorn. 1995. Violent marriages: Gender differences in levels of current violence and past abuse. *Journal of Family Violence* 10:159–76.

Lundgren, E. 1998. The hand that strikes and comforts: Gender construction and the tension between body and symbol. In *Rethinking violence against women,* edited by R. E. Dobash and R. P. Dobash. Thousand Oaks, CA: Sage.

Martin, Del. 1976. *Battered wives.* New York: Pocket Books.

McCaughey, M. 1998. The fighting spirit: Women's self-defense training and the discourse of sexed embodiment. *Gender & Society* 12:277–300.

Messerschmidt, J. 1993. *Masculinities and crime: A critique and reconceptualization of theory.* Lanham, MD: Rowman & Littlefield.

Messner, M. A. 1992. *Power at play: Sports and the problem of masculinity.* Boston: Beacon.

———. 1998. The limits of the "male sex role": An analysis of the men's liberation and men's rights movements' discourse. *Gender & Society* 12 (3): 255–76.

———. 2000. Barbie girls versus sea monsters: Children constructing gender. *Gender & Society* 14 (6): 765–84.

Pence, E., and M. Paymar. 1993. *Education groups for men who batter: The Duluth model.* New York: Springer.

Ptacek, J. 1990. Why do men batter their wives? In *Feminist perspectives on wife abuse,* edited by K. Yllö and M. Bograd. Newbury Park, CA: Sage.

Renzetti, C. M. 1992. *Violent betrayal: Partner abuse in lesbian relationships.* Newbury Park, CA: Sage

Savran, D. 1998. *Taking it like a man: White masculinity, masochism, and contemporary American culture.* Princeton, NJ: Princeton University Press.

Scully, D. 1990. *Understanding sexual violence: A study of convicted rapists.* Boston: Unwin Hyman.

Straton, J. C. 1994. The myth of the "battered husband syndrome." *Masculinities* 2:79–82.

Straus, M. A. 1993. Physical assaults by wives: A major social problem. In *Current controversies on family violence,* edited by R. J. Gelles and D. R. Loseke. Newbury Park, CA: Sage.

Walker, L. 1984. *The battered woman syndrome.* New York: Springer.

West, C., and S. Fenstermaker. 1995. Doing difference. *Gender & Society* 9:8–37.

Yllö, K. 1993. Through a feminist lens: Gender, power, and violence. In *Current controversies on family violence,* edited by R. J. Gelles and D. R. Loseke. Newbury Park, CA: Sage.

Introduction to Reading 45

Beth Quinn's research explores the accounts men and women give for what, on the surface, is a relatively benign form of sexual harassment—watching and rating women at work. To do so, she conducted forty-three interviews with men and women, lasting from one to three hours, asking about relationships at work. She then randomly sampled over half of the original sample (25) from one organization, and recruited the remaining individuals from a nearby community college and university. The accounts men give for sexual harassment indicate that they realize the effects of their girl watching on the women they work with and, if they were women, acknowledge that they would not like being watched this way. The interviews reveal that sexual harassment is instrumental in bonding men together and separating men from women in the workplace.

1. What is the role of objectification and (dis)empathy in maintaining masculinity?
2. How does sexual harassment work to enforce gender for men in the workplace, and what are the effects of such behavior for women and men?
3. What would have to change if "girl watching" were to disappear from the workplace?

Sexual Harassment and Masculinity

The Power and Meaning of "Girl Watching"

Beth A. Quinn

Confronted with complaints about sexual harassment or accounts in the media, some men claim that women are too sensitive or that they too often misinterpret men's intentions (Bernstein 1994; Buckwald 1993). In contrast, some women note with frustration that men just "don't get it" and lament the seeming inadequacy of sexual harassment policies (Conley 1991; Guccione 1992). Indeed, this ambiguity in defining acts of sexual harassment might be, as Cleveland and Kerst (1993) suggested, the most robust finding in sexual harassment research. . . .

This article focuses on the subjectivities of the perpetrators of a disputable form of sexual harassment, "girl watching." The term refers to the act of men's sexually evaluating women, often in the company of other men. It may take the form of a verbal or gestural message of "check it out," boasts of sexual prowess, or explicit comments about a woman's body or imagined sexual acts. The target may be an individual woman or group of women or simply a photograph or other representation. The woman may be a stranger, coworker, supervisor, employee, or client. For the present analysis, girl watching within the workplace is centered.

The analysis is grounded in the work of masculinity scholars such as Connell (1987, 1995) in that it attempts to explain the subject positions of the interviewed men—not the abstract and genderless subjects of patriarchy but the gendered and privileged subjects embedded in this system. Since I am attempting to delineate the gendered worldviews of the interviewed men, I employ the term "girl watching," a phrase that reflects their language ("they watch girls").

I have chosen to center the analysis on girl watching within the workplace for two reasons. First, it appears to be fairly prevalent. For example, a survey of federal civil employees (U.S. Merit Systems Protection Board 1988) found that in the previous 24 months, 28 percent of the women surveyed had experienced "unwanted sexual looks or gestures," and 35 percent had experienced "unwanted sexual teasing, jokes, remarks, or questions." Second, girl watching is still often normalized and trivialized as only play, or "boys will be boys." A man watching girls—even in his workplace—is frequently accepted as a natural and commonplace activity, especially if he is in the presence of other men.[1] Indeed, it may be required (Hearn 1985). Thus, girl watching sits on the blurry edge between fun and harm, joking and harassment. An understanding of the process of identifying behavior as sexual harassment, or of rejecting this label, may be built on this ambiguity.

Girl watching has various forms and functions, depending on the context and the men involved. For example, it may be used by men as a directed act of power against a particular woman or women. In this, girl watching—at least in the workplace—is most clearly identified as harassing by both men and women. I am most interested, however, in the form where it is characterized as only play. This type is more obliquely motivated and, as I will argue, functions as a game men play to build shared masculine identities and social relations.

Multiple and contradictory subject positions are also evidenced in girl watching, most notably that between the gazing man and the woman he watches. Drawing on Michael Schwalbe's (1992) analysis of empathy and the formation of masculine identities, I argue that girl watching is premised on the obfuscation of this multiplicity through the objectification of the woman watched and a suppression of empathy for her. In conclusion, the ways these elements operate to produce gender differences in interpreting sexual harassment and the implications for developing effective policies are discussed.

Previous Research

The question of how behavior is or is not labeled as sexual harassment has been primarily through experimental vignettes and surveys.[2] In both methods, participants evaluate either hypothetical scenarios or lists of behaviors, considering whether, for example, the behavior constitutes sexual harassment, which party is most at fault, and what consequences the act might engender. Researchers manipulate factors such as the level of "welcomeness" the target exhibits and the relationship of the actors (supervisor-employee, coworker-coworker).

Both methods consistently show that women are willing to define more acts as sexual harassment (Gutek, Morasch, and Cohen 1983; Padgitt and Padgitt 1986; Powell 1986;York 1989; but see Stockdale and Vaux 1993) and are more likely to see situations as coercive (Garcia, Milano, and Quijano 1989). When asked who is more to blame in a particular scenario, men are more likely to blame, and less likely to empathize with, the victim (Jensen and Gutek 1982; Kenig and Ryan 1986). In terms of actual behaviors like girl watching, the U.S. Merit Systems Protection Board (1988) survey found that 81 percent of the women surveyed considered "uninvited sexually suggestive looks or gestures" from a supervisor to be sexual harassment. While the majority of men (68 percent) also defined it as such, significantly more men were willing to dismiss such behavior. Similarly, while 40 percent of the men would not consider the same behavior from a coworker to be harassing, more than three-quarters of the women would.

The most common explanation offered for these differences is gender role socialization. This conclusion is supported by the consistent finding that the more men and women adhere to traditional gender roles, the more likely they are to deny the harm in sexual harassment and to consider the behavior acceptable or at least normal (Gutek and Koss 1993; Malovich and Stake 1990; Murrell and Dietz-Uhler 1993; Popovich et al. 1992; Pryor 1987; Tagri and Hayes 1997). Men who hold predatory ideas about sexuality, who are more likely to believe rape myths, and who are more likely to self-report that they would rape under certain circumstances are less likely to see behaviors as harassing (Murrell and Dietz-Uhler 1993; Pryor 1987; Reilly et al. 1992).

These findings do not, however, adequately address the between-group differences. The more one is socialized into traditional notions of [gender], the more likely it is for both men and women to view the behaviors as acceptable or at least unchangeable. The processes by which gender roles operate to produce these differences remain underexamined. Some theorists argue that men are more likely to discount the harassing aspects of their behavior because of a culturally conditioned tendency to misperceive women's intentions. For example, Stockdale (1993, 96) argued that "patriarchal norms create a sexually aggressive belief system in some people more than others, and this belief system can lead to the propensity to misperceive." Gender differences in interpreting sexual harassment, then, may be the outcome of the acceptance of normative ideas about women's inscrutability and indirectness and men's role as sexual aggressors. Men see harmless flirtation or sexual interest rather than harassment because they misperceive women's intent and responses.

Stockdale's (1993) theory is promising but limited. First, while it may apply to actions such as repeatedly asking for dates and quid pro quo harassment,[3] it does not effectively explain motivations for more indirect actions, such as displaying pornography and girl watching. Second, it does not explain why some men are more likely to operate from these discourses of sexual aggression contributing to a propensity to misperceive.

* * *

Findings: Girl Watching as "Hommosexuality"

> *[They] had a button on the computer that you pushed if there was a girl who came to the front counter. . . . It was a code and it said "BAFC"—Babe at Front Counter. . . . If the guy in the back looked up and saw a cute girl come in the station, he would hit this button for the other dispatcher to [come] see the cute girl.*
>
> —Paula, Police Officer

In its most serious form, girl watching operates as a targeted tactic of power. The men seem to want everyone—the targeted woman as well as coworkers, clients, and superiors—to know they are looking. The gaze demonstrates their right, as men, to sexually evaluate women. Through the gaze, the targeted woman is reduced to a sexual object, contradicting her other identities, such as that of competent worker or leader. This employment of the discourse of asymmetrical heterosexuality (i.e., the double standard) may trump a woman's formal organizational power, claims to professionalism, and organizational discourses of rationality (Collinson and Collinson 1989; Gardner 1995; Yount 1991).[4] As research on rape has demonstrated (Estrich 1987), calling attention to a woman's gendered sexuality can function to exclude recognition of her competence, rationality, trustworthiness, and even humanity. In contrast, the overt recognition of a man's (hetero)sexuality is normally compatible with other aspects of his identity; indeed, it is often required (Connell 1995; Hearn 1985). Thus, the power of sexuality is asymmetrical, in part, because being seen as sexual has different consequences for women and men.

But when they ogle, gawk, whistle and point, are men always so directly motivated to disempower their women colleagues? Is the target of the gaze also the intended audience? Consider, for example, this account told by Ed, a white, 29-year-old instrument technician.

> When a group of guys goes to a bar or a nightclub and they try to be manly. . . . A few of us always found [it] funny [when] a woman would walk by and a guy would be like, "I can have her." [pause] "Yeah, OK, we want to see it!" [laugh]

In his account—a fairly common one in men's discussions—the passing woman is simply a visual cue for their play. It seems clear that it is a game played by men for men; the woman's participation and awareness of her role seem fairly unimportant.

As Thorne (1993) reminded us, we should not be too quick to dismiss games as "only play." In her study of gender relations in elementary schools, Thorne found play to be a powerful form of gendered social action. One of its "clusters of meaning" most relevant here is that of "dramatic performance." In this, play functions as both a source of fun and a mechanism by which gendered identities, group boundaries, and power relations are (re)produced.

The metaphor of play was strong in Karl's comments. Karl, a white man in his early thirties who worked in a technical support role in the

Acme engineering department, hoped to earn a degree in engineering. His frustration with his slow progress—which he attributed to the burdens of marriage and fatherhood—was evident throughout the interview. Karl saw himself as an undeserved outsider in his department and he seemed to delight in telling on the engineers.

Girl watching came up as Karl considered the gender reversal question. Like many of the men I interviewed, his first reaction was to muse about premenstrual syndrome and clothes. When I inquired about the potential social effects of the transformation (by asking him, Would it "be easier dealing with the engineers or would it be harder?") he haltingly introduced the engineers' "game."

Karl:	Some of the engineers here are very [pause] they're not very, how shall we say? [pause] What's the way I want to put this? They're not very, uh [pause] what's the word? Um. It escapes me.
Researcher:	Give me a hint?
Karl:	They watch women but they're not very careful about getting caught.
Researcher:	Oh! Like they ogle?
Karl:	Ogle or gaze or [pause] stare even, or [pause] generate a commotion of an unusual nature.

His initial discomfort in discussing the issue (with me, I presume) is evident in his excruciatingly formal and hesitant language. The aspect of play, however, came through clearly when I pushed him to describe what generating a commotion looked like: "'Oh! There goes so-and-so. Come and take a look! She's wearing this great outfit today!' Just like a schoolboy. They'll rush out of their offices and [cranes his neck] and check things out." That this is as a form of play was evident in Karl's boisterous tone and in his reference to schoolboys. This is not a case of an aggressive sexual appraising of a woman coworker but a commotion created for the benefit of other men. . . .

Producing Masculinity

I suggest that girl watching in this form functions simultaneously as a form of play and as a potentially powerful site of gendered social action. Its social significance lies in its power to form identities and relationships based on these common practices for, as Cockburn (1983, 123) has noted, "patriarchy is as much about relations between man and man as it is about relations between men and women." Girl watching works similarly to the sexual joking that Johnson (1988) suggested is a common way for heterosexual men to establish intimacy among themselves.

In particular, girl watching works as a dramatic performance played to other men, a means by which a certain type of masculinity is produced and heterosexual desire displayed. It is a means by which men assert a masculine identity to other men, in an ironic "hommosexual" practice of heterosexuality (Butler 1990).[5] As Connell (1995) and others (Butler 1990; West and Zimmerman 1987) have aptly noted, masculinity is not a static identity but rather one that must constantly be reclaimed. The content of any performance—and there are multiple forms—is influenced by a hegemonic notion of masculinity. When asked what "being a man entailed, many of the men and women I interviewed triangulated toward notions of strength (if not in muscle, then in character and job performance), dominance, and a marked sexuality, overflowing and uncontrollable to some degree and natural to the male "species." Heterosexuality is required, for just as the label "girl" questions a man's claim to masculine power, so does the label "fag" (Hopkins 1992; Pronger 1992). I asked Karl, for example, if he would consider his sons "good men" if they were gay. His response was laced with ambivalence; he noted only that the question was "a tough one."

The practice of girl watching is just that—a practice—one rehearsed and performed in everyday settings. This aspect of rehearsal was evident in my interview with Mike, a self-employed house painter who used to work construction. In locating himself as a born-again Christian, Mike recounted the girl watching of his fellow construction workers with contempt. Mike was particularly disturbed by a man who brought his young son to the job site one day. The boy was explicitly taught to catcall, a practice that included identifying the proper targets: women and effeminate men.

Girl watching, however, can be somewhat tenuous as a masculine practice. In their acknowledgment (to other men) of their supposed desire lies the possibility that in being too interested in women the players will be seen as mere schoolboys giggling in the playground. Taken too far, the practice undermines rather than supports a masculine performance. . . . A man must be interested in women, but not too interested; they must show their (hetero)sexual interest, but not overly so, for this would be to admit that women have power over them.

The Role of Objectification and (Dis)Empathy

As a performance of heterosexuality among men, the targeted woman is primarily an object onto which men's homosocial sexuality is projected. The presence of a woman in any form—embodied, pictorial, or as an image conjured from words—is required, but her subjectivity and active participation is not. To be sure, given the ways the discourse of asymmetrical sexuality works, men's actions may result in similarly negative effects on the targeted woman as that of a more direct form of sexualization. The crucial difference is that the men's understanding of their actions differs. This difference is one key to understanding the ambiguity around interpreting harassing behavior.

When asked about the engineers' practice of neck craning, Robert grinned, saying nothing at first. After some initial discussion, I started to ask him if he thought women were aware of their game ("Do you think that the women who are walking by . . . ?")

Robert did not want to admit that women might not enjoy it ("that didn't come out right") but acknowledged that their feelings were irrelevant. Only subjects, not objects, take pleasure or are annoyed. If a woman did complain, Robert thought "the guys wouldn't know what to say." In her analysis of street harassment, Gardner (1995, 187) found a similar absence, in that "men's interpretations seldom mentioned a woman's reaction, either guessed at or observed."

The centrality of objectification was also apparent in comments made by José, a Hispanic man in his late 40s who worked in manufacturing. For José, the issue came up when he considered the topic of compliments. He initially claimed that women enjoy compliments more than men do. In reconsidering, he remembered girl watching and the importance of intent.

> There is [pause] a point where [pause] a woman can be admired by [pause] a pair of eyes, but we're talking about "that look." Where, you know, you're admiring her because she's dressed nice, she's got a nice figure, she's got nice legs. But then you also have the other side. You have an animal who just seems to undress you with his eyes and he's just [pause], there's those kind of people out there too.

What is most interesting about this statement is that in making the distinction between merely admiring and an animal look that ravages, José switched subject position. He spoke in the second person when describing both forms of looking, but his consistency in grammar belies a switch in subjectivity: you (as a man) admire, and you (as a woman) are undressed with his eyes. When considering an appropriate, complimentary gaze, José described it from a man's point of view; the subject who experiences the inappropriate, violating look, however, is a woman. Thus, as in Robert's account, José acknowledged that there are potentially different meanings in the act for men and women. In particular, to be admired in a certain way is potentially demeaning for a woman through its objectification. . . .

When asked to envision himself as a woman in his workplace, like many of the individuals I interviewed, Karl believed that he did not "know how to be a woman." Nonetheless, he produced an account that mirrored the stories of some of the women I interviewed. He knew the experience of girl watching could be quite different—in fact, threatening and potentially disempowering—for the woman who is its object. As such, the game was something to be avoided. In imagining themselves as women, the men remembered the practice of girl watching. None, however, were able to comfortably describe the game of girl watching from the perspective of a woman and maintain its (masculine) meaning as play.

In attempting to take up the subject position of a woman, these men are necessarily drawing on knowledge they already hold. If men simply "don't get it"—truly failing to see the harm in girl watching or other more serious acts of sexual harassment—then they should not be able to see this harm when envisioning themselves as women. What the interviews reveal is that many men—most of whom failed to see the harm of many acts that would constitute the hostile work environment form of sexual harassment—did in fact understand the harm of these acts when forced to consider the position of the targeted woman.

I suggest that the gender reversal scenario produced, in some men at least, a moment of empathy. Empathy, Schwalbe (1992) argued, requires two things. First, one must have some knowledge of the other's situation and feelings. Second, one must be motivated to take the position of the other. What the present research suggests is that gender differences in interpreting sexual harassment stem not so much from men's not getting it (a failure of the first element) but from a studied, often compulsory, lack of motivation to identify with women's experiences.

In his analysis of masculinity and empathy, Schwalbe (1992) argued that the requirements of masculinity necessitate a "narrowing of the moral self." Men learn that to effectively perform masculinity and to protect a masculine identity, they must, in many instances, ignore a woman's pain and obscure her viewpoint. Men fail to exhibit empathy with women because masculinity precludes them from taking the position of the feminine other, and men's moral stance vis-à-vis women is attenuated by this lack of empathy.

As a case study, Schwalbe (1992) considered the Thomas-Hill hearings, concluding that the examining senators maintained a masculinist stance that precluded them from giving serious consideration to Professor Hill's claims. A consequence of this masculine moral narrowing is that "charges of sexual harassment . . . are often seen as exaggerated or as fabricated out of misunderstanding or spite" (Schwalbe 1992, 46). Thus, gender differences in interpreting sexually harassing behaviors may stem more from acts of ignoring than states of ignorance.

The Problem with Getting Caught

But are women really the untroubled objects that girl watching—viewed through the eyes of men—suggests? Obviously not; the game may be premised on a denial of a woman's subjectivity, but an actual erasure is beyond men's power! It is in this multiplicity of subjectivities, as Butler (1990, ix) noted, where "trouble" lurks, provoked by "the unanticipated agency of a female 'object' who inexplicably returns the glance, reverses the gaze, and contests the place and authority of the masculine position." To face a returned gaze is to get caught, an act that has the power to undermine the logic of girl watching as simply a game among men. Karl, for example, noted that when caught, men are often flustered, a reaction suggesting that the boundaries of usual play have been disturbed.[6]

When a woman looks back, when she asks, "What are you looking at?" she speaks as a subject, and her status as mere object is disturbed. When the game is played as a form of hommosexuality, the confronted man may be baffled by her response. When she catches them looking, when she complains, the targeted woman speaks as a subject. The men, however, understand her primarily as an object, and objects do not object.

The radical potential of sexual harassment law is that it centers women's subjectivity, an aspect prompting Catharine MacKinnon's (1979) unusual hope for the law's potential as a remedy. For men engaged in girl watching, however, this subjectivity may be inconceivable.

From their viewpoint, acts such as girl watching are simply games played with objects: women's bodies. Similar to Schwalbe's (1992) insight into the senators' reaction to Professor Hill, the harm of sexual harassment may seem more the result of a woman's complaint (and law's "illegitimate" encroachment into the everyday work world) than men's acts of objectification. For example, in reflecting on the impact of sexual harassment policies in the workplace, José lamented that "back in the '70s, [it was] all peace and love then. Now as things turn around, men can't get away with as much as what they used to." Just whose peace and love are we talking about?

* * *

Conclusions

In this analysis, I have sought to unravel the social logic of girl watching and its relationship to the question of gender differences in the interpretation of sexual harassment. In the form analyzed here, girl watching functions simultaneously as only play and as a potent site where power is played. Through the objectification on which it is premised and in the nonempathetic masculinity it supports, this form of girl watching simultaneously produces both the harassment and the barriers to men's acknowledgment of its potential harm.

The implications these findings have for antisexual harassment training are profound. If we understand harassment to be the result of a simple lack of knowledge (of ignorance), then straightforward informational sexual harassment training may be effective. The present analysis suggests, however, that the etiology of some harassment lies elsewhere. While they might have quarreled with it, most of the men I interviewed had fairly good abstract understandings of the behaviors their companies' sexual harassment policies prohibited. At the same time, in relating stories of social relations in their workplaces, most failed to identify specific behaviors as sexual harassment when they matched the abstract definition. . . . [T]he source of this contradiction lies not so much in ignorance but in acts of ignoring. Traditional sexual harassment training programs address the former rather than the later. As such, their effectiveness against sexually harassing behaviors born out of social practices of masculinity like girl watching is questionable.

Ultimately, the project of challenging sexual harassment will be frustrated and our understanding distorted unless we interrogate hegemonic, patriarchal forms of masculinity and the practices by which they are (re)produced. We must continue to research the processes by which sexual harassment is produced and the gendered identities and subjectivities on which it poaches (Wood 1998). My study provides a first step toward a more process-oriented understanding of sexual harassment, the ways the social meanings of harassment are constructed, and ultimately, the potential success of antiharassment training programs.

Notes

1. For example, Maria, an administrative assistant I interviewed, simultaneously echoed and critiqued this understanding when she complained about her boss's girl watching in her presence: "If he wants to do that in front of other men . . . you know, that's what men do."

2. Recently, more researchers have turned to qualitative studies as a means to understand the process of labeling behavior as harassment. Of note are Collinson and Collinson (1996), Giuffre and Williams (1994), Quinn (2000), and Rogers and Henson (1997).

3. Quid pro quo ("this for that") sexual harassment occurs when a person with organizational power attempts to coerce an individual into sexual behavior by threatening adverse job actions.

4. I prefer the term "asymmetrical heterosexuality" over "double standard" because it directly references the dominance of heterosexuality and more accurately reflects the interconnected but different forms of acceptable sexuality for men and women. As Estrich (1987) argued, it is not simply that we hold men and women to different standards of sexuality but that these standards are (re)productive of women's disempowerment.

5. "Hommo" is a play on the French word for man, *homme*.

6. Men are not always concerned with getting caught, as the behavior of catcalling construction workers amply illustrates; that a woman hears is part of the thrill (Gardner 1995). The difference between the workplace and the street is the level of anonymity the men have vis-à-vis the woman and the complexity of social rules and the diversity of power sources an individual has at his or her disposal.

References

Bernstein, R. 1994. Guilty if charged. *New York Review of Books,* 13 January.

Buckwald, A. 1993. Compliment a woman, go to court. *Los Angeles Times.* 28 October.

Bumiller, K. 1988. *The civil rights society: The social construction of victims.* Baltimore: Johns Hopkins University Press.

Butler, J. 1990. *Gender trouble: Feminism and the subversion of identity.* New York: Routledge.

Cleveland, J. N., and M. E. Kerst. 1993. Sexual harassment and perceptions of power: An under-articulated relationship. *Journal of Vocational Behavior 42*(1): 49–67.

Cockburn, C. 1983. *Brothers: Male dominance and technological change.* London: Pluto Press.

Collinson, D. L., and M. Collinson. 1989. Sexuality in the workplace: The domination of men's sexuality. In *The sexuality of organizations,* edited by J. Hearn and D. L. Sheppard. Newbury Park, CA: Sage.

———. 1996. "It's only Dick": The sexual harassment of women managers in insurance sales. *Work, Employment & Society 10*(1): 29–56.

Conley, F. K. 1991. Why I'm leaving Stanford: I wanted my dignity back. *Los Angeles Times,* 9 June.

Conley, J., and W. O'Barr. 1998. *Just words.* Chicago: University of Chicago Press.

Connell, R. W. 1987. *Gender and power.* Stanford, CA: Stanford University Press.

———. 1995. *Masculinities.* Berkeley: University of California Press.

Estrich, S. 1987. *Real Rape.* Cambridge, MA: Harvard University Press.

Garcia, L., L. Milano, and A. Quijano. 1989. Perceptions of coercive sexual behavior by males and females. *Sex Roles 21*(9/10): 569–77.

Gardner, C. B. 1995. *Passing by: Gender and public harassment.* Berkeley: University of California Press.

Giuffre, P., and C. Williams. 1994. Boundary lines: Labeling sexual harassment in restaurants. *Gender & Society* 8:378–401.

Guccione, J. 1992. Women judges still fighting harassment. *Daily Journal,* 13 October, 1.

Gutek, B. A., and M. P. Koss. 1993. Changed women and changed organizations: Consequences of and coping with sexual harassment. *Journal of Vocational Behavior 42*(1): 28–48.

Gutek, B. A., B. Morasch, and A. G. Cohen. 1983. Interpreting social-sexual behavior in a work setting. *Journal of Vocational Behavior 22*(1): 30–48.

Hearn, J. 1985. Men's sexuality at work. In *The sexuality of men,* edited by A. Metcalf and M. Humphries. London: Pluto Press.

Hopkins, P. 1992. Gender treachery: Homophobia, masculinity, and threatened identities. In *Rethinking masculinity: Philosophical explorations in light of feminism,* edited by L. May and R. Strikwerda. Lanham, MD: Littlefield, Adams.

Jensen, I. W., and B. A. Gutek. 1982. Attributions and assignment of responsibility in sexual harassment. *Journal of Social Issues 38*(4): 121–36.

Johnson, M. 1988. *Strong mothers, weak wives.* Berkeley: University of California Press.

Kenig, S., and J. Ryan. 1986. Sex differences in levels of tolerance and attribution of blame for sexual harassment on a university campus. *Sex Roles 15*(9/10): 535–49.

MacKinnon, C. A. 1979. *The sexual harassment of working women.* New Haven, CT: Yale University Press.

Malovich, N. J., and J. E. Stake. 1990. Sexual harassment on campus: Individual differences in attitudes and beliefs. *Psychology of Women Quarterly 14*(1): 63–81.

Murrell, A. J., and B. L. Dietz-Uhler. 1993. Gender identity and adversarial sexual beliefs as predictors of attitudes toward sexual harassment. *Psychology of Women Quarterly 17*(2): 169–75.

Padgitt, S. C., and J. S. Padgitt. 1986. Cognitive structure of sexual harassment: Implications for university policy. *Journal of College Student Personnel* 27:34–39.

Popovich, P. M., D. N. Gehlauf, J. A. Jolton, J. M. Somers, and R. M. Godinho. 1992. Perceptions of sexual harassment as a function of sex of rater and incident form and consequent. *Sex Roles* 27 (11/12): 609–25.

Powell, G. N. 1986. Effects of sex-role identity and sex on definitions of sexual harassment. *Sex Roles* 14:9–19.

Pronger, B. 1992. Gay jocks: A phenomenology of gay men in athletics. In *Rethinking masculinity: Philosophical explorations in light of feminism,* edited by L. May and R. Strikwerda. Lanham, MD: Littlefield Adams.

Pryor, J. B. 1987. Sexual harassment proclivities in men. *Sex Roles.* 17(5/6): 269–90.

Quinn, B. A. 2000. The paradox of complaining: Law, humor, and harassment in the everyday work world. *Law and Social Inquiry* 25(4): 1151–83.

Reilly, M. E., B. Lott, D. Caldwell, and L. DeLuca. 1992. Tolerance for sexual harassment related to self-reported sexual victimization. *Gender & Society* 6:122–38.

Rogers, J. K., and K. D. Henson. 1997. "Hey, why don't you wear a shorter skirt?" Structural vulnerability and the organization of sexual harassment in temporary clerical employment. *Gender & Society* 11:215–38.

Schwalbe, M. 1992. Male supremacy and the narrowing of the moral self. *Berkeley Journal of Sociology* 37:29–54.

Smith, D. 1990. *The conceptual practices of power: A feminist sociology of knowledge.* Boston: Northeastern University Press.

Stockdale, M. S. 1993. The role of sexual misperceptions of women's friendliness in an emerging theory of sexual harassment. *Journal of Vocational Behavior* 42(1): 84–101.

Stockdale, M. S., and A. Vaux. 1993. What sexual harassment experiences lead respondents to acknowledge being sexually harassed? A secondary analysis of a university survey. *Journal of Vocational Behavior* 43 (2):221–34.

Tagri, S., and S. M. Hayes. 1997. Theories of sexual harassment. In *Sexual harassment: Theory, research and treatment,* edited by W. O'Donohue. New York: Allyn & Bacon.

Thorne, B. 1993. *Gender Play: Girls and boys in school.* Buckingham, UK: Open University Press.

U.S. Merit Systems Protection Board. 1988. *Sexual harassment in the federal government: An update.* Washington, DC: Government Printing Office.

Welsh, S. 1999. Gender and sexual harassment. *Annual Review of Sociology* 1999:169–90.

West, C., and D. H. Zimmerman. 1987. Doing gender. *Gender & Society* 1:125–51.

Wood, J. T. 1998. Saying makes it so: The discursive construction of sexual harassment. In *Conceptualizing sexual harassment as discursive practice,* edited by S. G. Bingham. Westport, CT: Praeger.

York, K. M. 1989. Defining sexual harassment in workplaces: A policy-capturing approach. *Academy of Management Journal* 32:830–50.

Yount, K. R. 1991. Ladies, flirts, tomboys: Strategies for managing sexual harassment in an underground coal mine. *Journal of Contemporary Ethnography* 19:396–422.

⚜ Topics for Further Examination ⚜

- Visit http://www.mencanstoprape.org to see what men are doing to stop rape.
- Look up resources on trafficking in women, including the website http://www.humantrafficking.org/.
- Try to find the latest reports of campus rape for your campus and surrounding campuses. Does this surprise you? Try to find organizations that provide support to women who have been sexually abused on your campus.
- Check out the sexual harassment laws for your state. Are they easily accessible? How are they enforced?

PART III

POSSIBILITIES

10

Nothing Is Forever

The title of this chapter stands for the principle that change is inevitable. Like the ever-evolving patterns of the kaleidoscope, change is inherent in all of life's patterns. Anything can be changed and everything does change, from the cells in our bodies to global politics. There is no permanent pattern, no one way of experiencing or doing anything that lasts forever. This fact of life can be scary, but it can also be energizing. The mystery of life, like the wonder of the kaleidoscope, rests in not knowing precisely what will come next.

The readings in this chapter address the changing terrain of gender. If one only takes a snapshot of life, it may appear as though current gender arrangements are relatively fixed. However, an expanded view of gender, over time and across cultures, reveals the well-researched fact that gender meanings and practices are as dynamic as any other aspect of life. Patterns of gender continuously undergo change, and they do so at every level of experience, from the individual to the global. Michael Schwalbe (2001) observes that there is both chance and pattern in the lives of individuals and in the bigger arena of social institutions. He makes the point that no matter how many rules there might be and no matter how much we know about a particular person or situation, "social life remains a swirl of contingencies out of which can emerge events that no one expects" (p. 127). As a result, life, including its gendered dimensions, is full of possibilities.

Social constructionist theory is especially helpful in understanding the inevitability of change in the gender order. Recall that social constructionist research reveals the processes by which people create and maintain the institution of gender. It underscores the fact that gender is a human invention, not a biological absolute. Particular gender patterns keep going only as long as people share the same ideas about gender and keep doing masculinity and femininity in a routine, predictable fashion (Schwalbe, 2001; Johnson, 1997). Given that humans create gender, gender patterns can be altered by people who, individually and collectively, choose to invent and negotiate new ways of thinking about and doing gender.

At the micro level of daily interaction, individuals participate in destabilizing the binary, oppositional gender order. They do so by choosing to bend conventional gender rules or changing the rules altogether (Lorber, 1994). For example, women and men are creating new forms of partnership based on shared care work and housework roles (see Gerson, Chapter 6). Other individuals purposefully transgress the boundaries of sexual and gender identities by mixing appearance cues via makeup, clothing, hairstyle, and other modes of self-presentation (Lorber, 1994). Alan Johnson's chapter reading offers excellent advice about how each one of us can make a difference by taking small steps toward gender equity. He offers an array of activities and forms of both ordinary and creative resistance that people can engage in to "plant the seeds of change" in themselves and their communities. Johnson makes the point that change in our gendered ways of life can be deliberately mapped out. He argues that we have the power to choose to change ourselves and our world. This is especially so if we are willing to forge relationships with others who recognize a need for change and are ready to get involved in developing modes of living together that do not rely on inequalities and destructive or exploitive patterns of living.

TRENDS

At the macro level of the gender order, change comes about through large-scale forces and processes, both planned and unplanned. Trends are unplanned changes in patterns that are sustained over time. For example, Kivisto (1998) states that the Industrial Revolution is a trend, marking the transition from agricultural to industrial economies. This so-called revolution involves complex economic, technological, and related changes, such as urbanization, that profoundly alter the fabric of social life over time. Consider the impact of industrialization on gender in work and family life in the United States. Prior to industrialization, women's labor was essential to agricultural life. Women, men, and children worked side by side to grow crops, make clothing, raise animals, and otherwise contribute to the family economy (Lorber, 2001). That is, work and family were closely intertwined and the distinction between home and workplace did not exist (Wharton, 2005).

As the Industrial Revolution got underway, productive work moved from the home into factories, and work came to be defined as valuable only if it resulted in a paycheck. Although essential work was still done at home, it did not produce income. The negative outcome was that household labor was transformed into an invisible and devalued activity. Work and family came to be seen as distinct, firmly gendered domains of life, especially in the middle-class. Women and children were relegated to the home and women were expected to be full-time housewives and mothers, while men were ordained to follow wage work, embrace the breadwinner role, and participate in the political arena (Godwin & Risman, 2001; Wharton, 2005).

The profound changes in gender relations and the organization of work and family wrought by the Industrial Revolution continue to be a source of conflict for many women and men today in the United States. For example, although most married women with children now work outside the home, the doctrine of natural separate spheres—unpaid household work for women and paid work for men—continues to operate as an ideal against which "working women" who have children are often negatively evaluated.

Industrialization continues as a force for social change, one that is amplified by globalization. The term *globalization* refers to the increasing interconnectedness of social, political, and economic activities worldwide (Held et al., 1999). Transnational forces such as geopolitical conflicts, global markets, multinational corporations, transnational media, and the migration of labor now strongly influence what happens in specific countries and locales (Connell, 2000). For example, the international trading system—dominated by nations such as the United States—encompasses almost every country in the world, while films and television programs, especially those produced in the West, circulate the globe (Barber, 2002).

Offering a perspective on the impact of globalization in his reading in this chapter, R. W. Connell (2000) argues that it has created a worldwide gender order. This world gender order has several interacting dimensions: (1) a gender division of labor in a "global factory" in which poor women and children provide cheap labor for transnational corporations owned by businessmen from the major economic powers, (2) the marginalization of women in international politics, and (3) the dominance of Western gender symbolism in transnational media.

However, despite the order Connell posits, globalization is not monolithic. There are countervailing forces challenging the homogenizing and hegemonic aspects of globalization. For example, indigenous cultures interact with global cultures to produce new cultural forms of art and music. In addition, globalization has spawned transnational social movements, such as feminism, the slow food movement, and environmentalism, which address worldwide problems of Western hegemony, global inequality, and human rights. Connell's reading includes analysis of the links between local and global social action involving men in gender change. His discussion of what he calls "the broad cultural shift toward a historical consciousness about gender" provides insight into the complexities of globalization.

Social Movements

Large-scale change may also come about in a planned fashion. Social movements are prime examples of change that people deliberately and purposefully create. They are conscious, organized, collective efforts to work toward cultural and institutional change, and share distinctive features including organization, consciousness, noninstitutionalized strategies (such as boycotts and protest marches), and prolonged duration (Kuumba, 2001). The United States has a long history of people joining together in organizations and movements to bring about justice and equality. The labor union movement, socialist movement, civil rights movement, and Gay Lesbian Bisexual Transgender (GLBT) movement have been among the important vehicles for change that might not otherwise have happened.

One of the most durable and flexible social movements is feminism. Consider the fact that the feminist movement has already lasted for two centuries. At the opening of the nineteenth century, feminism emerged in the United States and Europe. By the early twentieth century, feminist organizations appeared in urban centers around the world. By the turn of the twenty-first century, feminism had grown into a transnational movement in which groups are working at local and global levels to address militarism, global capitalism, racism, poverty, violence against women, economic autonomy for women and other issues of justice, human rights, and peace (Shaw & Lee, 2001).

Two readings in this chapter, one by Barbara Ryan and the other by Andrea Smith, provide in-depth analyses of identity politics and the feminist movement. Ryan makes the point that gender is not an "automatic path of connection" among women. Prisms of difference and inequality such as race and social class can and do divide women and present a challenge for the women's movement both within the United States and at a global level. Smith adds to this discussion, focusing on the complex and varied relationships between Native American women who call themselves feminists and those who do not, as well as the extent to which Native feminists work in coalition with non-Native feminists.

The Complexity of Change

Not only is social change pervasive at micro and macro levels of life and a function of both planned and unplanned processes, it is also uneven and infinitely complex. Change doesn't unfold in a linear, predictable fashion, and it may be dramatically visible or it may take us by surprise. Consider the passage of the Nineteenth Amendment to the U.S. Constitution in 1920 that guaranteed women the right to vote. This one historic moment uplifted the public status of women and did so in a visible fashion. But more often, change consists of alterations in the fabric of gender relations that are not immediately visible to us, both in their determinants and their consequences.

For instance, we now know that a complex set of factors facilitated the entry of large numbers of single and married women into the paid workforce and higher education in the second half of the twentieth century. Those factors included very broad economic, political, and technological developments that transformed the United States into an urban, industrial capitalist nation (Stone & McKee, 1998). Yet, no one predicted the extent of change in gender attitudes and relations that would follow the entry of women into the workforce. It is only "after the fact" that the implications have been identified and assessed. For example, marital relationships in the United States have moved toward greater equity in response to the reality that most married women are no longer wholly dependent on their husbands' earnings. As married women have increasingly embraced paid work, their husbands have reconceptualized and rearranged their priorities so that they can devote more attention to parenthood (Goldscheider & Rogers, 2001).

Finally, it is important for us to recognize that change occurs even under oppressive social conditions. Research has demonstrated how patriarchal traditions in seemingly rigid social institutions can be altered. For example, studies of the "forced" integration of women into previously all-male military academies, such as West Point and the Citadel, show that although women struggled against a powerful wall of male resistance, they have in the end demonstrated that they can "do military masculinity" (Kimmel, 2000).

Prisms of Gender and Change

Returning briefly to the metaphor of the kaleidoscope, let us recall that the prism of gender interacts with a complex array of social prisms of difference and inequality, such as race and sexual orientation. The prisms produce ever-changing patterns at micro and macro levels of life. Our metaphor points to yet another important principle of dynamic gender arrangements. We can link gender change to alterations in other structural dimensions of society, such as race, class, and age. For example, as Americans have moved toward greater consciousness and enactment of gender equality, they have also come to greater consciousness about the roles that heterosexism (i.e., the institutionalization of heterosexuality as the only legitimate form of sexual expression) and homophobia (i.e., the fear and hatred of homosexuality) play in reinforcing rigid gender stereotypes and relationships (see Chapter 6). It has become clear to many seeking gender justice that the justice sought after cannot be achieved without eliminating homophobia and the heterosexist framework of social institutions such as family and work.

Additionally, gender transformation in the United States is inextricably tied to race. This is true both historically and today. The first wave of feminism was an outgrowth of the antislavery movement, and it was the politics of racial justice that led to the second wave of feminism (Freedman, 2002). Racism, as well as ageism, classism, and other forms of oppression, had to be addressed by feminists because the struggle to achieve equal worth for women had to include all women (see Collins, Chapter 2). Anything less would mean failure.

The Inevitability of Change

Collectively, the articles in this chapter invite the reader to ask, "Why should I care about or get involved in promoting change in the gender status quo?" That is a good question. After all, why should one go to the trouble of departing from the standard package of gender practices and relationships? Change requires effort and entails risk. On the other hand, the cost of "going with the flow" can be high. There are no safe places to hide from change. Even if we choose "not to rock the boat" by closing ourselves off to inner and outer awareness, change will find us. There are two reasons for this fact of life. First, we cannot live in society without affecting others and in turn being affected by them. Each individual life intertwines with the lives of many other people, and our words and actions do have consequences, both helpful and harmful. Every step we take and every choice we make affect the quality of life for a multitude of people. If we choose to wear blinders to our connections to others, we run the risk of inadvertently diminishing their chances, and our own chances, of living fulfilling lives (Schwalbe, 2001). For example, when a person tells a demeaning joke about women, he or she may intend no harm. However, the (unintended) consequences are harmful. The joke reinforces negative stereotypes, and telling the joke gives other people permission to be disrespectful to women (Schwalbe, 2001).

Second, we can't escape broad, societal changes in gender relations. By definition, institutional and societal level change wraps its arms around all of us. Think about the widespread impact of laws such as the Equal Pay Act and Title VII, outlawing discrimination against women and people of color, or consider how sexual harassment legislation has redefined and altered relationships in a wide array of organizational settings. Reflect on the enormous impact of the large numbers of women who have entered the workforce since the latter half of the twentieth century. The cumulative effect of the sheer numbers of women in the workforce has been revolutionary in its affect on gender relations in family, work, education, law, and other institutions and societal structures.

Given the inevitability of change in gender practices and relationships, it makes good sense to cultivate awareness of who we are and what our responsibilities to one another are. Without awareness, we cannot exercise control over our actions and their impact on others. Social forces shape us, but those forces change. Every transformation in societal patterns reverberates through our lives. Developing the "social literacy" to make sense of the changing links between our personal experience and the dynamics of social patterns can aid us in making informed, responsible choices (O'Brien, 1999; Schwalbe, 2001).

References

Barber, B. R. (2002). Jihad vs. McWorld. In G. Ritzer (Ed.), *McDonaldization: The reader* (pp. 191–198). Thousand Oaks, CA: Pine Forge Press.

Connell, R. W. (2000). *The men and the boys.* Berkeley: University of California Press.

Freedman, E. (2002). *No turning back: The history of feminism and the future of women.* New York: Ballantine Books.

Godwin, F. K., & Risman, B. J. (2001). Twentieth-century changes in economic work and family. In D. Vannoy (Ed.), *Gender mosaic* (pp. 134–144). Los Angeles: Roxbury Publishing Company.

Goldscheider, F. K., & Rogers, M. L. (2001). Gender and demographic reality. In D. Vannoy (Ed.), *Gender mosaics* (pp. 124–133). Los Angeles: Roxbury Publishing Company.

Held, D., McGrew, A., Goldblatt, D., & Perraton, J. (1999). *Global transformations.* Stanford, CA: Stanford University Press.

Johnson, A. (1997). *The gender knot.* Philadelphia: Temple University Press.

Kimmel, M. (2000). Saving the males: The sociological implications of the Virginia Military Institute and the Citadel. *Gender & Society, 14*(4), 494–516.

Kivisto, P. (1998). *Key ideas in sociology.* Thousand Oaks, CA: Pine Forge Press.

Kuumba, M. B. (2001). *Gender and social movements.* Walnut Creek, CA: Altamira Press.

Lorber, J. (1994). *Paradoxes of gender.* New Haven, CT: Yale University Press.

Lorber, J. (2001). *Gender inequality: Feminist theories and politics.* Los Angeles: Roxbury Publishing Company.

O'Brien, J. (1999). *Social prisms.* Thousand Oaks, CA: Pine Forge Press.

Schwalbe, M. (2001). *The sociologically examined life.* Mountain View, CA: Mayfield Publishing Company.

Shaw, S. M., & Lee, J. (2001). *Women's voices, feminist visions.* Mountain View, CA: Mayfield Publishing Company.

Stone, L., & McKee, N. P. (1998). *Gender & culture in America.* Upper Saddle River, NJ: Prentice Hall.

Wharton, A. S. (2005). *The sociology of gender.* Malden, MA: Blackwell.

Worley, J. C., & Vannoy, D. (2001). The challenge of integrating work and family life. In D. Vannoy (Ed.), *Gender mosaics* (pp. 165–173). Los Angeles: Roxbury Publishing Company.

Introduction to Reading 46

Barbara Ryan, a sociologist, has studied social movements and feminism for over twenty years and has authored three books on the topic. This reading is the introduction to her latest book of the same title. She describes different perspectives, comparing postmodernists' emphasis on uniqueness to that of identity politics, which focuses on similarities across individuals' experiences. Activism is a central part of feminist thought; however, as we noted in the introduction to Chapter 1, activism for women's rights and equality is not a simple or straightforward issue. Ryan's discussion of the women's movement reflects the argument that gender is a complex concept representing multiple, socially constructed identities. It is these multiple social identities that have shaped and divided the women's movement.

1. What does Ryan mean by "identity politics" and how does this differ from a postmodernist perspective?
2. Are identity politics an impediment or facilitator for efforts to gain rights for women?
3. How are multiple social identities incorporated into the women's movement?

Identity Politics in the Women's Movement

Barbara Ryan

What is identity? Is identity recognition of a shared characteristic that enables a solidarity with members of a group? And does it conversely entail distance from those who lack this common feature? Or, is identity a social construction that ebbs and flows, is always in process, multilayered, and fragmented? What is gender? As a defining identity, does it fit the former or the latter conceptualization?

These questions are at the heart of understanding how identity politics affects social movements and the women's movement. The women's movement is a gender-focused movement. Yet, within the category of women there are other identities that work to keep women from recognizing gender commonalties. For instance, living in a largely segregated society, women of color feel a bonding with men of color that they do not usually feel with white women (Bell-Scott 1994; Collins 1990; Dill 1983; Fleming 1993; McKay 1993). Likewise, studies of lesbian feminist communities reveal the positive aspects of joining together to find acceptance and emotional support where it is lacking in the straight world (Franzen 1993; Kreiger 1982, 1983; Taylor and Whittier 1992).

Proponents of identity politics believe it important to affiliate with those who confront similar experiences based on social group characteristics. Members of an oppressed group may organize to change their situation, as well as their feelings of self-worth and place in the social structure. Hence, social characteristics that have been used to exclude certain groups have led to social movements organized by those groups to change their condition.

Critics of identity politics assert that it leads to further marginalization and that it prevents uniting with those who are working on similar issues but who differ in physical/social features (Gitlin 1993). Class, too, is left out of this analysis, as are differences within groups, which may have everyday practical consequences (Allison 1993). Indeed, Hall (1996:4–5) argues that unities based on essentialist identities are constructed within the dynamics of power and exclusion, "and thus are more the product of the making of difference and exclusion, than they are the sign of a naturally constituted unity." And problematically, as Grossberg (1996:88) points out, groups organized around their own model of repression often lack the capability of creating alliances with others.

Critiques

Postmodern thought sees identity as a process rather than a fact or deterministic force. Yet, recognizable identity traits continue to draw people together and to provide them with support for attempting social change. This means identity and identity politics are serious contenders in the political process and social movement arenas.

Eric Hobsbawm (1996) points out that in the late 1960s the *International Encyclopedia of the Social Sciences* had no entry under identity. Thus, Hobsbawm sees identity politics as a recent phenomenon. He also sees it as a problematic category. First, he argues that a collective identity is defined against others and is based not on what their members have in common but, rather, on differences between them when, in fact, "we" may have little in common except not being the "others." Second, no one has only a single identity; yet, identity politics leads one to disclaim other identities. And finally, identities are not fixed—they depend on contexts, which can change. . . .

Todd Gitlin (1995), like Hobsbawm, calls for a Left politics based on class position. However, his analysis is somewhat different, fashioned on a contemporary and North American model. According to Gitlin, the most serious and negative identity politics is white men who fear identity gains will come at their expense. But he is also opposed to identity politics in general because he feels it cultivates unity only within special groups, and there is an obsession with difference leading to the "borders identity politics draws" (Severson and Stanhope 1998).

Rather than organizing to reduce inequalities between rich and poor, Gitlin argues that identity politics struggles to change the color of inequality. And, what we need to be doing, instead, is to tend mutualities. Identity politics, according to Gitlin, (1995:236) has failed to tend and, even worse, has left the centers of power uncontested. Gitlin calls for a Left politics that includes everyone, a common—a cause of all. Although his argument is compelling in many ways, he barely mentions gender. Socialist feminists in the 1960s and 1970s explicitly pointed out how the inclusion of "all" in the Left of their day did not include them. When leftist writers in the late 1990s have little to say specifically to gender, are we to believe them? Does commonality leave women's "difference" out? Is leftist universalism like postmodern and deconstructionist analysis, wiping away all difference, even denying there is difference because there is no reality? Does identity politics as we know it in the women's movement leave the centers of power uncontested? These questions raised by leftist scholars present serious critiques of identity politics.

Feminist analysis also contains critiques of identity politics. Daphne Patai (1992), for instance, discusses the zealousness of feminist adherents to control thought and appearance, what she calls "ideological policing." Patai objects to the assumption that one's racial/ethnic identity is the same as one's views. Even more, she believes there has been a reversal of privilege, now residing with women of color, in which no white person can challenge their version of reality. These inclinations have led to concern for the "dogmatic turn" identity politics has presented to women's studies. In her experience, *Eurocentric* became a slur and teaching courses on other racial/ethnic groups was not accepted of a North American white.

Further, Patai (1992:B3) questions the ways identity politics gets used in a scarce job market, calling it "the fraud that accompanies familiar old ambitions dressed up in appropriate ideology."

Patai feels distress that these tendencies have arisen and, even more, that they are not discussed. Instead, identity politics has led to silencing. She considers her writing on these issues to be a defense of feminism.

Others also point to problems within feminism, particularly the focus on personal experience, which may have isolated the women's movement from more general social change struggles. Often, rather than oppression's being fought in the wider society, struggles are being fought on local levels (Adams 1989). L. A. Kauffman (1990) takes a more nuanced look at identity politics, dividing it into political and nonpolitical frameworks. Kauffman dates the beginning of identity politics not with black women's challenge to sisterhood but to the civil rights movement of Martin Luther King and the Black Power movement, where activists called for a new collective identity to offset white imperialism. In turning to the women's movement, Kauffman credits Kate Millett's *Sexual Politics* (1970) with defining gendered power as politics—structured relationships whereby one group controls another. In the 1980s and 1990s, though, she fears identity politics has evolved into fragments where "the notion of solidarity, so central to any progressive politics" is lost (Kauffman 1990:76).

Kauffman (1990:78) makes the point that the increasing movement of self-transformation (as political change) leads to thinking that problems are attitudes rather than power differentials and vested interests. Like Patai, she see this leading to an emphasis on lifestyle (who one reads, what one eats or wears) rather than on the actions one takes. Still, Kauffman calls for using identity as entry to challenging institutions of power, and as politics intent upon both social and individual transformation.

A central issue of importance is whether difference has displaced inequality as a central concern of social movements. As Anne Phillips (1997) points out, an *injustice* perspective seeks to eliminate differentiation used against powerless groups, and *difference* perspectives are intent on highlighting these identities. She cites the dilemma between strategies that are meant to diminish the significance of gender and strategies that focus on the intrinsic worth of one's sex.

Hazel Cathy (1990) adds another perspective when she questions whether the emphasis on diversity in feminist thought and practices is a way to avoid the politics of race, even as it appears that race is being confronted. Similarly, the disjunction between inclusive feminism and the reality of the organizations that make up the women's movement raises the crucial point "on whose behalf " inclusive ideologies are meant (Leldner 2001).

Identity Claims

In spite of critical questions associated with identity politics, there are important rationales for the development of a politics of identity, beginning with *The Second Sex,* Simone de Beauvoir's (1953) classic work. Her introduction sets the tone, unveiling a gendered identity politics by calling women "the other." By this, she means that women have failed to identify themselves as a group because they are considered a part of man (the subject) and, thus, are not segregated into their own group, as are some racial and ethnic groups. They have no history or religion that is particularly their own. She calls for women to see themselves as a group in order to change their situation.

De Beauvoir's writings inspired a collective conscience of women, which laid the foundation for the reemergence in the 1960s of women's activism in their own behalf. Yet, by the early 1980s, writings by women of color spoke to the need to claim an identity of their own. They formulated a base for organizing around that identity, even if it separated women from one another. The Combahee River Collective, a group of black feminists and forerunner to this claim, began meeting in 1974. They issued the first statement on black feminism, twelve years after the contemporary women's movement emerged and many more years after the U.S. publication of *The Second Sex.* The statement combined gender and race identity. Black women proclaimed the task of combating simultaneous oppressions as theirs because other movements failed to acknowledge their specific oppression. They named what they were doing "identity politics" based on their conclusion that "the only people who care enough about us to work consistently for our liberation are us" (Combahee River Collective 1978:275).

How did white feminists react to this challenge? Some were angry or dismissive. Some had already reached this awareness. Others welcomed it. Still others struggled with their past and worked to become multicultural in their feminist thought and actions, even as this became a painful process of stripping away their own identity, deciding what to keep, what to eliminate, what to change (Pratt 1984). This difficult process involved the acknowledgment of another's existence while not denying one's own. For instance, Minnie Bruce Pratt (1984:73) describes her fears as she tries to understand herself in "relation to folks different from me, when there are discussions, conflicts about anti-Semitism and racism among women, criticisms, criticisms of me; when, for instance in a group discussion about race and class, I say I feel we have talked too much about race, not enough about class, and a woman of color asks me in anger and pain if I don't think her skin has something to do with class."

Part of the problem in understanding "other worlds" is that women grow up learning different gender roles. For example, in many Native American groups women are strong and valued (Allen 1995), whereas other women have had to work at developing that consciousness. And having acquired an ideology of strength and independence, what happens if you become disabled or when you grow old (Klein 1992)?

Multiple Identities and Changing Identities

An obvious complexity within the field of identity politics is the reality of multiple identities, including those we are in the process of becoming or losing. Audre Lorde (1984a:41), who called upon women to speak—"your silence will not protect you"—used her life as an example of how we can rid ourselves of others' distortions by reclaiming all our identities so we can define them for ourselves.

As a forty-nine-year-old black lesbian socialist feminist, who was also a mother of two and part of an interracial couple, Lorde discussed her many group identities, including acquired identities that did not fit into acceptable society. This makes life difficult, and yet, she notes it is oppressed people who are expected to bridge the gap between their differences with more privileged groups. Lorde asserted that it is not the differences among us that separate us; it is the refusal to recognize the differences. An example she cites is the idea of "sisterhood." In a famous quote, she tells us: "Some problems we share as women, some we do not. You fear your children will grow up to join the patriarchy and testify against you, we fear our children will be dragged from a car and shot down in the street, and you will turn your backs upon the reasons they are dying" (Lorde 1984b:119).

In applying a wide lens, Lorde also talked about differences within black communities. Where racism is a living reality, differences within groups seem dangerous and suspect. The need for unity is often misnamed as a need for homogeneity, and a black feminist vision mistaken for betrayal. There is a refusal of some black women to recognize and protest against their oppression as women within the black community and of heterosexual women against lesbians, particularly among black women. She urged women to identify with one another and develop new ways of being in this world and new ways for this world to be. For, as she tells us, "the master's tools will never dismantle the master's house" (Lorde 1984b:123). In line with Lorde's analysis of divisions within groups, Marilyn Frye (1992) makes clear that even in what appears to be a cohesive commonality (in this case a lesbian community) there are substantial differences that must be acknowledged and worked through.

The necessity for claiming more than one identity is also true for Chicanas (Nieto 1997), Asians (Shah 1994), and women in developing countries. Like African American women, members of these racial/cultural groups often find U.S. feminism's focus on male/female relations alienating because they, too, are working against multiple oppressions of gender, class, race, and nationalism. But they add a difference to the experiences of African

American women, where much of identity politics has been focused, and that is invisibility, which, for them, is another form of oppression (Friedman 1995). Moreover, not all Third World women are women of color and not all women of color are Third World or poor. Hispanic women have reported experiencing racism through the rejection of black sisters because of being light skinned (Quintanales 1983), and class is a dividing agent among women of all nationalities and races.

Regardless of the identity issues that may divide them, feminists are concerned that in the rush to acknowledge and celebrate difference, the relations of power that create that difference are often ignored. Women of color, in particular, find the current popularity of diversity rhetoric all too often offers a decontextualized politics of difference, which turns out to be another way of preserving stratified social arrangements (Aguilar 1995).

In another vein, Carol Queen (1997) voices an unwelcome (at least for some) claim to feminist identity. Queen finds that sex radicals (regardless of race, class, ethnicity, or sexual orientation) have been silenced in the women's movement. She questions what it is that separates women who are opposed to sex work from those who do it, and why feminism does not take a more thoughtful look at this divide. She asks feminists to confront their "whorephobia" and agree to a dialogue, for she believes women have much to learn from sex workers.

The complexity of multiple identities is poignantly voiced by June Jordan (1985). From a vacation experience in the Bahamas she finds that, compared to the Bahamian people, she is a rich American woman. She is dismayed to find herself, as well as other black Americans (and whites), arguing prices on handmade items. Jordan uses this story to raise awareness of the complex interplay of race, class, and gender identity. She notes that she and the women workers are engaged in interactions that preclude seeing themselves as a united group of women. Jordan wonders how women are to connect with such different life circumstances, particularly when many women do not feel poor women's issues of poverty and crime are theirs. She asks, "Why aren't they everyone's?" Jordan's story shows that race, class, and gender are not automatic paths of connection; there are differences within identities that have been imposed.

Voices of African American Women

It was African American women in the early 1980s, more than any other group, who confronted the women's movement on identity politics issues. There was a desire for a more pluralistic approach to "sisterhood" that recognized similarities and differences among women (Dill 1983). Bernice McNair Barnett (1995:207) makes the interesting and telling remark that the barring of black women from the League of Women Voters in Montgomery, Alabama, showed that "it was white women, rather than black women, who placed their primary emphasis on race over gender." She also points out that the 1940s and 1950s were not a period of "doldrums" for women activists, as has been claimed for the women's movement (see Rupp and Taylor 1987). These were years of activism for black women in the civil rights movement, a movement dedicated to issues of freedom and equality.

Taking a different approach, Barbara Smith (1983) succinctly discusses the reasons feminism frightens black and Third World men and why they resist it. In her introduction to *Home Girls,* an early contribution to writings by black women, Smith shows why black women need a movement of their own. Revealing another perspective some ten years later, Ann duCille (1994) wonders if the effort to promote black women's lives has not gone too far. As a black woman who has long studied black women, she acknowledges having mixed feelings about this rise of "the occult of black womanhood." For instance, she questions the career-enhancing path women academics, white and black, have gained by claiming a "new" specialness for women of color and those who focus on them.

New questions are raised, such as looking at white middle-class women who are "housewives" to their husbands and the black working-class domestics they employ (Kaplan 1995). Both groups of women are in roles of serving others, but the white women exist with race and class privilege by means of their domestics. One conclusion, which can be drawn from this relationship, is that white women collude with the patriarchal/capitalist system that oppresses women. Another conclusion is that domestic workers enable white middle-class woman to avoid confronting their spouses about sharing household duties (Kaplan 1995:81). Moreover, household help releases the middle-class woman to become a woman of leisure or to have a career. The ways that women treat other women (using domestics as an example) may help explain why many black women stay away from the (white) feminist movement.

Sexuality and Sexualities

A second area of identity contestation centers on questions of sexual orientation and preference. One of the onerous aspects of heterosexual society is the normative expectation of appropriate sexual behavior that excludes homosexuality, bisexuality, sadomasochism, or transsexuality. Dichotomous thinking, rather than a continuum model of sexual identity, had long been critiqued within the feminist movement, yet it arose in the 1980s among lesbian feminists. These divisions revealed that there are exclusions and antagonisms among gays that differ from the full acceptance of sexual expression found in queer theory or, in the past, in the lesbian concept of the "woman-identified woman" (Radicalesbians 1970).

One challenge to agreed-upon thought was the deconstruction of commonly held views of sex workers, that is, to see them as workers deserving of workers' rights. Women in unions, armed with feminist ideologies, concretely address many of the issues of the women's movement—sexual harassment, maternity benefits, parental leave, and comparable worth (Chernow and Moir 1995). Yet, in the debates over prostitution and pornography, feminists who have argued for other women workers have not taken up these workers' cause (Alexander 1997).

There are differences within lesbianism, within feminism, and even within radical feminism. Eileen Bresnahan (2001) humorously relates an incident where the "original" radical feminists (with roots in the Left) collided with newer radical feminists, who were called cultural feminists. She laments the shift away from political process to lifestyle affirmation that she saw occurring in the mid-1970s. For her, this shift left an ambiguous meaning of radical feminism and was also a departure from the past, when being a radical feminist meant that one accepted definite agreed-upon principles of radical politics. Bresnahan explains her distrust of cultural feminism as the end product of her seriousness about feminist identity. She states that because "I'm a working-class woman who grew up in the 1950s and 1960s, the women's movement was the first time I took myself seriously and the first time I was taken seriously by others whom I could also respect." Thus, the challenge to agreed-upon thought was unwelcome in her mind and in her radical feminist circle.

What is a lesbian—who counts—is a continuing theme of sexuality inquiry. Divisions are found among lesbian feminists based on bisexuality, dress, associations, s/m practices, gender roles, and transgendered people. The 1970s woman-identified women represented a sisterhood against the patriarchy; 1990s lesbians aligned with gay men. Young lesbians have focused more on sex than political theory and often call themselves queer or "bad girls" rather than lesbians. These generational differences have led to clashes between lesbian feminism and queer ideologies. Problematically, these clashes have also raised charges of who is a real or fake lesbian, for example, what if you have a heterosexual past? The essentialism (true lesbian) of the 1970s is now confronted with more than one model for lesbian behavior. And even though this may seem confusing, as Vera Whisman (1993:58) says, "[T]he truth is, most of us sometimes feel incredibly queer, at other times indelibly female." One highly contested issue that has created division among lesbian feminists is sadomasochism. Shane Phelan considers it a mistake for activists to get involved in arguments of this kind, an issue that is rooted in the identity politics of what feminism is. By this, she does not mean that identity politics should be abandoned rather, she asks that we be more careful in distinguishing "the sorts of identity issues that are vital to our growth and freedom from those that are not" (Phelan 1989:133).

Other, more "acceptable" divisions among lesbians have been identified as class, age, and ideology. Trisha Franzen discovered that lesbian feminist university students considered the butch/fem roles played by many working-class bar lesbians to be tainted with heterosexuality. Thus, she argues that "sexuality is a problematic basis for political solidarity among women" (Franzen 1993:903). A similar dispute occurred at the Michigan Womyn's Music Festival over the admittance of transsexuals. In researching this issue, Joshua Garrison (1997:183) argues that identity requires difference and that building collective identities requires not simply pointing out commonalties but also marking off "who we are not." He finds these acts to be the boundary patrol of identity politics.

From these examples we can see that sexual identities and political affiliations often shift and are always contingent (Whisman 1993:58). Hence, we can no longer assume what the foundation of identical politics presumes; that is, the idea that identity groups, in this case lesbians, share an identity and therefore a politics. Even more pointed at the turn of the twenty-first century is to recognize that sexual identity, indeed all identities, are more provisional than most people realize. As Arlene Stein (1997) discussed in her research on ex-lesbians, there was a restructuring of the identity process based on situational factors that some feminists went through as they moved into lesbianism in the 1970s and out of it by the 1980s.

More Diversities—More Identities

While major divisions have arisen over issues of race and sexual diversity, there are other gender-plus identities that confront the women's movement. There is the issue of Jewish feminists and their place in the movement (Beck 1988), of age from the older woman's perspective (Macdonald 1995), as well as of the younger feminist viewpoint (Dietzel 1999; Heywood and Drake 1997; Looser and Kaplan 1997;Walker 1992). Inclusion itself has been questioned. For instance, Rosa María Pegueros (2001) reports on her own experience as a Latina activist in the National Organization for Woman (NOW). Achieving a high, visible position, she questions what that means for her. Is she a token, a traitor to her group, or an accepted member of a feminist elite (and does she want this)? What she has to say raises questions about the sincerity of inclusion some groups are promoting.

And, what of men? How does being a male feminist affect one's identity? There is the possibility that male feminism may be seen as a traitorous identity, indeed, traitorous perhaps to both men and women (Bettie 2001). Although it is self-evident that not all men are powerful, there is a danger in pointing out how men, too, are oppressed. This can seen as a denial of the history, and meaning of gender power relations. Deessentializing identity categories may be a necessary corrective to the conventional application of identity politics, but we must be careful not to become, then, an identity skeptic, refusing to recognize gender, race, and other identities (Bettie 2001).

Cutting across identity concerns is always the issue of class and class transformation. Moving from the working class to the middle class does not mean one has left all vestiges of one's background behind. Class has not been adequately explored in identity politics, perhaps because of the simultaneous desire to both reject and to retain this cultural identity. The challenge is to maintain a vigilant awareness of the inherent power these relations present while guarding against incorrect parallelisms, which can erase the political histories of difference (Bettie 2001).

The history of division within women's studies is legendary. Indeed, one could say that women's studies is itself identity politics (Perry 1995).Yet, in spite of the contentious debates over the category of women in the academy, the term *woman* has not been so starkly problematized in the larger society. Women of all races, ethnicity, sexual orientation (and preferences) are disadvantaged in a society that does not value women. There are other divisions among women . . . Women in revolutions, prisons, and armed services, and those living in rural areas . . . What is their relationship to other women and to feminism? What about differences between single and married women? Or those with or without children?

Divisions are often magnified when we begin to talk of global feminism and organizing transnationally to unite women from around the world. In Yemen in 2000, the Women's Studies Program at San'a University was closed down and the director fled the country because of the use of the word *gender* (Abu-nasr 2000). In Kuwait, women continue to be told they are not to be allowed the vote. How can women join together in India, where women are divided by caste? In what ways can Muslim women organize when ideas of a constructed gender identity are considered a Western concept? There is fear that a transnational unity of women might raise a counterargument, and perhaps repression, from conservative and religious forces that have used biological determinism in order to maintain gender segregation. At the same time, within countries, it must be recognized that the ideological and political realities women face limit the kinds of issues that can be raised. . . .

References

Abu-nasr, Donna. 2000. "Gender Controversy Shuts Down Women's Studies Program in Yemen." *Philadelphia Inquirer,* May 21, p. A14.

Adams, Mary Louise. 1989. "There's No Place Like Home: On the Place of Identity in Feminist Politics." *Feminist Review* 31 (Spring):22–33.

Aguilar, Delia D. 1995. "What's Wrong with the 'F' Word?" In *Frontline Feminism,* edited by Karen Kahn. San Francisco: Aunt Lute Books.

Alexander, Priscilla. 1997. "Feminism, Sex Workers, and Human Rights." In *Whores and Other Feminists,* edited by Jill Nagle. New York: Routledge.

Allen, Paula Gunn. 1995. "Where I Come From God Is a Grandmother." In *Frontline Feminism,* edited by Karen Kahn. San Francisco: Aunt Lute Books.

Allison, Dorothy. 1993. "A Question of Class." Pp. 46–60 in *Sisters, Sexperts, Queers: Beyond the Lesbian Nation,* edited by Arlene Stein. New York: Plume Books.

Barnett, Bernice McNair. 1995. "Black Women's Collectivist Movement Organizations: Their Struggles during the 'Doldrums.'" In *Feminist Organizations: Harvest of the New Women's Movement,* edited by Myra Marx Ferree and Patricia Yancy Martin. Philadelphia: Temple University Press.

Beck, Evelyn Torton. 1988. "The Politics of Jewish Invisibility." *NWSA Journal* 1:93–102.

Bell-Scott, Patricia, ed. 1994. *Life Notes: Personal Writings by Contemporary Black Women.* New York: Norton.

Bettie, Julie. 2001. "Changing the Subject: Male Feminism, Class Identity, and the Politics of Location." Chap. 15 in *Identity Politics in the Women's Movement* by Barbara Ryan. New York: New York University Press.

Bresnahan, Eileen. 2001. "The Strange Case of Jackie East: When Identities Collide." Chap. 21 in *Identity Politics in the Women's Movement* by Barbara Ryan. New York: New York University Press.

Introduction to Reading 47

This reading addresses the complex and varied Native American women activists' theories about feminism, the struggle against sexism within Native communities and society-at-large, and the importance of developing coalitions with non-Native feminists. The reading "brings to life" the major points made by Barbara Ryan in her article on identity politics. In addition, Andrea Smith analyzes current Native feminist sovereignty projects that address both colonialism and sexism through an intersectional framework.

1. Discuss links between Barbara Ryan's reading and Andrea Smith's analysis of Native American feminism.
2. Why is gender justice integral to issues of survival for indigenous people?
3. How does the boarding school project reveal connections between interpersonal gender violence and state violence?

Native American Feminism, Sovereignty, and Social Change

Andrea Smith

When I worked as a rape crisis counselor, every Native client I saw said to me at one point, "I wish I wasn't Indian." My training in the mainstream antiviolence movement did not prepare me to address what I was seeing—that sexual violence in Native communities was inextricably linked to processes of genocide and colonization. Through my involvement in organizations such as Women of All Red Nations (WARN, Chicago), Incite! Women of Color against Violence (www.incite-national.org), and various other projects, I have come to see the importance of developing organizing theories and practices that focus on the intersections of state and colonial violence and gender violence. In my ongoing research projects on Native American critical race feminisms, I focus on documenting and analyzing the theories produced by Native women activists that intervene both in sovereignty and feminist struggles.[1] These analyses serve to complicate the generally simplistic manner in which Native women's activism is often articulated within scholarly and activist circles.

Native Women and Feminism

One of the most prominent writings on Native American women and feminism is Annette Jaimes's (Guerrero) early 1990s article, "American Indian Women: At the Center of Indigenous Resistance in North America." Here, she argues that Native women activists, except those who are "assimilated," do not consider themselves feminists. Feminism, according to Jaimes, is an imperial project that assumes the giveness of U.S. colonial stranglehold on indigenous nations. Thus, to support sovereignty Native women activists reject feminist politics:

> Those who have most openly identified themselves [as feminists] have tended to be among the more assimilated of Indian women activists, generally accepting of the colonialist ideology that indigenous nations are now legitimate sub-parts of the U.S. geopolitical corpus rather than separate nations, that Indian people are now a minority with the overall population rather than the citizenry of their own distinct nations. Such Indian women activists are therefore usually more devoted to "civil rights" than to liberation per se. . . . Native American women who are more genuinely sovereigntist in their outlook have proven themselves far more dubious about the potentials offered by feminist politics and alliances.[2] According to Jaimes, the message from Native women is the same, as typified by these quotes from one of the founders of WARN, Lorelei DeCora Means:
>
> We are *American Indian* women, in that order. We are oppressed, first and foremost, as American Indians, as peoples colonized by the United States

From Smith, A., "Native American Feminism, Sovereignty, and Social Change" *Feminist Studies 31*(1), copyright © 2002. Reprinted with permission.

of America, *not* as women. As Indians, we can never forget that. Our survival, the survival of every one of us—man, woman and child—as *Indians* depends on it. Decolonization is the agenda, the whole agenda, and until it is accomplished, it is the *only* agenda that counts for American Indians.

The critique and rejection of the label of feminism made by Jaimes is important and shared by many Native women activists. However, it fails to tell the whole story. Consider, for instance, this quote from Madonna Thunder Hawk, who cofounded WARN with Means:

> Feminism means to me, putting a word on the women's world. It has to be done because of the modern day. Looking at it again, and I can only talk about the reservation society, because that's where I live and that's the only thing I know. I can't talk about the outside. How I relate to that term feminist, I like the word.
>
> When I first heard, I liked it. I related to it right away. But I'm not the average Indian woman; I'm not the average Indian activist woman, because I refuse to limit my world. I don't like that. . . . How could we limit ourselves? "I don't like that term; it's a white term." Pssshhh. Why limit yourself? But that's me.

My point is not to set Thunder Hawk in opposition to Means: both talk of the centrality of land and decolonization in Native women's struggle. Although Thunder Hawk supports many of the positions typically regarded as "feminist," such as abortion rights, she contends that Native struggles for land and survival continue to take precedence over these other issues. Rather, my argument is that Native women activists' theories about feminism, about the struggle against sexism both within Native communities and the society at large, and about the importance of working in coalition with non-Native women are complex and varied. These theories are not monolithic and cannot simply be reduced to the dichotomy of feminist versus nonfeminist. Furthermore, there is not necessarily a relationship between the extent to which Native women call themselves feminists, the extent to which they work in coalition with non-Native feminists or value those coalitions, whether they are urban or reservation-based, and the extent to which they are "genuinely sovereigntist." In addition, the very simplified manner in which Native women's activism is theorized straightjackets Native women from articulating political projects that both address sexism and promote indigenous sovereignty simultaneously.

Central to developing a Native feminist politic around sovereignty is a more critical analysis of Native activist responses to feminism and sexism in Native communities. Many narratives of Native women's organizing mirrors Jaimes's analysis—that sexism is not a primary factor in Native women's organizing. However, Janet McCloud recounts how the sexism in the Native rights movement contributed to the founding of the Indigenous Women's Network in 1985:

> I was down in Boulder, Colorado and Winona LaDuke and Nilak Butler were there and some others. They were telling me about the different kinds of sexism they were meeting up with in the movement with the men, who were really bad, and a lot of these women were really the backbone of everything, doing a lot of the kind of work that the movement needed. I thought they were getting discouraged and getting ready to pull out and I thought, "wow, we can't lose these women because they have a lot to offer." So, we talked about organizing a women's conference to discuss all the different problems. . . . Marsha Gomez and others decided to formally organize. I agreed to stay with them as a kind of a buffer because the men were saying the "Indignant Women's Organization" and blah, blah, blah. They felt kind of threatened by the women organizing.[3]

My interviews with Native women activists also indicate that sexism in Native communities is a central concern:

> Guys think they've got the big one, man. Like when [name of Native woman in the community] had to go over there and she went to these Indians because they thought they were a bunch of swinging dicks and stuff, and she just let them have it.

She just read them out. What else can you do? That's pretty brave. She was nice, she could have laid one of them out. Like you know, [name of Native man in the community], well of course this was more extreme, because I laid him out! He's way bigger than me. He's probably 5'11," I'm five feet tall. When he was younger, and I was younger, I don't even know what he said to me, it was something really awful. I didn't say nothing because he was bigger than me, I just laid him out. Otherwise you could get hurt. So I kicked him right in his little nut, and he fell down on the floor—"I'm going to kill you! You bitch!" But then he said, you're the man! If you be equal on a gut and juice level, on the street, they don't think of you as a woman anymore, and therefore they can be your friend, and they don't hate you. But then they go telling stuff like "You're the man!" And then what I said back to him, was "I've got it swinging!"

And although many Native women do not call themselves feminists for many well thought-out reasons, including but not limited to the reasons Jaimes outlines, it is important to note that many not only call themselves feminist but also argue that it is important for Native women to call themselves feminists. And many activists argue that feminist, far from being a "white" concept, is actually an indigenous concept white women borrowed from Native women.

(Interviewee 1)

I think one of the reasons why women don't call themselves feminists is because they don't want to make enemies of men, and I just say, go forth and offend without inhibition. That's generally why I see women hold back, who don't want to be seen as strident. I don't want to be seen as a man-hater, but I think if we have enough man-haters, we might actually have the men change for once. . . . I think men, in this particular case, I think men are very, very good at avoiding responsibility and avoiding accountability and avoiding justice. And not calling yourself a feminist, that's one way they do that. Well, feminism, that's for white women. Oh feminists, they're not Indian, They're counterrevolutionary. They're all man-haters. They're all ball-busters. They've gotten out of order. No, first of all that presumes that Native women weren't active in shaping our identity before white women came along. And that abusive male behavior is somehow traditional, and it's absolutely not. So I reject that. That's a claim against sovereignty. I think that's a claim against Native peoples. I think it's an utter act of racism and white supremacy. And I do think it's important that we say we're feminists without apology.

(Interviewee 2)

[On Native women rejecting the term "feminist"] I think that's giving that concept to someone else, which I think is ridiculous. It's something that there has to be more discussion about what that means. I always considered, they took that from us, in a way. That's the way I've seen it. So I can't see it as a bad thing, because I think the origins are from people who had empowered women a long time ago.

This reversal of the typical claim that "feminism" is white then suggests that Native feminist politics is not necessarily similar to the feminist politics of other communities or that Native feminists necessarily see themselves in alliance with white feminists. In addition, the binary between feminist versus nonfeminist politics is false because Native activists have multiple and varied perspectives on this concept. For instance, consider one woman's use of "strategic" feminism with another women's affirmation of feminist politics coupled with her rejection of the term "feminist." These women are not neatly categorized as feminists versus nonfeminists.

Native Feminism and Sovereignty

If we successfully decolonize, the argument goes, then we will necessarily eliminate problems of sexism as well. This sentiment can be found in the words of Ward Churchill. He contends that all struggles against sexism are of secondary importance because, traditionally, sexism did not exist in Indian nations. Churchill asks whether sexism exists in Indian country after Native peoples have attained sovereignty? His reply, "Ask Wilma Mankiller," former principal chief of the Cherokee

Nation.[4] Well, let's ask Mankiller. She says of her election campaign for deputy chief that she thought people might be bothered by her progressive politics and her activist background. "But I was wrong," she says:

> No one challenged me on the issues, not once. Instead, I was challenged mostly because of one fact—I am female. The election became an issue of gender. It was one of the first times I had ever really encountered overt sexism . . . (people) said having a female run our tribe would make the Cherokees the laughing stock of the tribal world.[5]

Regardless of its origins in Native communities, then, sexism operates with full force today and requires strategies that directly address it. Before Native peoples fight for the future of their nations, they must ask themselves, who is included in the nation? It is often the case that gender justice is often articulated as being a separate issue from issues of survival for indigenous peoples. Such an understanding presupposes that we could actually decolonize without addressing sexism, which ignores the fact that it has been precisely through gender violence that we have lost our lands in the first place.[6] In my activist work, I have often heard the sentiment expressed in Indian country: we do not have time to address sexual/domestic violence in our communities because we have to work on "survival" issues first. However, Indian women suffer death rates because of domestic violence twice as high as any other group of women in this country.[7] They are clearly not surviving as long as issues of gender violence go unaddressed. Scholarly analyses of the impact of colonization on Native communities often minimize the histories of oppression of Native women. In fact, many scholars argue that men were disproportionately affected by colonization because the economic systems imposed on Native nations deprived men of their economic roles in the communities more so than women.[8] By narrowing our analyses solely to the explicitly economic realm of society, we fail to account for the multiple ways women have disproportionately suffered under colonization—from sexual violence to forced sterilization. As Paula Gunn Allen argues:

> Many people believe that Indian men have suffered more damage to their traditional status than have Indian women, but I think that belief is more a reflection of colonial attitudes toward the primacy of male experience than of historical fact. While women still play the traditional role of housekeeper, childbearer, and nurturer, they no longer enjoy the unquestioned positions of power, respect, and decision making on local and international levels that were not so long ago their accustomed functions.[9]

This tendency to separate the health and well-being of women from the health and well-being of our nations is critiqued in Winona LaDuke's 1994 call to not "cheapen sovereignty." She discusses attempts by men in her community to use the rhetoric of "sovereignty" to avoid paying child support payments.

> What is the point of an Indian Child Welfare Act when there is so much disregard for the rights and well being of the children? Some of these guys from White Earth are saying the state has no jurisdiction to exact child support payments from them. Traditionally, Native men took care of their own. Do they pay their own to these women? I don't think so. I know better. How does that equation better the lives of our children? How is that (real) sovereignty?
>
> The U.S government is so hypocritical about recognizing sovereignty. And we, the Native community, fall into the same hypocrisy. I would argue the Feds only recognize Indian sovereignty when a first Nation has a casino or a waste dump, not when a tribal government seeks to preserve ground water from pesticide contamination, exercise jurisdiction over air quality, or stop clear- cutting or say no to a nuclear dump. "Sovereignty" has become a politicized term used for some of the most demeaning purposes.[10]

Beatrice Medicine similarly critiques the manner in which women's status is often pitted against sovereignty, as exemplified in the 1978 *Santa Clara Pueblo v. Martinez* case. Julia Martinez sued her tribe for sex discrimination under the Indian Civil Rights Act because the tribe had dictated that children born from female tribal members who married outside the tribe lost tribal status whereas children born from male tribal members who married outside the tribe did not. The Supreme Court ruled that the federal government could not intervene in this situation because the determination of tribal membership was the sovereign right of the tribe. On the one hand, many white feminists criticized the Supreme Court decision without considering how the Court's affirmation of the right of the federal government to determine tribal membership would constitute a significant attack against tribal sovereignty.[11] On the other hand, as Medicine notes, many tribes take this decision as a signal to institute gender-discriminatory practices under the name of sovereignty.[12] For these difficult issues, it is perhaps helpful to consider how they could be addressed if we put American Indian women at the center of analysis. Is it possible to simultaneously affirm tribal sovereignty and challenge tribes to consider how the impact of colonization and Europeanization may impact the decisions they make and programs they pursue in a manner which may ultimately undermine their sovereignty in the long term? Rather than adopt the strategy of fighting for sovereignty first and then improving Native women's status second, as Jaimes suggests, we must understand that attacks on Native women's status are themselves attacks on Native sovereignty. Lee Maracle illustrates the relationship between colonization and gender violence in Native communities in her groundbreaking work, *I Am Woman* (1988):

> If the State won't kill us
>
> we will have to kill ourselves.
>
> It is no longer good etiquette to
> head hunt savages.
>
> We'll just have to do it ourselves.
>
> It's not polite to violate "squaws"
>
> We'll have to find an Indian to
> oblige us.
>
> It's poor form to starve an Indian
>
> We'll have to deprive our young
> ourselves
>
> Blinded by niceties and polite
> liberality
>
> We can't see our enemy,
>
> so, we'll just have to kill each
> other.[13]

It has been through sexual violence and through the imposition of European gender relationships on Native communities that Europeans were able to colonize Native peoples in the first place. If we maintain these patriarchal gender systems in place, we are then unable to decolonize and fully assert our sovereignty.

Native Feminist Sovereignty Projects

Despite the political and theoretical straightjacket in which Native women often find themselves, there are several groundbreaking projects today that address both colonialism and sexism through an intersectional framework. One such attempt to tie indigenous sovereignty with the well-being of Native women is evident in the materials produced by the Sacred Circle, a national American Indian resource center for domestic and sexual violence based in South Dakota. Their brochure *Sovereign Women Strengthen Sovereign Nations* reads:

Another such project is the Boarding School Healing Project, which seeks to build a movement to demand reparations for U.S. boarding school abuses. This project, founded in 2002, is a coalition of indigenous groups across the United

Tribal Sovereignty *All Tribal Nations Have an Inherent Right to:*	*Native Women's Sovereignty* *All Native Women Have an Inherent Right to:*
1) A land base: possession and control is unquestioned and honored by other nations. To exist without fear, but with freedom.	1) Their body and path in life: the possession and control is unquestioned and honored by others. To exist without fear, but without freedom.
2) Self-governance: the ability and authority to make decisions regarding all matters concerning the Tribe without the approval or agreement of others. This includes the ways and methods of decision-making in social, political and other areas of life.	2) Self-governance: the ability and authority to make decisions regarding all matters concerning themselves, without others' approval or agreement. This includes the ways and methods of decision-making in social, political and other areas of life.
3) An economic base and resources: the control, use and development of resources, businesses or industries the Tribe chooses. This includes resources that support the Tribal life way, including the practice of spiritual ways.	3) An economic base and resources: the control, use and development of resources, businesses or industries that Native women choose. This includes resources that support individual Native women's chosen life ways, including the practice of spiritual ways.
4) A distinct language and historical and cultural identity: Each tribe defines and describes its history, including the impact of colonization and racism, tribal culture, worldview and traditions.	4) A distinct identity, history and culture: Each Native woman defines and describes her history, including the impact of colonization, racism and sexism, tribal women's culture, worldview and traditions.
******	******
Colonization and violence against Native people means that power and control over Native people's life way and land have been stolen. As Native people, we have the right and responsibility to advocate for ourselves and our relatives in supporting our right to power and control over our tribal life way and land tribal sovereignty.	*Violence against women, and victimization in general, means that power and control over an individual's life and body have been stolen. As relatives of women who have been victimized, it is our right and responsibility to be advocates supporting every woman's right to power and control over her body and life—personal sovereignty.*

States, such as the American Indian Law Alliance, Incite! Women of Color against Violence, Indigenous Women's Network, and Native Women of Sovereign Nations of the South Dakota Coalition against Domestic Violence and Sexual Assault. In Canada, Native peoples have been able to document the abuses of the residential school system and demand accountability from the Canadian government and churches. The same level of documentation has not taken place in the United States. The Boarding School Healing Project is documenting these abuses to build a movement for reparations and accountability. However, the strategy of this project is not to seek remedies on the individual level, but to demand collective remedy by developing links with other reparations struggles that fundamentally challenge the colonial and capitalist status quo. In addition, the strategy of this project is to organize around boarding schools as a way to address gender violence in Native communities.

That is, one of the harms suffered by Native peoples through state policy was sexual violence perpetrated by boarding school officials. The continuing effect of this human rights violation has been the internalization of sexual and other forms of gender violence *within* Native American communities. Thus, the question is, how can we form a demand around reparations for these types of continuing effects of human rights violations that are evidenced by violence *within* communities, but are nonetheless colonial legacies. In addition, this project attempts to organize against interpersonal gender violence *and* state violence simultaneously by framing gender violence as a continuing effect of human rights violations perpetrated by state policy. Consequently, this project challenges the mainstream anti-domestic/sexual violence movement to conceptualize state-sponsored sexual violence as central to its work. As I have argued elsewhere, the mainstream antiviolence movement has relied on the apparatus of state violence (in the form of the criminal justice system) to address domestic and sexual violence without considering how the state itself is a primary perpetrator of violence.[14] The issue of boarding schools forces us to see the connections between state violence and interpersonal violence. It is through boarding schools that gender violence in our communities was largely introduced. Before colonization, Native societies were, for the most part, not male dominated. Women served as spiritual, political, and military leaders. Many societies were matrilineal and matrilocal. Violence against women and children was infrequent or unheard of in many tribes.[15] Native peoples did not use corporal punishment against their children. Although there existed a division of labor between women and men, women's and men's labor was accorded similar status.[16] In boarding schools, by contrast, sexual/physical/emotional violence proliferated. Particularly brutalizing to Native children was the manner in which school officials involved children in punishing other children. For instance, in some schools, children were forced to hit other children with the threat that if they did not hit hard enough, they themselves would be severely beaten. Sometimes perpetrators of the violence were held accountable, but generally speaking, even when teachers were charged with abuse, boarding schools refused to investigate. In the case of just one teacher, John Boone at the Hopi school, FBI investigations in 1987 found that he had sexually abused more than 142 boys, but that the principal of that school had not investigated any allegations of abuse.[17] Despite the epidemic of sexual abuse in boarding schools, the Bureau of Indian Affairs did not issue a policy on reporting sexual abuse until 1987 and did not issue a policy to strengthen the background checks of potential teachers until 1989. Although not all Native peoples see their boarding school experiences as negative, it is generally the case that much if not most of the current dysfunctionality in Native communities can be traced to the boarding school era.

The effects of boarding school abuses linger today because these abuses have not been acknowledged by the larger society. As a result, silence continues within Native communities, preventing Native peoples from seeking support and healing as a result of the intergenerational trauma. Because boarding school policies are not acknowledged as human rights violations, Native peoples individualize the trauma they have suffered, thus contributing to increased shame and self-blame. If both boarding school policies and the continuing effects from these policies were recognized as human rights violations, then it might take away the shame from talking about these issues and thus provide an opportunity for communities to begin healing.

Unfortunately, we continue to perpetuate this colonial violence through domestic/sexual violence, child abuse, and homophobia. No amount of reparations will be successful if we do not address the oppressive behaviors we have internalized. Women of color have for too long been presented with the choices of either prioritizing racial justice or gender justice. This dualistic analysis fails to recognize that it is precisely through sexism and gender violence that colonialism and white supremacy have been successful. A question to ask ourselves then is, what would true reparations really look like for women of color who suffer state violence and interpersonal gender violence simultaneously? The Boarding School Healing Project provides an opportunity to organize around the connections between interpersonal gender violence and state violence that could serve as a model for the broader antiviolence movement.

In addition, this project makes important contributions to the struggle for reparations as a whole. That is, a reparations struggle is not necessarily radical if its demands do not call into question the capitalist and colonial status quo. What is at the heart of the issue is that no matter how much financial compensation the United States may give, such compensation does not ultimately end the colonial relationship between the United States and indigenous nations. What is at the heart of the struggle for native sovereignty is control over land and resources rather than financial compensation for past and continuing wrongs. If we think about reparations less in terms of financial compensation for social oppression and more about a movement to transform the neocolonial economic relationships between the United States and people of color, indigenous peoples, and Third World countries, we see how critical this movement could be to all of us. The articulation of reparations as a movement to cancel the Third World debt, for instance, is instructive in thinking of strategies that could fundamentally alter these relations.

Native Feminism and the Nation State

Native feminist theory and activism make a critical contribution to feminist politics as a whole by questioning the legitimacy of the United States specifically and the nation-state as the appropriate form of governance generally. Progressive activists and scholars, although prepared to make critiques of the U.S. government, are often not prepared to question its legitimacy. A case in point is the strategy of many racial justice organizations in the United States to rally against hate crimes resulting from the attacks of 9/11 under the banner, "We're American too." However, what the analysis of Native women activists suggests is that this implicit allegiance to "America" legitimizes the genocide and colonization of Native peoples, as there could be no "America" without this genocide. Thus by making anticolonial struggle central to feminist politics, Native women make central to their organizing the question of what is the appropriate form of governance for the world in general. Does self-determination for indigenous peoples equal aspirations for a nation-state, or are there other forms of governance we can create that are not based on domination and control?

Questioning the United States, in particular, and questioning the nation-state as the appropriate form of governance for the world, in general, allow us to free our political imagination to begin thinking of how we can begin to build a world we would actually want to live in. Such a political project is particularly important for colonized peoples seeking national liberation because it allows us to differentiate "nation" from "nation-state." Helpful in this project of imagination is the work of Native women activists who have begun articulating notions of nation and sovereignty that are separate from nation-states. Whereas nation-states are governed through domination and coercion, indigenous sovereignty and nationhood is predicated on interrelatedness and responsibility. As Crystal Ecohawk states:

> Sovereignty is an active, living process within this knot of human, material and spiritual relationships bound together by mutual responsibilities and obligations. From that knot of relationships is born our histories, our identity, the traditional ways in which we govern ourselves, our beliefs, our relationship to the land, and how we feed, clothe, house and take care of our families, communities and Nations.[18]

It is interesting to me . . . how often non-Indians presume that if Native people regained their landbases, that they would necessarily call for the expulsion of non-Indians from those landbases. Yet, it is striking that a much more inclusive vision of sovereignty is articulated by Native women activists. For instance, this activist describes how indigenous sovereignty is based on freedom for all peoples:

> If it doesn't work for one of us, it doesn't work for any of us. The definition of sovereignty [means that] . . . none of us are free unless all of us are free. We can't, we won't turn anyone away. We've been there. I would hear stories about the Japanese internment camps . . . and I could relate to it because

> it happened to us. Or with Africans with the violence and rape, we've been there too. So how could we ever leave anyone behind.

This analysis mirrors much of the work currently going on in women of color organizing in the United States and in other countries. Such models rely on this dual strategy of what Sista II Sista (Brooklyn) describes as "taking power" and "making power."[19] That is, it is necessary to engage in oppositional politics to corporate and state power ("taking power"). However, if we only engage in the politics of taking power, we will have a tendency to replicate the hierarchical structures in our movements. Consequently, it is also important to "make power" by creating those structures within our organizations, movements, and communities that model the world we are trying to create. Many groups in the United States often try to create separatist communities based on egalitarian ideals. However, if we "make power" without also trying to "take power" then we ultimately support the political status quo by failing to dismantle those structures of oppression that will undermine all our attempts to make power. The project of creating a new world governed by an alternative system not based on domination, coercion, and control does not depend on an unrealistic goal of being able to fully describe a utopian society for all at this point in time. From our position of growing up in a patriarchal, colonial, and white supremacist world, we cannot even fully imagine how a world not based on structures of oppression could operate. Nevertheless, we can be part of a collective, creative process that can bring us closer to a society not based on domination. To quote Jean Ziegler from the 2003 World Social Forum held in Porto Alegre, Brazil: "We know what we don't want, but the new world belongs to the liberated freedom of human beings. 'There is no way; you make the way as you walk.' History doesn't fall from heaven; we make history."

NOTES

1. Quotes that are not cited come from interviews conducted in Rapid City, New York City, Santa Cruz, Minneapolis, and Bemidji in 2001. These interviews are derived primarily from women involved in Women of All Red Nations (WARN) and the American Indian Movement (AIM). All are activists today.

2. M. Annette Jaimes and Theresa Halsey, "American Indian Women: At the Center of Indigenous Resistance in North America," in *State of Native America,* ed. M. Annette Jaimes (Boston: South End Press, 1992), 330–31.

3. Janet McCloud, "The Backbone of Everything," *Indigenous Woman* 1, no. 3 (n.d.): 50.

4. Ward Churchill, *Struggle for the Land* (Monroe, Maine: Common Courage Press, 1993), 419.

5. Wilma Mankiller, *Mankiller* (New York: St. Martin's Press, 1993), 241.

6. Andrea Smith, "Sexual Violence and American Indian Genocide," in *Remembering Conquest: Feminist/Womanist Perspectives on Religion, Colonization, and Sexual Violence*, ed. Nantawan Lewis and Marie Fortune (Binghamton, N.Y.: Haworth Press, 1999), 31–52.

7. Callie Rennison, "Violent Victimization and Race, 1993–1998" (Washington, D.C.: Bureau of Justice Statistics, 2001).

8. Lucy Eldersveld Murphy, "Autonomy and the Economic Roles of Indian Women of the Fox-Wisconsin Riverway Region, 1763–1832," in *Negotiators of Change: Historical Perspectives on Native American Women,* ed. Nancy Shoemaker (New York: Routledge Press, 1995), 72–89; Theda Purdue, "Women, Men, and American Indian Policy: The Cherokee Response to "Civilization," in *Negotiators of Change,* 90–114.

9. Paula Gunn Allen, *The Sacred Hoop* (Boston: Beacon Press, 1986), 202.

10. Winona LaDuke, "Don't Cheapen Sovereignty," *American Eagle* 4 (May 1996): n.d. www.alphacdc.com/eagle/op0596.html.

11. Catherine MacKinnon, *Feminism Unmodified* (Cambridge: Harvard University Press, 1987), 63–69.

12. Beatrice Medicine, "North American Indigenous Women and Cultural Domination," *American Indian Culture and Research Journal* 17, no. 3 (1993): 121–30.

13. Lee Maracle, *I Am Woman* (North Vancouver: Write-On Press Publishers, 1988).

14. Smith, "Sexual Violence and American Indian Genocide," 31–52.

15. Paula Gunn Allen, "Violence and the American Indian Woman," *The Speaking Profit*

Us (Seattle: Center for the Prevention of Sexual and Domestic Violence, n.d.), 5–7. See also *A Sharing: Traditional Lakota Thought and Philosophy Regarding Domestic Violence* (South Dakota: Sacred Shawl Women's Society, n.d.); and *Sexual Assault Is Not an Indian Tradition* (Minneapolis: Division of Indian Work Sexual Assault Project, n.d.).

16. See Jaimes and Halsey, "American Indian Women," 311–44; and Allen, *The Sacred Hoop.*

17. "Hello New Federalism, Goodbye BIA," *American Eagle* 4, no. 6(1994):19.

18. Crystal Echohawk, "Reflections on Sovereignty," *Indigenous Woman* 3, no. 1(1999):21–22.

19. Personal conversations with Sista II Sista members, ongoing from 2001–2005.

Introduction to Reading 48

R. W. Connell's article traces the emergence of a worldwide discussion of men and gender-equality reform and assesses the prospects of reform strategies involving men. Connell does so by locating recent policy discussions in the wider context of the cultural problematization of men and boys, the politics of "men's movements," the divided interests of men and boys in gender relations, and the increasing research evidence about the changing and conflict-ridden social construction of masculinities. Connell's analysis ranges from local to global, but the primary concern is with the global nature of debate about the role of men and boys in relations to gender equality.

1. Why is research on diverse social constructions of masculinity critical to worldwide efforts on behalf of achieving gender equality?
2. Connell states that men have a lot to lose from pursuing gender equality. However, Connell argues that men's advantages are conditions for the price they pay for their benefits. Discuss this notion.
3. What are the "grounds for optimism" and the "grounds for pessimism" set out in this reading?

Change Among the Gatekeepers

Men, Masculinities, and Gender Equality in the Global Arena

R. W. Connell

Equality between women and men has been a doctrine well recognized in international law since the adoption of the 1948 *Universal Declaration of Human Rights* (United Nations 1958), and as a principle it enjoys popular support in many countries. The idea of gender equal rights has provided the formal basis for the international discussion of the position of women since the 1975–85 UN Decade for Women, which has been a key element in the story of global feminism

Connell, R. W. 2005. "Change among the gatekeepers: Men, masculinities, and gender equality in the global arena" *Signs: Journal of Women in Culture and Society 30*(3). Reprinted with permission of The University of Chicago Press.

(Bulbeck 1988). The idea that men might have a specific role in relation to this principle has emerged only recently.

The issue of gender equality was placed on the policy agenda by women. The reason is obvious: it is women who are disadvantaged by the main patterns of gender inequality and who therefore have the claim for redress. Men are, however, necessarily involved in gender-equality reform. Gender inequalities are embedded in a multidimensional structure of relationships between women and men, which, as the modern sociology of gender shows, operates at every level of human experience, from economic arrangements, culture, and the state to interpersonal relationships and individual emotions (Holter 1997; Walby 1997; Connell 2002). Moving toward a gender-equal society involves profound institutional change as well as change in everyday life and personal conduct. To move far in this direction requires widespread social support, including significant support from men and boys.

Further, the very gender inequalities in economic assets, political power, and cultural authority, as well as the means of coercion, that gender reforms intend to change, currently mean that men (often specific groups of men) control most of the resources required to implement women's claims for justice. Men and boys are thus in significant ways gatekeepers for gender equality. Whether they are willing to open the gates for major reforms is an important strategic question.

In this article, I will trace the emergence of a worldwide discussion of men and gender-equality reform and will try to assess the prospects of reform strategies involving men. To make such an assessment, it is necessary to set recent policy discussions in the wider context of the cultural problematization of men and boys, the politics of "men's movements," the divided interests of men and boys in gender relations, and the growing research evidence about the changing and conflict-ridden social construction of masculinities.

In an article of this scope, it is not possible to address particular national agendas in detail. I will refer to a number of texts where these stories can be found. Because my primary concern is with the global character of the debate, I will give particular attention to policy discussions in UN forums. These discussions culminated in the 2004 meeting of the UN Commission on the Status of Women, which produced the first world-level policy document on the role of men and boys in relation to gender equality (UN Commission on the Status of Women 2004.)

Men and Masculinities in the World Gender Order

In the last fifteen years, in the "developed" countries of the global metropole, there has been a great deal of popular concern with issues about men and boys. Readers in the United States may recall a volume by the poet Robert Bly, *Iron John: A Book about Men* (1990), which became a huge best seller in the early 1990s, setting off a wave of imitations. This book became popular because it offered, in prophetic language, simple solutions to problems that were increasingly troubling the culture. A therapeutic movement was then developing in the United States, mainly though not exclusively among middle-class men, addressing problems in relationships, sexuality, and identity (Kupers 1993; Schwalbe 1996).

More specific issues about men and boys have also attracted public attention in the developed countries. Men's responses to feminism, and to gender-equality measures taken by government, have long been the subject of debate in Germany and Scandinavia (Metz-Göckel and Müller 1985; Holter 2003). In anglophone countries there has been much discussion of "the new fatherhood" and of supposed changes in men's involvement in families (McMahon 1999). There has been public agonizing about boys' "failure" in school, and in Australia there are many proposals for special programs for boys (Kenway 1997; Lingard 2003). Men's violence toward women has been the subject of practical interventions and extensive debate (Hearn 1998). There has also been increasing debate about men's health and illness from a gender perspective (Hurrelmann and Kolip 2002).

Accompanying these debates has been a remarkable growth of research about men's gender identities and practices, masculinities and the social processes by which they are constructed, cultural and media images of men, and related matters. Academic journals have been founded for specialized research on men and masculinities, there have been many research conferences, and there is a rapidly growing international literature. We now have a far more sophisticated and detailed scientific understanding of issues about men, masculinities, and gender than ever before (Connell 2003a).

This set of concerns, though first articulated in the developed countries, can now be found worldwide (Connell 2000; Pease and Pringle 2001). Debates on violence, patriarchy, and ways of changing men's conduct have occurred in countries as diverse as Germany, Canada, and South Africa (Hagemann-White 1992; Kaufman 1993; Morrell 2001a). Issues about masculine sexuality and fatherhood have been debated and researched in Brazil, Mexico, and many other countries (Arilha, Unbehaum Ridenti, and Medrado 1998; Lerner 1998). A men's center with a reform agenda has been established in Japan, where conferences have been held and media debates about traditional patterns of masculinity and family life continue (Menzu Senta 1997; Roberson and Suzuki 2003). A "traveling seminar" discussing issues about men, masculinities, and gender equality has recently been touring in India (Roy 2003). Debates about boys' education, men's identities, and gender change are active from New Zealand to Denmark (Law, Campbell, and Dolan 1999; Reinicke 2002). Debates about men's sexuality, and changing sexual identities, are also international (Altman 2001).

The research effort is also worldwide. Documentation of the diverse social constructions of masculinity has been undertaken in countries as far apart as Peru (Fuller 2001), Japan (Taga 2001), and Turkey (Sinclair-Webb 2000). The first large-scale comparative study of men and gender relations has recently been completed in ten European countries (Hearn et al. 2002). The first global synthesis, in the form of a world handbook of research on men and masculinities, has now appeared (Kimmel, Hearn, and Connell 2005).

The rapid internationalization of these debates reflects the fact—increasingly recognized in feminist thought (Bulbeck 1998; Marchand and Runyan 2000)—that gender relations themselves have an international dimension. Each of the substructures of gender relations can be shown to have a global dimension, growing out of the history of imperialism and seen in the contemporary process of globalization (Connell 2002). Change in gender relations occurs on a world scale, though not always in the same direction or at the same pace.

The complexity of the patterns follows from the fact that gender change occurs in several different modes. Most dramatic is the direct colonization of the gender order of regions beyond the metropole. There has also been a more gradual recomposition of gender orders, both those of the colonizing society and the colonized, in the process of colonial interaction. The hybrid gender identities and sexualities now much discussed in the context of postcolonial societies are neither unusual nor new. They are a feature of the whole history of imperialism and are visible in many contemporary studies (e.g., Valdés and Olavarría 1998).

Imperialism and globalization change the conditions of existence for gender orders. For instance, the linking of previously separate production systems changes the flow of goods and services in the gendered division of labor, as seen in the impact of industrially produced foods and textiles on household economies. Colonialism itself often confronted local patriarchies with colonizing patriarchies, producing a turbulent and sometimes very violent aftermath, as in southern Africa (Morrell 1998). Pressure from contemporary Western commercial culture has destabilized gender arrangements, and models of masculinity, in Japan (Ito 1992), the Arab world (Ghoussoub 2000), and elsewhere.

Finally, the emergence of new arenas of social relationship on a world scale creates new patterns of gender relations. Transnational corporations, international communications systems, global mass media, and international state structures (from the United Nations to the European Union) are such arenas. These institutions have their own gender regimes and may form the basis for new configurations of masculinity, as has recently been argued for transnational business (Connell 2000) and the international relations system (Hooper 2001). Local gender orders now interact not only with the gender orders of other local societies but also with the gender order of the global arena.

The dynamics of the world gender order affect men as profoundly as they do women, though this fact has been less discussed. The best contemporary research on men and masculinity, such as Matthew C. Gutmann's (2002) ethnographic work in Mexico, shows in fine detail how the lives of particular groups of men are shaped by globally acting economic and political dynamics.

Different groups of men are positioned very differently in such processes. There is no single formula that accounts for men and globalization. There is, indeed, a growing polarization among men on a world scale. Studies of the "super-rich" (Haseler 2000) show a privileged minority reaching astonishing heights of wealth and power while much larger numbers face poverty, cultural dislocation, disruption of family relationships, and forced renegotiation of the meanings of masculinity.

Masculinities, as socially constructed configurations of gender practice, are also created through a historical process with a global dimension. The old-style ethnographic research that located gender patterns purely in a local context is inadequate to the reality. Historical research, such as Robert Morrell's (2001b) study of the masculinities of the colonizers in South Africa and T. Dunbar Moodie's (1994) study of the colonized, shows how a gendered culture is created and transformed in relation to the international economy and the political system of empire. There is every reason to think this principle holds for contemporary masculinities.

Shifting Ground: Men and Boys in Gender-Equality Debates

Because of the way they came onto the agenda of public debate, gender issues have been widely regarded as women's business and of little concern to men and boys. In almost all policy discussions, to adopt a gender perspective substantially means to address women's concerns.

In both national and international policy documents concerned with gender equality, women are the subjects of the policy discourse. The agencies or meetings that formulate, implement, or monitor gender policies usually have names referring to women, such as Department for Women, Women's Equity Bureau, Prefectural Women's Centre, or Commission on the Status of Women. Such bodies have a clear mandate to act for women. They do not have an equally clear mandate to act with respect to men. The major policy documents concerned with gender equality, such as the UN *Convention on the Elimination of all Forms of Discrimination against Women* (United Nations [1979] 1989), often do not name men as a group and rarely discuss men in concrete terms.

However, men are present as background throughout these documents. In every statement about women's disadvantages, there is an implied comparison with men as the advantaged group. In the discussions of violence against women, men are implied, and sometimes named, as the perpetrators. In discussions of gender and HIV/AIDS, men are commonly construed as being "the problem," the agents of infection. In discussions of women's exclusion from power and decision making, men are implicitly present as the power holders.

When men are present only as a background category in a policy discourse about women, it is difficult to raise issues about men's and boys' interests, problems, or differences. This could be done only by falling into a backlash posture and affirming "men's rights" or by moving outside a gender framework altogether.

The structure of gender-equality policy, therefore, created an opportunity for antifeminist politics. Opponents of feminism have now found issues about boys and men to be fertile ground. This is most clearly seen in the United States, where authors such as Warren Farrell (1993) and Christina Hoff Sommers (2000), purporting to speak on behalf of men and boys, bitterly accuse feminism of injustice. Men and boys, they argue, are the truly disadvantaged group and need supportive programs in education and health, in situations of family breakup, and so forth. These ideas have not stimulated a social movement, with the exception of a small-scale (though active and sometimes violent) "father's rights" movement in relation to divorce. The arguments have, however, strongly appealed to the neoconservative mass media, which have given them international circulation. They now form part of the broad neoconservative repertoire of opposition to "political correctness" and to social justice measures.

Some policy makers have attempted to straddle this divide by restructuring gender-equality policy in the form of parallel policies for women and men. For instance, some recent health policy initiatives in Australia have added a "men's health" document to a "women's health" document (Schofield 2004). Similarly, in some school systems a "boys' education" strategy has been added to a "girls' education" strategy (Lingard 2003).

This approach acknowledges the wider scope of gender issues. But it also risks weakening the equality rationale of the original policy. It forgets the relational character of gender and therefore tends to redefine women and men, or girls and boys, simply as different market segments for some service. Ironically, the result may be to promote more gender segregation, not less. This has certainly happened in education, where some privileged boys' schools have jumped on the "gender equality" bandwagon and now market themselves as experts in catering to the special needs of boys.

On the other hand, bringing men's problems into an existing framework of policies for women may weaken the authority that women have so far gathered in that policy area. In the field of gender and development, for instance, some specialists argue that "bringing men in"—given the larger context in which men still control most of the wealth and institutional authority—may undermine, not help, the drive for gender equality (White 2000).

The role of men and boys in relation to gender equality emerged as an issue in international discussions during the 1990s. This development crystallized at the Fourth World Conference on Women, held in Beijing in 1995. Paragraph 25 of the *Beijing Declaration* committed participating governments to "encourage men to participate fully in all actions towards equality" (United Nations 2001). The detailed "Platform for Action" that accompanied the declaration prominently restated the principle of shared power and responsibility between men and women and argued that women's concerns could be addressed only "in partnership with men" toward gender equality (2001, pars. 1, 3). The "Platform for Action" went on to specify areas where action involving men and boys was needed and was possible: in education, socialization of children, child care and housework, sexual health, gender-based violence, and the balancing of work and family responsibilities (2001, pars. 40, 72, 83b, 107c, 108e, 120, 179).

Participating member states followed a similar approach in the twenty-third special session of the UN General Assembly in the year 2000, which was intended to review the situation five years after the Beijing conference. The "Political Declaration" of this session made an even stronger statement on men's responsibility: "[Member states of the United Nations] emphasise that men must involve themselves and take joint responsibility with women for the promotion of gender equality" (United Nations 2001, par. 6). It still remained the case, in this and the accompanying "Outcome Document," that men were present on the margins of a policy discourse concerned with women.

The role of men and boys has also been addressed in other recent international meetings.

These include the 1995 World Summit on Social Development, its review session in 2000, and the special session of the General Assembly on HIV/AIDS in 2001. In 1997 the UN Educational, Scientific, and Cultural Organization (UNESCO) convened an expert group meeting about "Male Roles and Masculinities in the Perspective of a Culture of Peace," which met in Oslo and produced studies on the links among personal violence, war, and the construction of masculinities (Breines, Connell, and Eide 2000).

International meetings outside the UN system have addressed similar issues. In 1997 the Nordic Council of Ministers adopted the *Nordic Action Plan for Men and Gender Equality.* In the same year the Council of Europe conducted a seminar on equality as a common issue for men and women and made the role of men in promoting equality a theme at a ministerial conference. In 1998 the Latin American Federation of Social Science (FLACSO) began a series of conferences about masculinities, boys, and men across Latin America and the Caribbean. The first conference in this series had the specific theme of gender equity (Valdés and Olavarría 1998). The European Commission has recently funded a research network on men and masculinities.

Divided Interests: Support and Resistance

There is something surprising about the worldwide problematizing of men and masculinities, because in many ways the position of men has not greatly changed. For instance, men remain a very large majority of corporate executives, top professionals, and holders of public office. Worldwide, men hold nine out of ten cabinet-level posts in national governments, nearly as many of the parliamentary seats, and most top positions in international agencies. Men, collectively, receive approximately twice the income that women receive and also receive the benefits of a great deal of unpaid household labor, not to mention emotional support, from women (Gierycz 1999; Godenzi 2000; Inter-Parliamentary Union 2003).

The UN Development Program (2003) now regularly incorporates a selection of such statistics into its annual report on world human development, combining them into a "gender-related development index" and a "gender empowerment measure." This produces a dramatic outcome, a league table of countries ranked in terms of gender equality, which shows most countries in the world to be far from gender-equal. It is clear that, globally, men have a lot to lose from pursuing gender equality because men, collectively, continue to receive a patriarchal dividend.

But this way of picturing inequality may conceal as much as it reveals. There are multiple dimensions in gender relations, and the patterns of inequality in these dimensions may be qualitatively different. If we look separately at each of the substructures of gender, we find a pattern of advantages for men but also a linked pattern of disadvantages or toxicity (Connell 2003c).

For instance, in relation to the gender division of labor, men collectively receive the bulk of income in the money economy and occupy most of the managerial positions. But men also provide the workforce for the most dangerous occupations, suffer most industrial injuries, pay most of the taxation, and are under heavier social pressure to remain employed. In the domain of power men collectively control the institutions of coercion and the means of violence (e.g., weapons). But men are also the main targets of military violence and criminal assault, and many more men than women are imprisoned or executed. Men's authority receives more social recognition (e.g., in religion), but men and boys are underrepresented in important learning experiences (e.g., in humanistic studies) and important dimensions of human relations (e.g., with young children).

One could draw up a balance sheet of the costs and benefits to men from the current gender order. But this balance sheet would not be like a corporate accounting exercise where there is a bottom line, subtracting costs from income. The disadvantages listed above are, broadly speaking, the conditions of the advantages. For instance, men cannot hold state power without some men becoming the agents of violence. Men cannot be the beneficiaries of women's domestic labor and "emotion work" without many of them losing intimate connections, for instance, with young children.

Equally important, the men who receive most of the benefits and the men who pay most of the costs are not the same individuals. As the old saying puts it, generals die in bed. On a global scale, the men who benefit from corporate wealth, physical security, and expensive health care are a very different group from the men who provide the workforce of developing countries. Class, race, national, regional, and generational differences cross-cut the category "men," spreading the gains and costs of gender relations very unevenly among men. There are many situations where groups of men may see their interest as more closely aligned with the women in their communities than with other men. It is not surprising that men respond very diversely to gender-equality politics.

There is, in fact, a considerable history of support for gender equality among men. There is certainly a tradition of advocacy by male intellectuals. In Europe, well before modern gender-equality documents were written, the British philosopher John Stuart Mill published "The Subjection of Women" (1912), which established the presumption of equal rights; and the Norwegian dramatist Henrik Ibsen, in plays like *A Doll's House* ([1923] 1995), made gender oppression an important cultural theme. In the following generation, the pioneering Austrian psychoanalyst Alfred Adler established a powerful psychological argument for gender equality (Connell 1995). A similar tradition of men's advocacy exists in the United States (Kimmel and Mosmiller 1992).

Many of the historic gains by women's advocates have been won in alliance with men who held organizational or political authority at the time. For instance, the introduction of equal employment opportunity measures in New South Wales, Australia, occurred with the strong support of the premier and the head of a reform inquiry into the public sector, both men (Eisenstein 1991). Sometimes men's support for gender equality takes the form of campaigning and organizing among men. The most prominent example is the U.S. National Organization of Men against Sexism (NOMAS), which has existed for more than twenty years (Cohen 1991). Men's groups concerned with reforming masculinity, publications advocating change, and campaigns among men against violence toward women are found widely, for instance, in the United Kingdom, Mexico, and South Africa (Seidler 1991; Zingoni 1998; Peacock 2003).

Men have also been active in creating educational programs for boys and young men intended to support gender reform. Similar strategies have been developed for adult men, sometimes in a religious and sometimes in a health or therapeutic context. There is a strong tradition of such work in Germany, with programs that combine the search for self-knowledge with the learning of antisexist behavior (Brandes and Bullinger 1996). Work of the same kind has developed in Brazil, the United States, and other countries (Denborough 1996; Lyra and Medrado 2001).

These initiatives are widespread, but they are also mostly small-scale. What of the wider state of opinion? European survey research has shown no consensus among men either for or against gender equality. Sometimes a third/third/third pattern appears, with about one-third of men supporting change toward equality, about one-third opposing it, and one-third undecided or intermediate (Holter 1997, 131–34). Nevertheless, examinations of the survey evidence from the United States, Germany, and Japan have shown a long-term trend of growing support for change, that is, a movement away from traditional gender roles, especially among members of the younger generation (Thornton 1989; Zulehner and Volz 1998; Mohwald 2002).

There is, however, also significant evidence of men's and boys' resistance to change in gender relations. The survey research reveals substantial levels of doubt and opposition, especially among older men. Research on workplaces and on corporate management has documented many cases where men maintain an organizational culture that is heavily masculinized and unwelcoming to women. In some cases there is active opposition to gender-equality measures or quiet undermining of them (Cockburn 1991; Collinson and Hearn 1996). Research on schools has also found cases where boys assert control of informal social life and direct hostility against girls and against boys perceived as being different. The status quo can be defended even in the details of classroom life, for instance, when a particular group of boys used misogynist language to resist study of a poem that questioned Australian gender stereotypes (Kenworthy 1994; Holland et al. 1998).

Some men accept change in principle but in practice still act in ways that sustain men's dominance of the public sphere and assign domestic labor and child care to women. In strongly gender segregated societies, it may be difficult for men to recognize alternatives or to understand women's experiences (Kandiyoti 1994; Fuller 2001; Meuser 2003). Another type of opposition to reform, more common among men in business and government, rejects gender-equality measures because it rejects all government action in support of equality, in favor of the unfettered action of the market.

The reasons for men's resistance include the patriarchal dividend discussed above and threats to identity that occur with change. If social definitions of masculinity include being the breadwinner and being "strong," then men may be offended by women's professional progress because it makes men seem less worthy of respect. Resistance may also reflect ideological defense of male supremacy. Research on domestic violence suggests that male batterers often hold very conservative views of women's role in the family (Ptacek 1988). In many parts of the world, there exist ideologies that justify men's supremacy on grounds of religion, biology, cultural tradition, or organizational mission (e.g., in the military). It is a mistake to regard these ideas as simply outmoded. They may be actively modernized and renewed.

Grounds for Optimism: Capacities for Equality and Reasons for Change

The public debates about men and boys have often been inconclusive. But they have gone a long way, together with the research, to shatter one widespread belief that has hindered gender reform. This obstacle is the belief that men *cannot* change their ways, that "boys will be boys," that rape, war, sexism, domestic violence, aggression, and self-centeredness are natural to men.

We now have many documented examples of the diversity of masculinities and of men's and boys' capacity for equality. For instance, life-history research in Chile has shown that there is no unitary Chilean masculinity, despite the cultural homogeneity of the country. While a hegemonic model is widely diffused across social strata, there are many men who depart from it, and there is significant discontent with traditional roles (Valdés and Olavarriá 1998). Though groups of boys in schools often have a dominant or hegemonic pattern of masculinity, there are usually also other patterns present, some of which involve more equal and respectful relations with girls.

Research in Britain, for instance, shows how boys encounter and explore alternative models of masculinity as they grow up (Mac an Ghaill 1994; O'Donnell and Sharpe 2000). Psychological and educational research shows personal flexibility in the face of gender stereotypes. Men and boys can vary, or strategically use, conventional definitions of masculinity. It is even possible to teach boys (and girls) how to do this in school, as experiments in Australian classrooms have shown (Davies 1993; Wetherell and Edley 1999).

Changes have occurred in men's practices within certain families, where there has been a conscious shift toward more equal sharing of housework and child care. The sociologist Barbara J. Risman (1998), who has documented such cases in one region of the United States, calls them "fair families." It is clear from her research that the change has required a challenge to traditional models of masculinity. In the Shanghai region of China, there is an established local tradition of relative gender equality, and men are demonstrably willing to be involved in domestic work. Research by Da Wei Wei (Da 2004) shows this tradition persisting among

Shanghai men even after migration to another country.

Perhaps the most extensive social action involving men in gender change has occurred in Scandinavia. This includes provisions for paternity leave that have had high rates of take-up, among the most dramatic of all demonstrations of men's willingness to change gender practices. Øystein Holter sums up the research and practical experience: "The Nordic 'experiment' has shown that a *majority* of men can change their practice when circumstances are favorable. . . . When reforms or support policies are well-designed and targeted towards an on-going cultural process of change, men's active support for gender-equal status increases" (1997, 126). Many groups of men, it is clear, have a capacity for equality and for gender change. But what reasons for change are men likely to see?

Early statements often assumed that men had the same interest as women in escaping from restrictive sex roles (e.g., Palme 1972). Later experience has not confirmed this view. Yet men and boys often do have substantial reasons to support change, which can readily be listed.

First, men are not isolated individuals. Men and boys live in social relationships, many with women and girls: wives, partners, mothers, aunts, daughters, nieces, friends, classmates, workmates, professional colleagues, neighbors, and so on. The quality of every man's life depends to a large extent on the quality of those relationships. We may therefore speak of men's relational interests in gender equality. For instance, very large numbers of men are fathers, and about half of their children are girls. Some men are sole parents and are then deeply involved in caregiving—an important demonstration of men's capacity for care (Risman 1986). Even in intact partnerships with women, many men have close relationships with their children, and psychological research shows the importance of these relationships (Kindler 2002). In several parts of the world, young men are exploring more engaged patterns of fatherhood (Olavarría 2001). To make sure that daughters grow up in a world that offers young women security, freedom, and opportunities to fulfill their talents is a powerful reason for many men to support gender equality.

Second, men may wish to avoid the toxic effects that the gender order has for them. James Harrison long ago issued a "Warning: The Male Sex Role May Be Dangerous to Your Health" (1978). Since then health research has documented specific problems for men and boys. Among them are premature death from accident, homicide, and suicide; occupational injury; higher levels of drug abuse, especially of alcohol and tobacco; and in some countries at least, a relative unwillingness by men to seek medical help when it is needed. Attempts to assert a tough and dominant masculinity sustain some of these patterns (Sabo and Gordon 1995; Hurrelmann and Kolip 2002).

Social and economic pressures on men to compete in the workplace, to increase their hours of paid work, and sometimes to take second jobs are among the most powerful constraints on gender reform. Desire for a better balance between work and life is widespread among employed men. On the other hand, where unemployment is high the lack of a paid job can be a damaging pressure on men who have grown up with the expectation of being breadwinners. This is, for instance, an important gender issue in post-Apartheid South Africa. Opening alternative economic paths and moving toward what German discussions have called "multioptional masculinities" may do much to improve men's well-being (*Widersprüche* 1998; Morrell 2001a).

Third, men may support gender change because they see its relevance to the well-being of the community they live in. In situations of mass poverty and underemployment, for instance in cities in developing countries, flexibility in the gender division of labor may be crucial to a household that requires women's earnings as well as men's. Reducing the rigidity of masculinities may also yield benefits in security. Civil and international violence is strongly associated with dominating patterns of masculinity and with marked gender inequality in the state. Movement away from these patterns makes it easier for men to adopt historically "feminine" styles of nonviolent negotiation and conflict resolution (Zalewski and Parpart 1998; Breines, Connell, and Eide 2000; Cockburn 2003). This may also reduce the toxic effects of policing and incarceration (Sabo, Kupers, and London 2001).

Finally, men may support gender reform because gender equality follows from their political or ethical principles. These may be religious, socialist, or broad democratic beliefs. Mill argued a case based on classical liberal principles a century and a half ago, and the idea of equal human rights still has purchase among large groups of men.

Grounds for Pessimism: The Shape of Masculinity Politics

The diversity among men and masculinities is reflected in a diversity of men's movements in the developed countries. A study of the United States found multiple movements, with different agendas for the remaking of masculinity. They operated on the varying terrains of gender equality, men's rights, and ethnic or religious identities (Messner 1997). There is no unified political position for men and no authoritative representative of men's interests.

Men's movements specifically concerned with gender equality exist in a number of countries. A well-known example is the White Ribbon Campaign, dedicated to mobilizing public opinion and educating men and boys for the prevention of men's violence against women. Originating in Canada, in response to the massacre of women in Montreal in 1989, the White Ribbon Campaign achieved very high visibility in that country, with support from political and community leaders and considerable outreach in schools and mass media. More recently, it has spread to other countries. Groups concerned with violence prevention have appeared in other countries, such as Men against Sexual Assault in Australia and Men Overcoming Violence (MOVE) in the United States. These have not achieved the visibility of the White Ribbon Campaign but have built up a valuable body of knowledge about the successes and difficulties of organizing among men (Lichterman 1989; Pease 1997; Kaufman 1999).

The most extensive experience of any group of men organizing around issues of gender and sexual politics is that of homosexual men, in antidiscrimination campaigns, the gay liberation movement, and community responses to the HIV/AIDS pandemic. Gay men have pioneered in areas such as community care for the sick, community education for responsible sexual practices, representation in the public sector, and overcoming social exclusion, which are important for all groups of men concerned with gender equality (Kippax et al. 1993; Altman 1994).

Explicit backlash movements also exist but have not generally had a great deal of influence. Men mobilizing as men to oppose women tend to be seen as cranks or fanatics. They constantly exaggerate women's power. And by defining men's interests in opposition to women's, they get into cultural difficulties, since they have to violate a main tenet of modern patriarchal ideology—the idea that "opposites attract" and that men's and women's needs, interests, and choices are complementary.

Much more important for the defense of gender inequality are movements in which men's interests are a side effect—nationalist, ethnic, religious, and economic movements. Of these, the most influential on a world scale is contemporary neoliberalism—the political and cultural promotion of free-market principles and individualism and the rejection of state control.

Neoliberalism is in principle gender neutral. The "individual" has no gender, and the market delivers advantage to the smartest entrepreneur, not to men or women as such. But neoliberalism does not pursue social justice in relation to gender. In Eastern Europe, the restoration of capitalism and the arrival of neoliberal politics have been followed by a sharp deterioration in the position of women. In rich Western countries, neoliberalism from the 1980s on has attacked the welfare state, on which far more women than men depend; supported deregulation of labor markets, resulting in increased casualization of women workers; shrunk public sector employment, the sector of the economy where women predominate; lowered rates of personal taxation, the main basis of tax transfers to women; and squeezed public education, the key pathway to labor market advancement for women. However, the same period saw an expansion of the human-rights agenda, which is, on the whole, an asset for gender equality.

The contemporary version of neoliberalism, known as neoconservatism in the United States, also has some gender complexities. George W. Bush was the first U.S. president to place a woman in the very heart of the state security apparatus, as national security adviser to the president. And some of the regime's actions, such as the attack on the Taliban regime in Afghanistan, were defended as a means of emancipating women.

Yet neoconservatism and state power in the United States and its satellites such as Australia remain overwhelmingly the province of men—indeed, men of a particular character: power oriented and ruthless, restrained by little more than calculations of likely opposition. There has been a sharp remasculinization of political rhetoric and a turn to the use of force as a primary instrument in policy. The human rights discourse is muted and sometimes completely abandoned (as in the U.S. prison camp for Muslim captives at Guantanamo Bay and the Australian prison camps for refugees in the central desert and Pacific islands).

Neoliberalism can function as a form of masculinity politics largely because of the powerful role of the state in the gender order. The state constitutes gender relations in multiple ways, and all of its gender policies affect men. Many mainstream policies (e.g., in economic and security affairs) are substantially about men without acknowledging this fact (Nagel 1998; O'Connor, Orloff, and Shaver 1999; Connell 2003b).

This points to a realm of institutional politics where men's and women's interests are very much at stake, without the publicity created by social movements. Public-sector agencies (Jensen 1998; Mackay and Bilton 2000; Schofield, forthcoming), private-sector corporations (Marchand and Runyan 2000; Hearn and Parkin 2001), and unions (Corman et al. 1993; Franzway 2001) are all sites of masculinized power and struggles for gender equality. In each of these sites, some men can be found with a commitment to gender equality, but in each case that is an embattled position. For gender-equality outcomes, it is important to have support from men in the top organizational levels, but this is not often reliably forthcoming.

One reason for the difficulty in expanding men's opposition to sexism is the role of highly conservative men as cultural authorities and managers. Major religious organizations, in Christianity, Islam, and Buddhism, are controlled by men who sometimes completely exclude women, and these organizations have often been used to oppose the emancipation of women. Transnational media organizations such as Rupert Murdoch's conglomerate are equally active in promoting conservative gender ideology.

A specific address to men is found in the growing institutional, media, and business complex of commercial sports. With its overwhelming focus on male athletes; its celebration of force, domination, and competitive success; its valorization of male commentators and executives; and its marginalization and frequent ridicule of women, the sports/business complex has become an increasingly important site for representing and defining gender. This is not traditional patriarchy. It is something new, welding exemplary bodies to entrepreneurial culture. Michael Messner (2002), one of the leading analysts of contemporary sports, formulates the effect well by saying that commercial sports define the renewed centrality of men and of a particular version of masculinity.

On a world scale, explicit backlash movements are of limited importance, but very large numbers of men are nevertheless engaged in preserving gender inequality. Patriarchy is defended diffusely. There is support for change from equally large numbers of men, but it is an uphill battle to articulate that support. That is the political context with which new gender-equality initiatives have to deal.

Ways Forward: Toward a Global Framework

Inviting men to end men's privileges, and to remake masculinities to sustain gender equality, strikes many people as a strange or utopian project. Yet this project is already under way. Many men around the world are engaged in gender reforms, for the good reasons discussed above.

The diversity of masculinities complicates the process but is also an important asset. As this diversity becomes better known, men and boys can more easily see a range of possibilities for their own lives, and both men and women are less likely to think of gender inequality as unchangeable. It also becomes possible to identify specific groups of men who might engage in alliances for change.

The international policy documents discussed above rely on the concept of an alliance between men and women for achieving equality. Since the growth of an autonomous women's movement, the main impetus for reform has been located in women's groups. Some groups within the women's movement, especially those concerned with men's violence, are reluctant to work with men or are deeply skeptical of men's willingness to change. Other feminists argue that alliances between women and men are possible, even crucial. In some social movements, for instance, environmentalism, there is a strong ideology of gender equality and a favorable environment for men to support gender change (Connell 1995; Segal 1997).

In local and central government, practical alliances between women and men have been important in achieving equal-opportunity measures and other gender-equality reforms. Even in the field of men's violence against women, there has been cooperation between women's groups and men's groups, for instance, in prevention work. This cooperation can be an inspiration to grassroots workers and a powerful demonstration of women and men's common interest in a peaceful and equal society (Pease 1997; Schofield, forthcoming). The concept of alliance is itself important, in preserving autonomy for women's groups, in preempting a tendency for any one group to speak for others, and in defining a political role for men that has some dignity and might attract widespread support.

Given the spectrum of masculinity politics, we cannot expect worldwide consensus for gender equality. What is possible is that support for gender equality might become hegemonic among men. In that case it would be groups supporting equality that provide the agenda for public discussion about men's lives and patterns of masculinity.

There is already a broad cultural shift toward a historical consciousness about gender, an awareness that gender customs came into existence at specific moments in time and can always be transformed by social action (Connell 1995). What is needed now is a widespread sense of agency among men, a sense that this transformation is something they can actually share in as a practical proposition. This is precisely what was presupposed in the "joint responsibility" of men invoked by the General Assembly declaration of the year 2000.[1]

From this point of view, the recent meeting of the UN Commission on the Status of Women (CSW) is profoundly interesting. The CSW is one of the oldest of UN agencies, dating from the 1940s. Effectively a standing committee of the General Assembly, it meets annually, and its current practice is to consider two main themes at each meeting. For the 2004 meeting, one of the defined themes was "the role of men and boys in achieving gender equality." The section of the UN secretariat that supports the CSW, the Division for the Advancement of Women, undertook background work. The division held, in June–July 2003, a worldwide online seminar on the role of men and boys, and in October 2003 it convened an international expert group meeting in Brasilia on the topic.

At the CSW meetings, several processes occur and (it is to be hoped) interact. There is a presentation of the division's background work, and delegations of the forty-five current member countries, UN agencies, and many of the nongovernmental organizations (NGOs) attending make initial statements. There is a busy schedule of side events, mainly organized by NGOs but some conducted by delegations or UN agencies, ranging from strategy debates to practical workshops. And there is a diplomatic process in which the official delegations negotiate over a draft document in the light of discussions in the CSW and their governments' stances on gender issues.

This is a politicized process, inevitably, and it can break down. In 2003 the CSW discussion on the issue of violence against women reached deadlock. In 2004 it was clear that some participating NGOs were not happy with the focus on men and boys, some holding to a discourse representing men exclusively as perpetrators of violence. Over the two weeks of negotiations, however, the delegations did reach consensus on a statement of "Agreed Conclusions."

Balancing a reaffirmation of commitment to women's equality with a recognition of men's and boys' potential for action, this document makes specific recommendations across a spectrum of policy fields, including education, parenthood, media, the labor market, sexuality, violence, and conflict prevention. These proposals have no force in international law—the document is essentially a set of recommendations to governments and other organizations. Nevertheless, it is the first international agreement of its kind, treating men systematically as agents in gender-equality processes, and it creates a standard for future gender-equality discussions. Most important, the CSW's "Agreed Conclusions" change the logic of the representation of men in gender policy. So far as the international discourse of gender-equality policy is concerned, this document begins the substantive presentation of gender equality as a positive project for men.

Here the UN process connects with the social and cultural possibilities that have emerged from the last three decades of gender politics among men. Gender equality is an undertaking for men that can be creative and joyful. It is a project that realizes high principles of social justice, produces better lives for the women whom men care about, and will produce better lives for the majority of men in the long run. This can and should be a project that generates energy, that finds expression in everyday life and the arts as well as in formal policies, and that can illuminate all aspects of men's lives.

Note

1. Twenty-third special session, UN General Assembly, "Political Declaration," par. 6.

References

Altman, Dennis. 1994. *Power and Community: Organizational and Cultural Responses to AIDS.* London: Taylor & Francis.

———.2001. *Global Sex.* Chicago: University of Chicago Press.

Arilha, Margareth, Sandra G. Unbehaum Ridenti, and Benedito Medrado, eds. 1998. *Homens e Masculinidades: Outras Palavras.* Sao Paulo: ECOS/Editora 34.

Bly, Robert. 1990. *Iron John: A Book about Men.* Reading, MA: Addison-Wesley.

Brandes, Holger, and Hermann Bullinger, eds. 1996. *Handbuch Männerarbeit.* Weinheim, Germany: Psychologic Verlags Union.

Breines, Ingeborg, Robert Connell, and Ingrid Eide, eds. 2000. *Male Roles, Masculinities and Violence: A Culture of Peace Perspective.* Paris: UNESCO.

Bulbeck, Chilla. 1988. *One World Women's Movement.* London: Pluto.

———. 1998. *Re-orienting Western Feminisms: Women's Diversity in a Postcolonial World.* Cambridge: Cambridge University Press.

Cockburn, Cynthia. 1991. *In the Way of Women: Men's Resistance to Sex Equality in Organizations.* Ithaca, NY: ILR Press.

———. 2003. *The Line: Women, Partition and the Gender Order in Cyprus.* London: Zed.

Cohen, Jon. 1991. "NOMAS: Challenging Male Supremacy." *Changing Men* (Winter/Spring): 45–46.

Collinson, David L., and Jeff Hearn, eds. 1996. *Men as Managers, Managers as Men: Critical Perspectives on Men, Masculinities and Managements.* London: Sage.

Connell, R. W. 1995. *Masculinities.* Berkeley: University of California Press.

———. 2000. *The Men and the Boys.* Sydney: Allen & Unwin Australia.

———. 2002. *Gender.* Cambridge: Polity.

———. 2003a. "Masculinities, Change and Conflict in Global Society: Thinking about the Future of Men's Studies." *Journal of Men's Studies* 11(3):249–66.

———. 2003b. "Men, Gender and the State." In *Among Men: Moulding Masculinities,* ed. Søren Ervø and Thomas Johansson, 15–28. Aldershot: Ashgate.

———. 2003c. "Scrambling in the Ruins of Patriarchy: Neo-liberalism and Men's Divided Interests in Gender Change." In *Gender—from Costs to Benefits,* ed. Ursula Pasero, 58–69. Wiesbaden: Westdeutscher.

Corman, June, Meg Luxton, D. W. Livingstone, and Wally Seccombe. 1993. *Recasting Steel Labour: The Stelco Story.* Halifax: Fernwood.

Da Wei Wei. 2004. "A Regional Tradition of Gender Equity: Shanghai Men in Sydney." *Journal of Men's Studies* 12(2):133–49.

Davies, Bronwyn. 1993. *Shards of Glass: Children Reading and Writing beyond Gender Identities.* Sydney: Allen & Unwin Australia.

Denborough, David. 1996. "Step by Step: Developing Respectful and Effective Ways of Working with Young Men to Reduce Violence." In *Men's Ways of Being,* ed. Chris McLean, Maggie Carey, and Cheryl White, 91–115. Boulder, CO: Westview.

Eisenstein, Hester. 1991. *Gender Shock: Practising Feminism on Two Continents.* Sydney: Allen & Unwin Australia.

Farrell, Warren. 1993. *The Myth of Male Power: Why Men Are the Disposable Sex.* New York: Simon & Schuster.

Franzway, Suzanne. 2001. *Sexual Politics and Greedy Institutions.* Sydney: Pluto.

Fuller, Norma. 2001. "The Social Constitution of Gender Identity among Peruvian Men." *Men and Masculinities* 3(3):316–31.

Ghoussoub, Mai. 2000. "Chewing Gum, Insatiable Women and Foreign Enemies: Male Fears and the Arab Media." In *Imagined Masculinities: Male Identity and Culture in the Middle East,* ed. Mai Ghoussoub and Emma Sinclair-Webb, 227–35. London: Saqi.

Gierycz, Dorota. 1999. "Women in Decision-Making: Can We Change the Status Quo?" In *Towards a Women's Agenda for a Culture of Peace,* ed. Ingeborg Breines, Dorota Gierycz, and Betty A. Reardon, 19–30. Paris: UNESCO.

Godenzi, Alberto. 2000. "Determinants of Culture: Men and Economic Power." In Breines, Connell, and Eide 2000, 35–51. Paris: UNESCO.

Gutmann, Matthew C. 2002. *The Romance of Democracy: Compliant Defiance in Contemporary Mexico.* Berkeley: University of California Press.

Hagemann-White, Carol. 1992. *Strategien gegen Gewalt im Geschlechterverhältnis: Bestandsanalyse und Perspektiven.* Pfaffenweiler, Ger.: Centaurus.

Harrison, James. 1978. "Warning: The Male Sex Role May Be Dangerous to Your Health." *Journal of Social Issues* 34(1):65–86.

Haseler, Stephen. 2000. *The Super-Rich: The Unjust New World of Global Capitalism.* London: Macmillan.

Hearn, Jeff. 1998. *The Violences of Men: How Men Talk about and How Agencies Respond to Men's Violence to Women.* Thousand Oaks, CA: Sage.

Hearn, Jeff, and Wendy Parkin. 2001. *Gender, Sexuality, and Violence in Organizations: The Unspoken of Organization Violations.* Thousand Oaks, CA: Sage.

Hearn, Jeff, Keith Pringle, Ursula Müller, Elzbeieta Oleksy, Emmi Lattu, Janna Chernova, Harry Ferguson, et al. 2002. "Critical Studies on Men in Ten European Countries: (1) The State of Academic Research." *Men and Masculinities* 4(4):380–408.

Holland, Janet, Caroline Ramazanoğlu, Sue Sharpe, and Rachel Thomson. 1998. *The Male in the Head: Young People, Heterosexuality and Power.* London: Tufnell.

Holter, Øystein Gullvåg. 1997. *Gender, Patriarchy and Capitalism: A Social Forms Analysis.* Oslo: Work Research Institute.

———. 2003. *Can Men Do It? Men and Gender Equality—the Nordic Experience.* Copenhagen: Nordic Council of Ministers.

Hooper, Charlotte. 2001. *Manly States: Masculinities, International Relations, and Gender Politics.* New York: Columbia University Press.

Hurrelmann, Klaus, and Petra Kolip, eds. 2002. *Geschlecht, Gesundheit und Krankheit: Männer und Frauen im Vergleich.* Bern: Hans Huber.

Ibsen, Henrik. (1923) 1995. *A Doll's House.* Cambridge: Cambridge University Press.

Inter-Parliamentary Union. 2003. "Women in National Parliaments: Situation at 30 December 2003." Available online at http://www.ipu.org/wmn–e/world.htm.

Ito, Kimio, 1992. "Cultural Change and Gender Identity Trends in the 1970s and 1980s." *International Journal of Japanese Sociology* 1 (1):79–98.

Jensen, Hanne Naxø. 1998. "Gender as the Dynamo: When Public Organizations Have to Change." In *Is There a Nordic Feminism? Nordic Feminist Thought on Culture and Society,* ed. Drude von der Fehr, Bente Rosenberg, and Anna G. Jóasdóttir, 160–75. London: UCL Press.

Kandiyoti, Deniz. 1994. "The Paradoxes of Masculinity: Some Thoughts on Segregated Societies." In *Dislocating Masculinity: Comparative Ethnographies,* ed. Andrea Cornwall and Nancy Lindisfarne, 197–213. London: Routledge.

Kaufman, Michael. 1993. *Cracking the Armour: Power, Pain and the Lives of Men.* Toronto: Viking.

———, ed. 1999. "Men and Violence." Special issue, *International Association for Studies of Men Newsletter* 6, no. 2.

Kenway, Jane, ed. 1997. *Will Boys Be Boys? Boys' Education in the Context of Gender Reform.* Canberra: Australian Curriculum Studies Association.

Kenworthy, Colin. 1994. "'We want to resist your resistant readings': Masculinity and Discourse in the English Classroom." *Interpretations* 27(2):74–95.

Kimmel, Michael S., Jeff Hearn, and R. W. Connell, eds. 2005. *Handbook of Studies on Men and Masculinities.* Thousand Oaks, CA: Sage.

Kimmel, Michael S., and Thomas E. Mosmiller. 1992. *Against the Tide: Pro-feminist Men in the United States, 1776–1990: A Documentary History.* Boston: Beacon.

Kindler, Heinz. 2002. *Väter und Kinder.* Weinheim, Germany: Juventa.

Kippax, Susan, R. W. Connell, G. W. Dowsett, and June Crawford. 1993. *Sustaining Safe Sex: Gay Communities Respond to AIDS.* London: Falmer.

Kupers, Terry. 1993. *Revisioning Men's Lives: Gender, Intimacy, and Power.* New York: Guilford.

Law, Robin, Hugh Campbell, and John Dolan, eds. 1999. *Masculinities in Aotearo/New Zealand.* Palmerston North, NZ: Dunmore.

Lerner, Susana, ed. 1998. *Varones, sexualidad y reproducción: Diversas perspectivas teórico-metodológicas y hallazgos de investigación.* El Colegio de México, México.

Lichterman, Paul. 1989. "Making a Politics of Masculinity." *Comparative Social Research* 11:185–208.

Lingard, Bob. 2003. "Where to in Gender Policy in Education after Recuperative Masculinity Politics?" *International Journal of Inclusive Education* 7(1):33–56.

Lyra, Jorge, and Benedito Medrado. 2001. "Constructing an Adolescent Father in Brazil." Paper presented at the Third International Fatherhood Conference, Atlanta, May 28–30.

Mac an Ghaill, Mairtin. 1994. *The Making of Men: Masculinities, Sexualities and Schooling.* Buckingham: Open University Press.

Mackay, Fiona, and Kate Bilton. 2000. *Learning from Experience: Lessons in Mainstreaming Equal Opportunities.* Edinburgh: Governance of Scotland Forum.

Marchand, Marianne H., and Anne Sisson Runyan, eds. 2000. *Gender and Global Restructuring: Sightings, Sites and Resistances.* London: Routledge.

McMahon, Anthony. 1999. *Taking Care of Men: Sexual Politics in the Public Mind.* Cambridge: Cambridge University Press.

Menzu Senta (Men's Center Japan). 1997. *Otokotachi no watashisagashi* (How are men seeking their new selves?). Kyoto: Kamogawa.

Messner, Michael A. 1997. *The Politics of Masculinities: Men in Movements.* Thousand Oaks, CA: Sage.

———. 2002. *Taking the Field: Women, Men and Sports.* Minneapolis: University of Minnesota Press.

Metz–Göckel, Sigrid, and Ursula Müller. 1985. *Der Mann: Die Brigitte–Studie.* Hamburg: Beltz.

Meuser, Michael. 2003. "Modernized Masculinities? Continuities, Challenges, and Changes in Men's Lives." In *Among Men: Moulding Masculinities,* vol. 1, ed. Søren Ervø and Thomas Johansson, 127–48. Aldershot: Ashgate.

Mill, John Stuart. 1912. "The Subjection of Women." In his *On Liberty; Representative Government; The Subjugation of Women: Three Essays,* 427–548. London: Oxford University Press.

Mohwald, Ulrich. 2002. *Changing Attitudes towards Gender Equality in Japan and Germany.* Munich: Iudicium.

Moodie, T. Dunbar. 1994. *Going for Gold: Men, Mines and Migration.* Johannesburg: Witwatersrand University Press.

Morrell, Robert. 1998. "Of Boys and Men: Masculinity and Gender in Southern African Studies." *Journal of Southern African Studies 24*(4):605–30.

———, ed. 2001a. *Changing Men in Southern Africa.* Pietermaritzburg, S.A.: University of Natal Press.

———. 2001b. *From Boys to Gentlemen: Settler Masculinity in Colonial Natal, 1880–1920.* Pretoria: University of South Africa Press.

Nagel, Joane. 1998. "Masculinity and Nationalism: Gender and Sexuality in the Making of Nations." *Ethnic and Racial Studies 21*(2):242–69.

Nordic Council of Ministers. 1997. *Nordic Action Plan for Men and Gender Equality, 1997–2000.* Copenhagen: Nordic Council of Ministers.

O'Connor, Julia S., Ann Shola Orloff, and Sheila Shaver. 1999. *States, Markets, Families: Gender, Liberalism and Social Policy in Australia, Canada, Great Britain, and the United States.* Cambridge: Cambridge University Press.

O'Donnell, Mike, and Sue Sharpe. 2000. *Uncertain Masculinities: Youth, Ethnicity and Class in Contemporary Britain.* London: Routledge.

Olavarría, José. 2001. *Y todos querian ser (buenos) padres: Varones de Santiago de Chile en conflicto.* Santiago: FLACSO-Chile.

Palme, Olof. 1972. "The Emancipation of Man." *Journal of Social Issues 28*(2): 237–46.

Peacock, Dean. 2003. "Building on a Legacy of Social Justice Activism: Enlisting Men as Gender Justice Activists in South Africa." *Men and Masculinities 5*(3): 325–28.

Pease, Bob. 1997. *Men and Sexual Politics: Towards a Profeminist Practice.* Adelaide: Dulwich Centre.

Pease, Bob, and Keith Pringle, eds. 2001. *A Man's World? Changing Mens in a Globalized World.* London: Zed.

Ptacek, James. 1988. "Why Do Men Batter Their Wives?" In *Feminist Perspectives on Wife Abuse,* ed. Kersti Yllö and Michele Bograd, 133–57. Newbury Park, CA: Sage.

Reinicke, Kenneth. 2002. *Den Hele Mand: Manderollen i forandring.* Aarhus, Denmark: Schønberg.

Risman, Barbara J. 1986. "Can Men 'Mother'? Life as a Single Father." *Family Relations 35*(1):95–102.

———. 1998. *Gender Vertigo: American Families in Transition.* New Haven, CT: Yale University Press.

Roberson, James E., and Nobue Suzuki, eds. 2003. *Men and Masculinities in Contemporary Japan: Dislocating the Salaryman Doxa.* London: Routledge.

Roy, Rahul. 2003. "Exploring Masculinities—a Travelling Seminar." Unpublished manuscript.

Sabo, Donald, and David Frederick Gordon, eds. 1995. *Men's Health and Illness: Gender, Power, and the Body.* Thousand Oaks, CA: Sage.

Sabo, Donald, Terry A. Kupers, and Willie London, eds. 2001. *Prison Masculinities.* Philadelphia: Temple University Press.

Schofield, Toni. 2004. *Boutique Health? Gender and Equity in Health Policy.* Sydney: Australian Health Policy Institute.

———. Forthcoming. "Gender Regimes in Public Policy Making." Unpublished manuscript, Faculty of Health Sciences, University of Sydney.

Schwalbe, Michael. 1996. *Unlocking the Iron Cage: The Men's Movement, Gender, Politics, and American Culture.* New York: Oxford University Press.

Segal, Lynne. 1997. *Slow Motion: Changing Masculinities, Changing Men.* 2nd ed. London: Virago.

Seidler, Victor T., ed. 1991. *The Achilles Heel Reader: Men, Sexual Politics and Socialism.* London: Routledge.

Sinclair-Webb, Emma. 2000. "'Our bülent is now a commando': Military Service and Manhood in Turkey." In *Imagined Masculinities: Male Identity and Culture in the Modern Middle East,* ed. Mai Ghoussoub and Emma Sinclair-Webb, 65–92. London: Saqi.

Sommers, Christina Hoff. 2000. *The War against Boys: How Misguided Feminism Is Harming Our Young Men.* New York: Simon & Schuster.

Taga, Futoshi. 2001. *Dansei no Jendâ Keisei: "Otoko-Rashisa" no Yuragi no Naka de* (The gender formation of men: Uncertain masculinity). Tokyo: Tôyôkan Shuppan-sha.

Thornton, Arland. 1989. "Changing Attitudes toward Family Issues in the United States." *Journal of Marriage and the Family 51*(4):873–93.

United Nations. 1958. *Universal Declaration of Human Rights.* New York: Department of Public Information, United Nations.

———. (1979) 1989. *Convention on the Elimination of All Forms of Discrimination against Women.* New York: Department of Public Information, United Nations.

———. 2001. *Beijing Declaration and Platform for Action, with the Beijing +5 Political Declaration and Outcome Document.* New York: Department of Public Information, United Nations.

United Nations Commission on the Status of Women. 2004. *The Role of Men and Boys in Achieving Gender Equality: Agreed Conclusions.* Available online at http://www.un.org/womenwatch/daw/csw/csw48/ac-men-auv.pdf.

United Nations Development Program (UNDP). 2003. *Human Development Report 2003.* New York: UNDP and Oxford University Press.

Valdés, Teresa, and José Olavarría. 1998. "Ser hombre en Santiago de Chile: A pesar de todo, un mismo modelo." In their *Masculinidades y equidad de género en América Latina,* 12–36. Santiago: FLACSO/UNFPA.

Walby, Sylvia. 1997. *Gender Transformations.* London: Routledge.

Wetherell, Margaret, and Nigel Edley. 1999. "Negotiating Hegemonic Masculinity: Imaginary Positions and Psycho-Discursive Practices." *Feminism and Psychology* 9(3):335–56.

White, Sara C. 2000. "Did the Earth Move? The Hazards of Bringing Men and Masculinities into Gender and Development." *IDS Bulletin 31*(2):33–41.

Widersprüche. 1998. "Multioptionale Männlichkeiten?" Special issue, no. 67.

Zalewski, Marysia, and Jane Parpart, eds. 1998. *The "Man" Question in International Relations.* Boulder, CO: Westview.

Zingoni, Eduardo Liendro. 1998. "Masculinidades y violencia desde un programa de accíon en México." In *Masculinidades y equidad de género en America Latina,* ed. Teresa Valdés and José Olavarría, 130–36. Santiago: FLACSO/UNFPA.

Zulehner, Paul M., and Rainer Volz. 1998. *Männer im Aufbruch: Wie Deutschlands Männer sich Selbst und wie Frauen Sie Sehen.* Ostfildern, Ger.: Schwabenverlag.

Introduction to Reading 49

This reading is valuable for the insight that Allan Johnson offers into: (1) the intellectual and emotional obstacles that stand in the way of individuals understanding positive social change, and (2) how individuals can overcome those obstacles in order to contribute to change. He offers specific suggestions for becoming involved in unraveling the gender knot and moving our own lives, and the lives of those around us, toward equal worth and justice for all people. Johnson also addresses the value of finding the courage and taking the risks to plant the seeds of change.

1. What are the two myths about social change that get in the way of individuals understanding change and participating in it?
2. Why do oppressive systems, such as patriarchy, often seem stable?
3. Which of Johnson's suggestions for how to participate in positive change appeal to you and why?

UNRAVELING THE GENDER KNOT

Allan Johnson

What is the knot we want to unravel? In one sense, it is the complexity of patriarchy as a system—the tree, from its roots to the smallest outlying twig. It is misogyny and sexist ideology that keep women in their place. It is the organization of social life around core patriarchal principles of control and domination. It is the powerful dynamic of fear and control that keeps the patriarchal engine going. But the knot is also about our individual and collective paralysis around gender issues. It is everything that prevents us from seeing patriarchy and our participation in it clearly, from the denial that patriarchy even exists to false gender parallels, individualistic thinking, and cycles of blame and guilt. Stuck in this paralysis, we can't think or act to help undo the legacy of oppression.

To undo the patriarchal knot we have to undo the knot of our paralysis in the face of it. A good place to begin is with two powerful myths about how change happens and how we can contribute to it.

Myth #1: "It's Always Been This Way, and It Always Will Be"

Given thousands of years of patriarchal history, it's easy to slide into the belief that things have always been this way. Even thousands of years, however, are a far cry from what "always" implies unless we ignore the more than 90 percent of humanity's time on Earth that preceded it. Given all the archaeological evidence pointing to the existence of goddess-based civilizations and the lack of evidence for perpetual patriarchy, there are plenty of reasons to doubt that life has always been organized around male dominance or any other form of oppression. . . . So, when it comes to human social life, the smart money should be on the idea that nothing has always been this way or any other.

This should suggest that nothing *will* be this way or any other, contrary to the notion that patriarchy is here to stay. If the only thing we can count on is change, then it's hard to see why we should believe for a minute that patriarchy or any other kind of social system is permanent. Reality is always in motion. Things may appear to stand still, but that's only because we have short attention spans, limited especially by the length of a human life. If we take the long view—the *really* long view—we can see that everything is in process all the time. Some would argue that everything *is* process, the space between one point and another, the movement from one thing toward another. What we may see as permanent end points—world capitalism, Western civilization, advanced technology, and so on—are actually temporary states on the way to other temporary states. Even ecologists, who used to talk about ecological balance, now speak of ecosystems as inherently unstable. Instead of always returning to some steady state after a period of disruption, ecosystems are, by nature, a continuing process of change from one arrangement to another and never go back to just where they were.

Social systems are also fluid. A society isn't some hulking *thing* that sits there forever as it is. Because a system only happens as people participate in it, it can't help but *be* a dynamic process of creation and recreation from one moment to the next. In something as simple as a man following the path of least resistance toward controlling conversations (and a woman letting him do it), the reality of patriarchy in that moment comes into being. This is how we *do* patriarchy, bit by bit, moment by moment. It is also how individuals can contribute to change—by choosing paths of *greater* resistance, as when men resist the urge toward control and women resist their own subordination. Since we can always choose paths of greater resistance or create new ones entirely, systems can only be as stable as the flow of human choice and creativity, which certainly isn't a recipe for permanence. In the short run, patriarchy may look stable and unchangeable. But the relentless process of social life never produces the exact same result twice in a row, because it's impossible for everyone to participate in any system in an unvarying and uniform way. Added to this are the dynamic interactions that go on among systems—between capitalism and the state, for example, or between families and the economy—that also produce powerful and unavoidable tensions, contradictions, and other currents of change. Ultimately, systems can't help but change, whether we see it or not.

Oppressive systems often *seem* stable because they limit our lives and imaginations so much that we can't see beyond them. But this masks a fundamental long-term instability caused by the dynamics of oppression itself. Any system organized around control is a losing proposition because it contradicts the essentially uncontrollable nature of reality and does such violence to basic human needs and values. As the last two centuries of feminist thought and action have begun to challenge the violence and break down the denial, patriarchy has become increasingly vulnerable. This is one reason why male resistance, backlash, and defensiveness are now so intense. . . .

Patriarchy is also destabilized as the illusion of masculine control breaks down. Corporate leaders alternate between arrogant optimism and panic, while governments lurch from one crisis to another, barely managing to stay in office, much less solving major social problems such as poverty, violence, health care, middle-class angst, and the excesses of global capitalism. Computer technology supposedly makes life and work more efficient, but it does so by chaining people to an escalating pace of work and giving them less rather than more control over their lives. The loss of control in pursuit of control is happening on a larger level, as well. As the patriarchal obsession with control deepens its grip on everything from governments and corporations to schools and religion, the overall degree of control actually becomes less, not more. The scale on which systems are out of control simply increases. The stakes are higher and the capacity for harm is greater, and together they fuel an upward spiral of worry, anxiety, and fear.

As the illusion of control becomes more apparent, men start doubting their ability to measure up to patriarchal standards of manhood. We have been here before. At the turn of the twentieth century, there was widespread white male panic in the United States about the "feminization" of society and the need to preserve masculine toughness. From the creation of the Boy Scouts to Teddy Roosevelt's Rough Riders, a public campaign tried to revitalize masculinity as a cultural basis for revitalizing a male-identified society and, with it, male privilege. A century later, the masculine backlash is again in full bloom. The warrior image has re-emerged as a dominant masculine ideal, from *Rambo, Diehard,* and *Under Siege* to right-wing militia groups to corporate takeovers to regional militarism to New Age Jungian archetypes in the new men's movement.[1]

Neither patriarchy nor any other system will last forever. Patriarchy is riddled with internal contradiction and strain. It is based on the false and self-defeating assumption that control is the answer to everything and that the pursuit of more control is always better than contenting ourselves with less. The transformation of patriarchy has been unfolding ever since it emerged seven thousand years ago, and it is going on still. We can't know what will replace it, but we can be confident that patriarchy will go, that it *is* going at every moment. It's only a matter of how quickly, by what means, and toward what alternatives, and whether each of us will do our part to make it happen sooner rather than later and with less rather than more human suffering in the process.

Myth #2: The Myth of No Effect and Gandhi's Paradox

Whether we help change patriarchy depends on how we handle the belief that nothing we do can make a difference, that the system is too big and powerful for us to affect it. In one sense the complaint is valid: if we look at patriarchy as a whole, it's true that we aren't going to make it go away in our lifetime. But if changing the entire system through our own efforts is the standard against which we measure the ability to do something, then we've set ourselves up to feel powerless. It's not unreasonable to want to make a difference, but if we have to see the final result of what we do, then we can't be part of change that's too gradual and long term to allow that. We also can't be part of change that's so complex that we can't sort out our contribution from countless others that combine in ways we can never grasp. Problems like patriarchy are of just that sort, requiring complex and long-term change coupled with short-term work to soften some of its worst consequences. This means that if we're going to be part of the solution to such problems, we have to let go of the idea that change doesn't happen unless we're around to see it happen and that what we do matters only if we make it happen. In other words, if we free ourselves of the expectation of being in control of things, we free ourselves to act and participate in the kind of fundamental change that transforms social life.

To get free of the paralyzing myth that we cannot, individually, be effective, we have to change how we see ourselves in relation to a long-term, complex process of change. This begins by changing how we relate to time. Many changes can come about quickly enough for us to see them happen. When I was in college, for example, there was little talk about gender inequality as a social problem, whereas now there are women's studies programs all over the country. But a goal like ending gender oppression takes more than this and far more time than our short lives can encompass. If we're going to see ourselves as part of that kind of change, we can't use the human life span as a significant standard against which to measure progress. . . .

[W]e need to get clear about how our choices matter and how they don't. Gandhi once said that nothing we do as individuals matters, but that it's vitally important that we do it anyway. This touches on a powerful paradox in the relationship between society and individuals. In terms of the patriarchy-as-tree metaphor, no individual leaf on the tree matters; whether it lives or dies has no effect on much of anything. But collectively, the leaves are essential to the whole tree because they photosynthesize the sugar that feeds it. Without leaves, the tree dies. So, leaves both matter and they don't, just as we matter and we don't. What each of us does may not seem like much, because in important ways, it *isn't* much. But when many people do this work together, they can form a critical mass that is anything but insignificant, especially in the long run. If we're going to be part of a larger change process, we have to learn to live with this sometimes uncomfortable paradox rather than going back and forth between momentary illusions of potency and control and feelings of helpless despair and insignificance.

A related paradox is that we have to be willing to travel without knowing where we're going. We need faith to do what seems right without necessarily knowing the effect that will have. We have to think like pioneers who may know the *direction* they want to move in or what they would like to find, without knowing where they will wind up.

Because they are going where they've never been before, they can't know whether they will ever arrive at anything they might consider a destination, much less what they had in mind when they first set out. If pioneers had to know their destination from the beginning, they would never go anywhere or discover anything. In similar ways, to seek out alternatives to patriarchy, it has to be enough to move *away* from social life organized around dominance and control and to move *toward* the certainty that alternatives are possible, even though we may not have a clear idea of what those are or ever experience them ourselves. It has to be enough to question how we think about and experience different forms of power, for example, how we see ourselves as gendered people, how oppression works and how we participate in it, and then open ourselves to experience what happens next. When we dare ask core questions about who we are and how the world works, things happen that we can't foresee; but they don't happen unless we *move,* if only in our minds. As pioneers, we discover what's possible only by first putting ourselves in motion, because we have to move in order to change our position—and hence our perspective—on where we are, where we've been, and where we *might* go. This is how alternatives begin to appear: to imagine how things might be, we first have to get past the idea that things will always be the way they are.

In relation to Gandhi's paradox, the myth of no effect obscures the role we can play in the long-term transformation of patriarchy. But the myth also blinds us to our own power in relation to other people. We may cling to the belief that there is nothing we can do precisely because we know how much power we do have and are afraid to use it because people may not like it. If we deny our power to affect people, then we don't have to worry about taking responsibility for how we use it or, more significant, how we don't. This reluctance to acknowledge and use power comes up in the simplest everyday situations, as when a group of friends starts laughing at a sexist joke and we have to decide whether to go along. It's a moment in a sea of countless such moments that constitutes the fabric of all kinds of oppressive systems. It is a crucial moment, because the group's seamless response to the joke reaffirms the normalcy and unproblematic nature of it and the sexism behind it. It takes only one person to tear the fabric of collusion and apparent consensus. . . .

Our power to affect other people isn't simply about making them feel uncomfortable. Systems shape the choices that people make primarily by providing paths of least resistance. We typically follow those paths because alternatives offer greater resistance or because we aren't even aware that alternatives exist. Whenever we openly choose a different path, however, we make it possible for people to see both the path of least resistance they're following and the possibility of choosing something else. This is both radical and simple. When most people get on an elevator, for example, they turn and face front without ever thinking why. We might think it's for purely practical reasons—the floor indicators and the door we'll exit through are at the front. But there's more going on than that, as we'd discover if we simply walked to the rear wall and stood facing it while everyone else faced front. The oddness of what we were doing would immediately be apparent to everyone, and would draw their attention and perhaps make them uncomfortable as they tried to figure out why we were doing that. Part of the discomfort is simply calling attention to the fact that we make choices when we enter social situations and that there are alternatives, something that paths of least resistance discourage us from considering. If the possibility of alternatives in situations as simple as where to stand in elevator cars can make people feel uncomfortable, imagine the potential for discomfort when the stakes are higher, as they certainly are when it comes to how people participate in oppressive systems like patriarchy.

If we choose different paths, we usually won't know if we affect other people, but it's safe to assume that we do. When people know that alternatives exist and witness other people choosing them, things become possible that weren't before. When we openly pass up a path of least resistance, we *increase* resistance for other people around that path because now they must reconcile their choice with what they've seen us do, something they didn't have to deal with before. There's no way to predict how this will play out in the long run, and certainly no good reason to think it won't make a difference.

The simple fact is that we affect one another all the time without knowing it. . . . This suggests that the simplest way to help others make different choices is to make them myself, and to do it openly so they can see what I'm doing. As I shift the patterns of my own participation in patriarchy, I make it easier for others to do so as well, *and harder for them not to.* Simply by setting an example—rather than trying to change them—I create the possibility of their participating in change in their own time and in their own way. In this way I can widen the circle of change without provoking the kind of defensiveness that perpetuates paths of least resistance and the oppressive systems they serve.

It's important to see that in doing this kind of work we don't have to go after people to change their minds. In fact, changing people's minds may play a relatively small part in changing systems like patriarchy. We won't succeed in turning diehard misogynists into practicing feminists. At most, we can shift the odds in favor of new paths that contradict core patriarchal values. We can introduce so many exceptions to patriarchal rules that the children or grandchildren of diehard misogynists will start to change their perception of which paths offer the least resistance. Research on men's changing attitudes toward the male provider role, for example, shows that most of the shift occurs *between* generations, not within them.[2] This suggests that rather than trying to change people, the most important thing we can do is contribute to the slow sea change of entire cultures so that patriarchal forms and values begin to lose their "obvious" legitimacy and normalcy and new forms emerge to challenge their privileged place in social life.

In science, this is how one paradigm replaces another.[3] For hundreds of years, for example, Europeans believed that the stars, planets, and sun revolved around Earth. But scientists such as Copernicus and Galileo found that too many of their astronomical observations were anomalies that didn't fit the prevailing paradigm: if the sun and planets revolved around Earth, then they wouldn't move as they did. As such observations accumulated, they made it increasingly difficult to hang on to an Earth-centered paradigm. Eventually the anomalies became so numerous that Copernicus offered a new paradigm, for which he, and later Galileo, were persecuted as heretics. Eventually, however, the evidence was so overwhelming that a new paradigm replaced the old one.

In similar ways, we can think of patriarchy as a system based on a paradigm that shapes how we think about gender and how we organize social life in relation to it. The patriarchal paradigm has been under attack for several centuries and the defense has been vigorous, with feminists widely regarded as heretics who practice the blasphemy of "male bashing." The patriarchal paradigm weakens in the face of mounting evidence that it doesn't work, and that it produces unacceptable consequences not only for women but, increasingly, for men as well. We help to weaken it by openly choosing alternative paths in our everyday lives and thereby providing living anomalies that don't fit the prevailing paradigm. By our example, we contradict patriarchal assumptions and their legitimacy over and over again. We add our choices and our lives to tip the scales toward new paradigms that don't revolve around control and oppression. We can't tip the scales overnight or by ourselves, and in that sense we don't amount to much. But on the other side of Gandhi's paradox, it is crucial where we "choose to place the stubborn ounces of [our] weight."[4] It is in such small and humble choices that patriarchy and the movement toward something better actually happen.

Stubborn Ounces: What Can We Do?

* * *

What can we do about patriarchy that will make a difference? I don't have the answers, but I do have some suggestions.

Acknowledge that Patriarchy Exists

A key to the continued existence of every oppressive system is people being unaware of what's going on, because oppression contradicts so many basic human values that it invariably arouses opposition when people know about it. The Soviet Union and its East European satellites, for example, were riddled with contradictions that were so widely known among their people that the oppressive regimes fell apart with barely a whimper when given half a chance. An awareness of oppression compels people to speak out, breaking the silence on which continued oppression depends. This is why most oppressive cultures mask the reality of oppression by denying its existence, trivializing it, calling it something else, blaming it on those most victimized by it, or drawing attention away from it to other things. . . .

It's one thing to become aware and quite another to stay that way. The greatest challenge when we first become aware of a critical perspective on the world is simply to hang on to it. Every system's paths of least resistance invariably lead *away* from critical awareness of how the system works. Therefore, the easiest thing to do after reading a book like this is to forget about it. Maintaining a critical consciousness takes commitment and work; awareness is something we either maintain in the moment or we don't. And the only way to hang on to an awareness of patriarchy is to make paying attention to it an ongoing part of our lives.

Pay Attention

Understanding how patriarchy works and how we participate in it is essential for change. It's easy to have opinions; it takes work to know what we're talking about. The easiest place to begin is by reading, and making reading about patriarchy part of our lives. Unless we have the luxury of a personal teacher, we can't understand patriarchy without reading, just as we need to read about a foreign country before we travel there for the first time, or about a car before we try to work under the hood. Many people assume they already know what they need to know about gender since everyone has a gender, but they're usually wrong. Just as the last thing a fish would discover is water, the last thing we'll discover is society itself and something as pervasive as gender dynamics. We have to be open to the idea that what we think we know about gender is, if not wrong, so deeply shaped by patriarchy that it misses most of the truth. This is why feminists talk with one another and spend time reading one another's work—seeing things clearly is tricky business and hard work. This is also why people who are critical of the status quo are so often self-critical as well: they know how complex and elusive the truth really is and what a challenge it is to work toward it. People working for change are often accused of being orthodox and rigid, but in practice they are typically among the most self-critical people around. . . .

Reading, though, is only a beginning. At some point we have to look at ourselves and the world to see if we can identify what we're reading about. Once the phrase "paths of least resistance" entered my active vocabulary, for example, I started seeing them all over the place. Among other things, I started to see how easily I'm drawn to asserting control as a path of least resistance in all kinds of situations. Ask me a question, for example, and the easiest thing for me to do is offer an answer whether or not I know what I'm talking about. "Answering" is a more comfortable mode, an easier path, than admitting I don't know or have nothing to say.[5] The more aware I am of how powerful this path is, the more I can decide whether to go down it each time it presents itself. As a result, I listen more, think more, and talk less than I used to. . . .

Little Risks: Do Something

The more we pay attention to what's going on, the more we will see opportunities to do something about it. We don't have to mount an expedition to find those opportunities; they're all over the place, beginning in ourselves. As I became aware of how I gravitated toward controlling conversations, for example, I also realized how easily men dominate group meetings by controlling the agenda and interrupting, without women objecting to it. This pattern is especially striking in groups that are mostly female but in which most of the talking nonetheless comes from a few men. I would find myself sitting in meetings and suddenly the preponderance of male voices would jump out at me, an unmistakable hallmark of male privilege in full bloom. As I've seen what's going on, I've had to decide what to do about this little path of least resistance and my relation to it that leads me to follow it so readily. With some effort, I've tried out new ways of listening more and talking less. At times it's felt contrived and artificial, like telling myself to shut up for a while or even counting slowly to ten (or more) to give others a chance to step into the space afforded by silence. With time and practice, new paths have become easier to follow and I spend less time monitoring myself. But awareness is never automatic or permanent, for patriarchal paths of least resistance will be there to choose or not as long as patriarchy exists.

As we see more of what's going on, questions come up about what goes on at work, in the media, in families, in communities, in religion, in government, on the street, and at school—in short, just about everywhere. The questions don't come all at once (for which we can be grateful), although they sometimes come in a rush that can feel overwhelming. If we remind ourselves that it isn't up to us to do it all, however, we can see plenty of situations in which we can make a difference, sometimes in surprisingly simple ways. Consider the following possibilities:

- *Make noise be seen.* Stand up, volunteer, speak out, write letters, sign petitions, show up. Like every oppressive system, patriarchy feeds on silence. Don't collude in silence. . . .

- *Find little ways to withdraw support from paths of least resistance and people's choices to follow them, starting with ourselves.* It can be as simple as not laughing at a sexist joke or saying we don't think it's funny; or writing a letter to the editor objecting to sexism in the media. . . .

- *Dare to make people feel uncomfortable, beginning with ourselves.* At the next local school board meeting, for example, we can ask why principals and other administrators are almost always men (unless your system is an exception that proves the rule), while the teachers they control are mostly women. Consider asking the same thing about church, workplaces, or local government. . . .

It may seem that such actions don't amount to much until we stop for a moment and feel our resistance to doing them—our worry, for example, about how easily we could make people feel uncomfortable, including ourselves. If we take that resistance to action as a measure of power, then our potential to make a difference is plain to see. The potential for people to feel uncomfortable is a measure of the power for change inherent in such simple acts of not going along with the status quo.

Some will say that it isn't "nice" to make people uncomfortable, but oppressive systems like patriarchy do a lot more than make people feel uncomfortable, and it certainly isn't "nice" to allow them to continue unchallenged. Besides, discomfort is an unavoidable part of any meaningful process of education. We can't grow without being willing to challenge our assumptions and take ourselves to the edge of our competencies, where we're bound to feel uncomfortable. If we can't tolerate ambiguity, uncertainty, and discomfort, then we'll never go beneath the superficial appearance of things or learn or change anything of much value, including ourselves.

- *Openly choose and model alternative paths.* As we identify paths of least resistance—such as women being held responsible for child care and other domestic work—we can identify alternatives and then follow them openly so that other people can see what we're doing. Patriarchal paths become more visible when people choose alternatives, just as rules become more visible when someone breaks them. Modeling new paths creates tension in a system, which moves toward resolution. . . .

- *Actively promote change in how systems are organized around patriarchal values and male privilege.* There are almost endless possibilities here because social life is complicated and patriarchy is everywhere. We can, for example,

—Speak out for equality in the workplace.

—Promote diversity awareness and training.

—Support equal pay and promotion for women.

—Oppose the devaluing of women and the work they do, from the dead-end jobs most women are stuck in to the glass ceilings that keep women out of top positions.

—Support the well-being of mothers and children and defend women's right to control their bodies and their lives.

—Object to the punitive dismantling of welfare and attempts to limit women's access to reproductive health services.

—Speak out against violence and harassment against women wherever they occur, whether at home, at work, or on the street.

—Support government and private support services for women who are victimized by male violence.

—Volunteer at the local rape crisis center or battered women's shelter.

—Call for and support clear and effective sexual harassment policies in workplaces, unions, schools, professional associations, churches, and political parties, as well as public spaces such as parks, sidewalks, and malls.

—Join and support groups that intervene with and counsel violent men.

—Object to theaters and video stores that carry violent pornography. . . .

—Ask questions about how work, education, religion, family, and other areas of family life are shaped by core patriarchal values and principles. . . .

• *Because the persecution of gays and lesbians is a linchpin of patriarchy, support the right of women and men to love whomever they choose.* Raise awareness of homophobia and heterosexism. . . .

• *Because patriarchy is rooted in principles of domination and control, pay attention to racism and other forms of oppression that draw from those same roots. . . .*

[P]atriarchy isn't problematic just because it emphasizes *male* dominance, but because it promotes dominance and control as ends in themselves. In that sense, all forms of oppression draw support from common roots, and whatever we do that draws attention to those roots undermines *all* forms of oppression. If working against patriarchy is seen simply as enabling some women to get a bigger piece of the pie, then some women probably will "succeed" at the expense of others who are disadvantaged by race, class, ethnicity, and other characteristics. . . . [I]f we identify the core problem as *any* society organized around principles of control and domination, then changing *that* requires us to pay attention to all of the forms of oppression those principles promote. Whether we begin with race or gender or ethnicity or class, if we name the problem correctly, we'll wind up going in the same general direction.

• *Work with other people.* This is one of the most important principles of participating in social change. From expanding consciousness to taking risks, it makes all the difference in the world to be in the company of people who support what we are trying to do. We can read and talk about books and issues and just plain hang out with other people who want to understand and do something about patriarchy. Remember that the modern women's movement's roots were in consciousness-raising groups in which women did little more than sit around and talk about themselves and their lives and try to figure out what that had to do with living in patriarchy. It may not have looked like much at the time, but it laid the foundation for huge social movements. One way down this path is to share a book like this one with someone and then talk about it. Or ask around about local groups and organizations that focus on gender issues, and go find out what they're about and meet other people. . . . Make contact; connect to other people engaged in the same work; do whatever reminds us that we aren't alone in this.

• *Don't keep it to ourselves.* A corollary of looking for company is not to restrict our focus to the tight little circle of our own lives. It isn't enough to work out private solutions to social problems like patriarchy and other forms of oppression and keep them to ourselves. It isn't enough to clean up our own acts and then walk away, to find ways to avoid the worst consequences of patriarchy at home and inside ourselves and think that's taking responsibility. Patriarchy and oppression aren't personal problems and they can't be solved through personal solutions. At some point, taking responsibility means acting in a larger context, even if that means just letting one other person know what we're doing. It makes sense to start with ourselves; but it's equally important not to *end* with ourselves.

If all of this sounds overwhelming, remember again that we don't have to deal with everything. We don't have to set ourselves the impossible task of letting go of everything or transforming patriarchy or even ourselves. All we can do is what *we* can *manage* to do, secure in the knowledge that we're making it easier for other people—now and in the future—to see and do what *they* can do. So, rather than defeat ourselves before we start:

- *Think small, humble, and doable rather than large, heroic, and impossible.* Don't paralyze yourself with impossible expectations. It takes very little to make a difference. . . .
- *Don't let other people set the standard for us.*
- *Start where we are and work from there. . . . set reasonable goals* ("What small risk for change will I take *today?*"). As we get more experienced at taking risks, we can move up our lists. . . .

In the end, taking responsibility doesn't have to be about guilt and blame, about letting someone off the hook or being on the hook ourselves. It is simply to acknowledge our obligation to make a contribution to finding a way out of patriarchy, and to find constructive ways to act on that obligation. We don't have to do anything dramatic or Earth-shaking to help change happen. As powerful as patriarchy is, like all oppressive systems, it cannot stand the strain of lots of people doing something about it, beginning with the simplest act of speaking its name out loud.

* * *

Notes

1. See James William Gibson, *Warrior Dreams: Violence and Manhood in Post-Vietnam America* (New York: Hill and Wang, 1994).
2. J. R. Wilkie, "Changes in U.S. Men's Attitudes Towards the Family Provider Role, 1972–1989." *Gender & Society* 7, no. 2 (1993): 261–279.
3. The classic statement of how this happens is by Thomas S. Kuhn, *The Structure of Scientific Revolutions* (Chicago: University of Chicago Press, 1970).
4. This is a line from a poem by Bonaro Overstreet that was given to me by a student many years ago. I have not been able to locate the source.
5. Or, as someone once said to me about a major corporation that valued creative thinking, "It's not OK to say you don't know the answer to a question here."

Topics for Further Examination

- Look up research on men who participated in the first and second waves of feminism in the United States. Check out the following Web sites: http://www.feminist.com/men.htm and http://www.nomas.org.
- Browse Web sites on women's organizations and gender issues such as: http://www.iwpr.org, http://www.feminist.org, http://www.un.org/wom-enwatch/, and http://www.wilpf.org.
- Locate articles on the impact of globalization on human rights. Go to the following Web sites: www.amnesty.org/ and www.un.org/rights/.

EPILOGUE

Possibilities

This book began with the metaphor of the kaleidoscope to aid in understanding the complex and dynamic nature of gender. Viewed kaleidoscopically, gender is not static. Gender patterns are social constructions, reconstructed as they intersect with multiple and changing social prisms such as race, ethnicity, culture, class, and sexuality. In concluding, we want to emphasize the dynamic nature of gender, underscoring the possibilities that the future holds. No one can predict the future; therefore, we will illustrate changes in the institution of gender using stories of how changing gender patterns shaped our lives and lives of many other people.

Reflecting on the course of our lives, we are struck by the depth and breadth of changes that have occurred in American culture, institutions, and social relationships since we were young girls in the 1950s. Many of these changes have been positive; patterns of oppression were reduced and the opportunity and power to participate meaningfully in America's social institutions were extended to more people. In the last fifty years, we have benefited from, and participated in, bringing about changes in the genderscape.

The cultural climate of the 1950s forged our early lives. Though romanticized in film and TV, that decade was in fact a deeply troubled time of blatant racism, sexism, and other forms of oppression. Civil rights had not yet been extended to people of color, women's rights were negligible, gay and lesbian Americans were largely closeted, poverty was ignored, political dissent was strongly discouraged, child abuse went unacknowledged or hidden, and an atomic war seemed ready to break out at any moment.

We didn't learn about the women's movement then, even though there had been significant organized social movements for gender change beginning almost one hundred years before we were born, culminating in the right to vote in 1920. The second wave of feminism and subsequent women's movements began when we were very young, after World War II, inspired by books such as Simone de Beauvoir's *The Second Sex* (1952), first published in France in 1949 and Betty Friedan's *The Feminine Mystique* (1963). Although not widely recognized, the United Nations Charter of 1945 affirmed the equal rights of men and women and established the U.N. Commission on the Status of Women a few years later (Schneir, 1994). However, in many ways, women were still second-class citizens in the 1950s.

We grew up in families similar to those of many White Americans of the 1950s. Our parents held traditional views of proper roles for women and men. They tried to live up to those roles, and yet, like many, they often failed or fell short. They suffered with their failures in silence, behind the closed doors of the nuclear family of that era. For men and women, marriage and children came with the gender territory. Social sanctions in society maintained this territory. For example, people called women who didn't marry "old maids" pejoratively, and looked at couples who didn't bear children suspiciously. Also, there were limited reproductive control options, and abortion, although widespread, was illegal. This left most women and many men with few options except for getting married and having children.

Jessie Bernard (1975) described men and women's roles in White, middle-class families of that time as being destructive to adults and children alike. She spoke of "the work intoxicated father" and the "pathogenic mother" as the end result of the efforts to fulfill the cultural imperative for a traditional family. Bernard observed that middle-class, White family roles at that time were mostly stressful and unsatisfying. Men detached from their families as they struggled to earn enough for the household, while women shouldered the sole responsibility of raising perfect children and keeping a perfect home. Family and gender researchers (e.g., Bernard, 1972; Coontz, 1992 and 1997; Rapp, 1983; Schwartz, 1994) found that this arrangement of distinct and separate roles did not foster full and loving relationships between men and women or parents and children. And, of course, by now you can guess that most people could not achieve this "perfect family."

However, television still reflected the image of the happy family behind the white picket fence and for decades it stood as an ideal, even for those who failed to meet it. Marriages were supposed to be happy, but often were not. Getting out of a conflict-ridden or abusive marriage was very difficult. There were divorces, but the courts only granted these if they decided there were "appropriate violations" of the marriage contract (Weitzman, 1985). As a result, many marriages persisted, even when there was unbearable alienation, violence, and abuse.

Intensifying these struggles further, there was little recognition of domestic violence. Marital rape and rape in general were not taken seriously, legally or socially. For example, the police, when called to a domestic conflict, often ignored pleas of beaten or raped women, and the rules and procedures that guided police work made interventions almost meaningless. There was a great deal of resistance to changes in legislation relative to domestic violence and rape, including marital rape. The words of one U.S. Senator captured the general attitude at that time; he said, "if you can't rape your wife, who can you rape" (Russell, 1982). Thus, domestic violence and rape reforms took many years to implement.

As you can imagine, women in our mothers' cohort had few choices and opportunities. The "best" occupations most women could aspire to were limited to clerical or secretarial jobs, teaching, and nursing. Salaries were low and women had a difficult time living independently, both financially and socially. There was no such legal concept as workplace sexual harassment, and equal pay for equal work was rarely considered. Married women who did work outside the home often had fragmented work lives, defined by their primary responsibility of caring for children and husbands.

The situation for White women was bad, but it was much worse for women of color and immigrant women. For example, most African American women worked outside the home, but were confined to the lowest paying and most degrading jobs such as domestic work. These jobs had even fewer protections against sexual harassment and workplace inequalities. Men of racial minority groups and immigrant men also endured considerable inequalities in the workplace as well as other domains of life.

Post–World War II saw a considerable increase in the number of people entering college; however, almost all new students were White men taking advantage of the G.I. Bill. Considerable gender segregation in higher education persisted throughout the 1950s, with many colleges and universities denying admittance to women and racial minorities. The proportion of women in higher education increased only slightly during the 1950s, from 31.6 percent in 1950 to only 35.9 percent in 1959 (U.S. Department of Education, 2005). And, some of the most exclusive colleges and universities maintained gender segregated spaces even after women were admitted. For example, although women were admitted to Harvard Law School in 1950, they were denied access to the only eating space at the law school until 1970 (Deckard, 1979). Harvard Law School even limited the days on which women could ask questions in classes (Deckard, 1979). Money was another resource denied to many women; there were few scholarships for females because administrators and faculty felt "men needed the money more" (Deckard, 1979, p. 130).

The decades of the 1960s and 1970s brought about the awakening of political consciousness and action by Americans from many walks of life. Early in this period, the civil rights movement resulted in the dismantling of many legal barriers to participation in American life for African Americans. In the 1960s, the second wave of feminism picked up steam, spawning several organized political movements around gender and race, ethnicity, nationality, and/or sexuality as described in Chapter 10. These social movements focused attention on the social disadvantages faced by women and brought about considerable change, including greater legal, economic, political, educational, and familial equality. Other social movements emerged out of this culture of change: gay and lesbian rights, antiwar and peace, environmental, and children's rights movements.

We were fortunate to enter early adulthood during this time of positive social change. For example, Title IX, 1972, opened up avenues in education previously closed to our mothers; attempts at pay equity made our labor somewhat more valuable than that of our mothers' labor; the naming and litigation of sexual harassment, marital rape, and other forms of gender violence made our lives somewhat safer than our mothers' lives; and the women's movement empowered us and helped us to understand how we could contribute to a more just world. Our lives took us down different pathways, yet many of the same social change forces touched us deeply, and eventually our lives intersected.

Kay joined the ranks of one of the first waves of college-bound baby boomers in the mid-1960s. Her undergraduate years coincided with the civil rights movement, the height of the Vietnam War, the development of a strong political left, and the emergence of countercultural life styles. She found her intellectual passion in sociology, a home for her experimental self in the counterculture, and a political focus in the antiwar movement. With B.A. in hand, Kay entered the work world at the very moment that the first of a series of economic recessions set in. She bounced from one unsatisfying, low-paying, female-type job to another and quickly found herself at an intellectual and emotional crossroads. Kay then chose a pathway that few women had gone down—graduate school.

In 1971, Kay joined the ranks of a rapidly growing number of women graduate students in departments of sociology across the United States. Although dominated by White men, this cohort contained more working-class White women and people of color than earlier generations. They moved through graduate school at the same time that the second wave of feminism spawned organized movements for gender equality around the world. With these changes, Kay's consciousness expanded to embrace feminism. She took one of the first gender courses, Sex Roles, to be offered in any American institution of higher education. Feminism opened up a world of choices never before available to her. Empowered and exhilarated, Kay chose a nontraditional life course, as did many of the women in her graduate school cohort. Women postponed marriage and children. Others chose singlehood or cohabitating relationships. Still others chose child-free marriages. All pursued careers.

Of course, change is never smooth and even-handed. Although the women's movement was well under way during Kay's graduate school years, women students and professors had regular encounters with sexism both in and out of the classroom. Sexual harassment was built into the everyday educational experiences of women who pursued a Ph.D. Also, as the women of Kay's cohort entered the professional world of teaching and research, barriers to hiring, tenure, and promotion would prove to be part of their ongoing struggle for respect, security, and equality.

Meanwhile, unlike Kay, Joan was a "good girl." She went to a secretarial school and pursued a gender-appropriate job as a secretary. She became dissatisfied, though, because she was not receiving raises or being paid as well as others who had two-year college degrees, so she went to a community college to earn that degree. One course in sociology was all Joan needed to become a student of understanding social processes. She married and had two children, but continued on in school, moving from community college to university-level education.

In trying to understand the patterns of daily life, gender became a major explanatory framework for Joan. The isolation of women during childrearing years and the lack of institutions to support raising children developed into major interests, along with the effects of work patterns on men and women. Despite this lack of support for women who chose to pursue family and profession, Joan pursued and received her Ph.D. while rearing children. Her consciousness raising occurred more informally than Kay's, via books and one-on-one conversations with friends, often while caring for children.

Entry into professional sociology, while at first off-putting given the White, male dominance of the field, became a path to connect ideas with action and to form meaningful relationships with an array of women. It was through a relatively new organization, Sociologists for Women in Society, that we met one another and many others who taught and worked to improve the situation for women in society.

Together with another sociologist, Martha Cornwell, we started an upstate New York chapter of Sociologists for Women in Society. We became involved in Women's Studies on our respective campuses, and conducted research on women in the arts, work, family, education, and on women's bodily and emotional experiences. Today, we continue the journey toward a world that is more just and humane, a world in which people can achieve their potential unimpeded by the social prisms of difference and inequality described in this book.

Although it may seem that most of the work toward gender equality has been accomplished, as individuals and as members of a global society, we have more work to do. For example, women still do not have equal pay for equal work; glass ceilings and sticky floors continue to keep women from high-paying jobs; only a few countries offer gender-equitable parental leaves; most women work a double day, at paid labor and in the home; violence against women remains a serious problem; heterosexist and homophobic beliefs and behaviors maintain restrictive gender patterns while oppressing gay men and lesbians; racism continues to degrade and diminish the lives of women and men of color; and hegemonic masculinity limits the life experiences of most men.

Although there is still more to be accomplished, the good news is that in a very short period of time—our lifetimes—much positive change has occurred in gender relations. A third wave of feminism has now emerged among young people (Baumgardner & Richards, 2000) and many of the movements that Barbara Ryan discussed in Chapter 10 remain strong. Clearly, more change is on the way. What possibilities for change do *you* see in your future? How might you make a difference?

References

Baumgardner, J., & Richards, A. (2000). *Manifesta: Young women, feminism, and the future.* New York: Farrar, Straus and Giroux.

Bernard, J. (1972). *The future of marriage.* New York: World Publishing.

Bernard, J. (1975). *Women, wives, mothers.* Chicago: Aldine Publishing Company.

Coontz, S. (1992). *The way we never were: Americans and the nostalgia trap.* New York: Basic Books.

Coontz, S. (1997). *The way we really are: Coming to terms with America's changing families.* New York: Basic Books.

de Beauvoir, S. (1952). *The second sex.* New York: Knopf.

Deckard, B. S. (1979). *The women's movement: Political, socioeconomic, and psychological issues.* New York: Harper & Row.

Friedan, B. (1963). *The feminine mystique.* New York: W. W. Norton.

Rapp, R. (1982). Family and class in contemporary America. In B. Thorne, with M. Yalom (Eds.), *Rethinking the family: Some feminist questions* (pp. 168–187). New York: Longman.

Russell, D. E. H. (1982). *Rape in marriage.* New York: Macmillan.

Schneir, M. (1994). *Feminism in our time: The essential writings, World War II to the present.* New York: Vintage Books.

Schwartz, P. (1994). *Peer marriage: How love between equals really works.* New York: Free Press.

U. S. Dept. of Education. (2005). *Digest of Education Statistics.* Table 170. Retrieved July 13, 2007, from http://nces.ed.gov/programs/digest/d05/tables/dt05_170.asp

Weitzman, L. J. (1985). *The divorce revolution: The unexpected social and economic consequences for women and children in America.* New York: Free Press.

INDEX

About the Editors

Joan Z. Spade is professor and chair of sociology at the State University of New York at Brockport. She received her Ph.D. from State University of New York at Buffalo and her B.A. from State University of New York at Geneseo. In addition to gender, Joan teaches courses on education, family, research methods, and statistics. She has published articles on rape culture in college fraternities and work and family, including women and men's orientations toward work. She has also coedited two books on education and has published articles on education, including research on tracking and gender and education. Joan is active in Sociologists for Women in Society, Eastern Sociological Society, and the American Sociological Association. Joan's two children and their spouses have four (soon-to-be five) grandchildren. In addition to visiting her children and grandchildren, she enjoys the arts and being outdoors, including a two-week trip hiking in Montana during the summer of 2007.

Catherine (Kay) G. Valentine is professor of sociology at Nazareth College in Rochester, New York. She received her Ph.D. from Syracuse University and her B.A. from the State University of New York at Albany. Kay teaches a wide range of courses, such as sociology of gender, senior seminar in sociology, sociology of bodies and emotions, sociology of consumerism, and human sexuality. Her publications include articles on teaching sociology, women's bodies and emotions, gender and qualitative research, and the sociology of art museums. She is the founding director of women's studies at Nazareth College and long-time member of Sociologists for Women in Society. She has also served as president of the New York State Sociological Association. Kay and her spouse, Paul J. Burgett, University of Rochester vice president and professor of music, are devotees of the arts and world travel.